Building Performance Simulation for Design and Operation

Effective building performance simulation can reduce the environmental impact of the built environment, improve indoor quality and productivity, and facilitate future innovation and technological progress in construction. It draws on many disciplines, including physics, mathematics, material science, biophysics and human behavioural, environmental and computational sciences. The discipline itself is continuously evolving and maturing, and improvements in model robustness and fidelity are constantly being made. This has sparked a new agenda focusing on the effectiveness of simulation in building life-cycle processes.

Building Performance Simulation for Design and Operation begins with an introduction to the concepts of performance indicators and targets, followed by a discussion on the role of building simulation in performance-based building design and operation. This sets the ground for in-depth discussion of performance prediction for energy demand, indoor environmental quality (including thermal, visual, indoor air quality and moisture phenomena), HVAC and renewable system performance, urban level modelling, building operational optimization and automation.

Produced in cooperation with the International Building Performance Simulation Association (IBPSA), this book provides a unique and comprehensive overview of building performance simulation for the complete building life-cycle from conception to demolition. It is primarily intended for advanced students in building services engineering, and in architectural, environmental or mechanical engineering; and will be useful for building and systems designers and operators.

Jan L.M. Hensen (Ph.D. & M.S., Eindhoven University of Technology) has his background in building physics and mechanical engineering. His professional interest is performance-based design in the interdisciplinary area of building physics, indoor environment and building systems. His teaching and research focuses on the development and application of computational building performance modelling and simulation for high performance.

Roberto Lamberts is a Professor in Construction at the Department of Civil Engineering of the Federal University of Santa Catarina, Brazil. He is also currently a board member of the IBPSA, Vice-President of the Brazilian Session and Counsellor of the Brazilian Council for Sustainable Buildings.

Building Performance Simulation for Design and Operation

Edited by Jan L.M. Hensen and Roberto Lamberts

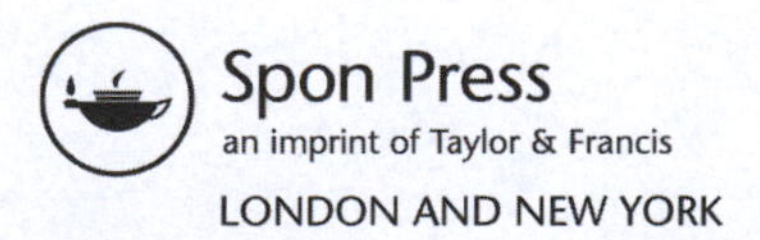

Published 2011 by Spon Press
2 Park Square, Milton Park, Abingdon, Oxon OX14 4RN

Simultaneously published in the USA and Canada by Spon Press
270 Madison Avenue, New York, NY 10016, USA

Spon Press is an imprint of the Taylor & Francis Group, an informa business

Typeset in Goudy by Swales & Willis Ltd, Exeter, Devon
Printed and bound in Great Britain by CPI Antony Rowe, Chippenham, Wiltshire

British Library Cataloguing in Publication Data
A catalogue record for this book is available
from the British Library

Library of Congress Cataloging in Publication Data
Building performance simulation for design and operation/edited Jan L.M. Hensen and Roberto Lamberts.
p. cm.
Includes bibliographical references.
1. Buildings—Performance—Computer simulation. I. Hensen, Jan L.M.
II. Lamberts, Roberto.
TH453.B875 2011
690'.22—dc22
2010021030

ISBN13: 978–0–415–47414–6(hbk)
ISBN13: 978–0–203–89161–2(ebk)

Contents

List of figures

List of tables

List of contributors

Godfried Augenbroe
Professor Godfried Augenbroe has a 35-year track record in the modelling and simulation of buildings, first at TU Delft in the Netherlands, and since 1997, at the Georgia Institute of Technology in Atlanta, USA.

Charles S. Barnaby
Charles S. Barnaby is Vice-President of Research at Wrightsoft Corporation. He focuses on the implementation of loads and energy simulation aspects of Wrightsoft software products. He also leads Wrightsoft's research projects funded by ASHRAE, the US Department of Energy and other agencies.

Ian Beausoleil-Morrison
Dr Ian Beausoleil-Morrison is Associate Professor at Carleton University in Ottawa, Canada, where he holds the Canada Research Chair in the Modelling and Simulation of Innovative Energy Systems for Residential Buildings. Currently he is President of the International Building Performance Simulation Association (IBPSA) and co-editor of the Journal of Building Performance Simulation.

Bert Blocken
Bert Blocken is Associate Professor at the Unit Building Physics and Systems (BPS) at Eindhoven University of Technology in the Netherlands. He was awarded his PhD in 2004 at the Katholieke Universiteit Leuven, Belgium, with the thesis "Wind-driven rain on buildings – measurements, numerical modelling and applications".

Jan Carmeliet
Professor Jan Carmeliet has been Chair of Building Physics at ETH Zürich and Head of the Laboratory of Building Technology of EMPA, Dübendorf (Swiss Federal Laboratories for Materials Testing and Technology) in Switzerland since June 2008. His research interests concern mainly physical processes in multi-scale (porous) materials, poromechanics, particle flow, flow at urban scale, materials for energy technology, computational modelling.

David E. Claridge
David E. Claridge (MS and PhD, Stanford University) is the Leland Jordan Professor of Mechanical Engineering at Texas A&M University and is Director of the Energy Systems Laboratory. He is a Fellow of ASME and ASHRAE and recipient of the E.K. Campbell Award from ASHRAE, a Faculty Distinguished Achievement in Research Award from Texas A&M University and has received three Best Paper Awards from ASME.

Drury B. Crawley
Dr Crawley is Director, Building Performance Products at Bentley Systems, Inc. He leads a group developing tools for building performance and sustainability. Previously he led the US Department of Energy's Commercial Building Initiative with a goal of market-ready net-zero energy commercial buildings by 2025.

Thijs Defraeye
Thijs Defraeye is a PhD student at the Laboratory of Building Physics at the Katholieke Universiteit Leuven, Belgium. His PhD research, which he started in 2006, mainly focuses on numerical modelling of convective heat and moisture transfer at exterior building surfaces due to air flow in the atmospheric boundary layer.

Dominique Derome
Since June 2008, Dominique Derome has been a Senior Scientist and Leader of the Multiscale modelling group in the Wood Laboratory of EMPA, Swiss Federal Laboratories for Materials Testing and Research. Her research interests include multi-scale modelling of coupled mechanical and hygrothermal behaviour of wood, transport of liquid in wood, determination of air-wood boundary conditions, and large-scale experimental investigation of hygrothermal behaviour of building assemblies.

Jan L.M. Hensen
Jan L.M. Hensen is Professor in Building Performance Simulation at Eindhoven University of Technology. His teaching and research focuses on development and application of computational modeling and simulation for high performance buildings while considering building physics, indoor environmental quality and building energy systems.

Gregor P. Henze
Gregor P. Henze is Professor of Architectural Engineering at the University of Colorado, holding a PhD in Civil Engineering as well as the Diplom-Ingenieur and MS in Mechanical Engineering. He has authored 70 peer-reviewed technical papers and is Associate Editor of the ASME's *Journal of Solar Energy Engineering*.

Roberto Lamberts
Roberto Lamberts is Professor of Construction at the Department of Civil Engineering of the Federal University of Santa Catarina. He is also currently a board member of the International Building Performance Simulation Association (IBPSA), Vice-President of the Brazilian Session, and Counsellor of the Brazilian Council for Sustainable Buildings.

Ardeshir Mahdavi
Professor Dr Ardeshir Mahdavi is Director of the Department of Building Physics and Building Ecology at the Vienna University of Technology in Austria. In the years 2005, 2006 and 2008, Professor Mahdavi has been consecutively awarded the "Austrian Building Award" in the research projects category.

Christian Neumann
Christian Neumann is one of the co-founders of the engineering company Solares Bauen GmbH, located in Freiburg, Germany (www.solares-bauen.de). Since 2005 he has been working as project engineer in the department of Thermal Systems and Buildings at Fraunhofer Institute for Solar Energy Systems in Freiburg, Germany. His main focus is the design and monitoring of energy-efficient buildings.

Christoph Reinhart
Christoph Reinhart is an Associate Professor of Architectural Technology at Harvard University, Graduate School of Design. His research expertise is in daylighting, passive climatization and the influence of occupant behaviour on building energy use. He is working on new design workflows and performance metrics that accommodate the complementary use of rules-of-thumb and simulations during building design.

Darren Robinson
A building physicist by training, Dr Robinson has been working for over ten years now in urban energy and environmental modelling. He is currently group leader of sustainable urban development with the Solar Energy and Building Physics Laboratory (LESO) at the Ecole

Polytechnique Fédérale de Lausanne (EPFL) in Switzerland. Darren has published over 50 scientific papers on the subject of urban modelling and was awarded the Napier-Shaw Medal by the CIBSE for one of these.

Jeffrey D. Spitler
Jeffrey D. Spitler is Regents Professor and C.M. Leonard Professor in the School of Mechanical and Aerospace Engineering at Oklahoma State University, where he teaches classes and performs research in the areas of heat transfer, thermal systems, building simulation, design cooling load calculations, HVAC systems, snow melting systems and ground source heat pump systems.

Jelena Srebric
Dr Srebric is an Associate Professor of Architectural Engineering and an Adjunct Professor of Mechanical and Nuclear Engineering at the Pennsylvania State University. She conducts research and teaches in the field of building energy consumption, air quality, and ventilation methods. She is a recipient of both NSF (National Science Foundation) and NIOSH (National Institute of Occupational Safety and Health) career awards.

Christoph van Treeck
Christoph van Treeck (Dr.-Ing. habil.) is Head of the Simulation Group of the Department of Indoor Environment at the Fraunhofer Institute for Building Physics in Germany. He is involved in teaching activities as Associate Professor at the Technische Universität München. He holds the *venia legendi* for the subject of computational building physics and is a board member of the International Building Performance Simulation Association.

Michael Wetter
Michael Wetter is a computational scientist in the Simulation Research Group at Lawrence Berkeley National Laboratory (LBNL). His research includes integrating building performance simulation tools into the research process, as well as the design and operation of buildings. He is a recipient of the IBPSA Outstanding Young Contributor Award, the Vice-President of IBPSA-USA and a member of ASHRAE.

Jonathan Wright
Jonathan Wright is Professor of Building Optimization at Loughborough University in the UK. He has 25 years of research experience in the field of building performance simulation and its application to the optimum design and operation of buildings. He has published widely on the theme of model-based building optimization, and is a member of the IBPSA Board of Directors. He is currently leader of the Building Services Engineering group at Loughborough University.

Foreword

The fossil fuels are entering their tertiary stage and steps are being taken in many countries to kick-start the transition to an alternative energy infrastructure. A pressing question is how this transition can best be managed, negative impacts mitigated, and the various technology options blended over time: fossil fuel de-carbonisation and sequestration in the short term, the deployment of energy efficiency measures, the switch to new and renewable source of energy, and the removal of barriers confronting new nuclear plant. A key aspect of any future energy infrastructure will be real-time demand management to facilitate the matching of demand with supply – especially where the latter comprises significant inputs from stochastic, distributed renewable energy sources. Because a large portion of a country's energy demand is associated with the built environment, it is here that productive action can be taken to reduce energy consumption whilst ensuring that expectations relating to human comfort/health and environmental protection are met.

The built environment is inherently complex and as a consequence conflicts abound, proffered solutions are often polarised and consensus is difficult to attain. This situation gives rise to three fundamental engineering challenges: how to consider energy systems in a holistic manner in order to address the inherent complexity; how to include environmental and social considerations in the assessment of cost-performance in order to ensure sustainable solutions; and how to embrace inter-disciplinary working in order to derive benefit from the innovative approaches to be found at the interface between the disciplines. In short, energy systems require an integrated approach to design: will the widespread deployment of micro-CHP within the urban environment be acceptable if the global carbon emission reduction to result is attained at the expense of reduced local air quality and increased maintenance cost?

Integrated building performance simulation has emerged as an apt means of addressing the above challenges while allowing collaborating practitioners to identify the action combinations that will be most effective in providing acceptable overall performance as a function of the unique climate, design and operational parameters defining specific buildings and communities, planned or existing. IBPS does this by modelling the heat, air, moisture, light, electricity, pollutant and control signal flows within building/plant systems and, thereby, nurturing performance improvement by design. The benefits of the power and universal applicability of the approach comes at a price however: application requires an understanding of design hypothesis abstraction, computer model building, multiple domain simulation, performance trade-offs, and the translation of outcomes to design evolution.

This book presents the complementary views of distinguished researchers in the field, arranged in a progressive format that covers the myriad issues underpinning the application of modelling and simulation when used to support decision-making relating to building performance and operation. In addition to the wide scope of topics covered, the book provides useful examples of the practical application of building simulation to formulate design and operational solutions that are acceptable in terms of performance criteria relating to indoor air quality, thermal/visual/acoustic comfort, operational/embodied energy, carbon emissions, and capital/running cost. A unique feature of the book is the balance between theory and practice on the one hand, and between the issues at the individual building and community level on the other. The book is essential reading to those practitioners and researchers who seek to understand and apply building simulation in a professional manner.

Joe Clarke
Energy Systems Research Unit
University of Strathclyde
Glasgow
December 2009

Preface

The rise of technology over the past decades has been something of a mixed blessing. On the one hand it has increased our freedom to move and communicate and has provided us with more comfort. On the other hand, it is widely understood that the energy use currently required to drive our modern way of living has led to critical environmental problems. These problems have been highlighted to such a degree through the explosion of research and news coverage over recent years, that it is now common knowledge that our lifestyle is unsustainable. In modern terminology, to slow down and hopefully reverse the manmade damage, we need to develop a sustainable and zero net energy built environment. This will involve not only the design of net energy producing new 'green' buildings, but also the optimization of energy use of existing buildings.

In line with the rise of technology, buildings and the systems within them have become exponentially more complex in recent times. The modern built environment is populated by a variety of building types with highly demanding performance and user requirements. The difficulties involved in optimizing energy use in buildings have been recognized for quite some time. The complexity of the task arises from the number of variables from a wide range of fields that must be considered. Many professionals and researchers, including ourselves, concluded that solving such a complex problem requires two things: interdisciplinary research involving a wide variety of disciplines, and well-developed technological tools to make the problem manageable.

In 1986 a group of like-minded individuals established the International Building Performance Simulation Association, IBPSA (www.ibpsa.org), a non-profit society of building performance simulation researchers, developers and practitioners dedicated to improving the built environment. IBPSA provides a forum for researchers, developers and practitioners to review building model developments, facilitate evaluation, encourage the use of software programs, address standardization, and accelerate integration and technology transfer.

IBPSA covers broad areas of building environmental and building services engineering. Typical topics include building physics (including heat, air and moisture flow, electric and day lighting, acoustics, smoke transport); heating, ventilation and air-conditioning systems; energy supply systems (including renewable energy systems, thermal storage systems, district heating and cooling, combined heating and power systems); human factors (including health, productivity, thermal comfort, visual comfort, acoustical comfort, indoor air quality); building services; and advancements and developments in modeling and simulation such as coupling with CAD, product modeling, software interoperability, user interface issues, validation and calibration techniques. All these topics may be addressed at different levels of resolution (from microscopic to the urban scale), and for different stages in the building life cycle (from early sketch design, via detailed design to construction, commissioning, operation, control and maintenance) of new and existing buildings worldwide.

In essence, IBPSA has two key objectives: to use computer simulation to (a) provide better support for the design of buildings; and (b) provide better support for building operation and management in the use phase of buildings. These two objectives have informed our own research over the last decades. This book aims to give the reader a thorough understanding of the recent progress made in building simulation and the key challenges that still need to be overcome.

The main motivation for developing this book is that at the time of writing no comprehensive text book on the subject was available even though building performance simulation has become an essential technology for architectural and engineering design and consultancy practices which aim to provide innovative solutions for their clients.

This book sets out to fill this gap by providing unique insight into the techniques of building performance modelling and simulation and their application to performance-based design and operation of buildings and the systems which service them. It provides readers with the essential concepts of computational support of performance based design and operation – all in one book. It provides examples of how to use building simulation techniques for practical design, management and operation, and highlights their limitations and suggests future research directions.

This book provides a comprehensive overview of building performance simulation for the complete building life-cycle from conception to demolition. It addresses theory, development, quality assurance and use in practice of building performance simulation. The book is therefore both theoretical and practical, and as such will be of interest to those concerned with modelling issues (universities, research organizations and government agencies) and real world applications (architects, engineers, control bodies, building operators). The book is primarily intended for (future) building and systems designers and operators of a postgraduate level. However, due to the interdisciplinary nature of research into the built environment, the book should also prove useful for a variety of connected fields.

The interdisciplinary nature of the research becomes clear when it is understood that building performance simulation draws its underlying theories from many disciplines including: physics; mathematics; material science; biophysics; human behavioral, environmental and computational sciences. The book would lend itself to adaption for multidisciplinary courses, for example AEC related university courses which address building performance prediction and operational issues. Other courses might include it on their recommended reading list, especially at the postgraduate level.

The book begins by introducing and describing the key features of building performance simulation and sets the scene for the rest of the book. The concepts of performance indicators and targets are discussed, followed by a discussion of the current and future role of building simulation in performance based building design and operation. This will lay the foundations for in-depth discussions of performance prediction for key aspects such as energy demand, indoor environmental quality (including thermal, visual, indoor air quality and moisture phenomena), HVAC and renewable system performance, urban level modelling, building operational optimization and automation. The book ends with a discussion of future directions for building performance simulation research and applications in practice. The book aims to show that when used appropriately, building performance simulation is a very powerful technique capable of helping us achieve a sustainable built environment, and at the same time improving indoor quality and productivity, as well as stimulating future innovation and technological progress in the architecture, engineering and construction (AEC) industry.

We believe this book to be long overdue. We have been contemplating the idea of writing it for many years. However, due to the interdisciplinary nature of the subject, writing such a book required the cooperation of many individuals. Toward the end of 2006 the idea became more concrete, and in early 2007 the co-authors and the publisher enthusiastically joined the adventure. Despite our busy schedules, in 2008 a symposium was organized in Brazil to bring co-authors together and allow them to present the content of their chapters. We would like to thank Eletrobrás and Conselho Nacional de Desenvolvimento Científico e Tecnológico (CNPq) for their financial support of this event. The event proved to be an important catalyst that allowed the book to progress to its finished state, even if it took a fair bit longer to finish the book than originally anticipated. In the end, we are really very pleased with the results, and hope you will enjoy reading the book too.

This book is the result of cooperation and dedication of many individuals, in particular of course all co-authors. We would like to take the opportunity to also acknowledge the support of our universities: Eindhoven University of Technology, The Netherlands, and Universidade Federal de Santa Catarina, Brazil. Last but not least, we wish to express our gratitude to Duncan Harkness, Roel Loonen, Ana Paula Melo, Martin Ordenes Mizgier, Jikke Reinten and Marija Trcka for their editorial and practical support.

Jan L.M. Hensen, Roberto Lamberts, March 2010

1 Introduction to building performance simulation

Jan L.M. Hensen and Roberto Lamberts

Scope

The aim of this chapter is to provide a general view of the background and current state of building performance simulation as an introduction to the following chapters in this book.

Learning objectives

To appreciate the context, background, current state, challenges and potential of building performance simulation.

Key words

Context/state-of-the-art/future needs/quality assurance

Introduction

Designing sustainable buildings that also fulfill all operational requirements of the users is an unprecedented challenge for our times. Researchers, practitioners and other stakeholders are faced with enormous challenges due to the need to recognize and take account of various dynamic processes around us, such as: global climate change; depletion of fossil fuel stocks; increasing flexibility of organizations; growing occupant needs and comfort expectations; increasing awareness of the relation between indoor environment and the health and wellbeing of the occupants, and consequently their productivity. Managing all of these aspects in order to achieve robust building and system solutions which will be able to withstand future demands requires an integrated approach of the subsystems, shown in Figure 1.1.

In response to the sustainability challenges that we are facing, the European Union has issued the rather strict 20–20–20 initiative (relative to 1999, by 2020 20% reduction of energy consumption, 20% reduction of CO_2, and 20% introduction of renewable energy. Similar and even stricter guidelines have also been developed by various countries from all over the world. Although strict, these initiatives and guidelines should be regarded as merely a first step towards the much more ambitious and demanding long-term goals as discussed in, e.g., Lund and Mathiesen (2009). Achieving these ambitious sustainability targets requires the development of net energy producing buildings or sites. For this we need models and tools which allow the consideration of interoperating domains such as transportation and large scale energy grids. Only then can the global optimization of energy production and consumption in the built environment be achieved.

In addition to these higher sustainability requirements, future buildings should also deliver considerable improvements to indoor environment quality. Rather than the current practice of merely complying with minimum standards for environmental parameters such as temperature, air quality, lighting and acoustical levels, future buildings should provide a positive indoor

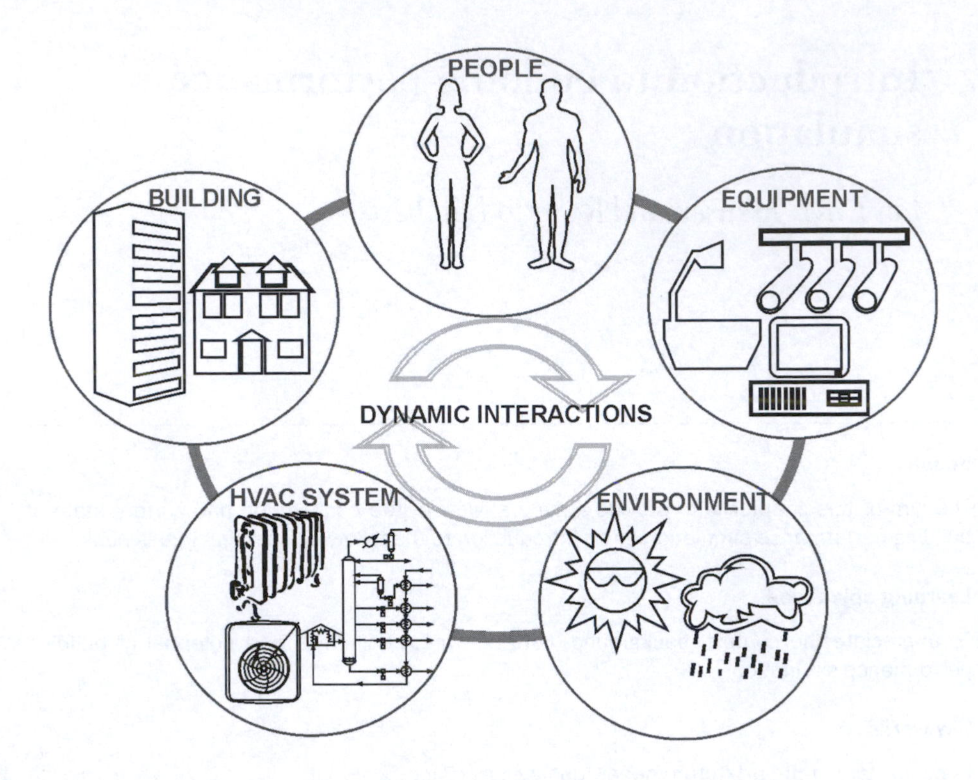

Figure 1.1 Dynamic interactions of (continuously changing) sub-systems in buildings.

environment that is stimulating, healing or relaxing, depending on the function. This will then result in truly high performance buildings (Green 2009).

At present, however, the focus – even in high performance buildings – is still very much on reducing energy demand. One recent, interesting approach is to engineer a building in such a way that it can adapt itself to the actual outdoor conditions. Loonen (2010) provides a comprehensive overview of climate adaptive building shell concepts. An illustration of a recent building with an adaptive building shell is given in Figure 1.2.

Given the relatively low volume of new building projects (in Europe only about 10% per year of the total building stock), it is evident that in order to reach the sustainability targets in time, a huge amount of work is needed in terms of refurbishment of existing buildings (see, e.g., Petersdorff *et al.* 2006). An interesting project in this context is "Cost Effective", which aims to convert facades of existing "high-rise" buildings into multifunctional, energy gaining components (Fraunhofer 2009).

One of the buildings considered is shown in Figure 1.3. In a holistic approach Deutsche Bank has been creating one of the most eco-friendly high-rise buildings in the world. As one of the manifold measures the bank replaced the complete facade of an existing building with new, super-insulating triple paned windows and improved insulation. As every second window can be opened in the future, less air needs to be moved through mechanical ventilation, thanks to the natural air circulation. The project shows just how much potential for optimization and sustainable energy efficiency there is for existing buildings as well as how a "green building" approach can be worthwhile in a wide variety of ways, even as an investment in existing properties. Energy consumption and CO_2 emissions of the so-called Greentowers building is expected to be reduced by 89 percent (DB Greentowers 2009).

So – both new and refurbishment – future projects face huge challenges that seem too complex for most current tools and approaches.

Figure 1.2 Virginia Tech Lumenhaus, Solar Decathlon 2009 hosted by the U.S. Department of Energy, National Mall, Washington, DC.

Source: Flickr 2009.

We feel that traditional engineering design tools are largely unsuitable for addressing the above challenges for the following reasons: they are typically mono-disciplinary, solution-oriented and very restricted in scope. They assume static (usually only extreme) boundary conditions, and are often based on analytical methods, which to a large extent can be characterized as aiming to provide an exact solution of a very simplified view (model) of reality).

Building performance simulation

Computational building performance modeling and simulation[1] on the other hand is multi-disciplinary, problem-oriented and wide(r) in scope. It assumes dynamic (and continuous in time) boundary conditions, and is normally based on numerical methods that aim to provide an approximate solution of a realistic model of complexity in the real world.

Computational simulation is one of the most powerful analysis/analytic tools in our world today – it is used to simulate everything from games to economic growth to engineering problems. However, it is very important to recognize that simulation does not provide solutions or answers, and that very often it is difficult to ensure the quality of simulation results. No matter what field they are from, those who have worked with simulation know that it is a complex task. Bellinger (2004), from the field of operational research, remarked:

> After having been involved in numerous modeling and simulation efforts, which produced far less than the desired results, the nagging question becomes; Why? . . . The answer lies in two areas. First, we must admit that we simply don't understand. And, second, we must pursue understanding. Not answers but understanding.

Figure 1.3 Deutsche Bank headquarters in Frankfurt am Main during refurbishment.

Source: DB Greentowers 2009.

As will become clear from reading the following chapters, both the power and the complexity of building performance modeling and simulation arise from its use of many underlying theories from diverse disciplines, mainly from physics, mathematics, material science, biophysics, human behavioral, environmental and computational sciences. It will also become clear that many theoretical and practical challenges still need to be overcome before the full potential of building modeling and simulation can be realized.

A short historical overview of the use of simulation for building design can be found in an editorial by Spitler (2006): "Building performance simulation: the now and the not yet". He explains that:

simulation of building thermal performance using digital computers has been an active area of investigation since the 1960s, with much of the early work (see e.g. Kusuda 1999) focusing on load calculations and energy analysis. Over time, the simulation domain has grown richer and more integrated, with available tools integrating simulation of heat and mass transfer in the building fabric, airflow in and through the building, daylighting, and a vast array of system types and components. At the same time, graphical user interfaces that facilitate use of these complex tools have become more and more powerful and more and more widely used.

Like many other technological developments, building performance simulation also experienced a so-called hype cycle (Fenn and Raskino 2008), as shown in Figure 1.4. The "recognition" took place in the early 1970s, with the peak of inflated expectations in the 1980s, followed by the trough of disillusionment. It seems fair to state that building performance simulation in general has been on an upward slope of productivity for almost two decades now. However, as will become clear from the following chapters, this does not appear to be the case yet for related developments such as object oriented programming, computational fluid dynamics, and building information technology.

In this context, the important role of the International Building Performance Simulation Association (IBPSA – www.ibpsa.org) should be acknowledged, since one of its most important goals is to increase awareness of building performance simulation while avoiding both inflated expectations and disillusionment.

The following chapters demonstrate that the building simulation discipline is continuously evolving and maturing and that improvements are continually being made to model robustness and fidelity. As a result much of the discussion has shifted from the old agenda focusing on software features, to a new agenda that focuses on the effectiveness of building performance simulation in building life cycle processes.

The development, evaluation, use in practice, and standardization, of the models and programs is therefore of growing importance. This is evidenced in, for example, green building

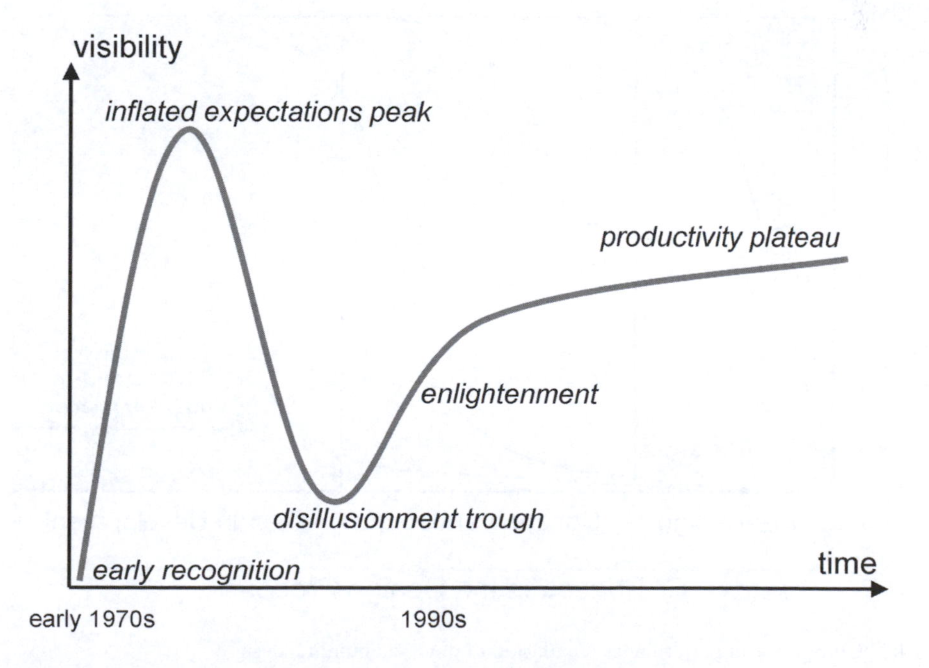

Figure 1.4 Technology hype cycle for building performance simulation.

rating systems currently being promoted around the world, such as LEED (Leadership in Energy and Environmental Design) and BREEAM (Building Research Establishment Environmental Assessment Method), in incentive programs such as the US EPAct (Energy Policy Act) and also in legislation such as the European EPBD (Energy Performance of Buildings Directive).

Current use in practice

It is widely recognized that predicting and analyzing future building behavior in advance is far more efficient and economical than fixing problems when the building is in the use phase. Nevertheless, the uptake of building performance simulation in current building design practice is surprisingly limited. The actual application is generally restricted to the final phases in building design as indicated in Figure 1.5.

At present, in addition to its relatively low adoption, the use of building performance simulation is largely restricted to a few key areas: for building envelope design; to predict the risk of overheating during the summer (see Figure 1.6 shows a typical example) and/or to calculate maximum cooling loads in view of equipment sizing (see Figures 1.7 and 1.8 for examples).

Although Figure 1.8 is from a study where simulation was used for mechanical engineering design, in reality these sorts of studies are rare. It is still much more common to see traditional design approaches being used for this. Thomas (2006) clearly demonstrates this point:

> Out of a typical large mechanical-electrical (M-E) design project consisting of 50,000 HVAC[2] labor-hours, about 100 hours is spent on energy analysis. Another 200 hours might be spent on loads calculations over the course of the project. This is about the total extent of HVAC engineering design using computer programs today. The remaining 99.9% of computer use is for drafting, word processing and spreadsheets for organizing information. The same is true for electrical, lighting, plumbing and fire protection design.

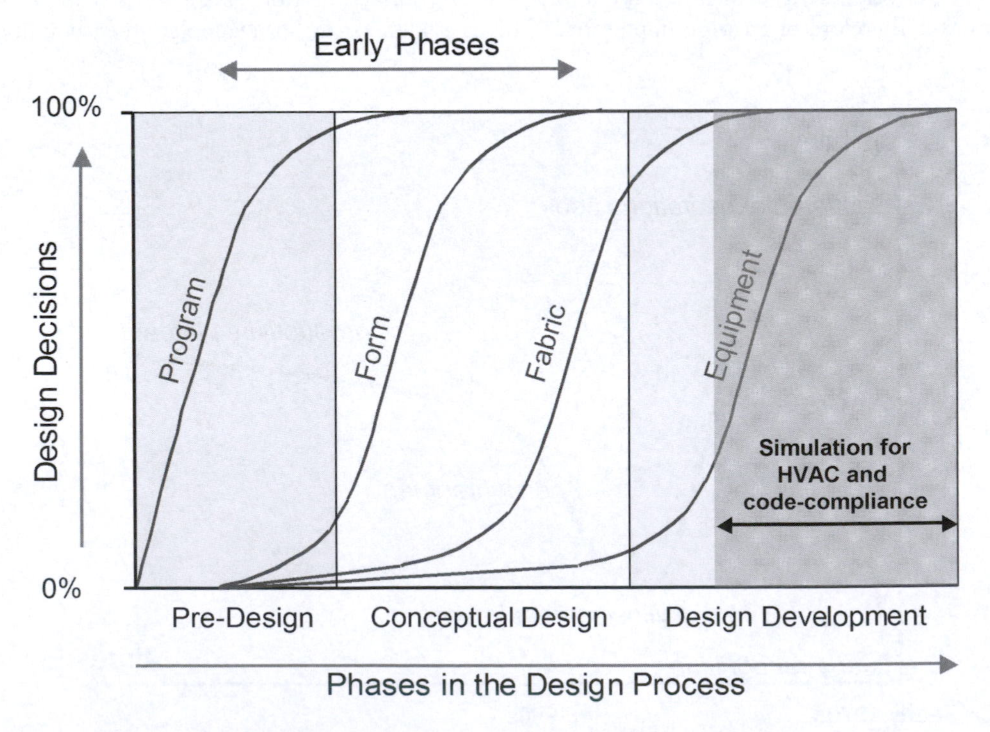

Figure 1.5 Current use of performance simulation in practical building design.

Source: Torcellini and Ellis 2006.

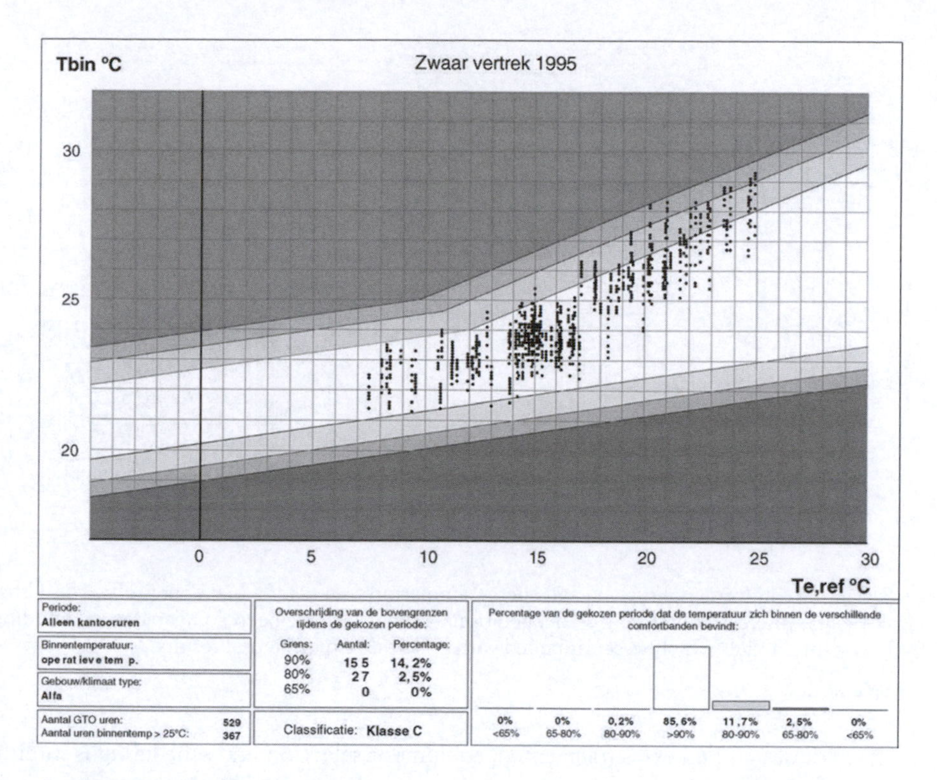

Figure 1.6 Output visualization of adaptive thermal comfort predictions for a medium-heavy office building in The Netherlands during the period from May to end September 1995.

Source: Linden *et al.* 2006.

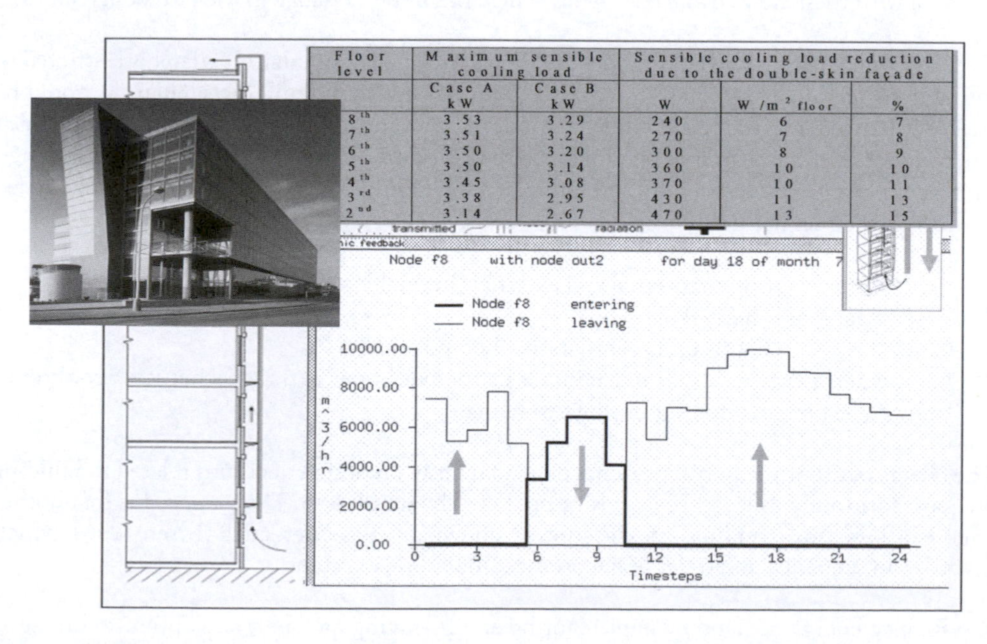

Floor level	Maximum sensible cooling load		Sensible cooling load reduction due to the double-skin façade		
	Case A kW	Case B kW	W	W /m² floor	%
8th	3.53	3.29	240	6	7
7th	3.51	3.24	270	7	8
6th	3.50	3.20	300	8	9
5th	3.50	3.14	360	10	10
4th	3.45	3.08	370	10	11
3rd	3.38	2.95	430	11	13
2nd	3.14	2.67	470	13	15

Figure 1.7 Sample airflow and cooling load simulation results for an office development in Prague with double-skin façade.

Source: Hensen *et al.* 2002.

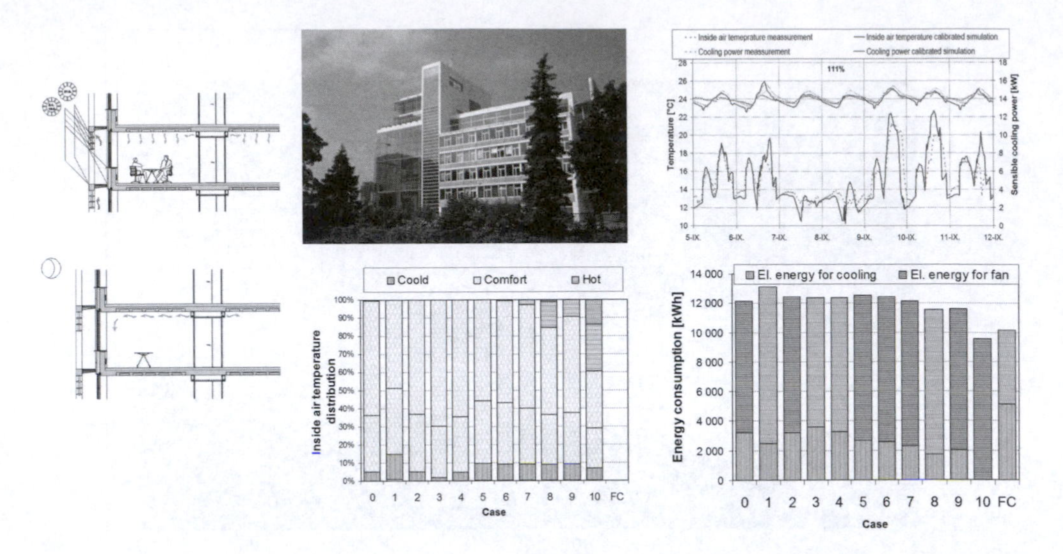

Figure 1.8 Sample comfort, cooling load and energy simulation results for a low-energy office building in Prague where 10 different system operational scenarios have been compared using building + systems models which were calibrated with real scale experimental results.

Source: Lain *et al.* 2005.

> The M-E design process is fragmented, equipment selection and scheduling is intermittent and the process consists of frequent revisions and continuous exchange of fragmented information between specialized architectural and engineering personnel. A-E design documents from schematic to construction are issued in 2-D whereas the new automated systems are in 3-D. The present organizational structure and specialized staff and tasks will not work well with these new advanced systems. There has to be a change in the A-E design culture.

To recap/summarize: in current practice building performance simulation is largely restricted to the analysis of a single design solution. The potential impact of building simulation would be greatly enhanced if its use was extended to (multiple variant) design optimization and included much earlier in the design process. To illustrate this point, consider the CIBSE (2005) design strategy for environment friendly and future proof building design, which can be summarized in the following sequential steps:

1. Switch off – relating to internal and external thermal loads;
2. Spread out – use thermal mass;
3. Blow away – apply (natural) ventilation when possible;
4. Cool when necessary – do not hesitate to include some extra (mechanical) cooling in order to be prepared for future climate change.

The effects of the three first approaches depend mainly on design decisions related to building program, form and fabric, and can only be predicted by simulation. The same is true for another – more or less corresponding – design strategy known as Trias Energetica (Lysen 1996), which involves, in sequence and order of importance, the following steps:

1. Reduce energy demand by implementing energy-saving measures and optimize the use of solar and casual gains;
2. Use sustainable sources of energy instead of finite fossil fuels;
3. Use fossil energy as efficiently as possible.

Again, the decisions related to the first step are taken primarily during the early design phases, where simulation is needed but currently not used.

Moving beyond the design phases, there exists a considerable and rapidly increasing interest – in practice and research – in the use of simulation for post-construction activities such as commissioning, operation and management. The uptake in current practice is still very limited, but we expect that the next decade will see a strong growth in application of building performance simulation for such activities. The two main reasons for this are (1) the current (considerable) discrepancy between predicted and actual energy consumption in buildings, and (2) the emergence of new business models driven by whole life time building (energy) performance.

Although a large number of building performance simulation tools exist (DOE 2010), it is also clear (e.g. from Crawley *et al.* 2008) that there is a huge overlap in functionality amongst many of these tools.

In various fields, including building and system design, people can be classified according to their innovativeness (Figure 1.9). From interviews with building and system designers (Hopfe *et al.* 2006) it was found that someone can be classified as an innovator in one aspect (say use of simulation tools) while belonging to the "late majority" in another aspect (say in terms of design team integration). A major requirement of an innovator would be that the simulation tool can be used to model and simulate all sorts of (not yet existing) building systems and applications. For that type of person it would not matter much if the tool would be quite hard to learn and use as long as it is very flexible and expandable. For the "late majority", on the other hand, it is essential that the tool appears intuitive and easy to use. This variety in needs combined with the relatively low potential number of software copies to be sold, remains a huge challenge for commercial software developers.

Quality assurance of simulation based decisions

Quality assurance is a very important and ongoing issue addressed throughout this book. The quality of simulation results depends, of course, on the physical correctness of the model. Although Robinson (1999) is not related to building simulation, its conclusion that it is not

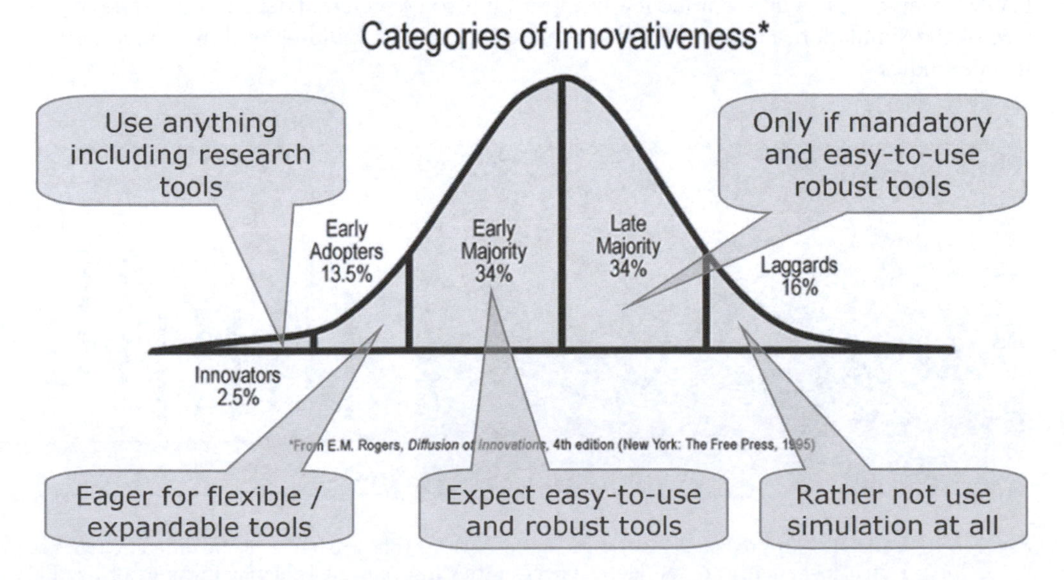

Figure 1.9 Attitude and expectations regarding simulation tools in relation to the (design) innovativeness of the user.

Source: Adapted from Rogers (1995).

possible to validate a model and its results, but only to increase the level of confidence that is placed in them, seems to be equally true for our domain. The ongoing BESTEST initiative (e.g. Judkoff and Neymark 1995; Neymark *et al.* 2001) represents a major international effort within the building domain to increase the confidence of simulation results. Its progress is reflected in its first footholds in professional standards such as the American Society of Heating, Refrigerating and Air-Conditioning Engineers (ASHRAE) Standard Method of Test 140.

It is worth noting that it is still common practice not to report confidence levels for simulation results. This is interesting because it is well known that, for example, real and predicted energy consumption of low-energy buildings is extremely dependent on uncertainties in occupant behavior; as illustrated in Figure 1.10.

User and use related aspects are very often under-appreciated in building performance simulation. In terms of application, for example, simulation is much more effective when used to predict the relative performance of design alternatives, than when used to predict the absolute performance of a single design solution.

In practice it can also be commonly observed that complex high resolution modeling approaches (such as computational fluid dynamics (CFD)) are used for applications where a lower resolution method would be quite sufficient and much more efficient. There is also a wide-spread misconception that increasing the model complexity will decrease the uncertainty of the results. As indicated in Figure 1.11, in reality, deviation from the optimum to either lower or higher complexity increases the potential error in the simulation results.

The above discussion is an element of conceptual modeling, i.e. the process of abstracting a model of a real or proposed system. Robinson (2008) states that:

> All simulation models are simplifications of reality . . . The issue in conceptual modelling is to abstract an appropriate simplification of reality . . . The overarching requirement is the need to avoid the development of an overly complex model. In general, the aim should be: to keep the model as simple as possible to meet the objectives of the simulation study.

The implication of this is that for the same physical artifact (e.g. a building, a façade or an HVAC component) a different modeling approach is to be preferred depending on the objective of the simulation. Hensen (2004) elaborates on this for building airflow related performance studies.

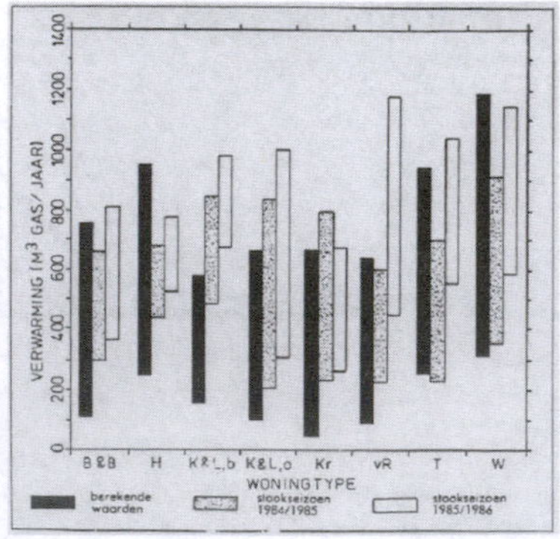

Figure 1.10 Variability in predicted gas use for space heating (black bars) in eight different types of Dutch low-energy houses due to uncertainties in occupant behavior in terms of heating set-point, casual gains and infiltration rates, compared to measured values for 1984–1986 heating seasons.

Source: Hensen 1987.

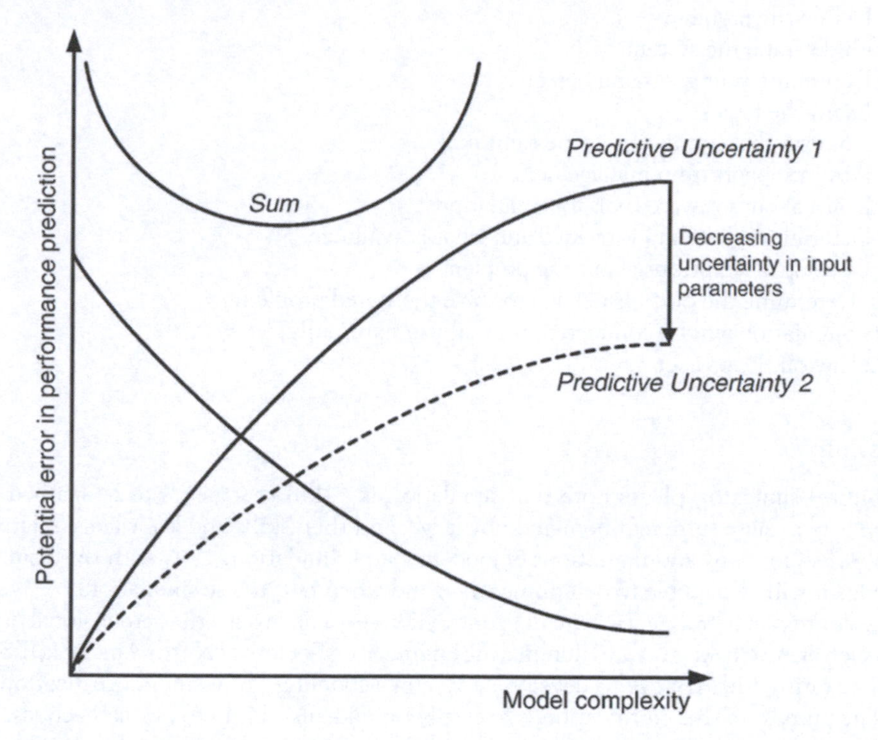

Figure 1.11 Potential error in performance prediction versus model complexity/level of detail.
Source: Trcka and Hensen 2010.

From the above it must be clear that the primary and paramount requirement for quality assurance is sufficient domain knowledge by the user. According to Becker and Parker (2009), we should nevertheless appreciate the distinction between being a subject matter expert in something, being able to describe that thing so it can be simulated, and actually implementing and testing the simulation.

There is a lot to be learned from modeling and simulation in other domains, especially with respect to methodological aspects. Banks and Gibson (1997), for example, relates to electrical engineering, but it is remarkable that there are many cases in our domain in which one or more of their rules "when not to simulate" are applicable. In summary, they say that simulation is not appropriate when:

1. The problem can be solved using "common sense analysis";
2. The problem can be solved analytically (using a closed form);
3. It's easier to change or perform direct experiments on the real thing;
4. The cost of the simulation exceeds possible savings;
5. There aren't proper resources available for the project;
6. There isn't enough time for the model results to be useful;
7. There is no data – not even estimates;
8. The model can't be verified or validated;
9. Project expectations can't be met;
10. System behavior is too complex, or can't be defined.

It also seems worthwhile to consider the application methodology from Banks and Gibson (1996) which involves the following sequential steps:

Step 1: Define the problem
Step 2: Understand the system
Step 3: Determine your goals and objectives
Step 4: Learn the basics
Step 5: Confirm that simulation is the right tool
Step 6: Attain support from management
Step 7: Learn about software tools for simulation
Step 8: Determine what data is needed and what is available
Step 9: Develop assumptions about the problem
Step 10: Determine the outputs needed to solve the stated problem
Step 11: Simulation will be conducted internally or externally?
Step 12: Kickoff the project

Discussion

As a (future) simulator, please note that simulation is a skill that needs to be learned. The first step is to acquire sufficient domain knowledge, and then skills and knowledge relating to principles, assumptions and limitations of modeling and simulation. Only with this combined knowledge it will be possible to determine when and when not, to use simulation.

In the context of user aspects of quality control it is very good to see that professional organizations such as ASHRAE and the Illuminating Engineering Society of North America (IESNA) are collaborating with IBPSA to develop an Energy Modeling Professional certification program. The purpose of this certification is to certify individuals' ability to evaluate, choose, use, calibrate, and interpret the results of energy modeling software when applied to building and systems energy performance and economics and to certify individuals' competence to model new and existing buildings and systems with their full range of physics.

Synopsis

This chapter demonstrates that building performance simulation has the potential to deliver, directly or indirectly, substantial benefits to building stakeholders and to the environment. It also explains that the building simulation community faces many challenges for the future. These challenges can be categorized in two main groupings:

- Provide better design support; issues here include early phase design support, multi-scale approaches (from construction detail to district level), uncertainty and sensitivity analysis, robustness analysis (employing use and environmental change scenarios), optimization under uncertainty, inverse approach (to address "how to" instead of being able to answer "what if" questions), multi-physics (particularly inclusion of electrical power flow modeling), and integration in the construction process (using building information modeling (BIM), process modeling, etc.);
- Provide support for building operation and management, with issues such as accurate in-use energy consumption prediction, whole building (total energy) performance analysis, model predictive (supervisory multi-input multi-output control).

Many (but not all) of these issues are addressed in the following chapters.

Notes

1 Quoting Becker and Parker (2009): "It is common to see the words *simulation* and *modeling* used as synonyms, but they are not really the same thing; at least, not to those in the field bearing those words in its name. To be precise in terminology, a **simulation** enacts, or implements, or instantiates, a **model**.

A model is a description of some **system** that is to be simulated, and that model is often a mathematical one. A **system** contains objects of some sort that interact with each other. A model describes the system in such a way that it can be understood by anyone who can read the description and it describes a system at a particular level of abstraction to be used."

2 Heating, Ventilation and Air-Conditioning.

References

Banks, J. and Gibson, R.R. (1996) "Getting Started in Simulation Modeling", *IIE Solutions*, November 1996, Institute of Industrial Engineers, IIE Solutions.

Banks, J. and Gibson, R.R. (1997) "Don't Simulate When: Ten Rules for Determining When Simulation is Not Appropriate", *IIE Solutions*, September 1997, Institute of Industrial Engineers, IIE Solutions.

Becker K. and Parker, J.R. (2009) "A simulation primer", in Gibson, D. and Baek, Y. (eds), *Digital Simulations for Improving Education: Learning through Artificial Teaching Environments*, Hershey, PA: IGI Global.

Bellinger G. (2004) "Simulation is Not the Answer", Online. Available at: http://www.systems-thinking.org/simulation/simnotta.htm (accessed 15.2.2010).

CIBSE (2005) "Climate Change and the Indoor Environment: Impacts and Adaptation", Chartered Institute of Building Services Engineers, publication M36, London, UK.

Crawley, D.B., Hand, J.W., Kummert, M. and Griffith, B.T. (2008) "Contrasting the Capabilities of Building Energy Performance Simulation Programs", *Building and Environment* 43(4): 661–673.

DB Greentowers (2009) "Deutsche Bank – Banking on Green", Online. Available at: http://www.greentowers.de (accessed 15.2.2010).

DOE (2010) "Building Energy Software Tools Directory", Online. Available at: http://apps1.eere.energy.gov/buildings/tools_directory/ (accessed 15.2.2010).

Fenn, J. and Raskino, M. (2008) *Mastering the Hype Cycle: How to Choose the Right Innovation at the Right Time*, Harvard Business Press.

Flickr (2009) "Lumenhaus 2", Online. Available at: http://www.flickr.com/photos/afagen/4006727335/ (accessed 1.3.2010).

Fraunhofer (2009) "EU-project 'Cost-Effective': Converting Facades of Existing High-rise Buildings into Multifunctional, Energy Gaining Components", presentation by Fraunhofer ISE, ECTP-conference, Brussels, Belgium. Online. Available at: http://www.ectp.org/documentation/Conference2009/D2-03-ResourceandCost-effectiveintegrationofrenewablesinexistinghigh-risebuildings.pdf (accessed 15.2.2010).

Green, H.L. (2009) "High-performance Buildings", *Innovations: Technology, Governance, Globalization* 4(4): 235–239.

Hensen, J.L.M. (1987) "Energieproeftuin: resultaten na twee stookseizoenen", *Bouw* 42(3): 38–41.

Hensen, J.L.M. Bartak, M., and Drakal, F. (2002) "Modeling and Simulation of Double-Skin Facade Systems", *ASHRAE Transactions* 108(2).

Hensen, J.L.M. (2004) "Integrated Building Airflow Simulation", in Malkawi, A. and Augenbroe, G. (eds) *Advanced Building Simulation*, New York: Spon Press, pp. 87–118.

Hopfe, C.J., Struck, C., Ulukavak Harputlugil, G. and Hensen, J. (2006) "Computational Tools for Building Services Design – Professional's Practice and Wishes", in *Proceedings of the 17th Int. Air-conditioning and Ventilation Conference*, Society of Environmental Engineering, Prague, Czech Republic.

Kusuda, T. (1999) "Early History and Future Prospects of Building System Simulation", in *Proceedings of the 6th International IBPSA Conference*, Kyoto, Japan, pp. 3–15.

Lain, M., Bartak, M., Drkal, F., and Hensen, J. (2005) "Computer Simulation and Measurements of a Building with Top-cooling", in *Proceedings of the 9th International IBPSA Conference*, Montreal, Canada, pp. 579–586.

Linden, A.C. van der, Boerstra, A.C., Raue, A.K., Kurvers, S.R. and de Dear, R.J. (2006) "Adaptive Temperature Limits: A New Guideline in The Netherlands: A New Approach for the Assessment of Building Performance with respect to Thermal Indoor Climate", *Energy and Buildings* 38(1): 8–17.

Loonen, R.C.G.M. (2010) "Overview of 100 Climate Adaptive Building Shell Concepts", MSc project report, Eindhoven University of Technology, The Netherlands.

Lund, H. and Mathiesen, B.V. (2009) "Energy System Analysis of 100% Renewable Energy Systems—The Case of Denmark in Years 2030 and 2050", *Energy* 34(5): 524–531.

Lysen, E.H. (1996) "The Trias Energica: Solar Energy Strategies for Developing Countries", in *Proceedings of the Eurosun Conference*, Freiburg, Germany.
Petersdorff, C., Boermans, T. and Harnisch, J. (2006) "Mitigation of CO2 Emissions from the EU-15 Building Stock. Beyond the EU Directive on the Energy Performance of Buildings", *Environmental Science and Pollution Research* 13(5): 350–358.
Robinson, S. (1999) "Simulation Verification, Validation and Confidence: A Tutorial", *Transactions of the Society for Computer Simulation International* 16(2): 63–69.
Robinson, S. (2008) "Conceptual Modelling for Simulation Part I: Definition and Requirements", *Journal of the Operational Research Society*. 59(3): 278–290.
Rogers, E.M. (1995) *Diffusion of Innovations*, New York: The Free Press.
Spitler, J.D. (2006) "Building Performance Simulation: The Now and the Not Yet", *HVAC&R Research* 12(3a): 549–551.
Thomas, V.C. (2006) "Using M-E Design Programs (some reasons for the lack of progress)", Presentation at IBPSA-USA Winter Meeting 2006, Chicago, IL.
Torcellini, P.A. and Ellis, P.G. (2006) "Early-phase design methods", Center for Buildings and Thermal Systems at National Renewable Energy Laboratory (NREL).
Trcka, M. and Hensen, J.L.M. (2010) "Overview of HVAC system simulation", *Automation in Construction* 19(2): 93–99.

Recommended reading

We recommend that you read the rest of this book, but not necessarily from the beginning to the end. Read the chapters according to your personal interest. This book aims to provide a comprehensive and in-depth overview of various aspects of building performance modelling and simulation, such as the role of simulation in design, outdoor and indoor boundary conditions, thermal modelling, airflow modelling, thermal comfort, acoustics, daylight, moisture, HVAC systems, micro-cogeneration systems, building simulation in operational optimization and in building automation, urban level modeling and simulation, building simulation for policy support, and finally a view on future building system modelling and simulation.

The structure of each chapter is the same. It starts with identifying the scope and learning objectives. This is followed by introduction, scientific foundation, computational methods, application examples, discussion, and synopsis. As we are limited by space constraints, each chapter includes recommendations for further reading. In order to practice what has been learned, each chapter finishes with some suggested activities or assignments.

Activities

After reading (parts of) this book, please email your feedback to j.hensen@tue.nl and lamberts@ecv.ufsc.br.

2 The role of simulation in performance based building

Godfried Augenbroe

Scope

This chapter explores how simulation supports performance based building through different lifecycle stages, with a focus on the design stage. As design evolves, a building project invokes many intermittent and recurring dialogues between stakeholders that engage in the definition and subsequent negotiation of performance requirements and their fulfillment. This requires, among others, an agreement about how to quantify requirements, i.e. how to choose adequate "measurement" methods. Measurement methods not only quantify performance criteria but also de facto define them (through their measures). The aim of this chapter is to show how an objective specification of performance measures and measurements can be derived without any subjective bias in their interpretation. In any real life design case, criteria span many types of performance aspects. As these need to be dealt with simultaneously, the design decision-making requires the careful balancing of different, and sometimes conflicting, performance targets. It is shown that a rigorous, system-theoretic definition of performance indicators is necessary to prepare a rational decision process. The treatment in this chapter explores the role of building simulation in the quantification of the performance criteria.

Learning objectives

Simulation does not exist in a vacuum. It has many purposes and one particular purpose, the support of design evolution, is discussed in this chapter. If simulation is used in performance based building it is essential to position it in a decision making framework. In doing so, this chapter has the following learning objectives:

- Understand the essence of building simulation as a "virtual experiment" and recognize the importance of the role of the experiment in studying and quantifying building behavior of two competing design options.
- Understand the need to define quantifiable expressions of performance to objectively compare design options; "raw" output data of simulations is typically not adequate to serve this purpose.
- Understand the basics of a system framework that identifies Aspect Systems as building systems that respond to a particular requirement, and which are hence subjected to a particular simulation in order to study building behavior in relation to the stated requirement.
- Recognize that every real life decision requires a multi criterion approach where some criteria may need simulation for their quantification, while others may rely on other methods.
- Be able to analyze a design problem, determine the important criteria and prepare multi-criterion decision making.
- Recognize the potential role of uncertainties in the outcome of building simulation and have general knowledge about how it may influence decision making.

Key words

Performance criteria/performance indicators/decision making/design evolution/system framework

Introduction

This chapter provides an engineering perspective on performance based building design. Statements of performance play a major role in expectations expressed by owners and future occupants. Traditionally, design solutions were driven by prescriptive terms, i.e. specifying properties of the solution rather than the expected performance of the solution. Building codes and regulations have long contributed to this by adopting prescriptive specifications. This is no longer the case as many countries are moving parts of their regulations and supporting standards to the performance domain (Foliente 2000), and recent research networking efforts have produced the roadmap towards their broad introduction in the industry (http://www.pebbu.nl/resources/allreports/). The key ingredient of performance based building is the up-front formulation of statements of performance requirements, and the subsequent management of a process that guarantees their fulfillment, i.e. through dialogues between designers, engineers and building managers. Designers seek to fulfill the client's performance requirements, but without a proper framework for the definition of measures and measurement methods, the dialogues will suffer from many disconnects between demand (client) and supply (design engineering service providers) side. In current practice, these disconnects are rampant throughout the whole building delivery process especially where the expression of expectations and the expressions of fulfillment have no formal basis. Systematic procedures for matching fulfillments with expectations is therefore considered an important target of the building industry to become more client-driven and provide better customer satisfaction and better value overall. The need to move the industry in this direction has fostered the introduction of performance based building methods (Foliente *et al.* 1998). Building teams are also adopting integrated project delivery, an approach which involves a greater number of stakeholders making design decisions earlier in the building procurement process (Barrett 2008). A performance based approach is a key enabler of rational decision-making across many stakeholders and based on a large set of performance criteria. Building simulation is the key tool to quantify performance criteria that inform decisions. Building simulation models and observes the building's behavior under a specific usage scenario (Augenbroe 2004) as depicted in Figure 2.1. A piece of reality is translated into a model, which is then studied in a variety of experiments (simulation runs) in an "experiment box" (the simulation tool). The experiment is set up to generate observable states that reveal something relevant about the behavior that contributes to the performance under study. It should be noted that the design of the experiment is far from trivial. It must be such that the observed states generated by the experiment reveal indeed the meaningful behavior that aggregates into the measure that defines the performance criterion and its quantification. It is obvious that this requires a deep understanding of the physical domain, the performance measure and the experiment box.

As Figure 2.1 shows, the performance criterion under study is at the heart of the experimental set-up: it demands a specific model, controls the set-up of the experiment and determines the choice of performance quantification. Looked at it this way, simulation is no longer the art of performing high fidelity simulations, but rather the art of performing the right type of virtual experiment with the right model/tool. In fact building performance simulation is no longer confined to the computational tool, but covers a process that starts much earlier and ends later. The main steps in this process are (1) agreement about performance criteria, (2) agreement about ways to measure them in order to quantify required and fulfilled levels of performance, (3) making rational design decisions that consider client preferences and trade-offs between potentially conflicting or hard to meet performance targets within the imposed time and budget limits. There are no well established step-by-step methods to guide client and design teams through this process, although progress has been made in supporting the second stage of the process in recent years (Becker 2008). Some authors have specifically dealt with the first stage of the process from different perspectives. Huovila *et al.* (2005) report the development of EcoProp as a specific tool for client involvement in the early stage; whereas Verheij (2005)

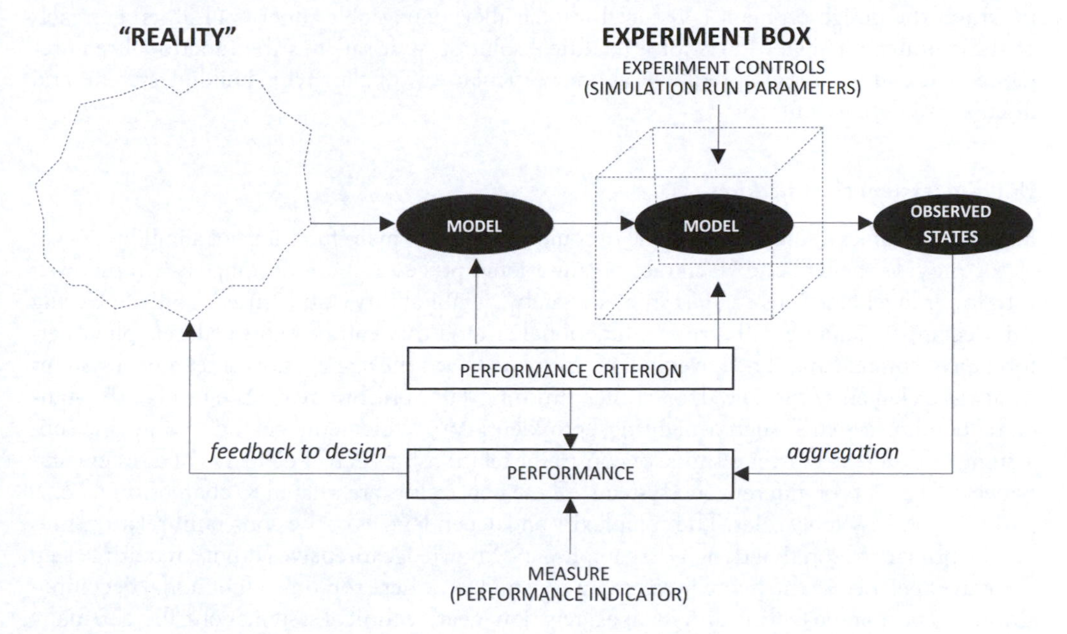

Figure 2.1 Performance testing requires a (virtual) experiment.

focuses on dedicated dialogue support between stakeholders in the discovery and agreement finding on performance criteria and related issues and conflicts to be considered. He asserts that the asymmetry in knowledge between client and design team members requires a rich platform that can structure the dialogue in a goal-driven way, reaching consensus about criteria and measurement. Ozsariyildiz (2006) advocates tools that remove the barrier between client and design team altogether thus blurring the distinction between demand and supply side, but in many situations this will not be practical or feasible. Things are negatively affected by the lack of transparent demand-supply roles in the building process (Beard *et al.* 2001) which explains many of the ineffective dialogues between clients and design teams in current practice. We argue that this is in part due to the lack of a framework for the categorization of functionality and their mapping into sets of performance criteria with well defined measures. The third stage of the process has received a lot of attention as multiple-stakeholder, multi-criteria decision making has become a fashionable research topic. Tools like analytical hierarchy process (AHP) (Saaty 1996), and utility theory (Quiggin 1997) are now routinely used to arrive at an optimal decision. Hazelrigg (2003) offers an interesting view on the adequacy of these approaches for rational decision making.

Scientific foundation

Dialogues about performance between stakeholders require the systematic introduction of functional requirements, their specialization in terms of performance criteria, and their translation into individual performance requirements. Each individual requirement must be accompanied by a measure and measurement method. The actual measurement outcome is represented by a "performance indicator" or PI. There is no established theory (e.g. formal classification) for the definition of performance criteria at whole building scale although several general schemes have been developed mostly linked to research networking activities initiated by the International Council for Building (CIB); the outcomes of the Pebbu networking effort can be found on http://www.pebbu.nl/resources/allreports/. Some attempts at a fundamental approach are worth mentioning. Inspired by design theories of Herbert Simon and his followers, Wim Gielingh in his General AEC Reference Model (Gielingh 1988) hypothesizes a design process in which the design problem is reduced into smaller manageable functional units. For each of these units, a designer looks for a technical solution that satisfies the functional requirements of the unit. This chapter looks at one particular way of classifying building systems and discusses its deployment.

Building system classification

Figure 2.2 assumes a similar functional decomposition. The main functions of a building (such as to "provide shelter" and "facilitate organizational processes") are decomposed (top-down) into lower level functions such as "provide safety", "habitability", and "sustainability". During this decomposition one will arrive at functional criteria that can be expressed as explicit performance requirements. The lower half of the figure shows the aggregation of technical systems from basic elements (window, door, chiller, finishing, etc.) until we reach the level of the standard "building systems" such as lighting, sprinkler, HVAC. Elements belong to only one subsystem, but there exists a multitude of constraint relationships across elements of a system and between elements of different sub-systems. These constraints are related to composition, compatibility, and assembly rules. The complexity and dependencies of the constraint relationships are the primary reason that design is a complex and knowledge intensive activity, requiring deep domain expertise. As the figure indicates, in the middle, where top-down functional decomposition and bottom-up technical systems aggregation meet, technical systems contribute to many building functions and building functions are supported by many technical systems.

The mapping of a design program into functional requirements at increasing levels of detail is a complex job, not supported by a well-established methodology. The closest methodologies come from the systems engineering world, where requirements engineering has been developed to an advanced level. Gilb (2005) for example presents one of the most insightful contributions to the field, although it must be stated that the application of his thinking requires re-invention and re-discovery in every project only supported by experience. It should be noted that building requirements engineering, and fundamental thinking about criteria and their management, is not taught, at least not a the fundamental level, in building design schools. For the bottom half of the picture, the situation is quite different. Building design students are trained

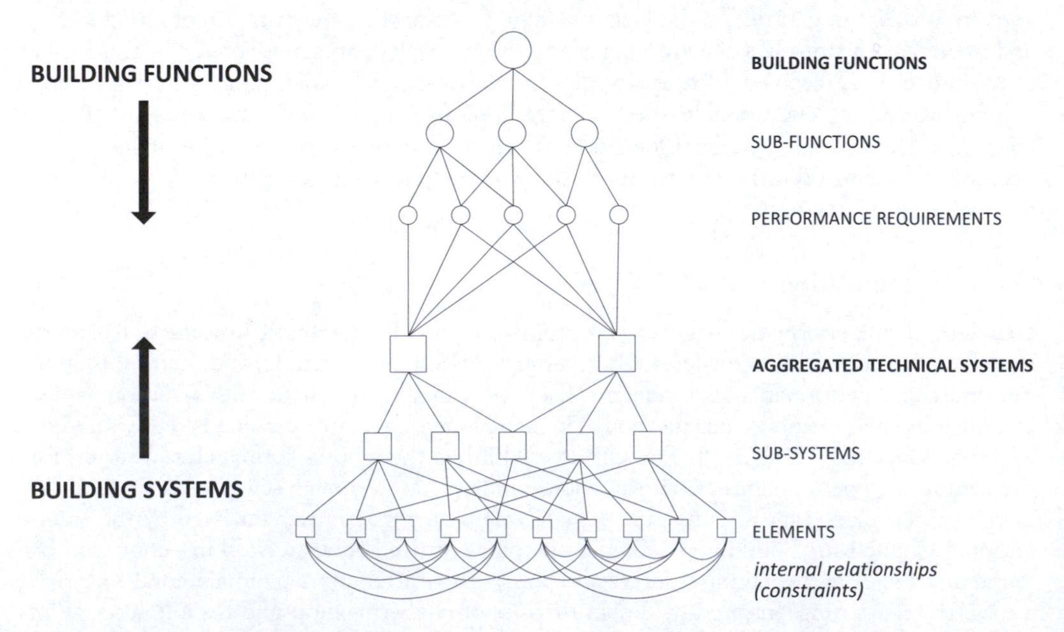

Figure 2.2 Top-down functional decomposition and bottom-up assembly of building systems.

to develop integral design solutions from existing part solutions. Most of the attention goes into recognizing the complexity of system integration, and knowing when to develop new solutions, i.e. when it is necessary to depart from the existing systems. Again, there are few authors that have tried to establish a systems integration theory that can be applied to practical design problems. The systems integration handbook (Rush 1986) and studies by Habraken (1998) are good guiding rods for practical and theoretical work in systems integration that over time could support the lower half of the figure better.

In spite of the lack of a workable framework for both halves of Figure 2.2, it is still useful to use the figure to understand and discuss the process of matching of performance requirements with design proposals. In fact, the center of the figure depicts where demand and supply traditionally interface in building projects. It is therefore relevant to use this figure as a template to explore the role that simulation plays in this interface.

First, it is necessary to understand performance requirements, i.e. what can and should be formulated as performance statements? The goal of specifying required performance, rather than prescribing properties, is to guarantee that a building system is capable of fulfilling the functional requirements at a defined level of performance. There are seven main concepts in the performance approach (Foliente *et al.* 1998; Vanier *et al.* 1996), as identified in Figure 2.3:

1. Goal (occupancy, or use) is usually a qualitative statement that addresses the needs of the user or consumer and determines the required level of performance.
2. Functional requirements are the mandatory requirements that must be fulfilled to ensure users are satisfied with the facility.
3. Performance requirements are the user requirements expressed in terms of the performance of a product. Performance requirements are "measured" by PIs.
4. PIs are quantifiable indicators that adequately represent a particular performance requirement. A PI, by definition, is an agreed-upon indicator that can be quantified specifying an agreed-upon (and sometimes standardized) measurement method.
5. Verification methods are used to evaluate whether performance requirements have been met. Verification methods can be experiments (tests), calculations, or a combination of both. Each PI has its own measurement method. A number of different measurement methods may exist for the same performance criterion, i.e. one criterion may have alternative PIs attached to it. Indeed many different ways to measure a certain performance can be introduced, each with its own merit and each requiring a different experimental set up or different model or different aggregation method. An elaboration of this discussion can be found in (Augenbroe *et al.* 2004).
6. Functional ("Aspect") Systems are parts of a building that fulfil a function. Elements of Aspect Systems can be any building element.
7. Agents alter the ability of functional elements to satisfy the functional requirements.

Figure 2.3 shows how the seven concepts are related to each other. The small filled circles indicate the cardinality of the relationship, where the cardinality could be specified as one-to-one or one-to-many. The latter detail is not shown in the figure, but most cardinalities can be assumed to be one-to-many. For example, *one* Performance Requirement can be measured by *many* Performance Indicators, whereas every Performance Indicator is related to exactly *one* Verification Method for its quantification.

In order to determine whether a particular design satisfies a performance requirement, PIs are used to quantify the performance requirement. Context information needed for the verification method is provided by the Aspect System that represents all the properties and all the relationships of all the elements in that system. Although this framework provides a useful model for the classification of performance related concepts, it is not a generic classification of building functions. With the infinite variability that characterizes the current style of

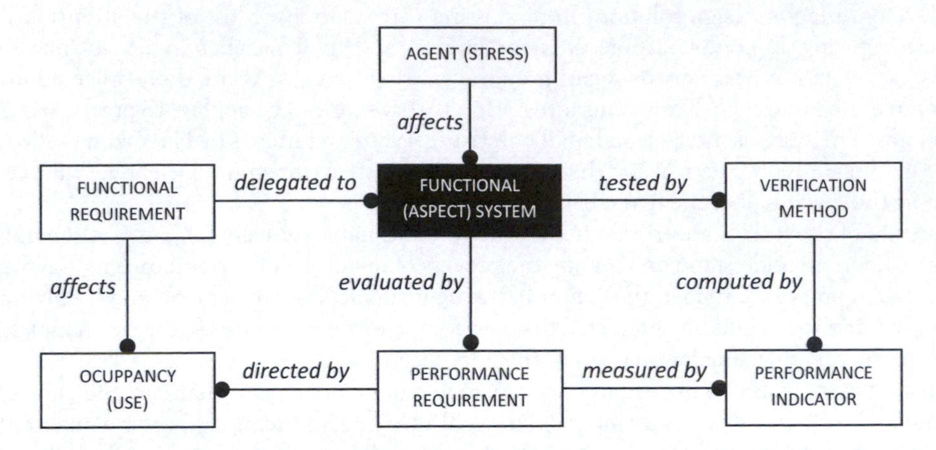

Figure 2.3 Relationship among concepts in the performance approach.

Source: Adapted from Vanier *et al.* (1996)

building design, it is doubtful whether such a functional classification can be made with sufficient generality, and moreover one should ask whether this would be of much practical use. In current practice, performance dialogues are typically generated in a "middle-out" fashion. Indeed, performance requirements are defined when certain design issues come into focus, i.e. when pertinent issues need to be resolved and decisions between discrete design variants need to be made. At that point, one is dealing with a small subset of functional requirements and a limited (but not small) subset of systems of Figure 2.2. At the interface of requirements and systems, all the concrete stakeholder dialogues with respect to expected and fulfilled performances take place. It is important to support these dialogues with quantified expressions of performance, and to identify the role of simulation in the quantification methods.

The role of simulation in the system framework

A closer look at Figure 2.2 reveals that the role of simulation is still not well defined. In fact, it is not clear at all *how* at a certain level of aggregation, a technical system relates to a functional requirement. The many-to-many relationship at the interface between performance requirement and technical systems is too vague to identify the appropriate virtual experiment. The example of visual comfort will clarify this. Figure 2.4 shows a particular decomposition until the level is reached where the visual comfort criterion is addressed. Understanding the physics of the problem (i.e. the building behavioral aspects of visual comfort) the following technical systems can be identified as having a role in producing the desired level of visual comfort. In parenthesis we state the most important elements or parameters of these systems impacting on the realization of visual comfort:

- external enclosure system (elevation: type, size and location of windows, shading devices, shading control)
- neighborhood system (nearby buildings, vegetation)
- electric lighting system (type of fixtures, location, control)
- internal enclosure (surface finishes (color, reflectivity))
- organization system (workplace location and orientation, internal partitions height and finishes).

The implication is that a collection of elements and element parameters, from diverse technical systems, is responsible for visual comfort. Together they form a functional system (Figure 2.4),

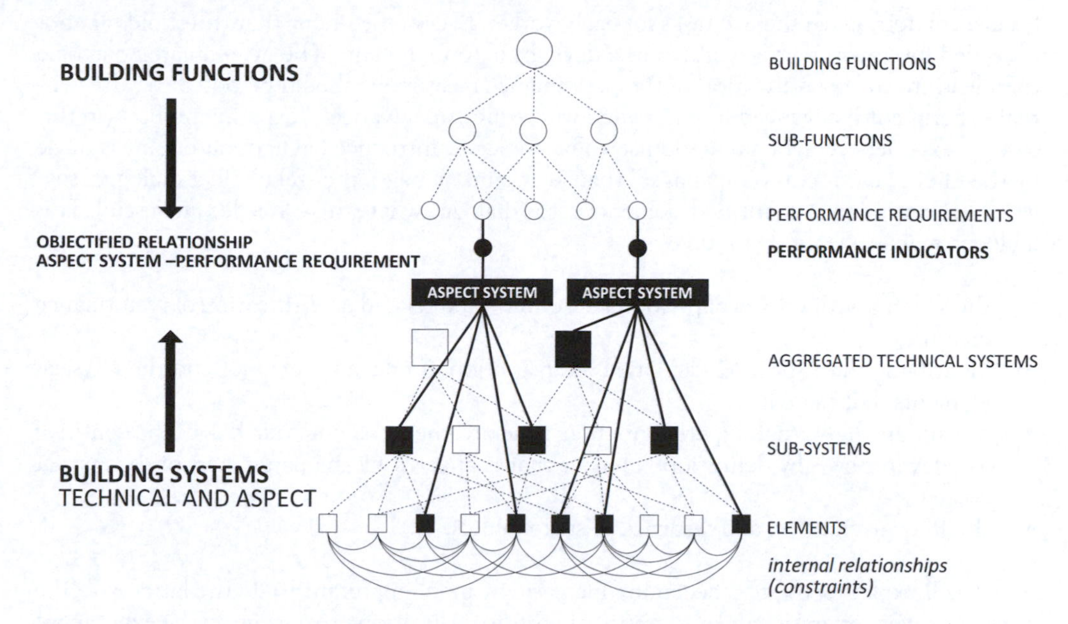

Figure 2.4 Aspects systems are performance criterion specific aggregations over technical systems.

which can be defined as a collection of elements from diverse technical (sub)systems. The choice of elements is determined by checking whether they contribute to building behavior that is relevant for a certain function. As introduced above, we denote this functional system as "Aspect System". The implication is that a quantification method (in most cases a simulation) for a performance requirement operates on an Aspect System that we compose in the way indicated above. In principle there are as many types of simulations (operating on a corresponding Aspect Systems) as there are performance criteria to be fulfilled. Figure 2.4 shows that each functional requirement can be linked to, per definition, exactly one Aspect System. The relationship between a functional requirement and the Aspect System can be "objectified" into the PI that we already introduced. The quantification of a PI is delegated to a virtual experiment as explained above, i.e. involving a simulation, or normative calculation, or any type of objective quantification method. To continue the example: assume that among the performance requirements linked to visual comfort is daylight autonomy, made quantifiable through a PI, one of many PIs related to visual comfort. As long as there is no standard available that defines the measurement method for daylight autonomy, each project will have to negotiate an accepted measurement method between involved stakeholders. Let's assume that all parties have agreed on the following PI: "the average percentage of office hours in the month of March that all workers (at their workstations) in a given space get sufficient daylight to perform required eye tasks with certain ease and without the need for artificial lighting". This implies that the unit of the measure is [%]. The quantification can be done with a variety of lighting simulation tools (for an overview of tools, see http://apps1.eere.energy.gov/buildings/tools_directory/) none of which offer a daylight autonomy measure as the outcome of a "standardized" verification method, i.e. a verification method that leaves no room for interpretation of how the measure is defined and how its quantification should be conducted. One can select a lighting simulation tool that takes the effect of all elements of the Aspect System into account, run hourly calculations for local average solar load and sky conditions, take shading configurations, surrounding buildings, vegetation, and ground reflection into account, to calculate the incident solar shortwave radiation on the fenestration. The next step is to calculate luminance, on desks and/or computer screens (given the workstation arrangements and worker tasks expressed in the office system as well as the reflectance of internal surfaces), and use accepted thresholds for eye comfort, given the eye tasks for each worker. Every office-hour that thresholds are not exceeded for any worker is counted as a daylight autonomy hour. The accumulated outcome then leads to the quantification of the performance measure. It should be noted that this virtual experiment is advanced in nature and will require an advanced "experimentalist" (in this case an experienced light simulationist). The choice to introduce daylight autonomy is made by the client and a "criteria manager" that steers and manages the stakeholder dialogue, suggesting the infusion of quantified elements in the dialogue where this is needed and useful. This dialogue management role involves:

- the systematic discovery of performance criteria and related quantification of performance measures
- identifying the Aspect Model based on inspection of relevant behavior and the physical elements that cause it
- setting up the virtual experiments that generate the behavior that has been identified as relevant, by is by definition behavior that is affected by the parameters of the Aspect System
- the PI quantification and feedback to stakeholders.

As we will explain soon, it is becoming increasingly more important to do the latter with the disclosure of uncertainties and assumptions in the quantification process, i.e. in the experiment that leads to the quantified PI.

Computational methods

The art of modeling and simulation is leaving things out that don't affect the answer (Sokolowski and Banks 2009). Translating this to performance simulation in the light of the previous section leads to a number of observations that are explored below.

Observation 1: Design decisions are never dealing with one single performance aspect; design is always inherently about satisfying many performance requirements at the same time. This implies that multiple stakeholders are engaged in the dialogue, and that performance verification deals with many performance criteria simultaneously. Informing the decision-making while recognizing trade-offs between different criteria is the primary objective. Delivering the most thorough and most detailed simulation in itself is not. Attempting the "best" simulation may in fact be counterproductive as it will often result in over-engineered models and time consuming simulation runs. There are three issues that determine the choice of simulation level: the complexity of the performance measure, the sophistication of the physical model, and the extent/scope of the Aspect Model. The elaboration of the daylight autonomy example can serve as explanation:

- One could attempt a simpler performance measure, such as the office time averaged illuminence level on a standard work plane in the center of the space; this would represent a particular case of a *simplified calculation* based on the consideration that this simpler measure would be equally well suited to support the pertinent design decisions.
- One could use a cruder lighting model that does not deal explicitly with the effect of internal reflections; for instance, one could calculate the hourly radiation coming into the room (Q_{tot}) and use an empirical formula producing the average illuminence value as a function of Q_{tot}, Depth of the office space, average reflectance. This approach represents a *normative approach* because the calculated outcome may not necessarily represent a physical (experimental) outcome, but is rather used as a surrogate for a true outcome. This approach is defendable when one can prove or ascertain that the value of the surrogate outcome has a strong correlation with the true outcome that quantifies the performance. Standardization bodies typically choose this route because normative calculations are

transparent and introduce no bias related to the individual expert or hidden assumptions in the experiment simulation tool. The development of a normative performance toolkit is described by Augenbroe and Park (2005).

- One could reduce the extent/scope of the Aspect Model, for example by ignoring surrounding buildings, or by ignoring the office system parameters. This is different from the previous two choices as a reduction of the Aspect Model, usually leads to *conditional statements of performance compliance*, e.g. of the type "PI (x | empty room | no surrounding buildings) meets $PI_{required}$" which should be read as "the performance for given design x, under the condition of empty room and no surrounding buildings, meets the performance requirement". In reality one will usually condition the outcome on a worst case approach, i.e. "PI (x | fully furnished room | high urban density of surrounding buildings) meets $PI_{required}$". This approach represents a safety factor as we usually apply to limit-state performance requirements, for example in structural safety requirements conditioned on worst case load scenarios.

Each of these choices leaves something out and thereby reduces the complexity of simulation or calculation that is used for the performance quantification. This can only be justified if there is proof that outcomes based on these reductions still inform design decisions adequately. In many cases simulation only needs to be adequate for comparative analysis of design variants. In such cases there is no high demand for absolute accuracy but good reflection of changes of design parameters in the outcomes is paramount. All of the above reduces the demand on the sophistication of the simulation tool and the needed expertise of the user.

Observation 2: performance quantification and comparisons take place in various stages of design, i.e. from conceptual design when the elements of the Aspect Model are not well developed yet, to late design when most design choices have been made and hence most of the elements and parameters of the Aspect Model are known. Research has generated a variety of tools that claim to support early design decisions, but very few if any are used in practice. Practice usually regards them as naïve, or not integrated, or too simplistic. It could be argued from a theoretical perspective that these tools have a fundamental flaw. To elucidate this, it should be understood that performance quantification in the conceptual design stage is by definition a conditional statement as explained above. Indeed, by definition the calculated performance is conditioned on the fact that certain assumptions about as yet unknown system parameters remain valid throughout the design process. Although this is in no way an eye opener, there are few researchers that realize that none of the early design tools attempt to express this properly (i.e. as probability distributions) and the user is hence left to guess what the results really mean. Sanguinetti *et al.* (2009) show in a façade design study that explicit statements about conditional performance can be formulated if one uses conditional probabilities of ensuing design variations, assuming one can make a reasonable guess about the uncertainties of as yet unknowns. The paper shows that such statements express the true nature of uncertainty in early design decisions. It is also shown that this uncertainty more often than not overwhelms the separation in performance between competing design options in such a way that the results of the early design simulation are close to irrelevant when it comes to make early design decisions.

Observation 3: irrespective of the design stage, every time a performance verification uses simulation to approximate (and forward-project) the real behavior of a building, one has to confront the fact that no building ever behaves as its model *(which is de facto a design idealization)* predicts. Many uncertainties play a role when constructing the model. Unfortunately, the state of the art of current simulation is that uncertainties are rarely dealt with explicitly, hence parameter values are based on best guesses or (if some statistical data is available), expected mean averages at best. The reality is that best guesses are just that, and fraught with assumptions and expert bias and in some cases informed by bad or incomplete science. In the actual realization and operation of a building, many things will not perform as expected. Compounding the effect of model uncertainties, unexpected operation, physical parameter variations, unpredictable deteriorations of systems etc, any outcome generated by a deterministic simulation has to be regarded as an arbitrary sample from a large statistical distribution. Figure 2.5(a) shows the example of a deterministic result for design options A and B. Figure 2.5 (b) shows the result of the same simulation model with uncertainty analysis for the same design options. It needs little explanation that the deterministic preference for A over B, has only limited validity when the full range of possible outcomes under uncertainty is known. Indeed as the figure shows there is a considerable chance that A underperforms B. A corollary of this treatment is that the original performance requirement may have no meaning. After all, requiring that a PI for a proposed design (x) should surpass a minimum value PI_{min} makes little sense when PI(x) is a stochastic variable.

The natural extension is therefore to formulate risk-oriented performance requirements, for example in the following form, $P(PI(x) < PI_{min}) < 5\%$ which expresses that the probability that the design performance does not meet the minimum required value shall be less than 5%. It is fairly straightforward to propagate the uncertainty ranges of parameters through a simulation model using a Monte Carlo approach as recent studies have shown (De Wit 2001; De Wit and Augenbroe 2002; Hopfe 2009; Hu and Augenbroe 2009; MacDonald 2002; Moon 2005). More general forms of stochastic simulation should replace the MC approach where appropriate. The bottleneck in these studies is the a priori estimates of uncertainties and there is clearly a need for more fundamental research in this area. The uncertainty analysis of performance should contain the screening of dominant parameters (Saltelli *et al.* 2008), and an iterative improvement of the a priori estimates is necessary in those cases where estimated uncertainty ranges in certain parameters are shown to have a dominant effect on the resulting under-performance risk. It can be said that the limited knowledge about causes and magnitude of uncertainties and deviations from expected behavior, causes of anomalistic behavior of building systems, and lack of anticipation of the large variability of operating conditions pose a severe handicap to performing risk-based performance assessment and verification on a large scale.

Observation 4: The Aspect System view can be cast in a conceptual data model (Aspect Model) of all the elements and element properties that contribute to a certain performance criterion. It is, in other words, a formal representation of the input model for the associated performance quantification method. If this quantification requires simulation, the Aspect Model is in fact the formal representation of the input model supplied to the simulation tool to perform the experiment on which the quantification is based. This insight has a number of consequences:

- Testing the adequacy of simulation tools: the current generation of simulation tools were not developed from the theoretical perspective to quantify PIs. Instead, they were developed to carry out virtual experiments in the generic sense, i.e. with emphasis on the experiment box rather than on the experiment. The consequence is that simulation tools have defined their own representations of buildings, in most cases not explicitly specified.

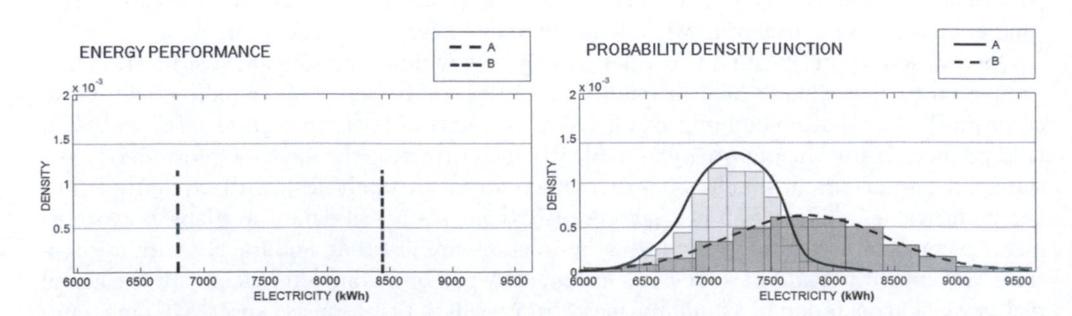

Figure 2.5 Deterministic versus probabilistic performance outcomes.

Future generations of building simulation should be developed with the explicit goal to support specific experiments and hence support explicitly defined Aspect Models. Current simulation tools could be pre-tested by checking a formal representation of their input model against the Aspect Model. This was the fundamental approach of the Design Analysis Integration project (Augenbroe and de Wilde 2003; Augenbroe *et al.* 2004) which set up simulation tools to respond to specific performance quantification requests. The key to this was the pre-definition of the Aspect Model and an automatic mapping from overall building model to local Aspect Model.

- Many recent efforts map building data in a Building Information Model (BIM) (http://www.iai-tech.org/) to the input models for simulation. In doing so they are mapping to the Aspect Models that we have defined in Figure 2.3. From the very nature of the Aspect Model, these mappings are hard to support in a generic way. The main reason is that currently the BIM does not support explicit performance criteria and associated Aspect Models. In order to make the mapping work, BIM interfaces are developed to "roam" technical building systems (they are well supported in the current BIM) and to pick and choose the elements and their properties that should be part of a particular performance assessment, i.e. quantification method. Much of this work would become obsolete if a future generation of the BIM would offer explicit support for Aspect Models. Given the vast amount of resources that would have to be spent on this, the more likely scenario is that future mappings will be driven by an explicit specification of Aspect Models and where interactive user augmentation of the mapping will be better supported.

In summary we can conclude that the systematic framework introduced on pages 17–22 enables us to identify the role of simulation in design performance assessment and moreover, identify the crucial choices that one has in defining different measures and the appropriate simulation method for their quantification. Moreover, the role of the Aspect System in identifying all information that describes the model for the virtual experiment cannot be underemphasized in the development of future Building Information Models.

Application examples

This section gives an account of the application of the performance dialogue process on real-world projects. The focus is to show how design problems are governed by performance criteria in multiple domains. Typically, only a subset of those can be quantified with existing simulation tools. Other criteria require the introduction of specific customized PIs for which often no established measurement method exists. The purpose of this section is to show that the role of simulation in performance design is still limited. The examples serve as an agenda to enlarge the scope of building simulation and become more universal in the support of performance dialogues in the design phase. Every example deals with the choice between two discrete design options. Our treatment aims at a comparison of the two options in such a way that a decision process that involves different stakeholders with different value systems could be effectively supported.

Case 1: Choice between granite stone and aluminum cladding

Case 1 background: this case is part of a range of design choices for an office building. The location of the project is in a city with high pollution and in a climate zone with severe winters and hot summers. The client wanted the building to be energy efficient, and to have a building cladding material that would be durable and requiring little maintenance. The client's needs focused on the durability of the material, i.e. the wall cladding should be able to withstand pollution and should not change in color, texture and tensile properties for at least a time period of 50 years.

Table 2.1 Case 1 Criteria and performance indicators (subset shown)

Performance Criteria	*PI #*	*PI Name/Description*	*Quantification Method*
Durability	1	Defacement because of weather and pollution	Assessment based on industry and manufacturer's average (based in part on durability experiments)
	2	Product's service life	Assessment based on industry and manufacturer's data; service life prediction.
Energy Performance	3	Energy required for heating and cooling	Building performance normative toolkit; produces accurate comparative energy consumption results
Constructability	4	Recycling capacity	Based on Eco-Impact calculator from ATHENA institute
	5	Waste production	Assessment based on industry data (normative)
Maintenance	6	Time taken to lay out stone	Calculation of time and people required to set up X m^2 of wall cladding (based on contractor survey)
	7	Replacement	Use of Markov Chain maintenance intervention matrix

Table 2.1 shows the criteria, PIs and quantification methods chosen in this case. The selected performance criteria are a reduction of an initially longer list offered initially by the client. The reduction took place in a dialogue with the designer to select those that were deemed most relevant in the choice between the two competing materials. It was left to the architectural engineer to come up with adequate PIs for the selected criteria and select the appropriate quantification methods. It is worth noting that no simulation was used in this case. The energy performance calculation was done using a normative calculation tool (Augenbroe and Park 2005), which was adequate for the comparative analysis. All other PI calculations were based on normative approaches as well.

Figure 2.6 shows the Aspect Systems for each of the performance requirements. The black boxes show the macro elements of the two competing technical systems. Three types of lines are used to show which element is part of which Aspect System. Figure 2.6 is used in the

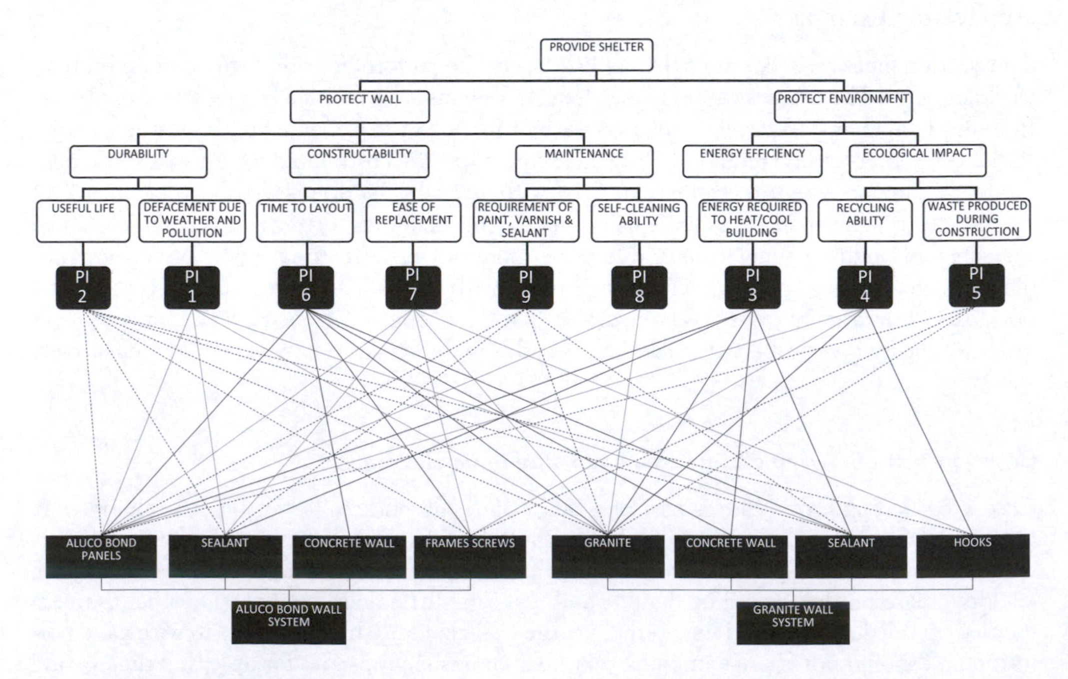

Figure 2.6 Aspect systems.

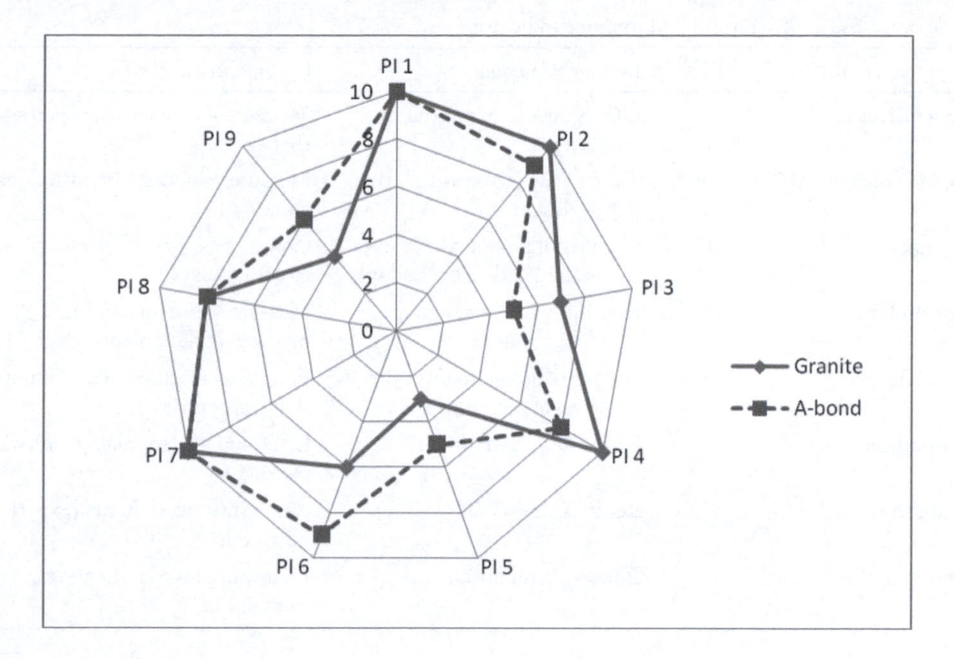

Figure 2.7 Normalized results.

quantification as schedule for the data (element properties) that need to be available to support the performance quantification. In practice this is a significant help in servicing the different consultants with the information needed to perform their assessment. As debated earlier, the explicit specification of the Aspect Models for the quantification method could also be used to drive constructive BIM interfaces, i.e. generating the input data from existing BIM data.

After normalization of the PI scales the radar chart in Figure 2.7 shows the outcomes.

Granite had been the first choice, not based on performance but merely on availability and cost. In the performance assessment, both products fare well on some performance indicators, but not all the criteria are met. Both the products are viable in terms of cost and availability and the skills required for installing the cladding systems. Both products have a service life of at least 50 years, and can last longer with routine cleaning and maintenance. In the assessment, the granite wall systems causes more air pollution during its lifecycle than Aluco Bond aluminum system, but it is saves more energy. The aluminum system is cheaper and quicker to install. It turns out that, based on a weighted average of the indicators in Figure 2.7 the aluminum cladding option offers the better choice.

Case 2: Choice between two apartment layouts

Case 2 background: the main design intention in this case was to offer a variety of apartment layouts within a single residential building. Because of the code regulations in the city where the project is located, the buildings' capacity (number of units) became constrained by the number of parking spaces provided. Two design options offering the highest number of car park units were evaluated, as point of departure, to develop the rest of the building, including the number of possible apartment types and façade variations with an apartment type.

Table 2.2 shows the criteria and their quantification method. For each performance criterion multiple PIs were defined representing alternative methods to quantify the criterion. Most of the resulting PIs within one criterion differed in their way of calculation, i.e. simulation or normative, low resolution or high resolution, first order approximation or high fidelity simulation. The last column shows the PI calculation method eventually used to inform the decision

Table 2.2 Case 2 Criteria and performance indicators

Performance criteria	*PI #*	*PI Name/description*	*Quantitative method*
Area efficiency	1–5	Utility of design in terms of area usage	Measure of geometric properties of shapes
Façade constructability	6–9	Time needed to construct the façade design	Measure provided by contractors and suppliers
Capacity	10–12	Number of tenants that can be accommodated in the building	Count of car park units, shops, and structural bays
Morphology	13–20	Flexibility of the design to allow for variation	Measure based on building matrices of design options
Fire Safety	21–23	Ability of design to meet fire safety regulation in the city	Egress calculation based on simple shape properties
Ventilation	24	Pedestrian wind hinder	CFD analysis (simulation software package)
Real estate	25	Lease/sale revenue	Real-estate consultants (expert judgment)
Energy performance	26	Energy consumption	Energy analysis (normative calculation)

making process. The column indicates three PIs that potentially need advanced simulation, i.e. fire safety/egress, ventilation/wind hinder and energy analysis. In the course of the project it was decided that egress and energy analysis could be based on normative calculations and only ventilation/wind hinder needed full blown simulation. These decisions were reached on the basis of time, convenience and the trust that normative methods were good enough for the comparative analysis. Figure 2.8 shows the sets of elements that compose the various Aspect Systems whereas Figure 2.9 shows the normalized results.

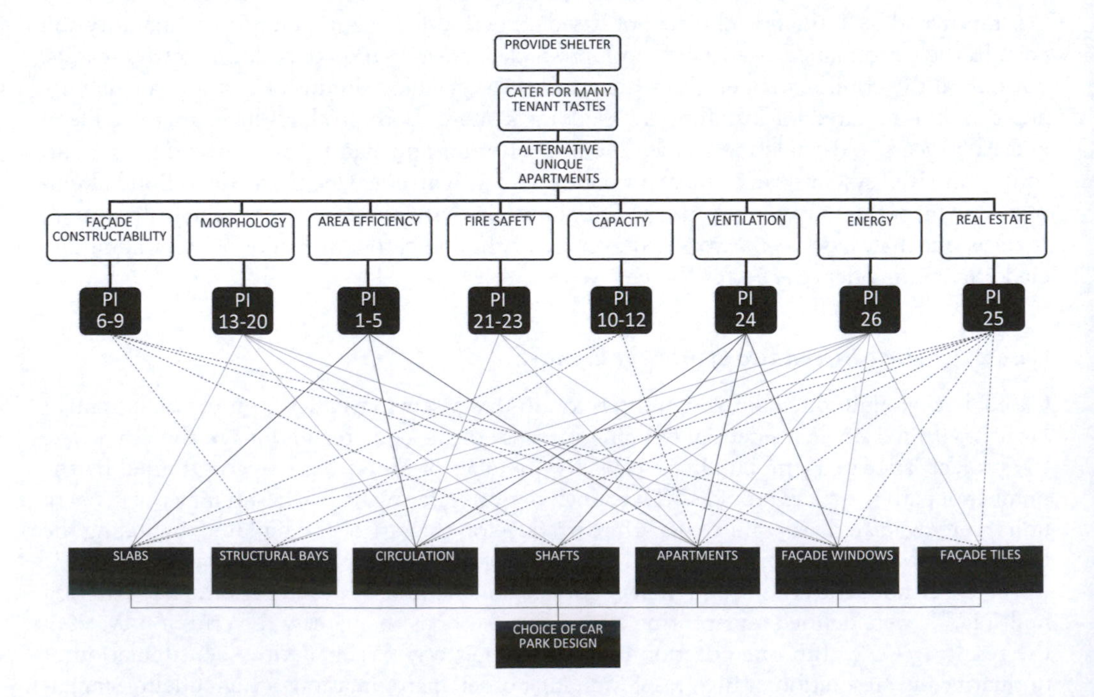

Figure 2.8 Aspect systems.

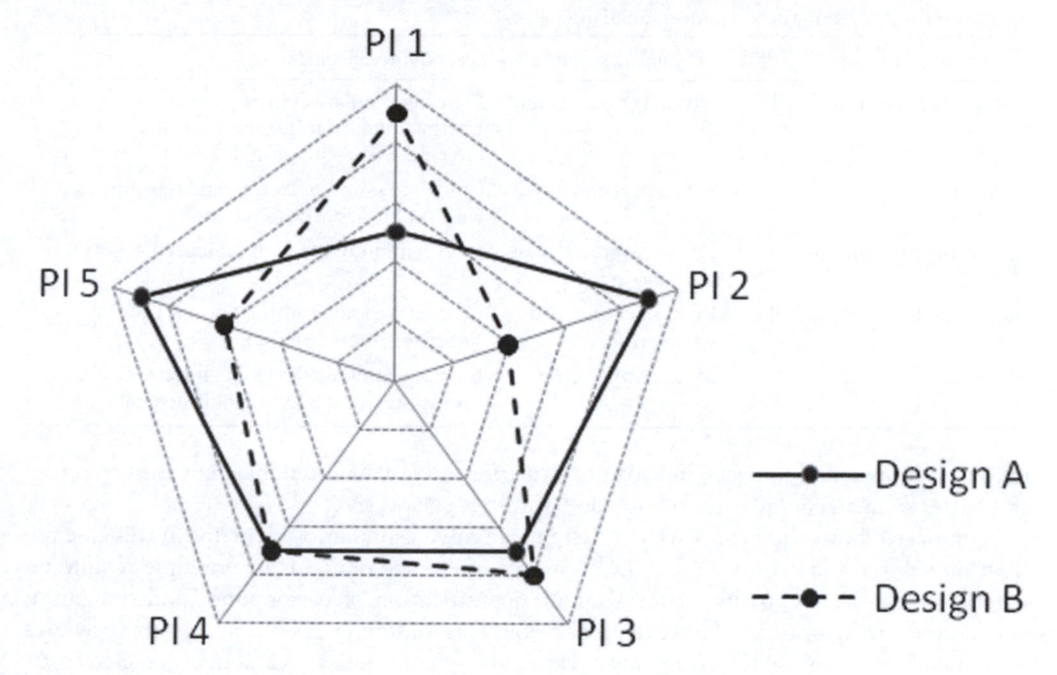

Figure 2.9 Normalized results.

As is obvious from Figure 2.9, showing only a selection of the major PIs, there is no clear winner. It will largely depend on the preferences of the stakeholders to arrive at a consensus "best" decision.

Case 3: Choice between off-grid PV panel system and green roof

Case 3 background: In this case, the owner provided a list of design goals and gave his preference on each goal, as given in Table 2.3. In this particular instance a SWOT analysis of the two options was used to select five main criteria and define one PI for each criterion as shown in Table 2.4. In the early analysis it was found that the PV panel solution performed well in terms

Table 2.3 Ranking of owner's needs

No.	*Goals*	*Weights*
1	STRENGTH OF CONCEPT	7
2	SIMPLICITY OF DESIGN	6
3	CREATIVITY OF DESIGN	7
4	CHALLENGING SOLUTION	4
5	CONNECTION OF INTERIOR & EXTERIOR	4
6	MATERIAL USE	6
7	QUALITY OF INDOOR SPACE	6
8	NATURAL LIGHTING	8
9	ENERGY SAVING	8
10	SUSTAINABILITY	8
11	INITIAL COST	4
12	MAINTENANCE & OPERATION COST	8
13	SHORT SCHEDULE	6
14	EFFICIENT USE OF SPACE ON SITE	5
15	PREFABRICATION	7
16	REUSABLE CONTENTS	8
	TOTAL	**100**

Table 2.4 Project E criteria and performance indicators

Performance criterion	*PI #*	*PI Name/description*	*Quantitative method*
Energy conservation	1	Annual energy saving	Lighting simulation in Ecotect Energy simulation (normative calculation) Green power estimation in PowerWatt
Acoustics	2	Acoustical comfort	Calculate sound reflection and transmission with simple acoustics simulation
Environmental impacts	3	Life cycle green house gas emission	LCA (simplified to GHG impact category)
Impact to structure	4	Extra structure load effect on cost	Calculate total weight of options (used as surrogate outcome for cost)
Maintenance	5	Maintenance costs	Estimate annual average maintenance and operation cost based on published data

of low environmental impacts, low load to structure and low maintenance, but the green roof was better from an energy conservation and acoustics perspective.

Figure 2.10 shows the macro Aspect Systems. Involved simulation domains in this case are lighting, energy and acoustics. The LCA and maintenance domains in principle require an extensive long term experiment for accurate quantification of performance and simulation would hence be appropriate. However, for obvious reason, standardized calculation procedures were used for to quantify life cycle impact (limited to greenhouse gas) and maintenance cost.

The role of simulation

The three cases show clearly that simulation is a totally integrated activity in the performance based design process. Performance simulation expertise is necessary from the start of the analysis of the options, through criteria definition and choice of quantification method. The actual execution of simulation is mostly defined by the earlier steps in the process. This defines the role of simulation in performance based design as an integral part of the preparation of decisions. The other remarkable fact is that the need for sophisticated simulation seems to be less than is often assumed. There are several reasons: (1) the multitude and integrality of the

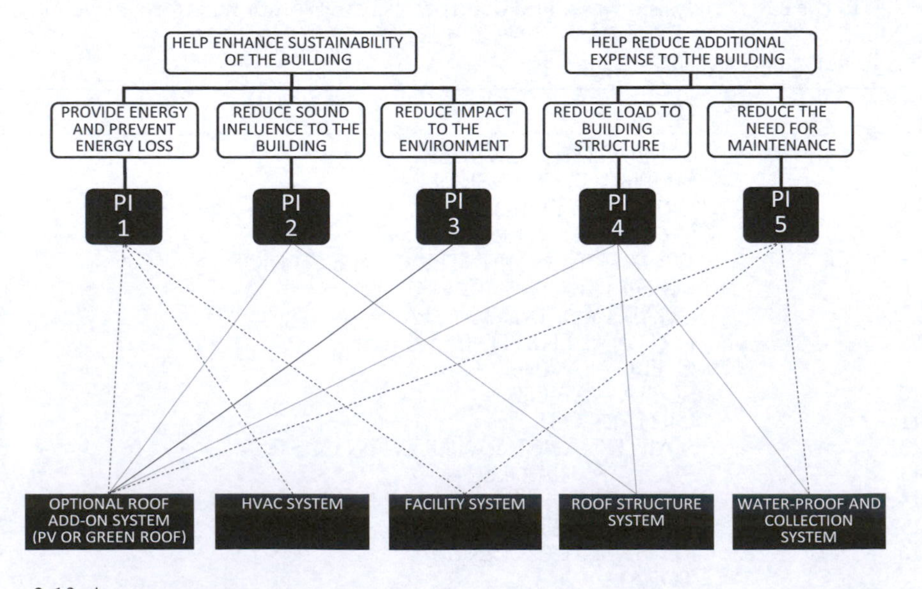

Figure 2.10 Aspect systems.

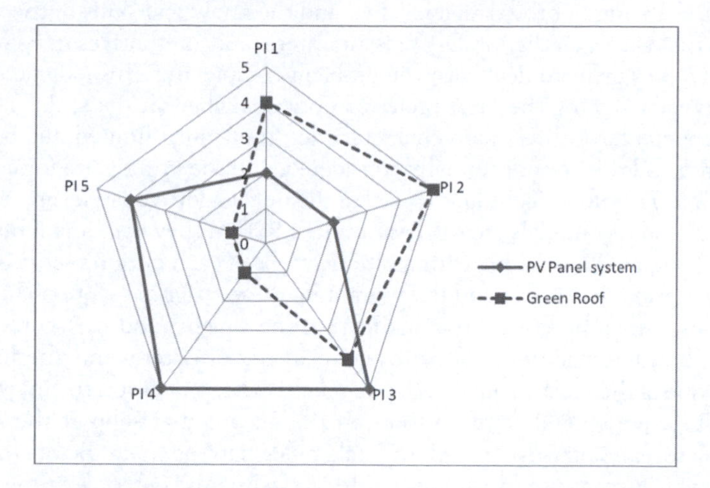

Figure 2.11 Weighted results.

combined criteria becomes the core of attention rather than any single criterion on its own, (2) normative comparative analysis and parametric fast turn around tools respond better to the dynamics of multi stakeholder decision making, (3) the systematic framing of the quantification in terms of Aspect Systems explains the role of simulation to the client and across different domains of expertise in the design team.

It is clear that the radar chart with normalized performance indicator values needs to be followed up by a method that helps stakeholders to reach a "best" decision, based on their individual ranking of criteria importance and individual weighting of the PI outcomes on each criterion. A brief overview is given in the next section.

Multi criterion decision making

Since Keeney and Raiffa (1976) proposed the Multi Criteria Decision Analysis (MCDA) approach in their seminal work, decision theory has steadily evolved. The analysis of the decision problem introduces and processes a formal subjective judgment during the alternatives' evaluation (Keeney 1982). Current MCDA procedures structure the decision problem in a sequence of steps that provide clarity for those that will make the decision. The MCDA starts when its aims and parties are identified. The parties are composed by three main players: the Decision Maker, who makes the decision and assumes the consequences, the Decision Analyst, who facilitates the process, and the Stakeholder, who will be affected by the consequences of the decision (Huovila *et al.* 2005). The next step is to develop the decision analysis system based on the exchange of information provided by the parties. Then, objectives and criteria, which express the means in which the alternatives create value, are identified (Dodgson *et al.* 2000). Once the parties agree on the decision problem organization, the subsequent step is to weight the criteria in relation to each alternative and to rank their importance in relation to the decision problem. Therefore, this process means weighting both the differences between options and "how much that difference matters" (Dodgson *et al.* 2000). Hence, a consistent weighting scale should be used in order to compare different criteria subjects. After the results are calculated, questions such as "what if . . ." are common and an analysis of it should be made in order to provide an understanding of the consequences of the decision. This can be made through a sensitivity analysis that aims to evaluate how the model is sensitive to changes and clarify conflicts of interests between stakeholders.

There are different decision techniques when involving multi criteria, and this will depend on the type of decision, how the technique deals with the data provided, the time available for the analysis, the nature of the data available, and the analytical skills of the stakeholders. All techniques that approach the MCDA present criteria and alternatives in its structure, and require subjective judgment to deal with the problem complexity. This judgment is mainly a numerical analysis by scoring the most preferred options higher on the scale. The cases presented in this chapter have three main characteristics: there are a limited number of alternatives, uncertainty and risk are not formally considered, and the criteria are formally independent of each other. Therefore, techniques based in the linear additive model are suitable to this approach. Linear additive models are the basis of the MCDA; they combine individual values into one overall output. This is done with multiplications of each option's score on each criterion by the weights of the criteria, and then summing these weighted scores. The well known AHP method also uses a linear additive model, but the weights and the scores obtained by alternatives are based, respectively, on paired comparisons of criteria and alternatives (Saaty 1994). AHP is one of the most frequently applied multi criteria decision techniques (Dodgson *et al.* 2000). There are several software tools on the market that support the AHP process which requires a several hour session with the stakeholders in the room. Before the session the criteria and participants are preloaded in the software, while the outcome of the performance assessments is distributed and explained to all stakeholders. One then proceeds using the stakeholders' ability to rank the criteria in relation to each alternative and to rank their importance in relation to the decision problem through a paired comparison. The tool then ranks the two options based on all inputs and allows a sensitivity analysis of the outcome. As stated before, the AHP methods foundation is sometimes criticized especially by those that prefer utility theory, with its strong theoretical foundation (Hazelrigg 2003).

After stipulating previously that uncertainty in the performance outcomes is unavoidable and necessary to consider in decision making, it should be observed that the treatment in this section has been purely based on deterministic results. Suffice it to say that the tools are increasingly expanded to deal with decision making under uncertainty (Hopfe 2009) but more work is still to be done in quantifying uncertainties in our models and experiments and above all, in conveying them to the stakeholders in such a way that the decision analysis can transparently deal with uncertainty and risk.

Discussion

In this chapter we have shown that the role of simulation in performance based design cannot be narrowed down to offering traditional simulation skills to the project. We argue that the simulationist will be an integral partner in the performance criteria management end ensuing dialogue concerning the definition of measures for the individual criteria and choice of appropriate quantification methods. In some cases this will call for advanced simulation, in other instances a simple calculation or expert judgment will be sufficient. From the perspective of quantification of performance, the role of these different methods is indeed interchangeable, and only determined by the special need and context of the decisions that the project team is facing.

A system theoretic approach has been shown to lead the way in defining the interface between physical building elements and performance requirements. A closer look at this interface reveals that the Aspect Model plays an important conceptual role. It is introduced as a composition of elements and parameter subsets of a variety of technical systems. Each Aspect Model is uniquely associated with a particular performance indicator. As such it is a representation of the physical model on which the experiment (simulation) is performed in order to quantify a particular performance indicator. One could imagine that this concept is used for the formal confirmation that a simulation software is suitable for the quantification of a given performance indicator. The explicit management of all Aspect Models for which a performance requirement has been specified would greatly enhance the integral data management across a design team. Future versions of BIM could explicitly support the Aspect Models and BIM driven interfaces would replace the current generation which haphazardly harvests elements from different technical systems to process in mapping software, for which there is no common ground across different interface development, leading to much unwarranted duplication of work.

Perhaps the biggest challenge of simulation in performance based design is to provide a variety of normative calculations when an advanced simulation cannot provide a more accurate answer, either because of the presence of uncertainties, the lack of available information, or the context of the decision demands it.

Synopsis

The chapter has placed building simulation in a design evolution context. It is shown that comparison of different design variants requires carefully crafted performance indicators. It is emphasized that a system theoretic approach is needed to define objective and universal performance indicators as the "outcome" of simulation. For this reason Aspect Systems are introduced, and it is shown that every simulation is in fact an operation on an Aspect System, the outcome of which can be expressed as a mathematical expression that only contains properties of the elements that compose the Aspect System. It is shown that the approach leads to unambiguous definitions of performance indicators that can be used to prepare multi criterion decision making. If all parties are satisfied with the definition of the measures and measurement methods, they will use them to express performance targets and to check whether design proposals meet these targets. It is shown that quantification, or any measurement method for that matter, can be supported by different types of building simulation, calculations, expert judgment, or other. Diverse examples have been presented to show that a well structured set of performance criteria can deliver the needed support for rational design decision making.

In addition, the explorations in this chapter have touched on issues of design evolution and decision-making, the anachronism between the importance of early design decisions and the deficiencies of current simulation tools to support them, the aggregation of simulation results into meaningful expressions of performance, the role of uncertainty, and the need to separate performance evaluation function from software execution.

Acknowledgements

Thanks to P. Sanguinetti for her valuable contributions to the paper as a whole, and to T. Chatterjee, M. El-Khaldi and F. Zhao for their contribution of the case studies used in this chapter.

References

Augenbroe, G. (2004) "Chapter 1 Introduction', in *Advanced Building Simulation*, Malkawi, A. and Augenbroe, G. (eds), Oxford: Taylor & Francis.

Augenbroe, G. and de Wilde, P. (2003) "Design Analysis Interface (DAI)', Final Report, Atlanta: Georgia Institute of Technology. Available from http: dcom.arch.gatech.edu/dai/

Augenbroe, G. and Park, C.S. (2005) "Quantification Methods of Technical Building Performance', *Building Research and Information* 33(2): 159–172.

Augenbroe, G., Malkawi, A. and de Wilde, P. (2004) "A Performance Based Language for Design Analysis Dialogues', *Journal of Architectural and Planning Research* 21(4): 321–330.

Barrett, P. (2008) *Revaluing Construction*, Blackwell.

Beard, J.L., Wundram, E.C. and Loulakis, M.C. (2001) *Design-Build: Planning Through Development*. New York: McGraw-Hill.

Becker, R. (2008) "Fundamentals of Performance-Based Building Design", *Building Simulation* 1(4): 356–371.

De Wit, M.S. (2001) "Uncertainty in Predictions of Thermal Comfort in Buildings", PhD thesis, Delft University of Technology, The Netherlands.

De Wit, S. and Augenbroe, G. (2002) "Analysis of Uncertainty in Building Design Evaluations and Its Implications", *Energy and Buildings* 34(9): 951–958.

Dodgson, J., Spackman, M., Pearman, A. and Phillips, L. (2000) *Multi-Criteria Analysis Manual*, London, UK: Department for Communities and Local Government.

Foliente, G.C. (2000) "Developments in Performance-Based Building Codes and Standards", *Forest Products Journal* 50(7/8): 12–21.

Foliente, G.C., Leicester, R.H. and Pham, L. (1998) "Development of the CIB Proactive Program on Performance Based Building Codes and Standards", BCE Doc 98/232. Australia: International Council for Research and Innovation in Building and Construction (CIB).

Gielingh, W. (1988) "General AEC Reference Model", ISO TC184/SC4/WG1 document 3.2.2.1 (also TNO report BI-88–150, October 1988).

Gilb, T. (2005) *Competitive Engineering*, Oxford: Elsevier.

Habraken, N.J. (1998) *Structure of the Ordinary; Form and Control in the Built Environment*, edited by Jonathan Teicher, Cambridge, MA: MIT Press.

Hazelrigg, G. (2003) "Validation of Engineering Design Alternative Selection Methods", *Journal of Engineering Optimization* 35(2): 103–120.

Hopfe, C. (2009) "Uncertainty and Sensitivity Analysis in Building Performance Simulation for Decision Support and Design Optimization", PhD Thesis, Eindhoven University of Technology, The Netherlands.

Hu, H. and Augenbroe, G. (2009). "Right Sizing an Off-grid Solar House", in *Proceedings of Building Simulation 2009*, Glasgow, Scotland, pp. 17–24.

Huovila, P., Porkka, J., Gray, C. and Al Bizri, S. (2005) "Conclusions and Recommendations on Decision Support Tools for Performance Based Building: Performance Based Building Thematic Network", Collaborative effort of VTT Building and Transport (Finland) and University of Reading (UK).

Keeney, R.L. (1982) "Decision Analysis: An Overview", *Operations Research* 30(5): 803–838.

Keeney, R.L. and Raiffa, H. (1976) *Decisions with Multiple Objectives: Preferences and Value Tradeoffs*, New York: John Wiley & Sons.

MacDonald, I.A. (2002) "Quantifying the Effects of Uncertainty in Building Simulation", PhD thesis, University of Strathclyde, Scotland.

Macdonald, I. and Strachan, P. (2001) "Practical Application of Uncertainty Analysis", *Energy and Buildings* 33(3): 219–227.

Moon, H.J. (2004) "Assessing Mould Risks in Buildings under Uncertainty", PhD thesis, Georgia Institute of Technology, Atlanta, USA.

Moon, H.J. and Augenbroe, G. (2007) "Application of the Probabilistic Simulation and Bayesian Decision Theory in the Selection of Mold Remediation Actions", in *Proceedings of Building Simulation 2007*, Beijing, China, pp. 912–918.

Ozsariyildiz, S. (2006) "Inception Support for Large-Scale Construction Projects. A knowledge Modeling Approach", PhD Thesis, TU Delft, The Netherlands.

Quiggin, J. (1997) *Generalized Expected Utility Theory: The Rank-Dependent Model*, Boston, MA: Kluwer Academic Publishers.

Rush, R.D. (ed.) (1986) *The Building Systems Integration Handbook*, AIA and Butterworth Heinemann.

Saaty, T.L. (1996) "Fundamentals of Decision Making and Priority Theory with the Analytic Hierarchy Process", *Analytic Hierarchy Process Series*; V. 6. 1st ed. Pittsburgh, PA: RWS Publications.

Saaty, T.L. (2004) "Fundamentals of the Analytic Network Process: Dependence and Feedback in Decision-Making with a Single Network", *Journal of Systems Science and Systems Engineering* 13 (June).

Saltelli, A., Ratto, S., Andres, T., Campolongo, F., Cariboni, J., Gatelli, D., Saisana, M. and Tarantola, S. (2008) *Global Sensitivity Analysis*. New York: Wiley & Sons.

Sanguinetti, P., Eastman, C. and Augenbroe, G. (2009) "Courthouse Energy Evaluation: BIM and Simulation Model Interoperability in Concept Design", in *Proceedings of Building Simulation 2009*, Glasgow, Scotland, 1922–1929.

Sokolowski, J.A. and Banks, C.M. (2009) *Principles of Modeling and Simulation. A Multi Disciplinary Approach*. New York: John Wiley & Sons.

Vanier, D.J., Lacasse, M.A. and Parsons, A. (1996) "Modeling of User Requirements Using Product Modeling", in *Proceedings of Third International Symposium – Applications of the Performance Concept in Building*, ed. Becker, R. and Paciuk, M. Tel-Aviv, Israel.

Verheij, J.M. (2005) "Collaborative Planning of AEC Projects and Partnerships", PhD thesis, Georgia Institute of Technology, USA.

Recommended reading

For an introduction in performance thinking in general, the author recommends the cited papers by Foliente and the 2005 paper by Augenbroe, as well as the documents found on the CIB-Pebbu web site mentioned in the text. A good overview from a broad perspective is given in the book "Assessing Building Performance", by Preiser and Vischer.

For follow-up reading on specific issues in performance simulation, decision making and standardization one should start with the excellent special issue of *Building Research and Information on Performance-based Building*, Vol. 33, March–April 2005.

Activities

A good exercise to train oneself in the use of the performance framework is to revisit and rationalize a design decision problem that one was part of. This should be a real problem where a real decision was made and enacted, preferably concerning a discrete decision case during building design, construction or operation. The following assignment challenges you to rationalize the original design decision problem.

Starting point: take a design, construction or operation context that you have been personally immersed in. You should single out a discrete decision problem that concerns a choice between design option A and design option B.

Now perform the following steps and answer the stated questions in each step:

Step 1: context

- describe the context of the case and the options
- describe the original decision; why was it taken; on what grounds?
- describe the two competing options that you are going to consider
- who (which stakeholders) is involved in the decision making?
- was risk analysis used in the original decision making?

Step 2: rationalization

- what functional requirements drive the decision? Use decomposition of requirements if needed until you arrive at the granularity that is adequate to quantify with a performance indicator
- match each functional requirement with a performance aspect
- choose and explain the performance indicators that you are going to use for each performance aspect.

Step 3: methodology

- describe the decision problem in the context of matching each functional requirement with an Aspect System (show this in a system diagram)
- describe every Aspect System in terms of aggregations over components and properties of technical subsystems (show in system diagram)
- explain the connection between Aspect Systems and functional requirements and locate your performance indicators on those connections.

Step 4: method

- develop the performance indicators and their quantification methods (at least five) and do a thorough quantification of at least three using building simulation tools.

Step 5: informing the decision making process

- produce a radar chart plotting options A and B.

Step 6: evaluation

- how will your approach affect the decision making process; is it likely that another outcome will result compared to the original decision
- how would the recognition of uncertainty affect your analysis.

Answers

The chapter treats three case studies that show the narrative as requested by the assignment. Read and study the case studies on pages 25–31 and apply to your own case.

3 Weather data for building performance simulation

Charles S. Barnaby and Drury B. Crawley

Scope

This chapter addresses the environmental boundary conditions that drive building simulation models. The qualitative short- and long-term behavior of weather data is described. Practical representation of this information in hourly weather files and typical weather years is covered. The extensive role of data models is emphasized. Important trends caused by urbanization and climate change are considered.

Learning objectives

The reader is to understand how highly varying weather phenomena are represented in practice for application in building performance simulation. In particular, the reader should understand the nature and limitations of the commonly used typical year hourly weather files. Further, the reader should see that boundary conditions imposed on building models are often themselves model results. Determining the appropriateness and uncertainty of any building simulation results needs to include consideration of the weather data used in the study.

Key words

Weather data/typical weather year/solar models

Introduction

Short-interval environmental data are required to drive building simulation models. These data provide the boundary conditions for the component models that are combined in simulation applications. Ideally, environmental data should be observed exactly at the project site, subject to all local conditions. Further, the observations should be made at intervals consistent with the modeling being undertaken. These ideals are virtually never realized. This chapter explores the strategies that have been developed to adapt available weather data sources to building simulation applications.

Historically, large-scale (national) weather data observation programs have supported various activities such as general forecasting, aviation, and agriculture. Requirements for engineering applications have received minimal consideration during development of observation procedures. Thus building simulation is often in the position of making do with what data are available. This situation is improving as technology lowers the cost and increases the sophistication of data observation, management, and modeling.

Hensen (1999) provides an overview of simulation-related weather data issues; most of that discussion remains applicable. An overriding point is that the type of simulation being conducted strongly determines weather data requirements. Traditionally, weather data for simulation have been represented in files of values for the 8760 hours of a year (or 8784 hours for a leap year). This form evolved for two reasons. First, until the 1990s, most weather observations were made manually; a one-hour measurement/reporting cycle was practical and sufficient for most applications. Second, a one-hour time step was (and still is) used in many building simulation models as a natural interval – short enough to capture behaviors of interest but long enough to allow practical execution times. Because of this history, it is common to think of weather data for building simulation as full-year hourly data. While hourly files remain important, it is essential to consider other forms as simulation applications become more sophisticated.

Two attributes of the modeled system determine weather data needs. The nature of the process defines the items required (temperature, solar radiation, and so forth) while its time constant suggests the appropriate observation interval. Building structural components generally have relatively long time constants (hours); this accounts for the success of the hourly simulation approach. However, many building elements respond much more quickly – the air within the building, many furnishings, HVAC equipment, and daylighting controls, to name a few. A common approach in hourly models is to represent such elements with quasi steady state models that calculate average performance assuming current conditions persist indefinitely. To the extent that actual processes are non-linear, this simplification can be seriously in error. Shorter time steps are needed using sub-hourly weather values.

In addition to needing the right values at the right interval, building simulation applications require weather data that are complete and self-consistent (colloquially, "clean"). Applications can behave unpredictably or fail completely when confronted by unexpected values such as non-zero beam solar when the sun is below the horizon or wet-bulb temperature above the dry-bulb temperature. Typical weather data, like any real measurements, contain erroneous or missing items. Thus, preprocessing, correction, and filling are important aspects of preparing weather data for simulation use.

The sections below expand on this introductory background. Much of the material concerns primarily software developers – the issues discussed are addressed within simulation applications or weather data processing utilities. However, the practitioner should generally understand how simulation boundary conditions are derived and the implications of those methods.

Characteristics of weather data

Selecting and applying weather data for building simulation should be informed by a qualitative understanding of the behavior of the driving phenomena. Figure 3.1 shows some example data observed at one-minute intervals for a typical day.

By inspection, it is seen that this day begins with rising temperature and wind speed accompanied by occasional clouds. At about 16:00 there is a sudden drop in temperature and solar radiation along with a rise in dew point – probably due to a rain shower (confirmed by precipitation data not shown). The figure shows significant variability within single hours, with the following characteristics:

- *Temperature*. Both dry-bulb and dew-point temperature vary relatively slowly, with only limited noise. Hour-to-hour changes are rarely more than a few degrees.
- *Solar*. Large, rapid changes occur over a matter of minutes.
- *Wind*. Wind velocity varies by a factor of two or more over time scales of a few minutes.

Representing this behavior with hourly values requires aggregation that necessarily omits a great deal of information. Conventional procedures for constructing hourly records use instantaneous values for meteorological data (such as temperatures, humidity, and wind speed) and totals over the prior hour for solar and illuminance data. Thus, for the Figure 3.1 example, the hourly record for 10:00 would show dry-bulb temperature of 25 °C and global solar radiation of about 705 Wh/m^2.

At the other end of the time scale, weather conditions also vary significantly from year to year. For many locations, there is no such thing as an average year – one or more anomalies

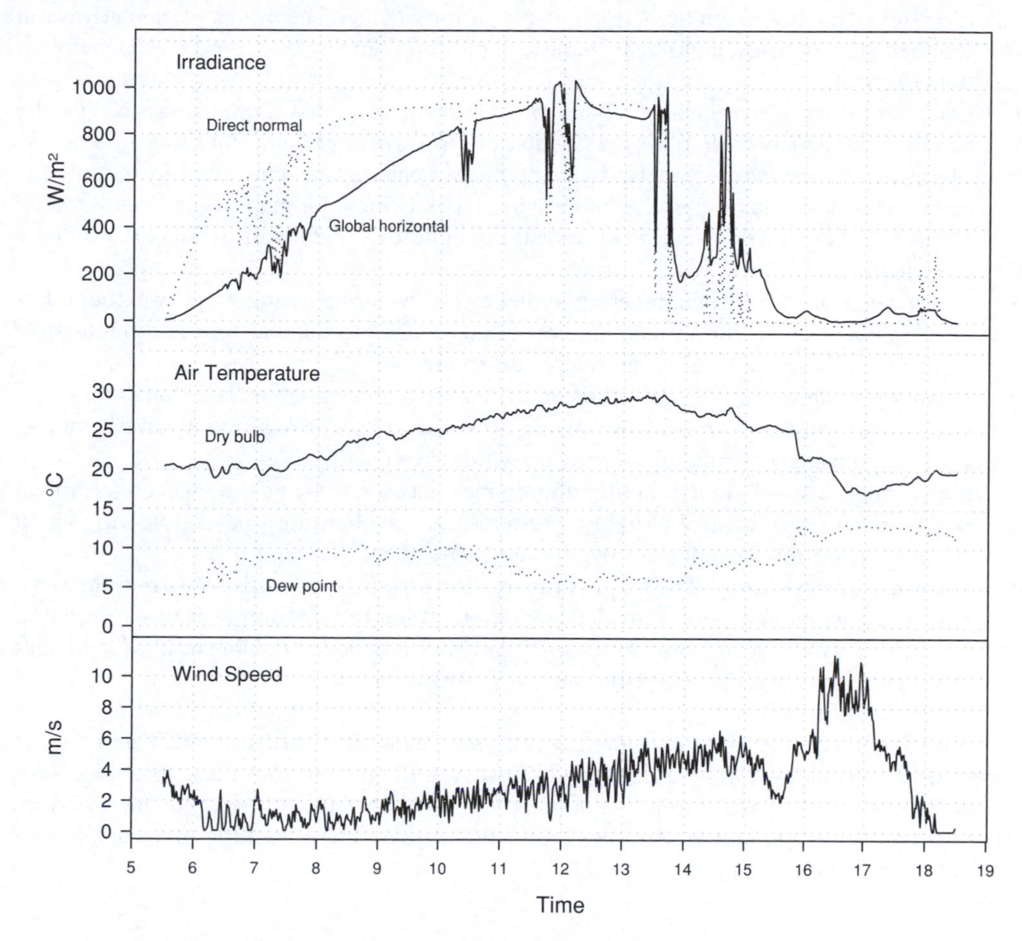

Figure 3.1 One minute temperature, wind, and solar radiation observations for 31 August 2008 at National Renewable Energy Laboratory, Golden, Colorado.

occur during each year. Five- or ten-year simulations could be needed to assess long-term building performance; although representative single-year analysis is often possible using fabricated typical weather years, discussed below.

Simulation applications and weather data requirements

The following are some common uses of simulation and associated implications for weather data characteristics.

- *Energy design and compliance analysis of fully conditioned buildings*. This is the "classic" simulation application. Energy savings information (generally annual) is derived by comparing results for design variants. In this application, the need for absolute accuracy is minimized – errors partially cancel, short-time inaccuracies are not significant, and the mechanical HVAC system will ensure actual building comfort. Similar baseline comparison procedures are used in energy code compliance and building labeling programs. Representative single-year hourly data are sufficient.
- *Performance of un- or semi-conditioned buildings*. Overheating analysis, passive solar design, and other "floating space temperature" problems involve investigation of performance under non-average conditions. For example, a comfort study for a building without

mechanical cooling might involve simulation using weather data from a particularly warm summer. Alternatively, multi-year studies are indicated. In any case, typical data are often not adequate.

- *Equipment sizing*. Determination of maximum demand on HVAC components is typically based on design-day or short period calculations using near-extreme conditions.
- *Model calibration, building troubleshooting, control optimization, and actual savings estimation*. Applications involving the performance of existing buildings require weather data observed during the study period at or near the building site. Historical data are generally of little use.
- *Engineering studies*. Many engineering studies can be performed using simple weather information. For example, the annual number of hours when mechanical ventilation is feasible can be readily estimated using bin temperature data.
- *Natural ventilation design*. Local wind conditions are highly variable; data from airport stations are notoriously unreliable for nearby sites. Great care is required when assembling wind data for study of buildings relying on wind-driven ventilation.
- *Daylighting studies*. Although hourly illuminance data are sufficient to estimate electrical savings from sensor-control of lighting systems, detailed lighting studies involving visual comfort or control dynamics can require sub-hourly data.
- *Renewable energy systems*. The output of solar electric systems depends on short-term variation and spectral make-up of incident radiation. Wind turbine output is proportional to the cube of wind velocity. Standard hourly data may produce unreliable results for systems that have non-linear characteristics such as these examples.

In short, the practitioner must consider the processes being simulated and take care to select appropriate weather data as boundary conditions. As can be seen from these examples, serious modeling errors can result from use of inappropriate data. Conversely, there are situations where simple or representative data are sufficient; in these cases, seeking more detailed or "accurate" information has little value.

Data items required for simulation

Table 3.1 enumerates common weather data items that have application in building simulation. Several deserve additional discussion. As noted, dry-bulb temperature and wind are subject to significant modification due to local (microclimate) surroundings, so weather station observations (often made at an open site such as an airport) often do not accurately represent even nearby site conditions. As discussed later in the chapter, dry-bulb temperature is can be reliably modified with relatively simple methods. Local wind direction and velocity, however, cannot be modeled with any simple procedures. Developing data to drive simulations of wind-sensitive designs often requires on-site measurements, scale model wind tunnel studies, computational fluid dynamics (CFD) modeling, and/or use of conservative synthetic data. Regarding solar and illuminance data, note that these important values are normally generated using models, as is extensively discussed later.

File formats

Representation and storage of short-term weather data have evolved along with the computer and information revolution of the last 40 years. Early automated data formats replicated punched card layouts formerly used for data entry by weather services. To save transcription costs, often data from only every third hour were made available. Weather data were expensive and were distributed on magnetic tape (one still occasionally hears the term "weather tape"). High cost, inconvenient formats, and omitted data made for difficult application to simulation.

Table 3.1 Weather data items of use for building simulation

Item	*Model use(s)*	*Availability and issues*
Dry-bulb air temperature	• Exterior surface convection • Infiltration/ventilation sensible heat transfer • Equipment (e.g. air-cooled condenser)	• Universally observed • Significant local effects (e.g. heat island)
Humidity (one of relative humidity, wet-bulb temperature, or dew-point temperature)	• Infiltration/ventilation latent heat transfer • Equipment (e.g. cooling tower)	• Commonly observed
Solar irradiance (direct and diffuse)	• Fenestration heat gain • Exterior surface heat balance • Solar thermal and photovoltaic systems	• Sparsely measured • If observed, often global only • Model sources widely used • Remote sensing opportunities
Solar illuminance (direct and diffuse)	• Daylight modeling	• Rarely measured (modeled from irradiance)
Sky temperature	• Exterior surface heat balance	• Rarely measured (modeled from temperature, humidity, and cloud cover)
Cloud cover/sky condition	• Sky models (e.g. for daylighting)	• Generally observed • Multiple data representation conventions • Evolution of automated instrumentation introduces uncertainties
Wind (velocity and direction)	• Exterior surface convection • Infiltration • Natural ventilation	• Generally observed • Local effects very significant for both velocity and direction • Low velocity observations unreliable
Ground temperature	Below-grade heat transfer	• Measured for agricultural purposes, limited exploitation of observed values for building simulation
Ground surface albedo	Reflected irradiance/illuminance	• Inferred from presence of snow
Weather conditions (e.g. rain)	Exterior surface wetting	• Generally measured; inconsistent reporting formats

Until recently, source files typically consisted of 8,760 fixed format records (values located in specific columns of each data line). Custom programming was often required to decode the records, although the formats were generally simple and well documented. Most simulation applications resorted to using external weather utilities that could decode source files, generate missing data, and write "packed" binary files that could be efficiently read during model execution. Additionally, this strategy saved precious disk space – the packed files are typically 50–200k bytes as opposed to over a megabyte for the ASCII source.

Current formats, such as TMY3 (Wilcox and Marion 2008) and EPW (EnergyPlus 2010b), are now more compatible with desktop spreadsheet tools. They are comma-separated and include metadata (such as site location) and column headings that make them somewhat self-documenting. In addition, many formats include field-level flags that indicate source and uncertainty. Figure 3.2 shows example data and illustrates that current files are straightforward in structure and admit of end-user manipulation. The EPW format is particularly rich, allowing inclusion of design conditions, extreme period sequences, and sub-hourly data. That is, EPW is

722280	BIRMINGHAM MUNICIPAL AP	AL	-6	33.567	-86.75	189								
Date (MM/DD/ YYYY)	Time (HH:MM)	ETR (W/m^2)	ETRN (W/m^2)	GHI (W/m^2)	GHI source	GHI uncert (%)	DNI (W/m^2)	DNI source	DNI uncert (%)	DHI (W/m^2)	DHI source	DHI uncert (%)	GH illum (lx)	GH illum source
1/1/1987	1:00	0	0	0	1	0	0	1	0	0	1	0	0	1
1/1/1987	2:00	0	0	0	1	0	0	1	0	0	1	0	0	1
1/1/1987	3:00	0	0	0	1	0	0	1	0	0	1	0	0	1
1/1/1987	4:00	0	0	0	1	0	0	1	0	0	1	0	0	1
1/1/1987	5:00	0	0	0	1	0	0	1	0	0	1	0	0	1
1/1/1987	6:00	0	0	0	1	0	0	1	0	0	1	0	0	1
1/1/1987	7:00	2	177	3	1	13	8	1	9	2	1	13	0	1
1/1/1987	8:00	153	1415	69	1	13	375	1	9	28	1	13	7041	1
1/1/1987	9:00	389	1415	236	1	9	683	1	9	47	1	13	24494	1
1/1/1987	10:00	580	1415	392	1	9	802	1	9	62	1	13	41498	1
1/1/1987	11:00	711	1415	507	1	9	862	1	9	72	1	13	52975	1
1/1/1987	12:00	773	1415	526	1	9	740	1	9	120	1	13	55519	1
1/1/1987	13:00	761	1415	513	1	9	779	1	9	93	1	13	53870	1
1/1/1987	14:00	677	1415	471	1	9	844	1	9	67	1	13	49464	1
1/1/1987	15:00	526	1415	346	1	9	779	1	9	57	1	13	36697	1
1/1/1987	16:00	319	1415	179	1	13	613	1	9	42	1	13	18587	1
1/1/1987	17:00	80	1144	40	1	13	197	1	9	20	1	13	3553	1
1/1/1987	18:00	0	0	0	1	0	0	1	0	0	1	0	0	1
1/1/1987	19:00	0	0	0	1	0	0	1	0	0	1	0	0	1
1/1/1987	20:00	0	0	0	1	0	0	1	0	0	1	0	0	1
1/1/1987	21:00	0	0	0	1	0	0	1	0	0	1	0	0	1
1/1/1987	22:00	0	0	0	1	0	0	1	0	0	1	0	0	1
1/1/1987	23:00	0	0	0	1	0	0	1	0	0	1	0	0	1
1/1/1987	24:00:00	0	0	0	1	0	0	1	0	0	1	0	0	1
1/2/1987	1:00	0	0	0	1	0	0	1	0	0	1	0	0	1
1/2/1987	2:00	0	0	0	1	0	0	1	0	0	1	0	0	1

Figure 3.2 Example TMY3 format data for Birmingham, Alabama, as displayed in a spreadsheet application. Only "upper left" portion of file is shown, full dimensions are 68 columns by 8762 rows.

truly organized to support building simulation, as opposed to being a format for distribution of meteorological data. EPW format can be read or translated by more than 20 building simulation programs.

In general, managing weather data for simulation is becoming simpler. A helpful trend is toward free availability of weather files on the Internet, at least in some countries (files remain expensive in Europe). For local studies, building automation systems now often record at least some weather parameters, often in an ad hoc (vendor specific) format. With spreadsheet manipulation, it is usually straightforward to adapt such data for simulation use. Thus tasks that were formerly cumbersome, expensive, and technically demanding are now manageable with desktop computing tools.

Typical weather years

The most common building modeling application is full-year hourly simulation to calculate energy use. Extensive effort has been applied to finding representative or typical weather years that allow estimation of long-term (multi-year) performance from a single year analysis.

Early work attempted to select a representative year from a multi-year set. For example, the Test Reference Year (TRY) data set was developed for 60 locations in the United States (NCDC 1976) using a process in which years in the period of record (~1948–1975) that had months with extremely high or low mean temperatures were progressively eliminated until only one year remained. This results in selection of mild years that exclude extreme conditions.

The lack of extreme conditions and solar data in the TRY files led to development of the Typical Meteorological Year (TMY) approach now in wide use. A TMY is a composite of typical months, not necessarily from the same year. A data record of N years contains N Januaries, N Februaries, and so forth. Various methods have been used to select the best representative

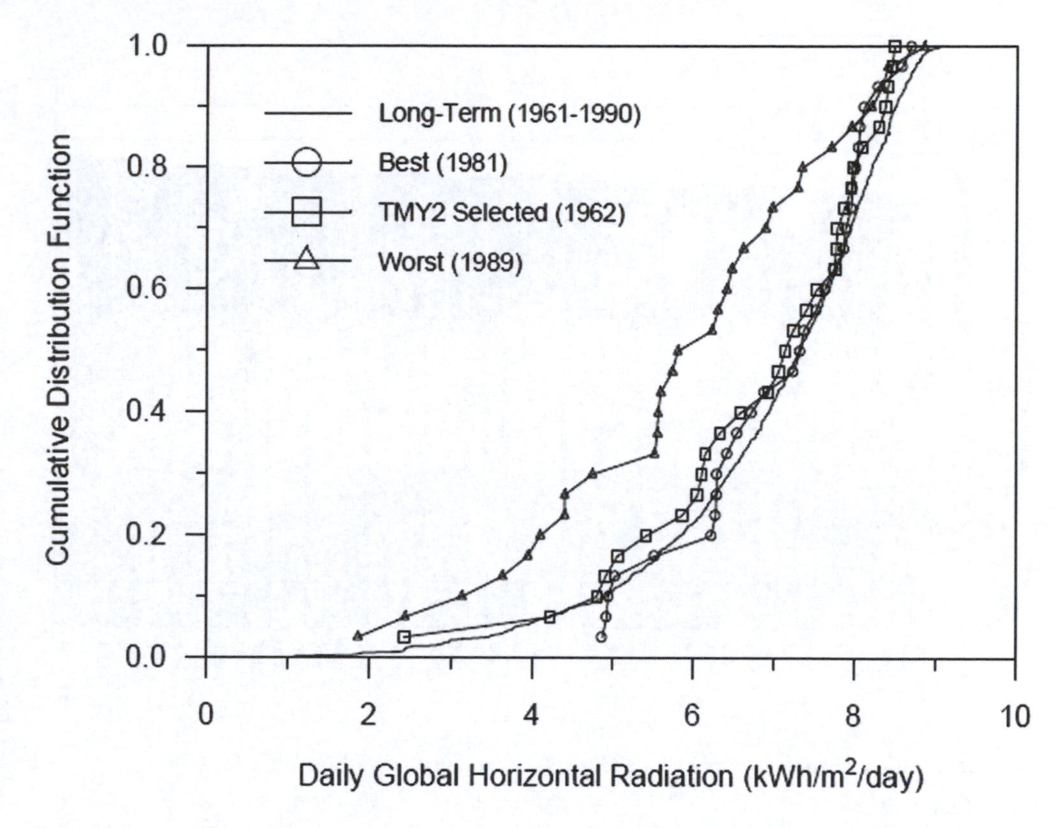

Figure 3.3 Cumulative distribution functions for June global horizontal solar radiation for Boulder, Colorado.

Source: Wilcox and Marion 2008.

for each of the 12. All rely on statistical measures of the similarity of the distributions of daily indices such as minimum, mean, and maximum dry-bulb temperature; minimum, mean, and maximum dew-point temperature; mean and maximum wind velocity; and daily total global and direct solar radiation. Weightings are used for the indices, influencing their importance. In some cases, final selections are made by hand, as illustrated in Figure 3.3. Wilcox and Marion (2008) provide a summary of methods used in the development of the TMY, TMY2, and TMY3 data sets developed for the United States.

The TMY procedure was used for the development of most of the current data sets widely available for building simulation. The Data Sources section (below) lists many of these. Note that in some countries, current files are designated TRY; this should not be confused with the older U.S. TRY format discussed above.

Studies have shown that TMY selection procedures produce weather years that yield average simulated energy use results for typical buildings, although predictions of peak energy demand may be less reliable. For example, Crawley (2008a) finds that office building end-use energy predictions made using typical years fall in the middle of the range of results from individual years; Figure 3.4 shows example results. Seo *et al.* (2009) showed that selection weightings have only moderate effect on energy use predictions for typical buildings.

In summary, research studies and field experience has shown that use of TMY type single year files is reasonably accurate and provides a practical route for estimating long-term building energy performance. However, since a single year cannot capture the full variability of the long-term record, peak energy predictions and uncertainty analyses based on TMYs are often not reliable.

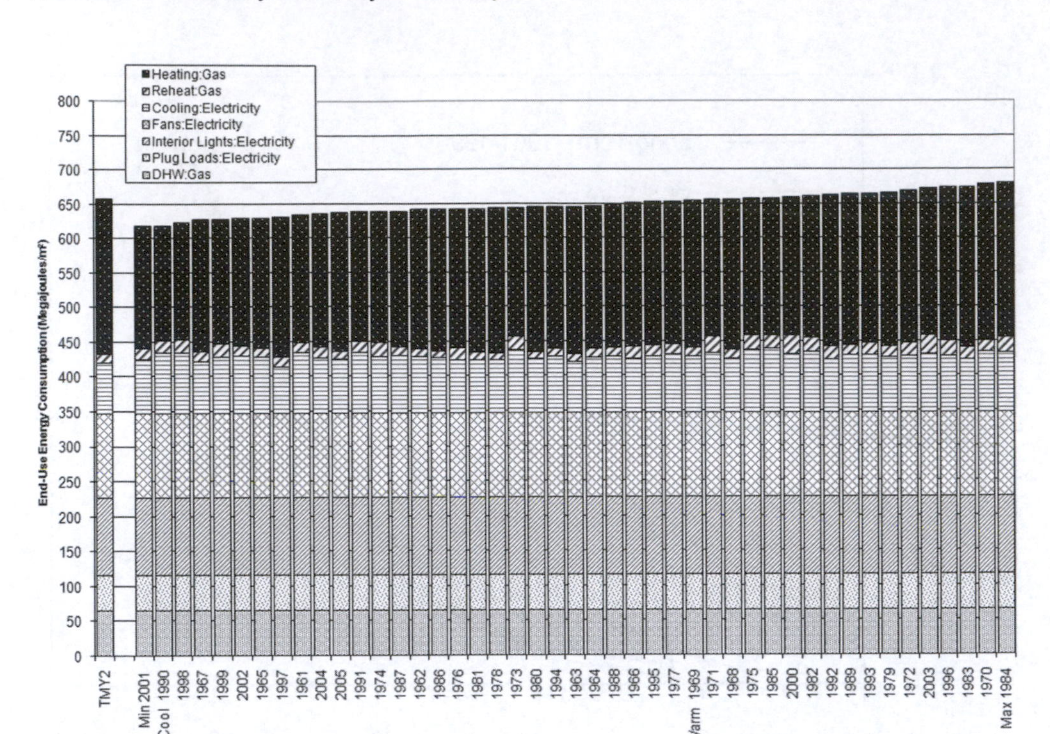

Figure 3.4 End-use energy predictions for 550-m² office building in Washington, DC.

Source: Crawley 2008a.

Modeled data

Weather data generated using many types of models are used extensively in the building simulation field. As discussed above, the weather sequences used by simulation applications must contain all required items for all time steps. It is common for data sources to not include required items and/or have erroneous or missing values. The following are examples of how models are used in adapting available information to simulation requirements.

Solar models

Modeling of solar and illuminance data are particularly important, since these values are significant (in some cases, dominant) drivers of building performance but are sparsely observed. With a few exceptions, all of the weather file collections listed in this chapter include modeled solar and illuminance values. For example, in the TMY3 data set, less than 40 of 1020 sites include measured solar values.

In general, for each hour, solar models derive a theoretically based cloudless sky global horizontal irradiation (GHI) and then apply empirical modifiers that depend on cloud conditions. Davies and McKay (1989) describe and compare a number of first generation procedures. Myers *et al.* (2005) analyze three additional models. Thevenard and Brunger (2002) compare daily modeled and measured values. These and other studies rely on data sets of high-quality measured values for a range of locations. Model quality is assessed using the Mean Bias Error (MBE) and the Root Mean Square Error (RMSE) calculated from monthly, daily, or hourly modeled and measured values. The MBE is zero when the average modeled value agrees with measured. RMSE represents the dispersion of the individual values and is zero only when every point is modeled perfectly. Current models produce very small MBEs (a few percent or less),

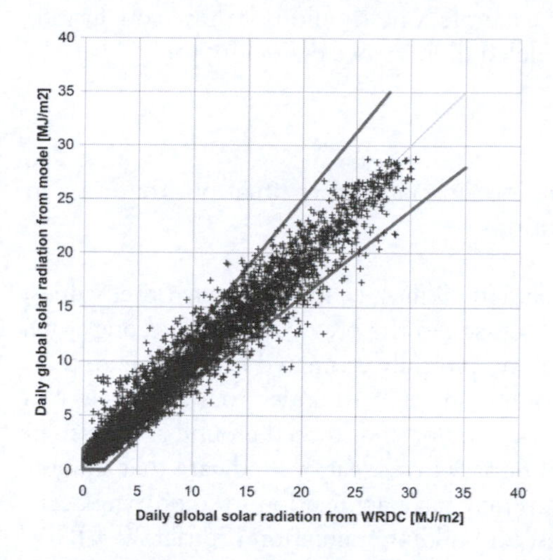

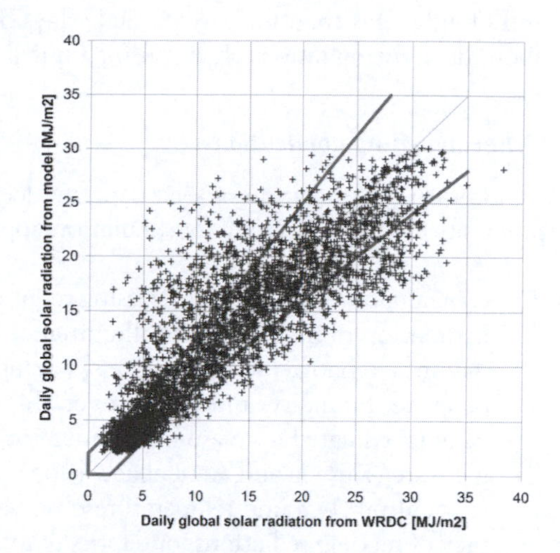

Figure 3.5 World Radiation Data Center (WRDC) measured daily global radiation vs. modeled for Dublin, Ireland (left) and Ulaangom, Mongolia (right).

Source: Thevenard and Brunger 2002, ©2002, ASHRAE, used with permission.

while RMSEs are much larger (perhaps 6% of monthly, 20% for daily, and 30% for hourly). Figure 3.5 shows typical results for daily values.

Recently, models have been enhanced to use satellite data to augment other observations. These models offer improved geographical coverage and have similar or somewhat better accuracy than station-based models. Perez *et al.* (2004) demonstrate that for any site more than 20–25 km from a measurement station, satellite-derived values are the most accurate available estimates of hourly global horizontal irradiance.

Thus far, this discussion of solar models has focused on global horizontal irradiance. However, building simulation requires separate direct and diffuse radiation components incident on each exterior surface. If daylighting is being modeled, illuminance values are needed as well. Partitioning models are used to estimate the direct and diffuse portions of the horizontal irradiance. Daylight models calculate the luminous fraction of the total radiation. Transposition models are used to derive the intensity on an arbitrary surface. The transposition step is necessarily done within simulation programs, since surface-specific information is required. Surface incident values depend on simple beam geometry, the distribution of sky brightness, reflections from other surfaces and surrounding objects, and shading due to other surfaces and surrounding objects. Highly detailed modeling is possible using computer graphics techniques, however, these are often too expensive computationally to be practical for full-year simulation. A common optimization is performing detailed solar geometry calculations for groups of days, as opposed to each day. Descriptions of many solar methods are found in Duffie and Beckman (2006), Clarke (2001), and EnergyPlus (2010a). Muneer (2004) provides comprehensive treatment of both theory and calculation methods, including sample FORTRAN implementations. Underlying research is documented in the technical literature; for example, see Hay and Davies (1978), Perez *et al.* (1990), and Gueymard (1987 and 2009). Citations in these sources will lead the interested reader to numerous publications going back to the 1960s.

In summary, calculation of surface irradiation is among the most complex aspects of building performance simulation and introduces large uncertainty in short-term results. The process starts with values that have significant short-term uncertainty and chains through several steps, each of which produces representative results based on limited inputs. Accurate long-term averages can be expected, but realistic hour-to-hour variability is not captured. This shortcoming is most pronounced for partly cloudy conditions – inter-day differences are smaller for clear and cloudy days than for partly cloudy days. For example, one should not draw conclusions about daylighting rate-of-change using typical modeled illuminance data as input.

Other modeling applications

In addition to generation of solar data, models are used in many other situations that arise in preparation of weather data for simulation applications.

- *Generation of required items*. Missing required data items must be inferred from other values. Estimation of irradiance and illuminance is discussed in the previous section. Long-wave radiant exchange models require sky temperature, typically estimated from dry-bulb temperature, humidity, and cloud cover. Temperature adjacent to underground surfaces can be approximated from annual temperature cycles and/or undisturbed ground surface temperature (Kusuda and Achenbach 1965). Preliminary investigations indicate that shallow (agricultural) ground temperature observations may have application in improving accuracy of model predictions for depths of interest for building simulation (Spitler 2009).
- *Filling gaps*. As in any set of observations, weather data sources can be incomplete or include obviously erroneous values. For example, if within a single record, wet-bulb temperature is greater than the dry-bulb temperature, one or the other is incorrect (or both). Errors can originate from instrumentation malfunctions, data-processing bugs, or transcription mistakes (for pre-automation hand-entered data), to name a few possibilities. For short gaps up to a few hours, linear or cubic spline interpolation is often satisfactory. Baltazar and Claridge (2006) compared 17 algorithms and found the least bias using linear interpolation for filling dry bulb and dew point temperature gaps up to 6 hours. Methods for filling longer gaps are a topic of active current research. Exploitation of mesoscale models and remote sensing techniques (discussed below) should soon allow accurate repair of nearly all data gaps.
- *Reconstruction of short time step data.* Simulation applications that utilize time steps shorter than an hour require corresponding short-interval weather data. Since hourly is the commonly available form, sub-hourly values must often be generated. As can be seen from Figure 3.1, linear interpolation can safely be used to estimate mid-hour temperature values. Reconstruction of plausible sub-hourly wind and especially solar data require more sophisticated models; for example, Meteonorm (Meteotest 2009) uses stochastic procedures to generate values for intervals as short as one minute. However, for general building energy estimation, simple linear methods are acceptable, since building mass damps short-term variation.
- *Generation of weather sequences*. A number of procedures have been developed that generate hourly weather sequences of up to a full year in length using longer term weather statistics as input. For example, see Meteotest (2009), Degelman (1991), Klein and Reindl (1998) and Malkin (2009). Degelman (2007) found that synthetic weather yields simulation results very similar to those derived with observed data for a case study building.
- *Adjustment of weather sequences*. A common requirement in practice is the need to adjust weather data to represent an alternative site, to account for urban effects, or to preview future climate conditions. Several methods and tools have been developed to modify data; among the earliest is Arens *et al.* (1980). A more recent tool is WeatherMaker (SBIC 2009). Additional methods are discussed in later sections on Urban Heat Island and Climate Change. Mesoscale models (discussed below) have potential use here also.
- *Time-shifting*. In some cases, weather data items must be manipulated to reconcile them with project site local time. This can be particularly important for solar values, since physically impossible conditions can arise if sun position is calculated relative to the project site and then applied to data observed elsewhere. Local time conventions, such as daylight savings, must also be correctly handled.

Current developments

On-going developments in measurement and modeling of environmental data are finding application in building performance simulation. The following are some examples of the advances that are expected to yield improvements in the quality and availability of weather data for simulation.

- *Mesoscale models*. Geographically aware regional models are now universally used for weather forecasting. It is possible to use these models to generate localized historical weather sequences by driving them with measured data from nearby stations. Malkin (2009) describes creation of a set of 55,000 files for the United States. As of 2009, mesoscale models remain too computationally intensive for routine use, but this may change given the rapid development of multi-processor computers. See Grell *et al.* (1995), Benjamin *et al.* (2004) and WRF (2009).
- *Reanalysis*. Meteorological services are continuously improving historical records in support of climate studies. For example, the North American Regional Reanalysis (NARR 2009) makes available data sets of numerous variables at 3-hour intervals for the period 1979–present on a 32 km grid. These resources remain to be exploited for building simulation.
- *Remote sensing*. Many years of satellite observations now exist. These data are beginning to contribute to weather data used in practice. Satellite-based data offer uniform worldwide coverage, a major advantage. The 2009 ASHRAE clear sky model uses parameters derived from satellite-based aerosol estimates, see Thevenard (2009a). Stackhouse *et al.* (2006) describe satellite-derived engineering data available from NASA.

Data sources

A growing body of simulation and engineering weather data is available. This section lists and characterizes some of the sources. However, Internet searching is recommended, since any published list is rapidly out of date. Many of these sources are from North America (where data files are often publicly available), however several of the data sets include files for other regions. Some of the files are available in EPW format at www.energyplus.gov/cfm/weather_data.cfm.

- *European Test Reference Years*. This set of files was created using composite TMY methodology for simulation of solar energy systems and energy consumption in buildings. Petrakis (1995) recommended revised procedures for generating files for Europe (Commission of the European Community 1985).
- *ASHRAE Design Weather Data*. WDView 4.0 provides design weather data for 5,564 worldwide locations, including monthly design temperatures, bin data, and heating or cooling degree days to any base (hourly data are not provided). In addition, clear sky solar radiation can be generated for all locations (ASHRAE 2009a, 2009b).
- *ASHRAE Weather Year for Energy Calculations (WYEC2)*. Typical year weather data for 77 locations, including modeled solar and illuminance data. See Stoffel and Rymes (1998). CD available from www.ashrae.org.
- *California Thermal Zones 2 (CTZ2)*. Second generation weather data for 16 California climate zones for use in demonstrating compliance with Title 24 with approved building energy simulation programs. Available from CEC (2009). A project was started in 2009 to update these files.
- *Canadian Weather for Energy Calculations (CWEC)*. The 80 CWEC files contain hourly weather observations representing an artificial one-year period specifically designed for building energy calculations. CWEC hourly files represent weather conditions that result in approximately average heating and cooling loads in buildings. The National Energy

Code of Canada requires the use of a CWEC file representative of a location when the performance path and customized design calculations are chosen as the means of building energy consumption compliance. The CWEC were derived from the Canadian Energy and Engineering Data Sets (CWEEDS) of hourly weather information for Canada from the 1953–1995 period of record. The CWEC are available from Environment Canada (Numerical Logics 1999).

- *Chartered Institution of Building Services Engineers (CIBSE)*. The CIBSE, in association with the (U.K.) Met Office has produced Test Reference Years and Design Summer Years for 14 U.K. locations for use with building energy simulation software. The data sets are available in several formats, including EnergyPlus/ESP-r (CIBSE 2009).
- *Chinese Standard Weather Data (CSWD)*. Developed for use in simulating building heating and air conditioning loads and energy use, and for calculating renewable energy utilization, this set of 270 typical hourly data weather files were developed by Dr. Jiang Yi, Department of Building Science and Technology at Tsinghua University and China Meteorological Bureau. The source data include annual design data, typical year data, and extreme years for maximum enthalpy, and maximum and minimum temperature and solar radiation (China Meteorological Bureau *et al.* 2005).
- *Chinese Typical Year Weather (CTYW)*. Developed for use in simulating building heating and air conditioning loads and energy use, and for calculating renewable energy utilization, this set of 57 weather files is based on a 1982–1997 period of record with data obtained from the U.S. National Climatic Data Center. Zhang Qingyuan of Tsukuba University Japan created the original data set, in collaboration with Joe Huang of Lawrence Berkeley National Laboratory. Zhang and Huang (2004) documented development of the original typical year weather files.
- *IMGW Weather data set for Poland*. The Polish Ministerstwo Infrastruktury prepared this set of 61 files for use in building energy calculations. The files are based on data from the Instytutu Meteorologii i Gospodarki Wodnej (IMGW 2009).
- *MSI Weather Data for Israel*. Weather data for Israeli locations developed by Faculty of Civil and Environmental Engineering, Technion – Israel Institute of Technology, Haifa, Israel, from data provided by the Meteorological Service of Israel (IMS 2009).
- *Indian Weather Data from the Indian Society of Heating, Refrigerating and Air-Conditioning Engineers (ISHRAE)*. Developed for use with building energy performance simulation programs, this set of 58 locations in India was created by the Indian Society of Heating, Refrigerating and Air-Conditioning Engineers (ISHRAE) in TMY2 format (ISHRAE 2009).
- *International Weather for Energy Calculations (IWEC)*. The IWEC data files are typical weather years suitable for use with building energy simulation programs for 227 locations outside the USA and Canada. The files are derived from up to 18 years of hourly weather data originally archived at the U.S. National Climatic Data Center. The weather data are supplemented by modeled solar radiation. Available from ASHRAE (2001). As of 2010, ASHRAE is nearing completion of the IWEC2 data set for nearly 3,000 locations outside the United States and Canada.
- *New Zealand National Institute of Water and Atmospheric Research Ltd (NIWA)*. The New Zealand Energy Efficiency and Conservation Authority (EECA) has developed a simulation-based Home Energy Rating Scheme (HERS) along with a supporting set of hourly weather data files for all climate zones of New Zealand. The files represent artificial years created from twelve representative months (Liley *et al.* 2007).
- *RMY Australia Representative Meteorological Year Climate Files*. Developed for the Australia Greenhouse Office for use in complying with Building Code of Australia, these data are for 66 locations throughout Australia, covering the climate regions in the Building Code of Australia (Energy Partners 2008).
- *Solar and Wind Energy Resource Assessment (SWERA)*. The SWERA project, funded by

the United Nations Environment Program, develops high quality information on solar and wind energy resources in 14 developing countries. Typical-year hourly data are available for 156 locations in Belize, Brazil, China, Cuba, El Salvador, Ethiopia, Ghana, Guatemala, Honduras, Kenya, Maldives, Nicaragua, and Sri Lanka. The data are available from the SWERA project web site (SWERA 2009).

- *Typical Meteorological Year 3 (TMY3)*. The TMY3 set includes files for 1020 locations in the United States including Guam, Puerto Rico, and the U.S. Virgin Islands, derived from a 1991–2005 period of record. The files contain hourly values of solar radiation and meteorological elements for a one-year period. Their intended use is for computer simulations of solar energy conversion systems and building systems to facilitate performance comparisons of different system types, configurations, and locations in the United States and its territories. Because they represent typical rather than extreme conditions, they are not suited for designing systems to meet the worst-case conditions occurring at a location. The files are available for download from the National Renewable Energy Laboratory for download (Wilcox and Marion 2008).

Real-time data

For some applications, notably those that use simulation models to forecast near-term building performance, very recent data are required. Example sources that make such data available (although not necessarily in ready-to-use form) include:

- *U.S. National Renewable Energy Laboratory*. NREL collates and makes available hourly data from about 4,000 worldwide sites. See Long (2006).
- *Oklahoma Mesonet*. In the U.S., the state of Oklahoma operates a network of 110 stations with high quality instruments (including pyranometers) recording observations at five-minute intervals. With about a one-hour lag, all data are available on line (Mesonet 2009).

Selecting the right weather file

A problem often encountered in practice is selection of a suitable weather data source when no weather station is close to the project site under study. Although there is no generally accepted procedure, the first step is to identify some candidate sources with approximately the same latitude and elevation as the site. Then comparison of monthly statistics derived from candidate files (using a simulation program weather utility) to climatological summaries for the project location will generally allow selection of an acceptable match. If no similar source is found, data synthesis or adjustment procedures should be considered.

Urban heat island

Conditions at urban building sites are modified by the surrounding environment and can differ significantly from those at nearby weather stations (typically at airports). Howard (1833) first described the altered meteorological conditions caused by pollution in London as "city fog," Howard also measured the temperature differences between the urban center and the countryside for a number of years, publishing his initial findings in 1820. In a footnote to his table of mean monthly temperature differences, Howard wrote: "night is 3.70° *warmer* and day 0.34° *cooler* in the city than in the country," recognizing what today we call the urban heat island (UHI) effect.

More recently, Mitchell (1953, 1961) measured the extent and intensity of the heat island phenomena. Oke (1988) and Runnalls and Oke (2000) were the first to develop diagrams to explain the diurnal and seasonal patterns of heat islands. Their diagrams were confirmed by the temperature measurements by Streuker (2003) and Morris and Simmonds (2000). Specifically, Streuker's measurements reinforced Oke's findings (1973) that heat island intensity depends on urban concentration (population density), vegetation, and surface albedo. Oke (2006) presents a broad overview of urban influences on local weather conditions and guidance about how to make meteorological measurements at urban sites. The U.S. Environmental Protection Agency's Heat Island Reduction Initiative estimates that the heat island effect is in the range of 2–10 °F (1–5°C) (U.S. EPA 2007). Additional discussion and references relating to UHI are found in Chapter 15. This extensive body of knowledge has seen only limited application in the building simulation field. Erell and Williamson (2006) describe an urban canyon air temperature (CAT) model. Williamson *et al.* (2009) and Crawley (2008b) offer two of the few studies to date on the impact on simulation results of using data modified to include urban influences. There is need for additional contributions in this area.

There is a clearly detectable warming trend in long-term climate records, probably caused primarily by urbanization, although global warming may be at work as well (and it certainly will be in the relatively near future, as discussed below). For example, Thevenard (2009b) analyzed building-related weather statistics for 1274 worldwide stations using inter-decade comparisons. Modest warming trends were found, somewhat more pronounced in winter than summer; an example comparison is shown in Figure 3.6.

Climate change

Over the past 15 years, the international scientific community as organized through the Intergovernmental Panel on Climate Change (IPCC) has focused significant effort to characterize the potential impacts of greenhouse gas emissions from human activities on the global climate. IPCC Working Group I focused on creating atmosphere-ocean general circulation models (GCMs), similar to models used to predict the weather, in which the physics of atmospheric motion are used to predict the future climate. The four major GCMs are

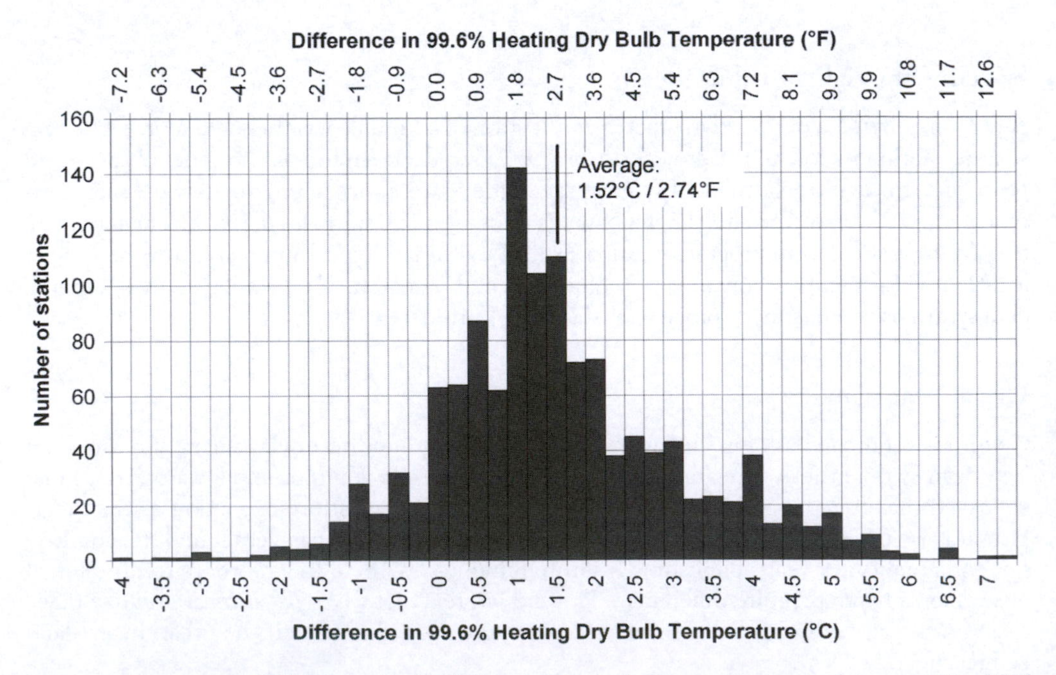

Figure 3.6 Histogram of differences in 99.6% heating dry bulb design temperature between 1997–2006 and 1977–1986.

Source: Thevenard 2009b, ©2009, ASHRAE, used with permission.

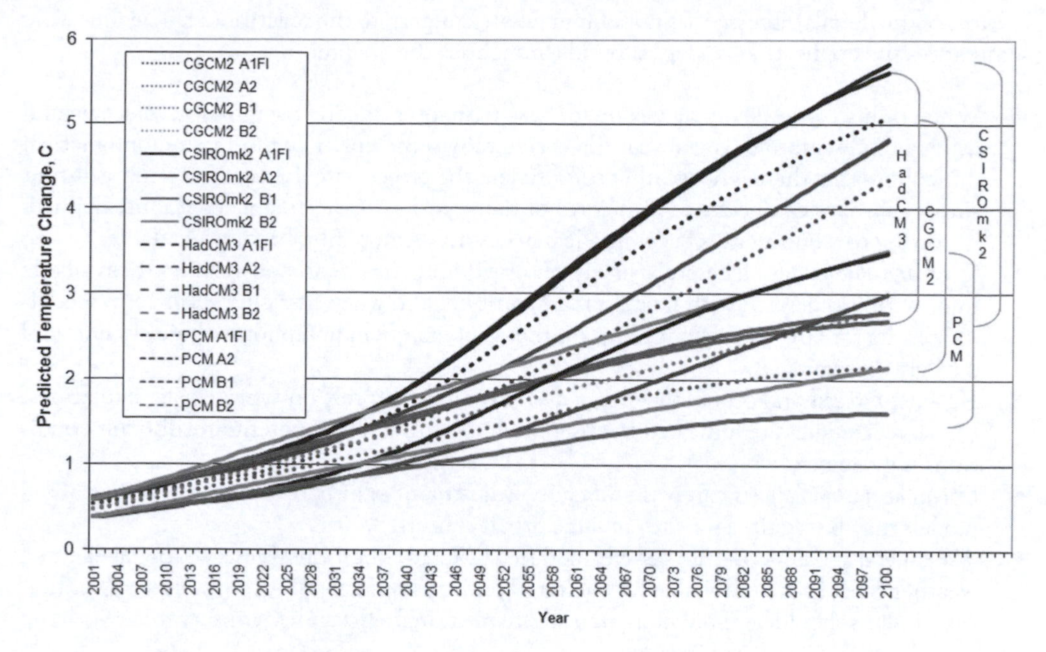

Figure 3.7 Change in annual average temperature relative to 1990 as predicted by several GCM models.

Source: Data from Intergovernmental Panel on Climate Change 2001.

HadCM3 (United Kingdom), CSIRO2 (Australia), CGCM2 (Canada), and PCM (USA) (IPCC 2001). The IPCC (2000) specified four scenarios representing a range of continued emission of carbon dioxide and other pollutants based on specific economic and political conditions. When these scenarios are simulated within the four major GCMs, they result in 16 combinations of scenario and emissions. The potential change in predicted annual average global temperature during the coming century is 1.5 to nearly 6 °C, as shown in Figure 3.7. It must be emphasized that the data shown are averages – some locations will experience larger changes.

The IPCC released its fourth assessment report (AR4) in 2007 (IPCC 2007), in which it identifies buildings as the sector with the highest economic mitigation potential of any other energy sector. Simulation methods will figure heavily in determining optimum mitigation strategies. To allow simulation of building behavior under projected conditions, techniques have been developed to adjust current weather files to reflect climate change scenarios. A "morphing" technique is proposed by Belcher *et al.* (2005) and used by Jentsch *et al.* (2008) in the publicly available tool CCWorldWeatherGen (2009). This method was used to produce weather years for 2020s, 2050s, and 2080s for 14 locations in the UK available from CIBSE (2009). Crawley (2008b) presents a methodology for generating predicted weather sequences for 2100. Guan (2009) characterizes proposed methods of preparing future weather data for simulation. For study of building design robustness, Struck *et al.* (2009) recommend using multi-year simulation with projected weather data, as opposed to single years projected from current reference years.

Synopsis

As can be seen from even the abbreviated summary presented in this chapter, an extensive body of knowledge and technology is employed in deriving the outdoor conditions that are used to simulate building performance. Some of the techniques are built into end-user simulation software while others are used during the preparation and distribution of data sets.

Most of the details discussed are not of immediate concern to the practitioner. The following summarizes the application guidance found throughout the chapter.

- When doing *typical hourly simulation* for design analysis or code compliance, selecting and using a representative typical year file is generally sufficient. The most common practical difficulty is that there is no weather station near the project site. In this case, a more distant station perhaps can be used (with care) or data synthesis/adjustment techniques applied. Resorting to adjustment is becoming less necessary as more files become available.
- *Solar and illuminance data* are sparsely observed and thus nearly all commonly available information is derived with models. Care must be used when assessing short-term modeling results for any system that relies on these inputs, including photovoltaic systems and daylighting strategies.
- *Net-zero projects* and other low-energy designs should not rely on typical year simulations. Multi-year studies are indicated to ensure that the full range of potentially difficult conditions is evaluated.
- *Urban heat island effects* can be significant. Adjustment of airport data or use of specialized models may be required for sites in large urban centers.
- *Wind* speed and direction are heavily modified by local conditions. Observations at exposed weather stations are unreliable for sheltered project sites, even if near by. Natural ventilation designs should be simulated using at-site measurements and/or conservative synthetic data.
- *Climate change* will have a small impact in the short term but is expected to cause significantly changed conditions in 50 to 100 years; this should be considered for long-life projects.
- *Model calibration* and other work with measured performance generally require contemporaneous weather data observed at the project site. Typical year data are not appropriate.

Because of the interaction of actual environmental conditions and shortcomings in observations (notably solar radiation), weather data remain a large source of uncertainty in building simulation predictions. This uncertainty may be trending lower, due to improvements such as remote sensing, enhanced models, and more observation stations. However, the practitioner must remain cognizant of his or her application and how model results are being used.

References

Arens, E.A., Flynn, L.E., Nall, D.N. and Ruberg, K. (1980) *Geographical Extrapolation of Typical Hourly Weather Data for Energy Calculation in Buildings*, National Bureau of Standards Building Science Series 126.

ASHRAE (2001) *International Weather for Energy Calculations (IWEC Weather Files) Users Manual and CD-ROM*, Atlanta: ASHRAE.

ASHRAE (2009a) *ASHRAE Handbook – Fundamentals*, Atlanta: ASHRAE.

ASHRAE (2009b) WDView 4.0 (CD), Atlanta: ASHRAE.

Baltazar, J.C. and Claridge, D.E. (2006) "Study of Cubic Splines and Fourier Series as Interpolation Techniques for Filling in Short Periods of Missing Building Energy Use and Weather Data," *Journal of Solar Energy Engineering* 128(2): 226–230.

Belcher, S.E., Hacker, J.N. and Powell, D.S. (2005) "Constructing Design Weather Data for Future Climates," *Building Services Engineering Research and Technology* 26(1): 49–61.

Benjamin, S.G., Devenyi, D.D., Weygandt, S.S., Brundage, K.J., Brown, J.M., Grell, G.A., Kim, D., Schwartz, B.E., Smirnova, T.G., Smith, T.L. and Manikin, G.S. (2004) "An Hourly Assimilation–Forecast Cycle: The RUC," *Monthly Weather Review* 132(2): 495–518.

CCWorldWeatherGen (2009) http://www.energy.soton.ac.uk/ccworldweathergen/index.html. Sustainable Energy Research Group, University of Southampton.

CEC (2009) California Energy Commission. http://www.energy.ca.gov/title24/.

China Meteorological Bureau, Climate Information Center, Climate Data Office and Tsinghua University, Department of Building Science and Technology (2005) *China Standard Weather Data for Analyzing Building Thermal Conditions*, Beijing: China Building Industry Publishing House.

CIBSE (2009) "Test Reference Years and Design Summer Years (current and future)," The Chartered Institution of Building Services Engineers. Online. Available at: www.cibse.org/index.cfm?go=publications.view&PubID=332&S1=y&L1=0&L2=0 (Accessed: October 18, 2010).

Clarke, J.A. (2001) *Energy Simulation in Building Design*, Oxford: Butterworth-Heinemann.

Commission of the European Community (1985) "Test Reference Years, Weather Data Sets for Computer Simulations of Solar Energy Systems and Energy Consumption in Buildings," CEC, DG XII, Brussels, Belgium: Commission of the European Community.

Crawley, D.B. (2008a) "Estimating the Impacts of Climate Change and Urbanization on Building Performance," *Journal of Building Performance Simulation* 1(2): 91–115.

Crawley, D.B. (2008b) "Building Performance Simulation: A Tool for Policymaking," PhD thesis. University of Strathclyde, Glasgow.

Davies, J.A. and McKay, D.C. (1989) "Evaluation of Selected Models for Estimating Solar Radiation on Horizontal Surfaces," *Solar Energy* 43(3): 153–168.

Degelman, L.O. (1991) "A Statistically-based Hourly Weather Data Generator for Driving Energy Simulation and Equipment Design Software for Buildings," in *Proceedings of Building Simulation 1991*, Nice, Sophia-Antipolis, France, pp. 592–599.

Degelman, L.O. (2007) "Testing the Accuracy of Synthetically-Generated Weather Data for Driving Building Simulation Models," in *Proceedings of Building Simulation 2007*, Beijing, China, pp. 1–8.

Duffie, J.A. and Beckman, W.A. (2006) *Solar Engineering of Thermal Processes*, 3rd ed., New York: Wiley.

Energy Partners (2008) *Australian Climate Data Bank: Weather Data Enhancement for Reference Meteorological Years*, Manuka, ACT, Australia. Available from www.environment.gov.au.

EnergyPlus (2010a) *EnergyPlus Engineering Reference*. http://energyplus.gov/pdfs/engineeringreference.pdf.

EnergyPlus (2010b) *EnergyPlus Weather Data Format Definition*. http://energyplus.gov/weatherdata_format_def.cfm

Erell, E. and Williamson, T. (2006) "Simulating Air Temperature in an Urban Street Canyon in All Weather Conditions Using Measured Data from a Reference Meteorological Station," *International Journal of Climatology* 26(12): 1671–1694.

Grell, G.A., Dudhia, J. and Stauffer, D.R. (1995) "A Description of the Fifth Generation Penn State/NCAR Mesoscale Model (MM5)," NCAR Technical Note NCAR/TN-398 + STR.

Guan, L. (2009) "Preparation of Future Weather Data to Study the Impact of Climate Change on Buildings," *Building and Environment* 44(4): 793–800.

Gueymard, C.A. (1987). "An Anisotropic Solar Irradiance Model for Tilted Surfaces and its Comparison with Selected Engineering Algorithms," *Solar Energy* 38(5): 367–386 and "Erratum," *Solar Energy* 40(2): 175.

Gueymard, C.A. (2009) "Direct and Indirect Uncertainties in the Prediction of Tilted Irradiance for Solar Engineering Applications," *Solar Energy* 83(3): 432–444.

Hay, J.E. and Davies, J.A. (1978) "Calculation of the Solar Radiation Incident on an Inclined Surface," in *Proceedings of the First Canadian Solar Radiation Data Workshop*, Toronto, Canada, pp. 59–72.

Hensen, J.L.M. (1999) "Simulation of Building Energy and Indoor Environmental Quality – Some Weather Data Issues," in *Proceedings of the International Workshop on Climate Data and Their Applications in Engineering*, Prague: Czech Hydrometeorological Institute.

Howard, L. (1833) *Climate of London Deduced from Meteorological Observations*, 3rd ed., London: Harvey & Darton.

IMGW (2009) http://www.mi.gov.pl/2-48203f1e24e2f-1787735.html (in Polish).

Intergovernmental Panel on Climate Change (2000) *Emissions Scenarios*, IPCC Special Report, Cambridge: Cambridge University Press.

Intergovernmental Panel on Climate Change (2001) *Climate Change 2001: Impacts, Adaptation and Vulnerability*, Cambridge: Cambridge University Press.

Intergovernmental Panel on Climate Change (2007) *Climate Change 2007: Synthesis Report*, Cambridge: Cambridge University Press.

ISHRAE (2009) www.ishrae.in.

Jentsch, M.F., Bahaj, S. and James, P.A.B. (2008) "Climate Change Future Proofing of Buildings: Generation and Assessment of Building Simulation Weather Files," *Energy and Buildings* 40(12): 2148–2168.

Klein, S.A. and Reindl, D.T. (1998) *Automated Generation of Hourly Design Sequences*, Final report, ASHRAE RP-962, Atlanta: ASHRAE.

Kusuda, T. and Achenbach, P.R. (1965) "Earth Temperature and Thermal Diffusivity at Selected Stations in the United States," *National Bureau of Standards Report* 8972.

Liley, J.B., Shiona, H., Sturman, J. and Wratt, D.S. (2007) *Typical Meteorological Years for the New Zealand Home Energy Rating Scheme*, Prepared for the Energy Efficiency and Conservation Authority, NIWA Client Report: LAU2007- 02-JBL. NIWA: Omakau, New Zealand.

Long, N. (2006) "Real-Time Weather Data Access Guide," NREL/BR-550=34303.

Malkin, S. (2009). "Weather Data for Building Energy Analysis," Autodesk Green Building Studio White Paper. Available at http://images.autodesk.com/adsk/files/weather_data_greenbuildingstudio_adsk_white_paper.pdf.

Mesonet (2009) Oklahoma Mesonet, www.mesonet.org

Meteotest (2009) Meteonorm 6.1, A Global Meteorological Database for Engineers, Planners, and Education, Bern, Switzerland, www.meteonorm.com.

Mitchell, J.M. (1953). "On the Causes of Instrumentally Observed Secular Temperature Trends," *Journal of Meteorology* 10(4): 244–261.

Mitchell, J.M. (1961) "The Temperature of Cities" *Weatherwise* 14: 224–229.

Morris, C.J.G. and Simmonds, I. (2000) "Associations between Varying Magnitudes of the Urban Heat Island and the Synoptic Climatology in Melbourne, Australia," *International Journal of Climatology* 20(15): 1931–1954.

MSI (2009) Meteorological Service of Israel.

Muneer, T. (2004) *Solar Radiation and Daylight Models*, 2nd ed., Oxford, UK: Butterworth-Heinemann.

Myers, D.R., Wilcox, S., Marion, W., George, R. and Anderberg, M. (2005) "Broadband Model Performance for an Updated National Solar Radiation Database in the USA," presented at ISES 2005 Solar World Congress, Orlando, FL, http://www.nrel.gov/docs/fy05osti/37699.pdf.

NARR (2009) North American Regional Reanalysis/Environmental Modeling Center. http://www.emc.ncep.noaa.gov.

NCDC (1976) *Test Reference Year (TRY)*, Tape Reference Manual, TD-9706, September 1976, Asheville, NC: National Climatic Data Center, U.S. Department of Commerce.

Numerical Logics (1999) *Canadian Weather for Energy Calculations*, Users Manual and CD-ROM, Downsview, Ontario: Environment Canada.

Oke, T.R. (1973) "City Size and the Urban Heat Island," *Atmospheric Environment* 7(8): 769–779.

Oke, T.R. (1988) *Boundary Layer Climates*, New York: Routledge.

Oke, T.R. (2006) *Initial Guidance to Obtain Representative Meteorological Observations at Urban Sites*. IOM Report No. 81, WMO/TD. No. 1250, Geneva, Switzerland: World Meteorological Organization.

Perez, R., Ineichen, P., Seals, R., Michalsky, J. and Stewart, R. (1990). "Modeling Daylight Availability and Irradiance Components from Direct and Global Irradiance," *Solar Energy* 44(5): 271–289.

Perez, R., Kmiecik, M., Moore, K., Wilcox, S., George, R., Renne, D., Vignola, F. and Ineichen, P. (2004) "Status of High Resolution Solar Irradiance Mapping from Satellite Data," in *Proceedings of the ASES Annual Meeting 2004*, Portland, OR, pp. 265–270.

Petrakis, M. (1995) "Test Meteorological Years: Procedures for Generation of TMY from Observed Data Sets," in *Natural Cooling Techniques, Research Final Report, PASCOOL*, Alvarez *et al.*, ed. Brussels, Belgium: Commission of the European Commission.

Runnalls, K.E. and Oke, T.R. (2000) "Dynamics and Controls of the Near-Surface Heat Island of Vancouver, British Columbia," *Physical Geography* 21(4): 283–304.

SBIC (2009) Energy10 (with WeatherMaker) available from Sustainable Buildings Industries Council. See www.sbicouncil.org.

Seo, D., Huang, J. and Krarti, M. (2009) "Evaluation of Typical Weather Year Selection Approaches for Energy Analysis of Buildings," *ASHRAE Transactions* 115(2): 654–667.

Spitler, J. D. (2009) Private communication.

Stackhouse, P.W., Whitlock, C.H., Chandler, W.S., Hoell, J.M., Zhang, T., Mikovitz, J.C., Leng, G.S. and Lilienthal, P. (2006) *Supporting Energy-Related Societal Applications Using NASA's Satellite and Modeling Data*, http://ntrs.nasa.gov.

Stoffel, T.L. and Rymes, M.D. (1998) "Production of the Weather Year for Energy Calculations version 2 (WYEC2) Data Files," 104(2) *ASHRAE Transactions*: 498–497.

Streuker, D.R. (2003) "A Study of the Urban Heat Island of Houston, Texas," PhD Thesis, Rice University, Houston, Texas.

Struck, S., de Wilde, P., Evers, J., Hensen, J. and Plokker, W. (2009) "On Selecting Weather Data Sets to Estimate a Building Design's Robustness to Climate Change," in *Proceedings of Building Simulation 2009*, Glasgow, Scotland. pp. 513–520.

SWERA (2009) Solar and Wind Resource Assessment. http://swera.unep.net.

Thevenard, D. (2009a) *Updating the ASHRAE Climatic Data for Design and Standards (RP-1453)*, ASHRAE Research Project, Final Report, Atlanta: ASHRAE.

Thevenard, D. (2009b) "Influence of Long-term Trends and Period of Record Selection on the Calculation of Climatic Design Conditions and Degree-Days" *ASHRAE Transactions* 115(2).

Thevenard, D. and Brunger, A. (2002) "The Development of Typical Weather Years for International Locations – Part II: Production," *ASHRAE Transactions* 108(2): 480–486.

U.S. Environmental Protection Agency (2007) *Heat Island Effect*. http://www.epa.gov/heatisland/.

Wilcox, S. and Marion, W. (2008) *User's Manual for TMY3 Data Sets*, NREL/TP-581-43156, April 2008, Golden, CO: National Renewable Energy Laboratory.

Williamson, T., Erell, J.E. and Soebarto, V. (2009) "Assessing the Error from Failure to Account for Urban Microclimate in Computer Simulation of Building Energy Performance," in *Proceedings of Building Simulation 2009*, Glasgow, Scotland. pp. 497–504.

WRF (2009) The Weather Research and Forecasting Model, http://www.wrf-model.org.

Zhang, Q. and Huang, J. (2004). *Chinese Typical Year Weather Data for Architectural Use* (in Chinese), Beijing: China Machine Press.

Recommended reading

TMY3 documentation (Wilcox and Marion 2008)
Meteonorm Handbook part II: Theory (Meteotest 2009)

Activities

Download a weather file for a location near where you live. Open it in a spreadsheet application and familiarize yourself with the data format.

Make a weather file suitable for use with a simulation program of your choice.

4 People in building performance simulation

Ardeshir Mahdavi

Scope

This chapter addresses the effects of people's presence and their actions on buildings' performance. It discusses the associated modeling approaches and information input options in view of a more reliable prediction of such effects in the course of building performance simulation studies.

Learning objectives

The reader is to become familiarized with a conceptual framework to systematically situate people in the context of building performance simulation. The reader is further expected to understand the mechanisms and corresponding models of how people's presence and interactions with buildings' environmental systems influence the outcome of building performance simulation studies.

Key words

Occupancy/control actions/building systems

Introduction

To conduct building performance simulation studies, various kinds of information are needed. When asked about the nature of such information, most professionals name primarily model ingredients such as building geometry, material properties, and boundary conditions (e.g., weather). Impact of people (building users, occupants) comes to mind often only as an afterthought, even though no simulation specialist would dispute the importance of the "people factor" in building performance simulation.

To highlight the critical role of the – occasionally underestimated – people factor, one needs only to consider the main purpose of building performance simulation, namely the provision of a solid basis for the evaluation of building designs in view of various performance criteria:

- What amount of energy will be needed to heat, cool, and ventilate the building?
- What kinds of indoor climate conditions (thermal parameter, indoor air quality) are to be expected in the building?
- What level of daylight availability can be expected in indoor environment under dynamically changing outdoor illuminance levels?
- Will the acoustical environment in indoor spaces provide the necessary conditions for communication and task performance?
- Can occupants be safely evacuated from the building in case of an emergency (e.g., outbreak of fire)?

Obviously, none of these questions can be reliably answered without considering the role of the people living and working in the buildings. People affect the performance of buildings, due to their presence and their actions. Energy and thermal performance of buildings is not only affected by the people's presence as a source of (sensible and latent) heat, but also due to their actions, including use of water, operation of appliances, and manipulation of building control devices for heating, cooling, ventilation, and lighting. User-based operation of luminaires and shading devices in a room affect the resultant light levels and visual comfort conditions. Presence of people in a room and the associated sound absorption influences the sound field and thus the acoustical performance of the room. Safety performance of a building cannot be evaluated without considering the behavior of people under emergency.

People's relevance to computational building performance modeling is, however, not restricted to the necessary input models for user presence and action models. Frequently, information on user-based requirements in view of thermal, visual, and acoustical comfort must be explicitly reflected in the settings of relevant simulation runs. For example, control settings for air temperatures and illuminance levels in architectural spaces (needed for the computation of heating, cooling, and lighting loads) must be defined in accordance with structured knowledge on people's needs and requirements. Moreover, building performance simulation results, i.e., values of performance variables, are expected to encapsulate information that is relevant to people's requirements and expectations (health, comfort, satisfaction). Appropriate selection and interpretation of building performance indicators requires thus that the relationship between buildings (occupied spaces) and their occupants are considered and understood.

Building performance simulation sciences and methods have made substantial advances in the last decades. Algorithmic domain models of underlying physical phenomena have become increasing sophisticated and versatile. Computational resources (speed and reliability) have grown in leaps and bounds. Representation of geometry (inclusive of topology) and building components, elements, and materials have experienced significant progress. Likewise, empirically grounded, well-structured, and high-resolution representations of dynamic boundary conditions have emerged. In comparison, however, practically applied models of people presence and actions are still rather simplistic. Moreover, there is a lack of shared methods and standards for representing people in the building simulation practice. This implies the necessity of sustained research and development efforts concerning advanced models of user presence and behavior (specifically control actions).

Toward this end, the present contribution mainly provides a conceptual framework to systematically situate people (users and occupants) in the context of building performance simulation. It deals with the mechanisms and corresponding models of how people's presence and interactions with buildings' environmental systems influence the outcome of performance simulations in terms of the values of relevant thermal and visual performance indicators. Specifically, possible approaches to the representation of occupants' presence and actions in terms of input information to simulation applications are discussed. Thereby, the focus will be on thermal and visual building performance simulation, even though many of the issues and approaches discussed are also relevant for other areas of building performance inquiry.

Effects on people

Broadly speaking, a useful distinction can be made between "passive" and "active" effects of users and occupants on buildings' performance. Passive effects of people on the hygro-thermal conditions in buildings denote those effects caused by the "mere" presence of people in the building. For instance, to correctly simulate hygro-thermal conditions and indoor air quality in architectural spaces, such passive people effects must be considered by the set up and solution of the systems of balance equations implemented in numeric simulation applications. Depending on their activity, people release not only various quantities of sensible and latent heat (see Figure 4.1), but also water vapor, carbon dioxide, and other execrations and odorous

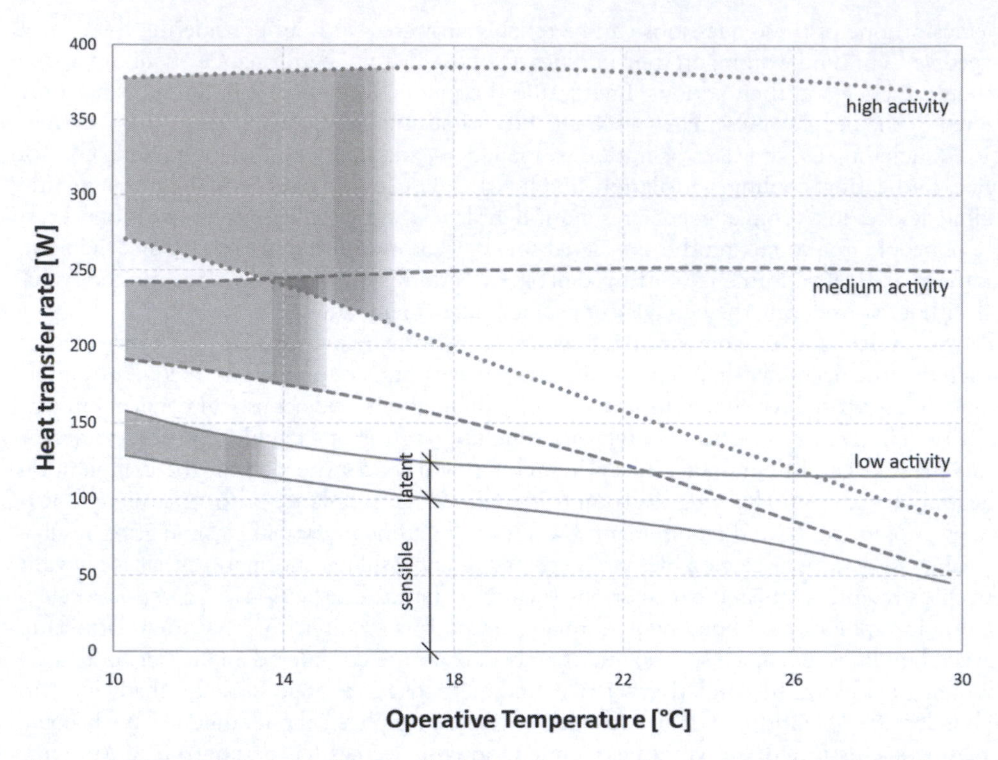

Figure 4.1 Heat transfer rate of human body as a function of activity and the ambient (operative) temperature.

Source: Adapted from Rietschel and Raiß 1968.

substances. Likewise, in the building and room acoustics domain, presence of people in a space has an effect on the sound field via introduction of additional sound absorption. To illustrate this point, Table 4.1 includes data showing the difference in the absorption coefficient of the audience seating area in a room under empty and occupied conditions. An example of the occupancy effect on the acoustical conditions is shown in Figure 4.2, which contrasts the reverberation time in a concert hall under empty and occupied conditions.

To model the passive effects of people's presence in buildings, simulation specialists typically rely on external sources of information such as occupancy load schedules (see, for example, Figure 4.3) derived from measurement results of people's metabolic rates. Provided such external information is available, the modeling process is as such straightforward, barring two possible complexities. Firstly, different levels of resolution are conceivable regarding temporal and spatial distribution of the passive effect in the model. For example, occupancy-based internal

Table 4.1 Sound absorption of unoccupied and occupied seating areas (based on data in Fasold and Veres 2003)

	Frequency [Hz]					
	125	*250*	*500*	*1000*	*2000*	*4000*
Unoccupied wooden seats	0.05	0.05	0.05	0.10	0.10	0.15
Audience on wooden seats	0.40	0.60	0.75	0.80	0.85	0.80
Unoccupied upholstered seats	0.30	0.55	0.55	0.65	0.75	0.75
Audience on upholstered seats	0.60	0.75	0.80	0.85	0.90	0.85

Source: Based on data in Fasold and Veres 2003.

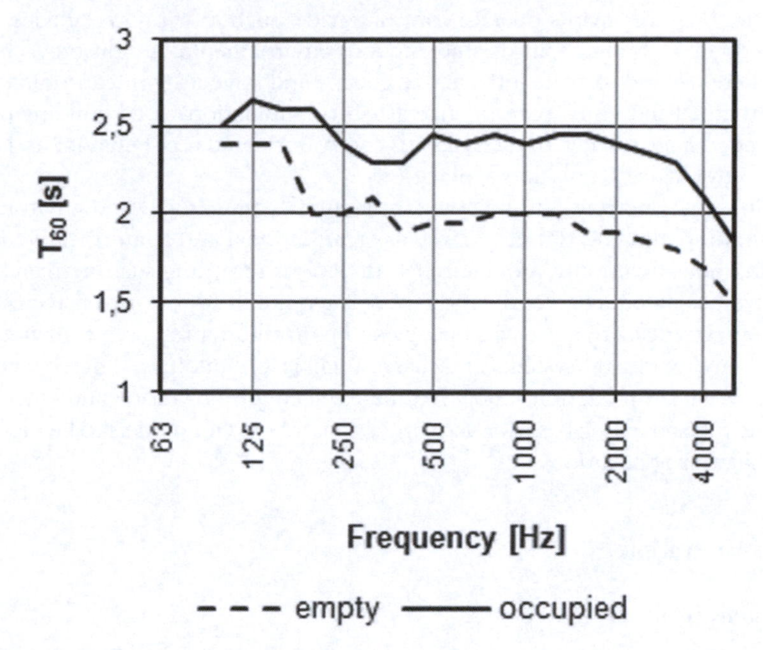

Figure 4.2 Measured reverberation times in a concert hall ("Berliner Philharmonie") under empty and occupied conditions.

loads may be modeled for a global occupancy schedule or, alternatively, in terms of autonomous agents representing individual occupants with distinct individual occupancy patterns. Secondly, the passive people effects such as heat emission (as illustrated in Figure 4.1) may depend on the context (e.g., thermal conditions in occupants' rooms). This interdependence would require – at least in case of highly detailed numeric simulation models – a dynamic coupling between the agent and its immediate environment.

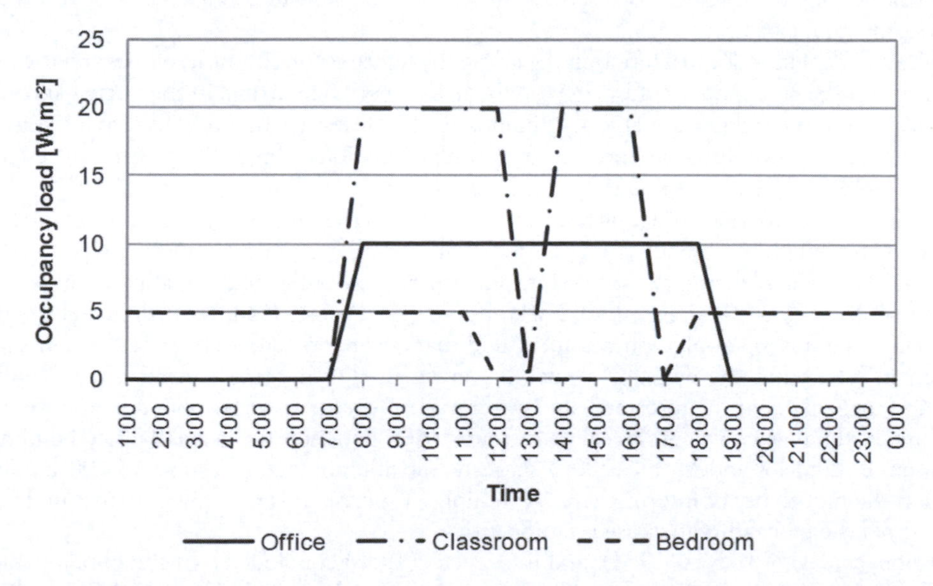

Figure 4.3 Example of an occupancy load model (sensible heat gain) as adapted in a commercially available simulation application.

Source: EDSL 2008.

In most buildings, occupants operate control devices such as windows, shades, luminaires, radiators, and fans to bring about desirable indoor environmental conditions. These control actions are here referred to as people's active effects, and have a significant impact on buildings' hygro-thermal and visual performance. Realistic simulation-based building performance predictions necessitate reliable models of such control-oriented user behavior and their incorporation in performance simulation applications.

General information about building type (residential, commercial) and environmental systems (free-running, air-conditioned) as well as organizational and administrative information (e.g., working hours) can only provide rough directions regarding such active effects. More representative people presence and action models require, however, extensive observational data based on empirical studies of occupancy and control-oriented user behavior (as related to buildings' environmental systems) in a large number of buildings. Thereby, possible relationships between control actions and environmental conditions inside and outside buildings could provide the underlying basis for derivation of user behavior models to be incorporated in building simulation applications.

Empirical observations

A brief summary of past research

A large number of studies have been conducted in the past decades to understand how building occupants interact with buildings' environmental control systems such as windows, blinds, and luminaires. A number of such studies are briefly described in the following. Note that this review is not claimed to be comprehensive. Rather, the objective is to provide an impression of the kinds and scope of the relevant research efforts. As such, there are significant differences between individual studies in this set in terms of size and type of buildings observed, relevant control devices (luminaires, shades, windows, etc.), duration of observation, measured environmental factors, as well as measurements' resolution and precision. Nonetheless, most of these studies share a common feature in that they attempt to establish a link between user control actions (or the state of user-controlled devices) and measurable indoor or outdoor environmental parameter.

Hunt (1979) introduced a function regarding the lighting conditions in offices and the probability that the occupants would switch on the lights upon their arrival in the office. According to this function (see Figure 4.4), only illuminance levels less than 100 lx lead to a significant increase of the "switching on" probability. Similar functions were subsequently suggested by Love (1998) and Reinhart (2001).

Pigg *et al.* (1996) found a strong relationship between the propensity of switching the lights off and the length of absence from the room, stating that people are more likely to switch off the light when leaving the office for longer periods. Similar relationships were found by other studies (Boyce 1980; Reinhart 2001). Boyce (1980) observed intermediate light switching actions in two open-plan offices and found that occupants tend to operate the lights more often in relation to the daylight availability given smaller lighting control zones. Reinhart (2004) suggested that the intermediate "switching on" events are more common at lower than at higher illuminance values. Based on a related study conducted in a small office building in Lausanne, Lindelöf and Morel (2006) suggested an illuminance threshold of 100 lx, above which the probability of intermediate "switching on" events was very low, whereas under this threshold the probability increased significantly.

Rubin *et al.* (1978), Rea (1984), and Inoue *et al.* (1988) concluded that the blind operation rates varied greatly in relation to building orientation. Lindsay and Littlefair (1992) conducted a study of five office buildings in UK and found a strong correlation between the operation of Venetian blinds and the solar radiation intensity (and sun position). Moreover, blinds were

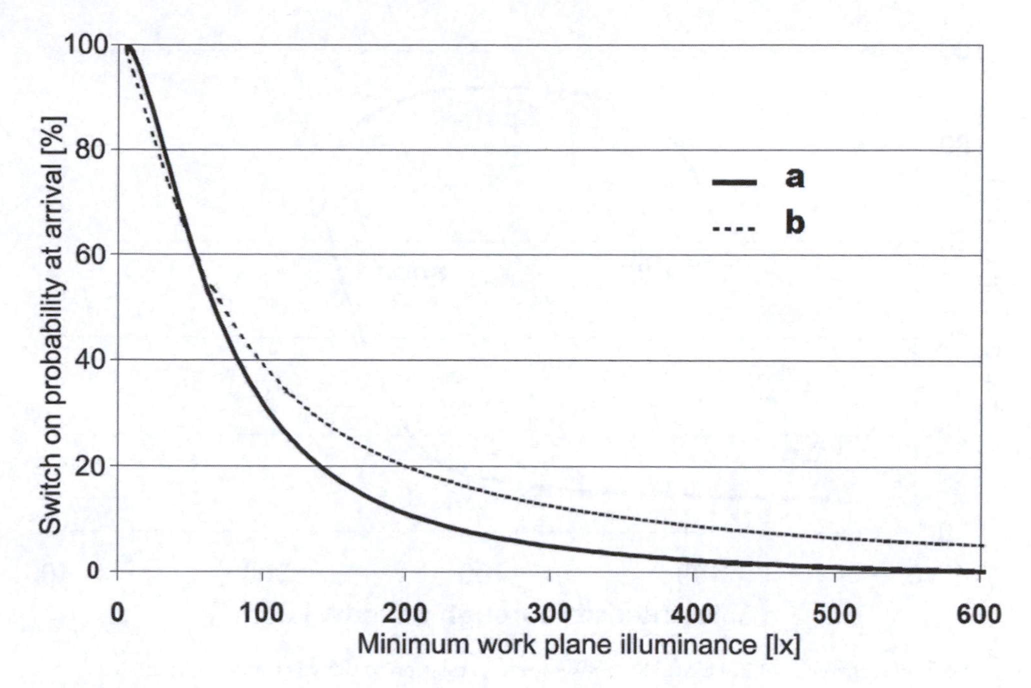

Figure 4.4 Probability of switching the lights on at arrival in the office.

Source: Based on (a) Hunt (1979) and (b) Reinhart (2001).

operated more frequently on the south façade. Rubin *et al.* (1978) suggested that occupants manipulate shades mainly to avoid direct sunlight and overheating. According to Inoue *et al.* (1988), above a certain threshold of vertical solar irradiance on a façade (50 W m^{-2}) the deployment level of shades is proportional to the depth of solar penetration into a room. This conjecture was corroborated by Reinhart (2001). Once closed, shades seem to remain deployed until the end of the working day or when visual conditions become intolerable. Rea (1984) observed a rather low rate of blinds operation throughout the day, implying that occupants' perception of solar irradiance is a long-term one. Inoue *et al.* (1988) observed a specific pattern (Figure 4.5, as adapted by Reinhart 2001) concerning the relation between blind operation and incident illumination on the façade. Inoue concluded that occupants largely ignore short-term irradiance dynamics.

Herkel *et al.* (2005) observed window operation in 21 south-facing single offices in Freiburg, Germany (with smaller and larger window units). Parameters such as window status, occupancy, indoor and outdoor temperatures, as well as solar radiation were regularly recorded. The analysis of the results revealed a strong seasonal pattern behind the window operation. In summer, 60 to 80% of the smaller windows were open, in contrast to 10% in winter. The frequency of window operation actions was observed to be higher in swing seasons spring and autumn. A strong correlation was found between the percentage of open windows and the outdoor temperature. Above 20 °C, 80% of the small windows were completely opened, whereas 60% of the large windows were tilted. The windows were opened and closed more frequently in the morning (9:00) and in the afternoon (15:00). Moreover, window operation occurred mostly when occupants arrived in or left their workplaces. At the end of the working day, most open windows were closed.

Exploring an stochastic simulation approach toward consideration of occupant behavior in buildings, Nicol (2001), as Hunt (1979) before him, used probit analysis (Finney 1947) to examine correlations between outdoor temperature and the use of windows, heating, and

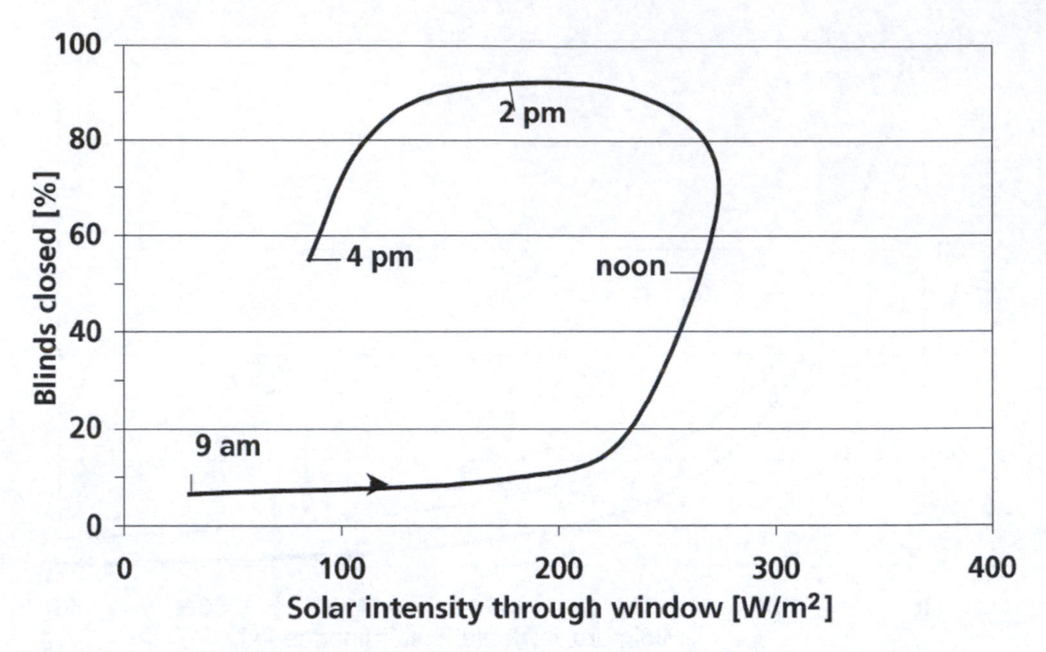

Figure 4.5 Percentage of blinds closed for SSW façade in relation to the vertical solar irradiance.

Source: Based on data from Inoue *et al.* (1988), as adapted in Reinhart (2001).

blinds. The study suggested that information on solar radiation intensity would be necessary to establish correlations pertaining to light and blind usage.

Reinhart (2004) developed LIGHTSWITCH 2002 using a dynamic stochastic algorithm. Based on an occupancy model and a dynamic daylight simulation application, predicted manual lighting and blind control actions provided the basis for the calculation of annual energy demand for electrical lighting. Page *et al.* (2007) hypothesized that the probability of occupancy at a given time step depends only on the state of occupancy at the previous time step. As suggested by Fritsch *et al.* (1990) in relation to window operation, Page *et al.* (2007) explored the use of Markov chains toward occupancy prediction. Most studies of user–system interactions are conducted for individual building systems (lighting, shading, etc.). Bourgeois (2005) attempted to bridge the gap between energy simulation and empirically based information on occupant behavior via a self-contained simulation module called SHOCC (Sub-Hourly Occupancy Control) that was integrated in ESP-r application (ESRU 2002).

Humphreys and Nicol (1998) introduced an adaptive approach to human thermal comfort stating that "people react in ways which tend to restore their comfort, if a change occurs such as to produce discomfort."

Conducting a field survey, Rijal *et al.* (2007) concentrated on window opening behavior in naturally ventilated buildings with regard to indoor (globe) temperature and outdoor air temperature as trigger parameters. The resulting "adaptive algorithm" (Humphreys and Nicol 1998) was implemented in ESP-r toward a more realistic thermal comfort and building performance assessment.

The above studies – and other similar ones – have provided a number of valuable insights into the circumstances and potential triggers of occupancy control actions in buildings. However, given the complexity of the domain, additional long-term and (geographically and culturally) broader studies are necessary to arrive at more dependable (representative) models of control-oriented user actions in buildings.

A recent case study

Objective and approach

Within the framework of a recent cross-section study performed by the author and his associates in Austria, an attempt was made to systematically collect a large consistent set of observational data regarding building occupants' presence and control action patterns (Mahdavi *et al.* 2008a, 2008b; Mahdavi and Pröglhöf 2008). General information regarding these offices is provided in Table 4.2.

The main intention of the study was to observe user control actions pertaining to lighting and shading systems (and in the case of one building, window operation) while considering the indoor and outdoor environmental conditions under which those actions occurred. The study – given its large-scale, long-term and, high-resolution nature – represents an appropriate case in point to demonstrate the potential, complexities, and challenges associated with the derivation of empirically grounded user presence and behavior models in buildings. In fact, one of the initial objectives was to conceive the research in terms of a model case, proposing and testing a general process for designing and conducting user behavior observations in buildings. Specifically, the study was conducted in five office buildings in Austria (Mahdavi *et al.* 2008a). These buildings are referred to henceforth as VC, ET, FH, UT, and HB. In some cases the data analyses for VC included a differentiation between office groups facing North and South-West. To denote this, the abbreviations VC-N and VC-S are used. Data collection was conducted on a long-term basis (9 to 14 months).

The parameters monitored in the course of the study are summarized in Table 4.3. The collected data were stored and processed in a database (see the respective data structure in Table 4.4) in terms of events (E) and states (S) for further analysis. This distinction was important toward organizing the monitored information in terms of a well-structured database that can be efficiently queried. In this taxonomy, events are either system-related (E_s) or occupancy-related (E_o). States can refer to systems (S_s), indoor environment (S_i), outdoor environment (S_e), and occupancy (S_o). The collected data was analyzed using standard approaches in descriptive statistics. Aside from temporal representations of states of occupancy and devices, occurrence of user actions were depicted (as relative frequency distributions or probability functions) in relationship to the value ranges of various indoor and outdoor environmental indicators.

Indoor environmental data (room temperature, relative humidity, and illuminance) were monitored using data loggers distributed across workstations. To obtain information regarding

Table 4.2 Summary information on selected office buildings

Code	*VC*	*FH*	*ET*	*UT*	*HB*
Location	Vienna	Vienna	Eisenstadt	Vienna	Hartberg
Function	International Organization	University	Telecom. services	Insurance	State government
Data collection	12 month	12 month	9 month	14 month	9 month
Work places observed	29	17	18	89	10
Orientation	N and SW	E	E	All	NW
Glazing to façade ratio	52%	34%	54%	89%	34%
Glazing to floor ratio	26%	18%	20%	51–80%	18%
Glazing transmittance	79%	65%	60%	65%	75%
External shades	—	Blinds (motorized)	Blinds (motorized)	Blinds (automated)	Blinds
Internal shades	Blinds	—	Vertical louvers	Indoor screens (motorized)	curtains
Windows	Not operable	Not operable	Operable	Operable	Operable
HVAC	Air-conditioned	Air-conditioned	Mix mode	Mix mode	Naturally ventilated

Table 4.3 Summary of the monitored indoor and outdoor parameters with the associated units

Monitored indoor parameters	*Monitored outdoor parameters*
Indoor air temperature [°C]	Outdoor air temperature [°C]
Indoor relative humidity [%]	Relative humidity [%]
Illuminance [lx]	Global horizontal irradiance [W m^{-2}]
Occupancy [occupied/non-occupied]	Wind speed [m s^{-1}]
State of luminaires [on/off]	
State of shade deployment [%]	
State of windows (opening) [%]	

Table 4.4 The structure of the collected data (data types and illustrative instances)

Data	*Type*	*Instances*
Events (E)	System-related (E_s)	Switching lights on/off Pulling shades up/down
	Occupancy-related (E_o)	Entering into (or leaving) an office
States (S)	System-related (S_s)	Lights on/off Position of shades/windows
	Indoor environment (S_i)	Air temperature Illuminance level
	Outdoor environment (S_e)	Outdoor air temperature Global irradiance
	Occupancy-related (S_o)	Office/workstation occupied/vacant

user presence and absence intervals, occupancy sensors were applied, which simultaneously monitored the state of the luminaries in the offices. The state of shading and windows were monitored via time-lapse digital photography: The degree of shade deployment for each office was derived based on regularly taken digital photographs of the façade. Shade deployment degree was expressed in percentage terms (0% denotes no shades deployed, whereas 100% denotes full shading). Measured outdoor environmental parameters included air temperature, relative humidity, and wind speed, as well as global horizontal irradiance. These parameters were monitored using weather stations, mounted either directly on the top of the building or the rooftop of a close-by building. With the exception of the shade and window states, which were monitored every 10 minutes, all the above data were monitored every five minutes. Vertical global irradiance incident on the façades was computationally derived based on measured horizontal global irradiance (see Mahdavi *et al.* 2006).

The measurement and associated error ranges of the sensors applied in the monitoring are listed in Table 4.5.

Table 4.5 Measuring range and accuracy of the sensors

Sensor	*Measuring range*	*Error range*
Solar radiation (weather station)	0 to 1280 W m^{-2}	± 5%
Outdoor temperature (weather station)	–40 to +75 °C	± 0.7 °C
Wind speed (weather station)	0 to 45 m s^{-1}	± 4%
Outdoor relative humidity (weather station)	0 to 100% (between 0 and 50 °C)	± 4%
Indoor temperature	–20 to +70 °C	± 0.4 °C
Indoor relative humidity	5 to 95%	± 3%
Indoor illuminance	50 to 20 000 lx	± 20%

Results

The above study generated an extensive quantity of information. Data analysis provided a number of noteworthy results, some of which are presented below. Thereby, the discussion of the results is structured in terms of occupancy, lighting operation, shades operation, and window operation.

OCCUPANCY

Figure 4.6 shows the mean occupancy level (i.e., presence in users' offices or at workstations) in VC, ET, FH, UT, and HB over the course of a reference day (averaged over the entire observation period). As this figure demonstrates, considerable differences exist amongst occupancy trends in different buildings. Moreover, as Figure 4.7 shows, different offices in the same building (in this case, FH) can also significantly differ in view of their occupancy patterns. Occupancy models in simulation applications must thus take into consideration the specific use types, functions, and associated presence hours of the respective occupants. As it will be discussed later, inter-individual differences in occupancy patterns can be important especially while simulating HVAC (heating, ventilating, and air conditioning) processes in buildings. To provide a statistically relevant sense of fluctuations in occupancy in an office, Figure 4.8 shows – for UT – the mean occupancy together with respective standard deviation ranges.

LIGHTING OPERATION

Figure 4.9 shows lighting operation (in observed offices in VC, FH, UT, and HB) in the course of a reference day expressed in terms of effective electrical lighting load. Figure 4.10 depicts (as regression lines), for all time intervals during the working hours in the observation period (in VC-N, VC-S, FH, and HB), the relationship between mean presence level (in percentage terms) and effective electrical lighting operation level (in percentages of the installed maximum lighting load). The information captured in this figure appears to imply a clear relationship between occupancy level and electrical light usage in the monitored offices. However, the

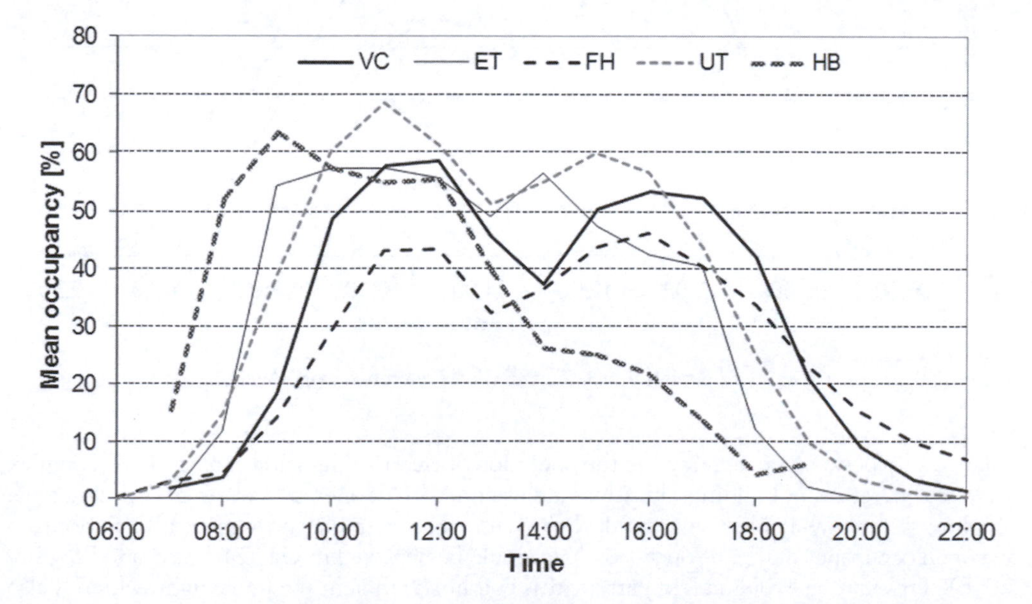

Figure 4.6 Mean occupancy level for a reference work day in VC, ET, FH, UT, and HB.

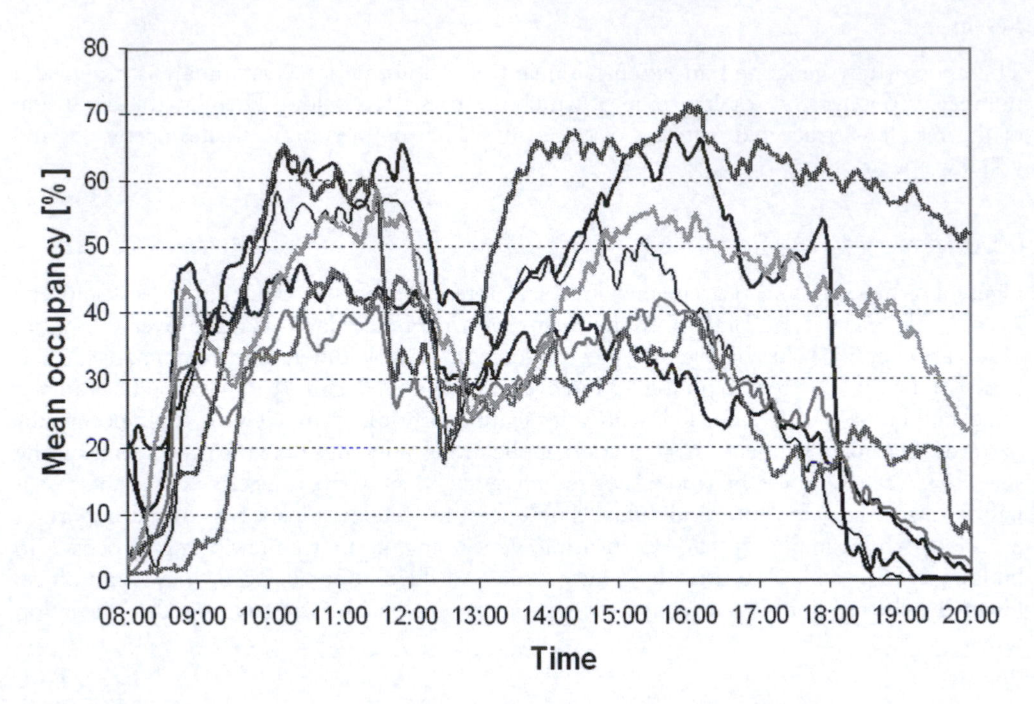

Figure 4.7 Observed occupancy levels in seven different offices in FH for a reference work day.

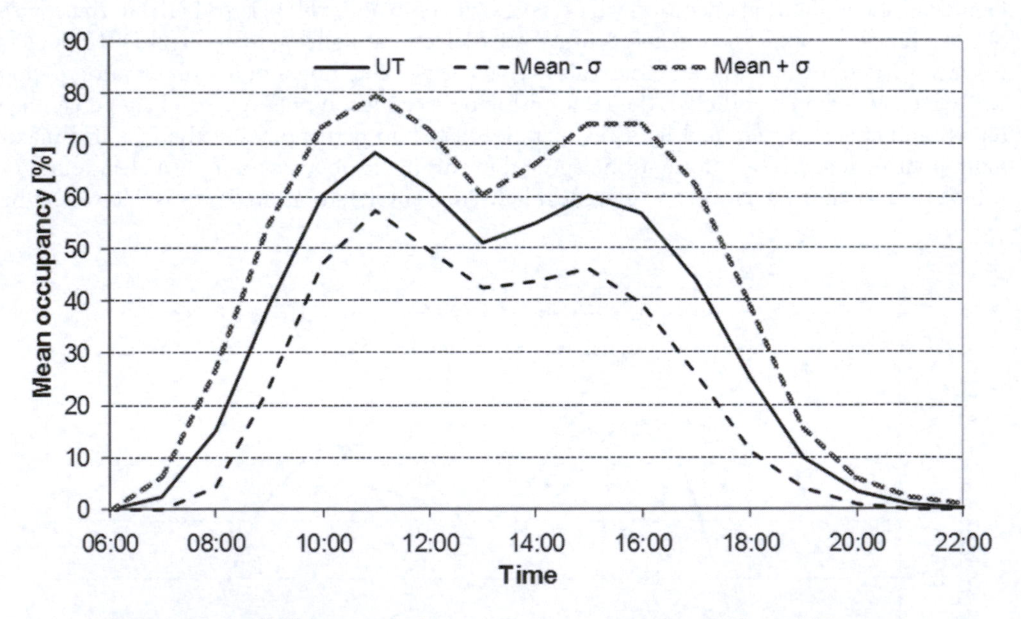

Figure 4.8 Mean and standard deviation of occupancy for a reference work day in UT.

relationship between occupancy and the operation of electrical lighting can be highly complex (due to differences in buildings' location and orientation, floor, window area and glazing type, shading system, available view and daylight, etc.). For example, no significant relationship between occupancy and light operation level could be observed in UT (Mahdavi and Pröglhöf 2008). This may be explained, in part, by this building's efficient use of daylight, which is also reflected both in the lower mean lighting load (see Figure 4.9) and the relationship between

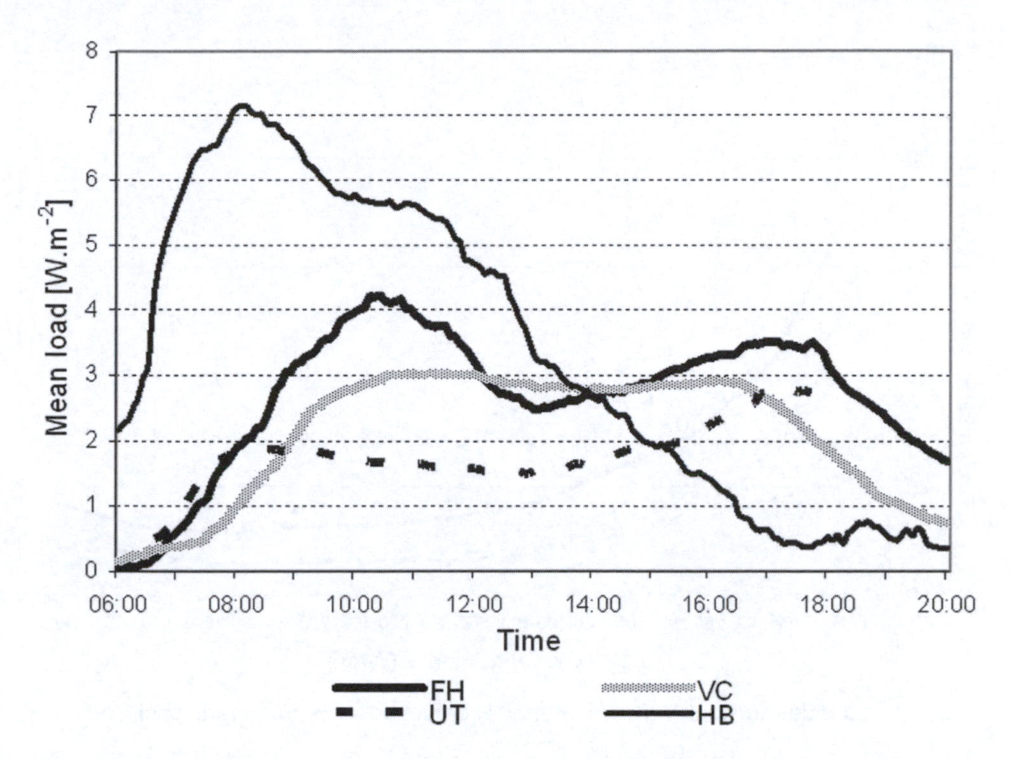

Figure 4.9 Lighting operation (in terms of mean effective lighting load) in VC, FH, UT, and HB offices for a reference work day.

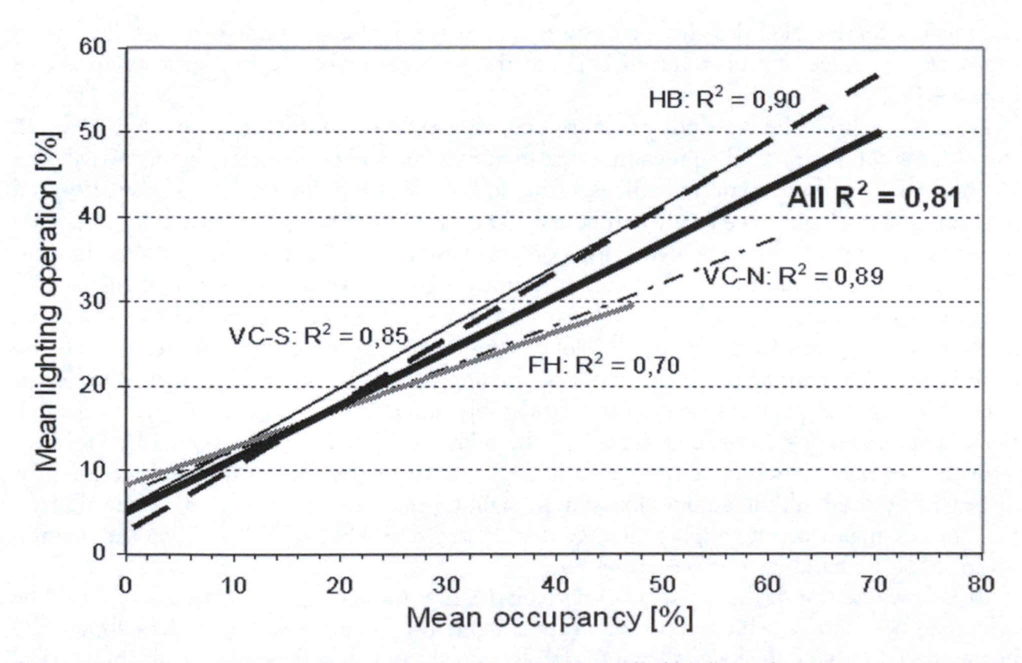

Figure 4.10 Lighting operation (in percentage of maximum installed load) in relation to mean occupancy for all time intervals of the working hours during the observation period in VC-N, VC-S, FH, and HB offices (shown is also the regression line for all observations).

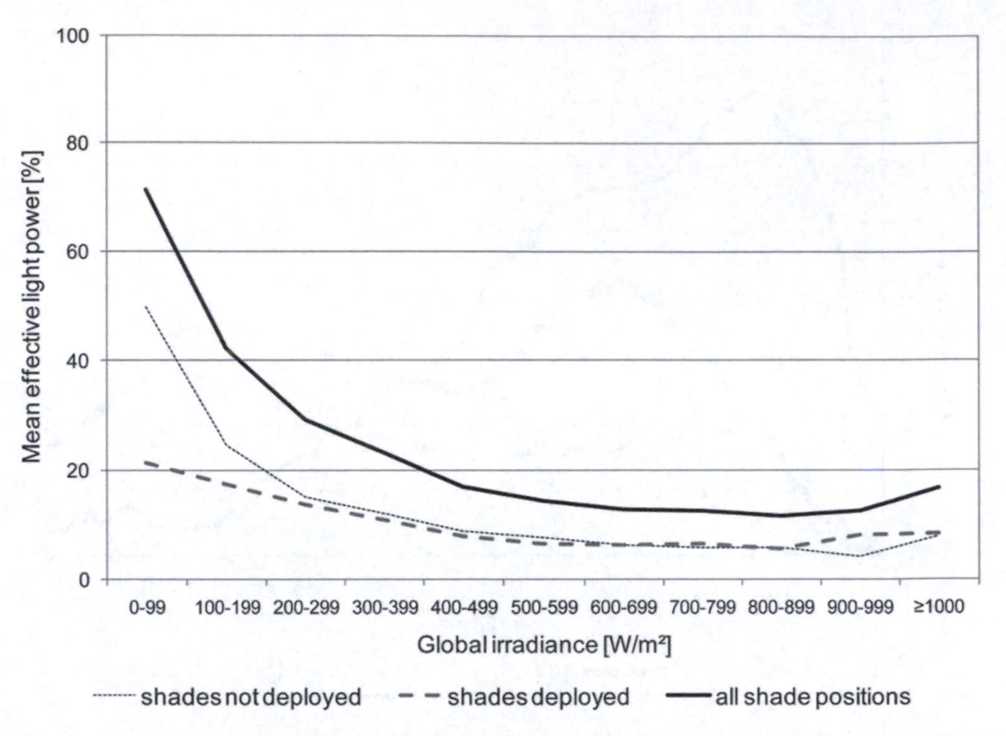

Figure 4.11 Mean effective electrical lighting power in UT (in percentage of maximum installed lighting power) averaged for all zones plotted against external global horizontal irradiance (time intervals with and without shade deployment are shown separately).

the mean effective electrical lighting power (expressed as the percentage of installed lighting power) averaged for all zones in UT and the external global horizontal irradiance (see Figure 4.11).

Figure 4.12 shows the probability that an occupant would switch the lights on upon arrival in the office as a function of the prevailing task illuminance level immediately before arrival (for VC and HB). In most monitored offices, rather low workstation illuminance levels (measured horizontal task illuminance levels well below 200 lx) appear to trigger a non-random increase in probability of switching the lights on upon occupants' arrival in their offices/workstations. This represents an interesting challenge in terms of the selection of appropriate (objective) visual performance criteria: Occupants' actions (in this case switching the lights on) is more likely to be triggered by the perceived general light conditions in the room, for which the horizontal task illuminance level may not be the appropriate indicator. The critical indicator value (for triggering actions) may shift considerably, if a different sensor position is selected. For example, Figure 4.13 illustrates, for UT, the relationship between the normalized relative frequency of light switch on actions and indoor light levels (horizontal illuminance levels as measured by the building automation system's ceiling-mounted light sensors). Note that, in UT, the occupants turn the lights (change the setting from 0 lx to 500 lx) on via the desktop interface of the building automation system.

In UT, where the daylight usage is relatively high, an unambiguous relationship could be discerned between switch on actions and the outside illuminance: As Figure 4.14 shows, the normalized relative frequency of "switch on" actions show a clear dependency on the vertical global irradiance incident on the façade measured for the orientation of the respective zones. For this analysis, only those time intervals are considered when all shades (internal and external) were open.

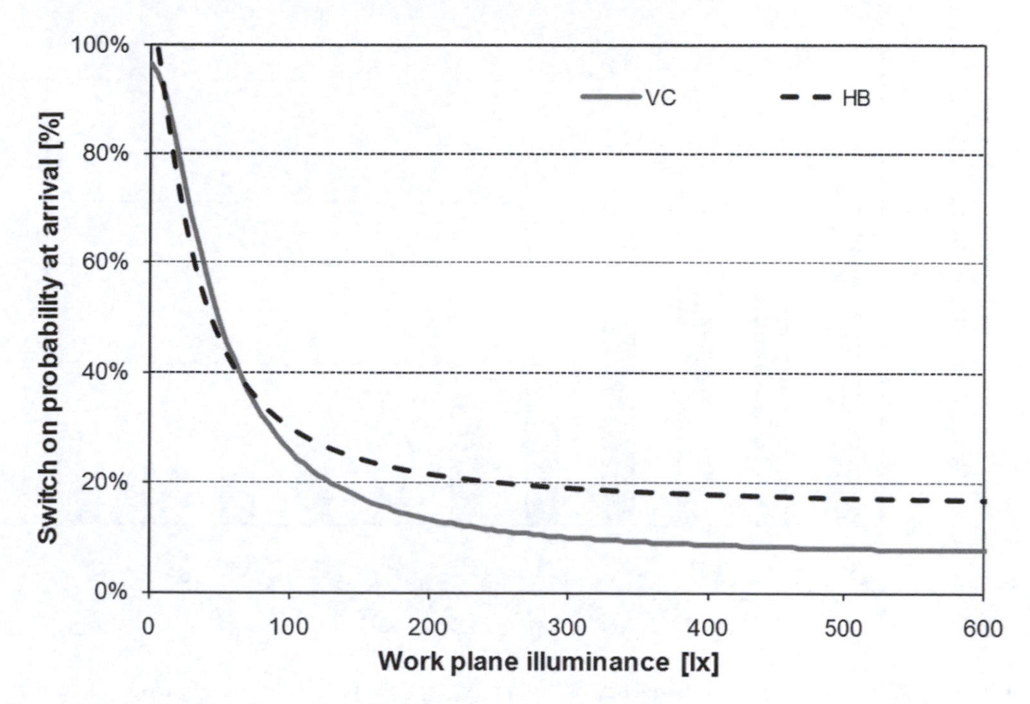

Figure 4.12 Probability of switching the lights on upon arrival in the office in VC, and HB as a function of the prevailing task illuminance level prior to an action.

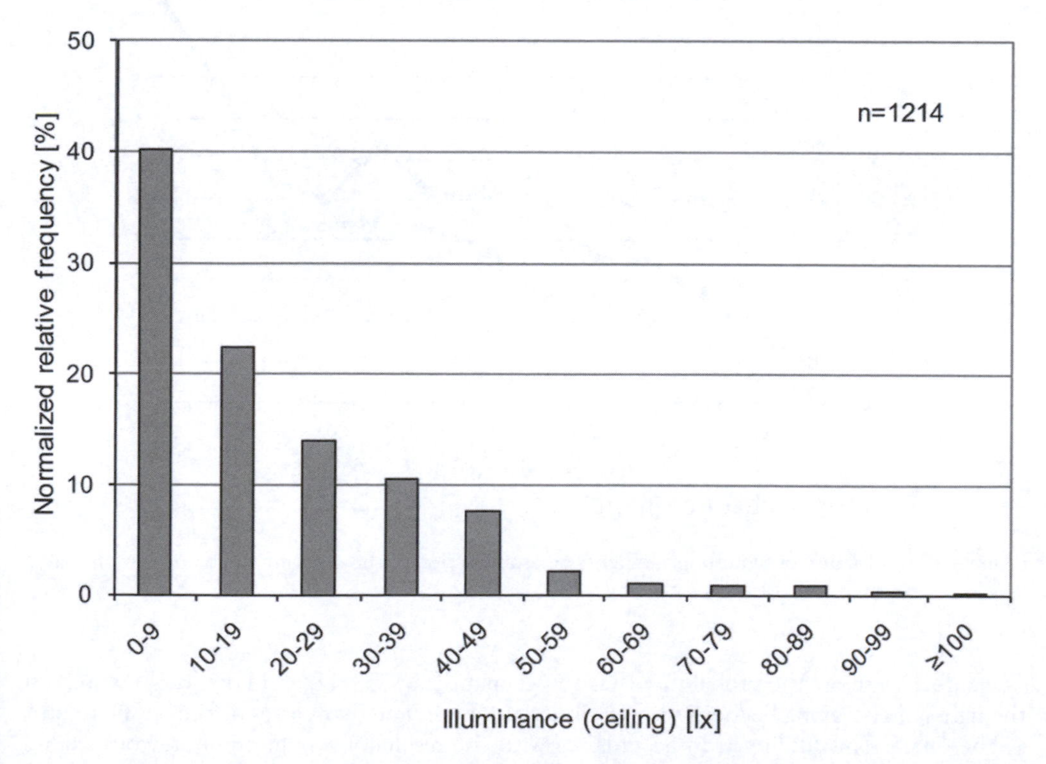

Figure 4.13 Normalized relative frequency of "switch on" actions (0–500 lx) in UT as a function of ceiling illuminance (6:00 to 18:00, all shades open).

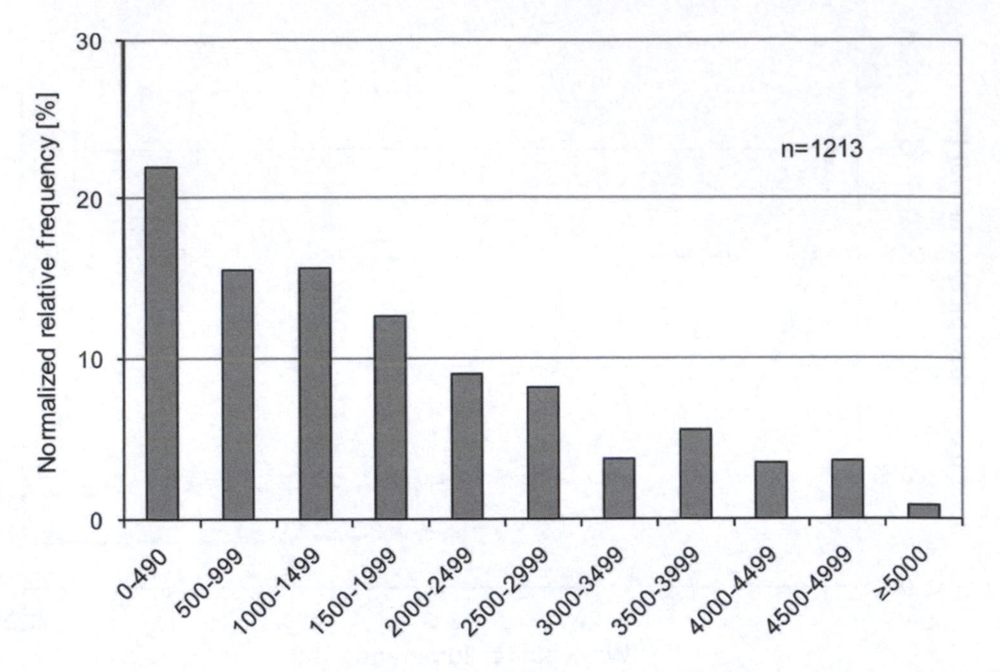

Figure 4.14 Normalized relative frequency of "switch on" actions (0–500 lx) as a function of vertical illuminance (intervals between 6:00 and 18:00 with shades open).

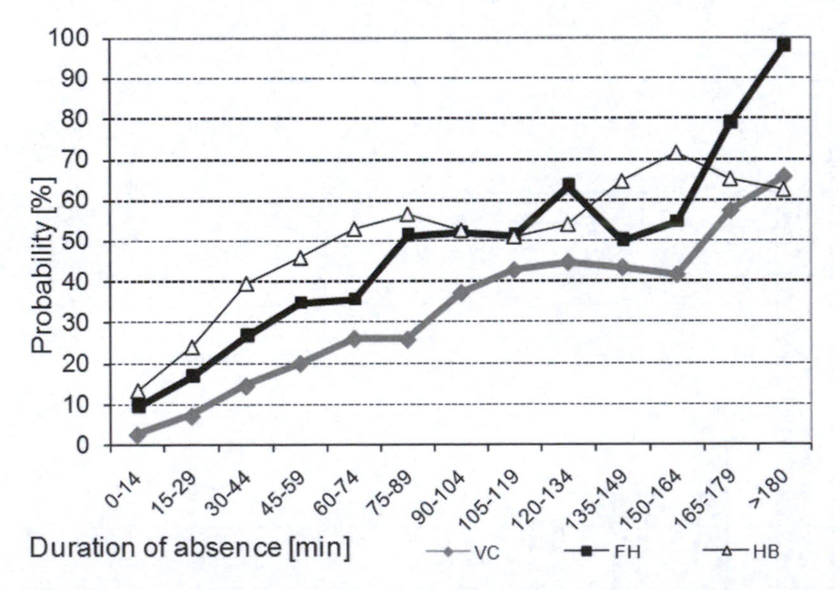

Figure 4.15 Probability of switching the lights off as a function of the duration of absence (in minutes) from the offices in VC, FH, and HB.

Figure 4.15 shows the probability that an occupant (in VC, FH, and HB) would switch off the lights upon leaving his/her office as a function of the time that passes before he/she returns to the office. This finding is in accordance with the results of a number of previous studies (Boyce 1980; Pigg *et al.* 1996) concerning the dependency of probability of light switch off actions by occupants who leave their workstations on the duration of the time they stay away.

SHADES OPERATION

The mean shade deployment levels (expressed as a percentage of window area occluded due to shade operation) differ from building to building and façade to façade (see Figure 4.16). In case of FH's east-facing façade, a relationship between shade deployment and the magnitude of solar radiation is observable. In case of VC-N and VC-S, the shade deployment level does not fluctuate much, but there is a significant difference in the overall shade deployment level between these two façades (approximately 75% in the case of south-west-facing façade, 10% in the case of the north-facing façade). Overall, our data suggests that the shade deployment level in buildings cannot be predicted reliably on the basis of incident irradiance alone.

WINDOW OPERATION

As to natural ventilation operation, observations (based on object HB) suggest that windows are opened somewhat more frequently early in the day, after the lunch hour, and towards the end of the working day (Mahdavi *et al.* 2008b). This is captured in Figure 4.17, which shows the occupancy-normalized relative frequency distribution of opening and closing windows actions for a reference day (representing the entire observation period). Closing actions are, however, observed at a significantly higher rate before the occupants leave their offices for the day. Window opening actions have a higher frequency during the colder months of the year, but this does not imply that the windows are open longer in these months. Rather, it is likely that, given lower outdoor temperatures, windows are not kept open too long and ventilation is achieved through more frequent – but brief – periods during which windows are kept open (see Figure 4.23). As Figure 4.18 suggests, considerably higher frequencies of window closing actions occur at lower indoor temperatures (below 22 °C). This result is consistent with the significant increase in window closing actions frequency for sub-zero outdoor temperatures (see Figure 4.19).

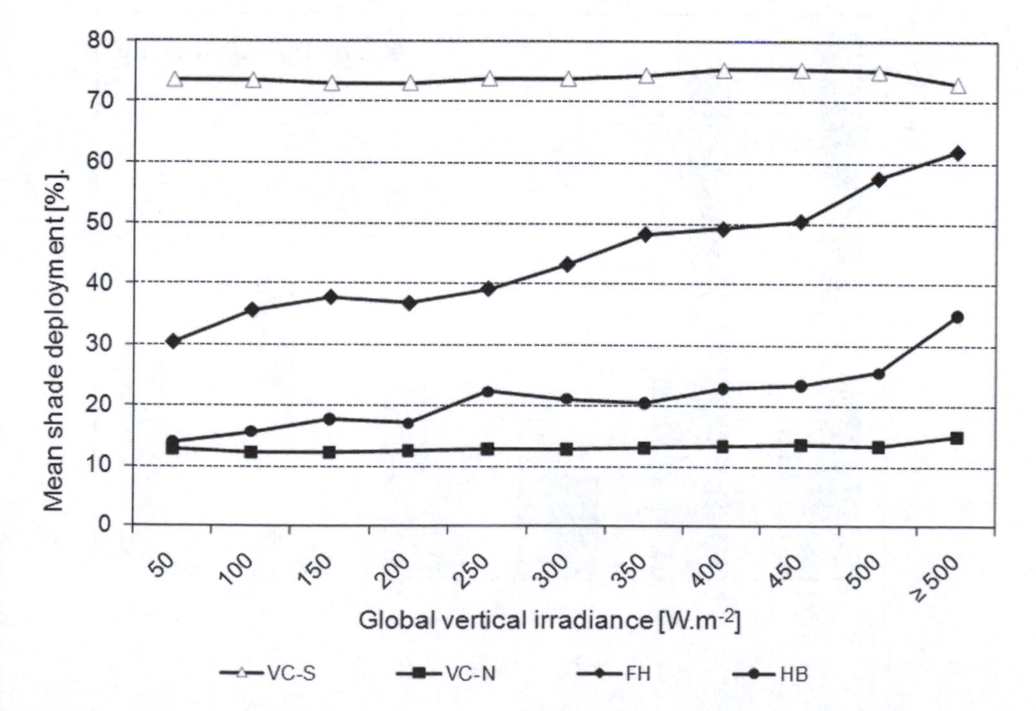

Figure 4.16 Mean shade deployment degree (VC-S, VC-N, FH, and HB) as a function of global vertical irradiance incident on the façade.

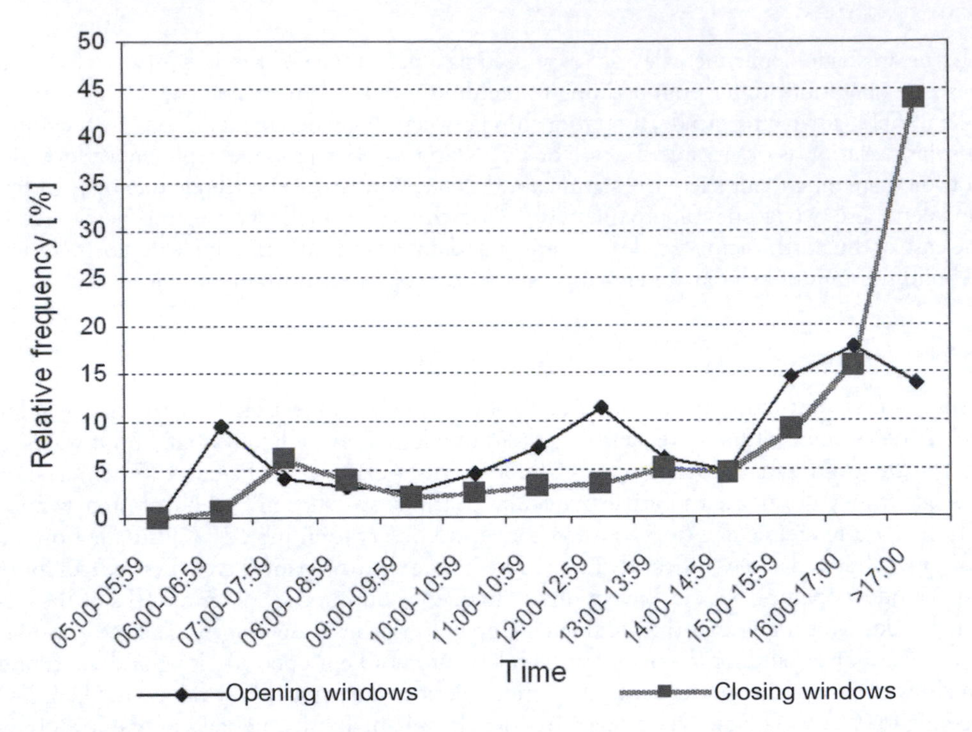

Figure 4.17 Normalized relative frequency of opening and closing actions over the course of a reference day.

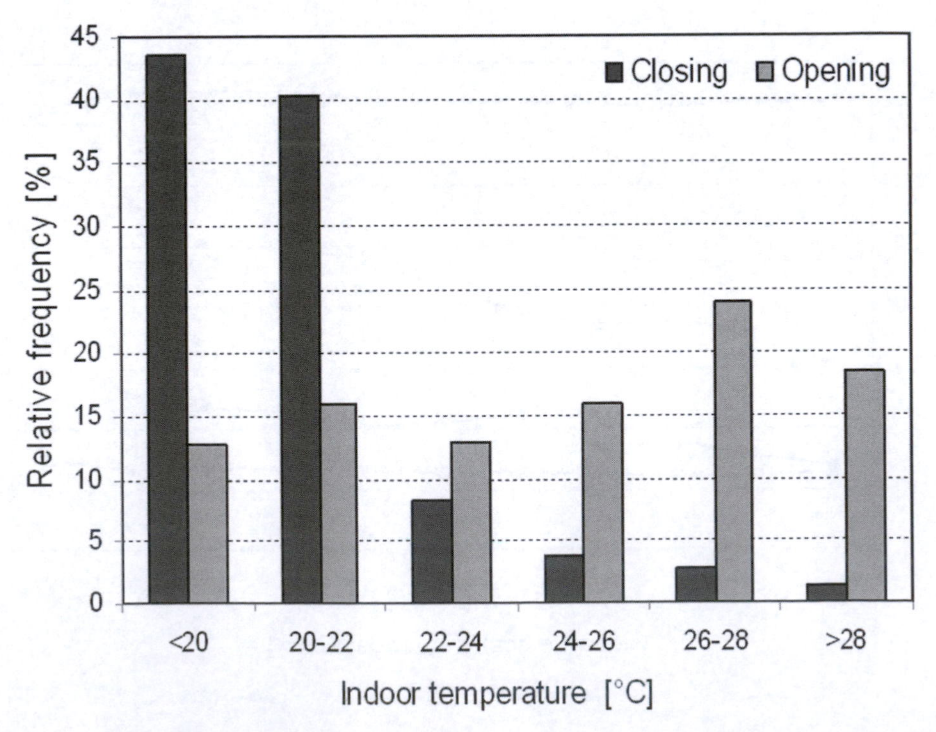

Figure 4.18 Normalized relative frequency of opening and closing windows (in HB) as a function of indoor temperature.

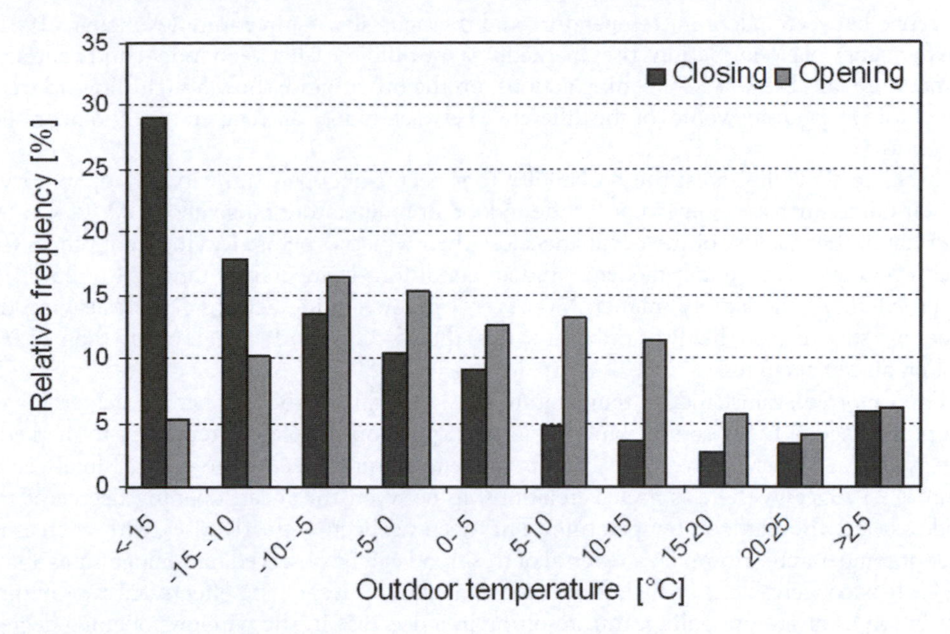

Figure 4.19 Normalized relative frequency of opening and closing windows (in HB) as a function of outdoor temperature.

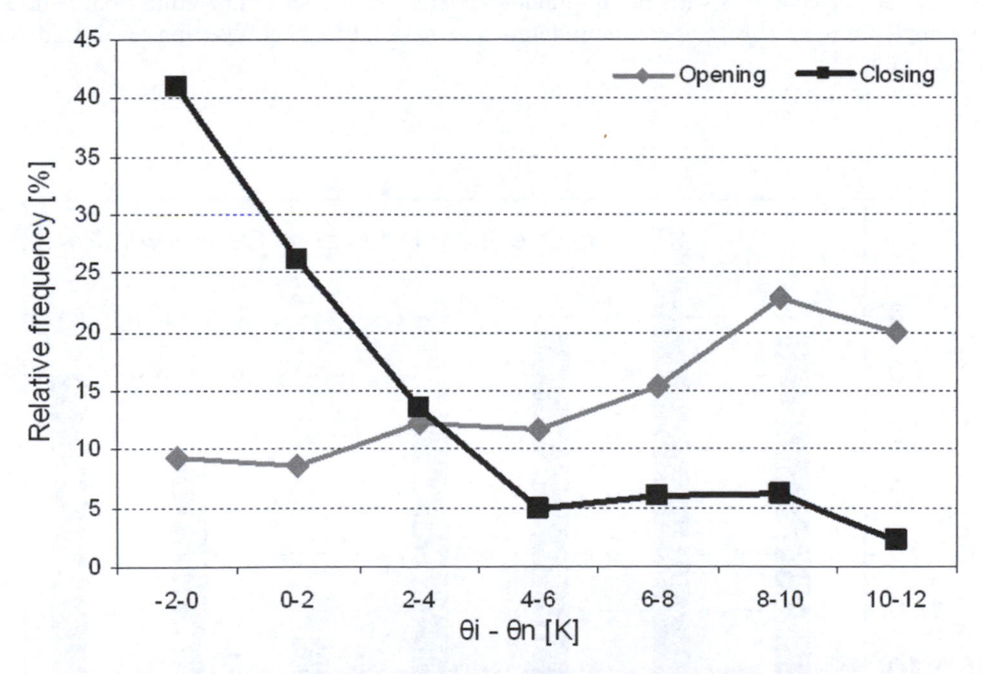

Figure 4.20 Relative frequency of window opening and closing actions (in HB) as a function of the difference between indoor air temperature and neutrality temperature.

It has been suggested that the deviation of indoor air temperature from the "comfort temperature" can trigger window opening actions (when the rooms appear warm or hot) and window closing actions (when the room appears cool or cold). This conjecture appears to be partially corroborated by the data collected in HB and illustrated in Figure 4.20. Thereby, the frequency distributions of window opening and closing actions are plotted against the difference between indoor air temperature and the neutrality temperature (Auliciems 1981). The frequency of closing actions rises markedly as the difference between indoor and neutrality temperature falls below 4 K. Opening actions, on the other hand, show a slight upward trend in according with rising values of the difference between indoor air temperature and neutrality temperature.

As Figure 4.21 illustrates, the probability that occupants open the windows upon arrival in their offices increases somewhat for the indoor air temperature range above 20 °C. On the other hand, the majority of the occupants close their windows before leaving their offices (for at least 3 hours). Only under moderate indoor conditions (temperature range 24 to 26 °C) is the probability considerably smaller. Moreover, window closing actions (before leaving the office) are slightly more likely at rather low (less than 5 °C) or rather high (more than 25 °C) outdoor air temperatures.

When plotted against indoor temperature bins, the mean window opening degree (given as a percentage of the observed windows in open position) displays a tendency to markedly increase for at higher (above 24 °C) room air temperatures (see Figure 4.22). Moreover, as Figure 4.23 suggests, there is a clear relationship between the mean opening degree of the windows and the outdoor temperature. The opening degree clearly rises with increasing outdoor temperature. However, a reversal of this trend can be observed for temperatures above 26 °C. It is conceivable that higher outdoor temperatures lessen the effectiveness of natural ventilation as means of cooling, thus resulting in a decrease in the windows opening degree. Finally, Figure 4.24 displays a clear relationship between window opening level and the difference between indoor temperature and neutrality temperature. The maximum mean window opening degree coincides with the minimum deviation of indoor temperature from neutrality temperature. As this temperature difference increases, the mean opening level decreases considerably.

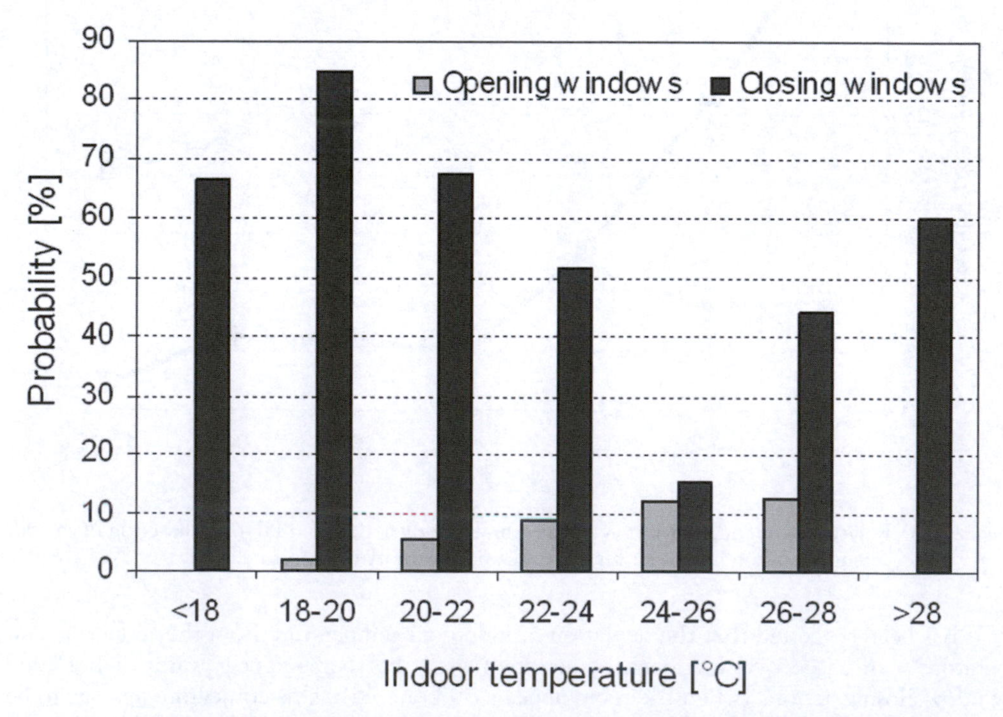

Figure 4.21 Probability of opening the windows (in HB) upon arrival in the office and closing windows upon leaving the office (for at least three hours) as a function of indoor temperature.

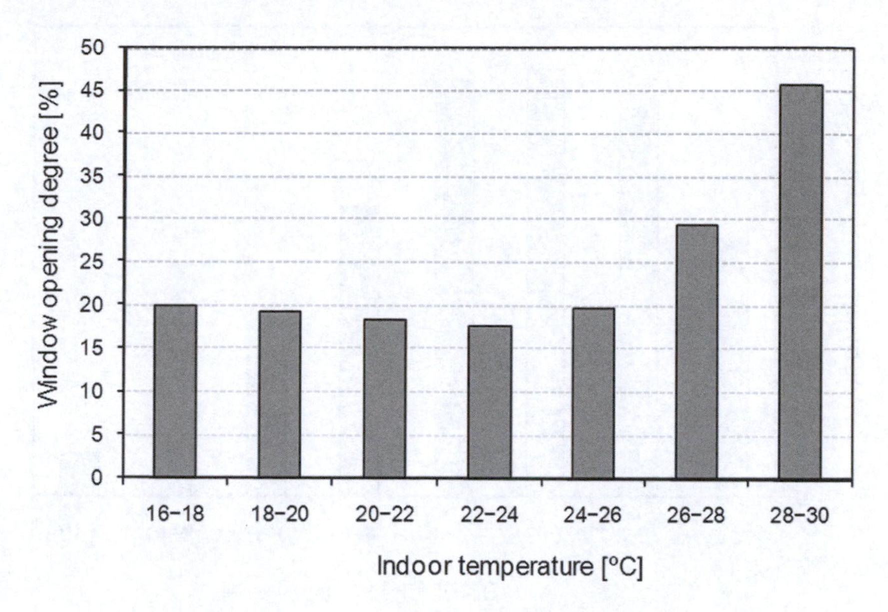

Figure 4.22 State of windows (in HB) plotted against indoor temperature.

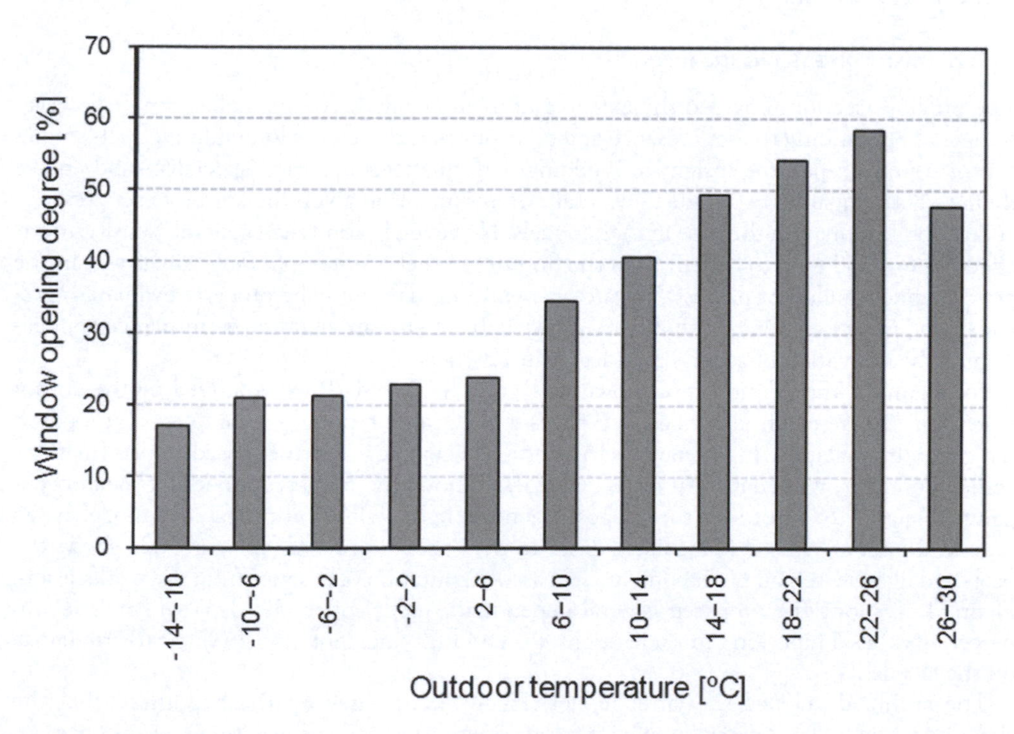

Figure 4.23 State of windows (in HB) plotted against outdoor temperature.

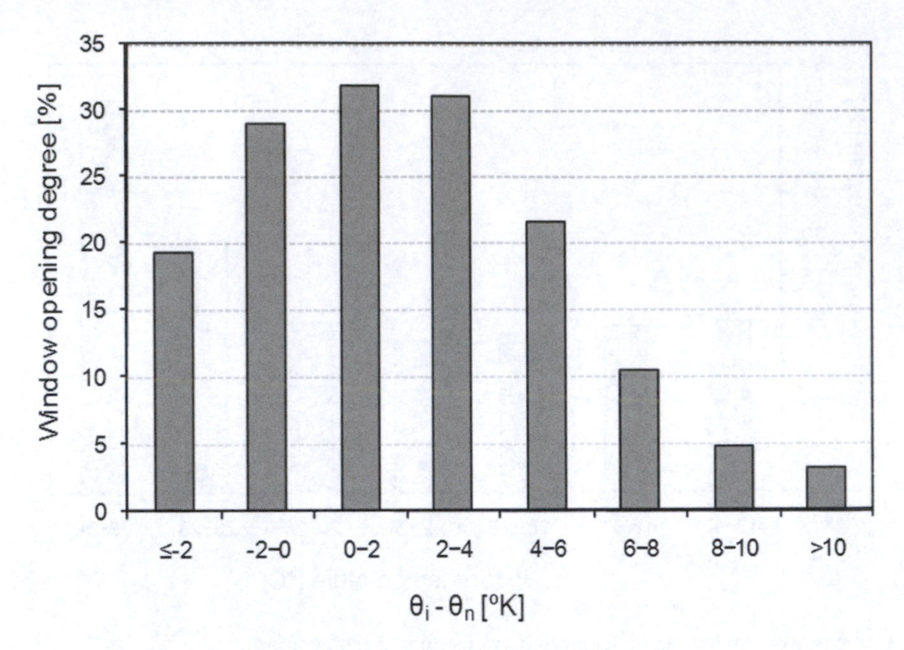

Figure 4.24 State of windows (in HB) as a function of the difference between indoor air temperature and neutrality temperature.

General reflections

Deterministic and stochastic methods

The previous section provided the example of a case study involving collection of observational data pertaining to user presence and control-oriented actions in buildings. Such studies can of course deepen the insights of building performance simulation specialists and enable them to better qualify the results they obtain from simulation given the kinds of user presence and action information they use in their models. However, beyond such general "sensitization" effect, one would wish to benefit from the findings of such studies in a more direct way in the performance simulation process. Toward this end, the data must be properly evaluated, prepared, and processed. One possibility to utilize such data toward building performance applications is the derivation of generalized (aggregate) models.

For example, the relationships shown in Figures 4.25 to 4.28 were derived by the author based on the research described on pages 63–76 and represent examples of proposals for general simulation input models to capture occupancy and user-based control actions (office buildings, work days, Austria). Figure 4.25 shows the proposed model for mean occupancy. Figure 4.26 illustrates the proposed general light switch on probability model based on task illuminance level immediately prior to the onset of occupancy. Figure 4.27 shows the proposed light switch off probability model based on duration of absence from the workstation. Figure 4.28 shows the proposed general dependency model of shades deployment level (in percentages) as a function of façade orientation and the incident global (vertical) irradiance on the façade.

The author developed the latter model (Figure 4.28) based on the hypothesis that the shade deployment level depends on both façade orientation and vertical irradiance intensity. A possible mathematical formulation of this dependency is expressed in Equation (4.1).

$$SD = a - b \cdot k_1 - c \cdot k_2 \cdot (d - E_v) \tag{4.1}$$

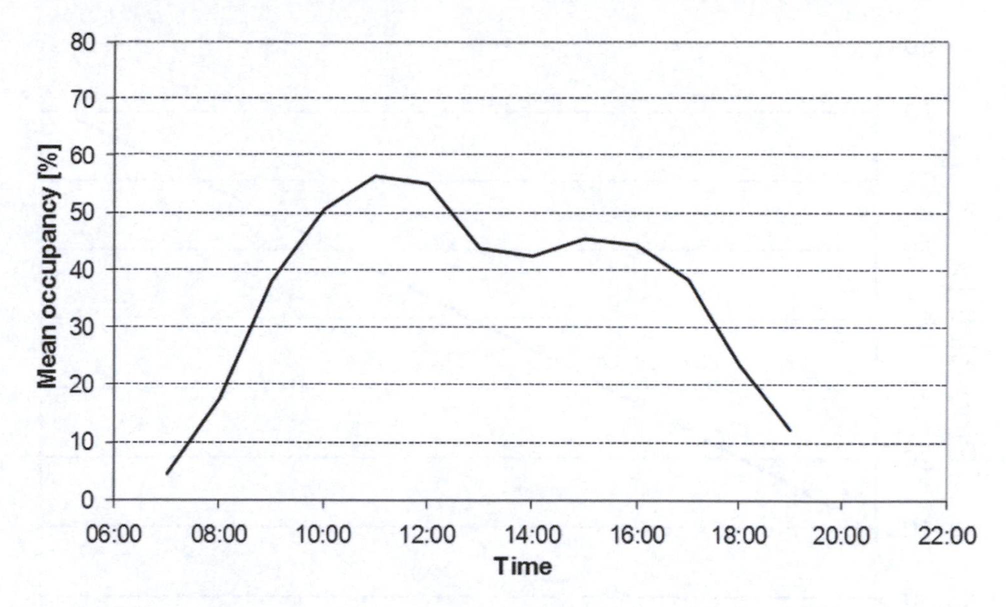

Figure 4.25 Proposed mean occupancy input model for building performance simulation applications (office buildings, work days, Austria).

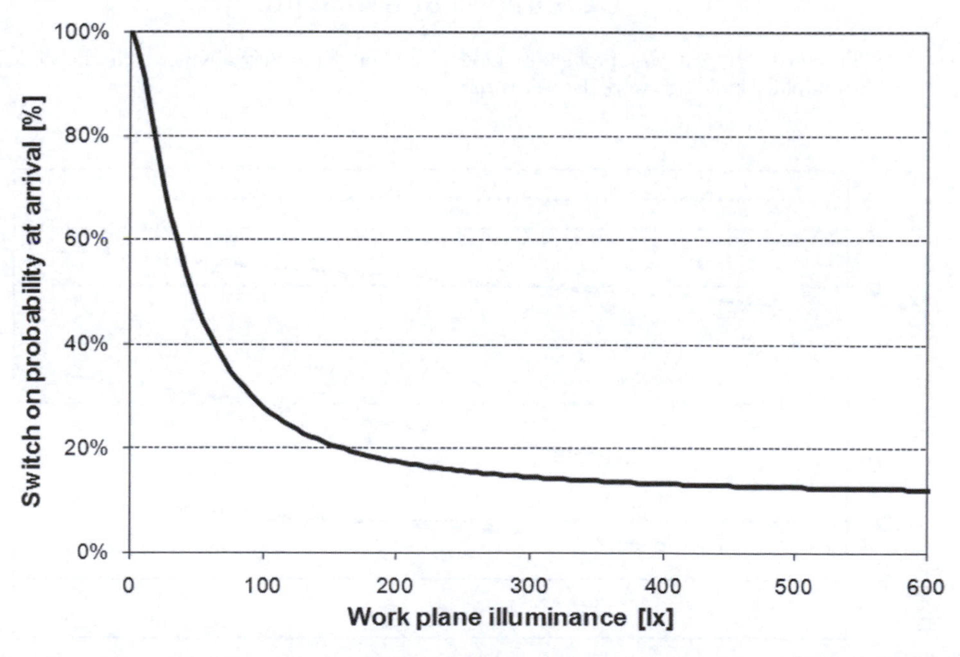

Figure 4.26 Proposed light switch on probability model based on task illuminance level immediately prior to the onset of occupancy (office buildings, work days, Austria).

Herein SD denotes the shade deployment level (in percentage) and E_v incident vertical irradiance (in W m^{-2}). The constants *a*, *b*, *c*, and *d* depend on location and building features and k_1 and k_2 are trigonometric functions of the façade normal azimuth α (in degrees, whereby the corresponding values for north, east, south, and west are 0, 90, 180, and 270 degrees respectively) as per the following equations:

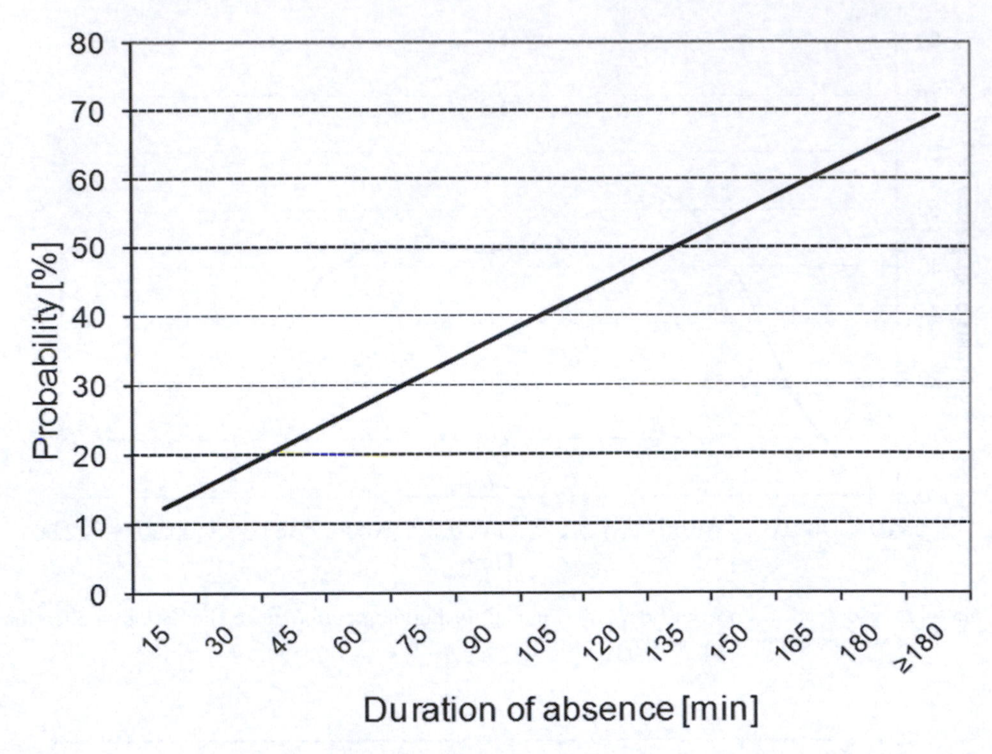

Figure 4.27 Proposed light switch off probability model based on duration of absence from the workstation (office buildings, work days, Austria).

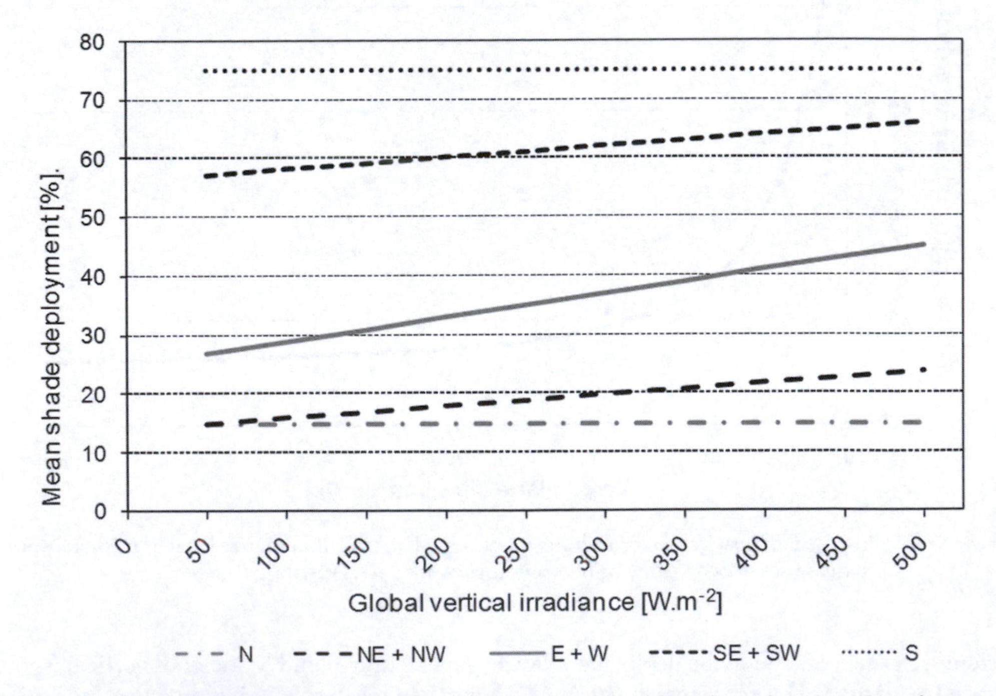

Figure 4.28 Proposed dependency model of shades deployment level (in %) as a function of façade orientation (S: South, NE: North-East, NW: North-West, E: East, W: West, SE: South-East, SW: South-West, S: South) and the incident global (vertical) irradiance on the façade (office buildings, work days, Austria).

$$k_1 = 0.5 \times (1 + \cos\alpha) \tag{4.2}$$

$$k_2 = 1 - \cos^2 \alpha \tag{4.3}$$

Figure 4.29 compares the predictions based on the Equation (4.1) with empirical findings documented in Figure 4.16. Thereby, the values for the constants were set as follows: $a = 75$, $b = 60$, $c = 0.04$, and $d = 500$. The actual façade normal azimuth values for the respective façades were applied for the computation and comparison with model predictions (206 degrees for VC-S, 103 degrees for FH, 51 degrees for HB, and 326 degrees for VC-N).

Such aggregate models represent, of course, general trends and not specific conditions in any specific building. Consequently, their relevance is limited to the context (country, climate, culture, building type, etc.) from which their underlying observational data originate (in this case various office buildings in Austria). Moreover, they do not capture the dynamism of actual processes and events in buildings. Nonetheless, they have a clear place and arguably fitting role in important areas of simulation-based building performance inquiry. Specifically, simulation-based assessment of design versions and design alternatives (a presumably common activity in the initial phases of the building delivery process) is often required to provide concise statements as to the overall performance of the building "hardware" under "standardized" conditions pertaining to both external climate and internal (occupancy-related) processes.

As indicated previously, differences in occupancy patterns over time and location can be very large (see, for example, Figure 4.7). Such differences can be important especially while simulating HVAC processes in buildings. Given sufficient observations, statistical methods can be used to generate individual occupancy patterns that, while unique – and realistic in their random fluctuations – could represent, in toto, the mean occupancy level associated with a building. Such models can be implemented in simulation applications in terms of autonomous agents with built-in methods (Chang and Mahdavi 2002) to generate stochastic behavior. However, a stochastic presence (and user action) model would generate, per definition,

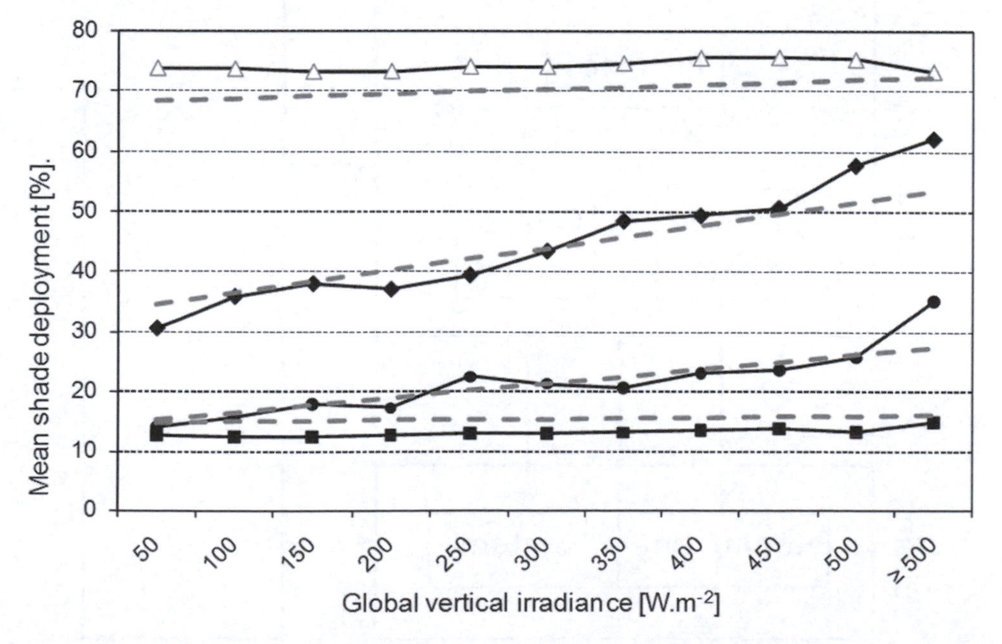

Figure 4.29 Comparison of the general shade deployment model's predictions (see Equation (4.1) and Figure 4.28) as shown in dashed lines with empirical observations (see Figure 4.16).

different input data at each simulation run. This could represent a problem for performance analyses of design alternatives, when the aim is to compare multiple (alternative) designs for the same contextual boundary conditions (weather) and for the same occupancy conditions. Thus, different approaches to representation of occupancy-related processes in building performance simulation may be appropriate given different scenarios.

Aggregate models of user presence and behavior would be rather problematic, if the detailed configuration of building service systems (for HVAC) is the main concern: For example, while dealing with the requirement of providing sufficient heating and cooling capacity to different zones of a building, the range and variability of required thermal loads at the local level must be reliably gauged. This cannot be based on spatially and temporally averaged occupancy assumptions. In such instances, application of realistic (stochastic) models may be critical. On the other hand, when the objective of a simulation-based inquiry is to parametrically compare (benchmark) basic design alternatives, parallel random variations of boundary conditions and internal processes may be counter-productive. Thus, to select the right kind of model (with the right level of complexity) for a specific line of performance inquiry may be in fact regarded as an inherent attribute of the "art of simulation".

The future of coupled models

High-resolution and dynamic simulation of environmental processes in buildings would have to include, ultimately, comprehensive multiple-coupled representations. To illustrate this vision, Figure 4.30 provides a schematic depiction of a multiple-coupled representation, including occupancy, building, and context models. In this scheme, the occupancy model, which can be implemented computationally in terms of a society of autonomous agents, is based on an underlying presence sub-model. This includes information about the presence of agents in the building. Given presence information, passive effects and control actions of occupants are derived. While passive effects may be computed based on physiological models of human body

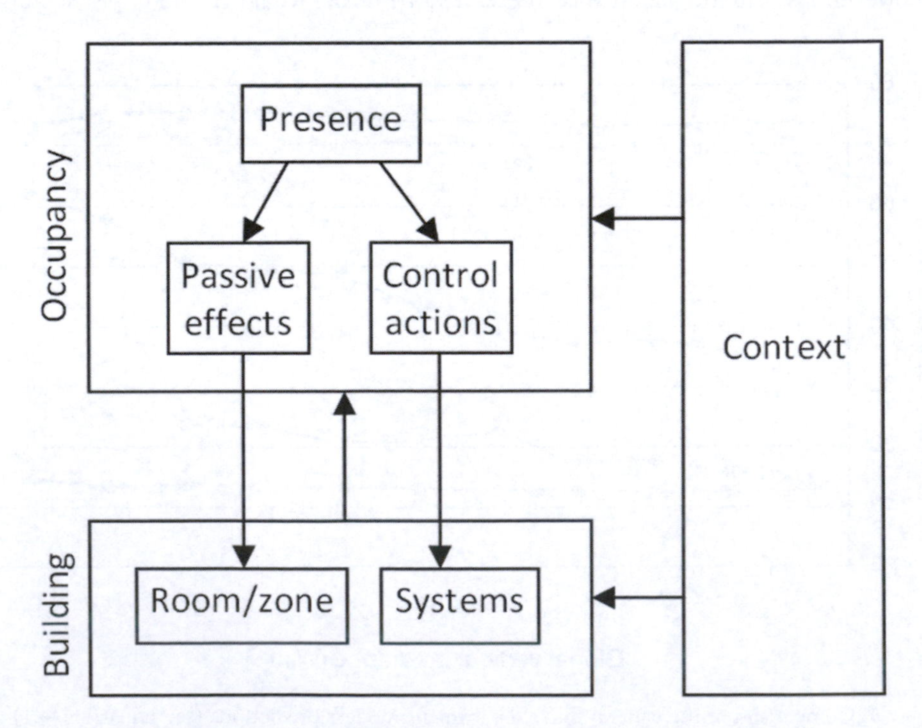

Figure 4.30 Schematic illustration of multiple-coupled models for occupancy, building, and context.

and metabolism, action probabilities would be generated based on both physiologically and psychologically based models (e.g., thermal comfort and thermal pleasantness models). Computed passive effects (e.g., sensible and latent heat generation) and predicted control action probabilities (e.g., operation of windows, luminaires, blinds) are provided to the room and system components of a coupled building model, which, in turn, provides boundary information (e.g., room air temperature and relative humidity) for the occupancy model. Both occupancy and building model are coupled with the context model, which supplies them with relevant information on external (weather) conditions around the building. The overall computational framework generates thus building performance data (thermal conditions indoors, heating and cooling loads, systems' energy use, etc.) under dynamic and concurrent (coupled) consideration of occupancy, building, and context states.

Conclusion

The reliability of results obtained from building performance simulation applications depends not only on the validity of computational algorithms, but also on the soundness of input assumptions. While there has been significant progress concerning methods and practices for specification of building geometry, material properties, and external (weather) conditions, the resolution of input information regarding occupancy (i.e., people's presence and behavior in buildings) is still rather low. However, the importance of people's passive and active effects on building performance (e.g., indoor conditions and energy performance) has been recognized for some time. Accordingly, many recent and ongoing research efforts attempt to construct models for passive and active occupancy effects on building performance. Thereby, physiological and psychological descriptions of occupancy as well as empirically based observational data provide the knowledge base. Specifically, long-term high-resolution empirical data on people's presence and control-oriented actions in buildings can support the generation of general patterns of user control behavior as a function of indoor and outdoor environmental parameters such as temperature, illuminance, and irradiance. These patterns can be expressed either as set of typologically differentiated standardized (aggregate) occupancy and control action models or realized in terms of emergent behavior of a society of computational agents with embedded stochastic features. Future developments in this area are expected to facilitate a detailed and dynamic simulation of environmental processes in buildings via comprehensive multiple-coupled representations that dynamically capture the states of occupancy, building, and context.

References

Auliciems, A. (1981) "Toward a Psycho-physiological Model of Thermal Perceptions", *International Journal of Biometeorology* 25(2): 109–122.

Bourgeois, D. (2005) "Detailed Occupancy Prediction, Occupancy-sensing Control and Advanced Behavioral Modeling within Whole-building Energy Simulation", PhD Thesis, Université Laval, Quebec, Canada.

Boyce, P. (1980) "Observations of the Manual Switching of Lighting", *Lighting Research & Technology* 12(4): 195–205.

Chang, S. and Mahdavi, A. (2002) "A Hybrid System for Daylight Responsive Lighting Control", *Journal of the Illuminating Engineering Society* 31(1): 147–157.

EDSL (2008) Tas Manager (Version 9.0.9e): *Tas Internal Condition Database*. Online. Available at: http://www.edsl.net (accessed 15.3.2009).

ESRU (2002) *The ESP-r System for Building Energy Simulation, User Guide Version 10 Series*. ESRU (Energy Systems Research Unit's) Manual U02/1, Glasgow: University of Strathclyde. Online. Available at: http://www.esru.strath.ac.uk (accessed 15.3.2009).

Fasold, W. and Veres, E. (2003) *Schallschutz und Raumakustik in der Praxis*, Berlin: Huss Medien/Verlag Bauwesen.

Finney, D. (1947) *Probit Analysis: A Statistical Treatment of the Sigmoid Response Curve*, London: Cambridge University Press, p. 333.
Fritsch, R., Kohler, A., Nygard-Ferguson, M. and Scartezzini, J.L. (1990) "A Stochastic Model of User Behaviour regarding Ventilation", *Building and Environment* 25(2): 173–181.
Herkel, S., Knapp, U. and Pfafferott, J. (2005) "A Preliminary Model of User Behavior regarding the Manual Control of Windows in Office Buildings", in *Proceedings of the Ninth International IBPSA Conference, Building Simulation 2005*, Montréal, Canada. pp. 403–410.
Humphreys, M.A. and Nicol J.F. (1998) "Understanding the Adaptive Approach to Thermal Comfort", *ASHRAE Transactions* 104(1): 991–1004.
Hunt, D. (1979) "The Use of Artificial Lighting in Relation to Daylight Levels and Occupancy", *Building and Environment* 14(1): 21–33.
Inoue, T., Kawase, T., Ibamoto, T., Takakusa, S. and Matsuo, Y. (1988) "The Development of an Optimal Control System for Window Shading Devices Based on Investigations in Office Buildings", *ASHRAE Transaction* 94(2): 1034–1049.
Lindelöf, D. and Morel, N. (2006) "A Field Investigation of the Intermediate Light Switching by Users", *Energy and Buildings* 38(7): 790–801.
Lindsay, C.T.R. and Littlefair, P.J. (1992) "Occupant Use of Venetian Blinds in Offices", *Building Research Establishment (BRE), Contract PD233/92*, Watford, UK: Garston Library.
Love, J.A. (1998) "Manual Switching Patterns Observed in Private Offices", *Lighting Research & Technology* 30(1): 45–50.
Mahdavi, A. and Pröglhöf, C. (2008) "User-System Interaction Models in the Context of Building Automation", in Zarli, A.S. and Scherer, R. (eds) *Proceedings of ECPPM 2008 – eWork and eBusiness in Architecture, Engineering and Construction*, Sophia-Antipolis, France, pp. 389–396.
Mahdavi, A., Dervishi, S. and Spasojevic, B. (2006) "Computational Derivation of Incident Irradiance on Building Facades Based on Measured Global Horizontal Irradiance Data", in *Proceedings of BauSIM 2006*, Munich, Germany, pp. 123–125.
Mahdavi, A., Mohammadi, A., Kabir, E. and Lambeva, L. (2008a) "Occupants' Operation of Lighting and Shading Systems in Office Buildings", *Journal of Building Performance Simulation* 1(1): 57–65.
Mahdavi, A., Kabir, E., Mohammadi, A. and Pröglhöf, C. (2008b) "User-based Window Operation in an Office Building", in Olesen, B., Strom-Tejsen, P. and Wargocki, P. (eds) *Proceedings of Indoor Air 2008 – The 11th International Conference on Indoor Air Quality and Climate*, Copenhagen, Denmark.
Nicol, J.F. (2001) "Characterising Occupant Behavior in Buildings: Towards a Stochastic Model of Occupant Use of Windows, Lights, Blinds, Heaters, and fans", in *Proceedings of the Seventh International IBPSA Conference, Building Simulation 2001*, Rio de Janeiro, Brazil, pp. 1073–1078.
Page, J., Robinson D., Morel N. and Scartezzini J.-L. (2007) "A Generalised Stochastic Model for the Simulation of Occupant Presence", *Energy and Buildings* 40(2): 83–98.
Pigg, S., Eilers, M. and Reed J. (1996) "Behavioral Aspects of Lighting and Occupancy Sensors in Private Offices: A Case Study of a University Office Building", in *ACEEE (American Council for an Energy Efficient Economy) 1996 Summer Study on Energy Efficiency in Buildings*, Pacific Grove, CA, USA. pp. 8, 161–168, 171.
Rea, M.S. (1984) "Window Blind Occlusion: A Pilot Study", *Building and Environment* 19(2): 133–137.
Reinhart, C. (2001) "Daylight Availability and Manual Lighting Control in Office Buildings – Simulation Studies and Analysis of Measurements", PhD thesis, University of Karlsruhe, Germany.
Reinhart, C. (2004) "LIGHTSWITCH-2002: A Model for Manual Control of Electric Lighting and Blinds", *Solar Energy* 77(1): 15–28.
Rietschel, H. and Raiß, W. (1968) *Lehrbuch der Heiz- und Lüftungstechnik* (15. Aufl.) Berlin/Göttingen/Heidelberg: Springer-Verlag.
Rijal, H.B., Tuohy, P., Humphreys, M.A., Nicol, J.F., Samuel, A. and Clarke, J. (2007) "Using Results from Field Surveys to Predict the Effect of Open Windows on Thermal Comfort and Energy Use in Buildings", *Energy and Buildings* 39(7): 823–836.
Rubin, A.I., Collins, B.L. and Tibbott, R.L. (1978) "Window Blinds as Potential Energy Saver – A Case Study", *NBS Building Science Series 112*, Gaithersburg, MD: National Institute for Standards and Technology.

Recommended reading

Systematic treatment of occupants' presence and actions in buildings and related models represent an emerging field of inquiry, resulting in a lack of standard textbooks and comprehensive review papers. The list of references includes nonetheless some valuable reading material.

Activities

Explicit assignments are not included for this chapter, given the specific character of the material covered (see also pages 56–57). However, various activities could deepen the appreciation of the challenges associated with accurate modelling of people's presence and actions in buildings. For example, readers might generate reports of their own presence and daily interactions with the building's environmental systems. Thereby, the general scheme captured in Table 4.4. can provide a conceptual framework. Moreover, such information could be used as input data for energy performance simulation runs. The results could then be compared to those obtained by using standard occupancy-related input models.

5 Thermal load and energy performance prediction

Jeffrey D. Spitler

Scope

This chapter covers heat transfer to, from, and within the building envelope; calculation of building heating and cooling loads; energy calculations related to the building envelope and simulation of rooms and zones to predict temperatures, loads, and required heat transfer to and from the building system.

Learning objectives

After studying this chapter, the student should be able to do the following:

(i) Understand how fundamental knowledge of heat transfer is applied to building simulation, (ii) Understand the basis for simulation of rooms and zones within buildings, and (iii) Apply the heat balance concept to building surfaces and zone air.

Key words

Load calculations/energy calculations/convection/radiation/conduction/heat transfer/infiltration

Introduction

Prediction of the building thermal load and thermal response is a fundamental to building performance simulation. Strictly speaking, the building thermal load is defined as the amount of heat that must be removed (cooling load) or added (heating load) to maintain a constant room air temperature. This quantity is usually used to determine design loads for sizing of ducts and equipment. In practice, a constant room air temperature is seldom maintained due to a variety of factors, e.g. setback thermostatic control of space temperature or use of natural ventilation for cooling. Accordingly, for prediction of energy performance, the actual heat extraction/addition rates to the space are of interest.

The building is generally divided into "zones" – portions of the building that can be considered to have an approximately uniform air temperature. This is most commonly a single room or group of rooms with similar thermal characteristics served by the same system, like a row of similar offices along the south side of the building. Each zone forms a control volume over which heat transfer into and out of the zone is analyzed.

This requires analysis of all three modes of heat transfer – conduction, convection, and radiation – within the building envelope and between the building envelope and its surroundings. (Here, the term building envelope refers to the walls, roofs, floors, and fenestrations that make up the building.)

The three modes of heat transfer all occur simultaneously, and it is the simultaneous solution of all three modes of heat transfer that complicates the analysis. In practice, this simultaneous solution is done with a building simulation program, though the very simplest forms of cooling and heating load calculation can be done by hand, relying on tabulated factors that were computed with a building simulation program.

Before concerning ourselves with the simultaneous solution, we will first consider each of the three modes independently. Then, after considering the three modes of heat transfer, various means for solving the entire problem will be considered. Each of the zone models utilizes a different approach for the simultaneous solution.

Conduction heat transfer

Conduction heat transfer through the building envelope, as well as conduction heat transfer into and out of the building envelope is an important contribution to the building load and the thermal response of the zone.

Steady-state conduction

Heat transfer through building walls and roofs is generally treated as a pure conduction heat transfer process, even though, for example, convection and radiation may be important in an internal air gap[1] in the wall. Conduction is the transfer of heat through a solid[2] via random atomic and molecular motion in response to a temperature gradient. Elements of the building envelope such as thermal bridges and corners distort the temperature gradients so that the heat flows in directions other than purely perpendicular to the envelope surfaces. Such heat flow is said to be multi-dimensional, and while it is possible to analyze multi-dimensional conduction heat transfer, for building load calculations, it is generally approximated as being one-dimensional; however the approximations do take into account the impact of thermal bridges. Heat loss from foundation elements is also multi-dimensional, but, again, approximations are made that simplify the calculation procedure.

One-dimensional steady-state heat conduction is described by the Fourier equation:

$$\dot{q} = -kA\frac{dT}{dx} \tag{5.1}$$

where:

$\dot{q}$ = heat transfer rate (W),
k = thermal conductivity (W/(m-K)),
A = area normal to the heat flow (m^2),
$\frac{dT}{dx}$ = temperature gradient (°C/m) or (K/m), which are equivalent.

For a single layer of a wall, with heat flow in the x direction, and with conductivity k and surface temperatures t_2 and t_1, as shown in Figure 5.1, Equation 5.1 can be integrated to give:

$$\dot{q} = -kA\frac{(T_2 - T_1)}{(x_2 - x_1)} \tag{5.2}$$

A convenient form of Equation 5.2 involves definition of the unit thermal resistance, R:

$$R = \frac{(x_2 - x_1)}{k} = \frac{\Delta x}{k} \tag{5.3}$$

where:

R = unit thermal resistance, (m^2K/W),
Δx = thickness of the layer; units are consistent with k: (m).

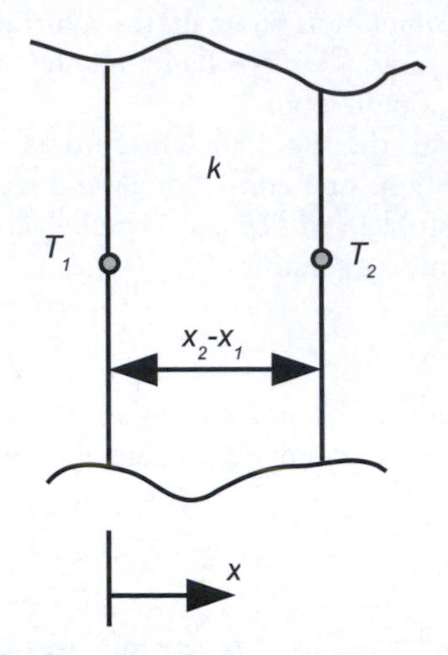

Figure 5.1 A single-layer plane wall.

Then, the conduction heat transfer rate through the wall is given by:

$$\dot{q} = A\frac{(T_1 - T_2)}{R} \tag{5.4}$$

Example 5.1

A 100 mm thick uninsulated wall of area 10 m^2, is made out of concrete with conductivity of 1.2 W/mK. What is the *R*-value of the wall? Under steady-state conditions, with the exterior surface temperature of the wall at –20 °C and the interior surface temperature of the wall at 15 °C, what is the total heat loss through the wall?

Solution:

The *R*-value is determined by dividing the thickness by the conductivity:

R= 0.1 m./1.2 W/mK= 0.083 m^{2}K/W.

The conduction heat transfer rate through the wall is given by Eqn. 5.4:

$$\dot{q} = 10\text{m}^2 \frac{(15^\circ\text{C} - (-20^\circ\text{C}))}{0.083\frac{\text{m}^2\text{K}}{\text{W}}} = 4,200\,\text{W}$$

End of solution.

A more common situation is that the wall has multiple layers, as shown in Figure 5.2. Here, the wall is made up of three layers with two exterior surfaces (1 and 4) and two interior interfaces (2 and 3). Each of the three layers is defined by a conductivity (e.g. k_{2-1} for the leftmost layer) and thickness (e.g. Δx_{2-1} for the leftmost layer).

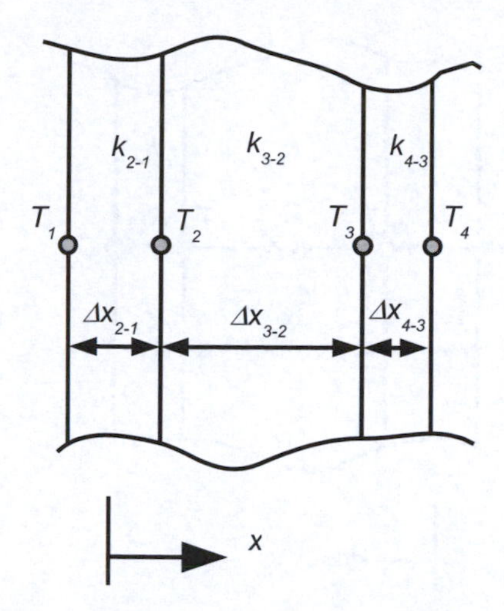

Figure 5.2 A multi-layer wall.

Individual unit thermal resistances are determined for each layer:

$$R_{2-1} = \frac{\Delta x_{2-1}}{k_{2-1}} \tag{5.5a}$$

$$R_{3-2} = \frac{\Delta x_{3-2}}{k_{3-21}} \tag{5.5b}$$

$$R_{4-3} = \frac{\Delta x_{4-3}}{k_{4-3}} \tag{5.5c}$$

There is an analogy between conduction heat transfer and electricity, as illustrated in Figure 5.3, where the unit thermal resistances (analogous to electrical resistances[3] per unit area) are shown connecting the surface temperatures (analogous to voltages). As a consequence, the total unit thermal resistance of the wall may be determined by simply adding the three individual unit thermal resistances (in series):

$$R_{4-1} = R_{2-1} + R_{3-2} + R_{4-3} \tag{5.6}$$

Then, the heat flux (analogous to current per unit area) may be determined as:

$$q'' = \frac{(T_1 - T_4)}{R_{4-1}} \tag{5.7a}$$

where:

q'' = heat flux rate (heat transfer rate per unit area), W/m^2.

This is also commonly expressed with an overall conductance, *U*, known as the *U*-factor or *U*-value for the wall:

$$q'' = U(T_1 - T_4) \tag{5.7b}$$

where:

$U = 1/R$, overall conductance, W/m^{2}K

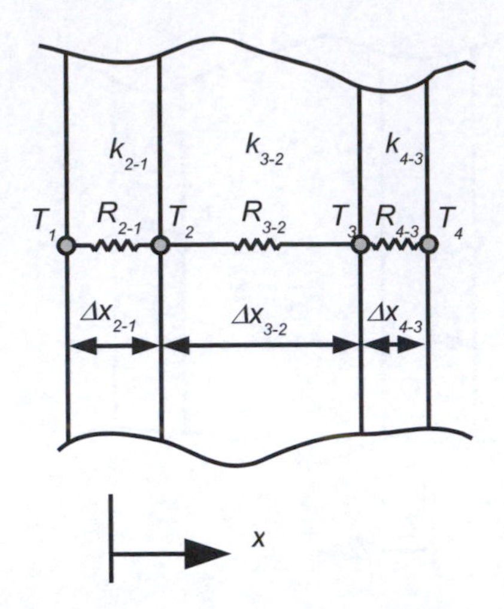

Figure 5.3 Multi-layer wall analysis based on electrical analogy.

Example 5.2

The concrete wall of Example 5.1 is improved by adding a 50 mm thick layer of expanded polystyrene, with an *R*-value of 1.8, and a layer of 12.7 mm thick drywall, with an *R*-value of 0.079. With these additional layers, what is the *R*-value of the wall? Under steady-state conditions, with the exterior surface temperature of the wall at –20 °C and the interior surface temperature of the wall at 15 °C, what is the total heat loss through the wall?

Solution:

Because the three layers are in series with each other, the total *R*-value is determined by adding the resistances of the three layers, as modeled in Equation 5.6:

$$R = 0.083 + 1.8 + 0.079 = 1.96\ \mathrm{m^2{\cdot}K/W}.$$

The conduction heat transfer rate through the wall is given by Eqn. 5.4:

$$\dot{q} = 10\mathrm{m}^2 \frac{\left(15^\circ\mathrm{C} - (-20^\circ\mathrm{C})\right)}{1.96\dfrac{\mathrm{m^2K}}{\mathrm{W}}} = 179\,\mathrm{W}$$

A dramatic reduction in heat loss can generally be obtained when insulation is added to uninsulated walls.
End of solution.

Generally, heating load calculations are performed for conditions that are assumed to be approximately steady-state: that is night-time, hours after the building occupants have gone home and the sun has gone down. In this case, the conduction heat transfer through the above-grade portion of the building envelope is determined by first calculating the overall thermal resistance and conductance (*U*-factor) of each surface; multiplying the *U*-factors by the surface areas; summing the *UA* values for each room, and multiplying the total *UA* for each room by the design temperature difference. For slab-on-grade floors, basement walls, and basement floors, a simple multi-dimensional analysis is used such that the solution is given in one-dimensional form.

Transient conduction

The discussion of conduction heat transfer in the previous section treated it as a steady-state phenomenon. While for heating load calculations this is generally sufficient, strong daily variations in incident solar radiation and outdoor air temperatures occurring under energy analysis and cooling load design conditions can not be ignored in most cases. This is particularly true as the thermal capacitance of the wall or roof increases. Thermal capacitance of a wall or roof element may be defined as the amount of energy required to raise the temperature by one degree:

$$C_{th} = Mc_p = \rho V c_p \qquad (5.8)$$

where:

C_{th} = thermal capacitance of a wall or roof layer (J/K),
M = mass of the wall or roof layer (kg),
c_p = specific heat of the wall or roof layer material (J/kg·K),
ρ = density of the wall or roof layer material (lb_m/ft^3) or (kg/m^3),
V = volume of the wall or roof layer (m^3).

The energy required to raise the temperature of the layer is then given by:

$$Q=C_{th}\,\Delta T =Mc_p\,\Delta T =\rho V c_p\,\Delta T \qquad (5.9)$$

where:

ΔT = temperature increase (°C).

Example 5.3 Thermal Storage

Consider the 100 mm thick layer of concrete in the 10 m^2 wall in Example 5.1. During a warm summer day, it is warmed from a temperature of 20 °C to a temperature of 30 °C. If the concrete has a density of 1600 kg/m^3 and a specific heat of 900 J/kgK, how much energy is absorbed in warming the concrete layer?

Solution:

The total volume of the concrete is 1 m^3; multiplying by the density gives the total mass of the concrete as 1600 kg. The required energy input is the mass multiplied by the specific heat multiplied by the rise in temperature:

$$Q= 1600 \text{ kg} \times 900 \text{ J/kgK} \times 10 \text{ °C} = 14{,}400 \text{ kJ}$$

End of solution.

Comparing the amount of energy required to raise the temperature of the layer by 10 °C to the expected heat transmission rate under steady-state conditions suggests that this heat storage effect will not be unimportant. However, because the thermal capacitance is distributed throughout each layer, there is no simple way to calculate the transient behavior. The distributed thermal capacitance is sometimes represented as shown in Figure 5.4, with the two parallel lines below each resistance representing the distributed capacitance.

In order to further illustrate the effects of thermal storage in the wall, consider what would happen if the three-layer (100 mm concrete, R-1.8 insulation, 12.7 mm drywall) wall described in Example 5.2 was placed in an office building in Atlanta, facing southwest. Under clear sky, high temperature conditions, the heat transmission through the wall may be calculated using two methods:

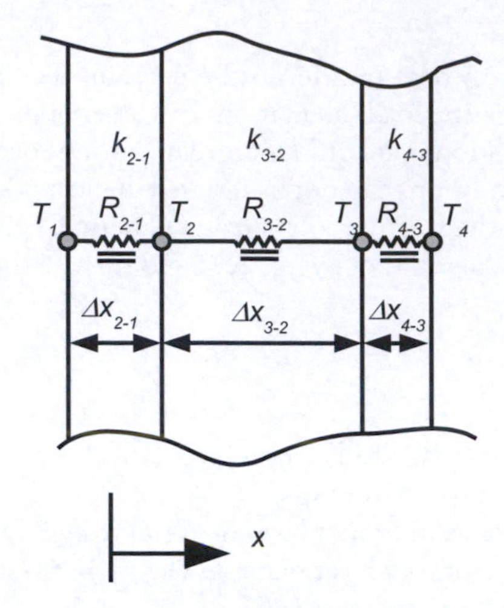

Figure 5.4 Electrical analogy with distributed thermal capacitance.

- A method that ignores the thermal mass in the wall, computing the conduction heat gain with Equation 5.4 at each hour of the day. In this case, the exterior temperature changing throughout the day gives conduction heat gains that change over the day. This method may be referred to as "Quasi-steady-state" because it treats each hour of the day as if the wall comes immediately to steady-state conditions. (Labeled "Quasi-SS" in Figure 5.5.)
- A method, described later in this chapter, which takes into account the thermal capacitance of each layer. (Labeled "Transient" in Figure 5.5.)

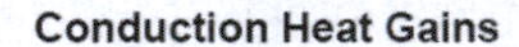

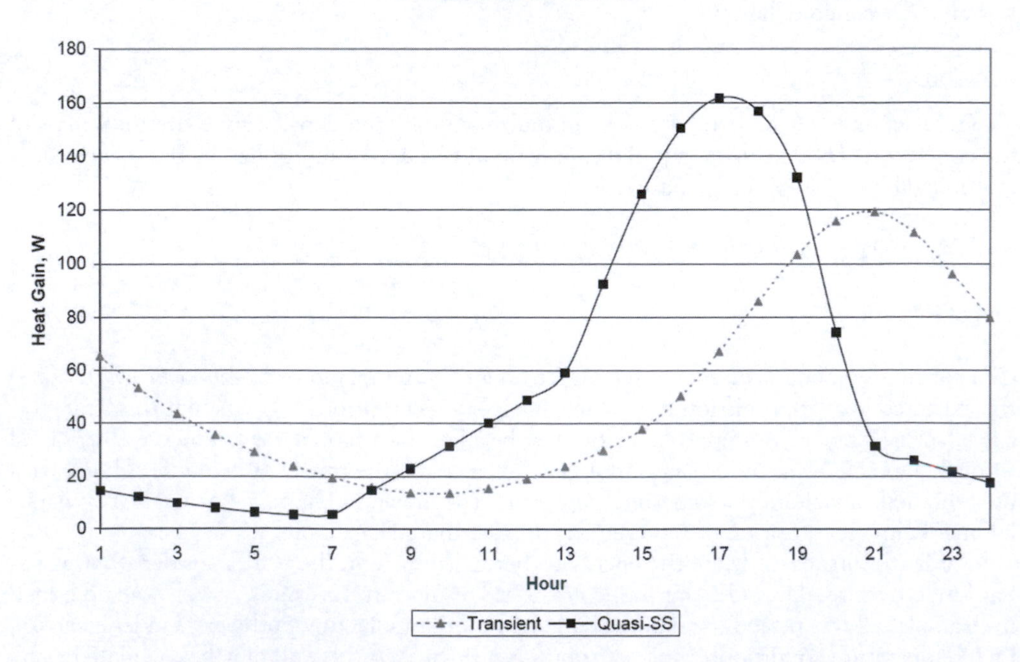

Figure 5.5 Comparison of transient and quasi-steady-state conduction heat gain calculations.

Conduction heat gains computed with each method are shown in Figure 5.5. In Figure 5.5, two effects of the thermal mass may be observed:

- The first may be referred to as "Time Delay" – the peak heat gain, when calculated with the quasi-steady-state method, occurs at 5 p.m. (hour 17). When the thermal capacitance is accounted for in the calculation, we can see that the peak heat gain actually occurs at 9 p.m. (hour 21). This four-hour delay in the peak heat gain may be important, especially if the peak heat gain is delayed such that it occurs after the occupants have gone home.
- The second effect may be referred to as "Damping" – the peak heat gain, when calculated with the transient method is about 30% lower than that calculated with the quasi-steady-state method.

Both effects are important, and this is why transient conduction calculations are typically performed as part of cooling load calculation procedures and energy analysis.

Transient one-dimensional conduction in a homogeneous slab is described by the heat diffusion equation:

$$\frac{k}{\rho c_p}\frac{\partial^2 T}{\partial x^2} = \frac{\partial T}{\partial t} \tag{5.10}$$

where:

k = thermal conductivity (W/mK),
ρ = density (kg/m^3),
c_p = specific heat (J/kgK),
T = temperature at any point in the slab (°C),
x = distance (m),
t = time (s).

Non-linear, time-dependent boundary conditions and walls and roofs that are not homogeneous preclude obtaining a simple solution of Equation 5.10. However, calculation of transient conduction heat transfer through walls and roofs may be performed with a number of different methods. These methods include:

1. Z-transform methods – methods based on Z-transform theory, including "response factors" and "conduction transfer functions". Z-transform methods have a high degree of accuracy and are computationally efficient. This class of methods are the most widely used for load calculations and building energy analysis.
2. Numerical methods – finite difference and finite element methods. Numerical methods also have a high degree of accuracy but typically are less computationally efficient than Z-transform methods. However, they offer significant increased flexibility for treating such phenomena as radiation across air gaps and phase change materials.
3. Lumped parameter methods – treating walls and roofs as discrete resistances and lumped capacitances. These methods are typically used in conjunction with lumped capacitance zone models.

Each of the methods is discussed below.

Transfer function and response factor methods

Due to their computational efficiency and accuracy, Z-transform methods are widely used in both design load calculations and building energy analysis applications. Z-transform methods have been utilized in building performance in one of three formulations, all of which utilize a fixed time step, with the most common time step being one hour:

- Response factors,
- Conduction Transfer Functions (CTF),
- Periodic Response Factors (PRF).

Response factors may be thought of as time series coefficients relating the current heat flux to past and present values of interior and exterior temperatures. (The interior and exterior temperatures may be air temperatures, sol-air temperatures, or surface temperatures, depending on the application.) Particularly for thermally massive constructions, large numbers of response factors may be required. Conduction transfer functions replace much of the required temperature history with heat flux history. In other words, many of the response factors are replaced with coefficients that multiply past values of heat flux. Periodic response factors are used for design day load calculations where the design day is assumed to repeat.

The conduction transfer function formulation for the inside heat flux is:

$$q_i''(\theta) = -Z_o T_{i,\theta} - \sum_{j=1}^{nz} Z_j T_{i,\theta-j\delta} + Y_o T_{o,\theta} + \sum_{j=1}^{nz} Y_j T_{o,\theta-j\delta} + \sum_{j=1}^{nq} \Phi_j q''_{ki,i-j\delta} \tag{5.11}$$

And, for the outside heat flux:

$$q_o''(t) = -Y_o T_{i,\theta} - \sum_{j=1}^{nz} Y_j T_{i,\theta-j\delta} + X_o T_{o,\theta} + \sum_{j=1}^{nz} X_j T_{o,\theta-j\delta} + \sum_{j=1}^{nq} \Phi_j q''_{ko,i,\theta-j\delta} \tag{5.12}$$

where:

X_j = Outside CTF coefficient, $j = 0,1,\ldots, nz$.
Y_j = Cross CTF coefficient, $j = 0,1,\ldots, nz$.
Z_j = Inside CTF coefficient, $j = 0,1,\ldots, nz$.
Φ_j = Flux CTF coefficient, $j = 0,1,\ldots, nq$.
T_i = Inside surface temperature (°C),
T_o = Outside surface temperature (°C),
q''_o = Conduction heat flux into wall at outside face (W/m^2),
q''_i = Conduction heat flux out of wall at inside face (W/m^2),
θ = Current time, seconds or hours,
δ = Length of time step, seconds or hours.

The subscript on each of the CTF terms indicates the time, relative to the current time, in terms of the number of time steps to which the term applies. e.g. X_1 is applied to the temperature one time step earlier, which is given in the temperature subscript as $\theta - 1\delta$. The first terms in the series (those with subscript "0") apply to the current time and have been extracted from the summation to facilitate solving for the current temperature in the solution scheme.

The two summation limits, nz and nq, depend on the wall construction and method used to calculate the CTFs. Generally, walls that are more thermally massive will require more terms. Response factors have a similar formulation, for which $nq = 0$. Theoretically nz is then infinite, but in practice, a large number of terms, perhaps 100 or more, may be sufficient. The CTF formulation is computationally more efficient.

Although use of either response factors or conduction transfer functions is relatively straightforward, a more difficult task is determining the response factors or conduction transfer function coefficients. A detailed explanation of an analytical procedure for determining response factors and conduction transfer function coefficients of multilayer slabs is given by Hittle

(1981). Spitler (1996) cites a number of other methods, of which the most common are the Laplace method (Mitalas and Arseneault 1970) and the state-space method (Seem *et al.* 1989). More recently, time-domain (Davies 1996) and frequency-domain regression (Wang and Chen 2003) methods have been developed. Although not usually done as a precursor to simulation, it is also possible to derive response factors or CTF coefficients experimentally, as demonstrated by Brown and Stephenson (1993).

Periodic response factors are used for design load calculations, where heat gains for a single day, assumed to be preceded by one or more identical days, are desired. Since periodic response factors are used in conjunction with simplifications to both the exterior and interior heat balances, heat gain is calculated based on an equivalent exterior temperature, the sol-air temperature, and the interior air temperature, which is presumed constant. Conductive heat gain with periodic response factors is calculated as:

$$q_{\theta} = A\sum_{j=0}^{23} Y_{Pj}(T_{e.\theta-j\delta} - T_{rc}) \tag{5.13}$$

where:

q_{θ} = hourly conductive heat gain for the surface (W),
A = surface area (m^2),
Y_{Pj} = *j*th periodic factor,
$T_{e,\theta-j\delta}$ = sol-air temperature, *j* hours ago (°C),
T_{rc} = presumed constant room air temperature (°C),
θ = the current hour (h),
δ = the time step (one hour) (h).

Periodic response factors can be computed with the same methods as conduction transfer functions, or directly from the values of the conduction transfer function coefficients, as described by Spitler and Fisher (1999).

Building simulation tools that use conduction transfer functions and periodic response factors will generally have internal routines for calculating the values based on the user's layer-by-layer description of the surface construction. Other options include the ASHRAE Load Calculations Toolkit (Pedersen *et al.* 2001) and the CTF/PRF Generator developed by Iu and Fisher (2004); the latter tool may be freely downloaded from www.hvac.okstate.edu.

Example 5.4

A wall is made up of five layers, as shown in Table 5.1, listed from outside to inside. Conduction Transfer Function coefficients for this wall have been calculated with a cooling load calculation program which uses the Laplace transform method; these values are shown in Table 5.2 (Mitalas and Arseneault 1970). For a day where the exterior and interior surface temperatures are as shown in Table 5.3, determine the interior surface heat flux. Assume the previous days had identical temperatures.

Solution:

Based on the provided CTF coefficients, nz=5 and nq=5. Because the calculation relies on past values of q'' to start the calculation, we must assume some values of q'' to get started. After starting the calculation, we will compute the same day over and over until a steady periodic solution is reached. Assuming values of zero will work, though with this very heavy wall, it may take a number of days to converge. Therefore, an improved estimate, which we will make use of here, is simply to find the steady-state heat transfer rate based on the average exterior and interior temperature.

Table 5.1 Layer descriptions for Example 5.4

Layer	*Thickness (mm)*	*Density (kg/m³)*	*Conductivity (W/mK)*	*Specific heat (J/kg K)*
Face brick	100	1920	0.87	800
Air gap	20	1	0.11	1005
Insulation	50	91	0.04	840
Solid concrete block	100	2096	1.63	920
Plaster	13	720	0.16	840

Table 5.2 Conduction transfer function coefficients for Example 5.4

j	X_j	Y_j	Z_j	Φ_j
0	2.171573E+01	3.375092E–05	1.033658E+01	
1	–4.514066E+01	5.617864E–03	–1.794393E+01	1.463783E+00
2	2.844604E+01	2.266381E–02	8.642595E+00	–5.568298E–01
3	–5.162262E+00	1.133217E–02	–1.021217E+00	2.488959E–02
4	1.826630E–01	8.227039E–04	2.664688E–02	–2.266001E–04
5	–1.023128E–03	8.775257E–06	–1.882663E–04	

Table 5.3 Exterior and interior surface temperatures for Example 5.4

Time	$T_{o,t}$ (°C)	$T_{i,t}$ (°C)
1	26.42	22.09
2	25.44	22.08
3	24.67	22.07
4	24.08	22.05
5	23.89	22.02
6	24.28	22.00
7	25.25	21.97
8	27.00	21.93
9	29.53	21.90
10	32.44	21.88
11	35.75	21.86
12	38.86	21.84
13	41.19	21.84
14	42.75	21.84
15	43.33	21.86
16	42.75	21.88
17	41.39	21.91
18	39.25	21.95
19	36.72	21.99
20	34.19	22.02
21	32.06	22.05
22	30.11	22.08
23	28.56	22.09
24	27.39	22.10

The *U*-value for the wall is 0.592 W/m²K and the average difference between the exterior and interior surface temperatures is 10.42 °C, yielding an estimate for the average heat flux of 6.16 W/m². Then applying Eqn. 5.11 for the first hour, we get:

$$q_i''(1) = -(10.336575 \times 22.09 - 17.943933 \times 22.10 + 8.642595 \times 22.09 - 1.021217 \times 22.08 + 0.026647 \times 22.05 - 0.000188 \times 22.02) + 0.000034 \times 26.42 + 0.005618 \times 27.39 + 0.022664 \times 28.56 + 0.011332 \times 30.11 + 0.000823 \times 32.06 + 0.000009 \times 34.19 + 1.463783 \times 6.16 - 0.556830 \times 6.16 + 0.024890 \times 6.16 - 0.000227 \times 6.16 = 6.07 \text{ W/m}^2$$

Applying it again for the second hour, we get:

$q_i''(2) = -(10.336575 \times 22.08 - 17.943933 \times 22.09 + 8.642595 \times 22.10 - 1.021217 \times 22.09 + 0.026647 \times 22.08 - 0.000188 \times 22.05) + 0.000034 \times 25.44 + 0.005618 \times 26.42 + 0.022664 \times 27.39 + 0.011332 \times 28.56 + 0.000823 \times 30.11 + 0.000009 \times 32.06 + 1.463783 \times 6.07 - 0.556830 \times 6.16 + 0.024890 \text{ x}06.16 - 0.000227 \times 6.16 = 5.90$ W/m^2

Equation 5.11 is applied for each hour of the day. Then, on the second day, the first day is used as history. For example, for the first hour of the second day:

$q_i''(1) = -(10.336575 \times 22.09 - 17.943933 \times 22.10 + 8.642595 \times 22.09 - 1.021217 \times 22.08 + 0.026647 \times 22.05 - 0.000188 \times 22.02) + 0.000034 \times 26.42 + 0.005618 \times 27.39 + 0.022664 \times 28.56 + 0.011332 \times 30.11 + 0.000823 \times 32.06 + 0.000009 \times 34.19 + 1.463783 \times 7.64 - 0.556830 \times 7.88 + 0.024890 \times 8.02 - 0.000227 \times 08.03 = 7.32$ W/m^2

Doing this for four days gives the results shown in Table 5.4. By the 4th day, the results are more-or-less converged.

Note that in an annual energy simulation, it is common to start with a short pre-conditioning period of one or a few days so as to develop reasonable historical values for past values of the heat fluxes prior to January 1.
End of Solution

Table 5.4 Heat fluxes at the interior surface

	Day			
Hour	*1*	*2*	*3*	*4*
1	6.07	7.32	7.34	7.34
2	5.90	6.96	6.97	6.97
3	5.69	6.57	6.58	6.58
4	5.44	6.17	6.18	6.18
5	5.16	5.77	5.78	5.78
6	4.88	5.38	5.39	5.39
7	4.60	5.02	5.02	5.02
8	4.36	4.70	4.71	4.71
9	4.17	4.46	4.46	4.46
10	4.07	4.31	4.31	4.31
11	4.09	4.28	4.28	4.28
12	4.24	4.40	4.40	4.40
13	4.53	4.66	4.67	4.67
14	4.96	5.07	5.07	5.07
15	5.48	5.57	5.57	5.57
16	6.06	6.13	6.13	6.13
17	6.64	6.70	6.70	6.70
18	7.17	7.22	7.22	7.22
19	7.59	7.64	7.64	7.64
20	7.89	7.92	7.92	7.92
21	8.03	8.06	8.06	8.06
22	8.02	8.05	8.05	8.05
23	7.88	7.90	7.90	7.90
24	7.64	7.66	7.66	7.66

Numerical methods

The literature on finite difference, finite volume and fine element methods applied to conduction heat transfer problems is voluminous. For a discussion of numerical methods applied to transient heat conduction in buildings, in the context of building simulation, see Clarke (2001), Underwood and Yik (2004), Waters and Wright (1985), Lewis and Alexander (1990), and Walton (1993). During the early years of building performance simulation, numerical methods were often considered too computationally demanding for practical application in hourly energy calculations. Increasing computational speed has rendered this a moot point for many applications and numerical methods are used in a number of building simulation programs. However, space precludes a detailed treatment here and the reader should consult the works cited above for more information.

Numerical methods treat the building envelope surface as being made up of discrete capacitances and resistances; this can be seen most clearly in the derivation of the finite volume method. For building walls modeled in one-dimension, the ESP-r program uses a default value of three discrete capacitances per layer. This gives reasonable computational speed for annual energy analysis. However, there are some applications for which further simplification achieved by using even fewer discrete capacitances is highly desirable. Methods that use only one to three discrete capacitances to model a wall are often called lumped parameter methods.

Lumped parameter methods

Lumped parameter methods treat the individual building envelope surfaces as being made up of a small number of discrete capacitances and resistances, typically only one or two. They are of particular interest for very short time-step (on the order of a few seconds) simulation of building systems. Transfer function models are unsuitable for such time steps because the required number of CTF coefficients becomes very large as the time steps become short. Numerical methods can be used, but they will typically require too much computational time at such short time steps. Furthermore, lumped parameter methods can be conveniently implemented within simulation environments such as MATLAB® that are commonly used for simulation of control systems. The history of lumped parameter methods is described by Gouda *et al.* (2002).

Typically, the total of the discrete resistance would be equivalent to the total wall resistance and the total of the discrete capacitances would be equal to the total wall capacitance. However, an approximate distribution of the discrete resistances and capacitances must be made. Gouda *et al.* (2002) develop an optimization-based method for tuning the parameters of a second-order (two discrete capacitances) model and demonstrate that it gives quite accurate results. Underwood and Yik (2004) give suggested distributions of resistances and capacitances for a range of wall and floor types.

Deviations from one-dimensional conduction-only heat transfer

For the most part, the models used for load calculations and energy calculations treat the wall heat transfer problem as a one-dimensional, transient conduction-only problem. Walls often include thermal bridges, for which one modeling approach is described by Burch *et al.* (1992), Seem *et al.* (1989) and Carpenter *et al.* (2003), describe development of CTF coefficients for two- and three-dimensional surfaces.

Karambakkam *et al.*(2005) describe a procedure for incorporating the effects of thermal bridges within the one-dimensional representation.

Heat transfer from the building foundation to and from the ground is also a multi-dimensional problem that is necessarily treated in a simplified manner. For purposes of design cooling load calculations, heat loss to the foundation is often neglected completely. For purposes of design heating load calculations, it is necessary to include the heat lost through the foundation and a range of approaches have been proposed. A simple approach for on-grade slabs, which utilizes a perimeter heat loss factor is given by ASHRAE (2005). For basements the same reference also gives a simplified approach based on the original work of Latta and Boileau (1969). For energy calculations, more general approaches are described by Beausoleil-Morrison and Mitalas (1997) and Krarti and Choi (1996). Beyond these approaches, numerical methods may be used, but they will, in most cases, require more computational time than the rest of the building simulation, perhaps by orders-of-magnitude.

Likewise, moisture transport into and out of the wall is most often neglected, despite its clear importance under some conditions. Mendes *et al.* (2003) give an overview of candidates for modeling moisture transport within a building simulation program. They also performed a study using the candidates to compute transient heat and mass conduction through four different walls in three climates while retaining an existing building simulation program to predict inside humidities. The effects of moisture transport on the conduction heat transfer were surprisingly high; one case showed a 200% overprediction in conduction heat gain if moisture transport was not taken into account. Barbosa and Mendes (2008) describe a more comprehensive study of a single building in Brazil. Compared to a detailed whole-building hygrothermal model, ignoring moisture effects resulted in a 13% overprediction of the peak cooling load and a 4% underprediction of annual energy consumption.

Other deviations from the standard one-dimensional transient conduction-only model include internal radiation (Goss and Miller 1989), diffuse infiltration (Liu and Claridge 1992; Virtanen *et al.* 1992) and internal convection (Park *et al.* 1987; Shipp 1983). Strand and Pedersen (1997) describe development of conduction transfer functions for walls with internal heat gain, e.g. radiant heating systems.

Convection heat transfer

Surface convection heat transfer

Thermal convection is the transport of energy in a fluid or gas by mixing in addition to conduction. For load calculations, we are primarily interested in convection between an envelope surface (wall, roof, etc.) and the indoor or outdoor air. The rate of convection heat transfer depends on the temperature difference, whether the flow is laminar or turbulent, and whether it is buoyancy-driven or driven by an external flow. The convection heat transfer rate is usually expressed[4] as:

$$\dot{q} = hA(T - Ts) \tag{5.14}$$

where:

h = convection coefficient (W/m^2K),
T = bulk temperature of the air (°C),
T_s = surface temperature (°C).

The convection coefficient, often referred to as the film coefficient, may be estimated with a convection correlation, or, for design conditions, may be read from a table.

Example 5.5 Convection

A 10 m^2 wall is heated by the sun to a temperature of 60 °C at a time when the outdoor air is at 20 °C, and a light breeze results in a convection coefficient of 10 $W/m^2{\cdot}K$. What is the heat lost by the surface due to convection?

Solution:

Equation 5.14 gives the heat transferred to the surface by convection. The heat transferred away from the surface by convection would be given by:

$$\dot{q} = hA(t_s - t) = 10\frac{\text{W}}{\text{m}^2\text{K}}(10\ \text{m}^2)(60°\text{C} - 20°\text{C}) = 4{,}000\ \text{W}$$

Note that the convection coefficient is typically given in units of W/m²K, but it could just as well be given in units of W/m² °C; h is always applied to temperature differences and a temperature difference is the same in either Kelvin or degrees Celsius; 40 K = 40 °C when we are referring to a temperature difference.
End of solution.

Exterior surface convection

Convection to exterior surfaces may be represented with a range of models, all of which involve the use of a convection coefficient:

$$q''_{convection,out,j,\theta} = h_c(T_o - T_{os,j,\theta}) \tag{5.15}$$

where:

h_c is the convection coefficient (W/m²K).

Palyvos (2008) gives a review of exterior convection correlations and also proposes a pair of linear convecton correlations, based on an empirical average of about 30 such correlations in the literature. Previous review papers include McClellan and Pedersen (1997) and Cole and Sturrock (1977). Given the very complex wind-driven and buoyancy-driven air flows around a building, a convective heat transfer model might be very complex and difficult to use. A correlation developed by Yazdanian and Klems (1994) seems to strike a reasonable balance between accuracy and ease of use for low-rise buildings. The correlation takes the form

$$h_c = \sqrt{\left[C_t(\Delta t)^{1/3}\right]^2 + \left[aV_o^b\right]^2} \tag{5.16}$$

where:

C_t = turbulent natural convection constant, given in Table 5.5,
ΔT = temperature difference between the exterior surface and the outside air (°C),
a, b = constants given in Table 5.5,
V_o = wind speed at standard conditions (m/s).

For high rise buildings, Loveday and Taki (1996) recommend the correlation:

$$h_c = CV_s^{0.5} \tag{5.17}$$

Table 5.5 Convection correlation coefficients for the Yazdanian and Klems (1994) model

Wind direction	C_t *(W/m²K⁴ᐟ³)*	*a* (W/m²K(m/s))	*b* (–)
Windward	0.84	2.38	0.89
Leeward	0.84	2.86	0.617

where:

$$C = 16.7\left(\frac{W}{m^2K}\left(\frac{m}{s}\right)^{-0.5}\right),$$

V_s = wind speed near surface (m/s).

The correlation was based on windspeeds between 0.2 m/s and 4 m/s. Loveday and Taki do not make a recommendation for windspeeds below 0.2 m/s, but a minimum convection coefficient of 7.5 W/m²K might be inferred from their measurements.

Interior surface convection

Interior convective heat transfer to and from room surfaces occurs under a wide range of regimes varying from natural convection to forced convection and from laminar flow to turbulent flow. At this time, there are no satisfactory models that cover the entire range of conditions. However, Beausoleil-Morrison and Mitalas (2001) have developed a model for rooms with ceiling diffusers that incorporates correlations (Alamdari and Hammond 1983; Fisher and Pedersen 1997) from a range of different flow regimes. Novoselac *et al.* (2006) give experimentally-based convection correlations for rooms with displacement ventilation. Similar models for other air supply or heat emitter configurations are still needed. Fortunately, for many buildings the cooling loads are only modestly sensitive to the interior convection coefficients. Buildings that are highly glazed are a notable exception.

A relatively simple model, strictly applicable for natural convection conditions, utilizes fixed convection coefficients extracted from the surface unit conductances (ASHRAE 2005). The surface unit conductances, which are combined convection-radiation coefficients, have a radiative component of about 5.1 W/(m²K). By subtracting the radiative component, we obtain the convective coefficients shown in Table 5.6. Once the convective coefficient is obtained, the convective heat flux from the wall to the zone air is

$$q''_{convection,in,j,\theta} = h_c(T_{is,j,\theta} - T_i) \tag{5.18}$$

Bulk convection – infiltration and ventilation

Infiltration and ventilation result in the introduction of outdoor air into the space. Methods used to estimate the quantity of infiltration and/or ventilation air are discussed in Chapter 6. Introduction of outdoor air results in both sensible and latent heat gain, which will be computed as follows:

$$\dot{q}_{infiltration,\theta} = \dot{m}_a c_p (T_o - T_i) = \frac{\dot{Q} c_p}{v_o}(T_o - T_i) \tag{5.19}$$

Table 5.6 Interior surface convection coefficients for use with the heat balance model

Position of Surface	*Direction of Heat Flow*	*h_c, (W/m²K)*
Horizontal	Upward	4.15
Sloping – 45°	Upward	3.98
Vertical	Horizontal	3.18
Sloping – 45°	Downward	2.39
Horizontal	Downward	1.02

$$\dot{q}_{infiltration,latent,\theta} = \dot{m}_a (W_o - W_i) i_{fg} = \frac{\dot{Q}}{v_o}(W_o - W_i) i_{fg} \tag{5.20}$$

where:

$\dot{q}_{infiltration,\theta}$ = sensible heat gain due to infiltration at time θ (W),
$\dot{q}_{infiltration,latent,\theta}$ = latent heat gain due to infiltration at time θ (W),
$\dot{m}_a$ = mass flow rate of infiltration on a dry-air basis (kg/s),
c_p = specific heat of the infiltration air (J/kg K),
T_o = outside air temperature (°C),
T_i = inside air temperature (°C),
$\dot{Q}$ = volume flow rate of air (m^3/s),
v_o = specific volume of the outside air (m^3/kg),
W_o = outside air humidity ratio (kg H_2O vapor/kg dry air),
W_i = inside air humidity ratio (kg H_2O vapor/kg dry air),
i_{fg} = phase change enthalpy (J/kg).

Bulk convection – air system

For rooms where heating or cooling is provided by an air system, the heat transfer to or from the space is affected by bulk convection. The air temperature and/or flow rate to the zone may be thermostatically controlled. It is possible to formulate the effects of the system in several different ways:

1. If the system is modeled simultaneously, and the mass flow rate and/or temperature of air entering the space is determined by the system simulation, it is possible to:

$$\dot{q}_{system,\theta} = \dot{m}_a c_p (T_{sys} - T_i) \tag{5.21}$$

2. If a design load calculation is being performed, a desired room air temperature can be set and the heat balance can be arranged to solve for the required system heat transfer rate.
3. If the system is not being simulated or it is being simulated sequentially, it is often convenient to represent the system heat transfer as a piecewise linear function of zone temperature, called a control profile:

$$\dot{q}_{system,\theta} = a + bT_i \tag{5.22}$$

where a and b are equation coefficients that apply over a specific range of temperatures.

Radiation heat transfer

Thermal radiation is the transfer of energy by electromagnetic waves. In building load calculations, thermal radiation is generally thought of as being a surface-to-surface phenomenon either between surfaces within the building, or between the surface of the sun and the building surfaces. Gases, aerosols, and particulates also emit and absorb radiation between surfaces. However, for analysis of radiation heat transfer between surfaces within a building, the radiation path lengths are short enough that emission and absorption of the indoor air can be neglected. For radiation heat transfer between the Sun and the building surfaces, emission and absorption in the Earth's atmosphere is accounted for in the models used to determine incident solar irradiation.

Thermal radiation emitted by any surface will have a range of wavelengths, depending upon the temperature of the emitting surface. The amount of thermal radiation absorbed, reflected, or emitted by any surface will depend on the wavelengths and direction in which the radiation is incident or emitted relative to the surface. Dependency on the wavelength is referred to as "spectral"; surfaces for which the properties are effectively independent of the wavelength are referred to as "gray" surfaces. Dependency on the direction is referred to as "specular"; surfaces for which the properties are effectively independent of the direction are referred to as "diffuse" surfaces. Properties of interest include:

- Absorptance,[5] α, the ratio of radiation absorbed by a surface to that incident on the surface,
- Emittance, ε, the ratio of radiation emitted by a surface to that emitted by an ideal "black" surface at the same temperature,
- Reflectance, ρ, the ratio of radiation reflected by a surface to that incident on the surface,
- Transmittance, τ, the ratio of radiation transmitted by a translucent surface to that incident on the surface.

Analysis of thermal radiation is greatly simplified when the surfaces may be treated as gray and diffuse. With two exceptions, surfaces in buildings are treated as both gray and diffuse in load calculation procedures.

The first exception is based on the fact that thermal radiation wavelength distributions prevalent in buildings may be approximately lumped into two categories – short wavelength radiation (solar radiation or visible radiation emitted by lighting) and long wavelength radiation (radiation emitted by surfaces, people, equipment, etc. that is all at relatively low temperatures compared to the sun.) Treatment of the two wavelength distributions separately in building load calculations might be referred to as a "two-band" model; but in practice this only means that surfaces may have different absorptances for short wavelength radiation and long wavelength radiation. As an example, consider that a surface painted white may have a short wavelength absorptance of 0.4 and a long wavelength absorptance of 0.9.

The second exception is for analysis of windows. Solar radiation is typically divided into specular ("direct" or "beam") and diffuse components. Since the window transmittance, absorptance, and reflectance tend to be moderately strong functions of the incidence angle, these properties are generally calculated for the specific incidence angle each hour.

A notable feature of thermal radiation is that the emission is proportional to the fourth power of the absolute temperature (i.e., Kelvin or Rankine.) As an example, consider a case with only two surfaces separated by a non-participating medium. The radiation heat transfer rate between surfaces 1 and 2 is given by:

$$\dot{q}_{1-2} = \frac{\sigma\left(T_1^4 - T_2^4\right)}{\dfrac{1-\varepsilon_1}{A_1\varepsilon_1} + \dfrac{1}{A_1F_{1-2}} + \dfrac{1-\varepsilon_2}{A_2\varepsilon_2}} \tag{5.23}$$

where:

σ = Stefan-Boltzmann constant (5.673×10^{-8} W/m^{2}K^4),
T_1 and T_2 = surface temperatures of surfaces 1 and 2 (K),
ε_1 *and* ε_2 = emittances of surfaces 1 and 2,
F_{1-2} = View factor from surface 1 to surface 2.

Example 5.5 Radiation

A 10 m^2 wall has an air gap separating a layer of brick and a layer of concrete block. At a time when the temperature of the brick surface adjacent to the air gap is 0 °C and the temperature of the concrete block surface adjacent to the air gap is 10 °C, what is the radiation heat transfer rate across the air gap?

Solution:

The radiation heat transfer rate can be calculated with Equation 5.23. In order to use Equation 5.23, there are a few terms that must first be evaluated. First, the temperatures T_1 and T_2 are needed in absolute temperature, Kelvin. Taking the concrete block as surface 1 and the brick as surface 2:

$$T_1 = 10 + 273.15 = 283.15\ \text{K}$$
$$T_2 = 0 + 273.15 = 273.15\ \text{K}$$

The surface emissivities are not given, but, in general for building materials that are not polished metals, a value of 0.9 is a reasonable assumption.

The view factor between two parallel surfaces facing each other is approximately one. Equation 5.23 can then be evaluated as:

$$\dot{q}_{1-2} = \frac{\sigma\left(283.15^4 - 273.15^4\right)}{\frac{1-0.9}{10 \cdot 0.9} + \frac{1}{10 \cdot 1} + \frac{1-0.9}{10 \cdot 0.9}} = 329\ W$$

If a low emissivity coating, e.g. a layer of aluminum foil, was applied to the concrete block layer and not compromised by being splattered with mortar or coated with dust during the bricklaying, the emittance might be as low as 0.05. (In practice, this would be nearly impossible, but just for purposes of example, we'll assume that it might be done.) The radiation heat transfer could then be calculated as:

$$\dot{q}_{1-2} = \frac{\sigma\left(283.15^4 - 273.15^4\right)}{\frac{1-0.05}{10 \cdot 0.05} + \frac{1}{10 \cdot 1} + \frac{1-0.9}{10 \cdot 0.9}} = 20\ W$$

End of solution.

In practice, rooms in buildings have more than two surfaces, and, in general, every surface exchanges radiant heat with every other surface. While it is possible to analyze the complete radiation network, building performance simulations typically adopt a simpler approximation. Several approximations are discussed in the section below entitled "Longwave radiation – exterior."

Shortwave radiation – exterior

Determination of incident solar radiation is covered in Chapter 3. For opaque surfaces, the next step is to determine the portion of the solar radiation that is absorbed. Although the absorptance of a surface to solar radiation will depend on the incidence angle, absorptance is often nearly constant until the incidence angle approaches 90°, at which point the solar irradiation is very low. Accordingly, the solar absorptance is reasonably approximated as a constant value and the absorbed solar irradiation is given by:

$$q''_{solar,out,j,\theta} = \alpha G_t \tag{5.24}$$

Table 5.7 Solar absorptances

Surface	*Absorptance*
Brick, red (Purdue)[a]	0.63
Paint, cardinal red[b]	0.63
Paint, matte black[b]	0.94
Paint, sandstone[b]	0.50
Paint, white acrylic[a]	0.26
Sheet metal, galvanized, new[a]	0.65
Sheet metal, galvanized, weathered[a]	0.80
Shingles, Aspen gray[b]	0.82
Shingles, autumn brown[b]	0.91
Shingles, onyx black[b]	0.97
Shingles, generic white[b]	0.75

Sources
a F. P. Incropera and D. P. DeWitt (1990) Fundamentals of Heat and Mass Transfer, 3rd ed., New York: Wiley.
b D. S. Parker, J. E. R. McIlvaine, S. F. Barkaszi, D. J. Beal and M. T. Anello (2000), Laboratory Testing of the Reflectance Properties of Roofing Material, FSEC-CR670-00. Florida Solar Energy Center, Cocoa, FL.

where:
α = solar absorptivity of the surface (–),
G_t = total solar irradiation incident on the surface (W/m^2).

Sample values of solar absorptances can be found in Table 5.7.

Shortwave radiation – windows

For transparent and semi-transparent surfaces, e.g. windows, the transmission of solar radiation through the window is of particular interest, as is the amount of solar radiation that is absorbed and then transmitted into the space in the form of longwave radiation or convection. For simple window systems, consisting of one, two, or three panes of homogeneous glass, analytical calculations based on ray tracing may be used. See Duffie and Beckman (1991) for an example. In current practice, there are a number of features that complicate the analysis:

- Modern windows often incorporate thin coatings of metallic oxides or other materials that are spectrally-selective. Short of an analysis which breaks down the spectrum into small wavelength bins, some approximation is needed.
- Furthermore, the properties of the thin coatings are often considered proprietary by window manufacturers, in which case only overall window properties are given rather than detailed layer-by-layer properties.
- Internal shading layers such as curtains and venetian blinds are difficult to analyze and needed data are rarely available. Actual positions of curtains and venetian blinds are controlled by occupants and seldom known accurately. A more detailed discussion of this subject can be found in Chapter 4.

Detailed model

A range of different approaches may be taken. A detailed model of such a window requires a method for determining the layer-by-layer absorption of solar radiation and transmission of solar radiation through all layers into the space. Direct (beam) solar radiation and indirect (diffuse) solar radiation are necessarily treated separately and the transmission and absorption of direct solar radiation is a function of the incidence angle. Models for this analysis are described by Klems (2002) and Wright and Kotey (2006).

As noted above, data (optical properties: transmittance, absorptance, and reflectance) supporting the models can be difficult to find; the WINDOW 5.2 program (LBNL 2009) developed by Lawrence Berkeley National Laboratory comes with data for over 10,000 different window layers, but no data are provided for commercially available glazing systems (i.e. if the window manufacturer were to disclose exactly which window layers were utilized in a specific window/glazing system, all data needed for a detailed analysis would then be available. But since that information is not generally available, the utility of the large data set of window layers is limited for building simulation practitioners).

Data that are actually provided by window manufacturers typically include:

- Number of layers – is the window single pane, double pane, triple pane?
- Some description of the glass type – coloration and whether or not it has a low-E coating, for example,
- U-factor (NFRC 2004b),
- Solar Heat Gain Coefficient (SHGC) at normal-incidence-angle (NFRC 2004a),
- Visual transmittance (NFRC 2004a).

With this information, and the WINDOW 5.2 program or, perhaps the WIS program (WinDat 2009) it may be possible, though tedious, to develop a layer-by-layer description that matches the overall performance. An alternative would be to consult a source such as Chapter 31 of the 2005 ASHRAE Handbook of Fundamentals, where a table of layer-by-layer properties for typical windows can be found. Using the manufacturer's description of the window, including the SHGC at normal incidence angle, a similar window can be sought.

Regardless of how the optical properties are obtained, for any given window, an overall solar transmittance (τ_{sol}) and absorptances (α_1 for outer pane, α_2 for inner pane) for each pane are the result, as shown in Table 5.8. These results, for a single pane window and two double pane windows give each of the properties of interest at intervals of 10 degrees. Properties for beam radiation would be found from the table, interpolating with the actual incidence angle. Properties for diffuse radiation would be taken from the "Hemispherical" column.

With these properties in hand, a detailed model of the window would then balance the heat at each pane, as described below in the section "Surface heat balances – semi-transparent surfaces."

Table 5.8 Window optical properties and SHGC

Angle	0	10	20	30	40	50	60	70	80	90	*Hemi-spherical*
Single pane, 3 mm thick, clear											
τ_{sol}	0.834	0.833	0.831	0.827	0.818	0.797	0.749	0.637	0.389	0	0.753
α_1	0.091	0.092	0.094	0.096	0.100	0.104	0.108	0.110	0.105	0	0.101
SHGC	0.859	0.859	0.857	0.854	0.845	0.825	0.779	0.667	0.418	0	0.781
Double pane, outer pane 3.2 mm thick with low e coating; inner pane 5.7 mm thick											
τ_{sol}	0.408	0.410	0.404	0.395	0.383	0.362	0.316	0.230	0.106	0	0.338
α_1	0.177	0.180	0.188	0.193	0.195	0.201	0.218	0.239	0.210	0	0.201
α_2	0.060	0.060	0.061	0.061	0.063	0.063	0.061	0.053	0.038	0	0.059
SHGC	0.469	0.472	0.466	0.459	0.448	0.428	0.382	0.291	0.152	0	0.400
Double pane, both panes 5.7 mm thick, clear											
τ_{sol}	0.607	0.606	0.601	0.593	0.577	0.546	0.483	0.362	0.165	0	0.510
α_1	0.167	0.168	0.170	0.175	0.182	0.190	0.200	0.209	0.202	0	0.185
α_2	0.113	0.113	0.115	0.116	0.118	0.119	0.115	0.101	0.067	0	0.111
SHGC	0.701	0.701	0.698	0.691	0.678	0.648	0.585	0.456	0.237	0	0.606

Note: Sample data generated with the WINDOW 5.2 program.

Simple model

Table 5.8 also shows, for each window, solar heat gain coefficients (SHGC), also as a function of incidence angle. A simpler model, appropriate for zone models that do not directly utilize a heat balance, e.g. the transfer function method, can utilize these values along with the manufacturer-specified U-factor to compute the heat gains due to fenestration. See Spitler (2009) for more in-depth coverage; a brief treatment follows here.

In this simple model, conduction heat gain is computed separately from the transmitted and absorbed solar heat gain. Because the thermal mass of the glass is very low, the conduction is approximately steady-state. Accordingly, for each hour, the conduction heat gain may be calculated as:

$$q_\theta = UA(T_{o,\theta} - T_{rc}) \tag{5.25}$$

where:

q_θ = hourly conductive heat gain (W) for the window,
U = overall heat transfer coefficient for the window (W/m²·K), as specified by the window manufacturer,
A = window area – including frame (m²),
$T_{o,\theta}$ = outdoor air temperature (°C),
T_{rc} = presumed constant room air temperature (°C),
θ = the current hour.

Transmitted and absorbed solar heat gains are calculated for each hour with a step-by-step procedure, as follows:

1. Compute incident angle, surface azimuth angle, incident direct (beam) irradiation and diffuse irradiation on window, as described in Chapter 3.
2. If exterior shading exists, determine sunlit area and shaded area, as described in Chapter 3.
3. For windows without interior shading, the beam, diffuse, and total transmitted and absorbed solar heat gains are given by:

$$q_{SHG,D} = E_D A_{sunlit} SHGC(\theta) \tag{5.26}$$

$$q_{SHG,d} = (E_d + E_r) A\, SHGC_{diffuse} \tag{5.27}$$

$$q_{SHG} = q_{SHG,D} + q_{SHG,d} \tag{5.28}$$

where:

$q_{SHG,d}$ = direct (beam) solar heat gain (W/m²),
$q_{SHG,d}$ = diffuse solar heat gain (W/m²),
q_{SHG} = total solar heat gain (W/m²),
E_D = incident direct (beam) irradiation (W/m²),
E_d = incident diffuse irradiation from sky (W/m²),
E_r = incident diffuse reflected irradiation (W/m²),
$SHGC(\theta)$ = angle dependent SHGC interpolated from Table 5.8,
$SHGC_{diffuse}$ = the SHGC for diffuse irradiation, taken from the “Hemispherical” column of Table 5.8,
A_{sunlit} = the unshaded area of the window (m²),
A = the total area of the window, including the frame (m²).

If there is no interior shading, the window heat gain calculation is complete.

INTERIOR SHADING

The detailed and simple models described above are both adequate for cases where the window unit is made up of one or more layers of glass or similar glazing material. The use of interior shading devices, such as Venetian blinds or draperies, through which air may pass further complicates the analysis. Likewise, insect screening can similarly complicate the analysis. For the detailed method, see Wright (2008) and Wright *et al.* (2009) for the most comprehensive treatment of the problem. For the simple model, an approximate method utilizing interior attenuation coefficients (IAC) is described by ASHRAE (2005) and Spitler (2009).

Example 5.6

On July 21 in Seattle, Washington (47.45° N latitude and 122.3° W longitude) the outdoor air temperature is 28.1 °C, and the room air temperature is 22 °C at 3 p.m. Pacific Daylight Savings Time. The southwest-facing double pane with low-e coating window shown in Table 5.8, is 1.2 m high and 1.2 m wide; the exterior surface is flush with the outside of the building. The rated U-factor for the window is 1.66 W/m^2K. At that time:

- the incident solar beam radiation is 486.1 W/m^2,
- the incidence angle is 55.88°,
- the incident solar diffuse radiation is 193.5 W/m^2 (106.8 W/m^2 diffuse from sky and 86.7 W/m^2 reflected).

Using the simple model, compute the solar heat gain for the window. Using the detailed model, determine the transmitted solar radiation and the solar radiation absorbed by both panes.

Solution:

First, with the simple model, the conduction heat gain may be determined. Then, the window properties including appropriate values of SHGC are determined. Lastly, the solar heat gains with and without internal shading are calculated.

Simple model – conduction heat gain

The U-factor is given by the manufacturer as 1.66 W/m^2K. The area of the window is 1.44 m^2. The air temperature at noon is 28.1 °C. From Equation 5.7b, the conduction heat gain at noon is:

$$q_{cond15} = 1.66 \times (1.44) \times (28.1 - 22.0) = 14.6 \text{ W}$$

Simple model – solar heat gain calculation

The incidence angle is 55.88°; interpolating in Table 5.8 gives a value of: *SHGC* (55.88) = 0.335

Then,

$$q_{SHG,D} = E_D A_{sunlit} SHGC(\theta) = 486.1 \times 1.44 \times 0.401 = 280.7\text{W},$$

$$q_{SHG,d} = (E_d + E_r)A\ SHGC_{diffuse} = 193.5 \times 1.44 \times 0.4 = 111.5\text{W},$$

$$q_{SHG} = q_{SHG,D} + q_{SHG,d} = 392\text{W}.$$

Combined with conduction, the total heat gain from the window is then 407 W.

Detailed model

With the detailed model, values of transmittance and absorptance for beam radiation are determined by interpolating in Table 5.8:

$$\tau_{sol}(55.88\,°) = 0.335,$$
$$\alpha_1(55.88\,°) = 0.211,$$
$$\alpha_2(55.88\,°) = 0.062.$$

Values for diffuse radiation are read from the last column:

$$\tau_{sol,d} = 0.338,$$
$$\alpha_{1,d} = 0.201,$$
$$\alpha_{2,d} = 0.059.$$

Then, the transmitted beam solar radiation is determined by multiplying the beam radiation by the beam transmittance and area:

$$q_{trans,D} = E_D A_{sunlit} \tau_{sol} = 486.1 \times 1.44 \times 0.335 = 234.5\text{W}$$

The transmitted diffuse solar radiation is determined by multiplying the diffuse radiation by the diffuse transmittance and area:

$$q_{trans,d} = (E_d + E_r) A \tau_{sol,d} = 193.5 \times 1.44 \times 0.338 = 94.2\text{W}$$

And the total transmitted solar radiation is:

$$q_{trans} = q_{trans,D} + q_{trans,d} = 329\text{ W}$$

In a similar manner, the absorbed solar radiation in each pane is determined.

$$q_{abs,1} = q_{abs,D,1} + q_{abs,d,1} = E_D A_{sunlit} \alpha_1 (55.88°) + E_d A \alpha_{1,d}$$
$$= 486.1 \times 1.44 \times 0.211 + 193.5 \times 1.44 \times 0.201 = 203.7\text{W}$$

$$q_{abs,2} = q_{abs,D,2} + q_{abs,d,2} = E_D A_{sunlit} \alpha_2 (55.88°) + E_d A \alpha_{2,d}$$
$$= 486.1 \times 1.44 \times 0.062 + 193.5 \times 1.44 \times 0.059 = 59.7\text{W}$$

Comparing the results from the simple model and the detailed model, we see that the simple model predicts a total heat gain through the window of 407 W. The detailed model predicts 329 W of transmitted solar radiation and a total absorbed radiation of 263 W. Of the 263 W, only a fraction will flow in to the room. The quantity that flows into the room will depend on the other outside and inside conditions and will be determined using a heat balance, described in the section "Surface heat balances – semi-transparent surfaces." Depending on those conditions, the detailed model may predict more or less heat gain to the space than the simple model.
End of solution.

Shortwave radiation – interior

Treatment of shortwave radiation, once it has been transmitted through the window, depends on the zone model described in Section 5.6, below. For transfer-function type models, a pre-computed transfer function or response function is applied to the solar heat gains determined with Equations 5.26–5.28. These heat gains include both transmitted and inward-flowing absorbed solar radiation.

For heat balance models and thermal network models, it is necessary to make an estimate of how the transmitted direct and diffuse solar radiation are distributed. In other words, the amount of transmitted solar radiation absorbed by each surface in the room must be determined. This can be (and sometimes is) analyzed in a very detailed manner, accounting for exactly where the radiation strikes each room surface, and then accounting for each reflection until it is all absorbed.

However, this level of detail is difficult to justify for most load calculation and energy calculation programs, as the amount of effort required to obtain the requisite input data is disproportionate to the user's ability to predict future configurations of furniture and window treatments. Therefore, a simpler model is often employed, assuming that some fraction of the transmitted beam radiation is absorbed by the floor. That fraction not absorbed by the floor is then assumed to be diffusely reflected and, along with the transmitted diffuse radiation, uniformly absorbed by all room surfaces.

If the total transmitted diffuse radiation and the reflected direct radiation (from the floor) are divided by the total interior surface area of the zone and distributed uniformly, then for all surfaces except the floor, the absorbed solar radiation flux is given by:

$$q''_{solar,in,j,\theta} = \frac{\sum_{i=1}^{M} \dot{q}_{trans,d,i} + (1-\alpha_{floor}) \sum_{i=1}^{M} \dot{q}_{trans,D,i}}{\sum_{j=1}^{N} A_j} \tag{5.29}$$

where:

$q''_{solar,in,j,\theta}$	=	solar radiation absorbed by interior surfaces other than the floor (W/m^2),
$\dot{q}_{trans,d,i}$	=	diffuse solar radiation transmitted by the *i*th window (W),
$\dot{q}_{trans,D,i}$	=	direct solar radiation transmitted by the *i*th window (W),
α_{floor}	=	the solar absorptance of the floor,
A_j	=	the area of the *j*th interior surface (m^2),
M	=	the number of windows,
N	=	the number of interior surfaces.

Then, the absorbed solar radiation flux (W/m^2) for the floor is given by:

$$q''_{solar,in,floor,\theta} = \frac{\sum_{i=1}^{M} \dot{q}_{trans,d,i} + (1-\alpha_{floor}) \sum_{i=1}^{M} \dot{q}_{trans,D,i}}{\sum_{j=1}^{N} A_j} + \frac{\alpha_{floor} \sum_{i=1}^{M} \dot{q}_{trans,D,i}}{A_{floor}} \tag{5.30}$$

This is a fairly simple model for distribution of transmitted solar heat gain. A number of improvements might be made, including determining which interior surfaces are actually sunlit by the direct solar radiation, and allowing for additional reflection of the beam radiation. Beyond that, more sophisticated algorithms are used for analysis of daylighting as described in Chapter 9 and might be adapted for cooling load calculation use.

Longwave radiation – exterior

Long wavelength (thermal) radiation to and from exterior surfaces is also a very complex phenomenon. The exterior surfaces radiate to and from the surrounding ground, vegetation, parking lots, sidewalks, other buildings, and the sky. In order to make the problem tractable, a number of assumptions are usually made:

- Each surface is assumed to be opaque, diffuse, and isothermal and to have uniform radiosity and irradiation.
- Each surface is assumed to be gray, having a single value of absorptivity and emissivity that applies over the thermal radiation spectrum. (The surface may have a different value of absorptivity that applies in the solar radiation spectrum.)
- Radiation to the sky, where the atmosphere is actually a participating medium, may be modeled as heat transfer to a surface with an effective sky temperature.
- Lacking any more detailed information regarding surrounding buildings, it is usually assumed that the building sits on a flat, featureless plane, so that a vertical wall has a view factor between the wall and the ground of 0.5, and between the wall and the sky of 0.5.
- Without a detailed model of the surrounding ground, it is usually assumed to have the same temperature as the air. Obviously, for a wall with a significant view to an asphalt parking lot, the ground temperature would be somewhat higher.

With these assumptions, the net long-wavelength radiation into the surface is given by

$$q''_{\text{radiation},out,j,\theta} = \varepsilon\sigma\left[F_{s-g}\left(T_g^4 - T_{os,j,\theta}^4\right) + F_{s-sky}\left(T_{sky}^4 - T_{os,j,\theta}^4\right)\right] \tag{5.31}$$

where:

ε = surface long wavelength emissivity,
σ = Stefan-Boltzmann constant = 5.67×10^{-8} (W/(m^2– K^4),
F_{s-g} = view factor from the surface to the ground,
F_{s-sky} = view factor from the surface to the sky,
T_g = ground temperature (K),
T_{sky} = effective sky temperature (K),
$T_{os,j,\theta}$ = The surface temperature (K).

Since it is usually assumed that the building sits on a featureless plain, the view factors are easy to determine:

$$F_{s-g} = \frac{1-\cos\alpha}{2} \tag{5.32}$$

$$F_{s-sky} = \frac{1+\cos\alpha}{2} \tag{5.33}$$

where α is the tilt angle of the surface from horizontal.

It is often convenient to linearize this equation by introducing radiation heat transfer coefficients:

$$h_{r,g} = \varepsilon\sigma\left[\frac{F_{s-g}(T_g^4 - T_{os,j,\theta}^4)}{T_g - T_{os,j,\theta}}\right] \tag{5.34}$$

$$h_{r,sky} = \varepsilon\sigma\left[\frac{F_{s-sky}(T_{sky}^4 - T_{os,j,\theta}^4)}{T_{sky} - T_{os,j,\theta}}\right] \tag{5.35}$$

Then Eqn. 5.31 reduces to

$$q''_{radiation,out,j,\theta} = h_{r,g}(T_g - T_{os,j,\theta}) + h_{r-sky}(T_{sky} - T_{os,j,\theta}) \tag{5.36}$$

If the radiation coefficients are determined simultaneously with the surface temperature, Eqn. 5.36 will give identical results to Eqn. 5.37.

A number of models are available (Cole 1976) for estimating the effective sky temperature seen by a horizontal surface under clear sky conditions. Perhaps the simplest is that used by the BLAST program (McClellan and Pedersen 1997), which simply assumes that the effective sky temperature is the outdoor dry bulb temperature minus 6 K.

For surfaces that are not horizontal, the effective sky temperature will be affected by the path length through the atmosphere. An approximate expression based on Walton's heuristic model (Walton 1983) is:

$$T_{sky,\alpha} = \left[\cos\frac{\alpha}{2}\right]T_{sky} + \left[1 - \cos\frac{\alpha}{2}\right]T_o \tag{5.37}$$

where:

$T_{sky,\alpha}$ = effective sky temperature for a tilted surface (°C) or (K),
T_{sky} = effective sky temperature for a horizontal surface (°C) or (K),
T_o = outdoor air dry bulb temperature (°C) or (K).

Longwave radiation – interior

From a building load calculation perspective, radiation is often divided into two categories: short wavelength and long wavelength. Short wavelength radiation is radiation with wavelengths near the visible spectrum, such as solar radiation and a portion of the radiation given off by lights. Long wavelength radiation is given off by sources near room temperature, such as people, equipment, walls, etc.

This section describes models for distribution of long wavelength radiation given off by surfaces, and heat gain sources in the room. The exchange of thermal radiation between surfaces in an enclosure is fairly well understood and is covered in any elementary heat transfer text book. However, the solution procedure is generally considered to require too much computational time for an hourly load calculation.

Furthermore, rooms are seldom empty and describing all of the interior surfaces and furnishings in detail is likely to be burdensome to the designer and to have little point, as the arrangement of the furnishings is not likely to remain constant over the life of the building. Therefore, simpler methods (Carroll 1980b; Davies 1988; Liesen and Pedersen 1997; Walton 1980) are often used for estimating radiation heat transfer. Many of these are based on defining a single temperature to which each surface radiates. Two additional simplifications are usually made when analyzing radiation heat transfer inside a room:

- Furnishings (e.g. desks, chairs, tables, shelves) are usually lumped into a single surface, sometimes called "internal mass".
- Radiation from equipment, lights, and people is usually treated separately. See the next section.

A reasonably simple model with acceptable accuracy is Carroll's *mean radiant temperature method* (Carroll 1980b). The model represents all of the surfaces in the room as a single fictitious surface, the so-called *mean radiant temperature* (MRT).

First, Carroll defines an MRT view factor, which is an approximate view factor that compensates for the self-weighting in the mean radiant temperature:

$$F_i = 1/(1 - A_i F_i / \sum_{i=1}^{N} A_i F_i) \tag{5.38}$$

where:

N = the number of surfaces in the zone.

Since the MRT view factors (F_i) appear on both sides of the equation, iteration is required to determine the values. Carroll (1980a) suggests setting F_i to one as an initial guess. The MRT view factors can be calculated once at the beginning of the simulation during the initialization phase. This procedure for calculating the MRT view factors is suitable for most zones. In the case of geometries with large but significantly different coplanar surfaces, Carroll suggests adjusting the procedure for estimating the MRT view factors.

Also before the simulation begins, a first estimate of the radiation coefficients for each surface to the mean radiant temperature should be made as follows:

$$h_{rad,ref,i} = \frac{4\sigma \overline{T}_{ref}^3}{1/F_i + (1-\varepsilon_i)/\varepsilon_i} \tag{5.39}$$

where:

$h_{rad,ref,i}$ = radiation coefficient for ith surface at reference temperature (W/m²K),

$\overline{T}_{ref}$ = mean reference temperature, taken to be 300 K,

F_i = the MRT view factor for ith surface (–),

ε_i = the longwave emissivity of ith surface (–).

Then, at each simulation time step and/or at each simulation iteration where radiation heat transfer is to be calculated, the estimates of the radiation coefficients are improved with a two-step process. First, given the actual surface temperatures, new estimates of the radiation coefficients ($h_{rad,1,j}$ – the 1 in the subscript refers to the first update) are calculated with this expression:

$$h_{rad,1,i} = (0.865 + T_i/200)h_{rad,ref,i} \tag{5.40}$$

Then the first estimate of the mean radiant temperature, $T_{MR,1}$, is calculated as the average surface temperature, weighted by the estimated radiative coefficients and surface areas:

$$T_{MR,1} = \sum_{i=1}^{N} A_i h_{rad,1,i} T_i / \sum_{i=1}^{N} A_i h_{rad,1,i} \tag{5.41}$$

A second, improved estimate of each radiation coefficient is made using the first estimate of the mean radiant temperature:

$$h_{rad,2,i} = (0.865 + T_{MR,1} / 200)h_{rad,1,i} \tag{5.42}$$

Then, a second improved estimate of the mean radiant temperature is made:

$$T_{MR,2} = \sum_{i=1}^{N} A_i h_{rad,1,i} T_i / \sum_{i=1}^{N} A_i h_{rad,2,i} \tag{5.43}$$

The net radiation flux in W/m² leaving each surface is then given by:

$$q''_{rad} = h_{rad,2,i}(T_i - T_{MR,2}) \tag{5.44}$$

And the net radiation in W leaving each surface is:

$$q_{rad} = h_{rad,2,i}A_i(T_i - T_{MR,2}) \tag{5.45}$$

Example 5.7

A room with a south-facing window is shown in Figure 5.6. The window is 5 m wide and 1.5 m high; its lower edge is 1 m above the floor and it is centered in the wall from east-to-west.

The surfaces are summarized in Table 5.9. On a cold winter's night, they have the temperatures shown in the last column.

Determine the net radiation leaving each of the seven surfaces.

Solution:

Based on the areas shown in Table 5.9, the MRT view factors are computed iteratively with Equation 5.38 and shown in the second column of Table 5.10. With the MRT view factors known, the reference values of the radiation coefficient are calculated with Equation 5.39. Using the known surface temperatures, Equation 5.40 is used to make the first estimate of the radiation coefficient, shown in the fourth column.

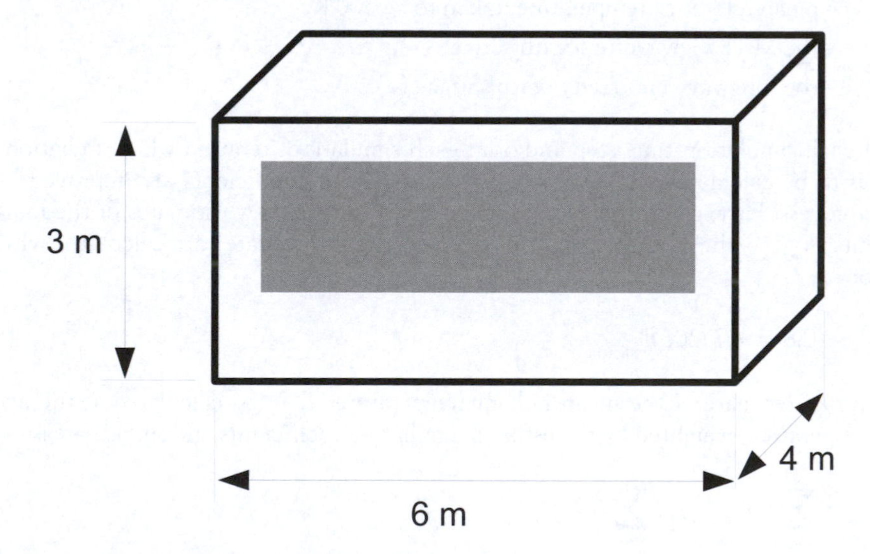

Figure 5.6 South-facing zone.

Table 5.9 Surfaces of room in Example 5.7

Number	*Name*	*Area (m²)*	*Emissivity*	*Temperature (°C)*
1	Floor	24	0.9	14
2	North wall	18	0.9	17
3	East wall	12	0.9	17
4	West wall	12	0.9	17
5	Ceiling	24	0.9	18
6	Window	7.5	0.9	15
7	South wall (surrounding window)	10.5	0.9	9

Table 5.10 Radiation example intermediate results

Number	F_i	$h_{rad,ref,i}$ (W/m²K)	$h_{rad,1,i}$ (W/m²K)	$h_{rad,2,i}$ (W/m²K)	q''_{rad} (W/m²)	q_{rad} (W)
1	1.318	7.04	6.58	6.22	–11.85	–284.48
2	1.197	6.47	6.14	5.80	6.35	114.27
3	1.114	6.07	5.76	5.44	5.96	71.47
4	1.114	6.07	5.76	5.44	5.96	71.47
5	1.318	7.04	6.72	6.35	13.30	319.17
6	1.065	5.42	4.93	4.66	–32.18	–241.36
7	1.096	5.98	5.63	5.31	–4.81	–50.55

The first estimate of the mean radiant temperature, $T_{MR,1}$, made with Equation 5.41, is 15.91 °C. Using this estimate of mean radiant temperature, the second estimates of the radiation coefficients, calculated with Equation 5.42, are given in the fifth column of Table 5.10. A second estimate of the mean radiant temperature, $T_{MR,2}$, made with Equation 5.43, is also 15.91 °C. Using Equations 5.44 and 5.45, the net radiative flux and net radiation leaving each surface are given in the last two columns of Table 5.10.
End of solution.

Zone models

To this point, all of the mechanisms for heat transfer into and out of the zone have been considered. In order to determine what actually happens in the zone, the individual mechanisms need to be analyzed simultaneously. A fundamental way of doing this is known as the heat balance model, though many versions of the model exist, so it might be more accurately referred to as a class of models. Heat balance models ensure that all energy flows in each zone are balanced and involve the solution of a set of energy balance equations for the zone air and the interior and exterior surfaces of each wall, roof, and floor. These energy balance equations are combined with equations for transient conduction heat transfer through walls and roofs and algorithms or data for weather conditions including outdoor air dry bulb temperature, wet bulb temperature, solar radiation, and so on.

A second class of models, known as thermal network models, are a form of heat balance model characterized by discretization of the building into a network of nodes, with interconnecting paths, through which energy flows; the use of numerical sub-models for conduction heat transfer (e.g. finite difference or finite volume); and provision for multiple zone air temperatures.

A third class of models use a heat balance model to precalculate zone responses to heat gains. This class of models utilizes transfer functions or response factors to relate current values of the cooling load or heating load to past values of heat gains, and, in most cases, past values of the cooling load or heating load.

The heat balance and thermal network models will balance all of the energy flows every hour, while the transfer function models may not. The heat balance method is also more flexible, able to incorporate such features as variable convection coefficients. However, the transfer function method generally runs faster than the heat balance method. Thermal network models are even more flexible than most heat balance models, while also requiring the most computation time.

The next three sections describe the three different models, with emphasis on the heat balance model.

Zone models: heat balance model

The heat balance model may be implemented with a wide variety of sub-models, e.g. different exterior convection models, different transient conduction models, and different interior radiation models may be used. What is described in this section is one implementation of the heat balance model made up of a particular combination of sub-models. Here, the sub-models are chosen primarily for simplicity in illustrating the model.

The heat balance model utilizes the heat balance concept at all interior and exterior zone surfaces, as well as applying a heat balance to the zone air.

Surface heat balances – opaque surfaces

To illustrate the heat balance model, consider a simple zone with seven surfaces: four walls, a window in one of the walls, a roof, and a floor. The zone has solar energy coming through the window, heat conducted through the exterior walls and roof and internal heat gains due to lights, equipment, and occupants. The heat balances on both the interior and exterior surfaces of a single wall or roof element are illustrated in Fig. 5.7. The heat balance on the *j*th exterior surface at time θ is represented conceptually by:

$$q''_{conduction,out,j,\theta} = q''_{solar,out,j,\theta} + q''_{convection,out,j,\theta} + q''_{radiation,out,j,\theta} \tag{5.46}$$

where:

$q''_{conduction,out,j,\theta}$ = conduction heat flux (W/m²),
$q''_{solar,out,j,\theta}$ = absorbed solar heat flux (W/m²),
$q''_{convection,out,j,\theta}$ = convection heat flux (W/m²),
$q''_{radiation,out,j,\theta}$ = thermal radiation heat flux (W/m²).

A few features of Fig. 5.7 that should be noted are:

- $q''_{conduction,out,j,\theta}$ is not equal to $q''_{conduction,in,j,\theta}$ unless steady-state heat transfer conditions prevail. This would be unusual for cooling load calculations.
- For the exterior surface, the wall is likely to radiate to both the sky and surroundings, and, perhaps to other buildings. For this figure, only one interchange is shown.
- For the interior surface, the wall actually radiates to all other surfaces. But, as described above, this is simplified by computing a mean radiant temperature for the zone.
- The solar radiation incident on the interior surface will have been transmitted through the window.

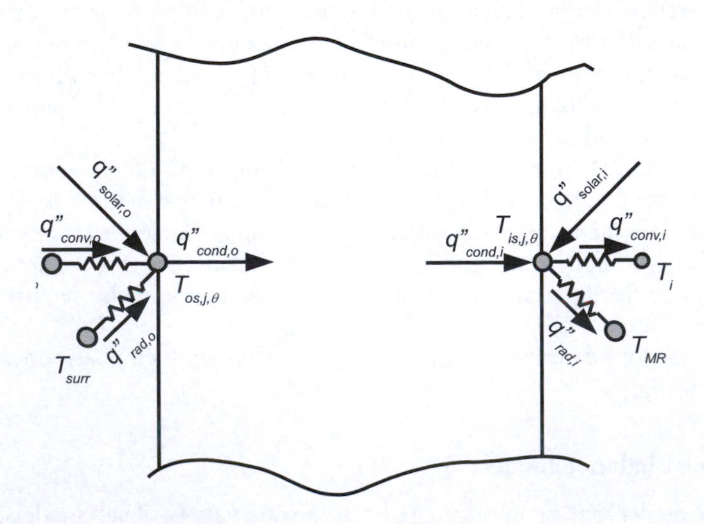

Figure 5.7 Opaque surface heat balance.

The interior surface heat balance on the *j*th surface at time θ may be represented conceptually as:

$$q''_{conduction,in,j,\theta} + q''_{solar,in,j,\theta} = q''_{convection,in,j,\theta} + q''_{radiation,in,j,\theta} \tag{5.47}$$

where:

$q''_{conduction,in,j,\theta}$ = conduction heat flux (W/m^2),
$q''_{solar,in,j,\theta}$ = absorbed solar heat flux (W/m^2),
$q''_{convection,in,j,\theta}$ = convection heat flux (W/m^2),
$q''_{radiation,in,j,\theta}$ = thermal radiation heat flux (W/m^2).

Exterior surface heat balance

We will first consider the exterior surface heat balance. Before deriving the full transient exterior surface heat balance equation, it may be helpful to consider a simpler case, based on a *steady-state* heat balance. Consider the exterior surface of a wall, as shown in Figure 5.8.

Here:

t_{es} = exterior surface temperature (°C),
t_{is} = interior surface temperature (°C),
t_o = outdoor air temperature (°C),
q''_{solar} = absorbed solar heat flux (W/m^2),
t_{surr} = temperature of the surroundings (°C).

Neglecting any transient effects, a heat balance on the exterior surface might be written as:

$$q''_{conduction} = q''_{solar} + q''_{convection} + q''_{radiation} \tag{5.48}$$

With each component of the heat transfer defined as follows:

$$q''_{conduction} = U_{s-s}(T_{es} - T_{is}) \tag{5.49}$$

$$q''_{solar} = \alpha G_t \tag{5.50}$$

$$q''_{convection} = h_c(T_o - T_{es}) \tag{5.51}$$

(5.52)

where:

U_{s-s} = surface-to-surface conductance (W/m^2),
α = absorptance of the surface to solar radiation,

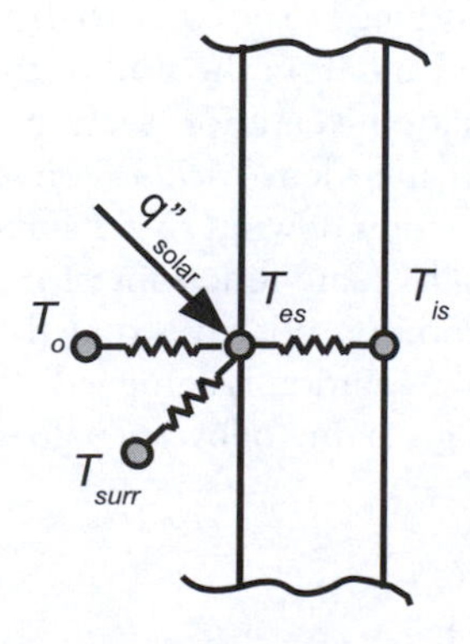

Figure 5.8 Exterior surface steady-state heat balance.

G_t = total incident solar radiation flux (W/m²),
h_c = exterior surface convection coefficient (W/m²),
h_r = exterior surface radiation coefficient (W/m²).

Here, the radiation heat transfer has been linearized, though in the heat balance solution procedure, the radiation heat transfer coefficient will be recalculated on an iterative basis. Substituting Equations 5.49–5.52 into 5.48, the exterior surface temperature can then be determined:

$$T_{es} = \frac{\alpha G_t + U_{s-s} T_{is} + h_c T_o + h_r T_{surr}}{U_{s-s} + h_c + h_r} \tag{5.53}$$

A similar procedure for the interior surface yields a similar equation for the interior surface temperature. The interior and exterior surface temperatures would then be solved simultaneously. When all interior surface temperatures are known, the cooling load can be calculated as the sum of the convective heat gains from the interior surfaces, lighting, equipment, people, and infiltration.

Example 5.8

For a very lightweight wall, a steady-state surface heat balance will give an approximate value of the exterior surface temperature and the conduction heat flux into the surface. Consider such a wall on a low rise building with a surface-to-surface *U*-factor of 0.3 W/m²K, a solar absorptivity of 0.8 and an emissivity of 0.9. The weather conditions include incident solar radiation is 680 W/m², the outdoor dry bulb temperature is 28.1 °C, the sky temperature is 22.1 °C and the windspeed is 3.6 m/s with the surface on the windward side of the building. At this time, the interior surface temperature of the wall is 21 °C.

Estimate the exterior surface temperature and the conduction heat flux into the surface.

Solution:

The first step is to establish the boundary conditions, where they are not explicitly stated. The sky temperature is given as 22.1 °C; because the wall does not face the sky (i.e. it is oriented vertically, not horizontally), the effective sky temperature should be calculated with Eqn. 5.37. The tilt angle, α, is 90° and the effective sky temperature is

$$T_{sky,\alpha} = \left[\cos\frac{90°}{2}\right] 22.1 + \left[1 - \cos\frac{90°}{2}\right] 28.1 = 23.9°\text{C}$$

The temperature of the ground will be assumed to be equal to the air temperature (22.1 °C) and view factors from the wall to the ground and from the wall to the sky will both be taken to be 0.5, as discussed in the "Longwave radiation – exterior" section.

For low-rise buildings, the Yazdanian and Klems (1994) convection correlation is appropriate. The convection coefficient is a function of wind speed and temperature difference. Similarly, the radiation coefficients, $h_{r,g}$ and $h_{r,sky}$, are functions of the surface temperature and sky temperature. Since the surface temperature is not known at the beginning, it is necessary to make an initial guess, then iterate until a solution is achieved.

Making an initial guess of t_{os}=50 °C, gives the following values for convection and radiation coefficients:

h_c = 7.76 W/m²K, from Eqn. 5.16,
$h_{r,sky}$ = 3.05 W/m²K, from Eqn. 5.34,
$h_{r,grd}$ = 3.11 W/m²K, from Eqn. 5.35.

Using these values in the surface heat balance, Eqn. 5.53,

$$T_{es} = \frac{\alpha G_t + U_{s-s} T_{is} + h_c T_o + h_{r,sky} t_{sky} + h_{r,g} T_g}{U_{s-s} + h_c + h_{r,sky} + h_{r,g}}$$

$$T_{es} = \frac{0.8(680) + 0.3(21) + 7.76(28.1) + 3.05(23.9) + 3.11(28.1)}{0.3 + 7.76 + 3.05 + 3.11} = 65.3°\text{C}$$

A second iteration with t_{os}=65.3 °C, gives the following values for convection and radiation coefficients:

h_c = 7.91 W/m²K, from Eqn. 5.16,
$h_{r,sky}$ = 3.29 W/m²K, from Eqn. 5.34,
$h_{r,grd}$ = 3.35 W/m²K, from Eqn. 5.35.

Using these values with the heat balance equation gives a new estimate of the surface temperature as 63.7 °C. A few additional iterations give slightly different values of the coefficients and a final estimate of the outside surface temperature as 63.81 °C. With this value of the outside surface temperature, the conductive heat flux, from Eqn. 5.9, is 12.8 W/m².

A few additional calculations illustrate the heat balance with the final values of the coefficients:

- The absorbed solar radiation is 0.8 × 680 = 544 W/m²,
- The heat leaving by convection is 7.90 × (63.81–28.1) = 282.0 W/m²,
- The heat radiating to the sky is 3.26 × (63.81–23.9)= 130.4 W/m²,
- The heat radiating to the ground is 3.33 × (63.81–28.1)= 118.8 W/m²,
- The heat conducted through the wall is 12.8 W/m².

A quick check of the results confirms that the heat transferred to the surface (544 W/m²) is equal to the sum of the heat being transferred away from the surface (282.0 + 130.4 + 118.8 + 12.8 = 544 W/m²).
End of solution

In practice, these calculations are further complicated by the transient conduction heat transfer and the need to analyze all modes of heat transfer simultaneously for all surfaces in a room.

The transient surface heat balance requires use of the conduction transfer function described above. For any given hour in the simulation, past values of the exterior surface temperature and conduction heat flux will be known or assumed. Therefore, all the historical terms from Eqn. 5.12 may be gathered into a single term, $H_{out,j,\theta}$,which can be calculated at the beginning of the time step.

$$H_{out,j,\theta} = -\sum_{n=1}^{N_y} Y_n T_{is,j,\theta-n\theta} + \sum_{n=1}^{N_x} X_n T_{os,j,\theta-n\delta} + \sum_{n=1}^{N_q} \Phi_n q''_{conduction,out,j,\theta-n\theta} \tag{5.54}$$

and Eqn. 5.12 may be represented as

$$q''_{conduction,j,\theta} = -Y_o T_{is,j,\theta} + X_o T_{os,j,\theta} + H_{out,j,\theta} \tag{5.55}$$

Then, by substituting the expressions for conduction heat flux (Eqn. 5.12), absorbed solar heat gain flux (Eqn. 5.24), convection heat flux (Eqn. 5.15), and radiation heat flux (Eqns.

5.34–5.36) into the exterior surface heat balance equation (Eqn. 5.46) and recasting the equation to solve for the exterior surface temperature, the following expression results:

$$T_{os,j,\theta} = \frac{Y_o T_{is,j,\theta} - H_{out,j,\theta} + \alpha G_t + h_c T_o + h_{r-g} T_g + h_{r-sky} T_{sky}}{X_o + h_c + h_{r-g} + h_{r-sky}} \tag{5.56}$$

Note that h_c, h_{r-g}, and h_{r-sky} all depend on the exterior surface temperature. While Eqn. 5.56 might be solved simultaneously with Eqns. 5.12, 5.34, 5.35, and 5.36 in a number of different ways, it is usually convenient to solve them by successive substitution. This involves assuming an initial value of the exterior surface temperature; then computing h_c, h_{r-g}, and h_{r-sky} with the assumed value; then solving Eqn. 5.56 for the exterior surface temperature; then computing h_c, h_{r-g}, and h_{r-sky} with the updated value of the exterior surface temperature; and so on until the value of the exterior surface temperature converges.

Also, the current value of the interior surface temperature appears in Eqn. 5.56. For thermally massive walls, Y_o will usually be zero. In this case, the exterior surface heat balance may be solved independently of the current hour's interior surface temperature. For thermally non-massive walls, the exterior surface heat balance must usually be solved simultaneously with the interior surface heat balance.

Example 5.9

For a day in September, the wall in Example 5.4 on a low-rise building is exposed to the weather conditions shown in the last four columns of Table 5.11. For hours 1–16, the surface temperatures and heat fluxes have already been calculated. At hour 17, the interior surface temperature has already been calculated and it is 21.91 °C.

The wall has an exterior solar absorptivity, α = 0.8 and an exterior thermal emissivity, ε = 0.9. The surface is on the windward side of the building.

Determine the exterior surface temperature and exterior conductive heat flux for hour 17.

Solution:

Like Example 5.8, the boundary conditions and appropriate correlations must be established first. The effective sky temperature for hour 17, calculated with Equation 5.37, is 24.79 °C. The Yazdanian and Klems correlation will again be used for the convection coefficient. The radiation coefficients are calculated in the same way as in Example 5.8.

Based on the previous hours, the history term for the conduction can be calculated with Equation 5.54:

$$H_{out,j,17} = -(0.005618 \times 21.88 + 0.022664 \times 21.86 + 0.011332 \times 21.84 + 0.000823 \times 21.84 + 0.000009 \times 21.84) - 45.140663 \times 61.82 + 28.446037 \times 58.04 - 5.162262 \times 51.75 + 0.182663 \times 43.92 - 0.001023 \times 35.65 + 1.463783 \times 226.59 - 0.556830 \times 280.05 + 0.024890 \times 292.69 - 0.000227 \times 264.88 = -1216.33$$

An initial guess for the hour 17 surface temperature must be made; the previous hour's surface temperature of 61.82 °C is a reasonable starting point. Using that value gives the following values for convection and radiation coefficients:

h_c = 7.91 W/m²K, from Eqn. 5.16,
$h_{r,sky}$ = 3.25 W/m²K, from Eqn. 5.34,
$h_{r,grd}$ = 2.95 W/m²K, from Eqn. 5.35.

Then, applying the heat balance, Eqn. 5.56:

Table 5.11 Weather conditions and previous hour

Time	$T_{os,\theta}$ (°C)	$T_{is,\theta}$ (°C)	q_o'' (W/m²)	q_i'' (W/m²)	*Incident Solar* (W/m²)	$T_{o,\theta}$ (°C)	*Wind speed* (m/s)	*Sky temperature* (°C)
1	27.39	22.09	−83.55	11.59	0.00	21.69	3.6	15.69
2	25.97	22.08	−71.15	10.86	0.00	21.22	3.6	15.22
3	24.75	22.07	−61.16	10.06	0.00	20.76	3.6	14.76
4	23.70	22.05	−52.59	9.23	0.00	20.38	3.6	14.38
5	22.83	22.02	−44.90	8.42	0.00	20.10	3.6	14.10
6	22.14	22.00	−37.46	7.62	0.00	20.00	3.6	14.00
7	21.67	21.97	−29.21	6.86	0.00	20.18	3.6	14.18
8	22.36	21.93	−0.30	6.13	40.00	20.63	3.6	14.63
9	23.51	21.90	21.58	5.45	72.59	21.47	3.6	15.47
10	24.97	21.88	39.18	4.83	99.05	22.66	3.6	16.66
11	28.42	21.86	91.80	4.29	198.22	24.06	3.6	18.06
12	35.65	21.84	196.58	3.86	423.81	25.63	3.6	19.63
13	43.92	21.84	264.88	3.61	627.37	27.14	3.6	21.14
14	51.75	21.84	292.69	3.68	784.36	28.25	3.6	22.25
15	58.04	21.86	280.05	4.19	873.26	29.01	3.6	23.01
16	61.82	21.88	226.59	5.17	873.85	29.30	3.6	23.30
17		21.91			760.91	29.04	3.6	23.04
18					475.74	28.38	3.6	22.38
19					0.06	27.38	3.6	21.38
20					0.00	26.16	3.6	20.16
21					0.00	24.96	3.6	18.96
22					0.00	23.92	3.6	17.92
23					0.00	23.00	3.6	17.00
24					0.00	22.24	3.6	16.24

$$T_{os,j,\theta} = \frac{Y_o T_{is,j,\theta} - H_{out,j,\theta} + \alpha G_t + h_c T_o + h_{r-g} T_g + h_{r-sky} T_{sky}}{X_o + h_c + h_{r-g} + h_{r-sky}}$$

$$T_{os,j,\theta} = \frac{3.375092 \cdot 10^{-5}(21.91) - (-1216.33) + 0.8(760.9) + 7.91(29) + 2.95(29) + 3.25(24.79)}{21.7157 + 7.91 + 2.95 + 3.25} = 62.00°\text{C}$$

The new estimate of surface temperature is so close to the original estimate that, to three significant figures, the convection and radiation coefficients do not change. Further iteration gives a final estimate of the surface temperature as 61.99 °C and the conduction flux can be calculated with Eqn. 5.55:

$$q''_{conduction,j,\theta} = -Y_o T_{is,j,\theta} + X_o T_{os,j,\theta} + H_{out,j,\theta} = 3.375092 \cdot 10^{-5}(21.91) + 21.7157(62.0) - 1216.33 = 129.9 \text{ W/m}^2.$$

End of solution.

Interior surface heat balance

Much like the outside surface heat balance, the inside surface heat balance insures that the heat transfer due to absorbed solar heat gain, convection, and long-wavelength radiation is balanced by the conduction heat transfer. Again, this comes about by solving for the surface temperature that results in a heat balance being achieved. This section will discuss how each heat transfer mechanism is modeled. For each mechanism, there are a number of possible models that could be used, ranging from very simple to very complex. We will again follow the approach of selecting a reasonably simple model for each heat transfer mechanism and referring the reader to other sources for more sophisticated and accurate models.

Like the exterior surface heat balance, the interior surface heat balance may be formulated to solve for a specific surface temperature. First, a history term that contains all of the historical terms for the interior CTF equation should be defined:

$$H_{in,j,\theta} = -\sum_{n=1}^{N_z} Z_n T_{is,j,\theta-n\delta} + \sum_{n=1}^{N_y} Y_n T_{os,j,\theta-n\delta} + \sum_{n=1}^{N_q} \Phi q''_{conduction,in,j,\theta-n\delta} \tag{5.57}$$

and then Eqn. 5.11 may be represented as

$$q''_{conduction,in,j,\theta} = -Z_o T_{is,j,\theta} + Y_o T_{os,j,\theta} + H_{in,j,\theta} \tag{5.58}$$

Furthermore, the net radiation leaving the surface is the surface-to-surface radiation minus the radiation due to internal heat gains:

$$q''_{radiation,in,j,\theta} = q''_{radiation-surf,in,j,\theta} - q''_{radiation,ihg,j,\theta} \tag{5.59}$$

Then, by substituting the expressions for conduction heat flux (Eqn. 5.11), convection heat flux (Eqn. 5.18), radiation heat flux (Eqns. 5.38–5.45), and absorbed solar heat gain (Eqn. 5.29 or 5.30) into the interior surface heat balance (Eqn. 5.47), and solving for the interior surface temperature, we obtain:

$$T_{is,j,\theta} = \frac{q''_{solar,in,j,\theta} + Y_o T_{os,j,\theta} + H_{in,j,\theta} + h_c T_i + h_{r,j} T_{f,j} + q''_{balance} + q''_{radiation-ihg,in,j,\theta}}{Z_o + h_c + h_{r,j}} \tag{5.60}$$

Note that $h_{r,j}$, $T_{f,j}$, and $q''_{balance}$ all depend on the other surface temperatures as well as on $T_{is,j,\theta}$. With a more sophisticted convection model, h_c might also depend on the surface temperature. As in the exterior heat balance, it is convenient to solve the equations iteratively with successive substitution.

Surface heat balances – transparent and semitransparent surfaces

The heat balance on windows must be treated differently than the heat balances on walls and roofs. The primary reason for this is that solar radiation may be absorbed throughout the window rather than just at the interior and exterior surfaces. This can lead to some rather arduous calculations, so we will make some simplifying assumptions:

- A window contains very little thermal mass, so we will assume that it behaves in a quasi-steady-state mode,
- Most of the overall thermal resistance of a window comes from the convective and radiative resistances at the interior and exterior surfaces and (if a multiple-pane window) between the panes. The conductive resistance of the glass or other glazing materials is quite small in comparison. Therefore, we will neglect the conductive resistance of the glass itself,
- Neglecting the conductive resistance causes each layer to have a uniform temperature. Therefore, there will be a single heat balance equation for each layer rather than an interior and an exterior surface heat balance equation for the entire window system,
- Layer-by-layer absorptance data may not generally be available for actual windows. If not, the engineer will have to make an educated guess as to the distribution of absorbed solar radiation in each layer.

Consider the thermal network for a double-pane window shown in Figure. 5.9. It has incident solar radiation from the outside, $q''_{solar,out,j,\theta}$ and solar radiation incident from the inside, $q''_{solar,in,j,\theta}$. The solar radiation incident from the inside was transmitted through a window and possibly reflected before striking the inside surface of the window. For both solar radiation fluxes, a certain amount is absorbed by both panes.

In addition to the heat transfer modes addressed earlier in this chapter, Fig. 5.9 also shows radiation and convection heat transfer between the panes. The radiative resistance may be estimated from a simple radiation network analysis as:

$$R_R = \frac{\left[\dfrac{1-\varepsilon_1}{\varepsilon_1} + 1 + \dfrac{1-\varepsilon_2}{\varepsilon_2}\right]}{4\sigma T_m^3} \tag{5.61}$$

where:

$\varepsilon_1, \varepsilon_2$ = emissivities of the two window surfaces, dimensionless,
σ = Stefan-Boltzmann constant, $5.67 \cdot 10^{-8}$ (W/m²K⁴),
T_m = mean temperature of the two window surfaces (K).

The convection resistance may be determined with a suitable correlation, such as that given by Wright (1996). Wright's correlation was developed for cavity height-to-width ratios greater than 40 and Rayleigh numbers less than 10^6, but it should be usable for any practical window configuration. The Nusselt number is defined based on the cavity width:

$$\mathrm{Nu} = h_c \cdot \frac{l}{k} \text{ or } h_c = \mathrm{Nu} \cdot \frac{k}{l} \tag{5.62}$$

where:

l = cavity width or spacing between the panes of glass (m),
k = thermal conductivity of the fill gas (W/mK),

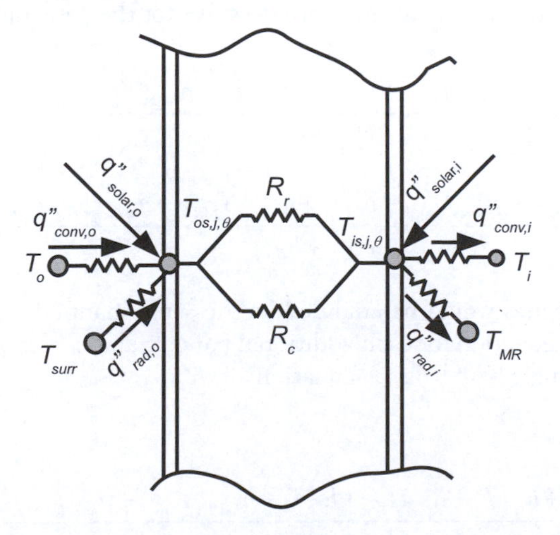

Figure 5.9 Heat balance for a double-pane window.

The Rayleigh number is also defined based on the cavity width:

$$\mathrm{Ra} = \frac{\rho^2 l^3 g c_p \Delta T}{\mu k T_m} \tag{5.63}$$

where:

ρ = fill-gas density (kg/m³),
c_ρ = fill-gas specific heat (J/kgK),
μ = fill-gas dynamic viscosity (Ns/m²),
T_m = mean temperature of fill gas (K),
g = acceleration due to gravity (m/s²).

Then, the Nusselt number is correlated with the following equations, depending on the value of the Rayleigh number:

$$\mathrm{Nu} = 0.0673838.\mathrm{Ra}^{(1/3)} \qquad \mathrm{Ra} > 5 \times 10^4 \tag{5.64a}$$

$$\mathrm{Nu} = 0.028154.\mathrm{Ra}^{0.4134} \qquad 10^4 < \mathrm{Ra} \le 5 \times 10^4 \tag{5.64b}$$

$$\mathrm{Nu} = 1 + 1.75967 \times 10^{-10}.\ \mathrm{Ra}^{2.2984755} \qquad \mathrm{Ra} \le 5 \times 10^4 \tag{5.64c}$$

Once the convection coefficient is found the convective resistance is:

$$R_C = \frac{1}{h_c} \tag{5.65}$$

Both the radiative resistance and convective resistance may be adjusted according to the surface temperatures. A *U*-factor may be defined:

$$U_{airspace} = \frac{1}{R_R} + \frac{1}{R_C} \tag{5.66}$$

As done previously with convection and resistance coefficients, the value of $U_{airspace}$ may be updated between iterations.

The heat balance for each pane may then be defined in a manner analogous to the heat balances previously developed for opaque exterior surfaces and opaque interior surfaces. The heat balance for each pane may then be cast in a form to solve for the pane temperatures:

$$T_{os,j,\theta} = \frac{q''_{absorbed,outer,j,\theta} + U_{airspace} T_{is,j,\theta} + h_{c,o} T_o + h_{r-g} T_g + h_{r-sky} T_{sky}}{U_{airspace} + h_{c,o} + h_{r-g} + h_{r-sky}} \tag{5.67}$$

$$T_{is,j,\theta} = \frac{q''_{absorbed,inner,j,\theta} + U_{airspace} T_{os,j,\theta} + h_{c,i} T_i + h_{rad,2,j} T_{MR,2} + q''_{radiation-ihg,in,j,\theta}}{U_{airspace} + h_{c,i} + h_{rad,2,j}} \tag{5.68}$$

A window with more panes would be analyzed in the same manner, but there would be an additional heat balance equation for each additional pane. Likewise, for a single pane window, there would only be a single heat balance equation:

$$T_{os,j,\theta} = T_{is,j,\theta} = \frac{q''_{absorbed,j,\theta} + h_{c,o} T_o + h_{r-g} T_g + h_{r-sky} t_{sky} + h_{c,i} T_i + h_{rad,2,j} T_{MR,2} + q''_{radiation-ihg,in,j,\theta}}{h_{c,o} + h_{r-g} + h_{r-sky} + h_{c,i} + h_{rad,2,j}} \tag{5.69}$$

Example 5.10

The 1.2 m high and 1.2 m wide window in Example 5.6, at 3 p.m. PDST on July 21 in Seattle, absorbed 141.5 W/m^2 in the outer pane and 41.5 W/m^2 in the inner pane. The rated U-factor for the window is 1.66 W/m^{2}K. At that time, the following conditions may be assumed to apply:

- outdoor air temperature and ground temperature is 28.1 °C,
- the effective sky temperature is 23.9 °C,
- the wind speed is 3.6 m/s,
- mean radiant temperature of zone is 21.8 °C,
- the air temperature of the zone is 21.5 °C,
- All of the window surfaces have emissivities of 0.84, except the inside of the outer pane, which has a low-e coating and an emissivity of 0.033,
- there are no internal heat gains (e.g. lighting, equipment, occupants),
- the interior convection coefficient from Table 5.6 is 3.18 W/m^{2}K,
- the interior radiation coefficient is 6.35 W/m^{2}K.

Note that several of the quantities would, in an actual simulation, be determined simultaneously with the window surface temperatures and heat flux.

Find the interior and exterior surface temperatures and convective and radiative heat gains from the window.

Solution:

Equations 5.67 and 5.68 represent the heat balance equations for the outside and inside panes, respectively. By stipulating values for the mean radiant temperature and air temperature in the zone, as well as the interior radiative and convective coefficients, the problem has been simplified so that these two heat balances can be considered independent of other surfaces and the zone air. As in most heat balance analyses, initial guesses must be made for surface temperatures; arbitrarily, we will choose 50 °C for the outer pane and 25 °C for the inner pane. Using these guesses gives the following values for convection and radiation coefficients:

h_c = 7.80 W/m^{2}K, from Eqn. 5.16,
$h_{r,sky}$ = 3.05 W/m^{2}K, from Eqn. 5.34,
$h_{r,grd}$ = 3.11 W/m^{2}K, from Eqn. 5.35,
$U_{airspace}$ = 2.44 W/m^{2}K, from Eqn. 5.66.

Then, Equation 5.67 gives the outer pane temperature:

$$T_{os,j,\theta} = \frac{q''_{absorbed,outer,j,\theta} + U_{airspace} T_{is,j,\theta} + h_{c,o} T_o + h_{r-g} T_g + h_{r-sky} T_{sky}}{U_{airspace} + h_{c,o} + h_{r-g} + h_{r-sky}}$$

$$= \frac{141.46 + 2.44(25) + 7.8(28.1) + 3.11(28.1) + 2.44(23.9)}{2.44 + 7.8 + 3.11 + 2.44} = 35.47°\mathrm{C}$$

and Equation 5.68 gives the inner pane temperature:

$$T_{is,j,\theta} = \frac{q''_{absorbed,inner,j,\theta} + U_{airspace} T_{os,j,\theta} + h_{c,i} T_i + h_{rad,2,j} T_{MR,2} + q''_{radiation-ihg,in,j,\theta}}{U_{airspace} + h_{c,i} + h_{rad,2,j}}$$

$$= \frac{41.46 + 2.44(50) + 3.18(21.5) + 6.35(21.8) + 0}{2.44 + 3.18 + 6.35} = 30.94°\text{C}$$

With the new estimates of the pane temperatures, the radiation and convection coefficients are:

h_c = 7.62 W/m²K,
$h_{r,sky}$ = 2.83 W/m²K,
$h_{r,grd}$ = 2.89 W/m²K,
$U_{airspace}$ = 2.26 W/m²K.

Solving Eqns. 5.67 and 5.68 again gives an outer pane temperature of 36.80 °C and an inner pane temperature of 27.86 °C. Then, with a few more iterations, the coefficients converge to:

h_c = 7.63 W/m²K,
$h_{r,sky}$ = 2.85 W/m²K,
$h_{r,grd}$ = 2.91 W/m²K,
$U_{airspace}$ = 2.26 W/m²K.

Final values of the outer and inner pane temperatures are 36.36 °C and 28.03 °C, respectively. A final check on the window heat balance shows the following:

- The incident solar radiation is 680 W/m².
- Of that, 228.5 W/m² are transmitted, 182.9 W/m² are absorbed by the two panes and 268.7 W/m² are reflected back outside.
- The 182.9 W/m² absorbed is distributed as follows –
  - Convected to the outside: 63.0 W/m²,
  - Radiated to the sky: 35.6 W/m²,
  - Radiated to the ground: 24.0 W/m²,
  - Convected to the zone air: 20.8 W/m²,
  - Radiated to the other zone surfaces 39.5 W/m²,
  - 63 + 35.6 + 24 + 20.8 + 39.5 = 182.9 W/m².

End of solution.

Zone air heat balance

Finally, with the assumption that the zone air has negligible thermal storage capacity, a heat balance on the zone air may be represented conceptually as:

$$\sum_{j=1}^{N} A_j q''_{convection,in,j,\theta} + q_{infiltration,\theta} + q_{system,\theta} + q_{internal,conv,\theta} = 0 \tag{5.70}$$

where:

A_j = area of the jth surface (m²),

$q_{infiltration,\theta}$ = heat gain due to infiltration (W),
$q_{system,\theta}$ = heat gain due to the heating/cooling system (W),
$q_{internal,conv,\theta}$ = convective portion of internal heat gains due to people, lights, or equipment (W).

The heat balance may be cast in several forms:

- Solving for the required system capacity to maintain a fixed zone air temperature,
- Solving for the zone temperature when the system is off,
- Solving for the zone temperature and system capacity with a system that does not maintain a fixed zone air temperature.

For the purposes of design cooling load calculations, the first formulation is usually of the most interest. The second formulation may be useful when modeling setback conditions or to help determine thermal comfort for naturally cooled buildings. The third formulation is the most general—with a fairly simple model of the system it is possible to model the first condition (by specifying a system with a very large capacity) or to model the second condition (by specifying a system with zero capacity). Also, while the first formulation is suitable for determining required system air flow rates and cooling coil capacities, it may be desirable to base the central plant equipment sizes on actual heat extraction rates. A more detailed discussion of the subject of HVAC performance prediction can be found in Chapter 11.

Before each formulation is covered, each of the heat transfer components will be briefly discussed.

Convection from surfaces

Convection from individual surfaces is described with Eqn. 5.18. The total convection heat transfer rate to the zone air is found by summing the contribution from each of the N surfaces:

$$\dot{q}_{convection,in,j,\theta} = \sum_{j=1}^{N} A_j q''_{convection,in,j,\theta} = \sum_{j=1}^{N} A_j h_{c,i,j}(T_{is,j,\theta} - T_i) \quad (5.71)$$

Convection from internal heat gains

Convection from internal heat gains is found by summing the convective portion of each individual internal heat gain:

$$\dot{q}_{internal,conv,\theta} = \sum_{j=1}^{M} \dot{q}_{j,\theta} F_{con,j} \quad (5.72)$$

where:

$\dot{q}_{internal,conv,\theta}$ = convective heat transfer to the zone air from internal heat gains (W),
$\dot{q}_{j,\theta}$ = heat gain for the jth internal heat gain element (W),
$F_{con,j}$ = convective fraction for the jth internal heat gain element.

Although the convective fraction for any heat gain element will vary with its surface temperature, the radiative environment, local airflow and other factors, it is often treated as a constant for specific heat gains. Typical values are tabulated by Spitler (2009) and ASHRAE (2005).

Heat gain from infiltration

Methods used to estimate the quantity of infiltration air are discussed in Chapter 6. The same methods apply to cooling load calculations. Both a sensible and latent heat gain will result and are computed as follows:

$$\dot{q}_{infiltration,\theta} = \dot{m}_{a,infiltration,\theta} c_p \left(T_o - T_i\right) = \frac{\dot{Q}c_p}{v_o}\left(T_o - T_i\right) \tag{5.73}$$

$$\dot{q}_{infiltration,latent,\theta} = \dot{m}_{a,infiltration,\theta} \left(W_o - W_i\right) i_{fg} = \frac{\dot{Q}}{v_o}\left(W_o - W_i\right) i_{fg} \tag{5.74}$$

Zone air heat balance formulations

The simplest formulation of the zone air heat balance is to determine the cooling load – i.e. for a fixed zone air temperature, determine the required system heat transfer. In this case, Eqns. 5.71, 5.72, and 5.73 can be substituted into Eqn. 5.70 to give:

$$\dot{q}_{system,\theta} = -\sum_{j=1}^{N} A_j h_{c,i,j} \left(T_{is,j,\theta} - T_i\right) - \dot{m}_{a,infiltration} c_p \left(T_o - T_i\right) - \dot{q}_{internal,conv,\theta} \tag{5.75}$$

Likewise, the zone air heat balance can be formulated to determine the instantaneous zone temperature when there is no system heat transfer. Setting the system heat transfer rate in Eqn. 5.75 equal to zero and solving for the zone air temperature gives:

$$T_i = \frac{\sum_{j=1}^{N} A_j h_{c,i,j} (T_{is,j,\theta}) + \dot{m}_{a,infiltration} c_p T_o + \dot{q}_{internal,conv,\theta}}{\sum_{j=1}^{N} A_j h_{c,i,j} + \dot{m}_{a,infiltration} c_p} \tag{5.76}$$

Finally, the zone air heat balance can be formulated to determine the zone temperature when there is system heat transfer. Substituting the piecewise linear expression for system capacity in Eqn. 5.22 into the zone air heat balance (Eqn. 5.70) and solving for the zone air temperature gives:

$$T_i = \frac{a + \sum_{j=1}^{N} A_j h_{c,i,j} \left(T_{is,j,\theta}\right) + \dot{m}_{a,infiltration} c_p T_o + \dot{q}_{internal,conv,\theta}}{-b + \sum_{j=1}^{N} A_j h_{c,i,j} + \dot{m}_{a,infiltration} c_p} \tag{5.77}$$

Note that the control profile coefficients depend on the value of the zone air temperature. Therefore, it is usually necessary to choose *a* and *b* based on an intelligent guess of the zone air temperature. Then, using those values of *a* and *b*, solve Eqn. 5.77 for T_i. If the value of T_i is not within the range for which *a* and *b* were chosen, then another iteration must be made.

Example 5.11

The zone shown in Figure 5.6 is conditioned by a variable air volume (VAV) system. The VAV system has a maximum flow of 140 L/s of air at 15 °C at zone air temperatures of 25 °C and higher. The minimum fraction the VAV terminal unit is 0.3, which occurs at zone air temperatures of 22 °C and lower. Between 22 °C and 25 °C, the airflow is modulated linearly.

Below 20 °C, reheat is provided. At a certain time, the outdoor air temperature is 28.1 °C, the infiltration rate is 6 L/s, the surface temperatures and convection coefficients are as shown in Table 5.12, and other heat gains to the space are summarized in Table 5.13, find the zone air temperature and cooling provided by the system.

Solution:

First, it might be noted that the surface temperatures in Table 5.12 are not generally known in advance, but rather are determined simultaneously with the zone air temperature. Then, the first step in solving the problem is to develop a piecewise linear profile for the system heat transfer rate, in the form shown in Eqn. 5.22. The zone air heat balance, Eqn. 5.70 is developed with the convention that heat transfer from the system to the zone air is positive when heat is being added to the zone.

The system heat transfer rate can be computed for three ranges:

- $T_i > 25$ °C, where the VAV terminal unit will be fully open,
- 25 °C $\geq T_i \geq$ 22 °C, where the VAV terminal unit is modulated between full open and 30% of full flow,
- 22 °C $> T_i \geq$ 20 °C, where the VAV terminal unit is at the minimum fraction.

As the supply air temperature is assumed to be held constant at 15 °C, regardless of zone air temperature, the system heat transfer may be computed at two points within each of the above three ranges, and then the coefficients of the linear equation giving system heat transfer rate as a function of zone air temperature may be determined. For each point, the heat transfer rate is determined with Eqn. 5.78:

$$\dot{q}_{system,\theta} = \dot{m}_a c_p (T_{SA} - T_i) \quad (5.78)$$

where T_{SA} is the supply air temperature, °C.

With the air density taken as 1.22 kg/m^3 and the specific heat of the air taken as 1006 J/kgK, the heat transfer rates shown in Table 5.14 can be computed.

Fitting a line for each of the temperature ranges, gives the set of *a* and *b* coefficients for the piecewise linear representation of the system heat transfer, often referred to as the system control profile, shown in Table 5.15.

Table 5.12 Zone surface temperatures and convection coefficients

Surface number	*Surface Name*	*Area (m²)*	*Temperature (°C)*	h_c *(W/m²K)*
1	Floor	24.0	22.3	4.15
2	North wall	18.0	24.1	3.18
3	East wall	12.0	23.7	3.18
4	West wall	12.0	23.8	3.18
5	Ceiling	24.0	25.3	1.02
6	Window	7.5	30.9	3.18
7	South wall (surrounding window)	10.5	24.3	3.18

Table 5.13 Other convective heat gains

Internal Heat Gain Type	*Convective Portion (W)*
Lighting	57
Equipment	207
People	117
$\dot{q}internal,conv,q$	381

Table 5.14 System heat transfer rate

System State	T_i(°C)	*Flow (L/s)*	$\dot{q}$ *system*, θ(W)
VAV TU fully open	30	140	–2579
VAV TU fully open	25	140	–1719
VAV TU min. frac	22	42	–361
Reheat comes on	20	42	–258

Table 5.15 System control profile

Temperature Range	*a*	*b*
$T_i > 25$°C	2579.12	–171.9411
25°C ≥ T_i ≥ 22°C	9600.04	–452.7782
22°C > T_i ≥ 20°C	773.73	–51.5823

Before performing the zone air heat balance, the infiltration mass flow rate and specific heat must be determined. Based on the outdoor air temperature, 28.1 °C, the density may be taken as 1.22 kg/m^3, giving a mass flow rate of 0.007 kg/s, and the specific heat is 1006 J/kgK.

The zone air heat balance, as formulated in Eqn. 5.77, has two summations which can be computed first. Both summations utilize values taken from Table 5.12. The first summation,

$$\sum_{j=1}^{N} A_j h_{c,i,j} T_{is,j,\theta} = (24 \times 4.15 \times 22.3) + (18 \times 3.18 \times 24.1) + \ldots = 7580.9$$

The second summation is similar to the first, but without the surface temperatures:

$$\sum_{j=1}^{N} A_j h_{c,i,j} = (24 \times 4.15) + (18 \times 3.18) + \ldots = 314.9$$

$$T_i = \frac{a + \sum_{j=1}^{N} A_j h_{c,i,j} \left(T_{is,j,\theta}\right) + \dot{m}_{a,infiltration} c_p T_o + \dot{q}_{internal,conv,\theta}}{-b + \sum_{j=1}^{N} A_j h_{c,i,j} + \dot{m}_{a,infiltration} c_p}$$

With numerical values determined for all of the terms in the zone air heat balance, as formulated in Equation 5.77, can now be applied to determine the zone air temperature. It is necessary to first guess which zone temperature range applies – an initial guess that the temperature is between 22 °C and 25 °C will be used. The zone air heat balance is then:

$$T_i = \frac{a + \sum_{j=1}^{N} A_j h_{c,i,j} \left(T_{is,j,\theta}\right) + \dot{m}_{a,infiltration} c_p T_o + \dot{q}_{internal,conv,\theta}}{-b + \sum_{j=1}^{N} A_j h_{c,i,j} + \dot{m}_{a,infiltration} c_p}$$

$$= \frac{9600.04 + 7580.9 + (0.007 \times 1006 \times 28.1) + 381}{-(-452.7782) + 314.9 + (0.007 \times 1006)}$$

= 22.92 °C So, our initial guess was correct; if it had not been, we would have tried the range suggested by the resulting temperature.

With a temperature of 22.92 °C, the zone load can be computed from Eqn. 5.70, or more simply from Eqn. 5.22,

$$\dot{q}_{system,\theta} = a + bT_i = 9600.04 - 452.7782(22.92) = -778 \text{ W}$$

The system will provide 778 W of cooling or remove 778 W from the space.
End of solution.

Zone models: thermal network models

Thermal network models (Clarke 2001; Lewis and Alexander 1990; Sowell 1990; Walton 1992) discretize the building into a network of nodes with interconnecting energy flow paths. The energy flow paths may include conduction, convection, radiation, and airflow. Thermal network models may be thought of as heat balance models with finer granularity. Where heat balance models generally have one node representing the zone air, a thermal network model may have several. Where heat balance models generally have a single exterior node and a single interior node per wall, thermal network models may have additional nodes. Where heat balance models generally distribute radiation from lights in a simple manner, thermal network models may model the lamp, ballast, and luminaire housing separately. Thermal network models are the most flexible of the three zone models discussed here. However, the added flexibility requires more computational time and, often, more user effort.

Furthermore, at the present time, most implementations of thermal network models are primarily research tools. One program that is more widely used is ESP-r (ESRU 2009), which has been under development by Prof. Joe Clarke and colleagues at the University of Strathclyde since the 1970s. A few of the features made possible by the thermal network approach include 2-D and 3-D conduction heat transfer (Strachan *et al.* 1995), integrated computational fluid dynamics analysis (Clarke *et al.* 1995) and phase-change materials (Heim and Clarke 2003).

Implementation of thermal network models in advanced simulation environments (Kalagasidis *et al.* 2007; Riederer 2005; Wetter 2006) remains an area of ongoing research.

Zone models: transfer function models

An alternative to the heat balance models and thermal network models are the transfer function models. Within this category, we may include models that use room transfer functions or room response factors to determine cooling loads from the heat gains. Examples of implementations include the DOE-2 (York and Cappiello 1981) program, the ASHRAE Transfer Function Method (TFM) (McQuiston and Spitler 1992) and Radiant Time Series Method (RTSM) (Spitler *et al.* 1997) cooling load calculation procedures

All of these models make certain simplifications to the heat balance model. These include:

- There is no exterior heat balance. Instead of modeling convection to the outdoor air, radiation to the ground and sky, and solar radiation separately, they are modeled as a single heat transfer between an "equivalent" temperature, known as the sol-air temperature, and the surface temperature. This allows the resistance between the sol-air temperature and the surface temperature to be included as a resistance in the transient conduction analysis, and it allows the exterior driving temperature for the transient conduction analysis to be determined prior to the load calculation. This has the limitation that a single fixed combined convection and radiation coefficient must be used, independent of the surface temperature, sky temperature, air temperature, wind speed, etc.,
- There is no interior surface heat balance. Instead, for radiation purposes, it is assumed that the other surfaces in the zone are effectively at the zone air temperature. Then, a single,

fixed value of the surface conductance is used, and folded into the transient conduction analysis,

- There is no zone air heat balance. Cooling loads are determined directly, but the zone air temperature is assumed to be constant. A second set of transfer functions can be used to estimate the heat extraction rate.
- The storage and release of energy by the walls, roofs, floors, and internal thermal mass are approximated with a predetermined zone response. Unlike the heat balance model, this phenomenon is considered independently of the through-the-wall conduction heat transfer. This has implications for the accuracy of the calculation. For cooling load calculations, it usually leads to a small overprediction of the cooling load. For zones with large quantities of high-conductance surfaces, it results in a significant overprediction. For example, the RTSM procedure will tend to overpredict (Rees *et al.* 1998) the peak cooling load for buildings with large amounts of glass.
- This also has implications for presentation of results, as it makes it possible to identify the individual heat gains and cooling loads on a component-by-component basis. For example, see Figure 5.10, where lighting heat gains for a conference room in an office building are plotted alongside the resulting cooling loads. The cooling loads show the delay and damping effects caused by some of the lighting heat gain being absorbed in the room structure and later released.

Each of the simplifications lead to common features of transfer function models – combined convection and radiation heat transfer, sol-air temperatures, and room transfer functions or response factors. Each of these features are discussed below, followed by a brief discussion of space air transfer functions, which allow the effects of system capacity and control to be included in a prediction of the actual room air temperature and actual heat extraction rate of the system.

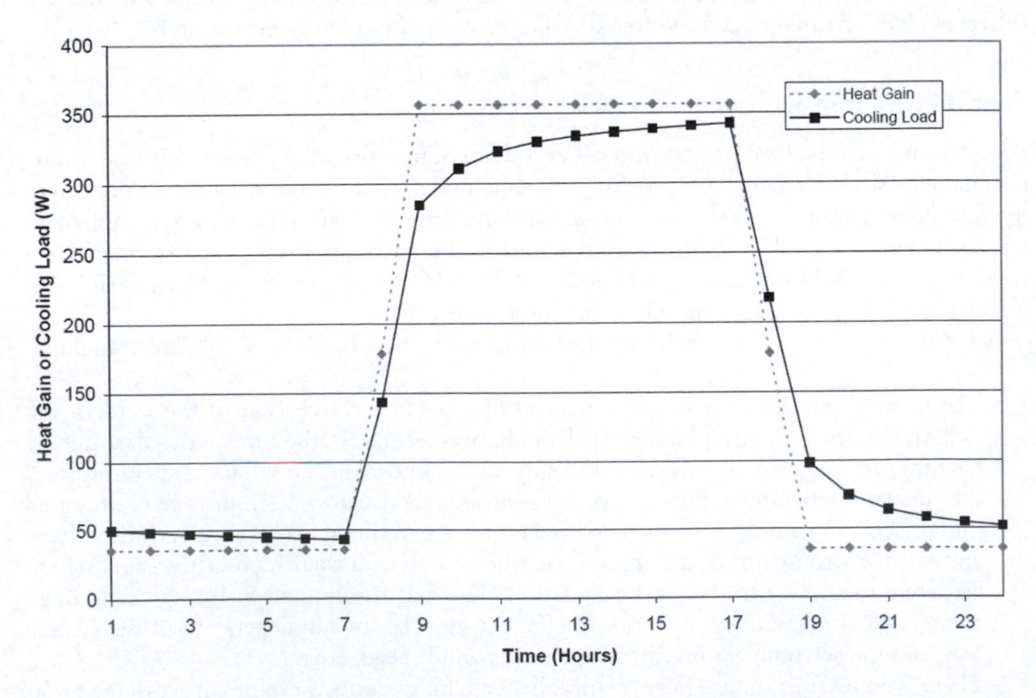

Figure 5.10 Lighting heat gain and the resulting cooling load for a conference room.

Combined convection and radiation heat transfer

An important feature of transfer function models is that convection and radiation analysis are often combined into a single surface conductance. This requires a linear approximation of the radiation heat transfer. Equation 5.23 quantifies the radiation heat transfer between two surfaces that only view each other. Considering that surface 2 in Equation 5.23 could represent, in aggregate, the surroundings, then approximately $A_2 >> A_1$ and Equation 5.23 can be simplified to:

$$\dot{q}_{1-2} = A_1 \varepsilon_1 \sigma \left(T_1^4 - T_2^4 \right) = A_1 h_r (T_1 - T_2) \tag{5.79}$$

$$h_r = \varepsilon_1 \sigma (T_1^2 + T_2^2)(T_1 + T_2) \approx 4\varepsilon_1 \sigma \bar{T}^3 \tag{5.80}$$

where:
h_r = radiation coefficient (W/m^2K).

Note that while the linearized radiation coefficient is calculated with absolute temperatures, it is applied to a temperature difference, which is the same whether it is taken in Kelvin or degrees Celsius:

$$\dot{q}_r = h_r A \left(T - T_w \right) \tag{5.81}$$

A surface conductance, which combines the radiation and convection coefficient, can then be defined as:

$$h_o = h_c + h_r \tag{5.82}$$

This has the advantage of allowing the combined convection and radiation heat transfer rate to be expressed as:

$$\dot{q} = h_o A \left(T - T_w \right) \tag{5.83}$$

Or, the inverse of the surface conductance can be treated as a conductive resistance, simplifying the analysis so that it appears as simply a conduction analysis. For the interior conductance, the indoor air temperature is taken as a suitable approximation to the actual effective temperature with which convection and radiation are exchanged. For the exterior conductance, an equivalent temperature, the sol-air temperature, is utilized, as described in the next section.

Sol-air temperature

In the case of the exterior conductance, the problem is further complicated by the presence of solar radiation. For simplified load calculation procedures, an equivalent air temperature, the sol-air temperature, is defined that gives approximately the same heat flux to the surface as the combined effects of solar radiation, convection, and radiation to the surroundings. It is given by:

$$T_e = T_o + \alpha G_t / h_o - \varepsilon \delta R / h_o \tag{5.84}$$

where:
T_e = sol-air temperature (°C),

T_o = outdoor air temperature (°C),
α = absorptance of the surface to solar radiation,
G_t = total incident solar radiation flux (W/m²),
h_o = exterior surface conductance (W/m²K),
ε = exterior surface emittance,
δR = difference between the thermal radiation incident on the surface from the sky or surroundings and the radiation emitted by a blackbody at the outdoor air temperature, (W/m²K).

The last term, $\varepsilon\delta R/h_o$, is often taken as 4 °C for horizontal surfaces (i.e. facing the sky) and 0 °C for vertical surfaces.

Room transfer functions and response factors

Room transfer functions and response factors relate the hourly heating and cooling loads due to individual types of heat gains to previous values of that type of heat gain and previous values of the cooling load due to that type of heat gain.

For example, consider a particular zone with equipment turned on for a certain fraction of the day. The equipment heat gain is partially radiant and partially convective. The convective portion immediately becomes space cooling load, the radiant portion is absorbed by thermal mass in the zone and later convected to the room air. If this zone is modeled with a set of heat balance equations and pulsed with a one-hour unit heat gain, the space cooling load each hour will be a fractional number indicating how much of the original pulse will be convected into the room air each hour. This set of fractional numbers is a set of thermal response factors (Mitalas and Stephenson 1967).

Room thermal response factors can be computed with a heat balance program. However, a more computationally efficient form is often desired and this involves computation of a set of room transfer functions, sometimes called weighting factors. One procedure for computation of weighting factors is described by the DOE-2 *Engineers Manual* (York and Cappiello 1981). For any zone, there are many possible combinations of heat gain terms and cooling load terms that may be used. A typical form is:

$$Q_\theta = v_0 q_\theta + v_1 q_{\theta-\delta} + v_2 q_{\theta-2\delta} - w_1 Q_{\theta-\delta} - w_2 Q_{\theta-2\delta} \tag{5.85}$$

where:
Q_θ = Cooling or heating load at hour θ (W),
q_θ = Heat gain at hour θ (W),
δ = time interval, typically one hour.

The terms v_0, $v_1 \ldots$, w_1, $w_2 \ldots$, are the coefficients of the Room Transfer Function:

$$K(z) = \frac{v_0 + v_1 z^{-1} + v_2 z^{-2}}{1 + w_1 z^{-1} + w_2 z^{-2}} \tag{5.86}$$

which relates the transform of the corresponding parts of the cooling load and of the heat gain. These coefficients depend of (i) the time interval of δ, (ii) the nature of the heat gain (fraction radiant and location in room), and (iii) the thermal characteristics of the room.

The RTSM cooling load calculation method is similar in many respects to the TFM, but it assumes that the design day cooling load calculation is for a single day, with the previous days having the same conditions. In other words, any energy stored in the building overnight will be consistent with the previous days having been identical in weather and internal heat gains. In the RTSM, the radiant portion of the heat gains is converted to cooling loads using radiant time factors, a 24-term periodic response factor series.

The radiant time series for a particular zone gives the time-dependent response of the zone to a single steady periodic pulse of radiant energy. The series shows the portion of the radiant pulse that is convected to the zone air for each hour. Thus, r_0 represents the fraction of the radiant pulse convected to the zone air in the current hour, r_1 in the last hour, and so on. The radiant time series thus generated is used to convert the radiant portion of hourly heat gains to hourly cooling loads according to Equation 5.87.

$$Q_\theta = r_0 q_\theta + r_1 q_{\theta-\delta} + r_2 q_{\theta-2\delta} + r_3 q_{\theta-3\delta} + \ldots + r_{23} q_{\theta-23\delta} \tag{5.87}$$

where:

Q_θ = cooling load (Q) for the current hour (θ),
q_θ = heat gain for the current hour,
$q_{\theta-n\delta}$ = heat gain n hours ago,
r_0,r_1, etc. = radiant time factors.

Space air transfer functions

To this point, the transfer function models have only considered determination of heating load or cooling load, i.e. the heat transfer rate required to maintain a constant room temperature. However, room temperature may vary for a number of reasons, and it is desirable to have a way of accounting for variations in room temperature, as well as predicting room temperature in some cases.

The heat extraction characteristics of the system are approximated by a simple linear expression, similar to a single segment of the control profile described for the heat balance model:

$$ER_\theta = W_\theta + (ST_{r\theta}) \tag{5.88}$$

where:

ER_θ = rate of heat removal from the space at time θ (W),
$T_{r\theta}$ = actual air temperature in the space at time θ (°C),
W = intercept characterizing the performance of cooling equipment (W),
S = slope characterizing the performance of cooling equipment (W/°C).

This linear relationship is expected to hold over the throttling range of the control system.

The heat extraction rate and the room air temperature are related by the Space Air Transfer Function:

$$\sum_{i=0}^{1} p_i \left(ER_{\theta-i\delta} - Q_{\theta-i\delta}\right) = \sum_{i=0}^{2} g_i \left(T_{rc} - T_{r\theta-i\delta}\right) \tag{5.89}$$

where:

gi and p_i = coefficients of the Space Air Transfer Function,
Q_θ = calculated cooling load for room at time θ, based on an assumed constant room temperature of T_{rc} (W).

Previous equations for heat extraction rate (5.88) and the Space Air Transfer Function (5.89) can be solved simultaneously to yield:

$$ER_\theta = (W_\theta g_0 + I_\theta S)/(S + g_0) \tag{5.90}$$

where:

$$I_\theta = T_{rc}\sum_{i=0}^{2} g_{i,\theta} - \sum_{i=1}^{2} g_{i,\theta} T_{r\theta-i\delta} + \sum_{i=0}^{1} p_i Q_{r\theta-i\delta} - \sum_{i=0}^{1} p_i ER_{\theta-i\delta} \tag{5.91}$$

If the value of ER_θ calculated by Equation 5.90 results in a value greater than ER_{max}, ER_θ is set equal to ER_{max}. If the value of ER_θ calculated by results in a value less than ER_{min}, ER_θ is set equal to ER_{min}. Then $T_{r\theta}$ is calculated from the expression:

$$T_{r\theta} = \left(I_\theta - ER_\theta \right) / g_{0,\theta} \qquad (5.92)$$

Note that Equation 5.90 requires some prior values of ER_θ and $T_{r\theta}$ which must be assumed to begin the computation process. The computation is then repeated until the results for successive days are the same. At that time, the results are independent of the values assumed initially.

Synopsis

This chapter has covered methods for determining zone loads for design load calculation or energy calculation purposes. The heat balance model has been covered in some detail, although for each of the needed sub-models (e.g. exterior convection on walls, transient conduction, interior radiation, etc.) only one of the simpler possibilities is covered. The number of possible variations on sub-models is enormous. The thermal network model, which might be characterized as a heat balance model, but with much more detailed sub-models for many of the phenomenon, has only been briefly described and the reader should consult the references given in the section for additional detail. Finally, the chapter has described the underlying basis for transfer function models, which, though limited in some respects, are widely used today for both energy calculations and load calculations.

Despite this chapter being by far the longest in the book, numerous topics have been omitted or given cursory treatment. The section "Recommended reading", below, gives some initial directions for further information on methods for modeling zones. Beyond those resources, the reader will find that ongoing developments are reported in the academic literature and, in particular, the conference proceedings of the International Building Performance Simulation Association (www.ibpsa.org) serve as an archive for much of the work done in the last twenty-five years.

The field continues to evolve for several reasons, including:

- In the nature of the case, buildings have essentially infinite variation in design. Phenomena that are unimportant for most buildings may be very important for some buildings. Therefore, methods are augmented to more accurately model a wider range of buildings.
- As building designs for common building types evolve to use less energy, the relative importance of heat transfer mechanisms shifts and methods are refined to improve accuracy.
- New developments in building components such as fenestration, lighting, and internal equipment require continuous updates to data and at least intermittent improvements in methods.
- Moore's Law improvements in computing hardware and corresponding developments in simulation environments and simulation methodologies used in the broader world continue to facilitate new possibilities for more accurate and flexible zone models.
- Experimental research continues to better inform models, both by experimental validation and by investigation of specific phenomena such as convection heat transfer leading to improved sub-models.

Finally, some reflection on experimental research and the current state-of-the-art does suggest that zone models will be undergoing improvement for many years in the future. Prediction of infiltration rates and accounting for the effects of imperfections introduced in the real-world construction process continue to be problematic. As another example, consider that, after many years of investigation of interior convective heat transfer with an eye towards improving building simulation, most load calculation design procedures still use convection coefficients that were experimentally measured in the 1930s. Though progress has been made in this area, an interior convection algorithm that covers all or most room conditions and room ventilation configurations is still not available. Similar limitations in modeling of other phenomena exist, indicating that further developments in zone modeling will continue indefinitely.

Notes

1 Even though heat transfer in an air gap is due to convection and radiation, it is approximated as being a conduction process, with a fixed thermal resistance that is independent of the temperatures of the gap surfaces.
2 Technically, conduction also occurs in liquids and gases, too, but here we are only concerned with conduction in solids.
3 The analogy is to discrete electrical resistors when expressed in thermal resistances ($R' = \Delta x/kA$) and heat transfer rates. For one-dimensional planar walls, it is often convenient to consider the analogy on a per unit area basis, in which case it may be expressed in terms of unit thermal resistances and heat fluxes (heat transfer rates per unit area.)
4 Also known as Newton's Law of Cooling.
5 Despite some attempts to reserve the endings "-ivity" for optically pure surfaces and "-tance" for real-world surfaces, the terms absorptivity, emissivity, reflectivity, and transmissivity are often used interchangeably for absorptance, emittance, reflectance and transmittance.

References

Alamdari, F. and Hammond, G. P. (1983) "Improved Data Correlations for Buoyancy-Driven Convection in Rooms", *Building Services Engineering Research and Technology* 4(3): 106–112.

ASHRAE (2005) *ASHRAE Handbook – Fundamentals Volume*, Atlanta: ASHRAE.

Barbosa, R. M. and Mendes, N. (2008) "Combined simulation of central HVAC systems with a whole-building hygrothermal model", *Energy and Buildings* 40(3): 276–288.

Beausoleil-Morrison, I. and Mitalas, G. (1997) "BASESIMP: A Residential-Foundation Heat-Loss Algorithm for Incorporating into Whole-Building Energy-Analysis Programs", *Building Simulation '97*, Prague, Czech Republic.

Brown, W. C. and Stephenson, D. G. (1993) "A Guarded Hot Box Procedure for Determining the Dynamic Response of Full-Scale Wall Specimens – Part 1", *ASHRAE Transactions* 99(1): 632–642.

Burch, D. M., Seem, J. E., Walton, G. N. and Licitra, B. A. (1992) "Dynamic Evaluation of Thermal Bridges in a Typical Office Building", *ASHRAE Transactions* 98(1): 291–304.

Carpenter, S. C., Kosny, J. and Kossecka, E. (2003) "Modeling Transient Performance of Two-Dimensional and Three-Dimensional Building Assemblies", *ASHRAE Transactions* 109(1): 561–566.

Carroll, J. A. (1980a) "An "MRT Method" of Computing Radiant Energy Exchange in Rooms", *Systems Simulation and Economic Analysis Conference*, San Diego, CA: 343–348.

Carroll, J. A. (1980b) "An 'MRT Method' of Computing Radiant Energy Exchange in Rooms", *Systems Simulation and Economic Analysis Conference*, San Diego, California.

Clarke, J. A. (2001) *Energy Simulation in Building Design*, Oxford: Butterworth-Heinemann.

Clarke, J. A., Dempster, W. M. and Negrao, C. (1995) "The Implementation of a Computational Fluid Dynamics Algorithm within the ESP-R System", in *Proceedings of Building Simulation 1995*, Madison, WI: 166–175.

Cole, R. J. (1976) "The Longwave Radiative Environment Around Buildings", *Building and Environment* 11(1): 3–13.

Cole, R. J. and Sturrock, N. S. (1977) "The Convective Heat Exchange at the External Surface of Buildings", *Building and Environment* 12(4): 207–214.

Davies, M. G. (1988) "Design Models to Handle Radiative and Convective Exchange in a Room", *ASHRAE Transactions* 94(2): 173–195.

Davies, M. G. (1996) "A Time-Domain Estimation of Wall Conduction Transfer Function Coefficients", *ASHRAE Transactions* 102(1): 328–343.

Davies, M. G. (2004) *Building Heat Transfer*, Chichester, West Sussex: Wiley.

Duffie, J. A. and Beckman, W. A. (1991) *Solar Engineering of Thermal Processes*, 2nd ed., New York: John Wiley & Sons.

ESRU (2009) "ESRU Home Page", from http://www.esru.strath.ac.uk/.

Fisher, D. E. and Pedersen, C. O. (1997) "Convective Heat Transfer in Building Energy and Thermal Load Calculations", *ASHRAE Transactions* 103(2).

Goss, W. P. and Miller, R. G. (1989) "Literature Review of Measurement and Predictions of Reflective Building Insulation System Performance: 1900–1989", *ASHRAE Transactions* 95(2): 651–664.

Gouda, M. M., Danaher, S. and Underwood, C. P. (2002) "Building Thermal Model Reduction Using Nonlinear Constrained Optimization", *Building and Environment* 37(12): 1255–1265.

Heim, D. and Clarke, J. A. (2003) "Numerical Modelling and Thermal Simulation of Phase Change Materials with ESP-R", in *Proceedings of 8th International IBPSA Conference*, Eindhoven, The Netherlands.

Hittle, D. C. (1981) "Calculating Building Heating and Cooling Loads Using the Frequency Response of Multilayered Slabs", University of Illinois at Urbana-Champaign.

Iu, I. and Fisher, D. E. (2004) "Application of Conduction Transfer Functions and Periodic Response Factors in Cooling Load Calculation Procedures", *ASHRAE Transactions* 109(2): 829–841.

Kalagasidis, A. S., Weitzmann, P., Nielsen, T. R., Peuhkuri, R.. Hagentoft, C. E. and Rode, C. (2007) "The International Building Physics Toolbox in Simulink", *Energy and Buildings* 39(6): 665–674.

Karambakkam, B. K., Nigusse, B. and Spitler, J. D. (2005) "A One-dimensional Approximation for Transient Multi-dimensional Conduction Heat Transfer in Building Envelopes", *The 7th Symposium on Building Physics in the Nordic Countries*, Reykjavik, Iceland: The Icelandic Building Research Institute.

Klems, J. H. (2002) "Solar Heat Gain through Fenestration Systems Containing Shading: Summary of Procedures for Estimating Performance from Minimal Data", *ASHRAE Transactions* 108(1): 512–524.

Krarti, M. and Choi, S. (1996) "Simplified Method for Foundation Heat Loss Calculation", *ASHRAE Transactions* 102(1): 140–152.

Latta, J. K. and Boileau, G. G. (1969) "Heat Losses from House Basements", *Canadian Building* 19(10): 39.

LBNL (2009) "LBNL Windows & Daylighting Software", retrieved 17 March 2009, from http://windows.lbl.gov/software/window/window.html.

Lewis, P. T. and Alexander, D. K. (1990) "HTB2: A Flexible Model for Dynamic Building Simulation", *Building and Environment* 25(1): 7–16.

Liesen, R. J. and Pedersen, C. O. (1997) "An Evaluation of Inside Surface Heat Balance Models for Cooling Load Calculations", *ASHRAE Transactions* 103(2): 485–502.

Liu, M. and Claridge, D. E. (1992) "The Energy Impact of Combined Solar Radiation/Infiltration/Conduction Effects in Walls and Attics", in *Proceedings of the ASHRAE/DOE/BTECC Conference, December 7–10, 1992, Thermal Performance of the Exterior Envelopes of Buildings V*, Clearwater Beach, FL.

Loveday, D. L. and A. H. Taki (1996) "Convective Heat Transfer Coefficients at a Plane Surface on a Full-scale Building Façade", *Int. J. Heat Mass Transfer* 39(8): 1729–1742.

McClellan, T. M. and Pedersen, C. O. (1997) "Investigation of Outside Heat Balance Models for Use in a Heat Balance Cooling Load Calculation Procedure", *ASHRAE Transactions* 103(2): 469–484.

McQuiston, F. C. and Spitler, J. D. (1992) "Cooling and Heating Load Calculation Manual", Atlanta: ASHRAE.

Mendes, N., Winkelmann, F. C., Lamberts, R. and Philippi, P. C. (2003) "Moisture effects on conduction loads", *Energy and Buildings* 35(7): 631–644.

Mitalas, G. P. and Arseneault, J. G. (1970) "Fortran IV Program to Calculate z-Transfer Functions for the Calculation of Transient Heat Transfer through Walls and Roofs", in *Proceedings of Use of Computers for Environmental Engineering related to Buildings*, Gaithersburg, MD.

Mitalas, G. P. and Stephenson, D. G. (1967) "Room Thermal Response Factors", *ASHRAE Transactions* 3(1): 2.1–2.10.

NFRC (2004a) "Procedure for Determining Fenestration Product Solar Heat Gain Coefficient and Visible Transmittance at Normal Incidence".

NFRC (2004b) "Procedure for Determining Fenestration Product U-Factors".

Novoselac, A., Burley, B. J. and Srebric, J. (2006) "Development of New and Validation of Existing Convection Correlations for Rooms with Displacement Ventilation Systems", *Energy and Buildings* 38: 163–173.

Palyvos, J. A. (2008) "A Survey of Wind Convection Coefficient Correlations for Building Envelope Energy Systems' Modeling", *Applied Thermal Engineering* 28(8–9): 801–808.

Park, J. E., Kirkpatrick, J. R., Tunstall, J. N. and Childs, K. W. (1987) "Building Wall Heat Flux Calculations", *ASHRAE Transactions* 93(2): 120–136.

Pedersen, C. O., Liesen, R. J., Fisher, D. E. and Strand, R. K. (2001) *Toolkit for Building Load Calculations (RP-987)*, Atlanta: ASHRAE.

Rees, S. J., Spitler, J. D. and Haves, P. (1998) "Quantitative Comparison of North American and U.K. Cooling Load Calculation Procedures – Results", *ASHRAE Transactions* 104(2): 47–61.

Riederer, P. (2005) "Matlab/Simulink for Building and HVAC Simulation – State of the Art", in *Proceedings of Building Simulation 2005*, Montreal, Canada: 1019–1026.

Seem, J. E., Klein, S. A., Beckman, W. A. and Mitchell, J. W. (1989) "Transfer Functions for Efficient Calculation of Multidimensional Transient Heat Transfer", *Journal of Heat Transfer* 111(11): 5–12.

Shipp, P. H. (1983) "Natural Convection Within Masonry Block Basement Walls", *ASHRAE Transactions* 89(1).

Sowell, E. F. (1990) "Lights: A Numerical Lighting/HVAC Test Cell", *ASHRAE Transactions* 96(2): 780–786.

Spitler, J. D. (1996) *Annotated Guide to Load Calculation Models and Algorithms*, Atlanta: ASHRAE.

Spitler, J. D. (2009) *Load Calculation Applications Manual*, Atlanta: ASHRAE.

Spitler, J. D. and Fisher, D. E. (1999) "On the Relationship between the Radiant Time Series and Transfer Function Methods for Design Cooling Load Calculations", *HVAC&R Research* 5(2): 125–138.

Spitler, J. D., Fisher, D. E. and Pedersen, C. O. (1997) "The Radiant Time Series Cooling Load Calculation Procedure", *ASHRAE Transactions* 103(2): 503–515.

Strachan, P., Nakhi, A. and Sanders, C. (1995) "Thermal Bridge Assessments", in *Proceedings Building Simulation* 95, Madison, WI, USA: 563–570.

Strand, R. K. and Pedersen, C. O. (1997) "Implementation of a Radiant Heating and Cooling Model into an Integrated Building Energy Analysis Program", *ASHRAE Transactions* 103(1): 949–958.

Underwood, C. P. and Yik, F. W. H. (2004) *Modelling Methods for Energy in Buildings*, Oxford: Blackwell Science.

Virtanen, M., Heimonen, I. and Kohonen, R. (1992) "Application of the Transfer Function Approach in the Thermal Analysis of Dynamic Wall Structures", *Thermal Performance of the Exterior Envelopes of Buildings V*, Clearwater Beach, FL: ASHRAE.

Walton, G. (1983) *Thermal Analysis Research Program Reference Manual*, National Bureau of Standards.

Walton, G. N. (1980) "A New Algorithm for Radiant Interchange in Room Loads Calculations", *ASHRAE Transactions* 86(2): 190–208.

Walton, G. N. (1992) *Computer Programs for Simulation of Lighting/HVAC Interactions*, National Institute of Standards and Technology.

Walton, G. N. (1993) *Computer Programs for Simulation of Lighting/HVAC Interactions*, National Institute of Standards and Technology.

Wang, S. and Chen, Y. (2003) "Transient Heat Flow Calculation For Multilayer Constructions Using A Frequency-Domain Regression Method", *Building and Environment* 38(1): 45–61.

Waters, J. R. and Wright, A. J. (1985) "Criteria for the Distribution of Nodes in Multilayer Walls in Finite-difference Thermal Modelling", *Building and Environment* 20(3): 151–162.

Wetter, M. (2006) "Multizone Building Model for Thermal Building Simulation in Modelica", *5th International Modelica Conference*, Vienna, Austria: 517–526.

WinDat (2009) "WIS Software", Retrieved 17 March 2009, from http://www.windat.org/wis/html/index.html .

Wright, J. L. (1996) "A Correlation to Quantify Convective Heat Transfer between Vertical Window Glazings", *ASHRAE Transactions* 102(1): 940–946.

Wright, J. L. (2008) "Calculating Center-Glass Performance Indices of Glazing Systems with Shading Devices", *ASHRAE Transactions* 113(2): 199–209.

Wright, J. L. and Kotey, N. A. (2006) "Solar Absorption by Each Element in a Glazing/Shading Layer Array", *ASHRAE Transactions* 112(2): 3–12.

Wright, J. L., Collins, M. R., Kotey, N. A. and Barnaby, C. S. (2009) *Improved Cooling Load Calculations for Fenestration with Shading Devices*, Final Report (ASHRAE 1311-RP), Atlanta: ASHRAE.

Yazdanian, M. and Klems, J. (1994) "Measurement of the Exterior Convective Film Coefficient for Windows in Low-Rise Buildings", *ASHRAE Transactions* 100(1): 1087–1096.

York, D. A. and Cappiello, C. C. (1981) *DOE-2 Engineers Manual (Version 2.1A)*, Lawrence Berkeley Laboratory and Los Alamos National Laboratory.

Recommended reading

The following references treat various aspects of room and zone simulation. A couple general works, not specific to a particular approach are Davies (2004), which gives a thorough treatment of the theory of building heat transfer, mainly from an analytical perspective; and Spitler (1996), which is an annotated bibliography of load calculation models and algorithms.

For the heat balance model, the most comprehensive treatment is probably the Pedersen *et al.* (2001) work, called a "toolkit" – it consists of a complete set of Fortran routines and a complete set of documentation, which, taken together form the basis of a heat balance model. Emphasis is given to clarity rather than speed. The heat balance model is also treated in some detail by McQuiston *et al.* (2005). Clarke (2001) gives an overview of building simulation, mostly informed by development of the ESP-r program, and so emphasizing the thermal network model approach. The text by Underwood and Yik (2004) treats several topics given scant coverage here, in particular numerical methods and lumped parameter methods for transient conduction.

For transfer function models, the York and Cappiello (1981) manual for the DOE-2 program is fairly comprehensive and Spitler (2009) describes the radiant time series method in detail.

Clarke, J. A. (2001) *Energy Simulation in Building Design*, Oxford: Butterworth-Heinemann.
Davies, M. G. (2004) *Building Heat Transfer*, Chichester, West Sussex: Wiley.
McQuiston, F. C., Parker, J. D. and Spitler, J. D. (2005) *Heating, Ventilating, and Air Conditioning Analysis and Design*, 6th ed., New York: John Wiley and Sons.
Pedersen, C. O., Liesen, R. J., Fisher, D. E. and Strand, R. K. (2001) *Toolkit for Building Load Calculations (RP-987)*, Atlanta: ASHRAE.
Spitler, J. D. (1996) *Annotated Guide to Load Calculation Models and Algorithms*, Atlanta: ASHRAE.
Spitler, J. D. (2009) *Load Calculation Applications Manual*, Atlanta: ASHRAE.
Underwood, C. P. and Yik, F. W. H. (2004) *Modelling Methods for Energy in Buildings*, Oxford: Blackwell Science.
York, D. A. and Cappiello, C. C. (1981) *DOE-2 Engineers Manual (Version 2.1A)*, Lawrence Berkeley Laboratory and Los Alamos National Laboratory.

Activities

5.1 A lightweight wall made with an exterior insulation finish system (EIFS) has the layers shown in Table 5.16. The resulting conduction transfer function coefficients are shown in Table 5.17. If the exterior wall surface temperatures are those given in Table 5.18 for Atlanta, and the interior surface temperature is held constant at 22 °C, compute the conduction heat flux at the interior surface for each hour of the day. Assume the day repeats.

5.2 A lightweight wall made with an exterior insulation finish system (EIFS) has the layers shown in Table 5.16. The resulting conduction transfer function coefficients are shown in Table 5.17. If the exterior wall surface temperatures are those given in Table 5.18 for Seattle, and the interior surface temperature is held constant at 22 °C, compute the conduction heat flux at the interior surface for each hour of the day. Assume the day repeats.

5.3 A lightweight wall made with an exterior insulation finish system (EIFS) has the layers shown in Table 5.16. The resulting conduction transfer function coefficients are shown in Table 5.17. If the exterior wall surface temperatures are those given in Table 5.18 for Denver, and the interior surface temperature is held constant at 22 °C, compute the conduction heat flux at the interior surface for each hour of the day. Assume the day repeats.

5.4 A heavyweight brick-and-block wall is made up of the layers shown in Table 5.1. The resulting conduction transfer function coefficients are shown in Table 5.2. If the exterior

Table 5.16 Layer descriptions for EIFS wall

Layer	*Thickness*	*Density*	*Conductivity*	*Specific heat*
	(mm)	*(kg/m³)*	*(W/mK)*	*(J/kg K)*
Stucco	12.7	1856	0.69	840
Extruded Polystyrene	50.0	32	0.03	1500
Equivalent Homogeneous Layer – Steel studs and insulation	89.0	65	0.09	550
Plaster board and skim	12.7	800	0.43	420

Table 5.17 Conduction transfer function coefficients for EIFS wall

j	X_j	Y_j	Z_j	$\square_j$
0	6.18619E+00	1.25614E–01	2.17049E+00	
1	–6.332701E+00	2.043432E–01	–1.949532E+00	8.691221E–02
2	4.842522E–01	7.784391E–03	1.167875E–01	

Table 5.18 Wall surface temperatures (°C)

Local Time	*Atlanta, July 21, SE-facing*	*Seattle, September 21, SW-facing*	*Denver, December 21, S-facing*
1	29.3	20.0	6.5
2	28.7	19.5	5.9
3	28.2	19.1	5.4
4	27.8	18.7	5.0
5	27.4	18.4	4.9
6	27.1	18.3	5.2
7	27.7	18.5	5.8
8	44.6	21.0	19.2
9	55.2	23.6	40.7
10	60.8	26.2	53.7
11	62.5	32.8	61.9
12	60.8	46.3	65.9
13	56.3	58.6	65.2
14	49.4	68.0	60.0
15	44.4	73.5	49.0
16	44.0	73.8	28.0
17	43.2	67.5	16.4
18	41.6	51.8	15.0
19	39.2	25.7	13.3
20	35.8	24.5	11.6
21	33.1	23.3	10.2
22	31.9	22.2	9.0
23	30.9	21.3	7.9
24	30.0	20.5	7.2

wall surface temperatures are those given in Table 5.18 for Atlanta, and the interior surface temperature is held constant at 22 °C, compute the conduction heat flux at the interior surface for each hour of the day. Assume the day repeats.

5.5 A heavyweight brick-and-block wall is made up of the layers shown in Table 5.1. The resulting conduction transfer function coefficients are shown in Table 5.2. If the exterior wall surface temperatures are those given in Table 5.18 for Seattle, and the interior surface temperature is held constant at 22 °C, compute the conduction heat flux at the interior surface for each hour of the day. Assume the day repeats.

5.6 A heavyweight brick-and-block wall is made up of the layers shown in Table 5.1. The resulting conduction transfer function coefficients are shown in Table 5.2. If the exterior wall surface temperatures are those given in Table 5.18 for Denver, and the interior surface temperature is held constant at 22 °C, compute the conduction heat flux at the interior surface for each hour of the day. Assume the day repeats.

5.7 For problem 5.1 and 5.4, compute the surface-to-surface U-factors and then compute the interior conduction heat flux at each hour. For each wall, quantify the delay and damping effects of the thermal mass.

5.8 Consider the zone shown in Figure 5.6, used for Example 5.7. If the south wall and window were coated with aluminum foil, the emissivity might be reduced to 0.1. If this were done, determine the radiative heat flux leaving each surface. How much of a reduction in heat loss might one expect, based solely on the inter-surface radiation?

5.9 Consider the zone shown in Figure 5.6, used for Example 5.7. If under hot summer conditions, the surface temperatures for surfaces 1–7 are 23.0, 23.2, 23.7, 22.5, 25.9, 30.1, and 28.2, respectively, what will the radiative heat fluxes leaving each surface be?

5.10 For a very lightweight wall, with negligible thermal mass, a steady-state surface heat balance will give an approximate value of the exterior surface temperature and the conduction heat flux into the surface. Consider the wall in Table 5.1 if it is painted such that the solar absorptivity is 0.9 and the emissivity is 0.9. The weather conditions for Atlanta are shown in Tables 5.19 (incident solar radiation on the wall) and 5.20 (hourly air temperature). Assume the sky temperature is 6 °C below the outdoor air temperature, the wind speed is a constant 4 m/s with the surface on the windward side of the building, and the building is a low-rise building. Also assume the interior surface temperature of the wall is constant at 22 °C. Find the exterior surface temperature and conductive heat flux at hour 15.

5.11 Find the exterior surface temperature and conduction heat flux at hour 15 for the situation described in Problem 5.10, except for the Seattle weather conditions.

Table 5.19 Incident solar radiation on vertical walls (W/m^2)

Local Time	*Atlanta, July 21, SE-facing*	*Seattle, September 21, SW-facing*	*Denver, December 21, S-facing*
1	0.0	0.0	0.0
2	0.0	0.0	0.0
3	0.0	0.0	0.0
4	0.0	0.0	0.0
5	0.0	0.0	0.0
6	0.0	0.0	0.0
7	10.7	0.0	0.0
8	326.4	40.0	230.8
9	514.7	72.6	606.8
10	602.0	99.0	815.0
11	609.7	198.2	930.0
12	549.4	423.8	966.2
13	432.6	627.4	925.8
14	273.8	784.4	806.3
15	158.6	873.3	592.0
16	139.1	873.9	201.8
17	122.1	760.9	0.0
18	99.3	475.7	0.0
19	69.5	0.1	0.0
20	25.9	0.0	0.0
21	0.0	0.0	0.0
22	0.0	0.0	0.0
23	0.0	0.0	0.0
24	0.0	0.0	0.0

5.12 Find the exterior surface temperature and conduction heat flux at hour 15 for the situation described in Problem 5.10, except for the Denver weather conditions.

5.13 Write a computer program or develop a spreadsheet to find the exterior surface temperature and conduction heat flux at all hours for the situation described in Problem 5.10.

5.14 Write a computer program or develop a spreadsheet to find the exterior surface temperature and conduction heat flux at all hours for the situation described in Problem 5.11.

5.15 Write a computer program or develop a spreadsheet to find the exterior surface temperature and conduction heat flux at all hours for the situation described in Problem 5.12.

5.16 Table 5.21 gives beam and diffuse solar radiation incident on a southeast facing vertical window in Atlanta on July 21 along with the corresponding incidence angle. For each hour of the day, find the transmitted and absorbed solar radiation in both panes of a double pane window with a low e coating – the second window in Table 5.8.

5.17 Table 5.21 gives beam and diffuse solar radiation incident on a southeast facing vertical window in Atlanta on July 21 along with the corresponding incidence angle. For each hour of the day, find the transmitted and absorbed solar radiation in both panes of a double pane window with clear panes – the third window in Table 5.8.

5.18 For the window of Problem 5.16, determine the outer pane and inner pane surface temperatures for each hour and the convective flux and radiative flux into the room for each hour. The outdoor air temperatures are those shown for Atlanta in Table 5.20. For purposes of solving this problem, you may assume that the distance between the panes is 13 mm; the sky temperature is 6 °C below the outdoor air temperature; the building is low-rise, so the Yazdanian and Klems (1994) model is applicable for exterior convection; the interior zone air temperature is fixed at 21.5 °C and the zone MRT is fixed at 21.8 °C; the interior convection coefficient is 4.18 W/m^{2}K; and all window panes have an emissivity of 0.84 except the low-e coated pane, which has an emissivity of 0.033 on the inside of the outer pane.

5.19 Work Problem 5.18, except use the window from Problem 5.17. In this case, you may assume that all window panes have an emissivity of 0.84.

Table 5.20 Outdoor air temperatures (°C)

Local Time	*Atlanta, July 21*	*Seattle, September 21*	*Denver, December 21*
1	29.3	22.3	6.5
2	28.7	21.8	5.9
3	28.2	21.3	5.4
4	27.8	20.9	5.0
5	27.4	20.6	4.9
6	27.1	20.4	5.2
7	27.1	20.5	5.8
8	27.4	21.0	7.0
9	28.0	21.8	8.7
10	29.0	23.0	10.6
11	30.3	24.5	12.8
12	31.8	26.2	14.8
13	33.5	27.9	16.3
14	34.9	29.2	17.3
15	36.0	30.1	17.7
16	36.7	30.5	17.3
17	36.8	30.3	16.4
18	36.4	29.7	15.0
19	35.6	28.6	13.3
20	34.4	27.3	11.6
21	33.1	26.0	10.2
22	31.9	24.8	9.0
23	30.9	23.8	7.9
24	30.0	23.0	7.2

Table 5.21 Solar radiation incident on a southeast-facing surface in Atlanta, July 21

Hour	*Beam (W/m²)*	*Diffuse (W/m²)*	*Incidence Angle (°)*
1	0.0	0.0	133.85
2	0.0	0.0	128.31
3	0.0	0.0	109.73
4	0.0	0.0	98.06
5	0.0	0.0	86.98
6	0.0	0.0	76.84
7	8.0	2.7	68.09
8	247.1	79.3	61.37
9	385.8	128.8	57.43
10	441.4	160.6	56.86
11	430.5	179.1	59.76
12	363.0	186.4	65.66
13	248.6	184.0	73.82
14	99.8	173.9	83.58
15	0.0	158.6	94.40
16	0.0	139.1	105.90
17	0.0	122.1	117.83
18	0.0	99.3	129.93
19	0.0	69.5	141.94
20	0.0	25.9	153.29
21	0.0	0.0	162.27
22	0.0	0.0	163.92
23	0.0	0.0	156.62
24	0.0	0.0	145.73

Answers

5.1 At hour 12, conduction heat flux is 14.71 W/m²; at hour 18, conduction heat flux is 7.67 W/m².

5.2 At hour 12, conduction heat flux is 5.50 W/m²; at hour 18, conduction heat flux is 15.05 W/m².

5.3 At hour 12, conduction heat flux is 14.98 W/m²; at hour 18, conduction heat flux is –1.86 W/m².

5.4 At hour 12, conduction heat flux is 8.34 W/m²; at hour 18, conduction heat flux is 15.06 W/m².

5.5 At hour 12, conduction heat flux is 0.54 W/m²; at hour 18, conduction heat flux is 11.93 W/m².

5.6 At hour 12, conduction heat flux is –3.75 W/m²; at hour 18, conduction heat flux is 11.06 W/m².

5.8 The radiation heat flux leaving the north wall is 3.60 W/m².

5.9 The radiation heat flux leaving the north wall is –8.44 W/m².

5.10 At hour 15, the exterior surface temperature is 43.30 °C and the conduction heat flux is 7.88 W/m².

5.11 At hour 15, the exterior surface temperature is 72.68 °C and the conduction heat flux is 18.75 W/m².

5.12 At hour 15, the exterior surface temperature is 48.99 °C and the conduction heat flux is 9.99 W/m².

5.16 At hour 11, the transmitted solar radiation is 197.05 W/m²; the absorbed solar radiation in the outer pane is 129.67 W/m² and the absorbed solar radiation in the inner pane is 36.85 W/m².

5.18 At hour 11, the temperature of the outer pane is 37.4 °C and the temperature of the inner pane is 27.8 °C. The inward longwave radiative heat flux is 38.38 W/m² and the inward convective heat flux is 20.17 W/m².

6 Ventilation performance prediction

Jelena Srebric

Scope

The chapter first presents an introduction to the state of the art in modeling and simulation of airflow, air temperature, and contaminant concentrations in buildings. Following the introduction of fundamental modeling concepts, the chapter continues with simulation examples for different ventilation systems, such as mechanical, natural, and hybrid ventilation. Overall, the entire chapter has outlined specific recommendations for potential new users to produce reliable ventilation performance predictions.

Learning objectives

To develop an understanding of the principles of ventilation performance predictions required for thermal comfort and indoor air quality studies.

Key words

Ventilation performance/air distribution/computational fluid dynamics/multi-zone airflow network models

Introduction

Ventilation systems are responsible for delivering air in air-conditioned spaces to accomplish the ventilation as well as heating/cooling tasks. As a result, ventilation systems have significant impact on thermal comfort and indoor air quality (IAQ). For example, increased ventilation is specified as one of the criteria in the LEED (Leadership in Energy and Environmental Design) certification process under indoor environmental quality credits (USGBC 2005). In the same group of credits, thermal comfort is also included, and ventilation provides thermal comfort to support the well-being of occupants. In the past decade, it has been found that the ventilation systems are associated with occupational health, perceived air quality including thermal comfort, and occupants' productivity (Fang *et al.* 1998; National Research Council 2006; NIOSH 1997;Wargocki *et al.* 2002). This central role of ventilation systems in achieving high indoor environmental quality requires reliable predictions of ventilation system performance. State-of-the-art prediction approaches include multi-zone airflow network and Computational Fluid Dynamics (CFD) simulation models. Both of these models are capable of calculating indoor air parameters such as airflow rates, temperature, and contaminant concentrations in buildings with different ventilation systems. Nevertheless, important fundamental differences between the two simulation techniques exist and define their accuracy and reliability in predicting ventilation performance. Knowledge of the strengths and weaknesses of both multi-zone airflow network and CFD models is the key for successful use of these simulation techniques for design and operation of ventilation systems.

Modeling approaches

Before simulation models, such as multi-zone airflow network and CFD, became widely accepted tools for ventilation performance predictions, semi-empirical equations were the standard design approach. These equations played an important historic role and the most successful ones are still in use. For example, equations for estimating infiltration flow rates are included in the multi-zone airflow network models. The use of semi-empirical equations, multi-zone airflow network or CFD models are complementary to each other because each of these models resolves the airflow patterns with different level of detail. More detailed models are always more time-consuming, and, therefore, appropriate models have to be chosen based on the particular ventilation design problem.

Overview of semi-empirical equations

The simplest models for ventilation performance prediction are the semi-empirical equations, which are also called single zone models. These equations represent a class of analytical models calibrated with empirical coefficients to predict bulk ventilation airflow rates through different types of openings. For openings at the building enclosure, such as windows or doors, the airflow is driven by the pressure difference between the indoor and outdoor environments. This pressure difference is caused by the wind airflow around a building, which is called the wind effect, as well as the temperature difference between indoor and outdoor environments, which is called the *stack effect*. It is important to notice that this assumption of combined *wind* and *stack effect* is strictly true only in the absence of heating, ventilating and air-conditioning (HVAC) systems, which cause indoor pressure changes depending on the system balancing and maintenance. As a result, this assumption is reliable for ventilation performance predictions through building openings in a natural ventilation regime.

Assuming the airflow through openings is proportional to the square root of the pressure differential across the openings, the total airflow rate due to the combined *wind* and *stack effect* can be approximated as following:

$$\dot{Q}_{tot} = \sqrt{\dot{Q}_w^2 + \dot{Q}_s^2}, \tag{6.1}$$

where $\dot{Q}_{tot}$ is the total volumetric flow rate [m^3/s], $\dot{Q}_w$ and $\dot{Q}_s$ are volumetric flow rates due to the *wind* and *stack* effect respectively [m^3/s]. For single-sided ventilation that represents natural airflow through a single opening for a single space in a building, the following simple correlations can be used (Allard 1998; CIBSE 2005):

$$\dot{Q}_w = 0.05AU_H, \tag{6.2}$$

$$\dot{Q}_s = 0.2A\sqrt{\frac{gh\Delta T}{T_{avg}}}, \tag{6.3}$$

where A is the opening area [m^2], U_H is the wind velocity at the building height H [m/s], g is the gravitational acceleration [m/s^2], h is the opening height [m], ΔT is the inside–outside temperature difference [K], and T_{avg} is the average of inside and outside temperature [K]. These simple equations are good first estimates for a single space (single zone) with bi-directional flow through a single opening. Nevertheless, natural ventilation scenarios for multiple spaces result in much more complex airflow patterns that require much more complex equations or use of tools such as multi-zone airflow network or CFD models.

Another airflow pattern that is often predicted with semi-empirical equations is building infiltration. One of the first widely used infiltration models was developed at the Lawrence

Berkeley National Laboratory (LBNL), and it is called LBNL model (Sherman and Grimsrud 1980). This model was similar to other formulations at that time because it used a power law relationship between the airflow rate and a pressure difference across the building enclosure. Nevertheless, the model was unique because it used the effective leakage area for building openings, which could be obtained from fan pressurization tests. The modern version of this model is known as the ASHRAE basic model (American Society of Heating, Refrigerating and Air-Conditioning Engineers (ASHRAE) 2005, Chapter 27):

$$\dot{Q}_{\text{inf}} = \frac{A_e}{1000}\sqrt{C_s \Delta T + C_w U^2}, \tag{6.4}$$

where $\dot{Q}_{inf}$ is the total volumetric flow rate of infiltration [m^3/s], A_e is the effective air leakage area [cm^2], ΔT is the average indoor–outdoor temperature difference for the time interval of calculation [K], U is the average wind speed measured at local weather station for the time interval of calculation [m/s], C_s is the stack coefficient [(L/s)^2/(cm^4·K)], and C_w is the wind coefficient [(L/s)^2/(cm^4 (m/s)^2)]. The stack and wind coefficients are available in tables (ASHRAE 2005, Chapter 27). The simplicity of this model has made it very popular, and important model assumptions are often overlooked. Among these assumptions are that the neutral pressure level is half of the building height, and the effects of wind directionality are averaged over the time period of interest. In real buildings, the neutral pressure level, where the indoor and outdoor pressures are equal, is often at different elevations depending on indoor and outdoor environmental conditions. In addition, the time averaging of wind directionality neglects the changes in leakage area available for infiltration as the wind impinges on different building surfaces.

The pressure created on the exterior surface of a building by wind has a very complex distribution that depends on many factors, such as wind direction, wind velocity, air density, building geometry, surface orientation, and nature of the surrounding environment (Etheridge and Sandberg 1996). Many of these factors are very hard to predict because of the stochastic nature of the wind itself. Usually, the wind pressures are positive on the windward surface and negative on the leeward surface of a building. The time averaged static pressures over the building surface are proportional to the velocity or dynamic pressure of undisturbed air stream. The pressure due to wind at a reference height (H) is calculated by the following expression:

$$p_v = \frac{1}{2} \cdot \rho_a \cdot U_H^2, \tag{6.5}$$

where U_H is the reference wind velocity at a reference height from the ground [m/s], ρa is the ambient air density [kg/m^3], and p_v is often referred to as the stagnation air pressure [Pa]. The reference height is typically 10 m above the ground. The difference between the pressure on a building surface and the atmospheric air pressure, at the same height for an otherwise undisturbed wind approaching the building, is defined as following:

$$p_w = C_p \cdot p_v = C_p \cdot \frac{\rho_a \cdot v^2}{2}, \tag{6.6}$$

where p_w is the equivalent static air pressure on the building surface [Pa], and C_p is the wind pressure coefficient [–]. This coefficient defines what portion of the wind kinetic energy is transformed to pressure energy on the vertical building surface. An overview of wall averaged wind pressure coefficients for different types of buildings is available in the form of databases (Blevis 1984; Hagentoft 2001). These databases with appropriate figures describing each case resulted from extensive literature reviews. As an example, Table 6.1 shows averaged values of the wind pressure coefficients for different surfaces of low-rise buildings. In this example, the incident wind angle, θ, at the windward surface 1, changes from 0° to 45°, and results in different wind pressure coefficients.

In addition to databases, wind pressure coefficients are also represented with correlations resulting from wind tunnel experiments. The literature on wind pressure coefficients is extensive, and the simplest correlations can be found in the ASHRAE Handbook of Fundamentals (2005, Chapter 16). These correlations are typically represented as a function of wind incident angle, such as the following equation for low-rise buildings (Swami and Chandra 1988):

$$C_p = \ln\left(\begin{array}{l} 1.248 - 0.703\sin\left(\dfrac{\theta}{2}\right) - 1.175\sin^2\theta + 0.131\sin^3(2\theta G) + \\ +0.769\cos\left(\dfrac{\theta}{2}\right) + 0.07G^2\sin^2\left(\dfrac{\theta}{2}\right) + 0.717\cos^2\left(\dfrac{\theta}{2}\right) \end{array}\right), \tag{6.7}$$

where θ is the incident wind angle [rad], and G is a shape factor [–], calculated by taking the natural logarithm of the ratio of the windward wall length to the length of the adjacent wall. This and other wind pressure coefficient equations can be used to calculate wind pressures on the building surfaces as input parameters for a multi-zone airflow network model.

Overall, even though the semi-empirical equation models incorporate numerous assumptions, they can provide good estimates of bulk airflow properties when the domain of interest such as a building can be treated as a single zone. For more complex situations, the semi-empirical models can be used for specifying boundary conditions, such as infiltration rates, for the multi-zone airflow network models.

Overview of multi-zone airflow network modeling approaches

The simplest mathematical models for calculating the distributions of indoor airflows, temperature, and contaminant concentrations are probably the multi-zone airflow network models. These models solve the conservation equation for mass, energy, and species concentration based on a number of approximations. The models divide a given space or a building into

Table 6.1 Wind pressure coefficients *Cp* for low-rise buildings up to three storeys in height

	WALL	
Location	*Wind angle: 0°*	*Wind angle: 45°*
Surface 1	0.4	0.1
Surface 2	–0.2	–0.35
Surface 3	–0.3	0.1
Surface 4	–0.3	–0.35
	ROOF, pitch angle – less than 10°	
Location	*Wind angle: 0°*	*Wind angle: 45°*
Front	–0.6	–0.5
Rear	–0.6	–0.5
	ROOF, pitch angle – between 10–30°	
Location	*Wind angle: 0°*	*Wind angle: 45°*
Front	–0.35	–0.45
Rear	–0.35	–0.45
	ROOF, pitch angle – greater than 30°	
Location	*Wind angle: 0°*	*Wind angle: 45°*
Front	0.3	–0.5
Rear	–0.5	–0.5

Source: Hagentoft 2001.

a number of volumes (zones). Then, the multi-zone airflow network models solve the mass, energy, and concentration equations for each zone to obtain a solution for specific boundary conditions and the perfect mixing assumption. Since the multi-zone airflow network models are simple and easy to understand, they are preferred models for design problems as long as modeling results are reliable.

A popular type of multi-zone airflow network models determines airflow within a space using Bernoulli's equation (Walton 1989). This multi-zone airflow network approach is effective for the simultaneous analyses of HVAC system, infiltration, and multi-room airflow problems. At present, the two most widely used multi-zone airflow network models are CONTAM and COMIS, which are developed by the National Institute of Standards and Technology (NIST) and Lawrence Berkeley National Laboratory (LBNL), respectively (Dols and Walton 2002; Feustel and Smith 1997). Both of these simulation tools are available on the shareware basis and have user-friendly interfaces accompanied with detailed user manuals. The comprehensive fundamentals are available in the manuals, and the airflow mass balance applied to each zone in a building for a transient problem is given in Equation (6.8):

$$\frac{\partial m_i}{\partial t} = \rho_i \frac{\partial V_i}{\partial t} + V_i \frac{\partial \rho}{\partial t} = \sum_j F_{j,i} + F_i, \tag{6.8}$$

where m_i is the mass of air in zone i [kg], t is time [s], ρ is the density [kg/m^3], V_i is the volume of zone i [m^3], $F_{j,i}$ is the airflow rate [kg/s] between zone j and zone i; positive value indicates flow from j to i and negative value indicates flow from i to j, and F_i is the airflow rate [kg/s] for non-flow processes that could add or remove significant quantities of air from the zone, usually sources and sinks specified as boundary conditions. Furthermore, flow paths (fluid flow components) are the connections between zones that allow airflow in one direction for small openings, and in both directions for large openings. The flow path equations represent fluid flow components such as doors, windows, and structural leakages that connect zones and allow airflow movements.

Figure 6.1(a) shows an office building that has a detailed multi-zone airflow network model. All four building floors are modeled, and Figure 6.1(b) presents the first floor in the graphical interface of a multi-zone airflow network model. Each separate space is a zone with appropriate flow paths, so the model is similar to the floor plan. The lines in the model define zone boundaries, while each dot represents either a flow path that connects zones or mechanical system air supplies and exhausts.

The airflow through a flow path, also known as a fluid flow component, is usually driven by pressure difference between connected zones. Flow path equations, based on one-dimensional fluid mechanics laws, are applied to quantify the airflow movement through the flow paths. For

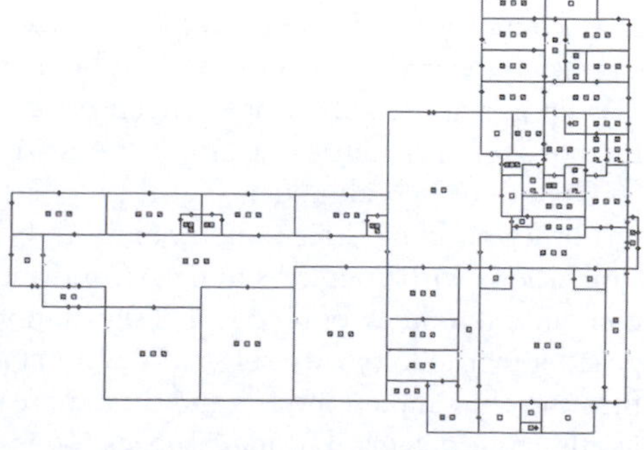

Figure 6.1 (a) Four-storey office building and (b) a representation of the first floor in a multi-zone airflow network simulation tool.

the pressure driven flow paths, the airflow rate from zone j to zone i, $F_{j,i}$ [kg/s], is a function of the pressure difference along the flow path, ΔP:

$$F_{j,i} = f(P_j - P_i) = f(\Delta P), \tag{6.9}$$

where P_j and P_i are the pressures at zone j and i, [Pa], which include static, dynamic and stack pressures for each zone. The following power-law equation is widely used to present the function $f(\Delta P)$ in Equation (6.9):

$$F_{j,i} = C \cdot (\Delta P)^n \tag{6.10}$$

where C is the flow coefficient (discharge coefficient) of the flow path[-], and n is the flow exponent of the flow path[-], usually 0.5 for large openings, 1.0 for narrow openings, and 0.6–0.7 for typical infiltration openings (Dols and Walton 2002).

The remaining key problems in effective use of multi-zone airflow network models are related to the perfect mixing assumption and flow coefficients of large openings. A recent experimental study has shown that the perfect mixing assumption for the temperature field is acceptable as long as the temperature gradient is relatively small, which is defined in the literature (Wang and Chen 2008). However, the method is unreliable, if applied to a single space containing several different sub-volumes with non-uniform indoor parameter distribution such as temperature field with displacement ventilation or contaminant concentration in a space with a point contaminant source (Srebric *et al.* 2008). Under these conditions, a multi-zone airflow network model can easily result in inaccurate predictions of contaminant flow rate intensity and direction from a point source.

Furthermore, for large openings, such as doors and windows, the airflow is typically bi-directional, and the fundamental nature of bi-directional airflow is based on two opposing flow forces creating unstable balance that changes dynamically. As a result, Equation (6.9) cannot accurately account for airflow through large openings because it relies on flow coefficients that do not have a general equation. Application of multi-zone airflow network models to a new complex geometry requires a more sophisticated simulation or experiment to determine new flow coefficients. Experimental and numerical methods are emerging to bridge the problems of multi-zone airflow network models because they are not only computationally inexpensive, but also relatively easy to set up when compared to CFD models (Srebric *et al.* 2008).

Hybridization of multi-zone airflow network models by coupling them with energy simulations and CFD is an important trend to improve the multi-zone airflow network model accuracy, but also to provide the necessary boundary conditions to energy simulation and CFD models (Clarke 2001; Hensen 1999; Novoselac 2004). The most widely used hybrid simulation tools are ESP-r and EnergyPlus developed by the University of Strathclyde, Glasgow and the U.S. Department of Energy, respectively (Clarke 2001; DOE 2008). EnergyPlus is available on the shareware basis, and can be obtained for implementation of custom simulation modules upon signing a developer's license agreement. Each of these hybrid simulation tools has included energy, multi-zone airflow network, and CFD simulation modules.

The use of hybrid models is still not a standard practice, but it appears to be a clear trend for future building design and operation. For example, a study demonstrated that the hybrid simulations with coupled simulation tools increase simulation accuracy, while using minimal computer resources for practical design applications (Djunaedy *et al.* 2004). The study also proposed a methodology for selection of an appropriate simulation strategy, which, in principle, first uses the simplest models and then increases the complexity of simulations to resolve building design problems. The main suggestion is to start with energy simulation models that assume uniform indoor parameters, then include multi-zone airflow network models, if necessary. Only use CFD as a last resort because it is the most computationally intensive simulation model. As the available computational power steadily increases, the problems associated with intense CFD computations will disappear. CFD will still require much more intense work on input and model development; therefore, multi-zone airflow network models will still be an important simulation option. Nevertheless, problems in determining reliable flow path equations for the multi-zone airflow network models stimulated the development of CFD simulation models for ventilation performance predictions.

Overview of CFD modeling approaches

CFD models can provide detailed indoor parameters of design interests such as velocities, temperatures, and contaminant concentrations in a reliable and fast manner. These simulation models are recommended as an inexpensive approach compared to the measurements and a more detailed approach compared to the multi-zone airflow network models. In principle, a CFD model subdivides the interior space into a number of cells. For each cell, the conservation of mass is satisfied so that the sum of mass flows into or out of a cell from all its neighbours is balanced to zero. Similarly the exchange of momentum from the flow into or out of a cell must be balanced in each direction with pressure, gravity, viscous shear, and energy transport by turbulent eddies. For transitional and fully turbulent flows, which normally exist in room flow, the challenge is to properly describe the transport of momentum and energy. The nature of turbulence is complex and not yet fully understood. The turbulent flow is three-dimensional and random with many vortices (turbulent eddies) that enhance mixing in the flow field. The mixing decreases velocity gradients and dissipates kinetic energy of the fluid stream. The dissipation is irreversible, and kinetic energy is transformed to the internal energy of the fluid. To study the impact of turbulent mixing on thermal comfort and indoor air quality, partial differential equations that govern turbulent flow must be solved.

Indoor airflow is typically modelled as turbulent incompressible flow of Newtonian fluid, and the following transport equations describe the flow in the tensor notation.

Mass continuity:

$$\frac{\partial \rho}{\partial t}+\frac{\partial \rho \tilde{u}_i}{\partial x_i}=0, \tag{6.4}$$

where ρ is the air density [kg/m^3], $\tilde{u}_i$ is the instantaneous velocity component in x_i-direction [m/s], x_i is the coordinate (for i=1, 2, 3, x_i corresponds to three perpendicular axes) [m], and t is the time [s].

Momentum conservation:

$$\frac{\partial \rho \tilde{u}_i}{\partial t}+\frac{\partial \rho \tilde{u}_i \tilde{u}_j}{\partial x_j}=-\frac{\partial \tilde{p}}{\partial x_i}+\frac{\partial}{\partial x_j}\left[\mu\left(\frac{\partial \tilde{u}_i}{\partial x_j}+\frac{\partial \tilde{u}_j}{\partial x_i}\right)\right]+\rho\beta\left(T_o-\tilde{T}\right)g_i, \tag{6.5}$$

where $\tilde{u}_j$ is the instantaneous velocity component in x_j-direction [m/s], $\tilde{p}$ is the instantaneous pressure [Pa], μ is the molecular viscosity [Pa s], β is the thermal expansion coefficient of air [1/K], To is the temperature of a reference point [K], $\tilde{T}$ is the instantaneous temperature [K], and g_i is the gravity acceleration in i-direction [m/s^2].

Energy conservation:

$$\frac{\partial \rho \tilde{T}}{\partial t}+\frac{\partial \rho \tilde{u}_j \tilde{T}}{\partial x_j}=\frac{\partial}{\partial x_j}\left(\frac{k}{c_p}\frac{\partial \tilde{T}}{\partial x_j}\right)+S_T, \tag{6.6}$$

where k is the thermal conductivity of air [W/(m K)], S_T is the thermal source, where S_T=(Heat flux for the source/cp) [W (kg K)/J], and c_p is the specific heat at constant pressure [J/(kg K)]. In addition, $\frac{c_p}{k}=\frac{1}{k/c_p}=\frac{\mu}{\mu k/c_p}=\frac{\mu}{\mathrm{Pr}}$, where *Pr* is the *Prandtl number*.

Species concentration conservation:

$$\frac{\partial \rho \tilde{c}}{\partial t}+\frac{\partial \rho \tilde{u}_j \tilde{c}}{\partial x_j}=\frac{\partial}{\partial x_j}\left(\rho D \frac{\partial \tilde{c}}{\partial x_j}\right)+S, \tag{6.7}$$

where $\tilde{c}$ is the instantaneous species concentration [m_c^3/m_{air}^3], D is the molecular diffusion coefficient for the species [m_c^2/s], and S is the species source [m_c^3/s]. In addition, $\rho D=\frac{1}{1/\rho D}=\frac{\mu}{\mu/\rho D}=\frac{\mu}{Sc}$, where *Sc* is the *Schmidt number*.

Ventilation performance predictions with CFD use the Boussinesq approximation for thermal buoyancy as shown in Equation (6.5). This approximation makes air density a constant in all the governing equations and considers the buoyancy influence on air movement by the temperature difference. The conservation equations (Equations (6.4)–(6.7)) must be solved numerically. The solution provides the field distributions of pressure, air temperature, velocity, and contaminant concentrations. Depending on the numerical solution procedure of the conservation equations, there are three CFD methods: (1) direct numerical simulation (DNS), (2) large-eddy simulation (LES), and (3) *Reynolds averaged Navier-Stokes* (RANS) equations with turbulence models. DNS solves the highly reliable *Navier-Stokes* equations without approximations. Therefore, DNS requires a grid resolution as fine as the *Kolmogorov microscale* of small eddies for which the inertia and viscous effects are of equal magnitude. Based on dimensional analyses, the ratio of smallest to largest length scales in turbulent flows is in order of $Re^{3/4}$, where *Re* is the *Reynolds number*. For three-dimensional flows, such as indoor airflow, the approximate number of cells or grid points becomes $Re^{9/4}$ (Versteeg and Malalasekera 2007). Since the *Reynolds number* for a typical indoor airflow pattern is approximately 10^5, the required grid number for solving the three-dimensional airflow pattern is approximately 10^{11} to 10^{12}. Current super-computers have an approximate grid resolution as fine as 10^8, and the new PetaFlops super-computers are expected to allow 10^9 and 10^{12} grid resolution in the near future (Nakahashi 2008). In addition, the DNS method requires very small time steps, where the ratio of smallest to largest time scales is approximated as $Re^{1/2}$ (Versteeg and Malalasekera 2007). Overall, current computer capacity is still far too small to solve ventilation airflows with the DNS method in design practice.

LES was developed in the early 1970s by Deardorff (1970) for meteorological applications. He separated turbulent motion into large eddies and small eddies. The separation of the two did not significantly affect the evolution of large eddies. LES solves the large-eddy motion by a set of filtered equations governing three-dimensional, time-dependent motion. Turbulent transport approximations are used for small eddies and the small eddies are modeled independently from the flow geometry. The success of LES stems from the fact that the main contribution to turbulent transport comes from large-eddy motion. LES is also a more realistic application of computational effort for building performance simulations than DNS because LES can be performed on a large and fast workstation or computer clusters. Nevertheless, LES is still too time consuming because it calculates time-dependent flow, although the time and space steps can be larger than those for DNS.

The RANS method is the fastest, but potentially the least accurate method. RANS solves ensemble-averaged *Navier-Stokes* equations by using turbulence modeling. In RANS, all unsteadiness is averaged and regarded as part of the turbulence (Ferziger and Peric 2001). The grid number used for the simulation is normally much less than that for LES or DNS. Most importantly, steady flow can be solved as time-independent; therefore, computing costs are less than those for DNS and LES. However, turbulence modeling introduces an additional error into the calculations. Nevertheless, calculations based on RANS are popular and widely used in many building design solutions because they give reasonably good quality results with an affordable computing effort. Even with more detailed calculations such as LES and DNS, the results should be averaged for design assessments of ventilation performance because the current design guidelines employ the steady-state assumption.

The first step in averaging the conservation equations is to decompose instantaneous variables into their mean and fluctuating parts, which is called *Reynolds decomposition*. The mean parts are denoted with capital letters and the fluctuating parts are denoted with a "′" superscript:

$$\tilde{u}_i = U_i + u_i', \qquad \tilde{p} = P + p', \qquad \tilde{T} = T + T', \qquad \tilde{c} = C + c'. \tag{6.8}$$

To illustrate typical mean and fluctuating parts of indoor airflow variables, Figure 6.2 shows instantaneous values of velocity and temperature measured in a full-scale environmental chamber with a square mixing diffuser at the ceiling (Srebric 2000). The results are for a period of 4.5 minutes sampled at 5 Hz. Two measurement points are selected to show how different airflow variables can easily exist within the same space. Both measurement points are at the same vertical line, but at different heights (h). The first measurement point is in the air jet region close

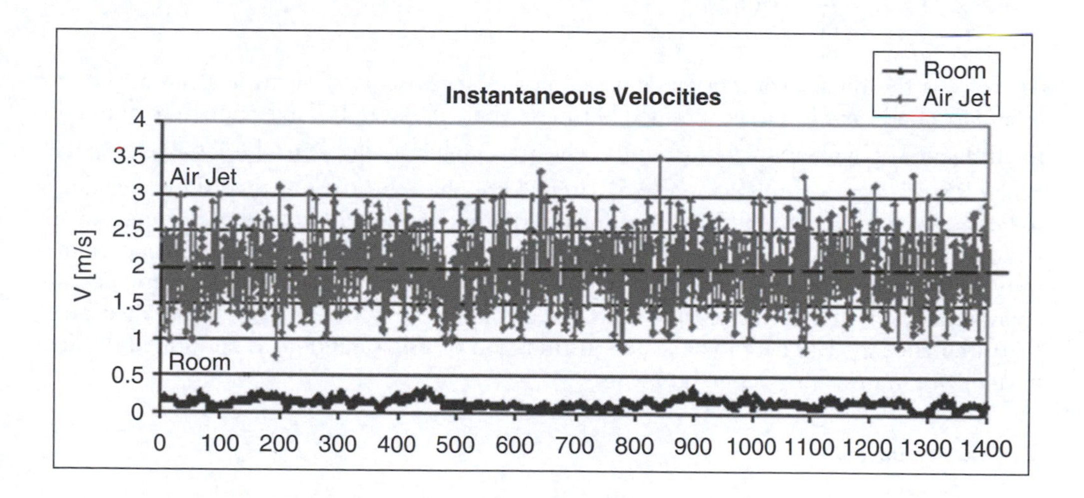

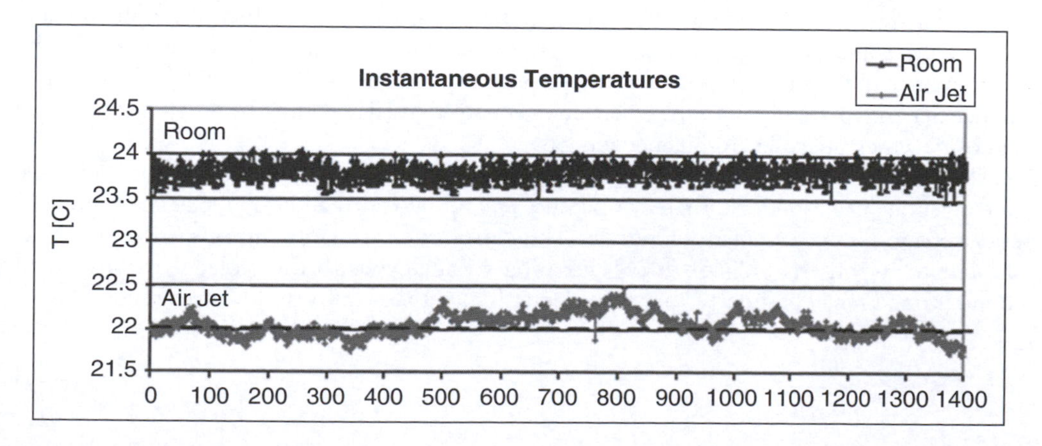

Figure 6.2 Instantaneous characteristics of air velocities (a) and temperatures (b) in a room with a square ceiling diffuser measured at two different heights (for room h=1.10m, and for jet h=2.40m).

to the ceiling (h=2.40m), and the second one is in the occupied zone (h=1.10m) in the room. The results show that the fluctuation and unsteadiness are much higher in the jet region than those in the occupied zone.

Since any instantaneous variable can be represented as the sum of the average value and fluctuation around the average, a general variable can be introduced:

$$\tilde{\varphi}(x_i,t)=\varphi(x_i)+\varphi'(x_i,t). \tag{6.9}$$

The averaging for statistically steady flow gives the mean value of the variables, and has the following integral form:

$$\varphi(x_i)=\lim_{T\to\infty}\frac{1}{T}\int_0^T\tilde{\varphi}(x_i,t)\,dt. \tag{6.10}$$

The averaging time, T, has to be much larger than the time scale of turbulence to obtain results independent of the integration interval. To evaluate the mean flow parameters for unsteady turbulent flow, ensemble averaging is used:

$$\varphi(\mathrm{x_i},\mathrm{t})=\lim_{\mathrm{N}\to\infty}\frac{1}{\mathrm{N}}\sum_{\mathrm{n=1}}^{\mathrm{N}}\tilde{\varphi}(\mathrm{x_i},\mathrm{t}), \tag{6.11}$$

where N is the number of independent samples. A measurement with the same initial and boundary conditions has to be repeated N times, where N has to be large enough to eliminate the influence of fluctuation on the results. The averaging introduces an additional term $\rho\overline{u_i'u_j'}$ called *Reynolds stresses* in Equation (6.5), and additional terms $\rho\overline{u_i'\varphi'}$ called the scalar fluxes in Equations (6.6) and (6.7). These nonlinear terms in the original equations are grouped with viscous diffusion terms to enable their modeling with eddy-viscosity turbulence approximation (Versteeg and Malalasekera 2007). Turbulence should be considered as an engineering approximation rather than scientific law (Ferziger and Peric 2001). Hence, the evaluation of the turbulence model's performance and justification of its assumptions is performed through its application in solving design problems.

Turbulence modeling

The modeling of eddy-viscosity and diffusivity is based on an analogy with laminar flow and molecular transport. Turbulent eddies are represented as molecules, which on a microscopic scale, collide and exchange energy. Although, the analogy is empirical and questionable, the performance of the model is good for many practical applications (Wilcox 2006). The performance depends particularly on the equation or equations used to define the distribution of eddy viscosity, which are called turbulence models.

The simplest eddy viscosity models are zero-equation turbulence models. These models have one algebraic equation for turbulent viscosity, μ_t, and no (zero) additional partial differential equations to resolve turbulent properties. The basis for zero-equation models is *Prandtl*'s mixing-length hypothesis (Wilcox 2006). Forming an analogy with the molecular transport of momentum, *Prandtl* postulated the mixing-length hypothesis in the following form:

$$\mu_t=\mu\, l_{mix}^2\left|\frac{dU}{dx_j}\right|. \tag{6.12}$$

where μ_t is the turbulent viscosity[Pa s], l_{mix} is the mixing length [m], ρ is the fluid density [kg/m^3], and U is the average velocity [m/s].

The mixing length is an empirical variable defined as a free path of small eddies (fluid lumps) over which their original momentum is preserved. The eddy-viscosity and mixing length are not physical properties of the fluid, but they are properties of the flow field. The product of the mixing length and the velocity gradient is called mixing velocity, which is also the property of the turbulent flow. Calibration of the mixing length is always needed for complex flows such as indoor airflows. A zero-equation turbulence model was developed and validated for indoor airflow CFD simulations (Srebric *et al.* 1999). Nevertheless, more typical turbulence modeling requires one or two additional partial differential equations to calculate parameters for the turbulent viscosity. The standard k-ε model is the most widely used two-equation model in practice (Launder and Spalding 1974). The two additional transport equations include the turbulent kinetic energy and turbulent energy dissipation rate as:

$$\frac{\partial \rho k}{\partial t}+\frac{\partial \rho U_j k}{\partial x_j}=\frac{\partial}{\partial x_j}\left(\frac{\mu_t}{\sigma_k}\frac{\partial k}{\partial x_j}\right)+2\mu_t S_{ij}\cdot S_{ji}-\rho\varepsilon, \tag{6.13}$$

$$\frac{\partial \rho \varepsilon}{\partial t}+\frac{\partial \rho U_j \varepsilon}{\partial x_j}=\frac{\partial}{\partial x_j}\left(\frac{\mu_t}{\sigma_\varepsilon}\frac{\partial \varepsilon}{\partial x_j}\right)+C_{1\varepsilon}\frac{\varepsilon}{k}2\mu_t S_{ij}\cdot S_{ji}-C_{2\varepsilon}\rho\frac{\varepsilon^2}{k}, \tag{6.14}$$

where k is the turbulent kinetic energy [J/kg], ε is the turbulent energy dissipation rate [J/(kg s)], S_{ij} is the strain rate tensor ($S_{ij}=\frac{1}{2}\left(\frac{\partial U_i}{\partial x_j}+\frac{\partial U_j}{\partial x_i}\right)$) [1/s], and the empirical model constants are: $\sigma_k = 1.0$, $\sigma_\varepsilon = 1.3$, $C_{1\varepsilon} = 1.44$, and $C_{2\varepsilon} = 1.92$.

The turbulent viscosity is then calculated based on k and ε without assumptions for length scale, which is difficult to estimate:

$$\mu_t=\rho C_\mu\frac{k^2}{\varepsilon}, \tag{6.15}$$

where $C_\mu = 0.09$ is the empirical constant.

With an eddy-viscosity turbulence model, such as the standard k-ε model, the transport equations for all indoor airflow variables, including turbulence properties, can be written in terms of the general variable ϕ, as the following time-averaged *Navier-Stokes* equation:

$$\frac{\partial \rho \varphi}{\partial t}+\frac{\partial \rho U_j \varphi}{\partial x_j}=\frac{\partial}{\partial x_j}\left(\Gamma_{\varphi,eff}\frac{\partial \varphi}{\partial x_j}\right)+S_\varphi, \tag{6.16}$$

where φ can be 1, u_i, T, c, k and ε to represent the mass, momentum, energy, species concentration, turbulent kinetic energy, and turbulent energy dissipation rate conservation laws respectively, $\Gamma_{\varphi,eff}$ is the effective diffusion coefficient in each conservation law, and S_φ is the source for a specific variable ϕ. Table 6.2 shows different variables with corresponding effective diffusion coefficients and source terms for Equation (6.16). Overall, Equation (6.16) has transient $\frac{\partial \rho\varphi}{\partial t}$, convection $\frac{\partial \rho U_j\varphi}{\partial x_j}$, diffusion $\frac{\mu_{eff}}{Pr_{eff}}=\frac{\mu}{Pr}+\frac{\mu_t}{Pr_t}$, and source S_φ terms as defined by different conservation laws. For ventilation performance predictions, a system of partial differential equations will typically have the mass conservation, momentum conservation for three velocity components, energy conservation, and two equations for turbulent properties, k and ε, which is a system of seven partial differential equations for each cell (grid point).

Table 6.2 The k-ε, turbulence model parameters

Equation	φ	$\Gamma_{\varphi eff}$	S_{φ}
Continuity	1	0	0
Momentum	U_i	$\mu + \mu_t$	$-\frac{\partial p}{\partial x_i} + \rho\beta(T_o - T)g_i$
Turbulent kinetic energy	k	$\mu + \frac{\mu_t}{\sigma_\kappa}$	$2\mu_t S_{ij} \cdot S_{ji} - \rho\varepsilon$
Dissipation rate of kinetic energy	ε	$\mu + \frac{\mu_t}{\sigma_\varepsilon}$	$C_{1\varepsilon}\frac{\varepsilon}{k}2\mu_t S_{ij} \cdot S_{ji} - C_{2\varepsilon}\rho\frac{\varepsilon^2}{k}$
Temperature	T	$\frac{\mu}{\mathrm{Pr}} + \mu_t$	S_T
Concentration	C	$\frac{\mu}{\mathrm{Sc}} + \mu_t$	S

The standard k-ε model introduces two additional differential equations for the turbulent transport processes. To solve the two additional equations requires considerable computing effort because of the highly non-linear characteristics of the equations. For example, the prediction of ventilation performance in an office requires several hours of computing time on a PC. If a designer wants to use the CFD method to vary the design, such as to change HVAC equipment location, the additional computing time renders this process inefficient, especially for conceptual designs. Therefore, a zero-equation model, referring to an algebraic turbulence model is more suitable for design, when the simplicity of the model does not lead to incorrect solutions of the transport flow equations (Srebric *et al.* 1999).

Due to the complexity of turbulent flow, it is difficult to obtain a universal turbulence model for all types of ventilation performance predictions. A general conclusion is that the turbulent models perform differently from one case to another, although all simulated flows are indoor airflows. In fact, for indoor airflow simulations, no turbulent models perform significantly better than the standard k-ε model. The standard k-ε model is stable, easily implemented, widely validated, and reasonably accurate for many indoor airflow applications. Nevertheless, the standard k-ε model is not universal, and has provided erroneous results in certain indoor airflow regimes. For example, the standard k-ε model has had problems in predicting strong buoyancy, separated flows, axisymmetric jets, swirl flows and heat transfer from a wall. Therefore, many modified k-ε models, such as the *RNG* k-ε model (Yokhot and Orszag 1986), have emerged to improve performance for certain classes of problems. Compared to the standard k-ε model, the *RNG* k-ε model has an additional source term, R, in the equation for the turbulent energy dissipation rate, ε. This term improves flow prediction for regions with large strain rates, and the term is negligible for small strain rates. Therefore, the model can better predict separate flows, which are commonly present in indoor airflow. Several studies have shown that the *RNG* k-ε model has better performance than the standard k-ε model for indoor airflow simulations (Chen 1995; Loomans 1998). However, the *RNG* model assumes high Reynolds numbers for the flow field, which is one of the assumptions that causes problems in modeling with the standard k-ε model.

CFD boundary conditions

Boundary conditions are necessary for the mathematical solution of the system of partial differential equations representing the conservation laws. It is important to notice that CFD takes boundary conditions as an additive source term on the left-hand side of Equation (6.16). This is a rather crude, but quite practical representation of ventilation transport process. As a result, CFD models should be a simplified version or real detailed geometry such as the example office space with six cubicles in Figure 6.3(a). The entire office was reduced to floor, ceiling,

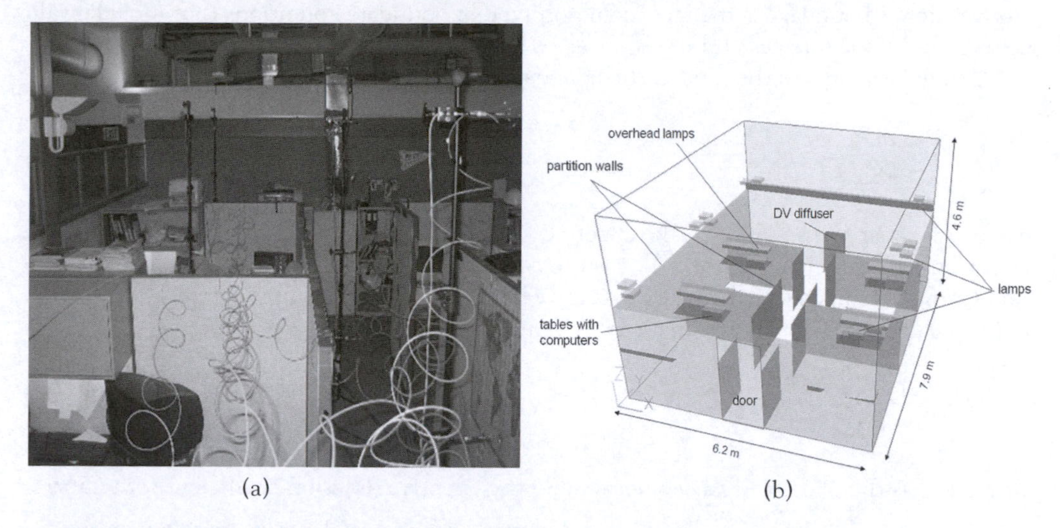

Figure 6.3 (a) An example office, and (b) its simplified model for CFD.

Source: Srebric *et al.* 2008.

four walls, cubicle partitions, tables, computers, lamps and displacement ventilation (DV) diffuser as shown in Figure 6.3(b). All of these items represent CFD boundary conditions with defined size, location, and source term intensity for the conservation laws. This is an appropriate level of detail for CFD models used for ventilation performance predictions, and additional items in the office can be neglected without sacrificing accuracy of CFD results. In principle, only important boundary conditions have to be introduced in CFD models. There are three types of boundaries of practical importance: free boundary, symmetry surface, and conventional boundary.

Free boundary is a boundary surface that is adjacent to an inviscid stream. Examples are the air supply and exhaust. For a supply, the boundary conditions are:

$$u_i = u_{supply}, \qquad T = T_{supply}, \qquad k = k_{supply}, \qquad \varepsilon = \varepsilon_{supply}, \text{ and } \qquad c = c_{supply}, \tag{6.17}$$

where subscripts "supply" are the parameter values at the supply opening. These equations apply only for a slot opening, while supply openings with diffusers require more complex specifications of boundary conditions, which are different for different types of ventilation methods discussed later in this chapter.

Pressure is normally given for an exhaust opening, and zero gradients normal to the surface are assumed for the other parameters:

$$p = p_{exhaust}, \quad \frac{\partial U_i}{\partial x_i} = 0, \quad \frac{\partial T}{\partial x_i} = 0, \quad \frac{\partial k}{\partial x_i} = 0, \quad \frac{\partial \varepsilon}{\partial x_i} = 0, \quad \frac{\partial C}{\partial x_i} = 0, \tag{6.18}$$

where $p_{exhaust}$ is the pressure at a return [Pa], and x_i is the coordinate normal to the surface of the exhaust opening or diffuser [m].

Symmetry surface boundaries are used to reduce the size of simulation domains that are symmetric. If the x_i coordinate is normal to the symmetry surface, the following equations describe the boundary conditions of the symmetry surface:

$$\frac{\partial U_i}{\partial x_i} = 0, \quad \frac{\partial T}{\partial x_i} = 0, \quad \frac{\partial k}{\partial x_i} = 0, \quad \frac{\partial \varepsilon}{\partial x_i} = 0, \quad \frac{\partial C}{\partial x_i} = 0, \tag{6.19}$$

Conventional boundary is the most common type of boundary conditions that include wall, ceiling, and floor surfaces and the surfaces of furniture, appliances, and occupants. If the x_i coordinate is parallel to the surface, the boundary conditions are:

$$\tau = \mu_{eff} \frac{\partial U_i}{\partial x_j}, \; \dot{q} = h(T_w - T) \tag{6.20}$$

where τ is shear stress [Pa], μ_{eff} is the effective viscosity, $\mu_{eff} = \mu + \mu_t$ [Pa s], h is the convective heat transfer coefficient [W/(m^2 K)], $\dot{q}$ is the convective heat transfer rate [W], and T_w is the wall (surface) temperature [K]. The convective heat transfer coefficient is determined from the following equation, which is similar to the *Reynolds analogy*:

$$h = \frac{\mu_{eff}}{\Pr_{eff}} \frac{c_p}{\Delta x_j}, \tag{6.21}$$

where $\Pr_{eff}$ is the effective *Prandtl number*, and the effective diffusion coefficient for temperature equation, $\frac{\mu_{eff}}{\Pr_{eff}} = \frac{\mu}{\Pr} + \frac{\mu_t}{\Pr_t}$, is based on laminar, *Pr*, and turbulent, Pr_t, *Prandtl numbers*.

Equations (6.20) and (6.21) can be used as wall boundary conditions for the laminar flows as well as for the turbulent flows. The zero-equation turbulence models apply directly to the specified equations. However, the standard *k-ε* or *RNG k-ε* model developed for high *Re* numbers needs adjustment for the near wall region where the *Re* number is very small. The *k-ε* models use additional equations for the near-wall region, called wall functions (Tennekes and Lumley 1972).

The wall functions are dimensionless velocity, temperature and concentration profiles in the wall boundary region. The wall boundary layer has three regions, the laminar sub-layer ($0 < y+ \leq 5$), buffer zone ($5 < y+ \leq 30$), and inertial sub-layer ($30 < y+ \leq 130$), where y+ is the dimensionless distance to the wall. The wall has a damping effect on the turbulent flow, and the layer closest to the wall has predominant viscous forces. In the buffer zone, viscous forces and *Reynolds stresses* have the same order of magnitude and neither can be neglected. Finally, the inertial sub-layer is predominately turbulent. The wall functions give expressions for laminar and turbulent sub-layers, while the buffer zone is neglected. A near wall flow is considered laminar if $y+ \leq 11.63$, and if $y+ > 11.63$ the flow is turbulent. The wall function equations for velocity and temperature are:

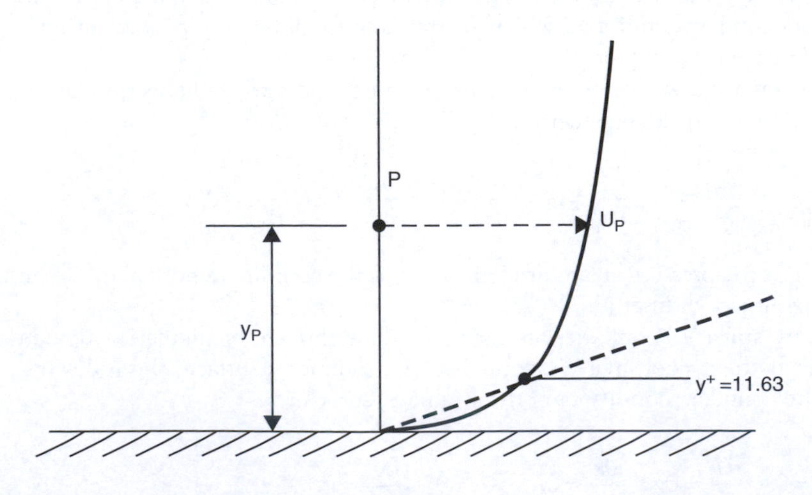

Figure 6.4 Velocity distribution at the near wall region.

$$u^{+} = y^{+},\ T^{+} = \Pr y^{+}, \text{ for } y+ \leq 11.63 \tag{6.22}$$

$$u^{+} = \frac{1}{\kappa}\ln\left(Ey^{+}\right),\ T^{+} = \Pr_t\left[u^{+} + E\left(\frac{\Pr}{\Pr_t} - 1\right)\left(\frac{\Pr_t}{\Pr}\right)^{1/4}\right], \text{ for } y+ > 11.63 \tag{6.23}$$

$$u^{+} = \frac{U_P}{u_\tau},\ u_\tau = \sqrt{\frac{\tau_w}{\rho}},\ y^{+} = \frac{y_P u_\tau \rho}{\mu},\ T^{+} = \frac{(T_w - T_P)u_\tau}{q} \tag{6.24}$$

where u^{+} is the dimensionless mean velocity [–], U_P is the velocity parallel to the wall [m/s], u_τ is the friction velocity [m/s], τ_w is the wall shear stress [Pa], $y+$ is the dimensionless wall distance [–], y_P is the distance to the wall [m], κ is the von Karman constant (κ=0.41) [–], E is the roughness parameter (E=9 for hydraulically smooth walls with constant shear stress) [–], T_w is the wall temperature [K], T_P is the air temperature at location denoted as P [K], and $\dot{q}$ is the convective heat transfer rate [W].

In the viscous sub-layer, the mean velocity profile is approximated as the linear function of wall distance as shown in Figure 6.4. At the wall surface (y=0), air velocity is zero ("no-slip" conditions) and air temperature is equal to the wall surface temperature. For the turbulent sub-layer, the velocity profile is the log-law function.

Assumptions used to derive the wall functions restrict the application of these formulae to a certain class of flows. The assumptions are the *Prandtl mixing hypothesis*, the *Boussinesq eddy-viscosity assumption*, a fully developed flow, and no pressure gradients or other momentum sources (constant shear stresses). For typical ventilation performance predictions, these assumptions are acceptable and wall functions are widely used. However, predictions of the heat transfer in the near-wall region tend to be incorrect, and dependent on the distance to the wall y_P. Improved calculation of heat transfer is possible with improved temperature profile equations or use of Equation (6.20) with a prescribed empirical value for the convective heat transfer coefficient h (Novoselac 2004). This temperature profile or heat flux can be obtained from coupling CFD with the energy simulation tools such as ESP-r or EnergyPlus (Hensen 1999). Overall, a correct prediction of the wall boundary conditions is crucial for accurate ventilation performance predictions.

CFD model performance

Professional organizations such as ASHRAE provide guidelines for quality assessment of CFD model predictions (ASHRAE 2005; Chen and Srebric 2002; Sorensen and Nielsen 2003). The quality assessment has three steps that include: (1) verification, (2) validation, and (3) reporting of results. The verification step determines whether a CFD model has capabilities to represent important physical phenomena in support of ventilation design decisions. The validation step establishes a CFD model and user capabilities to accurately simulate physical phenomena. Finally, the reporting of results provides instructions to enable quality assessment of CFD results with sufficient details describing verification and validation of ventilation performance predictions. These steps represent guidelines for best practices in CFD use for ventilation performance predictions, and cannot be extended into a standard design procedure because CFD models are still under intense development (AIAA 1998).

Verification of a CFD model starts with the assessment of necessary models to realistically represent ventilation transport processes and finishes with the evaluation of numerical simulation methods. These verifications assessments include selection of:

- Airflow and heat transfer models,
- Turbulence model, and
- Numerical solution methods.

For ventilation design, it is important to achieve thermal comfort, so both velocity and temperature fields need to be modeled. The airflow and heat transfer models are connecting convection, diffusion, and/or radiation in the simulation domain. Therefore, mass, momentum, and energy conservation laws have to be solved simultaneously with turbulence model. It is important to notice that most ventilation airflows are weakly turbulent because of the high *Rayleigh number*, *Ra*, and sometimes high *Reynolds number*, *Re*. Typical *Ra* ranges from 10^9 to 10^{12}, and *Re* from 10^4 to 10^7, while turbulence occurs when $Ra > 10^9$ and/or $Re > 10^4$ (ASHRAE 2005). For a new CFD user, discussion on different turbulence models is useful, but the starting point in gaining modeling experience should be a CFD model with the standard k-ε model (Equations 6.13 and 6.14). This model assumes fully developed turbulent flow, and calculation errors can occur in the simulation regions where this assumption is incorrect such as near wall boundaries. Other more complex turbulence models are recommended only for more experienced users because they can introduce additional partial differential equations that are hard to solve.

As discussed, with the standard k-ε model seven partial differential equations have to be solved simultaneously for each grid point in the simulation domain. The equations are discretized and solved with numerical solution procedures that start with an initial guess for calculated variable distributions. The solution procedure for the system of discretized transport equations is iterative because the equations are coupled and the convection terms in momentum equations are non-linear. The discretization introduces an error that can be reduced by refining the grid size. A commonly accepted method for grid refinement is to double the grid size until the two different grid resolutions give similar CFD results (Wilcox 2006). In this way, the discretization error is minimized, and other numerical and physical approximations, such as turbulence models, can be explored as potential sources of simulation errors.

Due to the non-linearity of convection terms in the transport equations, solution procedures can be unstable. The variables must be solved simultaneously, and the progress towards a solution has to be simultaneous. To control the rate of change for the calculated variables, relaxation factors are used. Relaxation factors serve to control the change in variable calculations and, hence, decrease the numerical instability. A general problem with relaxation factors is that they depend on a solved case, and therefore their values have to be estimated and corrected during calculations. The relaxation only influences the calculation procedure and the final solution of a problem is independent of relaxation factors. Overall, the numerical solution should be independent of grid distributions, initial variable distributions, and relaxation factors. The criterion for ending the numerical procedure is called the convergence criterion. Usually, the residual of the continuity equation must be smaller than a certain predefined value, such as:

$$\text{Residual for mass} = \frac{\text{Sum of absolute residuals in each cell}}{\text{Total mass inflow}} < 0.1\%, \tag{6.25}$$

$$\text{Residual for energy} = \frac{\text{Sum of absolute residuals in each cell}}{\text{Total heat gain}} < 1\%. \tag{6.26}$$

In the end, converged and grid-independent CFD results represent approximate performance of ventilation system in a building. The level of accuracy depends on the airflow and heat transfer models, turbulence model, and numerical solution methods selected in the verification step. The next step is the validation to quantitatively address the model accuracy.

Validation includes quality assessment of both the CFD tool and user in modeling a ventilation system in a building. This step is necessary for every new user who plans on using CFD as a design tool. Example simulation cases for ventilation systems are available in the literature or in the end of this chapter. It is important that the validation data are detailed and complete. Validation data have to include: (1) detailed geometry and boundary conditions of simulation domain, (2) detailed measurements of flow properties such as distributions of velocity, pressure, and temperature, and (3) complementary overall measurements such as total mass flow rate and heat fluxes (Versteeg and Malalasekera 2007).

It is important to notice that measured data also contain uncertainties and errors introduced by acquisition equipment, experimental set-up, and data collection process. For example, accuracy of low-velocity measurement probes is relatively low when collecting data in the occupied zone, where velocities are typically below 0.2m/s and many times even below 0.05m/s as shown in Figure 6.2(a) (Melikov *et al.* 2007). Depending on probe type and calibration, the inaccuracy of velocity measurements can be the same order of magnitude as the measured velocity. On the contrary, temperature probes can be highly accurate and depending on the manufacturer and price, a typical accuracy is in the range from ±0.1°C to ±0.5°C. As a result, errors in velocity or turbulence predictions can point to measurement as well as model uncertainties. Nevertheless, errors in temperature distribution, if not resulting from erroneous velocity distributions, can point to modeling problems. Overall, high accuracy of CFD results when compared to measured data is desirable, but not always feasible. The validation criteria should reflect CFD application requirements (ASHRAE 2005). Luckily, for ventilation design process, model changes are typically incremental and comparative analyses of results can be sufficient.

When creating a CFD model, the best practice is to start from a very simple geometry as well as boundary conditions, and only gradually to increase model complexity. This approach is recommendable for all users, experienced and inexperienced, when starting a new type of modeling exercise. The first CFD model should probably include just the simulation domain with supply and exhaust diffusers to calculate simple mass balance in the domain. The subsequent modeling efforts should gradually include other important boundary conditions, such as partitions, computers and occupants. Finally, the CFD model should also include temperature field and non-isothermal boundary conditions. The gradual introduction of complexity in the simulation domain minimizes probability of simple input errors, but also helps determine sources of potential convergence or instability problems. Overall, a successful validation process confirms that appropriate decisions were made with respect to airflow and heat transfer models, turbulence model, and numerical solution methods, but it also confirms user's knowledge and capabilities.

Reporting is the last step in quality assurance process for use of CFD models for ventilation performance predictions. A report should contain all of the verification and validation steps outlined with sufficient details that another user can reproduce CFD results (Chen and Srebric 2002). If the simulation case is not new, the report can reference another study with detailed case geometry, boundary conditions and measurement uncertainties/errors, and proceed to list the details of numerical model. Both quantitative and qualitative evaluations have to be included. For example, simple qualitative statements such as "bad" or "good" agreement between CFD results and measured data can be misleading if the comparison criteria are not specific. Finally, the report has to draw conclusions that can benefit the broader design community interested in this type of CFD simulation for ventilation performance predictions.

Multi-zone airflow network and CFD model selection

Both multi-zone airflow network and CFD models are appropriate for evaluations of the ventilation system performance. As outlined in the model overview, multi-zone airflow network is applicable to evaluations where perfect mixing assumption holds, and only bulk flow properties are needed. On the contrary, CFD is suitable for applications with known detailed boundary conditions such as locations, size and intensity of mass, momentum, heat and contaminant sources. Hybridization of these two models by their coupling can take advantage of both model strengths, while mitigating some of the weaknesses. In order to compare different models, experimental measurements were conducted in an interior cubicle office with displacement ventilation system as shown in Figure 6.3. Detailed measured and simulation data are available in the literature (Srebric *et al.* 2008). In this chapter, the focus is on practical aspects of efforts necessary for model set-up, execution and accuracy.

Three different models were developed for this cubicle office: CFD, coupled CFD-multi-zone airflow network, and multi-zone airflow network. Each model simulated the temperature field and contaminant distribution. A point source of SF_6 gas was placed in front of displacement diffuser, and the source was slightly shifted towards one side of the diffuser. The simplest model, which is the multi-zone airflow network model, included 9 zones that covered the lower level and another 9 zones that covered the upper level of the office. A total of 18 zones were sufficient to simulate this office with the multi-zone airflow network model. For the coupled model, CFD handled the area close to the diffuser, which was called the source zone with approximately 100,000 grid points. Additional 12 zones covered the space outside of the source zone with the multi-zone airflow network model. Finally, the stand-alone CFD model covered the entire space and used approximately 200,000 grid points. The computational time was: t_{mz} (couple of seconds) for multi-zone airflow network, $4500\times t_{mz}$ for coupled model, and $16000\times t_{mz}$ for CFD model.

Considering the computational time of each method, the coupled model is three and half times faster than the CFD model, and significantly slower than the multi-zone airflow network model. In addition, the time required to set up a coupled model is much shorter than the time needed to set up the CFD model because the computation domain covered by the multi-zone simulation required only associated source/sink intensities, rather than a detailed geometry, detailed space layout, and detailed source/sink locations. Further interesting observations come from the evaluation of simulation result accuracy as shown in Figure 6.5. The asymmetry of the concentration in this case was nicely captured by the coupled model, while the multi-zone airflow network model failed to predict the asymmetry due to its inherent assumption of the uniform concentrations in each simulation zone. Nevertheless, the multi-zone model performed well outside of the immediate vicinity of the source zone. Even the buoyancy driven transport of contaminants with displacement ventilation can be captured with the multi-zone simulations as long as the temperature field is correctly specified or calculated.

Overall, the justification for use of either of the three models, CFD, multi-zone airflow network or coupled models, is case dependent. At present, users of the simulation models are responsible for making this judgment based on the tradeoffs between required accuracy and available computational time. These tradeoffs become quite obvious for simulation domains that are much larger than a single cubicle office, such as an entire building shown in Figure 6.1. This building has four stories plus basement with an approximate area of 2000 m^2

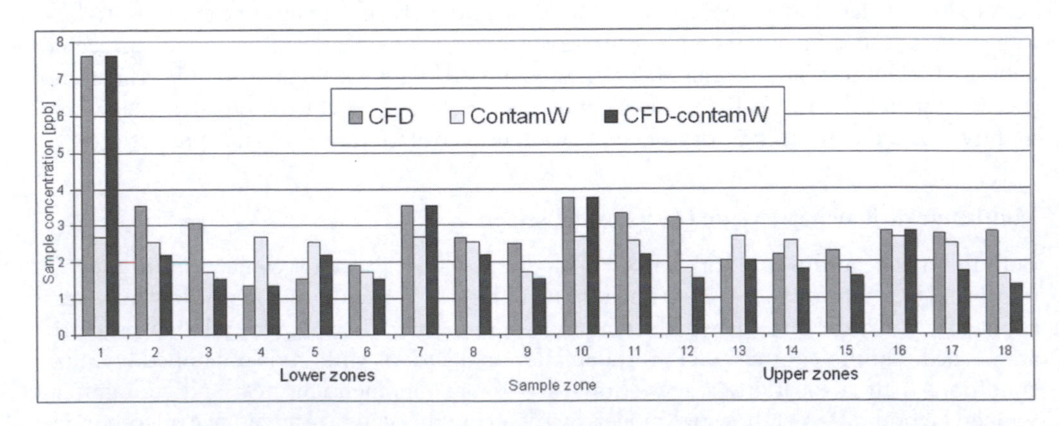

Figure 6.5 The comparison of CFD, multi-zone airflow network (ContamW) and coupled CFD-contamW model results.

Source: Srebric *et al.* 2008.

per floor. A multi-zone airflow network model was created for the entire building with a total of 250 zones, where each floor had roughly 50 zones. For each zone, we had at least 4 flow elements that included a window, door, air supply and exhaust. This resulted in at least 1000 flow elements. This model set-up took more than a week to finish, while the simulations were still very fast (measured in seconds). It was relatively easy to decide that the development and execution of a CFD model for the building shown in Figure 6.1 was too time consuming from both model set-up and computational time requirements. The assessments of ventilation performance in this building had to be accomplished with the multi-zone airflow network model.

Ventilation system performance

Ventilation systems in buildings are designed and operated to accomplish two primary functions: (1) to deliver fresh air to occupants while removing internally generated contaminants, and (2) to provide acceptable thermal comfort in the vicinity of occupants. During the operation of a ventilation system, occupants' focus is on the thermal comfort because perception of thermal comfort is immediate and thermal discomfort is intolerable. However, poor air quality is more difficult to notice, and, therefore, occupants' response time is longer. In addition, designers traditionally pay more attention to the thermal comfort than to the air quality because thermal comfort is better defined by standards. For the ventilation system design, the thermal comfort is typically defined by the ANSI/ASHRAE Standard 55 (ASHRAE 2004), and the indoor air quality requirements are specified in ANSI/ASHRAE Standard 62.1 (ASHRAE 2007). Chapter 7 discusses in more detail thermal comfort parameters, while the current chapter discussions will focus on indoor air quality parameters. It is important to notice that simulations of thermal comfort and indoor air quality parameters are based on the same system of equations, such as the equations in Table 6.2. Calculated airflow rates, velocities, turbulence parameters, temperatures, and contaminant concentrations are inputs into the thermal comfort and indoor air quality indicators defined by these two standards.

Indoor air quality (IAQ) indicators

To evaluate performance of ventilation systems, a large number of indoor air quality indicators are in use today such as air exchange rates, air exchange efficiency, ventilation effectiveness, and contaminant removal effectiveness. Each of these indicators is useful for a certain type of ventilation system analysis. However, the air exchange efficiency (ε_a) and the contaminant removal effectiveness (ε) appear to be the most suitable for use in comparative analysis of ventilation strategies because they are general and almost all other indicators are extensions of these two (Novoselac and Srebric 2003). These indicators are applicable to all types of ventilation strategies and airflow patterns. Also, they can be easily measured in the field or the laboratories, which qualify them for extensive use in design and standards.

The first developed IAQ indicator is the contaminant removal effectiveness ε (Yaglou and Witheridge 1937), which is based on the room/space average contaminant concentration $\langle C \rangle$, the contaminant concentration at the supply C_s, and the concentration at the exhaust C_e:

$$\varepsilon = \frac{C_e - C_s}{\langle C \rangle - C_s}. \tag{6.27}$$

On the contrary, air exchange efficiency (ε_a) is based on a local value of age of air (τ), and it was developed in chemical engineering. Later, ε_a was experimentally and theoretically evaluated for use in the indoor air quality field (Etheridge and Sandberg 1996). The contaminant removal effectiveness represents the ratio between shortest possible time needed for replacing the air in the room (τ_n) and the average time for air exchange (τ_{exe}):

$$\varepsilon_a = \frac{\tau_n}{\tau_{exe}} = \frac{\tau_n}{2\langle\tau\rangle}. \quad (6.28)$$

The average time for air exchange can be calculated as $\tau_{exe} = 2\langle\tau\rangle$, where $\langle\tau\rangle$ represents the average local values of age of air [s]. Also, the shortest possible time needed for replacing the air in the room (τ_n) can be calculated as a reciprocal value of the number of Air Changes per Hour (ACH) in the room (τ_n=1/ACH, and ACH= Volume flow rate/Room volume [1/hour]).

It is important to keep in mind that ε_a and ε are only qualitative indicators of indoor air quality and that additional parameters such as flow rates and supply contaminant concentration or local contaminant distribution should be taken into account in the design process. Detailed distributions of IAQ or thermal comfort indicators can be calculated with CFD based tools for different types of ventilation systems (see Chapter 7 as well).

As an example using CFD, the contaminant removal effectiveness (ε) and the air exchange efficiency (ε_a) were calculated in the same room, and results are presented in the middle vertical cross section, shown in Figure 6.6 (Novoselac and Srebric 2003). More specifically, Figure 6.6(a) presents two-dimensional airflow velocities with a bypass and a recirculation in the occupied zone. Figure 6.6(b) shows corresponding age of air normalized by the shortest time needed for the replacement of air (τ/τ_n). Figures 6.6(c) and 6.6(d) represent contaminant concentration normalized with the contaminant concentration at the exhaust (C/C_n) for two different positions of a contaminant source; in the room centre (Figure 6.6(c)) and within the air jet (Figure 6.6(d)). The comparison of Figure 6.6(b) with 6.6(c) and 6.6(d) show that, depending on the position of the contaminant source, the mean age of air and the contaminant concentration field might have a similar or completely different distribution. Consequently,

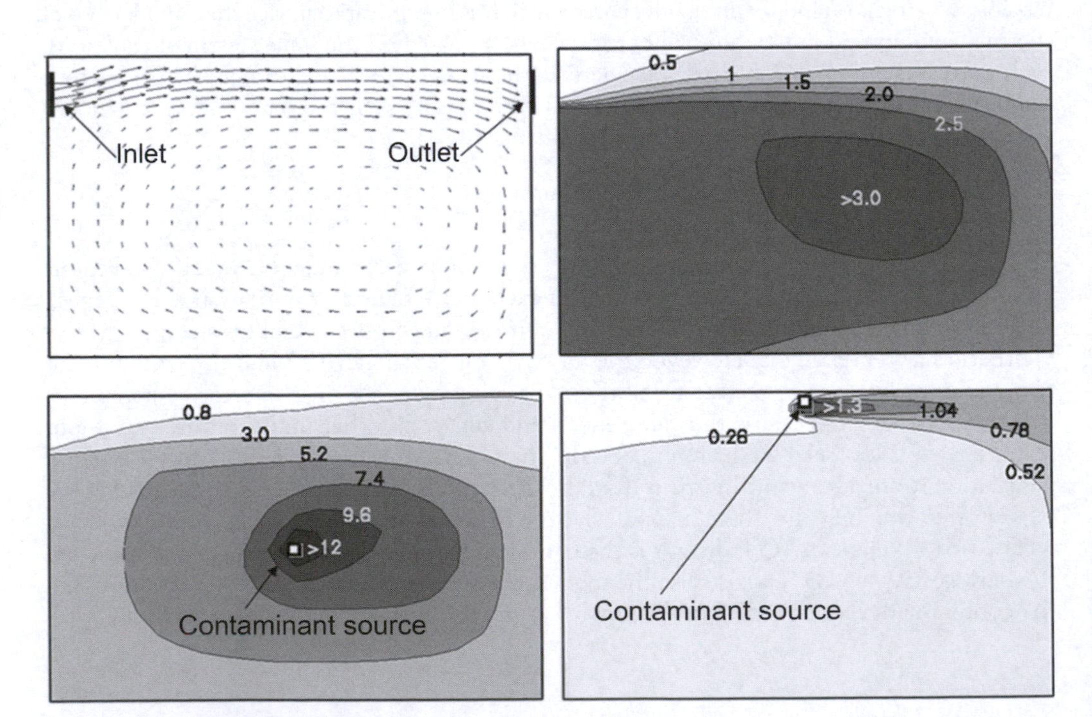

Figure 6.6 Normalized local age of air and contaminant concentration distributions for the same airflow field and different positions of the contaminant source (a) Airflow pattern, (b) Normalized age of air τ/τ_n (εa = 0.21 for this airflow pattern), (c) Normalized concentration C/C_e (εa = 0.21 and e = 0.19 for contaminant source at room centre), (d) Normalized concentration C/C_e (εa = 0.21 and ε = 2.20 for contaminant source in the supply air jet).

Source: Novoselac and Srebric 2003.

the air exchange efficiency (ε_a) and the contaminant removal effectiveness (ε) might have the same or conflicting values for perceived indoor air quality in the same space. Nevertheless, these two parameters are the general single number IAQ indicators actually evaluating performance of the air distribution system.

Overall, air exchange efficiency (ε_a) quantifies the quality of the airflow pattern. The airflow pattern depends on the inlet and outlet positions, the temperature difference between the supply jet and room air, the jet inertial force, and the intensity of buoyant flow created by cooling/heating loads. On the contrary, contaminant removal effectiveness (ε) is the indicator of the contamination level in a room. It depends not only on the airflow pattern, but also on the intensity, size, position, and, type of contaminant sources relative to this pattern (Novoselac and Srebric 2003). Overall, different ventilation systems result in different air quality levels that can be assessed with CFD.

Ventilation examples

A classification of ventilation systems based on the driving mechanism for air circulation includes: (1) mechanical, (2) hybrid, and (3) natural ventilation. Mechanical ventilation uses fans, ducts and diffusers, called air distribution systems, to move the air through building spaces. As an alternative to mechanical ventilation, natural ventilation takes advantage of pressure difference between indoor and outdoor environments due wind and stack effect to accomplish the same ventilation task. As a result, natural ventilation is one of the passive technologies that offer significant energy saving opportunities for buildings in mild climates. Nevertheless, natural ventilation is not as reliable as the mechanical ventilation because it depends on the building geometry, surrounding terrain, and stochastic nature of outdoor air temperature, humidity and wind velocity. In recent years, many building design projects explored opportunities to take advantage of the mechanical ventilation reliability and natural ventilation energy savings by combining them into one hybrid ventilation system. This hybrid system uses both natural and mechanical ventilation via control system that switches between or complements the two operational modes based on an optimization algorithm designed to minimize the HVAC energy consumption while satisfying thermal comfort and IAQ requirements. Overall, the mechanical ventilation is still the dominant system for building ventilation, while natural ventilation is becoming more widely used for its potential to save energy.

Natural and hybrid ventilation

Natural ventilation is the oldest form of ventilation, and it has been used as long as the buildings have been constructed. At present, natural ventilation is an interesting alternative to mechanical ventilation even in commercial buildings due to its potential to reduce building energy consumption, and to increase environmental quality such as thermal comfort, ventilation effectiveness, and indoor air quality (Allard 1998; CIBSE 2005). The potential benefits of natural ventilation are not always easy to take advantage of due to their dependence on weather conditions and surrounding environment. The local weather conditions including wind velocity, air temperature and humidity have to be suitable for natural ventilation. In addition, in urban environments, local air and noise pollution can dramatically reduce potential for natural ventilation, even if the local weather conditions are favorable. Other considerations in design of natural ventilation have to include fire and safety regulations as well as security issues (Allard 1998). Furthermore, Chapter 15 covers urban level performance predictions that can be directly used in design considerations of natural ventilation.

The design process of natural ventilation can be based on empirical correlations, multi-zone airflow network modeling, and/or CFD simulations. At present, there is no standard design procedure for natural ventilation, such as the established design procedures for mechanical ventilation. In principle, design of natural ventilation has to take into account different considerations. From the perspective of ventilation, design considerations have to include: (1) climate analysis and site planning, (2) layout and sizing of window openings at building envelope, and (3) floor planning and layout of internal partitions. These three steps are not independent of each other, and, as a result, the design process is typically iterative. As the final outcome, well designed natural ventilation should be able to satisfy requirements for thermal comfort and indoor air quality as specified in the standards (ASHRAE 2004; ASHRAE 2007).

The first step in natural ventilation design, climate analysis and site planning, relies on assessment of wind conditions in the building local microenvironment. Figure 6.7 shows an example of site planning for a residential neighbourhood in Thailand to promote natural ventilation for all of the residences (Tantasavasdi *et al.* 2001). It was demonstrated that even in this hot and humid climate, certain parts of the year can be suitable for natural ventilation, provided that front row houses did not block the incoming wind completely. Local weather data, described in Chapter 3, should be used to determine how many hours/days in a month/season a building would be able to take advantage of natural ventilation. Typically, outdoor air temperature and humidity are compared to the thermal comfort zone (ASHRAE 2004), and associated wind velocities are assessed for their potential to provide sufficient air change rate per hour (ACH) for a building. In principle, ACH can be correlated to the incoming wind velocity and temperature difference between indoor and outdoor environments because they are the two driving forces that create the pressure difference to establish airflow rates through open windows (Awbi 1991; Clarke 2001; Mahdavi and Pröglhöf 2008). The calibrated correlations can be found in the form of equations (Clarke 2001; Mahdavi and Pröglhöf 2008) or as graphically represented functions (Awbi 1991) with the following general form:

$$ACH = \text{Function}\ (\textit{wind velocity, temperature difference}). \tag{6.29}$$

This simple equation can be used only for rough, but useful estimates, especially if a particular building has an available calibrated function. The calculation of the actual airflow rates for different outdoor weather conditions is not simple because the actual airflow rates depend not only on the combined effect of *wind* and *stack*, but also the window distribution and size. In the design process for predicting these airflow rates, the starting point is determination of wind pressure on the external building surface. Semi-empirical equations (ASHRAE 2005, Chapter 16) or CFD can be used to calculate pressure coefficients for different site plans and wind pressures can be estimated for different building shapes and different locations of building openings. These CFD simulations need to incorporate several rows of buildings and an incoming wind profile is typically defined as the power law wind velocity profile (Awbi 1991 and Allard 1998):

$$U(z) = KU_H z^a, \tag{6.30}$$

where U_H is the reference wind velocity [m/s], z is the distance from the ground [m], while the coefficient K and the exponent a are constants dependent on the terrain surface roughness [–]. The values for these coefficients are given in Table 6.3. The reference wind velocity U_H is

Table 6.3 Typical values for terrain dependent parameters.

Terrain types	*Power-law coefficients*	
	K	a
Open flat country	0.68	0.17
Country with scattered wind breaks	0.52	0.20
Urban (industrial or forest areas)	0.35	0.25
Center of large city	0.21	0.33

Source: Allard 1998.

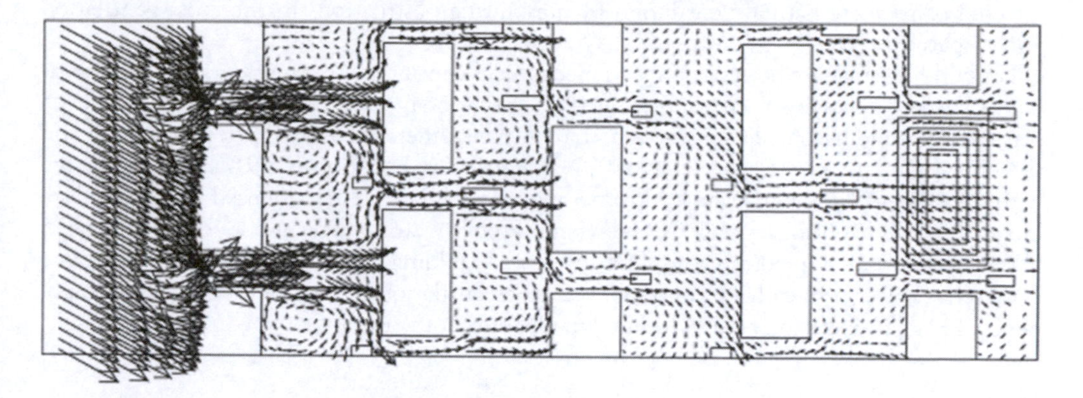

Figure 6.7 An example of CFD calculated wind velocities to support site planning and promote natural ventilation for all residences in an urban neighborhood.

Source: Adapted from Tantasavasdi *et al.* 2001.

measured at weather stations, and it is defined as the hourly averaged wind speed, i.e. the wind speed that is exceeded 50% of the time at a particular site (see Chapter 3).

Once the wind velocities and pressure distribution or pressure coefficients are established, the design process of natural ventilation focuses on the internal environment through calculations of airflow at the building openings and the internal environment. Empirical correlations for airflow through different types of openings are available in the literature, not only for the simple single-sided ventilation, but also for much more complex natural ventilation cases (Allard 1998; CIBSE 2005). In addition, CFD can be used to study the flow through building openings as well as internal environment to address the design issue of floor planning and layout of internal partitions. Figure 6.8 shows the calculated velocities with CFD simulations that can be used in design decisions of floor planning. In particular, the partitions are

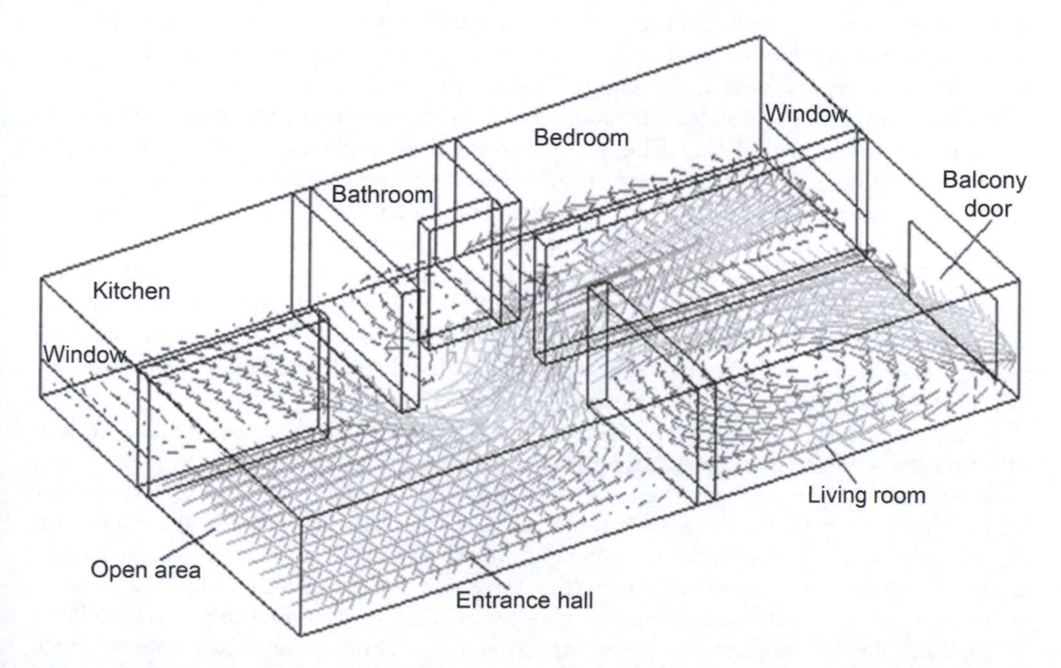

Figure 6.8 A CFD example showing air velocities used in floor planning and layout of internal partitions for a naturally ventilated apartment.

selected to promote natural ventilation by minimizing obstructions in the cross-ventilation airflow pattern.

For buildings that use both natural and mechanical ventilation, in the form of hybrid ventilation, the last step is the integration of the two ventilation modes. This integration is achieved through the building management system that provides different control options depending on the climate and season (Lomas *et al.* 2007). In particular, Equation (6.29) can be calibrated with experimental or multi-zone airflow network model data, which are used in the development of optimal control strategy (Mahdavi and Pröglhöf 2008). This model-based control of HVAC systems is an important trend for integrated building design (Hensen and Augenbroe 2004) that will dynamically optimize not only the ventilation systems, but also other components of HVAC operation as also outlined in Chapter 2, 13, and 14.

Mechanical ventilation

For mechanically ventilated rooms, thermal comfort and IAQ strongly depend on the performance of air supply diffusers. The prediction of diffuser flow characteristics is challenging because the jet flow from a diffuser is complicated due to the complex diffuser geometry and the effects of confined space. Jet formulae are commonly used to describe the flow and thermal characteristics of supply diffusers. The jet formulae are empirical expressions for jet velocity and temperature decay, profiles, and trajectory for inclined jets and separation distance for cold ceiling jets. Researchers have spent numerous decades developing the jet formulae, and today the approach is the most widely used to design mechanical ventilation systems.

Diffuser air jets are classified according to the most important parameters for jet diffusion, such as jet temperature and distance to the nearest enclosure surface. A jet can be isothermal if the jet temperature is the same as the room air temperature or non-isothermal if is the temperatures are different. With respect to the distance from enclosure surfaces, a jet can be free or attached. When an air diffuser is installed close to the side walls, floor or ceiling, the jet attaches to the surface due to the *Coanda effect*. The attachment occurs because the jet has a restricted entrainment on the wall side, and, as a result, a zone of lower air pressure forms and drags the jet towards the wall. On the contrary, a free jet has unrestricted momentum diffusion and room air entrainment. In practice, if the enclosure is big enough to have a minimal impact on the jet trajectory and spread, a jet can be treated as a free jet.

Normally, an air jet is divided into four zones: core, transitional, main, and terminal as shown in Figure 6.9 (Awbi 1991). The notation in the figure includes: the air supply velocity, U_o, the effective area of the supply opening, A_o, the centerline velocity, U_m, and the distance from the supply opening, x. In the core zone, the centerline velocity is equal to the initial jet velocity. The core region usually extends approximately four diameters or widths of the supply opening. The transitional region can be neglected for circular jets or plane jets when the aspect ratio (width/height of square opening) is smaller than 13.5 (Awbi 1991). For most commonly used supply diffusers, the first two zones are relatively small compared to the third or main jet zone. In the main jet zone, the flow is fully developed, and the velocity profiles are self-similar and can be described with non-dimensional jet profiles.

A fully developed jet flow is very important for ventilation design because diffuser jets usually enter the occupied space within their main zone (third zone). Therefore, the main zone is a subject of study by many researchers (Li *et al.* 1993). Actually, not only jet velocity profiles, but also jet temperature and concentration profiles tend to be self-similar in the main zone. Within a jet, heat and mass transfer are more intense than momentum transfer. Consequently, for the temperature and concentration profiles, the core region fades faster. The main zone for temperature and concentration develops faster than that for momentum (Grimitlyn and Pozin 1993). The fully developed region has many formulae to define the jet flow parameters such as velocity and temperature, which are key design parameters for the air distribution systems (Awbi 1991; Srebric 2000).

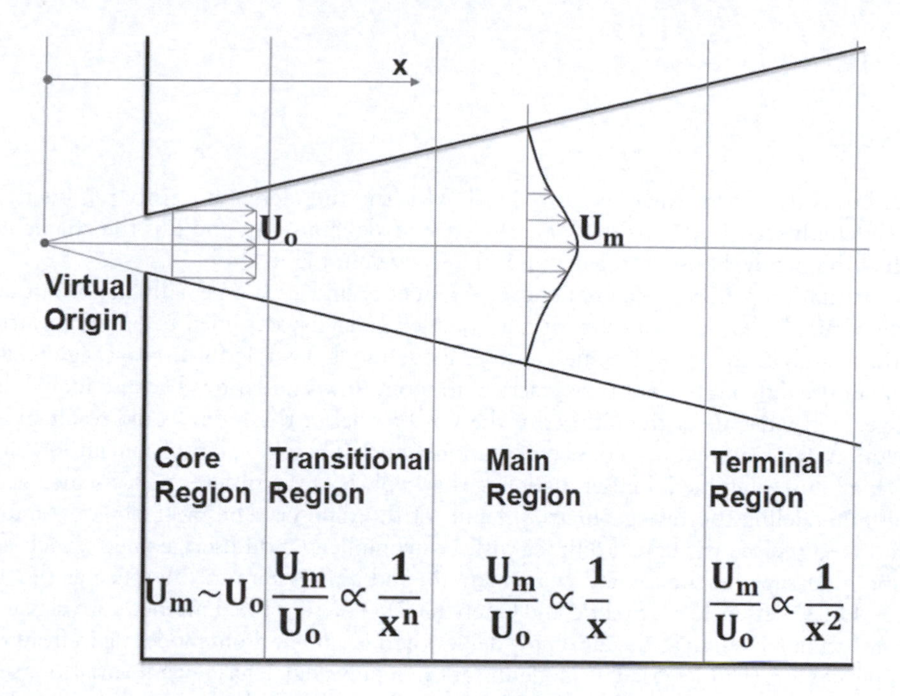

Figure 6.9 Jet zones and characteristic velocity profiles in log-log plot.
Source: Adapted from Awbi 1991.

Finally, in the terminal region, jet velocity decay is very rapid, and jet velocities become the same order as room airflow velocities. In the terminal region, jet separation may occur in a cold air distribution system, which can cause draft problems. The theory of the terminal jet region is still under development, and no formulae are available to describe jet degradation.

Some diffuser jets can be successfully predicted with the jet formulae, but not all of them. For example, Figure 6.10 shows the development of jet flow in front of the displacement diffuser (Srebric 2000). The jet drops immediately to the floor in the front of the diffuser because of the low air supply velocity and buoyancy effect. The jet then spreads over the floor and reaches the opposite wall. In front of the diffuser, the jet velocity profile changes along its trajectory. Close to the diffuser no jet formulae can be used since the jet is in a transition region. Only after 0.9 m does the jet form an attached jet. This result was obtained experimentally, but it can also be simulated with CFD.

The CFD approach is very popular due to its flexibility in testing different diffuser layouts and supply air jet parameters. Nevertheless, special attention has to be paid to boundary condition as the supply diffusers are usually dominant sources of momentum that create airflow pattern responsible for temperature and concentration distributions. The supply boundary condition may be specified as a supply velocity, pressure, or airflow rate. The airflow inlet condition requires specifying the mass flow rate and the temperature and contaminant concentration at the supply diffuser. The velocities and the pressure are calculated, by extrapolation, at the supply diffuser boundaries. In order to have an accurate CFD simulation for a turbulent flow, experimentally measured values of the turbulence parameters are also required for supply diffusers. For the k-ε turbulence models, these values are turbulent kinetic energy k and turbulent dissipation rate ε. When these values are not available from experimental data, they can be estimated from the following equations:

$$k = \frac{3}{2}(U_{ref}TI)^2, \tag{6.31}$$

$$\varepsilon = C_{\mu}^{3/4} \cdot \frac{k^{3/2}}{\ell}, \quad (6.32)$$

$$l = 0.07L, \quad (6.33)$$

where, U_{ref} is the mean stream velocity [m/s], *TI* is the turbulence intensity [–], *l* is the turbulence length scale [m], C_{μ} is the *k-ε* turbulence model constant, and *L* is the characteristic length of the supply diffuser (for a duct *L* is the equivalent radius) [m].

For ventilation performance predictions, the inlet boundary is especially important due to the use of supply diffusers with potentially complex geometry designed to produce particular performance characteristics. Detailed diffuser modeling is possible for limited regions in the vicinity of the diffuser, but not very practical in room-flow simulations because including the small geometric details of the diffuser in the CFD model of the room would result in a grid distribution that requires enormous computational time. Therefore, most room airflow simulations should use simplified diffuser modeling that replicates the diffuser performance without explicitly modeling the fine geometrics details of the diffuser. The most obvious simplified method is to replace the actual diffuser with less complicated diffuser geometry, such as slot opening, that supplies the same flow momentum and airflow rate to the space as the actual diffuser does (Nielsen 1992; Srebric and Chen 2002). The simplified methods are classified as jet momentum modeling at the air supply devices and jet momentum modeling in front of the air supply devices (Fan 1995). The modeling at the supply device has several variations such as basic, slot and momentum model, while the methods for the modeling in front of the diffuser are prescribed velocity, box, and diffuser specification model (Srebric and Chen 2002). The most widely used methods are:

- momentum,
- box, and
- prescribed velocity.

The momentum method decouples the momentum and mass boundary conditions for the diffuser in CFD simulations (Chen and Moser 1991). The diffuser is represented with an opening that has the same gross area, mass flux and momentum flux as a real supply diffuser. This model enables specification of the source terms in the conservation equations over the real diffuser area. The air supply velocity, U_o, for the momentum source term is calculated from the mass flow rate $\dot{m}$ and the diffuser effective area A_o:

$$U_0 = \frac{\dot{m}}{\rho A_0} \quad (6.34)$$

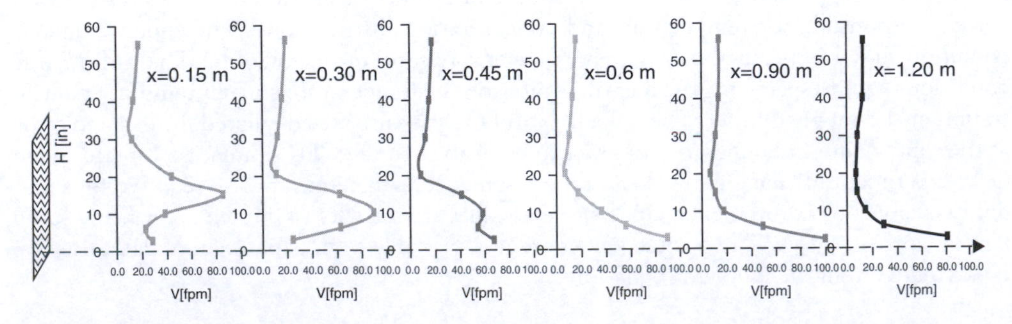

Figure 6.10 Development of the wall jet in front of the displacement diffuser (x is the distance from the diffuser).

The specification of boundary conditions with the momentum method is very simple, but the method might not work well for certain types of diffusers (Srebric 2000; Srebric and Chen 2002).

The box method is based on the wall jet flow generated close to the diffuser (Nielsen 1992; Srebric and Chen 2002). The details of the flow in the immediate vicinity of the supply opening are ignored, and the supplied jet is described along imaginary box surfaces in front of the supply diffuser. The data for velocity distribution in a wall jet (or a free jet) generated by different commercial diffusers can be obtained from diffuser catalogues or design guide books as ASHRAE Fundamentals and from other textbooks (Awbi 1991; Etheridge and Sandberg 1996; Rajaratnam 1976).

The prescribed velocity method has also been successfully used in the numerical prediction of room air movement. The inlet profiles are given as boundary conditions of a simplified slot diffuser, and they are represented only by a few grid points. All the variables, except the velocities, are predicted in a volume close to the diffuser as well as in the rest of the room. The velocities are prescribed in the volume in front of the diffuser as the fixed analytical values obtained for a wall jet from the diffuser, or they are given as measured values in front of the diffuser (Nielsen 1992).

More detailed description of available simplified methods and their applicability to commonly used supply diffusers can be found in the literature (Chen and Srebric 2002). Figure 6.11 shows how real diffusers can be simplified by the momentum method and box method in CFD simulations (Srebric 2000).

Mixing and displacement ventilation

Depending on the type of the supply diffusers, mechanical ventilation is classified as either mixing or displacement ventilation. Mixing ventilation relies on the forced airflow in the occupied space to remove air contaminants, bring fresh air for breathing, and preserve thermal comfort. These tasks are accomplished by completely mixing the air in the occupied zone with the supply air before it is exhausted. Different versions of this system exist and include supply with different ceiling diffusers, such as square or round ceiling diffusers. In contrast, displacement ventilation relies on stratified flow with minimal or no mixing, and several different variations of this system exist.

In displacement ventilation systems for cooling applications, conditioned air with a temperature slightly lower than the desired room air temperature in the occupied zone is supplied. The air supply diffusers are located at or near the floor level, and the supply air is directly introduced to the occupied zone. Returns are located at or close to the ceiling through which the warm room air is exhausted from the room as shown in Figure 6.12. The supply air is spread over the floor and then rises as it is heated by the heat sources in the occupied zone. Heat sources, such

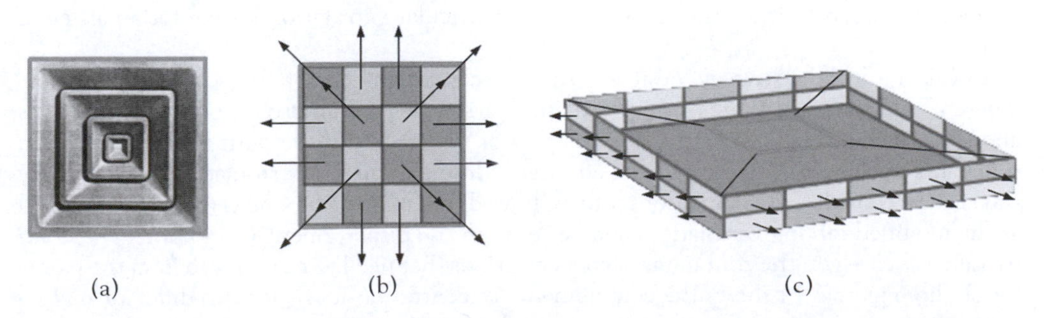

Figure 6.11 An example of simplified boundary condition for (a) supply (diffuser) modeling for square diffuser (b) diffuser modeling by momentum method, and (c) diffuser modeling by box method.

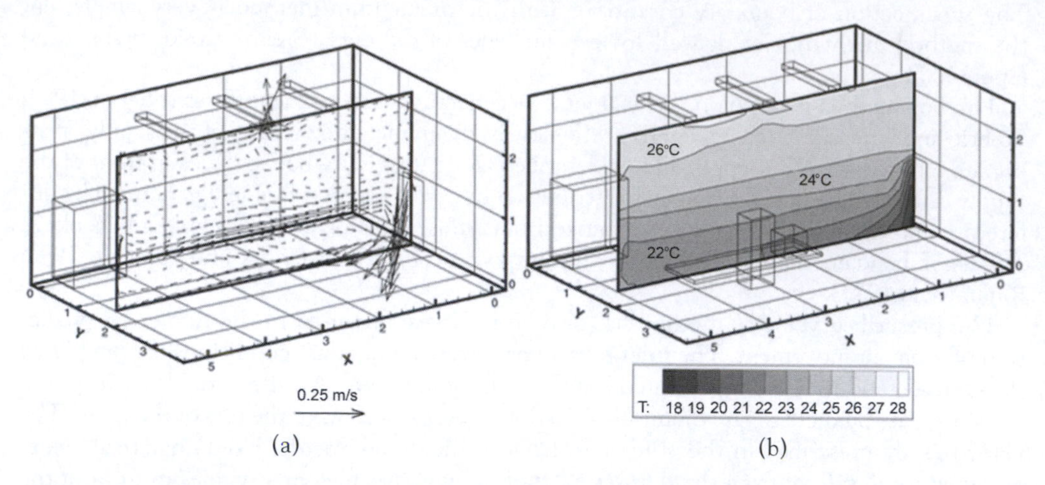

Figure 6.12 Computational fluid dynamics (CFD) simulation results for the displacement ventilation (a) velocities, and (b) temperatures in the vertical middle section.

as people and computers, in the occupied zone create upward convective flows in the form of thermal plumes. These plumes remove heat and contaminants that are less dense than air from the surrounding occupied zone.

Design example for mixing and displacement ventilation

An example of a practical CFD use in air distribution system design includes a decision on type of air distribution system with respect to targeted indoor air quality, thermal comfort, and building energy consumption. The objective of a practical study was to evaluate air distribution systems in a painting studio based on airflow pattern, air velocities, air temperatures, and contaminant concentrations. A properly designed ventilation system should provide better indoor air quality, such as lower contaminant concentrations, while maintaining good thermal comfort. Three different air distribution systems were evaluated, and one was selected based on the results of this study. The studied ventilation systems include existing mixing ventilation, modified mixing ventilation, and displacement ventilation, and one of the three systems was recommended for installation.

The mixing ventilation generates a strong jet flow, while the jet from displacement diffuser is weak. As a result, the mixing ventilation "steers" the air in the entire space and creates airflow with mostly uniform air parameters. On the contrary, the displacement diffuser makes a stratified flow for all indoor parameters. Nevertheless, these general descriptions are not sufficient to understand capabilities and deficiencies of a particular ventilation system for a particular space.

To select the best strategy, designers should account for use of the space and multiple different solutions, not only for the ventilations system type, but also for occupant activities and space layout. Figure 6.13 shows the layout of the studied space during art classes when occupants were creating paintings and afterwards drying them in the storage space. The study was performed with CFD simulations, which showed that existing mixing ventilation is as good as the modified mixing ventilation, and better than the displacement ventilation. The comparison was based on the contaminant concentrations that need to be removed from the painting studio. Figure 6.14 shows the contaminant concentration levels for two different mixing ventilation systems. As the final result of the study, the selected system was the existing mixing system with several modifications. The recommended modifications included decreasing the flow rate on the supply diffuser above the storage space, and adding an exhaust opening in the

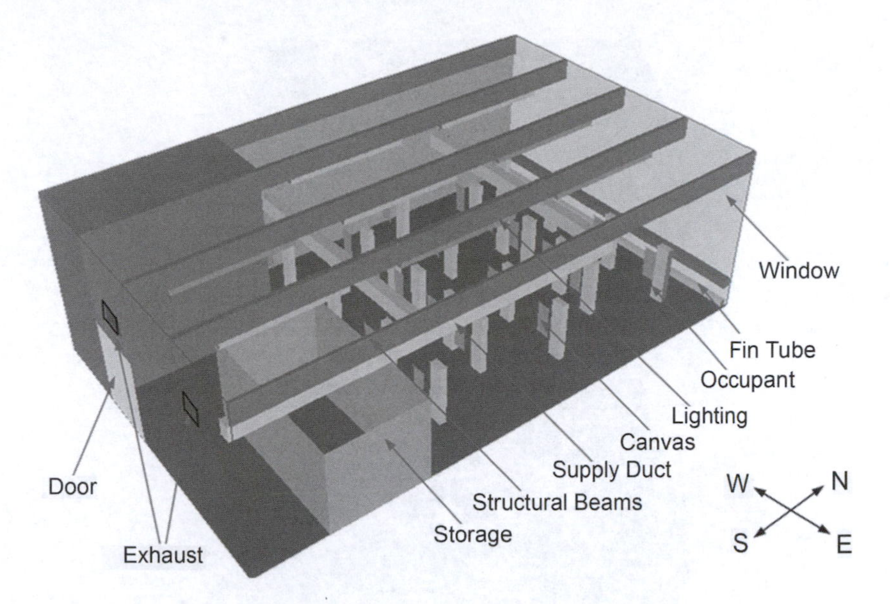

Figure 6.13 Layout of the painting studio during art classes.

corner of the room close to the storage area. The study concluded that these minor modifications are more effective on the improvement of the indoor air quality than the change of the air distribution system. In this way, costly and unnecessary modifications were avoided, while the goal of better ventilation was achieved.

The example of the ventilation in a painting studio is a case where only mechanical ventilation can be used to address removal of contaminants from internal sources. Nevertheless, there are many other practical applications where natural ventilation can be used by itself or as hybrid ventilation to include both mechanical and natural ventilation modes.

Discussion

This chapter briefly introduced different simulation tools available for performance prediction of mechanical and natural ventilation systems. The tools range from semi-empirical correlations, through simple multi-zone airflow network simulations, and complex CFD simulations. Each of the described methods has a set of implied assumptions such as perfect mixing and bulk flow for simpler models, and even much more complex models are based on numerous assumptions such as the eddy viscosity or the Boussinesq approximation. A responsible user of different models has to be aware of all of these assumptions, so that over a period of extensive use it becomes clear how these approximations make models either effective or ineffective in treating specific design and operation questions. It is easy to become excited about opportunities for simulations tools, but they have severe limitations that should be understood and addressed through future research efforts.

A good starting point in learning specific simulation tool for ventilation performance prediction is to reproduce an existing study with detailed published data. In particular, it is important to pick a study that has both simulation and experimental data to provide a new user with an opportunity to understand the accuracy of simulation tools as described in Chapter 1. For example, when comparing the simulation results to measured data, users can understand the instrumentation accuracy, and difference between computational and experimental results can be put into a perspective. If the simulation results and experimental data appear not to "disagree," but the difference is still smaller than the accuracy of the instruments used for measurements, then the agreement is actually "good," even though the graphical representation

(a)

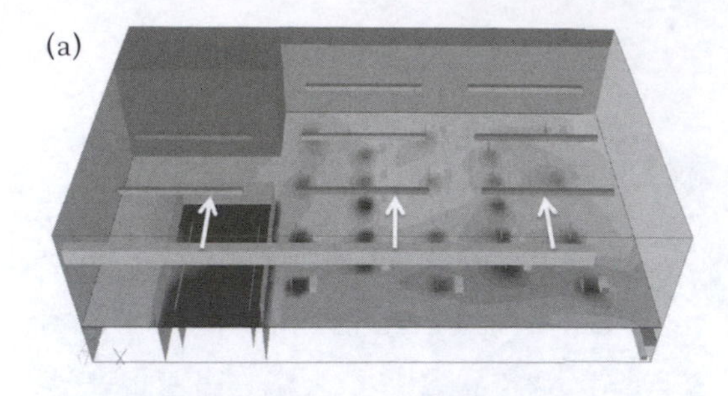

(b)

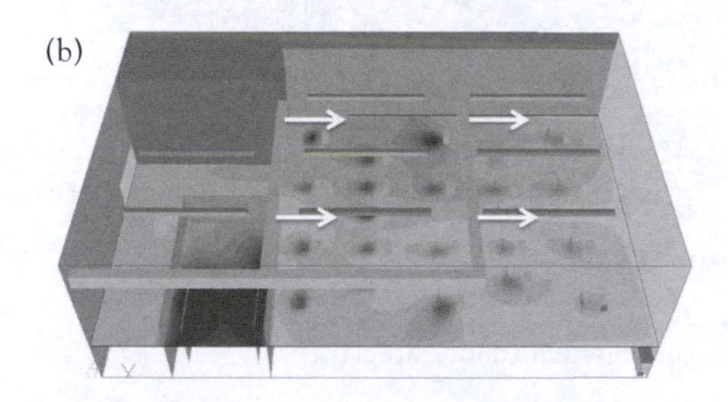

Figure 6.14 Contaminant concentrations in the breathing plane (1.5 m from the floor) for (a) the existing, and (b) modified mixing ventilation (darker shades indicate higher concentration levels).

appears to present discrepancies. Two case studies are presented as assignments in the end of this chapter, so that a new user can start building experience with multi-zone airflow network and CFD simulation tools.

The accuracy of simulation tools strongly depends on the appropriate setting of boundary conditions and numerical simulation parameters. As discussed, every CFD user with an intention of applying the results in design and operational practice is responsible for verification of the model capabilities, validation of calculated data, and appropriate reporting of not only the results, but also the simulation parameters. For new users, the verification process includes search for a similar previously published study as well as understanding of required physical models. The validation process does not have to be conducted on every single simulated case. In principle, the user should find an existing case as the close as possible to the new design case of interest, and first perform simulations and validation for the existing case. In this way, the user builds necessary experience, and can make an informed evaluation of new simulation results. Last, but not the least, a detailed report of CFD results that includes important simulation parameters, provides a quality assessment of conducted CFD study.

Synopsis

This chapter described a range of different analytical and simulation tools to predict ventilation performance in design and operation of buildings. The performance predictions used examples of mechanical, natural, and hybrid ventilation to illustrate capabilities of different simulation tools for ventilation design. Simulation tool users are strongly advised to expand the briefly introduced concepts by exploring the recommended readings. A clear future trend is identified to be represented by the coupled simulation tools that have available empirical equations, multi-zone airflow network models, CFD models, and energy simulations in a single platform. The users of coupled tools will be responsible to use the simplest, most applicable available tools in order to accurately resolve design and operational problems for ventilation systems.

References

AIAA (1998) "Guide for the Verification and Validation of Computational Fluid Dynamics Simulations", American Institute of Aeronautics and Astronautics (AIAA) Standard G-077–1998.

Allard, F. (1998) *Natural Ventilation in Buildings: A Design Handbook*, London: James & James Science Publishers Ltd.

ASHRAE (2004) "ANSI/ASHRAE Standard 55–2004: Thermal Environmental Conditions for Human Occupancy", American Society of Heating, Refrigerating and Air-conditioning Engineers (ASHRAE).

ASHRAE (2005) *Handbook of Fundamentals*, Chapters 16, 27, 33, and 34. American Society of Heating, Refrigerating, and Air-Conditioning Engineers (ASHRAE), Atlanta, Georgia.

ASHRAE (2007) "ANSI/ASHRAE Standard 62.1–2007: Ventilation for Acceptable Indoor Air Quality (IAQ)", American Society of Heating, Refrigerating, and Air-conditioning Engineers (ASHRAE).

Awbi, H.B. (1991) *Ventilation of Buildings*, London: Chapman & Hall.

Blevis, R.D. (1984) *Applied Fluid Dynamics Handbook*, New York: Van Nostrand Reinhold Company Inc.

Chen, Q. (1995) "Comparison of Different k-ε Models for Indoor Airflow Computation", *Numerical Heat Transfer, Part B, Fundamentals* 28(3): 353–369.

Chen, Q. and Moser, A. (1991) "Simulation of a Multiple-nozzle Diffuser", *Proceedings of the 12th AIVC Conference*, Ottawa, Canada, Vol. 2, pp. 1–14.

Chen, Q. and Srebric, J. (2002) "A Procedure for Verification, Validation, and Reporting of Indoor Environment CFD Analyses", *International Journal of HVAC&R Research* 8(2): 201–216.

CIBSE (2005) *Natural Ventilation in Non-Domestic Buildings*, London: Chartered Institution of Building Service Engineers.

Clarke, J.A. (2001) *Energy Simulation in Building Design*, 2nd ed., Oxford: Butterworth-Heinemann.

Deardorff, J.W. (1970) "A Numerical Study of Three-dimensional Turbulent Channel Flow at Large Reynolds Numbers", *Journal of Fluid Mechanics* 42: 453–480.

Djunaedy, E., Hensen, J.L.M. and Loomans, M.G.L.C. (2004) "Selecting an Appropriate Tool for Airflow Simulation in Buildings", *Building Services Engineering Research and Technology* 25(3): 269–278.

DOE (2008) "User Guides: Getting Started and Overview", Department of Energy, November 2008.

Dols, W.S. and Walton, G.N. (2002) "CONTAMW2.0 User Manual", National Institute of Standards and Technology (NIST), NISTIR 6921.

Etheridge, D. and Sandberg, M. (1996) *Building Ventilation – Theory and Measurement*, Chichester, West Sussex: John Wiley & Sons.

Fan, Y. (1995) "CFD Modeling of the Air and Contaminant Distribution in Rooms", *Energy and Buildings* 23(1): 33–39.

Fang, L., Clausen, G. and Fanger, P.O. (1998) "Impact of Temperature and Humidity on the Perception of Indoor Air Quality", *Indoor Air* 8(2): 80–90.

Ferziger, J.H. and Peric, M. (2001) *Computational Methods for Fluid Dynamics*, New York: Springer-Verlag.

Feustel, H.E. and Smith, B.V. (1997). *COMIS 3.0 – User's Guide*, Lawrence Berkeley National Laboratory.

Grimitlyn, M.I. and Pozin, G.M. (1993) "Fundamentals of Optimizing Air Distribution in Ventilated Spaces", *ASHRAE Transactions* 99(1): 1128–1138.

Hagentoft, C.E. (2001). *Introduction to Building Physics*, Lund, Sweden: Studentlitteratur.

Hensen, J. (1999) "A Comparison of Coupled and Decoupled Solutions for Temperature and Air Flow in a Building", *ASHRAE Transactions* 105(2): 962–969.

Hensen, J. and Augenbroe, G. (2004) "Performance Simulation for Better Building Design", *Energy and Buildings* 36(8): 735–736.

Launder, B.E. and Spalding, D.B. (1974) "The Numerical Computation of Turbulent Flows", *Computer Methods in Applied Mechanics and Energy* 3: 269–289.

Li, Z.H., Zhang, J.S., Zhivov, A.M. and Christianson, L.L. (1993) "Characteristics of Diffuser Air Jets and Airflow in the Occupied Regions of Mechanically Ventilated Rooms – A Literature Review," ASHRAE Transactions, 99(1): 1119–1127.

Lomas, K.J., Cook, M.J. and Fiala, D. (2007) "Low Energy Architecture for a Severe US Climate: Design and Evaluation of a Hybrid Ventilation Strategy", *Energy and Buildings* 39(1): 32–44.

Loomans, M. (1998) "The Measurement and Simulation of Indoor Air Flow", PhD Thesis, Eindhoven University of Technology, The Netherlands.

Mahdavi, A. and Pröglhöf, C. (2008) "A Model-based Approach to Natural Ventilation", *Building and Environment* 43(4): 620–627.

Melikov, A.K., Popiolek, Z., Silva, M.C.G., Care, I. and Sefker, T. (2007) "Accuracy Limitations for Low-velocity Measurements and Draft Assessment in Rooms", *International Journal of HVAC&R Research*, 13(6): 971–986.

Nakahashi, K. (2008) "Building-Cube Method: A CFD Approach for Near-future PetaFlops Computers", in *Proceedings of the 8th World Congress on Computational Mechanics (WCCM8)*, Venice, Italy.

National Research Council (2006) "Review and Assessment of the Health and Productivity Benefits of Green Schools: An Interim Report Committee to Review and Assess the Health and Productivity Benefits of Green Schools", National Research Council of the National Academies.

Nielsen, P.V. (1992) "Description of Supply Openings in Numerical Models for Room Air Distribution", *ASHRAE Transactions* 98(1): 963–971.

NIOSH (1997) "NIOSH Facts: Indoor Environmental Quality (IEQ)", National Institute for Occupational Safety and Health (NIOSH) Online. Available at: http://www.cdc.gov/niosh/topics/indoorenv/ (accessed 05.12.2009).

Novoselac, A. (2004) "Combined Airflow and Energy Simulation Program for Building Mechanical System Design", PhD Thesis, Pennsylvania State University.

Novoselac, A. and Srebric, J. (2003) "Comparison of Air Exchange Efficiency and Contaminant Removal Effectiveness as IAQ Indices", *ASHRAE Transactions* 109(2): 339–349.

Rajaratnam, N. 1976. *Turbulent jets*, Amsterdam: Elsevier Scientific Publishing Co.

Sherman, M. and Grimsrud, D.T. (1980) "Infiltration-Pressurization Correlation: Simplified Physical Modeling", *ASHRAE Transactions* 86(2): 778–807.

Sorensen, D.N. and Nielsen, P.V. (2003) "Quality Control of Computational Fluid Dynamics in Indoor Environments" *Indoor Air* 13(1): 2–17.

Srebric, J. (2000) "Simplified Methodology for Indoor Environment Design", PhD Thesis, Massachusetts Institute of Technology, Cambridge, MA.

Srebric, J. and Chen, Q. (2002) "Simplified Numerical Models for Complex Air Supply Diffusers", *International Journal of HVAC&R Research*, 8(3): 277–294.

Srebric, J., Chen, Q. and Glicksman, L.R. (1999) "Validation of a Zero-equation Turbulence Model for Complex Indoor Airflows", *ASHRAE Transactions* 105(2): 414–427.

Srebric, J., Yuan, J. and Novoselac, A. (2008) "In-Situ Experimental Validation of a Coupled Multi-zone and CFD Model for Building Contaminant Transport Simulations", *ASHRAE Transactions*, 114(1): 273–281.

Swami, M.V. and Chandra, S. (1988) "Correlations for Pressure Distribution on Buildings and Calculation of Natural-ventilation Airflow", *ASHRAE Transactions* 94(1): 243–266.

Tantasavasdi, C., Srebric, J. and Chen, Q. (2001) "Natural Ventilation Design for Houses in Thailand", *Energy and Buildings* 33(8): 815–824.

Tennekes, H. and Lumley, J.L. (1972) *A First Course in Turbulence*, Cambridge, MA: MIT Press.

USGBC (2005) "LEED for New Construction & Major Renovations – Version 2.2", U.S. Green Building Council (USGBC).

Versteeg, H.K. and Malalasekera, W. (2007) *An Introduction to Computational Fluid Dynamics*, Harlow: Longman.

Vukovic, V., Tabares-Velasco, P.C. and Srebric, J. (2010) "Real-Time Identification of Indoor Pollutant Source Positions Based on Neural Network Locator of Contaminant Sources (LOCS) and Optimized Sensor Networks", *Journal of the Air & Waste Management Association*, 60: 1034–1048.

Walton, G.N. (1989) "Airflow Network Models for Element-based Building Airflow Modeling", *ASHRAE Transactions* 95(2): 611–620.

Wang, L. and Chen, Q. (2008) "Evaluation of Some Assumptions Used in Multi-zone Airflow Network Models", *Building and Environment* 43(10): 1671–1677.
Wargocki, P., Sundell, J., Bischof, J., Brundrett, G., Fanger, P.O., Gyntelberg, F., Hanssen, O., Harrison, P., Pickering, A., Seppanen, O.A. and Wouters, P. (2002) "Ventilation and Health in Non-industrial Indoor Environments: Report from European Multidisciplinary Scientific Consensus Meeting (EUROVEN)". *Indoor Air* 12(2):113–128.
Wilcox, D.C. (2006) *Turbulence Modeling for CFD*, La Cañada, CA: DCW Industries, Inc.
Yaglou, C.P. and Witheridge, W.N. (1937) "Ventilation Requirements", *ASHRAE Transactions*, 42(2): 423–436.
Yokhot, V. and Orszag, S.A. (1986) "Renormalization Group Analysis of Turbulence: I. Basic theory", *Journal of Scientific Computing* 1(1): 3–51.
Yoshino, H., Yun, Z., Kobayashi, H. and Utsumi, Y. (1995) "Simulation and Measurement of Air Infiltration and Pollutant Transport Using a Passive Solar Test House", *ASHRAE Transactions* 101(1): 1091–1099.

Recommended reading

Clarke, J.A. (2001) *Energy Simulation in Building Design*, 2nd ed., Oxford, UK: Butterworth-Heinemann.
Etheridge, D. and Sandberg, M. (1996) *Building Ventilation – Theory and Measurement*, West Sussex, England: John Wiley & Sons.
Versteeg, H.K. and Malalasekera, W. (2007) *An Introduction to Computational Fluid Dynamics*, Harlow, Essex: Longman.

Activities and model answers

Two assignments are included to enable practice of multi-zone airflow network and CFD modeling. Both assignments are specified in such a way that users can conduct them with any of the available simulation tools. The first assignment is to model a house with a multi-zone airflow network tool, and the second assignment is to model an office with a CFD tool.

1 Multi-zone airflow network assignment

The objective of this assignment is to simulate contaminant dispersion due to infiltration in a small house. To correctly predict dispersion, the prediction of infiltration and airflow pattern has to be accurate. This particular case is selected because it has detailed measured data published in the literature (Yoshino *et al.* 1995). This study conducted measurements in a three-zone single story building as shown in Figure 6.15(a). Rooms 1, 2 and 3 had contaminant concentration sensors and were completely isolated from the unmarked room on the right-hand side in Figure 6.15(a). Each of the three rooms (zones) had a burst source releasing different tracer gases: $CHCl_2F$, N_2O and SF_6. This set-up allowed simultaneous tracer gas releases and concentration measurements within the three rooms (zones).

This assignment is focused on simulation of SF6 with a point source in room 3. Figure 6.15(b) presents constructed multi-zone airflow network model for three relevant rooms, defining the contaminant source location in room 3 (zone 3). The following model parameters were selected based on the experimental study (Yoshino *et al.* 1995; Vukovic *et al.* 2010):

The height of the floor is set to 2.8 m with zero elevation from the ground.

Table 6.4 specifies detailed properties of the three building zones.

For all flow paths between the zones, one-way flow elements are selected. The choice is based on multi-zone airflow network tool that allows users to define flow exponent values only for one-way flow elements. Consequently, to take into account two-way flows, two differently elevated elements are placed for each real flow path.

Additional flow elements define infiltration due to background leakage. Table 6.5 specifies the properties of flow path elements between the zones. Relative elevation of the flow elements defines the position of the centre of the leakage area of the element with respect to the

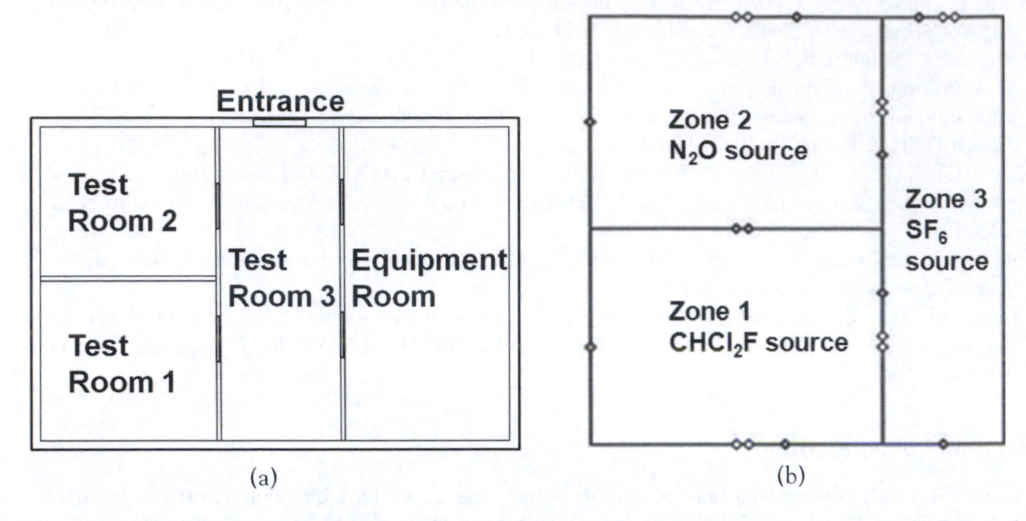

(a) (b)

Figure 6.15 (a) The floor plan of the test house and (b) multi-zone airflow network model (◊-flow paths symbol).

Sources: (a) Adapted from Yoshino *et al.* 1995; (b) Adapted from Vukovic *et al.* 2010.

Table 6.4 The summary of the zone properties

Zone number	*Zone name*	*Floor area*	*No. of flow paths* [m^2]	*Temperature* [°C]
1	Room 1	7.425	9	33.4
2	Room 2	7.425	9	26.0
3	Room 3	9.735	10	28.2

Table 6.5 The summary of airflow path properties

Connected zones	*Relative elevation* [m]	*Leakage area* [cm^2]
1–2	0.625	86
1–2	1.875	86
1–3	0.625	124
1–3	1.875	124
1–3	1.4	3
1–ext_s	0.85	26
1–ext_s	1.95	26
2–3	0.625	150
2–3	1.875	150
2–3	1.4	3
2–ext_n	1.725	21
2–ext_n	2.175	21
2–ext_n	1.4	6
2–ext_w	1.4	6
3–ext_n	0.625	141
3–ext_n	1.875	141
3–ext_n	1.4	172
3–ext_s	1.4	172

Legend: ext – exterior n – north s – south w – west

floor level. All selected flow elements have discharge coefficient 0.6 and pressure drop of 10 Pa during the measurements.

Outdoor conditions are specified as following: south-west wind of 1.06 m/s, normal atmospheric pressure, and outdoor temperature of 25.3 °C.

The initial uniform concentration of SF_6 in room 3 is 1 ppm, which is normalized value, so that the results are independent of source strength.

Model validation: All measured concentration data showed non-zero background contaminant concentrations. Background concentration levels also differ with respect to contaminant species. In order to compare calculated SF_6 to measured data, these background concentrations are subtracted from all measured concentration values. This procedure enables the assignment modeling task to be independent of the background concentration levels. Figure 6.16 shows a validation that compares simulation results to measured data for SF_6 concentrations in each of the three rooms. Overall, even though the model is simple, the exercise is challenging because it includes the validation.

2 *CFD assignment*

This assignment requires CFD modeling of a displacement ventilation case shown in Figure 6.17. In addition, Table 6.6 and Table 6.7 provide most important thermal, and flow boundary conditions to be used in any CFD simulation tool, and additional detailed geometry data are available in the literature (Srebric 2000). This particular set-up has also available detailed measured data in the occupied zone, so it presents an opportunity for users to validate and verify their simulation technique (Chen and Srebric 2002). Figure 6.12 shows the simulated velocities and temperatures that are plotted in the vertical middle section. The simulated results are in very good agreement with the observed flow pattern and measured velocities and

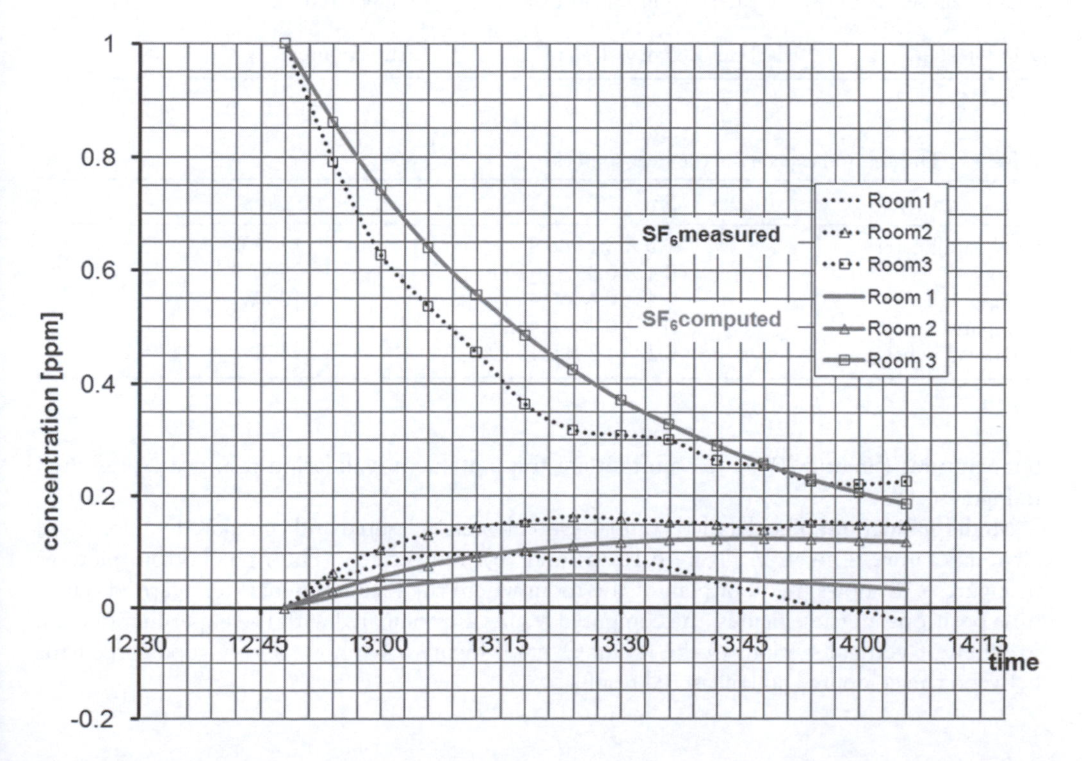

Figure 6.16 Measured vs. computed SF_6 concentrations.

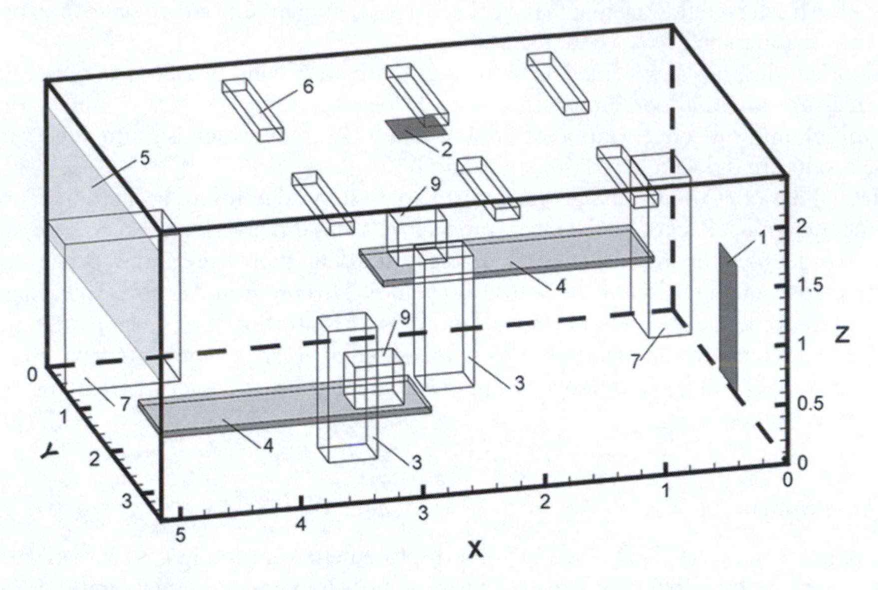

Figure 6.17 The layout of the displacement ventilation simulated and tested in the chamber (inlet –1, outlet –2, person –3, table –4, window –5, fluorescent lamps –6, cabinet –7, computer –9).

Table 6.6 The geometrical, thermal, and flow boundary conditions for the diffuser and window

	Displacement ventilation case		
Supply diffuser	Size: 0.53 m × 1.1 m	Temperature: 17.0 °C	Velocity: 0.086 m/s
Window	Size: 3.65 m × 1.16 m	Temperature: 27.7 °C	–

Table 6.7 The size and capacity of the heat sources

Heat source	*Size*	*Power*
Baseboard heater	$1.2 \times 0.1 \times 0.2$ m^3	1500 W
Person	$0.4 \times 0.35 \times 1.1$ m^3	75 W
Computer 1	$0.4 \times 0.4 \times 0.4$ m^3	108 W
Computer 2	$0.4 \times 0.4 \times 0.4$ m^3	173 W
Overhead lighting	$0.2 \times 1.2 \times 0.15$ m^3	34 W

temperatures (Srebric 2000). To substantiate this statement, validation results are presented in Figure 6.18.

Model validation: The detailed comparison between measured and calculated velocity and air temperature is given in Figure 6.18 (a) and (b), respectively. The right bottom pictures in Figure 6.18 shows the floor plan of the room where the measurements were carried out at nine positions. In these figures, the computed results are compared with the experimental data in different vertical sections in the room. Overall, Figure 6.18 shows a very good agreement between the computed and measured results.

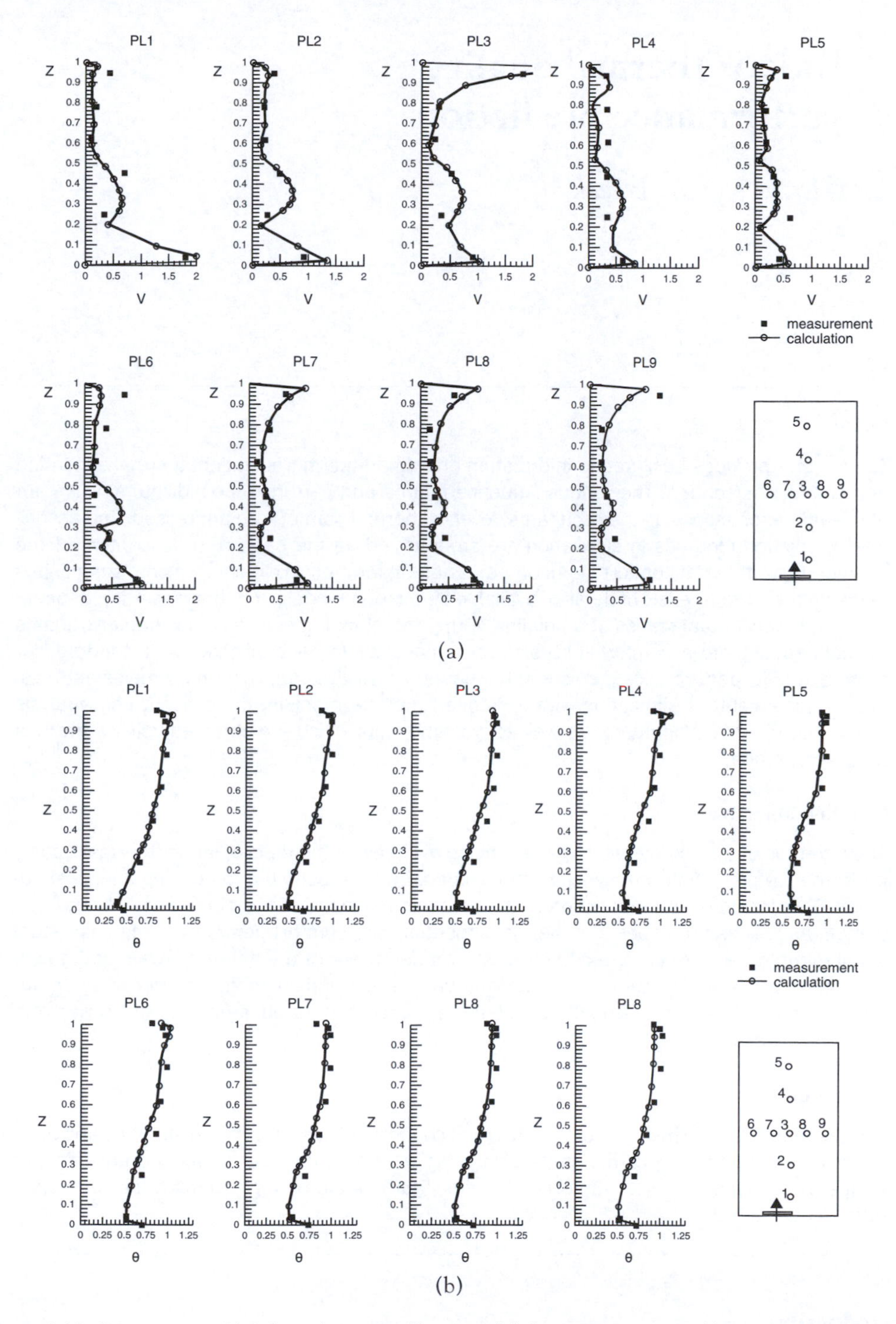

Figure 6.18 Comparison of (a) the velocity profiles, and (b) the temperature profiles at nine positions in the room between the simulated and measured data for the displacement ventilation case. Z=height/total room height (H), V=velocity/inlet velocity (Vin), θ=(T-Tin/Tout-Tin), H=2.43m, Vin=0.086m/s, Tin=17.0°C, and Tout=26.7°C.

7 Indoor thermal quality performance prediction

Christoph van Treeck

Scope

The chapter provides background information on human thermoregulation, thermal sensation and comfort perception. The various influence factors and performance indicators which are of relevance for assessing the occupants' level of thermal comfort are addressed. The associated prediction methods in simulation are emphasized on the basis of a description of the conditions for thermal comfort in uniform and non-uniform environments which distinguishes between the various levels of detail of available simulation models and their accuracy, depending on the conceptual stages of a building during the planning process. The chapter explains the concept of thermal comfort indices and summarizes international codes and standards for thermal quality performance prediction. The chapter also gives application examples addressing the local comfort simulation with a detailed thermal manikin, the draught risk analysis using a coupled CFD approach, as well as typical results which are obtained using a thermal multi-zone model.

Learning objectives

The intention of this chapter is to give a concise overview of the topic of indoor thermal quality performance prediction with special emphasis on, but not restricted to building energy simulation. This includes an introduction to the dynamic reactions of the human body to thermal stimuli, as well as the aspects of thermal sensation and comfort perception. The case studies presented here are designed to make the reader aware of the different levels of detail in simulation and the physical resolution of the various methods, enable him/her to select an appropriate assessment method in simulation practice and introduce him to the codes and standards available.

Key words

Factors that influence thermal comfort/thermal comfort indices/calculation methods and algorithms/other relevant industrial areas/design stage related levels of detail and uncertainty/temperature sensation vs. comfort perception/human thermoregulation/codes and standards/case studies

Introduction

From the human body to thermal quality performance prediction in buildings

It is well known from experiments that the temperature sensation and the perception of thermal comfort are related to the thermal state of the human body, as detected by thermoreceptors, that they depend on skin and core temperatures, respectively, and their variation over time (Fanger 1972; Gagge 1973; Stolwijk 1971). The efforts for regulating body temperature and moisture sensation on the skin are also of significance (Gagge 1973). The thermal state of the body (pages 183–187) is composed of several thermophysical and thermoregulatory processes with varying impacts on individual body parts. The body becomes acclimatized due to physiological adaptations. With these remarkable properties the human body is able to maintain its core temperature at a constant level for a wide range of ambient conditions.

Thermal sensation and comfort can be expressed using assessment scales (pages 189–190) and are usually regarded in terms of steady state, uniform conditions for the body as a whole (Fanger 1972) in relation to the indoor microclimate (pages 190–191), which is a rational approach for a large number of simulation applications in practice. Recently, so-called adaptive models (pages 195–196 and 198–201) also take into account the behavioral adaptation of people in naturally ventilated buildings and relate the acceptability of thermal conditions to the change in ambient climate over time.

The ergonomic requirements of the local thermal environment can be obtained by taking measurements and interviewing subjects exposed to ambient conditions, as indicated in Figure 7.1. A non-uniform environment is then correlated to a uniform environment using the concept of equivalent temperatures (see pages 188–189 and 192–194) for individual parts of the body respectively. Performing this type of experiment is, however, costly and time-consuming.

Building performance simulation offers a cost-effective tool for a thorough investigation of the aforementioned effects. Models are available for statistically correlating performance indicators with people's average assessment and thus for predicting the impact of ambient conditions. As previously shown in this book (cf. Chapter 5), simulation approaches range from steady-state monthly and annual balances (rough level) over dynamic multi-zone models up to

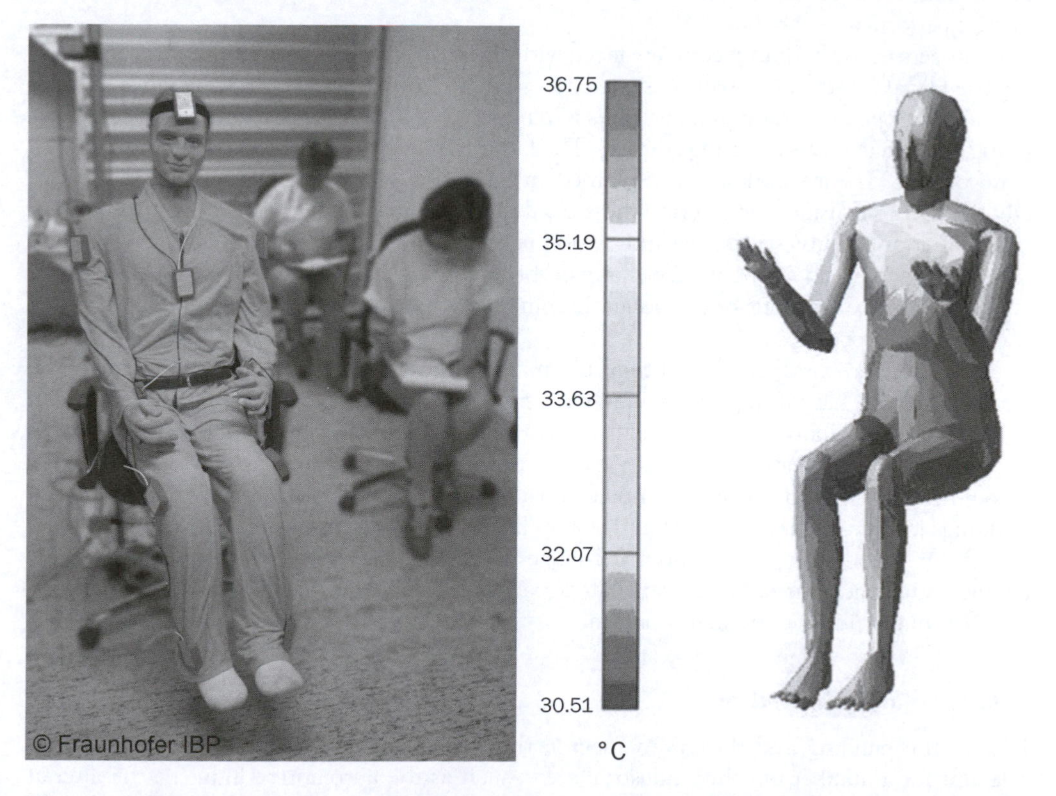

Figure 7.1 Thermal manikin measuring resultant surface temperature (RST) values (courtesy of Fraunhofer Institute for Building Physics, Germany) versus results of RST values obtained by numerical simulation.

Source: Adapted from van Treeck *et al.* 2007a.

computational methods modeling fluid flow and radiative heat transfer in detail (fine level). Combined approaches exist in order to bridge the gap between the multi-scale and multi-level problem (Beausoleil-Morrison 2000; van Treeck 2004).

For assessing the indoor thermal quality of a building in terms of comfort, descriptive integral quantities are required, such as the frequency of temperatures exceeding a certain limit during a defined period, or an even more sophisticated model for predicting the risk of overheating. The section on pages 197–202 introduces the relevant prediction methods. The required data can be acquired using dynamic simulation models which use the building geometry, the structure and materials of all components, the site specific conditions and the type of the energy concept as input data.

Researchers increasingly tend to concentrate on both highly refined and combined numerical models for assessing comfort. For such an integrated study, essentially a flow and a thermal radiation solver are needed, as well as a thermal manikin modeling the heat exchange between the body and the environment, including human thermoregulation. In addition, a model is required to relate the computed skin temperatures to local thermal sensation votes. The sections on pages 192–194 and 207–209 address this issue.

When predicting performance, a distinction is generally made between the aspects of indoor thermal quality and indoor air quality. The latter is dealt with in Chapter 6 of this book. With regard to health criteria, engineers are confronted with complex integrated relationships involving thermal, physical, chemical and microbiological phenomena, such as the sick building syndrome (Brager and de Dear 1998; Kröling 1998; Stolwijk 1991) or the tight building syndrome (Recknagel *et al.* 2007). The thermal indoor climate and air quality influence both health and productivity; a well-designed indoor thermal environment helps to increase the mental performance capability.

Engineers in the building sector are faced with the problem that the energy concept, including the HVAC system, is usually designed according to heating and cooling demands in order to maintain operative room temperatures at a reasonable level – and not primarily according to local thermal effects affecting comfort. The latter temperature level can be determined by a comfort zone. The methodology is explained on pages 197–198. Such local effects are not usually taken into consideration during the early design stages of a building project. It is particularly important to investigate the interaction between energy system, building envelope and (thermally active) components, especially in the energy efficient building design (Pfafferott *et al.* 2007), as modifications of the design become more expensive during subsequent stages of the planning process.

People who are dissatisfied with their thermal environment are likely to try to remedy the situation, which has an impact on energy consumption (de Dear *et al.* 2002; van der Linden *et al.* 2002). The indoor thermal quality accordingly influences the energy consumption as regards cooling and heating demands. As explained later in this chapter, several standards give recommendations for a reasonable choice of parameters affecting the thermal environment in buildings in terms of both comfort and system layout (see the sections on pages 202–209 and 209–211). Among the most significant changes in standards over the last few years is to take people's adjustment in naturally ventilated spaces in account as a result of conscious action aimed at improving the thermal situation.

Other relevant industrial areas

Besides the building and the HVAC sector, the assessment of thermal comfort is likewise relevant for a number of other industrial areas, such as the automotive industry, the aircraft industry and the railway/coach industries, as indicated in Figures 7.2 and 7.3. In vehicles, for example, many effects interact within a small space, such as short-wave radiation which is locally absorbed by parts of the body while the air-conditioning directs a cold stream of air to the thorax, for example, and areas of skin get wet because they have been in contact with

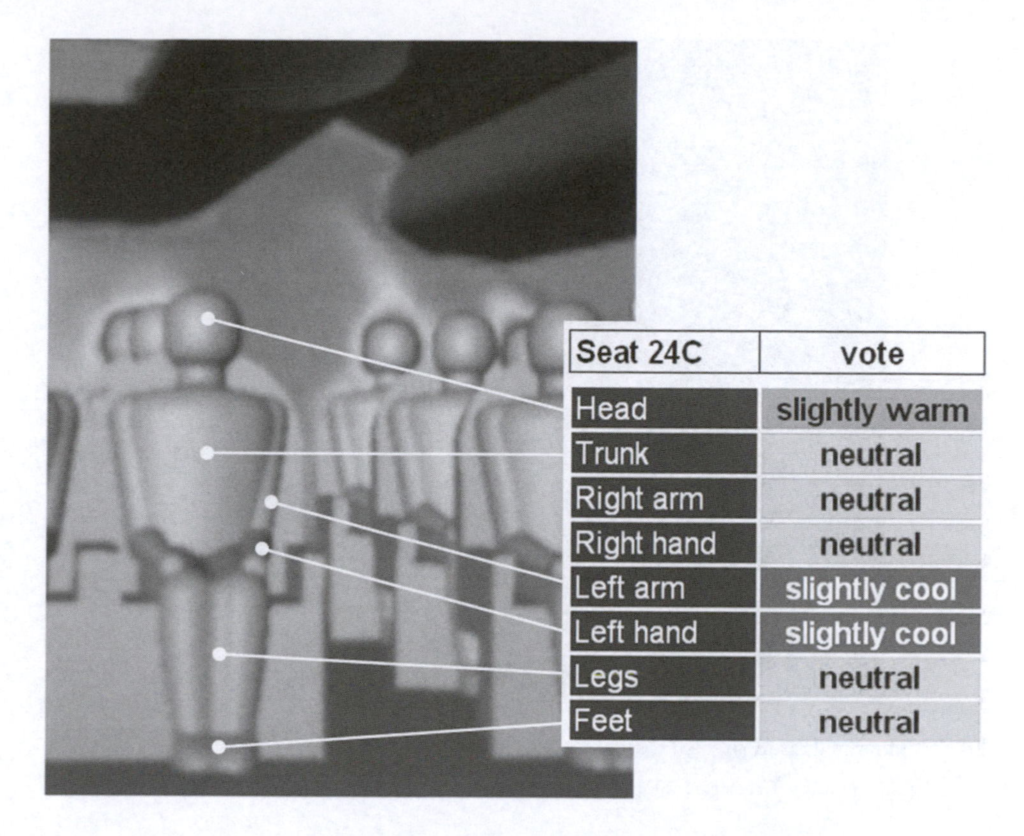

Figure 7.2 Example of a thermal flow analysis in an aeroplane passenger cabin using a lattice Boltzmann type CFD solver and the comfort assessment using equivalent temperatures. The cross-section shows the thermal plumes of the convective flow.

Source: Adapted from van Treeck *et al.* 2007b.

the fabric, which may cause discomfort. A single effect may accordingly dominate the overall individual perception.

In addition to the aspect of thermal quality, health, occupational safety and risk management may also become important issues in respect of the design. For example, operative temperatures in spaces equipped with machinery must be limited to moderate levels in order to ensure safe working conditions. Special performance indicators relating the heat stress are used in this context.

Human thermoregulation, energy balance and temperature sensation

Human thermoregulation

The human body is able to maintain its core temperature at approximately 37 °C through a wide range of ambient conditions. The thermal state of the body is composed of several thermophysical and thermoregulatory processes with various effects on different parts of the body. The loss of heat to the environment due to convective and radiative heat transfer, heat conduction through contact, evaporation and respiration is compensated by internal heat production due to conscious or unconscious muscle activity and by a basal metabolic rate, which together give the total metabolic rate.

Figure 7.3 Flow analysis in an open train carriage. The picture shows the average velocity field.

Source: Adapted from van Treeck *et al.* 2007a.

Human thermoregulation involves four essential autonomic regulation mechanisms in order to control the heat exchange with the environment. The skin blood flow can be changed by vasoconstriction (narrowing of blood vessels) and vasodilatation (widening of blood vessels). Evaporative heat loss is caused by sweating while shivering produces heat by increased metabolism in the muscles. Conscious reactions, such as changing clothing, performing exercises or changing one's position, likewise actively support the regulation process.

Thermoregulation is controlled by various stimuli. Skin temperature is detected by cutaneous thermoreceptors which are inhomogeneously distributed over the surface and which possess regionally varying sensitivities. Heat and cold receptors operate over different but overlapping ranges and cold receptors are more densely distributed than warm receptors; it is important to distinguish between receptors sensing warmth, cold and pain. The behavior of the heat and cold receptors makes it possible to use a so-called set-points concept in thermoregulation models. Both types show a dynamic overshoot in their response frequency if skin temperatures change over time (Hensel 1982). Concerning the body core, thermo-sensitive neurons are mainly concentrated within the hypothalamic region. The hypothalamus can also be stimulated by local warming or cooling which causes dilatation and sweating or constriction and shivering, respectively. It plays a central role in information processing.

To summarize, thermoregulatory reactions of the central nervous system are the response of multiple functions of signals from core and peripheral areas. Moreover, local changes in skin temperature also cause local reactions, such as modifications in the perspiration rate or local vasodilatation. Significant indicators are the average skin temperature and its variation over time, and the hypothalamus temperature. Calorimetric values only exert a slight influence. The indicators can be correlated with the autonomic responses described above to form a detailed thermoregulation model (Fiala *et al.* 1999; Gagge 1973; Gordon 1974; Holmér 2004; Huizenga *et al.* 2001; Stolwijk 1971; Tanabe *et al.* 2002; and Wissler 1985).

Thermal exchange with the environment

The thermal exchange between the human body and the environment can be described using the following energy balance equation (Fanger 1972).

$$\underbrace{M - W}_{\substack{\textit{net heat}\\ \textit{production}}} = \underbrace{(Q_{conv} + Q_{cond} + Q_{rad} + E_{skin})}_{\substack{\textit{skin}\\ \textit{heat losses}}} + \underbrace{(C_{resp} + E_{resp})}_{\substack{\textit{respirative}\\ \textit{heat losses}}} + \underbrace{S}_{\substack{\textit{heat}\\ \textit{storage}}} \tag{7.1}$$

The net heat production ($M - W$) comprises the total metabolic rate M minus the portion of generated heat which is converted into external mechanical power W. The metabolic activity is considered in the unit [met] in terms of the heat production per unit of skin area where 1met = $58 W/m^2$ for a resting adult person.[1] Tables indicating the metabolic heat generation for different levels of activity are summarized in standards, e.g., ISO 7730 or ASHRAE Standard 55.

On the right-hand side of Equation (7.1) the terms for heat loss and storage are accumulated. The sensitive heat losses from the skin involve convective and radiative heat transfer Q_{conv} and Q_{rad}, heat conduction through contact Q_{cond} (often neglected by using adiabatic boundary conditions in simulation), and latent heat loss by sweat evaporation and moisture diffusion E_{skin}. The respiratory system participates by means of sensitive heat flow C_{resp} and latent heat flow E_{resp}. The heat storage term S, i.e. the change in internal energy, can be described in terms of the thermal capacity and the time rate of change in transient conditions. The model can be refined by regarding the body as consisting of two thermal compartments, skin and core, using the two-node Gagge model (1973) or in a more detailed way with a multi-segment model (Stolwijk 1971), which includes models for the blood circulation and thermoregulation.

In simulation, the thermal exchange between the human body and the environment can be computed by numerical approximation of the governing partial differential equations or described with simplified algebraic expressions using linearized heat transfer coefficients (Bejan 1993; Modest 2003). The balance is usually established on the outer clothing or the exposed skin surface where the sensitive and latent heat losses can be expressed in terms of the parameters of skin temperature and skin wetness in relation to the independent environmental variables of air temperature, air velocity, mean radiant temperature and ambient water vapor pressure. For a detailed description of the terms in the heat balance equation, the reader is referred to Fanger (1972); Fiala *et al.* (2001); Stolwijk (1971) or ASHRAE (2005).

A short review of thermal sensation versus comfort

It is important to distinguish between the terms skin temperature, thermal sensation and thermal comfort (acceptability). As cutaneous thermoreceptors are inhomogeneously distributed over the skin and possess a different regional sensitivity (Guyton 2002; Hensel 1982), this asymmetry needs to be reflected by the formulae for calculating the average skin temperature. Details are given in Fiala (1998) and McIntyre (1980). Fiala (1998) also summarizes the interpretation of the various thermal stimuli from core and peripheral areas in terms of the autonomic thermoregulatory responses.

The autonomic responses of the human thermoregulation are correlated with the skin and core temperatures and their variation over time. With this dynamic property, for example, people feel colder when their skin temperature is actively falling compared to when the temperature remains at a constant but cold level (ASHRAE 2005). The moisture sensation on the skin is likewise significant, as the skin moisture correlates with warm discomfort (Gagge 1973, 1979) as well as the effort for regulating body temperatures (Berglund and Cunningham 1986).

Gagge (1973, 1986) proposed a simplified two-node thermoregulation model and derived two measures dedicated to thermal sensation and thermal discomfort. The indices are obtained from the quantities, such as (i) modified effective temperature (ET*, see page 189) and (ii) skin wetness and are based on experimental data under uniform and steady-state conditions (cf. Fanger 1972; Nevins *et al.* 1966; Rohles and Wallis 1979). Provided the operative temperature is within the comfort range, humidity fluctuations tend not to have an appreciable effect for relative humidities between 20 and 70 per cent (Gonzalez and Gagge 1973; Gonzalez and Berglund 1979; Nevins *et al.* 1975).

For uniform and steady state conditions (Nevins *et al.* 1966; Rohles and Wallis 1979) and moderate climates, Fanger (1972) found out that the level of comfort can be related to the disequilibrium in the energy balance of the human body (see pages 190–191). For most engineering applications it is therefore advisable to use a statistically correlated assessment of the comfort perception for the body as a whole, wherever details of the thermoregulation are not required. It is practical to directly relate environmental parameters to a single value in terms of comfort perception criteria (pages 190–191 and 191–192). However, a number of

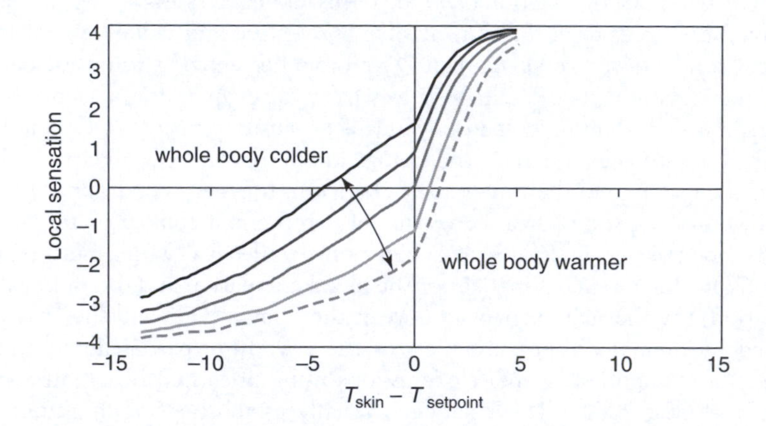

Figure 7.4(a) Local thermal sensation with respect to set-point deviation (from thermal neutrality) and thermal state of the whole body.

Source: Zhang 2003; Zhang *et al.* 2004. Reprinted with the author's permission.

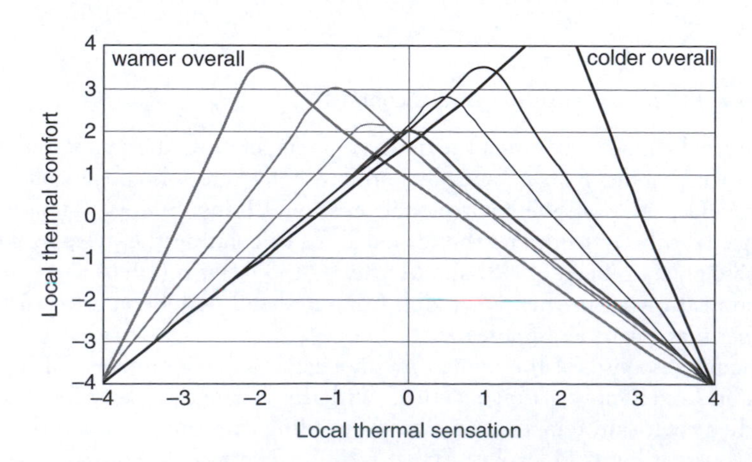

Figure 7.4(b) Local thermal comfort related to the set-point deviation.

Source: Zhang 2003; Zhang *et al.* 2004. Reprinted with the author's permission.

applications require additional predictions of local effects such as asymmetric conditions and/or transient behavior where the aforementioned approach is not feasible (see pages 192–194).

Mower (1976) showed that local thermal sensation is governed by skin temperatures. However, local (dis)comfort was found to depend on the general comfort perception (Cabanac and Hardy 1969; Issing and Hensel 1981). Zhang (2003) showed that temperature sensation and comfort perception depend on the overall thermal state of the body as captured by thermoreceptors, and she quantified the weighted influence of the individual body parts. Figure 7.4(a) shows the relationship between the skin temperatures' deviation from their set points at thermal neutrality and the local thermal sensation. Negative values correspond to perceptions of cold and positive values to warm ones. The graphs indicate an asymmetric behavior of the logistic function with respect to its response to higher and lower ambient temperatures as well as the influence of the thermal state of the whole body (Zhang 2003).

In Figure 7.4(b), the assessed local thermal sensation is plotted versus the comfort perception in terms of their acceptability. Again, the result depends on the overall thermal state. So the perceived sensation of optimum comfort shifts according to the overall feeling. If, for example, the core temperature is high, cold stimuli are perceived as more agreeable than warm ones (Zhang 2003; Zhang *et al.* 2004).

In asymmetrical environments, parts of the body may therefore feel comfortable, even if the body as a whole does not. So it is impossible to apply the concept of neutral thermal sensation without modification. This effect is even stronger in transient situations. In order to account for dynamic responses to transient conditions, for instance, a time derivative of, say, the skin temperature needs to be included in a model that corresponds to the transient sensations experienced. In this case, signals from thermoreceptors contain a static and a dynamic part (Hensel 1982; Ring and de Dear 1991).

Performance indicators and conditions for thermal comfort

This section summarizes the environmental and personal parameters affecting thermal comfort, explains the idea of operative and equivalent temperatures, and introduces thermal sensation scales and basic environmental indices. The concept of the predicted mean vote index is introduced and discussed with respect to people's behavioral adaptation. The section on pages 197–202 also details the issue of thermal comfort prediction in buildings from the point of view of comfort envelopes and the respective guidelines.

Definition of thermal comfort

Benzinger (1979) defines thermal comfort as "the absence of driving impulses" from cutaneous and hypothalamic receptors causing the body to counteract with physiological adaptations. A resting and unclothed individual perceives ideal thermal comfort at thermoneutral conditions at approximately 30 °C ambient temperature (Gagge *et al.* 1967) where conscious actions are not required for controlling the thermal state. The ASHRAE Standard 55 associates thermal comfort with the "condition of mind that expresses satisfaction with the thermal environment".

Parameters affecting thermal comfort

The primary environmental parameters which have influence on thermal sensation and comfort are:

- the air temperature (McNall and Schlege1968),
- the (relative) air velocity (Rohles *et al.* 1974),

- the mean radiant temperature of the surrounding surfaces, and
- the ambient water vapor pressure (Rohles *et al.* 1971).

In addition to affecting the heat transfer in the energy balance, these variables are "perceptible" if local effects are present such as asymmetrical radiation, draught or vertical temperature gradients which are significant indicators for describing discomfort (Fanger *et al.* 1985). The thermal comfort perception for people all over the world living in different climates and in cultural diversities appears to be statistically uniform if similar environmental and personal parameters are taken into account (ASHRAE 2005; Busch 1992; de Dear *et al.* 2002; Fanger 1972).

The air velocity has a significant influence if, for example, a cold air stream is directed towards part of the body. Furthermore, the degree of turbulence is relevant, i.e. if fluctuations perpendicular to the main flow direction occur (ASHRAE 2005). In summer, a high room air temperature may be compensated by an increase in the ventilation rate (EN ISO 7730). The influence of thermal radiation is particularly noticeable if the surrounding surfaces are colder than the environment (Hensen 1990); the effect of locally absorbed short-wave radiation has hardly been investigated to date.

Similarly, air humidity also plays an important role because it influences the intensity of evaporation. This is significant if perspiration occurs at hot room temperatures and if the limit of thermal comfort has already been attained (Recknagel *et al.* 2007). While cold discomfort is governed by the average skin temperature, heat discomfort is related to skin wetness and skin and core temperatures (ASHRAE 2005; Berglund and Cunningham 1986; Fanger 1972; Gagge 1979; Gonzalez and Gagge 1973; Nevins *et al.* 1975). *Skin wetness* expresses the fraction of skin covered with water and is a strong indicator of the level of thermal discomfort (Gagge 1979; Gonzalez and Gagge 1973). In winter, however, respiratory diseases frequently occur when humidity ratios are low, as the mucous membranes are drying out and the dust accumulation in rooms is further supported by the dry air (ASHRAE 2005; Recknagel *et al.* 2007).

The thermal response is further influenced by behavioral actions with respect to the personal parameters

- activity and
- clothing (insulation, moisture permeability).

The thermal resistance of clothing is usually expressed in [clo] units where 1clo = $0.155(m^2 \cdot K)/W$. Values of the insulation and vapor permeability of clothing ensembles can be obtained from measurements and are summarized in standards (EN ISO 9920, ASHRAE Standard 55). Similarly, metabolic rates for different levels of activity are given in standards in tabular form.

Operative/equivalent temperatures and environmental indices

Equivalent homogeneous temperatures (Bohm *et al.* 1990; Wyon *et al.* 1989) or equal thermal environments (Mayer and Schwab 1993) are a concept for combining parameters into a single value in order to transfer inhomogeneous ambient conditions into a uniform perspective. The idea is advantageous to relate thermal conditions to thermal sensation assessments perceived by subjects. Subjective responses are assumed to be similar, irrespective of the combination of the heat transfer mechanisms.

The *operative temperature* (T_{OP}) represents "the temperature of a uniform environment with radiant black enclosure that transfers dry heat by radiation and convection at the same rate as in the actual environment" (ASHRAE Standard 55, 2004). The operative temperature considers the mean radiant temperature and the air temperature at actual velocity and can accordingly be approximated as the average between both temperatures if both values are close to each other and if the air velocity is small.

Other regional expressions exist for assessing humidity, such as the *humid operative temperature*, which is the temperature in an environment at a relative humidity of 100 per cent that yields the same total heat loss from the skin as for the actual environment. The *effective temperature (ET*)* by Gagge (1973) is the same, except that it is for 50 per cent relative humidity (ASHRAE 2005). Both indices are defined in terms of operative temperatures and are useful indices for taking skin wetness into account.

Alongside the operative temperature, the *equivalent temperature* (T_{EQ}) also takes into account the non-evaporative heat loss from the human body and includes the effect of low air velocities. EN ISO Standard 14505–2 (2004) defines the equivalent temperature as "the uniform temperature of the imaginary enclosure with air velocity equal to zero in which a person will exchange the same dry heat by radiation and convection as in the actual non-uniform environment". The standard explains experimental methods and calculation procedures with respect to segment-wise and whole-body values of T_{EQ}. For a historical review the reader is referred to (Nilsson 2004).

Environmental indices are frequently used in a scope-dependent manner to express the physiological stress a thermal environment imposes. A distinction is made between rational and empirical indices (ASHRAE 2005; EN ISO 11399). In hot ambient conditions, such as industrial environments, the *wet bulb globe temperature index (WBGT)* and the *sweat rate* are also used for predicting comfort diagnostically or analytically, respectively (ISO 7243, ISO 7933). Another formulation is the *heat stress index (HSI)* which, according to ASHRAE (2005), expresses the ratio of the total evaporative heat loss required for thermal equilibrium to the maximum possible evaporative loss for specific ambient conditions. Under cold conditions, for example, the *wind chill index (WCI)* and the *required clothing insulation (IREQ)* are defined in (ASHRAE 2005, ISO/TR 11079). EN ISO Standard 11399 (2000) summarizes calculation procedures for the various situations.

T_{OP}, T_{EQ} and *ET** and the predicted mean vote index (to be explained on pages 190–191 below) are frequently used measures employed to define so-called comfort zones relating the indoor climate with a range of acceptable thermal conditions.

Sensation scales

The level of thermal comfort can be expressed using well-defined assessment scales. However, thermal comfort is a subjective measure and the wording for applying these scales is frequently not very precise. We distinguish between the following definitions:

The level of (dis)comfort can be expressed by the sequence *{comfortable, slightly uncomfortable, uncomfortable, very uncomfortable}*. Gagge *et al.* (1967) introduced the unpleasantness scale with the four terms *{pleasant, indifferent, slightly unpleasant, unpleasant}*. The so-called Bedford scale (Bedford 1936) uses mixed expressions in order to combine thermal sensation and comfort perception. The seven-point scale reads *{much too warm, too warm, comfortably warm, comfortable, comfortably cool, too cool, much too cool}*, with ratings from one to seven. Bedford showed a linear relationship between comfort assessment and equivalent temperatures in his scaling. In addition, he found out that the influence of humidity is insignificant for temperatures below 24 °C.

Thermal sensation as a rational experience was suggested by Rohles *et al.* (1971) with the seven-point ASHRAE scale

{cold, cool, slightly cool, neutral, slightly warm, warm, hot}

which is numerically correlated with the range *{–3, –2, –1, 0, +1, +2, +3}*, zero being the neutral element (ASHRAE Standard 55, 2004). In a large field study, Rohles and Nevins (1971) developed linear correlations between comfort level, dry bulb temperature, humidity, gender and length of exposure which are summarized in (ASHRAE 2005).

Another scale is the mean thermal vote (MTV) which is similar to the Bedford scale but counting from –3 to +3 for *{much too cold, too cold, cold but comfortable, neutral, hot but comfortable, too hot, much too hot}*, having again zero as the neutral element, like in the ASHRAE scale (Nilsson *et al.* 1997; Wyon *et al.* 1989). Unlike the ASHRAE scale, however, it clearly distinguishes between acceptable, i.e. comfortable ratings (–1, 0, +1), and ratings which are unacceptable.

It should be noted that differences in language and in subjective observations cause misinterpretations in experimental data, as well. The data transfer between experiments performed by different authors must accordingly be handled with care even if the same scale has been used.

Predicted mean vote and predicted percentage of dissatisfied

Most engineering applications are related to moderate climates. In this context, a common method for steady state and uniform conditions is to predict the average assessment (the predicted mean vote, PMV) of people's satisfaction with the thermal environment according to the well-known Fanger model (1972). Fanger statistically related the physiological responses of a large group of subjects to their relevant sensation assessment.

The idea of the model is that, close to thermal neutrality, the skin temperature and perspiration rate are the main parameters influencing the heat balance of the human body (ASHRAE 2005). Fanger reduced the correlations discovered through experimentation to a single equation. The empirical PMV model, thus expresses the disequilibrium between the energy balance equation for the body as a whole and the perceived level of comfort on the seven-point ASHRAE scale (see pages 189–190) for a large group of people exposed to the same thermal environment. Such a disequilibrium (i.e. the difference between the left and right-hand side of the balance equation defined on page 185) occurs if, for a specific activity and clothing, the actual heat flow differs from the flow which would be required for an optimum level of comfort (*PMV*=0).

Since the obtained *predicted mean vote index (PMV)* is a statistical measure, it can be further related to a *predicted percentage of dissatisfied (PPD)* people. Figure 7.5 shows correlations for estimating the PPD in terms of neutral assessments for the PMV. Fanger (1972) defines neutral conditions (bold line) for *PMV*=0. EN ISO 7730, for example, contains clo-met diagrams indicating isolines of optimum operative temperatures for PMV=0. In Fanger's correlation,

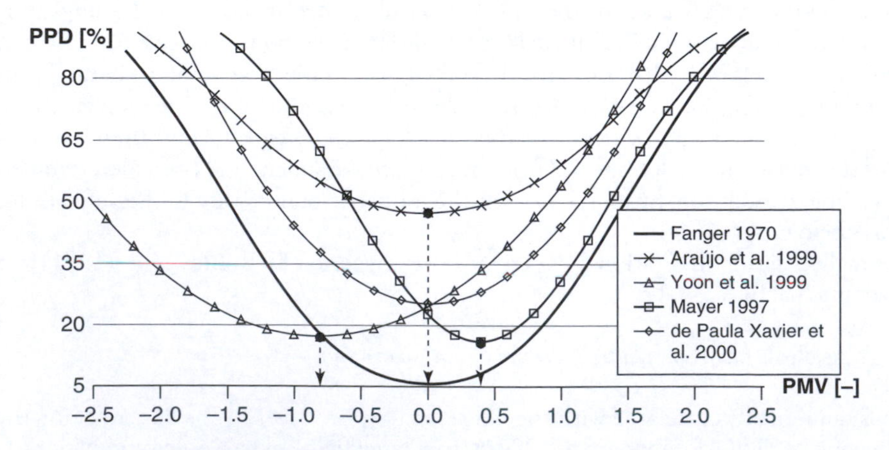

Figure 7.5 Different relationships for PMV and PPD.

Source: Redrawn from the formulae summarized in van Hoof 2008.

PMV=0 means that still 5 per cent of all people remain dissatisfied (*PPD*<5%). Mayer (1997) found the position at *PMV*=+0.4 (*PPD* < 16%) had shifted from the experiments instead. Different ratios are shown by Araújo and Araújo (1999), de Paula Xavier and Robert (2000) and Yoon *et al.* (1999), which are summarized in (van Hoof 2008).

Applying this approach to building simulation is straightforward as the six personal and environmental parameters, i.e. the average air temperature, mean radiant temperature, average air velocity (and degree of turbulence), relative humidity and the level of activity and clothing, are required as input data, respectively. The calculation procedure is defined in the EN ISO Standard 7730 (2005) and the ASHRAE Standard 55 (2004) in detail. Equations for the surface temperature and the convective heat transfer coefficient have to be solved iteratively. Clothing insulation parameters are given in EN ISO 9920, values for the metabolic rates during different activities are detailed in EN ISO 8996, or in the ASHRAE Standard 55, for example.

In practice, comfort envelopes, which are introduced on page 197, are often defined in terms of PMV ranges such as "–0.5 < *PMV* < +0.5". Assuming usual combinations of clothing and activity, a typical relative humidity and low local air velocities, a reasonable range of operative temperatures using the PMV model can thus be identified as comfort criterion for the whole body.

In other words, having limited the PPD to a certain value, for example 6 per cent, a permitted range of PMV values is fixed (–0.2 < *PMV* < +0.2). Choosing the said parameters in a case-dependent manner, for instance assuming people are performing sedentary work (1.2met), wearing winter (1.0clo) or summer (0.5clo) clothing, a valid range of operative temperatures can be obtained. In this case, the minimum requirements for the operative temperatures are 21 °C in winter and the maximum ones 25.5 °C in summer.

The accuracy of a comfort prediction depends on the resolution of the simulation model. With a thermal multi-zone model including a detailed radiation model, most of the information is available. However, a detailed assessment requires coupling with a CFD solver in order to replace average quantities with field values for the air temperature and the local flow velocity. An example is given on pages 206–207.

Local thermal comfort in terms of PPD

The PMV model is formulated for the body as a whole, expressing an average assessment which is statistically representative for a group of people. At the same time, local thermal conditions must be taken into account such as draught risk (Fanger 1964; Fanger *et al.* 1988; Heiselberg 1994; Mayer and Schwab 1990), asymmetrical radiation (Fanger 1977; Fanger *et al.* 1985; Glück 1994), warm and cold floor surfaces (Olesen 2002, 2008), or vertical temperature gradients (Olesen *et al.* 1979), which are significant indicators for describing discomfort (Fanger 1977). Local effects are especially present if people perform light, sedentary activities.

For thermal conditions close to neutral, the standards ISO 7730 and ASHRAE 55 provide additional formulae and diagrams in order to assess these issues in terms of PPD. The criteria should be applied collectively for the whole-body assessment using the PMV model to identify local deficiencies.

- *Draught risk*. Local fluctuations in the velocity influence the perception of draught, which occurs if parts of the body are cooled by a locally increased convective air flow. The local degree of turbulence also contributes to the average air temperature and the average velocity (Fanger *et al.* 1988). The standards provide an algebraic expression to assess the draught risk in terms of PPD. In order to apply the draught risk model in simulation, it is necessary to know the velocity and temperature fields, and these can be obtained by CFD. Fanger (1964), Heiselberg (1994) and Nielsen (1981) provide empirical formulae for assessing the risk of down draught from cold window surfaces. The methods are summarized in (Recknagel *et al.* 2007).

- *Compensation of high temperatures*. Increased flow velocities can, on the other hand, compensate higher ambient temperatures. The latter effect is used in terms of the behavioral adaptation if occupants open windows for natural ventilation. In its appendix, ISO 7730 quantifies the impact of increased air velocities with respect to raising temperatures above the respective comfort temperatures. However, for naturally ventilated buildings it is wise to use the adaptive guidelines.
- *Asymmetrical radiant temperature*. Hot ceilings or cold vertical surfaces, such as cold windows, may lead to uncomfortable situations (Fanger 1977). Both standards summarize PPD relations with respect to hot and cold vertical and horizontal surfaces. These effects can be postprocessed by thermal multi-zone models if a geometry based long wave radiation model is present.
- *Warm and cold floor surfaces*. The standards further show the relation between PPD and cold or hot floor surfaces, respectively. A minimum PPD of 6 per cent is obtained at approximately 23–24 °C, for example. Also, in this case, multi-zone models are appropriate if the short and long wave radiation is accounted for in detail. Radiant floor heating and cooling is discussed by Olesen (2002, 2008), for example.
- *Vertical temperature gradients*. ISO 7730 and ASHRAE 55 provide a diagram for assessing the PPD caused by a vertical temperature gradient between head and ankles, measured 1.1m and 0.1m above the floor surface. Assuming a well-distributed air temperature within a thermal zone, empirical formulae modeling the stratification or natural stack effect may be applied, where applicable. However, a full picture of the temperature distribution can only be obtained with CFD.

Inhomogeneous and transient conditions

The PMV model together with the said local comfort criteria is a straightforward, useful way of predicting discomfort in simulation, particularly with respect to the design of HVAC components. In many situations of practical interest, however, it is necessary to evaluate human responses to inhomogeneous conditions in steady or transient states, which makes the assessment more complex.

The section on pages 185–187 has already set out the basic correlations between local thermal sensation and comfort perception, which both depend on the overall thermal state of the body. Information on human responses to asymmetrical (Fanger 1977) or transient uniform conditions (Gagge *et al.* 1967; Griffiths and McIntyre 1974; Hensen 1990), or non-uniform environments is available in the literature. Only a limited number of publications address human responses under both non-uniform and transient conditions at the same time, as shown by Zhang *et al.* (2004).

Local ergonomic requirements of the thermal environment can be obtained through manikin measurements and by interviewing individuals who are exposed to ambient conditions. The real non-uniform environment is then correlated to a uniform environment using equivalent temperatures, as explained on pages 188–189. The procedure is repeated for each segment of the body (EN ISO 14505–2 2004; Wyon *et al.* 1989). Regressions usually tend to show linear relationships between the local thermal sensation votes and the local equivalent temperatures. The correlations that are obtained in this way are accordingly limited to (i) the range of ambient conditions and (ii) to the specific settings in terms of clothing and activity as given by the experiment. Reports on experience gained in measuring and modeling human responses under various conditions are provided in Bohm *et al.* (1990), Han *et al.* (2001), Holmér (1995), Mayer (2000), Nilsson (2004), Strøm-Tejsen *et al.* (2007) and Zhang *et al.* (2004), for example.

The comfort range obtained by the regression can be displayed in a chart with respect to

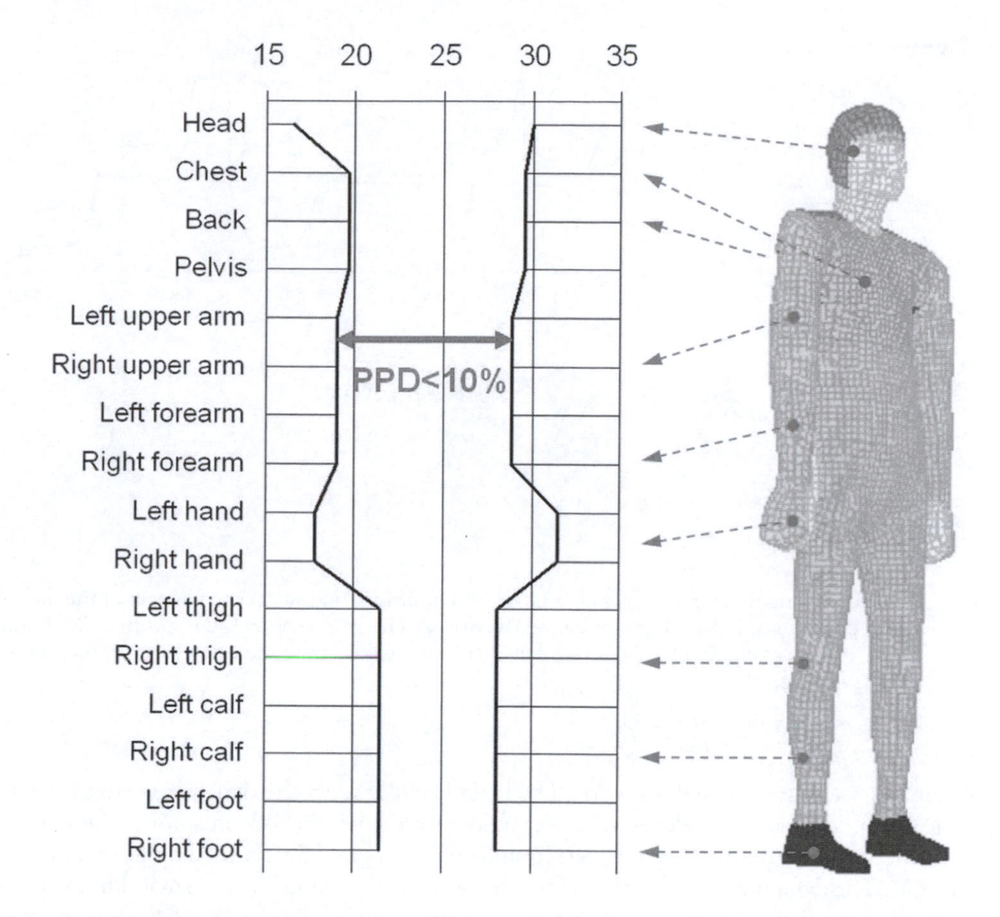

Figure 7.6 Example of a comfort zone for 90 per cent acceptability in terms of equivalent temperatures.
Source: Redrawn from the values presented in (Bohm *et al.* 1990; Han *et al.* 2001) for a car cabin.

the equivalent temperatures. The method is demonstrated in Figure 7.6, which has been adapted from the findings of Bohm *et al.* (1990) and Han *et al.* (2001) for the passenger cabin of a car. The limiting values shown in the figure are determined by the selected percentage of (dis)satisfied people, in this case for 90 per cent acceptability. Measured or simulated local equivalent temperatures (not shown in the figure) can then be visually compared with the comfort zone.

In order to address transient and non-uniform conditions, current research focuses on the integration of multi-segment thermal manikin models (Fiala *et al.* 1999 and 2001; Huizenga *et al.* 2001; McGuffin *et al.* 2002; Stolwijk 1971; Tanabe *et al.* 2002) into numerical simulation tools and their application in practice (Frijns *et al.* 2006; Kato *et al.* 2007; Kok *et al.* 2006; Nielsen *et al.* 2003; Nilsson 2004; Nilsson *et al.* 2007; Niu and Gao 2008; Paulke 2007; Rugh *et al.* 2004; Sano *et al.* 2008; Streblow *et al.* 2008; van Treeck *et al.* 2008, 2009; Zhang and Yang 2008; Zhu *et al.* 2007, 2008). A detailed numerical analysis calls for the coupling between a CFD and radiation solver with a thermal manikin and is accordingly fairly complicated in terms of mesh generation and the numerical coupling procedure. The section on pages 207–209 gives an example of such a numerical analysis using the detailed thermal manikin model of Fiala *et al.* (2001) taking physical and physiological properties, the blood circulation and the human thermoregulation system into consideration.

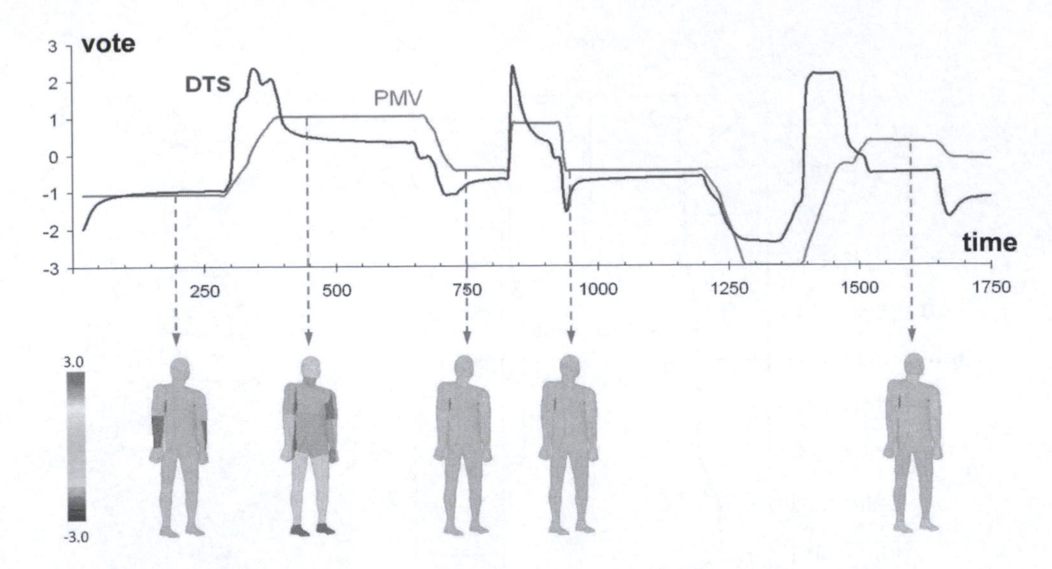

Figure 7.7 Virtual thermal manikin exposed to transient conditions and subject to changes of the ambient conditions and clothing insulation over time. The graph shows the respective PMV and DTS (Fiala *et al.* 2001) indices together with the local thermal sensation vote on the manikin's surface.

Source: Modified from van Treeck *et al.* 2009.

A dynamic sensation model is proposed by Fiala (1998). With the dynamic thermal sensation (DTS) index, Fiala introduces a comfort model for the whole body in a uniform environment, which is based on physiological effects and can be applied over a wide range of ambient conditions. The dynamic nature of the DTS index is shown in Figure 7.7 for a manikin exposed to changing boundary conditions and subject to changes in the clothing insulation over time. We wish to emphasize that the nature of DTS differs from PMV, as the latter is a measure of the ambient climatic conditions and does not account for the responses of the body as DTS does. The empirical DTS model (Fiala 1998) is formulated for the skin and core temperatures simulated using the thermoregulation model (Fiala *et al.* 2001) with a hyperbolic tangent function, which takes the change in skin temperature, the effect of the core temperature on the thermal sensation and the dynamic changes as input data.

From a series of experiments addressing the local and global temperature sensation and comfort perception of subjects, Zhang *et al.* (2004) derived mathematical models for both non-uniform and transient conditions. The researchers developed local sensation and comfort models for 19 individual body parts, as well as global sensation and comfort models, taking into account the weights of the different body parts. The models are formulated in terms of skin and core temperatures and their rates of change. Regression coefficients are detailed in (Zhang *et al.* 2004).

Some of the correlations between thermal sensation and comfort perception have already been mentioned on pages 185–187. Figure 7.4 shows the asymmetric nature of the static local sensation model, which is further extended by a term accounting for the rate of change of the skin temperature under transient conditions (Zhang *et al.* 2004).

Secondary factors affecting comfort

Besides the main environmental and personal parameters mentioned in the section on pages 187–188, the indoor air quality (air age, chemical and microbiological issues, contamination) and other effects also influence the comfort perception, health, productivity and mental performance capability. The influence of acoustic effects, daylight and the use of blinds, electromagnetic fields and other secondary effects are summarized in ASHRAE 2005 and Recknagel *et al.* 2007.

Olesen and Hellwig (Recknagel *et al.* 2007) conclude that secondary psycho-social effects are not yet completely understood. Day-to-day variations in the perception are negligible according to Fanger (1972). Differences in the age to the comfort perception are mainly due to a lower activity level among elderly people. Men and women show almost same preferences but women tend to have a lower metabolism.

For building performance simulation it is important to note that people's subjective thermal comfort perception seems to be statistically uniform if similar environmental and personal conditions are taken into consideration (ASHRAE 2005). In large experiments, Fanger (1972) found out that people do not prefer other thermal environments due to physiological adaptation and, accordingly, that the same comfort conditions apply for all people, independently of any cultural diversity (Fanger and Toftum 2002). However, several researchers showed that factors beyond the fields of physics and physiology influence occupants' expectations and thermal preferences in buildings (de Dear and Brager 2002). Taking into account psychological effects, clothing preferences and behavioral action leads to the formulation of adaptive comfort standards (see pages 195–196 and 198–201).

Acceptability, satisfaction and adaptation

Thermal sensation and comfort perception in terms of acceptability or satisfaction with the thermal environment is not only influenced by the aforementioned six environmental and personal factors. Behavioral adaptations and people's expectations are likewise of relevance (Brager and de Dear 1998; de Dear and Brager 2002).

Humphreys and Nicol (1998) state that in conditions of thermal discomfort people react in such a way as to restore their thermal comfort and consequently to reduce the physiological strain (see also Auliciems 1983). De Dear and Brager used extensive field studies (1998) to show that people evaluate the indoor thermal performance differently if the building design makes provisions for the individual control of the indoor climate. Such behavioral actions include the opening of windows, wearing appropriate clothes, if allowed by the dress code, adjusting the scheduling, closing roller blinds, installing local fans, or taking refreshments and drinks, for example. De Dear and Brager (2002) indicate that people living and working in naturally ventilated buildings accept a somewhat wider range of temperatures, by making appropriate adaptations, than predicted by the traditional static thermal comfort envelopes (EN ISO 7730, 2005; ASHRAE Standard 55, 2004).

In Figure 7.8, both cases reflect de Dear and Brager's regressions (1998) in the observed and PMV predicted indoor comfort temperatures versus the average outdoor temperatures: Figure 7.8(a) for controlled HVAC buildings and Figure 7.8(b) for naturally ventilated (NV) buildings. NV buildings show a much steeper gradient in the observed responses, i.e. their occupants seem to accept a wider range of ambient conditions. The PMV model in turn predicts HVAC buildings very accurately (de Dear and Brager 2002).

Another important factor alongside behavioral adaptation is what people expect of the indoor climate, based on the ambient thermal conditions and the time history of the preceding days (psychological adaptation). The section on pages 198–201 discusses adaptive comfort guidelines which are mainly formulated as a function of the running average outdoor temperature (van der Linden *et al.* 2006).

The PMV model does, to some extent, include behavioral actions, as the model generally accounts for the personal parameters of clothing and activity as well as the local ambient conditions. Fanger and Toftum (2002) enhanced the PMV model by multiplying PMV by a so-called expectancy factor, a function of region, season and indoor environment. The

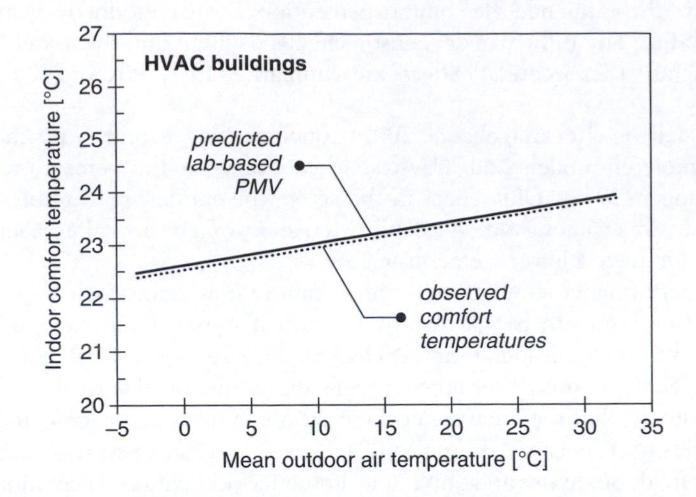

Figure 7.8(a) Qualitative behavior of the observed and predicted indoor comfort temperatures from a field study conducted by de Dear and Brager (1998) for HVAC buildings.

Source: Redrawn from de Dear and Brager 2002.

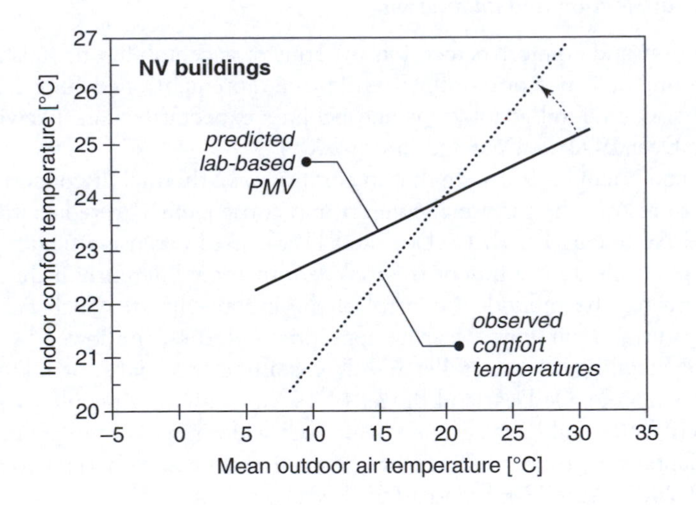

Figure 7.8(b) The same but for naturally ventilated (NV) buildings.

Source: Redrawn from de Dear and Brager 2002.

expectancy factor is a corrective term that extends the thermal sensation assessment range of people in non-air-conditioned buildings in warm climates (van Hoof 2008; van Hoof and Hensen 2007).

Prediction of thermal comfort in buildings

Comfort assessment in buildings

The assessment of thermal comfort in buildings can generally be achieved by (i) designing the HVAC system layout in such a way as to meet the heating and cooling power required for maintaining operative temperatures within their defined limits, by (ii) taking measurements and conducting surveys in existing buildings while they are in operation, or by (iii) simulation. As previously demonstrated in this book, simulation is a highly efficient method and provides details of the building performance and the indoor thermal quality in advance. For example, a dynamic simulation can supply hourly values of the room air temperature in summer in order to assess the thermal quality of a naturally ventilated building zone. If the air temperature is controlled by an HVAC system, the level of comfort is mainly determined by the respective set point temperature of the control device.

Professional designers often apply an annual or monthly static balancing approach for the system layout. If this approach is used, similar comfort categories should be applied for both the HVAC system layout and for calculating consumption. Where the system is designed on the basis of a dynamic simulation, the EN Standard 15251 (2007) recommends temperature ranges for heating and cooling with respect to a selected comfort category, type of clothing and level of activity, for example. The level of the indoor thermal quality strongly influences the heating and cooling demands of a building. The next section summarizes the respective comfort envelopes.

When designing HVAC systems, it is important to distinguish between mechanically ventilated (i.e., heated and/or cooled) buildings, naturally ventilated buildings, and "mixed-mode" buildings. The evaluation methods differ as occupants tend to show different expectations for naturally ventilated buildings. In naturally ventilated buildings it is assumed that people adapt to warmer conditions by changing clothes or by opening windows and doors (de Dear and Brager 2002). Adaptive comfort guidelines account for these effects by relating the individual expectation and local adaptation to a gradual change in the ambient climate over time.

Predicting the indoor thermal performance is also subject to a number of uncertainties (De Wit 2001). For a detailed review on the subject of user behavior the reader is referred to Chapter 4.

Comfort envelopes

A number of standards define comfort envelopes or categories from the point of view of the indoor thermal quality performance. The comfort criteria of these categories are related to a certain range of PMV or PPD with respect to a selected clothing insulation, level of activity and climate zone in terms of humidity. Some formulations make use of a psychrometric chart in order to express satisfaction with the thermal environment in terms of operative temperature and humidity.

For moderate climates, EN Standard 15251 defines four categories for (I) high, (II) normal, (III) moderate and (IV) low expectations. For category (II), for example, which is suggested for new or renovated buildings, it is recommended to keep $PPD < 10\%$ and thus to allow for $-0.5 < PMV < +0.5$ for mechanically ventilated buildings. For people working in an open-plan office with sedentary, light activity (1.2met) this would imply the following conditions: During the heating period, the minimum operative temperature (1clo) should be maintained above 20 °C, and during the cooling period (0.5clo) below 26 °C for category II. Other configurations and settings are defined in tables in EN 15251 (2007). Local thermal comfort criteria, such as draught risk or radiation asymmetries, have to be satisfied simultaneously in accordance with ISO 7730.

Both Standard EN ISO 7730 (2005) and the ASHRAE Standard 55 (2004) define "ergonomics of the thermal environment" and "thermal environmental conditions for human occupancy", respectively. ISO 7730 (2005) proposes three categories A, B and C for limiting the PPD (and accordingly the range of PMV), each with respect to the whole body and to the four major local comfort criteria (on pages 191–192), for steady-state conditions (cyclical temperature changes restricted to 1K and drifts or ramps limited to 2K per hour). The standard also summarizes the formulae and tables required to calculate PMV and to assess the local effects in terms of PPD, such as the draught risk (DR). As indicated on pages 190–191, these criteria make it possible to derive the permissible range of operative temperatures from estimated values for clothing and activity, humidity and local air velocity. Table 7.1 summarizes some of these ISO categories.

Similarly, the ASHRAE Standard 55 is based on the energy balance method for the whole body and provides formulae and calculation procedures, also with respect to the mentioned local conditions. Compared to the ISO, the ASHRAE Standard 55 uses a similar methodology but draws the comfort zone in terms of the range of operative temperatures for 80 per cent acceptability as well as accounting for limiting the wet-bulb temperatures. A psychrometric chart is used to define comfort envelopes for a summer zone (0.5clo) and a winter zone (1.0clo), respectively, based on a diagram which is spanned by the operative temperatures and the humidity ratio (or the dew point temperature). The limiting values for the operative temperatures are found by selecting $PPD < 10\%$ for general comfort plus an additional 10 per cent for local effects which determine the range of PMV, provided the activity levels (1 to 1.3met) and the air speed (<0.2m/s) have been set. The humidity is limited by an upper and lower bound as a high relative humidity causes discomfort due to skin wetness and increased skin friction, and low humidity is accompanied by dry skin, throat and eyes and may lead to respiratory disorders (ASHRAE 2005).

In contrast to ISO 7730, the ASHRAE Standard 55 also reflects the dependence of the occupants' comfort temperature to the outside conditions in free-running buildings by providing an alternative method. This issue is discussed in the next section.

Adaptive comfort guidelines

Adaptive thermal comfort (ATC) guidelines define comfort envelopes for free-running, non-air-conditioned buildings in terms of acceptable indoor temperatures which vary as a function of the running outdoor temperatures. Several methods have been proposed to average the outside temperatures. Humphreys (1978) suggests exponentially weighted running average values over a seven-day period, van der Linden *et al.* (2006) a four-day period, and de Dear and Brager (2002) take monthly average values of the dry-bulb outdoor temperature instead.

Examples are the EN ISO Standard 15251 (2007), the Dutch ISSO Standard 74 (2005) and the method described in the ASHRAE Standard 55 (2004). As the operative indoor and the outdoor temperatures are the only inputs, the use of these ATC procedures in simulation is straightforward. The application is restricted to office type buildings providing a high level of occupant control (Herkel *et al.* 2005) and where people mainly perform light, sedentary work.

Table 7.1 Categories for indoor thermal performance characterization according to EN ISO 7730 (2005). The standard also restricts local conditions with respect to vertical air temperature gradients, warm or cold floor surfaces and asymmetric radiation.

ISO 7730 Category	*Whole body-related discomfort*		*Local discomfort*
	Predicted percentage of dissatisfied	*Predicted mean vote*	*Percentage of dissatisfied in terms of draft risk*
	PPD [%]	PMV	DR [%]
A	< 6	−0.2 < PMV < +0.2	< 10
B	< 10	−0.5 < PMV < +0.5	< 20
C	< 15	−0.7 < PMV < +0.7	< 30

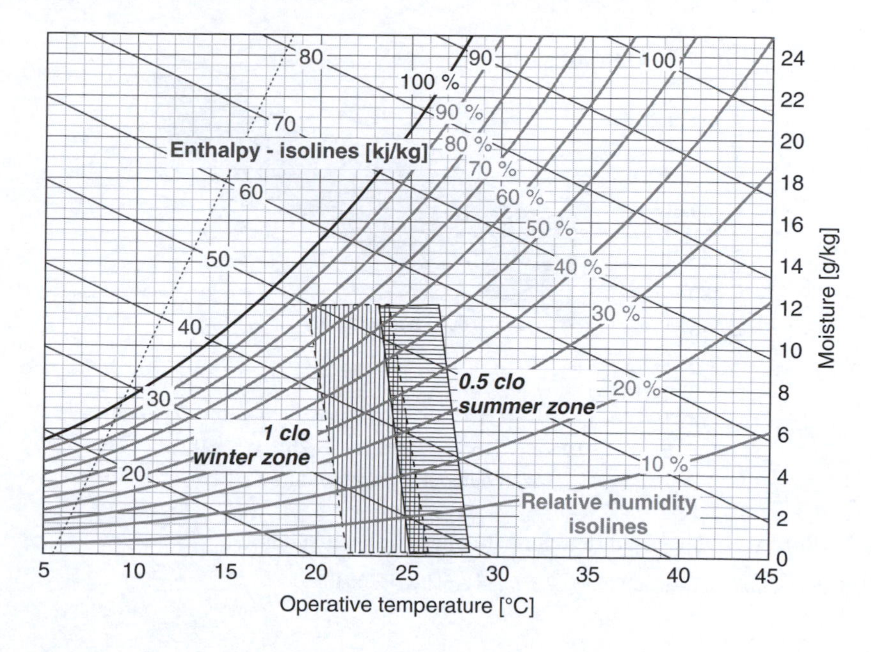

Figure 7.9 Psychrometric chart (Mollier chart) showing summer and winter comfort envelopes within ±0.5 PMV limits. The comfort zones have been adapted from the specifications of the ASHRAE Standard 55 (2004).

The range of acceptable temperatures is centred around the comfort temperature, which is defined by the ATC regression equation (see also pages 195–196). The comfort envelope is based on PMV ranges with respect to certain categories of acceptability. De Dear and Brager (2002) propose the modified equation $T_{comf} = 0.31\ T_{a,out} + 17.8$ for the comfort temperature T_{comf} and the average outdoor temperature $T_{a,out}$ and suggest a bandwidth of 5K and 7K for 90 and 80 per cent acceptability, respectively, as shown in Figure 7.10(a) for the ASHRAE Standard 55–2004.

EN ISO 15251 (2007) suggests three categories, see Figure 7.10(b), for naturally ventilated buildings without air-conditioning installed. The comfort temperature regression equation in this case is $T_{comf} = 0.33\ T_{rm} + 18.8$ using the running average of an exponentially weighted, daily outdoor temperature T_{rm} (EN ISO 15251 (2007)). The temperature bandwidths for the three categories I, II and III are 4K, 6K and 8K, respectively. The settings for mechanically ventilated buildings are applied outside the intervals indicated in the diagram.

Similarly, the 90, 80 and 65 per cent limits (class A, B or C) suggested by the Dutch ISSO Standard 74 (2005) are derived from de Dear and Brager's research (1998). The standard distinguishes between type Alpha buildings, offering a high degree of occupant control, and a type Beta for centrally controlled HVAC systems. During the heating period, the same operative temperature limits are applied for temperatures below 10–12 °C for the type Alpha buildings as for the Beta model. For details, the reader is referred to ISSO Standard 74 (2005).

Van der Linden *et al.* (2008) quantify the differences between the ATC and PMV approaches for moderate climates by investigating the range of input parameters for the PMV model in a moderate climate, compared with the ATC method. To this end, seasonal scenarios are selected and correlations between average outdoor temperatures and clothing habits for naturally ventilated buildings, discovered by De Carli *et al.* (2007), are used. Different humidities and air velocities are assumed for winter and summer months. These settings make it possible to obtain

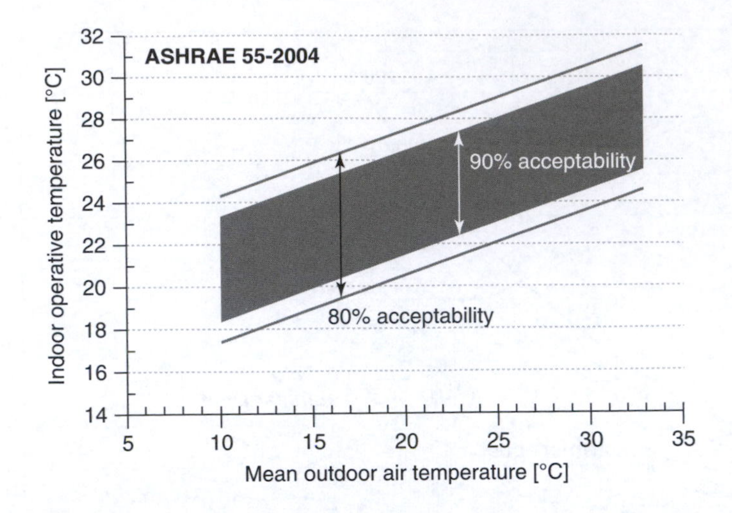

Figure 7.10(a) Acceptable conditions for naturally ventilated buildings.

Source: Redrawn from de Dear and Brager 2002 and ASHRAE 2004.

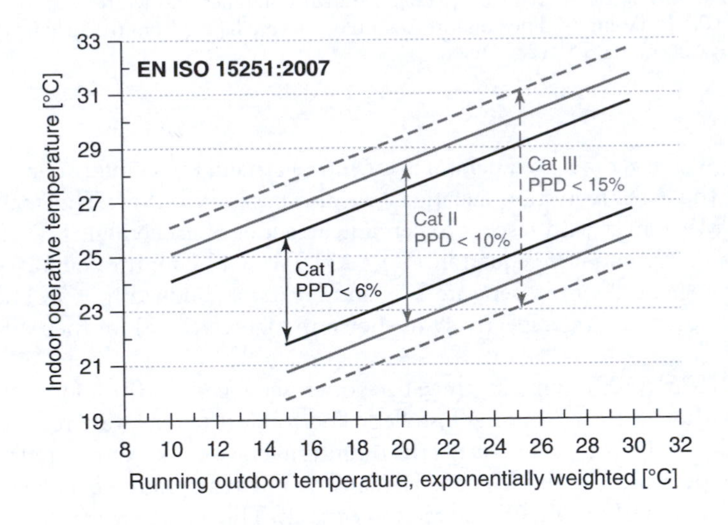

Figure 7.10(b) ATC design values for the three categories of EN ISO 15251 (2007).

acceptable ranges of operative temperatures, which are then compared with the Dutch ATC guideline. The results indicate that the adaptive ranges can be satisfactorily explained in terms of the PMV ranges up to 21 °C average outdoor temperatures.

In other words, the PMV model does, in fact, provide a flexible approach if parameters are adjusted according to people's adaptations and expectations. The ATC approach, however, defines a wider range of acceptable conditions, provided a naturally ventilated building is evaluated. The deficiencies of the balance method, namely ignoring the psychological effects (de Dear and Brager 2002), can be compensated to some extent by applying the expectancy factor concept (Fanger and Toftum 2002). Van Hoof and Hensen (2007) identified an energy reduction potential of the ATC method for type Alpha buildings of up to 10 per cent.

Robinson and Haldi (2008) suggest that the adaptive effects can be interpreted as an aggregation of individual adaptive processes which may be further isolated in terms of successive adaptive temperature increments which determine the slope of the adaptive curve of the envelope to some extent.

Long-term indicators

The discussion of the guidelines indicates that two types of indicators are needed for the assessment of simulation data such as an integrated measure for the annual thermal quality performance as well as indicators capturing the local thermal behavior subject to the proper functioning of the HVAC system. An additional, local analysis (see pages 191–192) calls for a refined model and is therefore applicable for selected critical periods of a reference year only.

For an all-year thermal simulation, ISO 7730 (2005) and EN 15251 (2007) propose several long-term indicators for predicting the frequency of occurrence based on penalty functions if the thermal quality is not maintained within the desired range of a comfort zone.

- *Hourly performance index*. The number of hours per year, or the percentage of the time when the comfort criteria are satisfied or not, can be summarized and thus quantified as a single indicator.
- *Degree hours*. Another descriptive performance indicator is obtained if the time is weighted by the number of degrees by which the temperature has been exceeded (degree hours), i.e. where the operative temperature is out of the defined comfort range.
  The hours are accordingly multiplied by the difference $w = T_{op} - T_{op,limit}$ of the operative temperature T_{op} and the respective upper or lower boundary value of the comfort zone $T_{op,limit}$, respectively. The product of the weighting factor w and the time is summed for characteristic periods, i.e., for the heating period if $T_{op}<T_{op,limit}$, or for the cooling period if $T_{op}>T_{op,limit}$. The unit of the performance index is expressed in degree hours.
- *PPD weighted performance index*. In a similar procedure the PMV index can be weighted and summed over characteristic periods. The weighting factor is hereby determined as the ratio between the PPD value of the actual PMV based on the operative temperature and the PPD, which corresponds to the PMV at the boundary of the comfort range if the actual PMV is out of this range. If the PMV is within the range, the weight is zero.
  The PMV-based weighting factor results in more hours than the index based on degree hours and is therefore a sharper criterion. The appendix of EN 15251 (2007) gives an example.

However, these long-term indicators are subject to several discussions in the literature. Van der Linden *et al.* (2006) compare different assessment methods in terms of the Dutch TO/GTO methods (TO is based on exceeding temperatures and GTO means weighted excess hours) and the ATC method (ISSO Standard 74 (2005)). The examples indicate that the ATC guideline is more rigorous than the degree hour method as a building's comfort category is already downgraded as soon as a single value drops out of a specified range of acceptability.

We therefore suggest using an integrated performance approach to provide a descriptive graphical evaluation by visualizing simulation results based on the sketches representing the comfort envelopes as well as indicating weighted excess hours or similar integral quantities. However, these performance indicators fail to accurately account for the accumulation of heat stress events that cause overheating.

Overheating risk assessment

Adaptive models assume that weather and seasons influence peoples' behavioral and psychological adaptations and consequently allow for a wider range of acceptable thermal conditions in naturally ventilated buildings (de Dear and Brager 2002). According to Baker and Standeven (1996), occupants only tolerate a limited number of occasions where the indoor temperature exceeds a certain limit. It is, however, not clear if and how these stresses – if counted or weighted using one of the methods described above – accumulate and cause overheating.

In order to account for the accumulation of heat stimuli, Robinson and Haldi (2008) propose a mathematical model to predict the overheating risk in buildings for moderate climates. The model hypothesizes an analogy between charging and discharging of humans' tolerance to overheating stimuli and the principle of charging an electrical capacitor. It is assumed that, during a warm period, the overheating tolerance is discharged and, during a series of cold days, it is recharged. The tolerance is accordingly the inverse of the probability of overheating. Robinson and Haldi (2008) formulate their overheating probability model analytically with empirical time constants for (dis)charging. This promising model was tested in a field survey which revealed that overheating is caused by an accumulation of heat stress events rather than single events (Robinson and Haldi 2008).

As the adaptability is implicitly given in the (dis)charging time constants of the model, Robinson and Haldi (2008) also propose an explicit model which eliminates the adaptive increments, i.e. the elevations of neutral temperature due to adaptation (Baker and Standeven 1996). Therefore probability functions for the adaptive actions, which are shown to be primarily correlated with indoor temperatures, are proposed. The elevation of neutral temperature is further quantified by explicitly relating comfort temperatures and their increments to actions and their conjugates.

Certification

In order to assess the whole building energy efficiency, the methods set out above are required for classifying the indoor thermal quality in terms of a single indicator. EN 15251 (2007) gives some recommendations according to the European Energy Performance of Buildings Directive (EPBD 2003). It is useful to treat the issues of thermal quality and air quality separately, for example, by indicating the percentage of the time periods where the indoor climate is within the limits, as defined by a certain comfort category (I, II, III, IV). For the different energy labels and country-specific settings, the reader is referred to national standards.

Application: level of detail-dependent prediction in simulation

In the introductory section, the range of simulation approaches was already discussed. Three typical examples for the indoor thermal performance prediction will now be briefly introduced. It shall be noted that with increasing resolution in space and time also the engineering costs are elevated in terms of the time which is needed for creating or coupling the respective models.

Comfort analysis with thermal multi-zone models

Dynamic building performance simulation (BPS) offers most of the data which are relevant for assessing thermal comfort. Care must be taken as BPS usually assumes a well-distributed or stratified air temperature in spaces. The operative temperatures, i.e. the average air and radiant temperatures of the surrounding surfaces as well as the air humidity, depending on the model, are predicted with a high resolution in time. (The average air speed may be estimated from the thermodynamic conditions of the respective space.) Quantities are usually evaluated at hourly intervals or less for the period of a whole year, i.e. 8760 time-lines or more are available for post-processing (1a = 365d = 8760h). For methods incorporating user behavior into these tools, the reader is referred to Chapter 4.

The next section looks at a specimen room (German VDI Standard 6020:2001) in a naturally ventilated office-building which is exposed to a moderate climate. A single rectangular space with an area of 17.5m² and a large, south-facing window with appropriate shading devices is considered for the simulations. The building is of a heavy construction. For the comfort assessment, the static thermal comfort envelope of the ISO Standard 7730 (2005) and the two adaptive guidelines ISSO Standard 74 (2005) and the European Standard EN ISO 15251 (2007) were applied.

Figures 7.11, 7.12 and 7.13 show the obtained results with respect to the comfort envelopes ISO 7730, EN ISO 15251 and ISSO 74, respectively. The abscissas are each based on the definitions of the standards. For the ISO 7730, the daily average ambient air temperature was used. As the ISO 7730 comfort envelope is static, for summer and winter different values for clothing (0.5clo and 1.0clo), local air speed and relative humidity are assumed (see also ISO 7730 (2005)). For the sake of better readability, only the daily min. and max. values are drawn.

The frequency of occurrence if the criteria of the comfort zones are met or violated are summarized in Table 7.2 (for all time steps). In this case, the building would be classified under ISO 7730 as "worse than category C", under ISO 15251 as "category III", and under ISSO 74 as "worse than category C" in terms of acceptability.

In order to relate the frequency of occurrence of exceeding temperatures with the respective time points of these occurrences, Figure 7.14 exemplarily plots the deviation of the daily maximum operative temperatures from the definitions of the comfort envelope using a calendar-type diagram. The Figure evaluates the results in accordance with ISO 7730 for values within the upper band of category B. (Such calendar-type layouts could be drawn for any kind or band of comfort envelope.) The depiction is descriptive for locating local hot spots if it is combined with the diagrams shown in Figures 7.11–7.13. The latter method merely indicates if temperatures go beyond the permitted range but not when this is obviously the case.

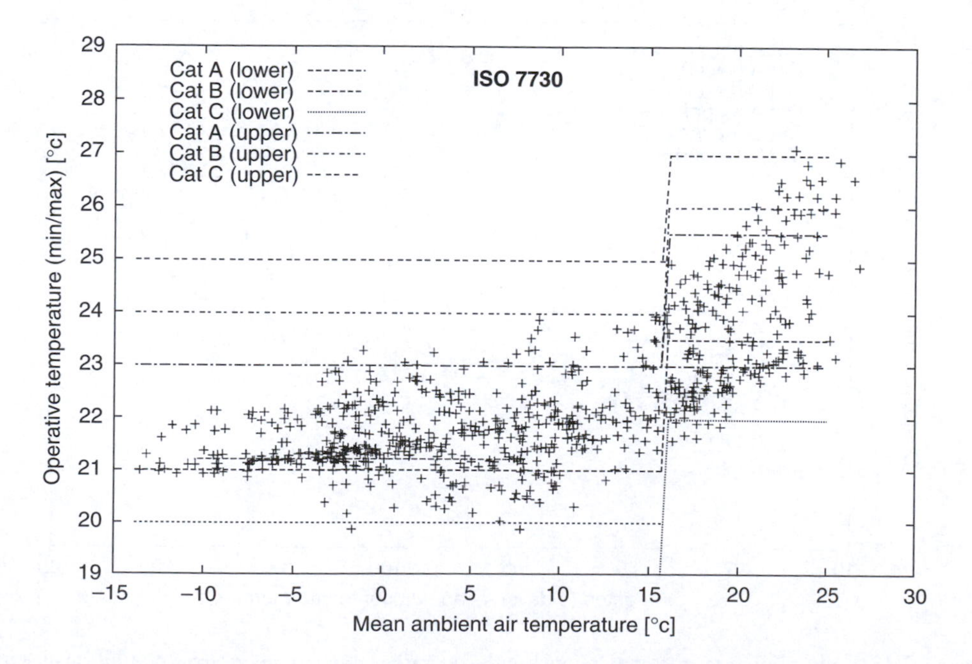

Figure 7.11 Daily minimum and maximum hourly operative room temperatures plotted against the daily average ambient air temperatures for the comfort envelope of the ISO 7730 (2005).

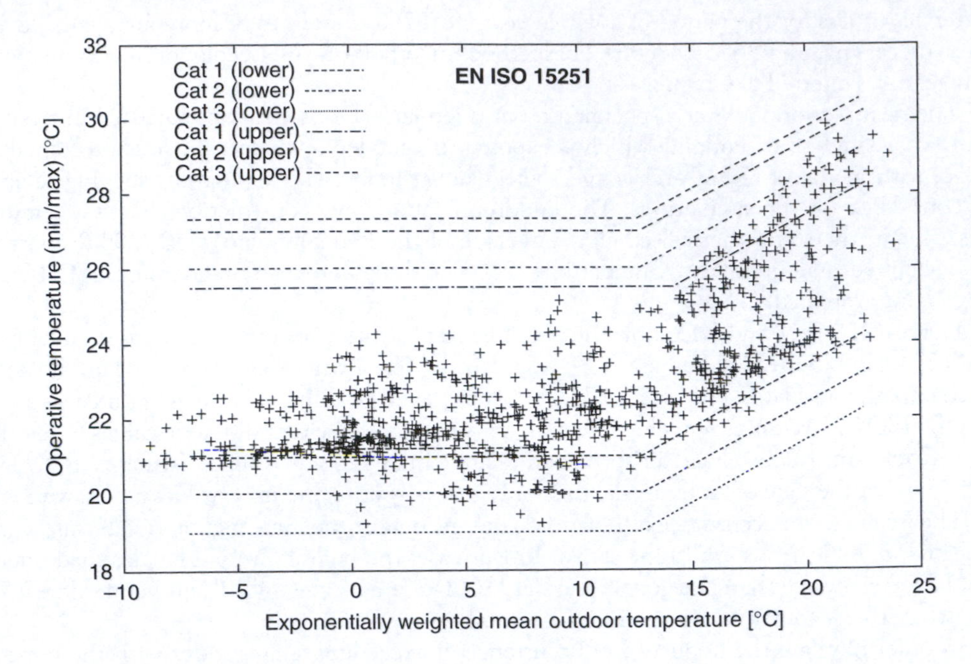

Figure 7.12 Daily minimum and maximum hourly operative room temperatures plotted against the exponentially weighted running average outdoor temperatures for the comfort envelope of the EN ISO 15251 (2007).

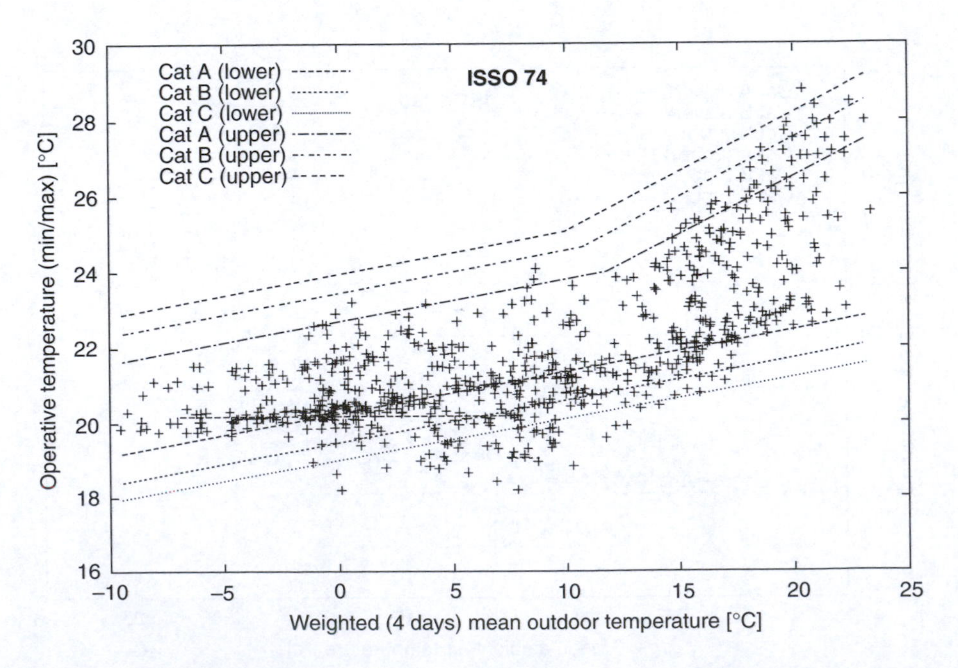

Figure 7.13 Daily minimum and maximum hourly operative room temperatures plotted against the weighted running average outdoor temperatures for the comfort envelope of the ISSO 74 (2005).

Table 7.2 Comparison of the different comfort classification schemes (categories)

ISO 7730	A	B	C	*else*
Acceptability	94%	90%	85%	<85%
Frequency of occurrence	75.9%	16.1%	6.5%	1.6%
EN ISO 15251	*I*	*II*	*III*	*IV*
Acceptability	94%	90%	85%	<85%
Frequency of occurrence	93.8%	6.0%	0.2%	0%
ISSO 74	A	B	C	*else*
Acceptability	90%	80%	65%	<65%
Frequency of occurrence	94.8%	4.4%	0.6%	0.1%

Maximum values (ISO 7730 / CAT A-B)

	Jan	Feb	Mar	Apr	May	Jun	Jul	Aug	Sep	Oct	Nov	Dec
1	0.0	0.0	0.0	0.0	0.0	0.0	0.4	2.4	2.1	0.0	0.0	0.0
2	0.0	0.0	0.0	0.4	0.3	0.0	0.0	1.1	2.9	0.0	0.0	0.0
3	0.0	0.0	0.0	0.0	0.0	0.0	2.0	0.2	2.3	0.0	0.7	0.0
4	0.0	0.0	0.0	0.0	0.0	0.0	1.9	1.7	3.0	0.0	0.7	0.0
5	0.0	0.0	0.0	0.0	0.0	0.0	1.0	1.7	2.5	0.0	1.1	0.0
6	0.0	0.0	0.0	0.0	0.5	0.0	1.0	1.5	0.3	0.0	1.5	0.0
7	0.0	0.0	0.0	0.0	0.8	0.6	0.0	3.2	0.2	0.1	1.3	0.0
8	0.0	0.0	0.0	0.0	0.0	0.0	0.1	0.8	1.1	0.0	0.0	0.0
9	0.0	0.0	0.0	0.0	0.0	0.0	1.1	0.0	0.8	0.0	0.0	0.0
10	0.0	0.0	0.0	0.0	0.0	0.0	0.7	2.6	3.1	0.0	0.5	0.0
11	0.0	0.0	0.0	0.0	0.0	1.3	0.5	1.8	1.4	0.0	0.3	0.0
12	0.0	0.0	0.0	0.0	0.0	2.0	0.0	2.3	2.9	0.0	0.7	0.0
13	0.0	0.0	0.0	0.0	0.0	0.0	0.3	1.7	0.0	0.0	0.4	0.0
14	0.0	0.0	0.0	0.0	0.0	0.0	2.5	2.1	0.0	1.4	0.4	0.0
15	0.0	0.0	0.0	0.0	1.2	0.0	2.2	1.9	0.4	0.6	0.0	0.0
16	0.0	0.0	0.0	0.0	1.3	0.1	2.0	0.2	0.0	0.0	0.0	0.0
17	0.0	0.0	0.0	0.0	–0.1	0.0	2.2	0.1	0.8	0.0	0.0	0.0
18	0.0	0.0	0.0	0.3	–0.2	0.4	2.5	1.1	1.8	0.0	0.0	0.0
19	0.0	0.2	0.0	0.0	0.0	0.8	0.6	1.7	1.6	0.0	0.0	0.0
20	0.0	0.7	0.0	0.0	1.6	0.2	0.1	0.9	0.3	0.0	0.0	0.0
21	0.0	0.5	0.0	0.0	0.0	1.6	1.2	0.0	0.6	0.0	0.0	0.0
22	0.0	0.0	0.0	0.0	0.0	1.8	1.0	1.3	0.0	0.0	0.0	0.0
23	0.0	0.0	0.0	0.4	0.0	0.0	1.5	0.3	0.0	0.0	0.0	0.0
24	0.0	0.0	0.0	0.0	0.8	1.3	0.9	0.0	0.0	0.0	0.0	0.0
25	0.0	0.0	0.0	0.0	0.7	2.6	1.0	0.5	0.0	0.0	0.0	0.0
26	0.0	0.0	0.0	0.0	0.0	0.2	2.4	0.4	0.0	0.0	0.0	0.0
27	0.0	0.0	0.0	0.0	0.0	0.0	2.3	0.0	0.0	0.0	0.0	0.0
28	0.0	0.2	0.0	0.0	1.1	1.3	0.6	0.0	0.0	0.0	0.0	0.0
29	0.0		0.0	0.0	0.0	1.2	0.9	0.0	0.0	0.0	0.0	0.0
30	0.0		0.0	0.0	0.8	0.0	1.4	0.0	0.0	0.0	0.0	0.0
31	0.0		0.0		–0.3		2.4	0.3		0.0		0.0

Figure 7.14 Deviation in [K] of the daily maximum operative temperatures from the upper band of category B of the comfort envelope of the ISO 7730 (2005), for example. Dark (red) colors indicate high deviations.

Source: Modified from van Treeck 2004.

Local draught risk analysis using CFD

The results shown in the previous subsection were related to the body as a whole in terms of acceptability using the PMV/PPD and adaptive assessment methods. As indicated on pages 191–192, local conditions are also of relevance. This section demonstrates the local draught risk analysis using the formulae provided in ISO Standard 7730 and ASHRAE-55, respectively. For the local draught risk assessment, field values of the average air temperature, the air velocity and the degree of turbulence are required, which can be obtained by CFD.

In computational fluid dynamics (CFD), the Navier-Stokes equations and the energy equation are solved numerically, which offers a detailed insight into the indoor air-flow behavior. The turbulent convective nature of the air flows calls for the application of turbulence models (Bejan 1993) and the refinement of the locations closest to the walls. A geometric model is required to set up a CFD simulation in order to create a surface and volume mesh. This is a very time-consuming procedure in terms of the engineering work involved, so it restricts the application of CFD in practice. Initial and boundary conditions, such as inflow and outflow conditions (velocity or pressure values), temperatures or heat fluxes at surfaces bounding the flow domain, are required. Realistic boundary conditions can be obtained from a preceding simulation or by coupling with a thermal multi-zone model, as shown by Beausoleil-Morrison (2000) or van Treeck *et al.* (2006).

The capabilities of such a high-resolution CFD analysis are demonstrated for the case of a cold down-draught from the cold glass façade of the atrium of a building (26 m high). The south-facing atrium underwent a number of investigations, which are detailed in (Schälin *et al.* 1996). Van Treeck (2004) compared available measurement data with simulations by realizing a coupling algorithm between a thermal simulation solver and a CFD model in order to support the CFD model with boundary conditions.

Based on the computed velocity and temperature fields for the winter scenario with closed vents, the local draught risk is calculated and depicted in Figure 7.15 on a horizontal plane 0.4 m above the floor surface. The Figure clearly shows that the down-draught from the cold façade (15 °C air temperature) produces high local velocities (up to 0.5 m/s). The inconvenience of this local situation in the floor area is expressed in terms of PPD on a 2D plane, the acceptance is less than 20 per cent. The situation can be remedied by installing local radiators in the façade area or by improving the glazing system from the point of view of the U value.

Local and dynamic comfort analysis using a thermal manikin

The preceding sections have demonstrated the comfort assessment for steady-state uniform conditions for the body as a whole and in terms of local effects, where the thermal body state

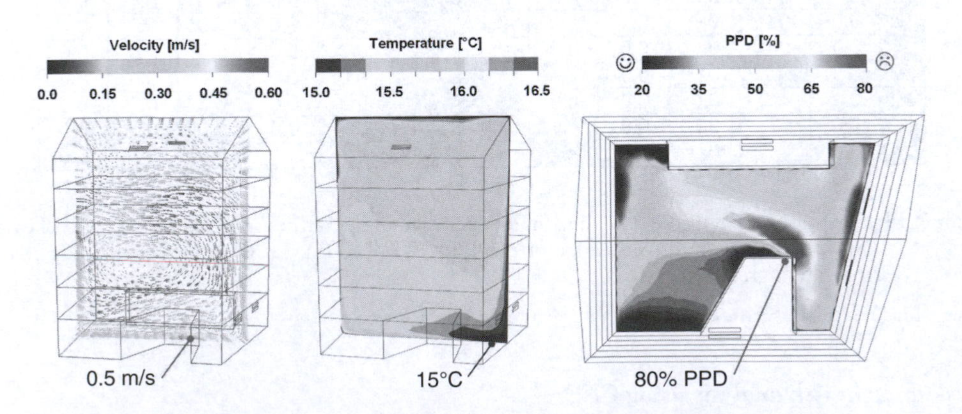

Figure 7.15 From left to right: averaged velocity field on a vertical plane, vertical temperature profile and draught risk in terms of PPD on a horizontal plane 0.4 m above the ground.

Sources: Data from van Treeck 2004. Reprinted from van Treeck *et al.* 2009.

was close to neutral. This section addresses the application of a multi-segment thermal manikin for a local and dynamic thermal comfort analysis. Surface, skin and core temperatures are computed in order to express local thermal sensations in inhomogeneous conditions, and the dynamic responses of the human body are displayed.

For a detailed simulation, both a flow and thermal radiation solver are required in order to compute the convective and radiative heat transfer between body and environment. To reduce the computational overhead, predefined view factors and heat transfer coefficients may be applied (Fiala *et al.* 2001; van Treeck and Frisch 2009). With a thermal manikin, the body heat exchange is modeled using a (passive) system taking physical and physiological properties, the blood circulation and the (active) human thermoregulation system into consideration (Fiala *et al.* 2001; Stolwijk 1971; Tanabe *et al.* 2002). The outcome covers the surface, skin and core temperatures for the respective body parts and their change over time. The surface temperatures can be converted into local equivalent temperatures which, in turn, can be related to experimentally obtained local thermal sensation votes, as described on pages 192–194.

The following section deals with the model devised by Fiala *et al.* (2001), which can be used to predict human thermal reaction to a wide range of environmental conditions. The results are computed by implementing Fiala's model (Paulke 2007) in THESEUS-FE (2008), taking physical and physiological properties, the blood circulation and the human thermoregulation system into consideration. The body is discretized sector-wise using spherical and cylindrical elements based on a finite element shell model (FIALA-FE 2008). In addition, the model is also coupled to other solvers using a parametric manikin model (van Treeck *et al.* 2009).

The skin and surface temperatures shown in Figure 7.16 are correlated with local thermal sensation votes using equivalent temperatures (see pages 188–189 and 192–194 for explanation) and regression equations, which have been obtained from an experimental study in a joint research project (van Treeck *et al.* 2009).

Figure 7.17 shows the graphical user interface of an integrated simulation tool (a "virtual climate chamber") by way of an example, which allows for monitoring the state of the manikin in terms of thermal and physiological properties that change over time. The tool can be used

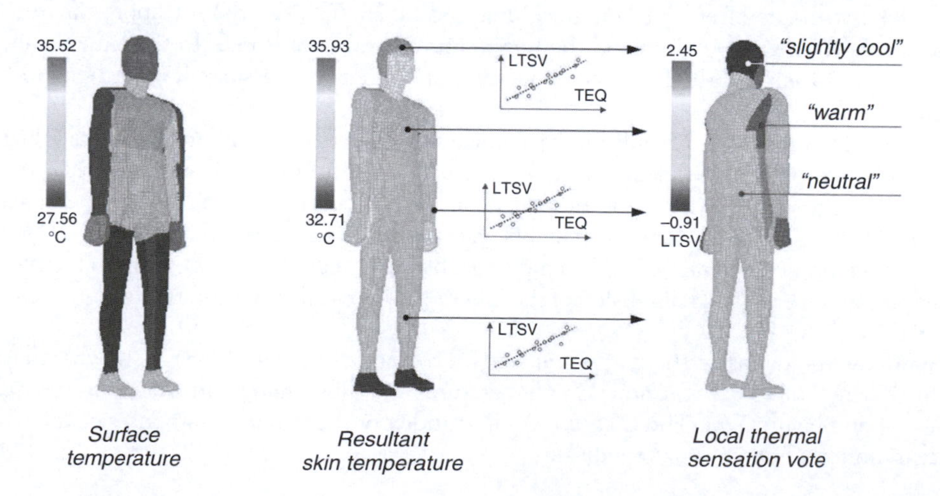

Figure 7.16 Skin and surface temperatures simulated using the human thermoregulation model FIALA-FE. Results are mapped to local thermal sensation assessments (LTSV) via regression and displayed on the artificial manikin skin.

Source: van Treeck *et al.* 2009.

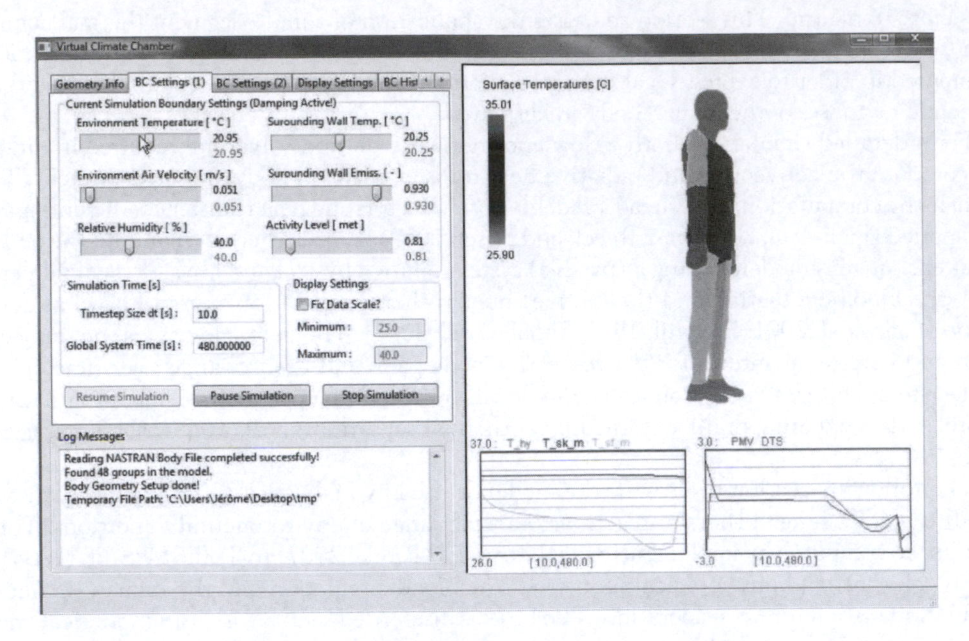

Figure 7.17 A virtual climate chamber displaying the change in the thermal and physiological properties over time.

Sources: Adapted from Fiala 1998, Paulke 2007, Raven and Horvath 1970, van Treeck *et al.* 2009.

in a stand-alone manner or coupled with other external solvers (Frisch 2008 and van Treeck *et al.* 2009).

The prediction of dynamic responses for transient conditions shall be demonstrated for the configuration of the experiment of Raven and Horvath (1970), which is also used for benchmarking in (Fiala *et al.* 2001) and Paulke (2007). During the test, eleven male people dressed in shorts (0.1clo) were exposed to temperatures of 5 °C (70%rh) and shivering for two hours after first being acclimatized at 28 °C (45%rh) for approximately 30 minutes. Both the vital values and the temperatures were monitored. In the simulation, an air speed of 0.1m/s and the basic metabolic rate of 0.8met were assumed, with the manikins unclothed.

Figure 7.18 compares the results of the average skin temperatures (a) and the metabolic rate (b) with the experiment conducted by Raven and Horvath (1970). The temperature results for the skin agree well. After 30 minutes, the ambient conditions are changed. The dynamic metabolic changes are predicted reasonably well where shivering is reproduced by the model with a deviation of several watts. The model is, however, calibrated according to an average adult person; for a detailed discussion of the differences in the metabolism the reader is referred to (Fiala *et al.* 2001).

Similarly, the dynamic thermal sensation (DTS) index of Fiala (1998) is shown in Figure 7.18(b), which takes the core and skin temperatures and their changes into account (see pages 192–194 and Figure 7.7). The DTS is a comfort index of the dynamic responses of the whole body in uniform but transient conditions.

Discussion: design-related level of detail of simulation

The previous section demonstrated a few examples of predicting the indoor thermal quality performance at fairly different resolutions in space and time. Building performance simulation

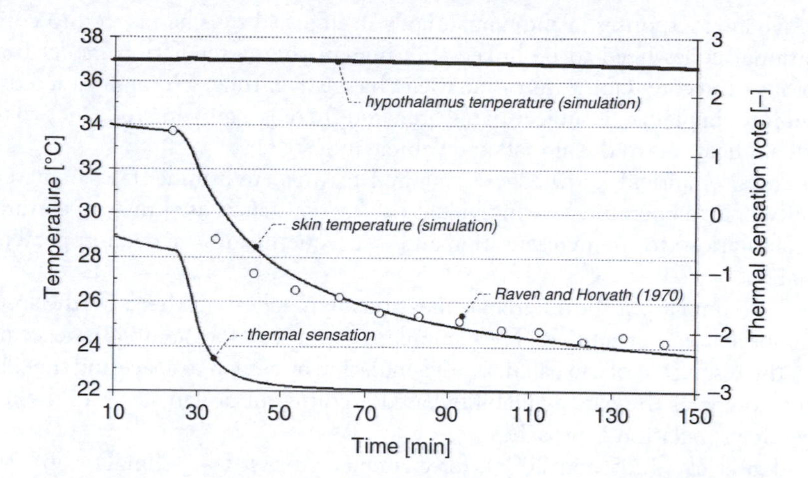

Figure 7.18(a) Comparison of simulation and experiment of dynamic thermal responses under transient conditions. Display of the temperature values.

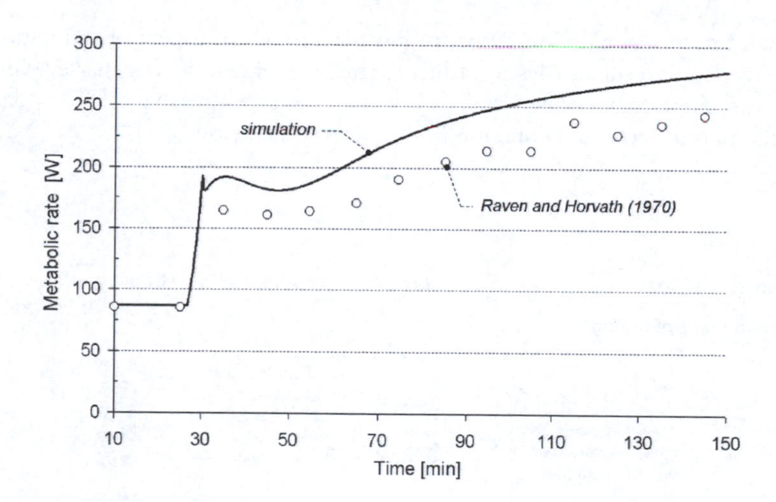

Figure 7.18(b) Display of the metabolic rate.

(BPS) can be identified as an effective tool that provides insight into the whole-year characteristics of the thermal quality of a building. The spatial resolution in spaces is restricted as a uniform distribution of the air temperature is assumed and as no further information on local air flow patterns is available. Integral measures are useful for detecting critical periods from the point of view of overheating occurrences and how frequently they occur. BPS is the recommended method for investigating the thermal quality performance, for exploring the thermal dynamics of a building and its components and, thus, for validating the overall conceptual design.

In order to capture local effects, a detailed CFD analysis is required, which introduces much more complexity in terms of both modeling and computation. The application of CFD is restricted to single spaces and selected critical time periods, which may have been identified via BPS before. Hence, CFD can support building simulation to detect and investigate local situations, such as the position of the air inlets of an HVAC system. CFD results can be "translated" into comfort data and depicted, using the PMV or DR model, for example.

If the dynamic responses of the human body itself are also to be taken into consideration, computational codes need to be linked to a human thermoregulation model that provides details of skin and core temperatures and their change over time. The application of such models in building simulation is subject to research and there is a growing tendency to use them in industrial applications in the aircraft and vehicle industries.

In all cases, empirical knowledge is required to link environmental parameters (such as air velocity and temperature) or individual parameters (such as skin temperature and moisture accumulation) to thermal sensation and, consequently, to the comfort perception of the individual.

Figure 7.19 summarizes the different simulation approaches together with the applicable performance indicators respectively. The "granularity" of a comfort analysis is accordingly determined by the resolution of the balancing or simulation approach in space and time. In addition, the Figure compares the data availability based on different design stages to the resolution of the respective simulation approaches.

Hausladen *et al.* (2005 and 2006), for example, proposed a "ClimaDesign" guideline for predicting indoor climate and energy efficiency of buildings during the very early design stages, in the absence of detailed planning data, by simply pointing out the interplay between individual design variables and their different sensitivities. Several other simplified tools are available which are applicable to early design stages, such as the MIT Design Advisor (Urban and Glicksman 2006).

The visualization of simulation results is another important aspect of simulation practice. The expressiveness of results varies according to the level of detail. At a coarse level, for example, significant details may not be visible, while at a fine level the sheer flood of data needs to be condensed in order to extract and interpret relevant quantities.

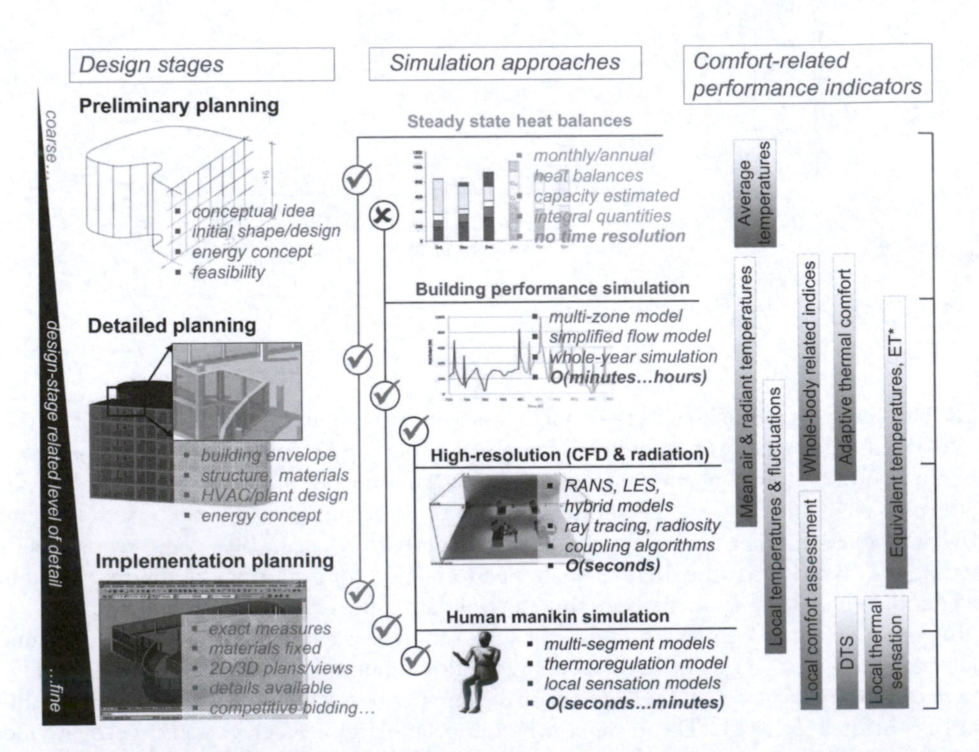

Figure 7.19 Simulation approaches and required resolution of relevant performance indicators with respect to the interplay between different design stages.

Synopsis

The indoor thermal quality performance and the energy efficiency of a building and its components, i.e. the envelope and the HVAC system, are inherently connected with one another. In the integrated building design, engineers are confronted with the interplay between aesthetical, functional and economical aspects, which interact with both the energy and the thermal quality performance. The conceptual planning stages are accompanied by different levels of detail of available simulation data and, consequently, different levels of accuracy.

This chapter served to summarize the relevant indicators and integral measures for predicting the indoor thermal quality performance in spaces and buildings. Background information was provided on human thermoregulation, thermal sensation of the individual and comfort assessment methods in uniform and non-uniform environments. The relevant codes and standards have been introduced as well as examples demonstrating the application of the various models and their relevant resolution.

When evaluating the thermal quality performance of buildings by means of simulation, special emphasis is placed on the post-processing and interpretation of the results obtained. Descriptive integrated views, such as calendar-type diagrams, or well-founded models that account for the accumulation of heat stress events, are therefore required.

The PMV/PPD model supported by the adaptive comfort guidelines are well-established and recommended methods employed in engineering practice. Their basic idea is to statistically relate peoples' assessment in terms of thermal sensation and acceptability directly to the said environmental and personal parameters. The relevant data can be acquired by building performance simulation, which makes the application of these models straightforward. However, in some situations, special attention needs to be paid to local effects. Some of these effects can be predicted with the help of empirical formulae, others require the additional application of refined models such as CFD, or even manikin models. Prediction methods that account for asymmetrical and transient conditions are still undergoing research and involve extensive experimental work with volunteer subjects.

International codes and standards

ASHRAE (2004) "Standard 55: Thermal Environmental Conditions for Human Occupancy", Atlanta: American Society of Heating, Refrigerating and Air-Conditioning Engineers.

DIN V 18599 (2005) "Energy efficiency of buildings. Calculation of the net, final and primary energy demand for heating, cooling, ventilation, domestic hot water and lighting", Beuth Verlag.

DIN 33403–2 (2001) "Klima am Arbeitsplatz und in der Arbeitsumgebung", Beuth Verlag.

ISSO 74 (2004) "Thermische behaaglijkheid; eisen voor de binnentemperatuur in gebouwen", publicatie 74, ISSO, Rotterdam.

ISO 7243 (1989) "Hot environments – Estimation of the heat stress on working man, based on the WBGT-index (wet bulb globe temperature)".

ISO 7726 (1998) "Ergonomics of the thermal environment – Instruments for measuring physical quantities".

ISO 7730 (2005) "Ergonomics of the thermal environment – Analytical determination and interpretation of thermal comfort using calculation of the PMV and PPD indices and local thermal comfort criteria".

ISO 7933 (2004) "Ergonomics of the thermal environment – Analytical determination and interpretation of heat stress using calculation of the predicted heat strain".

EN ISO 8996 (2004) "Ergonomics of the thermal environment – Determination of metabolic rate".

EN ISO 9920 (2007) "Ergonomics of the thermal environment – Estimation of thermal insulation and water vapor resistance of a clothing ensemble".

ISO 10551 (1995) "Ergonomics of the thermal environment – Assessment of the influence of the thermal environment using subjective judgement scales".

ISO 11079 (2007) "Ergonomics of the thermal environment – Determination and interpretation of cold stress when using required clothing insulation (IREQ) and local cooling effects".

ISO 11399 (1995) "Ergonomics of the thermal environment – Principles and application of relevant International Standards".
EN ISO 13731 (2001) "Ergonomics of the thermal environment – Vocabulary and symbols".
EN ISO 14505–2 (2004) "Ergonomics of the thermal environment – Evaluation of thermal environment in vehicles – Part 2: Determination of the Equivalent Temperature".
EN 15251 (2007) "Indoor environmental input parameters for design and assessment of energy performance of buildings addressing indoor air quality, thermal environment, lighting and acoustics".
VDI 6020 (2001) "Requirements for methods of building performance simulation", VDI Verein Deutscher Ingenieure, Beuth Verlag.

Note

1 Differences between ethnic and geographical groups (ASHRAE 2005).

References

Note: Codes and standards have been summarized in the previous section and are not repeated here.

Araújo, V.M.D. and Araújo E.H.S. (1999) "The Applicability of ISO 7730 for the Assessment of the Thermal Conditions of Users of the Buildings in Natal-Brazil", in *Proceedings of Indoor Air 1999*, Edinburgh, Scotland, pp. 148–153.
ASHRAE (2005) *Handbook – Fundamentals (SI)*, American Society of Heating, Refrigerating and Air-Conditioning Engineers.
Auliciems, A. (1983) "Psychophysical Criteria for Global Thermal Zones of Building Design", *International Journal of Biometeorology* 27: 69–86.
Baker, N. and Standeven, M. (1996) "Thermal Comfort for Free-running Buildings", *Energy and Buildings* 23(3): 175–182.
Beausoleil-Morrison, I. (2000) "The Adaptive Coupling of Heat and Air Flow Modeling within Dynamic Whole Building Simulation", PhD Thesis, University of Strathclyde, Glasgow.
Bedford, T. (1936) "The Warmth Factor in Comfort at Work", *MRC Industrial Health Board Report*, Vol. 76, London: HMSO.
Bejan, A. (1993) *Convection Heat Transfer*, Chichester: John Wiley & Sons, Inc.
Benzinger, T. (1979) "The Physiological Basis for Thermal Comfort", in Fanger, P.O. and O. Valbjorn, O. (eds), *Indoor Climate*, Danish Building Research Institute, Copenhagen, pp. 441–476.
Berglund, L.G. and Cunningham, D.J. (1986) "Parameters of Human Discomfort in Warm Environments", *ASHRAE Transactions* 92(2B): 732–746.
Bohm, M., Browen, M., Holmer, I., Nilsson, P.V. and Noren, O. (1990) "Evaluation of Vehicle Climate with a Thermal Manikin", Swedish Institute of Agricultural Engineering.
Brager, G.S. and de Dear, R.J. (1998) "Thermal Adaptation in the Built Environment: A Literature Review", *Energy and Buildings* 27(1): 83–96.
Busch, J.F. (1992) "A Tale of Two Populations: Thermal Comfort in Air-conditioned and Naturally Ventilated Offices in Thailand", *Energy and Buildings* 18(3–4): 235–249.
Cabanac, M. and Hardy, J.D. (1969) "Réponses unitaires et thermorégulatrices lors de réchauffements et réfroidissements localisés de la région préoptique et du mésencéphale chez le lapin", *Journal of Physiology* 61: 331–347.
de Carli, M., Olesen, B.W., Zarrella, A. and Zecchin, R. (2007) "People's Clothing Behaviour according to External Weather and Indoor Environment", *Building and Environment* 42(12): 3965–3973.
de Dear, R.J. and Brager, G.S. (1998) "Developing an Adaptive Model of Thermal Comfort and Preference", *ASHRAE Technical Data Bulletin* 14(1): 27–49.
de Dear, R.J. and Brager, G.S. (2002) "Thermal Comfort in Naturally Ventilated Buildings: Revisions to ASHRAE Standard 55", *Energy and Buildings* 34(6): 549–561.
de Paula Xavier, A.A. and Roberto, L. (2000) "Indices of Thermal Comfort Developed from Field Survey in Brazil", *ASHRAE Transactions* 106: 45–58.
de Wit, S. (2001) "Uncertainty in Predictions of Thermal Comfort in Buildings", PhD Thesis, TU Delft, The Netherlands.
EPBD (2003) "Directive 2002/91/EC of the European Parliament and of the Council of 16 December

2002 on the Energy Performance of Buildings", *Official Journal of the European Union* 1, 4.1.2003: 65–71.

Fanger, P.O. (1964) "Termiske luftströmninger langs vinduer og kolde vægge", *Ingeniören* 19.

Fanger, P.O. (1972) *Thermal Comfort*, New York: McGraw-Hill (repr. 1982, Malabar, FL: Robert E. Krieger).

Fanger, P.O. (1977) "Thermal Discomfort Caused by Radiant Asymmetry, Local Air Velocities, Warm and Cold Floors, and Vertical Air Temperature Gradients", in Durand, J. and Raynaud, J. (eds), 'Confort thermique: Aspects physiologiques et psychologiques' *INSERM* 75: 145–151.

Fanger, P.O. and Toftum, J. (2002) "Extension of the PMV Model to Non Airconditioned Buildings in Warm Climates", *Energy and Buildings* 34(6): 533–536.

Fanger, P.O., Ipsen, B.N., Langkilde, G., Olesen, B.W., Christensen, N.K. and Tanabe S. (1985) "Comfort Limits for Asymmetric Thermal Radiation", *Energy and Buildings* 8: 225–236.

Fanger, P.P., Melikov, A.K., Hanzawa, H. and Ring, J. (1988) "Air Turbulence and Sensation of Draught", *Energy and Buildings* 12(1): 21–39.

Fiala, D. (1998) "Dynamic Simulation of Human Thermoregulation and Thermal Comfort", PhD Thesis, De Montfort University, Leicester, UK.

Fiala, D., Lomas, K. and Stohrer, M. (1999) "A Computer Model of Human Thermoregulation for a Wide Range of Environmental Conditions: The Passive System", *Journal of Applied Physiology* 87(5): 1957–1972.

Fiala D., Lomas K. and Stohrer, M. (2001) "Computer Prediction of Human Thermoregulatory and Temperature Responses to a Wide Range of Environmental Conditions", *International Journal of Biometeorology* 45(3): 143–159.

FIALA-FE (2008) "Manikin Model Validation Manual, Release 2.0". Online. Available at: http://www.theseus-fe.de (accessed 07.12.2009).

Frijns, A., Limpens-Neilen, D., Schoffelen, M. and van Leeuwen, G. (2006) "Simulation of the Human Thermal Responses to Local Heating", in *Proceedings of IEA*, Maastricht, The Netherlands.

Frisch, J. (2008) "Parametrisches Modell zur interaktiven Simulation menschlicher Thermoregulation", MSc Thesis, TU München.

Gagge, A.P. (1973) "Rational Temperature Indices of Man's Thermal Environment and Their Use with a 2-node Model of His Temperature Regulation", *Federal Proceedings* 32(5): 1572–1582.

Gagge, A.P. (1979) "The Role of Humidity during Warm Discomfort", in Fanger, P.O., and Valbkorn, O. (eds), *Indoor Climate*, Copenhagen : Danish Building Research Institute, pp. 527–538.

Gagge, A.P. and Gonzalez, R.R. (1973) "Physiological Bases of Warm Discomfort for Sedentary Man", *Archives des sciences physiologiques* 27(4): 409–424.

Gagge, A.P., Stolwijk, J.A. and Hardy, J.D. (1967) "Comfort and Thermal Sensation and Associated Physiological Responses at Various Ambient Temperatures", *Environmental Research* 1(1): 1–20.

Gagge, A.P., Fobelets, A.P. and Berglund, P.E. (1986) "A Standard Predictive Index of Human Response to the Thermal Environment", *ASHRAE Transactions* 92(2B): 709–731.

Glück, B. (1994) "Zulässige Strahlungstemperatur-Asymmetrie", *Gesundheitsingenieur* 115(6): 285–344.

Gonzalez, R.R. and Berglund, L.G. (1979) "Efficacy of Temperature and Humidity Ramps in Energy Conservation", *ASHRAE Journal* 6: 34–41.

Gonzalez, R.R. and Gagge, A.P. (1973) "Magnitude Estimates of Thermal Discomfort during Transients of Humidity and Operative Temperature", *ASHRAE Transactions* 79(1): 88–96.

Gordon, R. (1974) "The Response of Human Thermoregulatory System in the Cold", PhD Thesis, University of California, Santa Barbara, CA.

Griffiths, I.D. and McIntyre, D.A. (1974) "Sensitivity to Temporal Variations in Thermal Conditions", *Ergonomics* 17(4): 499–507.

Guyton, C. (2002) *Textbook of Medical Physiology*, Philadelphia, London: W.B. Saunders.

Han, T., Huang, L., Kelly, S., Huizenga, C. and Zhang, H. (2001) "Virtual Thermal Comfort Engineering", *SAE Technical Paper Series* 2001–01–0588.

Hausladen, G., Saldanha, M., Liedl, P. and Sager, C. (2005) *ClimaDesign: Lösungen für Gebäude, die mit weniger Technik mehr können*, München: Verlag Callwey.

Hausladen, G., Saldanha, M. and Liedl, P. (2006) *ClimaSkin: Konzepte für Gebäudehüllen, die mit weniger Energie mehr leisten*, München: Verlag Callwey.

Heiselberg, P. (1994) "Stratified Flow in Rooms with a Cold Vertical Wall", *ASHRAE Transactions* 100(1): 1011–1023.

Hensel, H. (1982) *Thermal Sensation and Thermoreceptors in Man*, Springfield, IL: Charles C. Thomas.

Hensen, J.L.M. (1990) "Literature Review on Thermal Comfort in Transient Conditions", *Building and Environment* 25(4): 309–316.
Herkel, S., Knapp, U. and Pfafferott, J. (2005) "A Preliminary Model of User Behaviour Regarding the Manual Control of Windows in Office Buildings", in *Proceedings of Building Simulation 2005*, Montreal, Canada, pp. 403–410.
Holmér, I. (1995) "Heated Manikins as a Tool for Evaluating Clothing", *Annals of Occupational Hygiene* 39(6): 809–818.
Holmér, I. (2004) "Thermal Manikin History and applications", *European Journal of Applied Physiology* 92(6): 614–618.
van Hoof, J. (2008) "Forty Years of Fanger's Model of Thermal Comfort: Comfort for All?", *Indoor Air* 18(3): 182–201.
van Hoof, J. and Hensen, J.L.M. (2007) "Quantifying of relevance of Adaptive Thermal Comfort Models in Moderate Thermal Climate Zones", *Building and Environment* 42(1): 156–170.
Huizenga, C., Hui, Z. and Arens, E. (2001) "A Model of Human Physiology and Comfort for Assessing Complex Thermal Environments", *Building and Environment* 36(6): 691–699.
Humphreys, M.A. (1978) "Outdoor Temperatures and Comfort Indoors", *Building Research & Information* 6(2): 92–105.
Humphreys, M.A. and Nicol, J.F. (1998) "Understanding the Adaptive Approach to Thermal Comfort", *ASHRAE Transactions* 104(1): 991–1004.
Issing, K. and Hensel, H. (1981) "Static Thermal Sensation and Thermal Comfort", *Zeitschrift für Physikalische Medizin, Balneologie und Medizinische Klimatologie* 11: 354–365.
Kato, S., Nagano, H. and Yang, L. (2007) "A CFD Manikin with a Thermo Physiology Model", in *Proceedings of RoomVent 2007*, Helsinki, Finland.
Kok, J., van Muijden, J., Burgers, S., Dol, H. and Spekreijse, S. (2006) "Enhancement of Aircraft Cabin Comfort Studies by Coupling of Models for Human Thermoregulation, Internal Radiation, and Turbulent Flows", in Wesseling, P., Onate E. and Periaux J. (eds), *ECCOMAS CFD 06*, Egmond aan Zee, The Netherlands.
Kröling, P. (1998) "Sick Building Syndrom: Ursachen und Prophylaxe gebäudebedingter Gesundheitsstörungen", *Allergologie* 21(5): 180–191.
Van der Linden K., Boerstra A.C., Raue A.K. and Kurvers S.R. (2002) "Thermal Indoor Climate Building Performance Characterized by Human Comfort Response", *Energy and Buildings* 34(7): 737–744.
Van der Linden, A.C., Boerstra, A.C., Raue, A.K., Kurvers, S.R. and de Dear, R.J. (2006) "Adaptive Temperature Limits: A New Guideline in The Netherlands", *Energy and Buildings* 38(1): 8–17.
Van der Linden, W., Loomans, M. and Hensen, J. (2008) "Adaptive Thermal Comfort Explained by PMV", in *Proceedings of Indoor Air 2008*, Copenhagen, Denmark.
McGuffin, R., Burke, R., Huizenga, C., Zhang, H., Vlahinos, A. and Fu, G. (2002) "Human Thermal Comfort Model and Manikin", *Society of Automotive Engineers* Paper 2002–01–1955.
McIntyre, D.A. (1980) *Indoor Climate*, London: Applied Science Publishers.
McNall, P.E. and Schlegel, J.C. (1968) "The Relative Effect of Convective and Radiation Heat Transfer on Thermal Comfort", *ASHRAE Transactions* 74: 131–143.
Mayer, E. (1997) "A New Correlation between Predicted Mean Votes (PMV) and Predicted Percentages of Dissatisfied (PPD)", in Woods, J.E., Grimsrud, D.T. and Boschi N. (eds), *Proceedings of Healthy Buildings/IAQ 1997*, Washington DC, pp. 189–193.
Mayer, E. (2000) "Manfitted Measurement of Thermal Climate by a Dummy Representing Suit for Simulation of Human Heatloss (DRESSMAN). *Healthy Buildings* 2: 551–556.
Mayer, E. and Schwab, R. (1990) "Untersuchungen der physikalischen Ursachen von Zugluft", *Gesundheitsingenieur* 111(1): 17–30.
Mayer, E. and Schwab, R. (1993) "Evaluation of Heat Stress by an Artificial Skin", in *Proceedings of Indoor Air*, Helsinki, Finland, pp. 73–78.
Modest, M.F. (2003) *Radiative Heat Transfer*, 2nd ed., San Diego, CA: Academic Press.
Mower, G.D. (1976) "Perceived Intensity of Peripheral Thermal Stimuli Is Independent of Internal Body Temperature", *Journal of Comparative and Physiological Psychology* 90(12): 1152–1155.
Nevins, R.G., Rohles, F.G., Springer, W. and Feyerherm, A.M. (1966) "Temperature-humidity Chart for Thermal Comfort of Seated Persons", *ASHRAE Transactions* 72(1): 283–291.
Nevins, R.G., Gonzalez, R.R., Nishi, Y. and Gagge, A.P. (1975) "Effect of Changes in Ambient Temperature and Level of Humidity on Comfort and Thermal Sensations", *ASHRAE Transactions* 81(2): 169–182.

Nielsen, P.V. (1981) "Luftströmninger i ventilerede arbejdslokaler", *SBI-rapport 128*, Statens byggeforskningsinstitut.

Nielsen, P.V., Murakami, S., Kato, S., Topp, C. and Yang, J.H. (2003) "Benchmark Test for a Computer Simulated Person", Aalborg University. Online. Available at: http://www.cfd-benchmarks.com (accessed 07.12.2009).

Nilsson, H.O. (2004) "Comfort Climate Evaluation with Thermal Manikin Methods and Computer Simulation Models", PhD Thesis, Royal Institute of Technology, Sweden.

Nilsson, H.O., Holmér, I., Bohm, M. and Norén, O. (1997) "Equivalent Temperatures and Thermal Sensation, Comparison with Subjective Responses", in *Proceedings of ATA Conference: Comfort in the Automotive Industry*, Bologna, Italy, pp. 157–162.

Nilsson, H., Brohus, H. and Nielsen, P. (2007) "CFD Modeling of Thermal Manikin Heat Loss in a Comfort Evaluation Benchmark Test", in *Proceedings of Roomvent 2007*, Helsinki.

Niu, J. and Gao, N. (2008) "Personalized Ventilation for Commercial Aircraft Cabins", in *Proceedings of Indoor Air 2008*, Copenhagen, Denmark.

Olesen, B.W. (2002) "Radiant Floor Heating in Theory and Practice", *ASHRAE Journal* 44(7): 19–24.

Olesen, B.W. (2008) "Radiant Floor Cooling Systems", *ASHRAE Journal* 50(9): 16–22.

Olesen, B.W., Schöler, M. and Fanger, P.O. (1979) "Discomfort Caused by Vertical Air Temperature", in Fanger, P.O. and O. Valbjorn, O. (eds), *Indoor Climate*, Danish Building Research Institute, Copenhagen, pp. 561–579.

Paulke, S. (2007) "Finite Element-based Implementation of Fiala's Thermal Manikin in THESEUS-FE", in *Proceedings of EUROSIM 2007*.

Pfafferott, J., Herkel, S., Kalz, D. and Zeuscher, A. (2007) "Comparison of Low-energy Office Buildings in Summer Using Different Thermal Comfort Criteria", *Energy and Buildings* 39(7): 750–757.

Raven, P. and Horvath, S. (1970) "Variability of Physiological Parameters of Unacclimatized Males during a Two-hour Cold Stress of 5 °C", *International Journal of Biometeorology* 14(3): 309–320.

Recknagel, H., Sprenger, E. and Schramek, E.R. (2007) *Taschenbuch für Heizung, Klima und Technik*, 73th ed., München: Oldenbourg Industrieverlag.

Ring, J.W. and de Dear, R.J. (1991) "Temperature Transients: A Model for Heat Diffusion through the Skin, Thermoreceptor Response and Thermal Sensation", *Indoor Air* 1(4): 448–456.

Robinson, D. and Haldi, F. (2008) "An Integrated Adaptive Model for Overheating Risk Prediction", *Journal of Building Performance Simulation* 1(1): 43–55.

Rohles, F.H. and Nevins, R.G. (1971) "The Nature of Thermal Comfort for Sedentary Man", *ASHRAE Transactions* 77(1): 239–246.

Rohles, F.H. and Wallis, S.B. (1979) "Comfort Criteria for Air Conditioned Automotive Vehicles", *SAE Technical Paper Series 790122*, pp. 440–449.

Rohles, F.H., Woods, J.E. and Nevins, R.G. (1974) "The Effect of Air Movement and Temperature on the Thermal Sensations of Sedentary Man", *ASHRAE Transactions* 80(1): 101–118.

Rugh, J.P., Farrington, R.B., Bharathan, D., Vlahinos, A., Burke, R., Huizenga, C. and Zhang, H. (2004) "Predicting Human Thermal Comfort in a Transient Nonuniform Thermal Environment", *European Journal of Applied Physiology* 92(6): 721–727.

Sano, J., Tanabe, S., Murakami, K., Nagayama, H., Oi, H. and Itami, T. (2008) "Evaluation of Thermal Environments in Vehicles by CFD Coupled with Numerical Thermoregulation-model JOS", in *Proceedings of Indoor Air 2008*, Copenhagen, Denmark.

Schälin, A., Moser, A., van der Maas, J. and Aiulfi, D. (1996) "Application of Air Flow Models as Design Tools for Atria", in *Proceedings of Roomvent 1996*, Yokohama, Japan, pp. 171–178.

Stolwijk, J. (1971) "A Mathematical Model of Physiological Temperature Regulation in Man", CR-1855, NASA.

Stolwijk, J.A. (1991) "Sick-Building Syndrome", *Environmental Health Perspectives* 95: 99–100.

Streblow, R., Müller, D. and Gores, I. (2008) "A Coupled Simulation of the Thermal Environment and Thermal Comfort with an Adapted Tanabe Comfort Model", in *Proceedings of Indoor Air 2008*, Copenhagen, Denmark.

Strøm-Tejsen, P., Zukowska, D., Jama, A. and Wyon, D. (2007) "Assessment of the Thermal Environment in a Simulated Aircraft Cabin Using Thermal Manikin Exposure", in *Proceedings of RoomVent 2007*, Helsinki, Finland, pp. 227–234.

Tanabe, S.I., Kobayashi, K., Nakano, J., Ozeki, Y. and Konishi, K. (2002) "Evaluation of Thermal Comfort Using Combined Multi-node Thermoregulation (65MN) and Radiation Models and Computational Fluid Dynamics (CFD)", *Energy and Buildings* 34(6): 637–646.

THESEUS-FE (2008) Theory Manual, Release 2.0. Online. Available at: http://www.theseus-fe.de (accessed 07.12.2009).

van Treeck, C. (2004) "Gebäudemodell-basierte Simulation von Raumluftströmungen", PhD Thesis, Technische Universität München.

van Treeck, C. and Frisch, J. (2009) "Model-adaptive Analysis of Indoor Thermal Comfort", in *Proceedings of Building Simulation 2009*, Glasgow, Scotland, pp. 1374–1381.

van Treeck, C., Rank, E., Krafczyk, M., Tölke, J. and Nachtwey, B. (2006) "Extension of a Hybrid Thermal LBE Scheme for Large-Eddy Simulations of Turbulent Convective Flows", *Computers and Fluids* 35(8–9): 863–871.

van Treeck, C., Wenisch, P., Borrmann, A., Pfaffinger, M., Wenisch, O. and Rank, E. (2007a) "ComfSim – Interaktive Simulation des thermischen Komforts in Innenräumen auf Höchstleistungsrechnern", *Bauphysik* 29(1): 2–7.

van Treeck, C., Hillion, F., Cordier, A., Pfaffinger, M. and Rank, E. (2007b) "Flow and Thermal Sensation Analysis in an Aeroplane Passenger Cabin Using a Lattice Boltzmann Based Approach", in *Proceedings of ICMMES 2007*, Munich, Germany.

van Treeck, C., Pfaffinger, M., Wenisch, P., Frisch, J., Yue, Z., Egger, M. and Rank, E. (2008). "Towards Computational Steering of Thermal Comfort Assessment", in *Proceedings of Indoor Air 2008*, Copenhagen, Denmark.

van Treeck, C., Frisch, J., Pfaffinger, M., Rank, E., Paulke, S., Schweinfurth, I., Schwab, R., Hellwig, R. and Holm, A. (2009) "Integrated Thermal Comfort Analysis Using a Parametric Manikin Model for Interactive Real-time Simulation", *Journal of Building Performance Simulation* 2(4): 233–250.

Urban, B.J. and Glicksman, L.R. (2006) "The MIT Design Advisor – a Fast, Simple Building Design Tool", in *Proceedings of IBPSA-Germany Conference BauSIM2006*, pp. 203–204.

Wissler, E. (1985) "Mathematical Simulation of Human Thermal Behaviour Using Whole Body Models", *Heat Transfer in Medicine and Biology* 1(13): 325–373.

Wyon, D.P., Larsson, S., Forsgren, B. and Lundgren, I. (1989) "Standard Procedures for Assessing Vehicle Climate with a Thermal Manikin", *SAE Technical Paper* 890049, pp. 1–11.

Yoon, D.W., Sohn, J.Y. and Cho, K.H. (1999) "The Comparison on the Thermal Comfort Sensation between the Results of Questionnaire Survey and the Calculation of the PMV Values", in *Proceedings of Indoor Air 1999*, Edinburgh, Scotland, pp. 137–141.

Zhang, H. (2003) "Human Thermal Sensation and Comfort in Transient and Non-uniform Thermal Environments", PhD Thesis, University of California, Berkeley.

Zhang, H., Huizenga, C., Arens, E. and Wang, D. (2004) "Thermal Sensation and Comfort in Transient Non-uniform Thermal Environments", *European Journal of Applied Physiology* 92(6): 728–733.

Zhang, Y. and Yang, T. (2008) "Simulation of Human Thermal Responses in a Confined Space", in *Proceedings of Indoor Air 2008*, Copenhagen, Denmark.

Zhu, S.W., Kato, S., Ooka, R. and Sakoi, T. (2007) "Development of a Computational Thermal Manikin Applicable in a Non-uniform Thermal Environment (Part 1)", *ASHRAE HVAC&R Research* 13(4): 661–679.

Zhu, S.W., Kato, S., Ooka, R., Sakoi, T. and Tsuzuki K. (2008) "Development of a Computational Thermal Manikin Applicable in a Non-uniform Thermal Environment (Part 2)", *ASHRAE HVAC&R Research* 14(4): 545–564.

Activities and solution strategy

Choose a preferred building simulation tool. Create a suitable single-zone model of heavy weight type, with an appropriate insulation standard, and with a south-facing wall and window surface. All other walls are adiabatically connected with the rest of the building. Use weather data from a moderate climate zone and select typical internal heat sources.

Case A: Natural ventilation

Set up a standard convective heating system and define a reasonable tempurature threshold and a plant operation schedule for the winter period. Select a high infiltration rate during the day and omit mechanical cooling for the summer period.

- Compute the annual and monthly heating performance and validate your results using a spreadsheet tool.
- Compute the operative room temperatures for each hour of the year. Choose an adaptive thermal comfort guideline, compute the minimum, maximum and average operative temperatures for each day. Plot these values against the weighted running average outdoor temperature, as required by the standard. Indicate the comfort categories in your diagram.
- Classify the building in terms of the comfort category by counting the number of occurrences of the operative temperature values within each category (separately for the upper and lower bound).
- Locate critical time periods in terms of overheating. Try to optimize your model by changing the design (night ventilation, increase/decrease the thermal capacity of the components, reduce internal heat gains, increase the ventilation rate during the day, etc.).
- Try to find a method to account for the accumulation of heat stress events causing overheating.
- Add a radiant cooling system covering the ceiling. Again, compute the operative room temperatures and check the thermal quality performance of the model. Evaluate the impact of the cooling device in terms of local thermal comfort. Use the methods described in the ASHRAE Standard 55 or ISO Standard 7730.

Case B: Mechanical ventilation

Use the scenario of case A and add a convective cooling system. Similarly, add a plant operation schedule for the summer period. Select a low infiltration rate during set-back times.

- Again, compute the annual and monthly heating performance and validate your results with the help of a spreadsheet tool.
- Compute the operative room temperatures for each hour of the year. Choose a thermal comfort guideline based on the PMV/PPD model to evaluate the thermal quality performance.
- Design the cooling plant performance in such a way that the same comfort category is obtained from the point of view of PPD as with the natural ventilation case. Compare the annual energy consumption.

8 Room acoustics performance prediction

Ardeshir Mahdavi

Scope

This chapter covers the fundamentals of computational room acoustics. It discusses common approaches applied in room acoustics simulation applications to model the sound propagation phenomena in architectural spaces.

Learning objectives

The reader will gain insights into the computational methods adapted in room acoustics simulation applications as well as into the variety and purposes of numeric performance indicators in room acoustics. Moreover, the reader should become familiar with application potential and reliability issues of acoustical simulation and auralization in view of room acoustics design.

Key words

Room acoustics/sound propagation/absorption/ray tracing/image source method/auralization

Introduction

The subject of "Room acoustics" evokes associations with concert halls, opera houses, and theater spaces. Prediction and evaluation of the acoustical quality of such spaces is indeed one of the main concerns of room acoustics as a science and as a source of knowledge for design-support. However, good room acoustics is not important only regarding spaces for music and speech performances. More "ordinary" spaces such as a living room, an open-plan office, a library reading room, or the waiting room of a clinic may be experienced as possessing good or poor room acoustics.

A scientifically-based computationally-supported evaluation of architectural designs in view of their room acoustical quality must meet two challenges: Firstly, methods and tools are needed to correctly predict the properties of the sound field in a designed (an as yet not existing) space. Secondly, predicted properties of the sound field in a space must be analyzed and interpreted in view of relevant acoustical performance criteria. For a long time, there was a general belief that the acoustics of a room cannot be predicted, that room acoustics is rather an art, not a science. There are indeed still unsolved problems concerning both the exact prediction of the sound field in a space, and the relationship between objective parameters of the sound field and subjective perception of acoustical quality. Nonetheless, recent advances in room acoustics have reached a point so that satisfactory acoustics cannot only be defined rationally, but can be also predicted reliably. Such advances pertain to sophisticated mathematical methods, powerful computational implementations, and deeper insights concerning the relationship between objective (measureable) sound field parameters and subjective perception of room acoustics quality.

This chapter provides a brief introduction to computational room acoustics. Toward this end, pages 219–222 deal with computational methods adapted in room acoustics simulation applications. Pages 222–226 include an overview of a number of numeric performance indicators in room acoustics. Pages 226–231 address the issue of reliability as related to the application of simulation and auralization tools in the design process. Summary and conclusions are included on page 232.

Note that this chapter does not address questions pertaining to electronic equipment and systems for sound reinforcement and control in rooms. Not included are also discussions regarding general issues of noise control in buildings. The latter represents a second important application area of applied acoustics in the built environment domain, namely building acoustics. This discipline is concerned with those situations when exposure to sound events is not desired, necessitating thus various noise control strategies. Typical instances of noise control applications in the built environment concern, for example, traffic noise exposure, air-borne and impact sound transmission between rooms, and noise due to mechanical installations and equipment. While not addressed explicitly in the present contribution, proper solutions in building acoustics are a prerequisite of successful room acoustics. For example, musical performances in a concert hall require both adequate envelope sound insulation (in view of sound transmission from external environment and from adjacent rooms) and quietly operating building service systems.

Computational methods

Introduction

There is no absolute and unique way of modeling the sound propagation in rooms. Given a specific inquiry, criteria such as room size and features, and critical frequency range, computational capacity, and necessary (or appropriate) level of precision are amongst the factors that may give one method an advantage over another. Generally speaking, acoustical phenomena could be described either as waves or as particles. As such, numeric solutions can be applied to solve the wave equation. But while wave models can provide accurate results for single frequencies, their application for large rooms and higher frequencies are deemed less practical, given – amongst others – the computational problems in obtaining eigenvalues and associated normal modes. An alternative computational approach for describing the characteristics of sound fields in architectural spaces (suitable also for large complex rooms and for high frequencies) involves the notion of sound particles whose propagation paths in the space are abstracted in terms of sound rays. This idea goes back to the discipline of geometrical room acoustics, in which the concept of waves is replaced by that of sound rays. Analogous to light rays in geometrical optics, a sound ray denotes a small portion of a spherical wave with vanishing aperture which, originates from a specific point and follows a well-defined direction of propagation (Kuttruff 2000).

Two geometrically-based methods have been applied to simulate sound propagation in large rooms, namely the "Ray tracing method" and the "Image source method". As these methods, however, do not embody the wave property of sound phenomena, they imply high order reflections that, in terms of their accuracy, deviate from the actual behavior of sound waves. As a consequence, the assumption of purely geometrical behavior should be restricted to low order reflections, necessitating alternative (statistical) approaches for computational description of higher order reflections. In the following, the main features of the ray tracing method and the image source method, as well as combinations of those (hybrid solutions) are briefly discussed.

The ray tracing method

The Ray Tracing Method (Krokstad *et al.* 1968; Kulowski 1985; Lewers 1993; Mahdavi *et al.* 1997; Vian and van Maercke 1986) involves the emission of sound particle rays from a source in various directions. The collision of a particle with a surface results in:

(a) An energy loss whose magnitude is a function of the surface's absorption coefficient α (i.e., the ratio of non-reflected to incident sound intensity);
(b) The change of the direction of propagation, to be determined – in case of a specular reflection – according to Snell's law.

Certain conditions must be met in order to ensure that, in the course of modeling, neither false reflections are generated nor valid reflections are ignored. The probability that a ray will find a surface of area A after a travelling time of *t* is quite high, if the wave front area per ray is smaller than A/2. From this fact the required minimum number of rays *N* may be estimated using Equation (8.1).

$$N \geq 8\pi c^2 t^2 A^{-1} \tag{8.1}$$

Here, *c* denotes the speed of sound propagation in air.

The image source method

Particularly in those cases where enclosure boundaries of a space are composed of planar surfaces, the concept of image sources facilitates the efficient construction and pursuit of sound rays about the space. Given a point sound source in front of a plane surface element, each reflection from this surface can be imagined to originate from a virtual source located behind this surface and at the same distance from it as the real source. The concept of image sources can be extended to include image sources of higher order. To illustrate this, Figure 8.1 shows, using the example of a sound source (*S*) in a simple room, the image sources of first order (*S′*) and second order (*S″*).

The number of image sources within a radius of *ct* in a room (with volume *V*) can be estimated using Equation (8.2). It provides an approximate value for the number of reflections that arrive at the location of a receiver up to the time *t* after the emission of the sound signal.

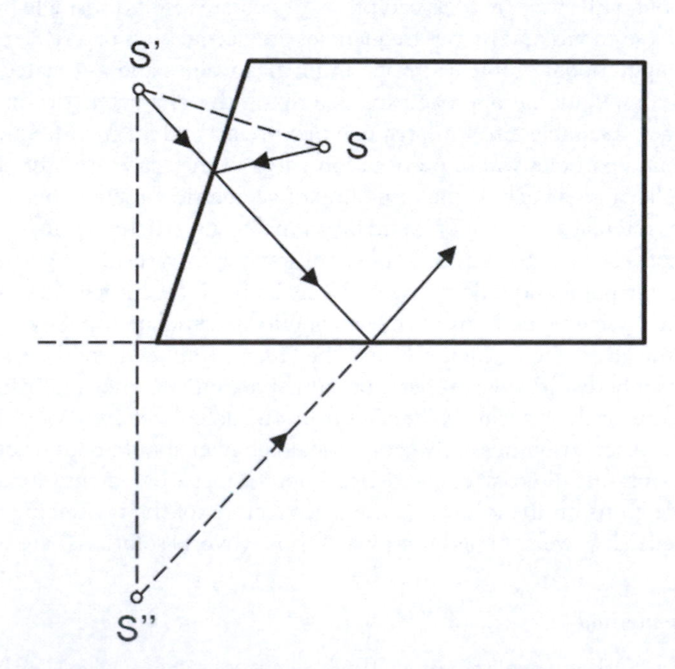

Figure 8.1 Schematic illustration of first and second order image sources.

Source: After Kuttruff 2000, modified.

$$N_{refl} = 4\pi c^3 t^3 (3V)^{-1} \tag{8.2}$$

Assuming a room with N (plane) surfaces and a corresponding number of first order source images (mirrored by each surface), result in $N(N-1)$ new second order images (Kuttruff 2000). This procedure can be repeated so that the total number of images of order up to i_0 is obtained as follows (see Equation (8.3)):

$$N(i_0) = N((N-1)^{i_0} - 1)(N-2) \tag{8.3}$$

The resulting number of possible image sources can thus rapidly increase, making image source methods practically applicable only in cases where it is sufficient to only consider reflections of low order and in case of simple rooms with regular geometry.

The hybrid methods

Hybrid methods attempt to combine the positive features of both Ray Tracing and Image Source methods (Naylor 1993; Vorländer 1989). Thereby, efficient solutions are worked out to determine if image sources are valid or not. Toward this end, rays are traced from the source until they hit a surface. The resulting reflections are then examined ("visibility test") to see if they are relevant for (energetically contribute to) the receiver location. This operation involves tracking back from the receiver toward the image source. Assuming an image source has been consequently deemed to be valid, the corresponding reflection's energy can be computed as the product of source energy (in the direction of radiation) and the reflection coefficient of the surfaces involved. Typically, an "image tree" is constructed in the process to keep track of the emerging early images, avoiding thus that each valid image is accounted for more than once. Otherwise, the ensuing reflectogram could include redundant reflections and become erroneous as a result (Rindel 2000).

The temporal length of a reflectogram itself represents a basic computational challenge: There is a limit to the number of rays that can be used in a computation. Thus, the sequence of reflections in a reflectogram can only be accurately predicted until a certain time limit, constrained by the finite number of the rays considered. To construct the "reverberation tail" in the reflectogram, other approaches are needed (Mahdavi *et al.* 1997; Naylor 1993). For example, the "*secondary source method*" (Naylor 1993) deploys different methods to treat early and late reflections. While early reflections are obtained via image source generation, late reflections are modeled in terms of secondary sources that emerge when a ray hits a surface. Secondary sources act as elemental area radiators whose energy is calculated based on the product of the energy of the primary source (per ray) and the reflection coefficient of the surfaces involved in the reflection sequence. The energy arrived at a receiver from a secondary source is proportional to the cosine of the angle between the surface normal at the location of the secondary source and the vector from the secondary source to the receiver point. Moreover, the energy level attenuates from the source to the receiver in accordance with the inverse square law. Given the total path length of a reflection (from the primary source through intermediate reflections to the secondary source and thereafter to the receiving point), the reflection arrival time can be determined.

According to Rindel (2000), a very high transition order (from early to late reflection) does not necessarily lead to more accurate simulation results: The likelihood that images would be visible from the receiver point decreases at higher reflection orders. Moreover, as compared to real circumstances, geometrical acoustics overestimates the strength of reflections from smaller surfaces (relevant for high order reflections). Thus, missing such reflections due to the application of secondary source method does not introduce significant errors in predictions. Ray numbers of up to 1,000 and transition orders of two or three have been claimed to be sufficient to provide reliable simulation results for most architectural spaces.

Specular and diffuse reflection

To characterize the reflective behavior of a surface in a simulation environment, the notion of a scattering coefficient(s) is applied. It denotes the ratio between reflected sound intensity in non-specular directions and the total reflected sound intensity. Thus, $s = 0$ denotes fully specular reflection, while $s = 1$ suggest complete diffusivity. To simulate the direction of a diffuse reflection, stochastic methods can be used, involving random numbers and a probability function based on Lambert's cosine law (Stephenson 1985). The scattering coefficient enables the allocation of weights to the diffuse and specular components of a reflection. There is currently a lack of detailed empirical data on the scattering coefficients of various architecturally relevant surfaces. In most instances scattering coefficient values of about 0.1 are used for large planar surfaces, and values of about 0.7 for irregular surface. Ongoing efforts to assemble comprehensive acoustical material properties in general and scattering coefficient information in particular, are expected to improve the predictive utility of room acoustical simulation tools.

Auralization

Analogous to visualization, auralization (Kleiner *et al.* 1993) allows, in theory, to circumvent – or at least to complement – the utility of quantitative (numeric) performance indicators, facilitating instead a "direct" perception of the acoustical quality already in the design stage. Computational simulation of room acoustics produces an impulse response model that can be used toward modeling the head-related transfer function of the listener. The binaural room impulse response (BRIR) can be computed as a pair of impulse responses for a listener's ears whose head is located at the receiver position. Thereby, the head related transfer function (HRTF) is used as obtained from an artificial head representing an "average" human head (Rindel 2000). Thus, all kinds of (anechoically recorded) signals such as music and speech passages could be implanted in the virtual room by combining the signal with the computed BRIR.

Performance indicators

Once the major physical phenomena concerning the interaction of sound waves and built structures are captured in a simulation model, it can be used to support evaluative processes pertaining to room acoustics. A common approach to evaluate the acoustics of rooms involves the computation of the values of relevant room acoustical performance indicators and their comparison with recommended values (Fasold and Veres 2003). Such indicators typically result from process, whereby a consensus is reached that a certain proposed computable derivative of the sound field is relevant to and correlates with subjective (perceived) quality assessments by experts and participants of controlled tests. An exhaustive list of such indicators cannot be presented here, as research in this area is ongoing. Table 8.1 provides examples of room acoustic indicators, divided into three broad groups, namely general, position-related, and space-related indicators. Included in this table are also (rough) recommended value ranges for each indicator.

The oldest and still widely applied indicator of room acoustic is the reverberation time (T). It is defined as the time that it takes for the mean sound pressure level in a room to drop 60 dB after the source of a sound in a room is turned off. Reverberation time in a room can be estimated, amongst others, via Equation (8.4) (Eyring 1930) and Equation (8.5) (Sabine 1927). Sabine equation is applicable only to rooms with rather low absorption coefficients ($\alpha < 0.3$).

$$T = 0{,}163 \frac{V}{-\ln(1-\bar{\alpha})S_{tot}} [\mathrm{s}] \tag{8.4}$$

Table 8.1 Examples of room acoustics performance indicators with recommendations for their values

Type	*Room acoustics indicators*	*recommended value for speech*	*music*
General	Reverberation time T [s]	ca. 1 s	ca. 2 s
	Early decay time EDT [s]		ca. 2,2 s
	Bass-ratio BR [-]	0,9–1,0	1,1–1,3
Position-related	Speech Clarity C_{50} [dB]	≥ –2 dB	
	Articulation loss of consonants Al_{cons} [%]	< 15 %	
	Clarity measure C_{80} [dB]		≥ –3 dB
	Direct sound measure C_7 [dB]		≥ –10 dB
	Speech transmission index STI [-]	> 0,5	
	Center time t_s [ms]	60–80 ms	70–150 ms
	Echo criterion EK_{Speech} [ms]	< 50 ms	
	Strength measure G [dB]	1–10 dB	1–10 dB
Space-related	Reverberance measure R [dB]		2–6 dB
	Lateral fraction LF [%]		25–40 %
	Echo criterion EK_{Music} [ms]		80–100 ms

$$T = 0{,}163 \frac{V}{A_{tot}} [s] \tag{8.5}$$

$$\bar{\alpha} = A_{tot} / S_{tot} [-] \tag{8.6}$$

where:
V Room volume [m³]
A_{tot} Total equivalent absorption area [m²]
S_{tot} Total room surface area [m²]

The early decay time (*EDT*) relates likewise to the sound pressure level decay in a room. However, it is based on the initial 10 dB sound pressure level drop (Jordan 1980). *EDT* is said to better correlate with the subjective perception of the duration of reverberation.

The bass-ratio (*BR*) further characterizes the frequency aspect of reverberation time as a single-number indicator (Beranek 1962). It is calculated using Equation (8.7), based on the reverberation time (*T*) values at four (octave-band) mid-frequencies (125, 250, 500, and 1,000 Hz).

$$BR = \frac{T_{125} + T_{250}}{T_{500} + T_{1000}} \tag{8.7}$$

Speech clarity (C_{50}) (Januska 1968; Lochner and Burger 1961; Niese 1965; Thiele 1953) provides a measure of speech intelligibility (see Equation (8.8)).

$$C_{50} = 10 \log \left(\frac{E_{0...50ms}}{E_{50ms...\infty}} \right) [dB] \tag{8.8}$$

$E_{0...50ms}$ Sound energy reaching the listener in the interval from 0 ms (direct sound) to 50 ms
$E_{50ms...\infty}$ Rest sound energy (reaching the listener after 50ms)

Articulation loss of consonants Al_{cons} (see Equation (8.9)) is an alternative to C_{50} toward objective assessment of speech intelligibility in rooms or with sound reinforcement systems (Peutz 1971).

$$Al_{cons} \approx 0,65\left(\frac{s}{r_H}\right)^2 T[\%] \tag{8.9}$$

where:

s distance between sound source and listener [m]

r_H Half-room diffuse-field distance [m] (see Equation (8.10))

$$r_H = 0,057\sqrt{\frac{V}{T}}[\mathrm{m}] \tag{8.10}$$

The clarity measure C_{80} (see Equation (8.11)) is relevant for temporal and register clarity of music (particularly musical passages of high tempo) (Abdel Alim 1973).

$$C_{80} = 10\log\left(\frac{E_{0...80ms}}{E_{80ms...\infty}}\right)[\mathrm{dB}] \tag{8.11}$$

where:

$E_{0...80ms}$ Sound energy reaching the listener in the interval from 0 ms (direct sound) to 80 ms

$E_{80ms...\infty}$ Rest sound energy (reaching the listener after 80 ms).

Direct Sound Measure C_7 (see Equation (8.12)) represents an objective correlate to the subjective perception of "nearness" and "directness" of the sound source (Ahnert and Schmidt 2007).

$$C_7 = 10\log\left(\frac{E_7}{E_\infty - E_7}\right)[\mathrm{dB}] \tag{8.12}$$

where:

E_7 Sound energy reaching the listener in the interval from 0 ms up to 7 ms (direct sound)

$E_\infty - E_7$ Sound energy reaching the listener after the direct sound (reverberations).

Speech transmission index (*STI*) (Houtgast and Steeneken 1985; Steeneken and Houtgast 1980) is computed based on the reduction of the signal modulation from sound source location to receiver with octave center frequencies of 125 Hz to 8,000 Hz (see Equations (8.13) to (8.15)). To ascertain this influence, the modulation transfer function (*MTF*) is employed. Thereby, the available useful signal *S* is related to the prevailing interfering noise *N*. The modulation reduction factor *m(F)* characterizes the interference with speech intelligibility: Modulation frequencies from 0.63 Hz to 12.5 Hz in third octaves are used for the calculation. In addition, the modulation transfer function is subjected to a frequency weighting (*WMTF* – weighted modulation transfer function). Thereby, the modulation transfer function is divided into seven frequency bands that are each modulated with the modulation frequency. This results in a matrix of 98 (7 times 14) modulation reduction factors m_i. Table 8.2 provides a key to the interpretation of *STI* values in view of the implied speech intelligibility.

$$STI = \frac{\bar{X}_i + 15dB}{30dB}[-] \tag{8.13}$$

$$X_i = 10\log\left(\frac{m_i}{1-m_i}\right)[\mathrm{dB}] \tag{8.14}$$

$$m(F) = \frac{1}{\sqrt{1+(2\pi F\cdot T/13,8)^2}}\cdot\frac{1}{1+10^{-\frac{S/N}{10dB}}} \tag{8.15}$$

F modulation frequency [Hz]

T Reverberation Time [s]

S/N Signal/Noise ratio [dB].

The center time (t_s) (Kürer 1971) is a quality criterion relevant for both speech intelligibility and musical clarity (see Equation (8.16)).

$$t_S = \frac{\int_0^\infty t \cdot p^2(t)dt}{\int_0^\infty p^2(t)dt} [\text{ms}] \tag{8.16}$$

where:
p sound pressure [Pa]
t time [ms].

Echo criterion EK_{Speech} (Dietsch and Kraak 1986) allows for the examination of the occurrence of high-energy late reflections (echoes). One requirement of good room acoustics implies that at all seats the reflection sequence is uniform and tight and that no high-energy late reflections occur. Strong reflections arriving later than 50 ms after the direct sound (not preceded by any or only few weaker reflections), are subjectively perceived as echoes.

Strength Measure (G) (Jamagushi 1972; Lehmann 1976) is a normalized measure of overall sound energy at a listener's position as related to the energy resulting from a power-equivalent omni-directional sound source in a distance of 10 m from the source (see Equation (8.17)).

$$G = 10 \log \frac{E_{tot}}{E_{tot10}} [\text{dB}] \tag{8.17}$$

where:
E_{tot} Total sound energy at the listener's location
E_{tot10} Total sound energy of a reference sound source (in free field) at 10 m distance from the source.

Reverberance measure R (see Equation (8.18)) is an indicator of the acoustic liveliness or "reverberance" of the acoustic impression in case of musical performances (Beranek and Schultz 1965).

$$R = 10 \log \frac{E_\infty - E_{0...50ms}}{E_{0...50ms}} [\text{dB}] \tag{8.18}$$

Lateral energy fraction LF (see Equation (8.19)) (Barron 1993) denotes the ratio between the laterally arriving sound energy components and those arriving from all sides (within an interval of 80 ms). It serves to gauge the subjective appraisal of the extension of a (musical) sound source. The higher the lateral efficiency, the acoustically broader the sound source appears.

$$LF = \frac{E_{5...80ms(lateral)}}{E_{0...80ms}} [\%] \tag{8.19}$$

Table 8.2 Perceived speech intelligibility for various ranges of STI-values

Syllable intelligibility	*STI*
poor	0 to 0,3
satisfactory	0,3 to 0,45
good	0,45 to 0,6
very good	0,6 to 0,75
excellent	0,75 to 1,0

Advanced room acoustic simulation applications allow for the computation of the values of most of the above-listed indicators. However, the judicious selection of such indicators for any given design task, as well as the interpretation of the results require a certain level of experience. Measurements of such parameters in existing spaces, combined with regular auditory experiences of the sound field in actual rooms can effectively enrich the proficiency of the users of room acoustic simulation applications toward proper construction of computational room models and correct interpretation of associated simulation results.

Validation and reliability

Round robin tests

A common way to explore the validity of a simulation tool is to conduct a round robin, whereby multiple developers simulate the same (real) object independently. Subsequently, all results are compared with the measurement results from the real object. Regarding room acoustical simulation programs, one round robin test (Vorländer 1995) involved an 1,800 m^3 auditorium for which eight acoustical parameters (ISO 1997) were computed by 16 participants. Simulations were compared with measurement results (average of multiple measurements as performed by seven participants). Three programs were found to produce accurate results: the deviations of these tools' predictions from the measurements were not larger than the difference between the multiple measurements.

In the course of a subsequent round robin test (Bork 2005), a number of room acoustical parameters (T_{30}, EDT, D_{50}, C_{80}, TS, G) were simulated for a 78 m^2 recording studio (V= 400 m^3). The room acoustical parameters were also measured in this room using two source positions and three different receiver positions. Simulations were performed using six different commercially available room acoustical simulation programs. Thereby, two different room models were considered. The first involved a simplified representation of the geometry of diffusers and absorbers as plane surfaces, whose absorption and scattering data were derived based on measurements. The second representation involved the detailed modeling of diffuser and wooden absorber geometries. The difference between the simulation results for these two representations was not found to be significant. The deviation of the simulated parameter values from the respective measurements were derived for all programs in terms of mean relative errors at 1,000 Hz (averaged over all six positions) with the dimension of "limen" (i.e., the smallest audible sensation) for each parameter. These are 0.05 s for T_{30} and EDT, 5% for D_{50}, 1 dB for C_{80} and G, and 10 ms for TS. The relative errors, as averaged over all frequencies, revealed similar magnitudes. The relative errors in prediction of T_{30} and EDT ranged from 2 to 5 %. Simulation errors regarding the other parameters, as calculated with four of the six programs, were found to be below the limit of auditory detectability. Certain errors (for example deviation of simulated reverberation times at 1,000 Hz) were suggested to be the consequence of input assumptions concerning absorption coefficients. Laboratory measurements of absorption coefficients are typically performed under conditions that favor diffuse reflection. The materials' reflective behavior in common architectural situations may turn out to be partially directional, leading to underestimation of the reverberation time in simulation. Moreover, locally varying parameter ranges at lower frequencies cannot be detected with these applications, as they do not consider the sound wave phases. Nonetheless, the results of this round robin test suggest that a number of commercially available simulation applications can, under controlled test conditions, and given reliable material information, provide results that show a good agreement with measurements.

Realistic simulation deployment scenarios

Application of simulation tools in the context of actual design processes differs from the above described controlled test scenario, as detailed design specifications (final geometry configuration, definitive data on material properties) may not be available. A comparable situation is faced, when acoustical simulations are performed for existing spaces, for which detailed documentations of acoustic properties (specifically of room surfaces) are not available. An examination of the reliability of room acoustic simulation in such circumstances may thus provide clues as to the reliability of simulation-based predictions of acoustical performance indicator values in the architectural design phase. To illustrate this point, data from recent acoustical measurements and simulations in 14 spaces (Mahdavi *et al.* 2008a, 2008b) are considered in the following discussion. These spaces include a lecture room and a ceremonial hall in a university, a small recording studio, five atria in different building complexes (museums, office buildings, etc.), as well as six bath buildings of different scales.

For these objects, frequency-dependent reverberation times and sound pressure levels (for various source and receiver positions) were measured and simulated. Simulations were performed using a commercially available room acoustical simulation and auralization tool (Christensen 2005). To compute the reverberation times and sound pressure levels, virtual loudspeakers were placed in the model in the same location as the real loudspeakers in the course of measurements. Likewise, the location of virtual microphone positions in the model matched those of the real microphones during the measurements.

The input data assumptions concerning the absorption coefficient data for surface finishes were based mostly on general literature and the simulation tool's material database. Material properties for simulation input were defined independent of (and unaffected by) the measurement results.

Figure 8.2 shows the mean relative error and respective standard deviations (in percent) of simulated frequency-dependent reverberation times. Figure 8.3 shows the correlation between measured and simulated reverberation times (all objects, all frequencies). Figure 8.4 shows the mean difference (and respective standard deviations in decibel) between measured and simulated frequency-dependent sound pressure levels for all objects. Figure 8.5 shows the correlation between measured and simulated sound pressure levels (all objects, all frequencies).

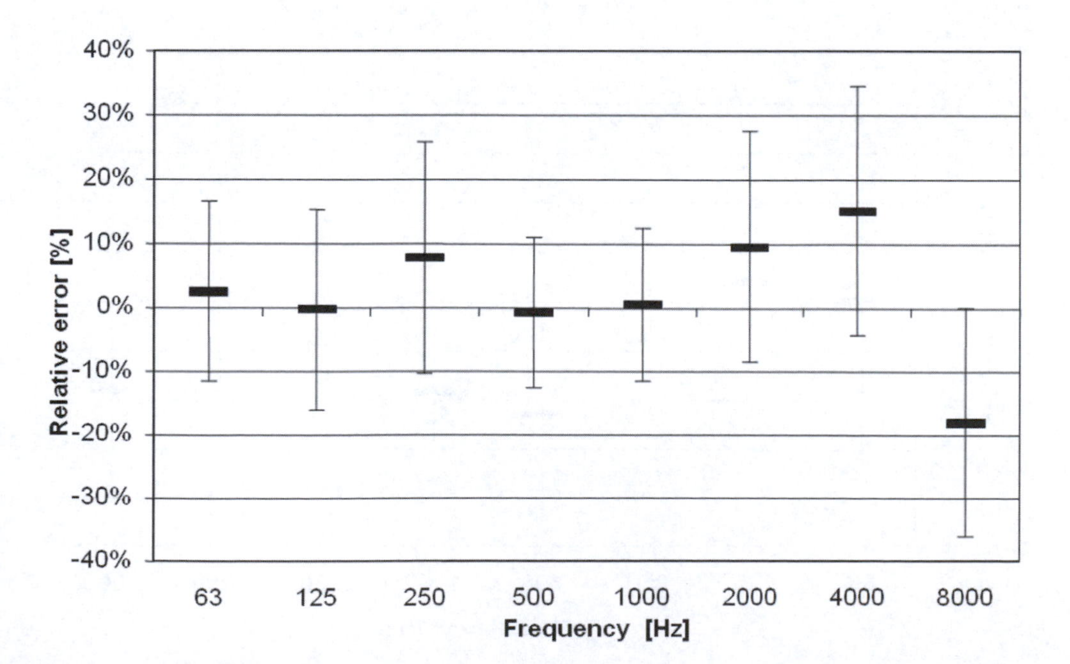

Figure 8.2 Mean relative error and respective standard deviations (in percent) of simulated frequency-dependent reverberation times.

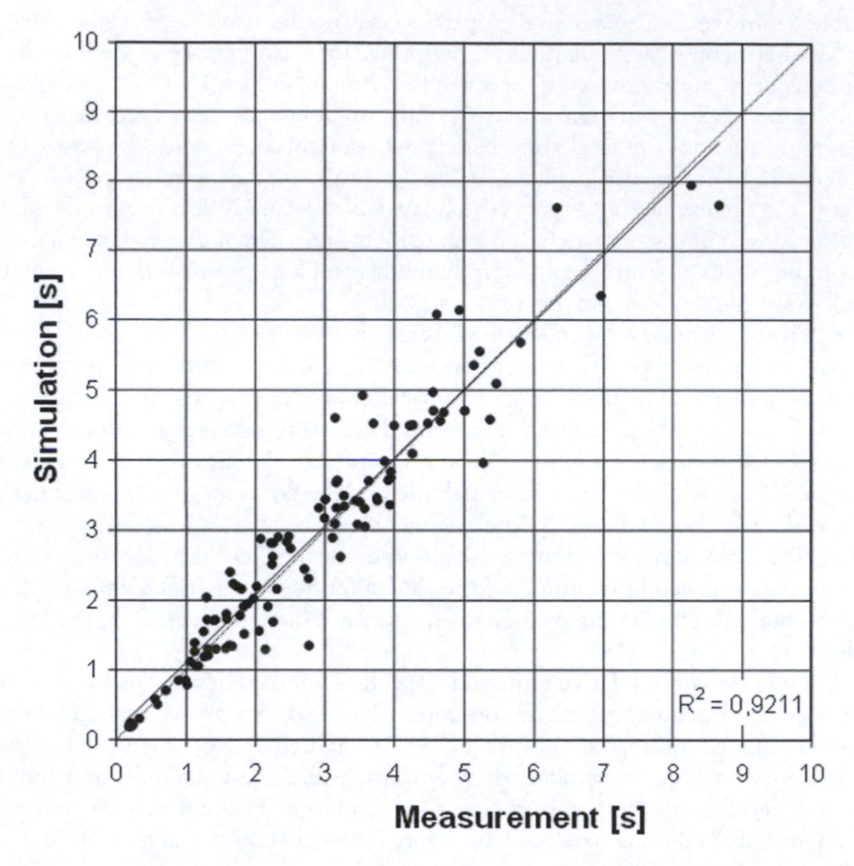

Figure 8.3 Measured vs. simulated reverberation times (all objects, all frequencies).

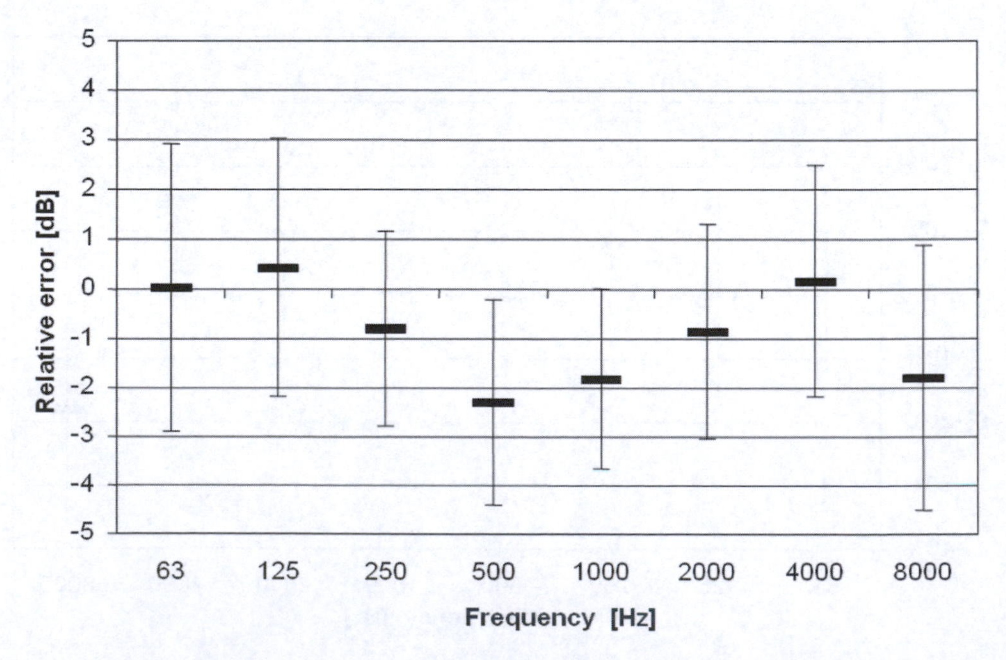

Figure 8.4 Mean differences (together with respective standard deviations) between measured and frequency-dependent sound pressure levels (all objects).

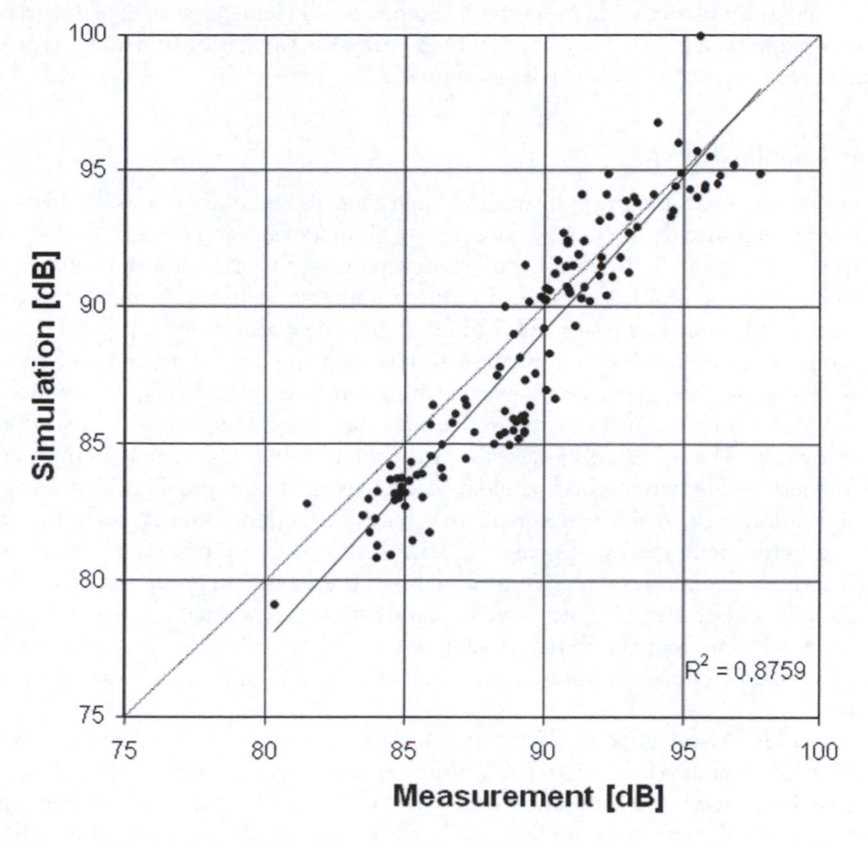

Figure 8.5 Measured versus simulated sound pressure levels (all objects, all frequencies).

These results imply, for the objects studied, a generally good agreement between measurements and simulations of reverberation times and sound pressure levels. The frequency-dependent mean relative error of reverberation time simulations over all objects did not exceed 20%. Individual relative errors (for different objects and frequencies) of reverberation time simulations were only in about 10% of the cases more than 20%. The mean relative error of the mean reverberation times (over 125, 500, and 2,000 Hz), as calculated for all objects, is 3% (with a standard deviation of 15%). The comparison of measured and simulated sound distribution levels displays likewise a reasonable agreement. The mean difference between the measured and simulated sound pressure levels (over 125, 500, and 2,000 Hz) aggregated over all objects is –0.9 dB (with a standard deviation of ±2.3 dB).

It is likely that the simulation errors are primarily the result of inaccuracies in modeling the material properties. In this study, similar to most application instances of simulation in the early stages of the design process, absorption coefficients were primarily obtained from general information in literature. Acoustical properties of building elements are indeed seldom specified in the pertinent technical design documents. Nonetheless, despite difficulties in accurately determining the absorption coefficients of surfaces, it was possible to reproduce measured reverberation times and sound distribution patterns with reasonable accuracy. Given adequately detailed models of room geometry and availability of relevant information on acoustical surface properties, simulation can effectively support the evaluation of design options and alternatives in view of room acoustical performance criteria. Moreover, it is important that performance requirements are specified early in design addressing the expected space use patterns. Architectural documentation of buildings must include detailed information on the acoustical properties of materials and components applied. Finally, more comprehensive empirically obtained databases of acoustical properties of architectural elements should be incorporated in acoustical simulation applications in order to expedite the simulation and analysis process and thus make it more effective toward design support.

Real and auralized rooms

The conventional scenario in application of room acoustics simulation runs, highly simplified, as follows: A proposed design (e.g., a concert hall) is modeled using a scientific room acoustic simulation application. This leads to the prediction of acoustical conditions in the modeled space. The predictions are expressed in terms of numeric values of a number of acoustical performance indicators (see pages 222–226 for an overview of such indicators). Designers and consultants typically compare such numeric results with applicable requirements (recommendations, standards, etc.) to decide if a particular design meets mandated performance criteria. If not, modifications are made in the design and the performance is iteratively tested until satisfactory results have been obtained. Recently, the possibility has emerged that such traditional numeric evaluation methods could be complemented by approaches that rely on scientific auralization tools, enabling the users to "directly" experience and evaluate the acoustical quality of architectural spaces (see page 222). In this context, an important research question arises: To which extent can subjective assessments of the acoustical properties of a real space be reproduced by assessments that are based on auralization (i.e., virtual acoustical replacements of real spaces)? The reproducibility of subjective acoustic evaluations of real spaces based on their virtual (auralized) counterparts is an open research question and subject to ongoing inquiries.

In one study (Mahdavi *et al.* 2005), previous research methods and results (Fasold and Winkler 1976; Mahdavi and Eissa 2002) were used to select a metric that captures certain subjective dimensions of the acoustical experience of spaces (see Table 8.3). Three rooms with differing acoustical attributes were selected for the study, including a ceremonial hall (aula) in a university, a mid-size lecture room, and a small recording studio. Reverberation time, sound level distribution, and ambient sound level were measured in all three spaces. A digital recording of a sample of acoustical events was prepared including speech, music, and various other sounds. The recording also included a reading of a set of logatoms (Fasold and Winkler 1976) for speech intelligibility test purposes. The acoustical properties of these rooms were evaluated by a group of 30 test participants using the above mentioned metric, as the pre-recorded sequence of acoustical events was being played in the rooms. Additionally, a speech intelligibility test was performed using the logatoms. Moreover, the participants were asked to specify the location of the sound source by specifying the prevailing direction from which they believed the sound was coming.

Auralizations of the spaces were generated using a room acoustics simulation tool (Christensen 2005) and the aforementioned pre-recorded sequence of acoustical events. The underlying simulation models were tested and found to correctly reproduce measured sound distribution pattern and reverberation times. The auralizations of the three spaces were presented (via head phones attached to computers) to and evaluated by a second demographically similar group of 75 test participants, who again made use of aforementioned acoustical quality metric and also located the sound source. As with the real spaces, a speech intelligibility test was performed based on a logatoms list as included in the sound sample recording.

The evaluation results of room acoustics in the real spaces were compared with those based on the virtual spaces. Figure 8.6 shows, for all three spaces, the correlation between the evaluation means of the real spaces versus the evaluation means of the auralizations for the eight categories considered. The results show a general agreement between acoustical evaluations based on real and auralized rooms, particularly for the ceremonial hall and the recording studio. However, the degree of correlation is different from space to space. A stronger congruence between the two evaluation methods was found in acoustically more "pronounced" cases. The ceremonial hall is rather large and reverberant, whereas the recording studio is very small and acoustically "dry". The lecture room, however, is moderate both in size and acoustical properties. This plausibly suggests that the computational reproduction of acoustical properties of real spaces is in view of evaluative utility (as expressed in terms of a quality metric) more effective in rooms with pronounced acoustical attributes. A closer look at the individual scales of the quality metric reveals different levels of "expressiveness": whereas the numeric values of some scales change noticeably from space to space, others change only marginally (see Figure 8.6). In this sense, the scales for reverberation, clarity, speech intelligibility, and general impression may be said to be more expressive. Sharpness, warmth, and brightness scales, however, are less expressive.

Auralization appears to provide an effective means to predict the speech intelligibility in spaces. The results of the auralization-based test were strikingly close to those from the real spaces. However, auralization did not provide a reliable reproduction of the perceived direction of the incoming sound. This latter failure is probably due to the difficulties inherent in the evocation of spatial acoustical orientation via headphones.

Table 8.3 Evaluation axes and corresponding bi-polar scales with associated points (from –2 to +2) for the evaluation of the acoustical quality of space

Evaluation axis	*Bi-polar terms*
Reverberation	Reverberant (+2) – dry (–2)
Loudness	Loud (+2) – quiet (–2)
Clarity	Clear (+2) – blurry (–2)
Sharpness	Shrill (+2) – dull (–2)
Warmth	Warm (+2) – cool (–2)
Brightness	Bright (+2) – dark (–2)
General impression	Pleasant (+2) – unpleasant (–2)
Speech intelligibility	Good (+2) – poor (–2)

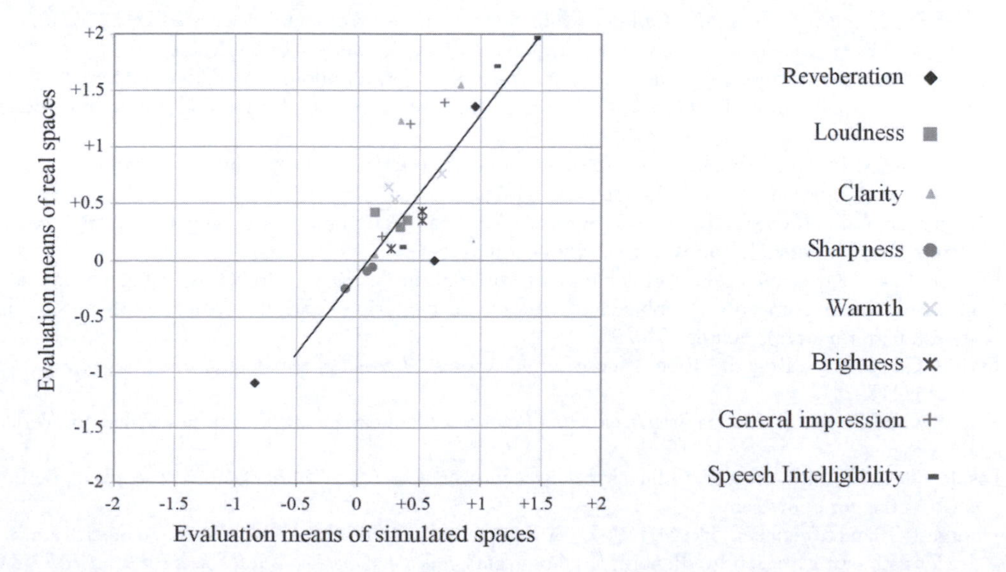

Figure 8.6 Evaluation means based on real and auralized spaces (regression analysis across all spaces).

Conclusion

The design of architectural spaces with good acoustical attributes has become, thanks in part to advances in computational methods and techniques, an accountable process. Using advanced simulation applications, it is now possible to predict sound behavior in rooms with a high degree of accuracy. Thus, numeric values of a comprehensive set of performance indicators can be computed and compared to desired (recommended) ranges. Moreover, sound propagation phenomena can be visualized not only in terms of sound level distribution maps (for various plans and sections), but also in terms of rays that, emanating from the sound source, travel throughout the room and bounce off from the surfaces. Using such representations and associated reflectograms, critical reflections and arrival times can be appraised and necessary adjustments in design can be initiated. At a minimum, the simulation-supported design process can ensure that essential performance requirements are fulfilled and design failures are avoided. Furthermore, the advent of powerful auralization tools opens new and exciting perspectives: proposed designs can be henceforth not only evaluated based on predicted values of numerically expressible performance indicators, but be made the direct objects of acoustical experience. This development can support the refinement of designs in view of numerically intangible yet subjectively evident (experiential) space qualities.

Needless to say, simulation applications must be further improved in the future. More elaborate algorithmic solutions will have to be implemented to better capture the wave nature of sound phenomena. Likewise, more detailed descriptions of the interactions between sound waves and room enclosure are needed, addressing, amongst others, the coincidence and resonance frequencies of surfaces. Moreover, comprehensive and reliable databases of acoustical material properties must be implemented as the integral feature of any effective simulation tool. Ultimately, the utility of simulation-based design support process is quintessentially dependent on the resolution, consistency, and fidelity of the input information, i.e., the design model itself.

References

Abdel Alim, O. (1973) "Abhängigkeit der Zeit- und Registerdurchsichtigkeit von raumakustischen Parametern bei Musikdarbietungen" (Dependence of time and register definition of room acoustical parameters with music performances), PhD Thesis, TU Dresden, Germany.

Ahnert, W. and Schmidt, W. (2007) "Fundamentals to Perform Acoustical Measurements", *Appendix to EASERA (version 1.1) Tutorial*. Online. Available at: http://www.easera.de (accessed 15.3.2009).

Barron, M. (1993) *Auditorium Acoustics and Architectural Design*, London: Spon Press.

Beranek, L.L. (1962) *Music, Acoustics and Architecture*, New York/London: John Wiley and Sons.

Beranek, L.L. and Schultz, T.J. (1965) "Some Recent Experiences in the Design and testing of Concert Halls with Suspended Panel Arrays", *Acustica* 15: 307.

Bork, I. (2005) *Report on the 3rd Round Robin on Room Acoustical Computer Simulation – Part II: Calculations, Acta Acustica* united with *Acustica* 91: 753–763.

Christensen, C.L. (2005) *ODEON Room Acoustics Program*. Version 8.0. User manual. Industrial, Auditorium and Combined Editions (http://www.odeon.dk).

Dietsch, L. and Kraak, W. (1986) "Ein objektives Kriterium zur Erfassung von Echostörungen bei Musik- und Sprachdarbietungen" (An objective criterion for capturing echo disturbances with music and speech performances), *Acustica* 60: 205.

Eyring, C.F. (1930) "Reverberation Time in Dead Rooms", *Journal of the Acoustical Society of America* 1(2A): 217–241.

Fasold W. and Veres E. (2003) *Schallschutz und Raumakustik in der Praxis*, Berlin: Huss-Medien, Verlag Bauwesen.

Fasold, W. and Winkler, H. (1976) *Bauphysikalische Entwurfslehre: Band 5 Raumakustik*, 1. Auflage, Berlin: VEB Verlag für Bauwesen.

Houtgast, T. and Steeneken, H.J.M. (1985) "A Review of the MTF Concept in Room Acoustics and Its Ise for Estimating Speech Intelligibility in Auditoria", *Journal of the Acoustical Society of America* 77(3): 1060–1077.

ISO (1997) *ISO 3382 "Measurement of the reverberation time of rooms with reference to other acoustical parameters"*.

Jamagushi, K. (1972) "Multivariate Analysis of Subjective and Physical Measures of Hall Acoustics", *Journal of the Acoustical Society of America* 52(5A): 1271–1279.

Januska, I. (1968) "Experimentally Stated Correlation between Objective Echogramm Evaluation and Speech Intelligibility", *Archivum Akustiki* 140.

Jordan, V.L. (1980) *Acoustical Design of Concert Halls and Theatres*, London: Applied Science Publishers Ltd.

Kleiner, M., Dalenbäck, B.I. and Svensson, P. (1993) "Auralization – An Overview", *Journal of the Audio Engineering Society* 41(11): 861–875.

Krokstad, A., Stroem, S. and Soersdal, S. (1968) "Calculating the Acoustical Room Response by the Use of a Ray Tracing Technique", *Journal of Sound and Vibration* 8(1): 118–125.

Kulowski, A. (1985) "Algorithmic Representation of the Ray Tracing Technique", *Applied Acoustics* 18: 449–469.

Kürer, R. (1971) "Einfaches Messverfahren zur Bestimmung der 'Schwerpunktzeit' raumakustischer Impulsantworten" (A simple measuring procedure for determining the "center time" of room acoustical impulse responses), in *Proceedings of the 7th International Congress on Acoustics*, Budapest, Hungary.

Kuttruff, H. (2000) *Room Acoustics*, London: Spon Press, Taylor & Francis Group.

Lehmann, P. (1976) "Über die Ermittlung raumakustischer Kriterien und deren Zusammenhang mit subjektiven Beurteilungen der Hörsamkeit", PhD Thesis,Technische Universität Berlin, Germany.

Lewers, T. (1993) "A Combined Beam Tracing and Radiant Exchange Computer Model of Room Acoustics", *Applied Acoustics* 38(2–4): 161–178.

Lochner, J.P.A. and Burger, J.F. (1961) "The Intelligibility of Speech under Reverberant Conditions" *Acustica* 11: 195.

Mahdavi, A. and Eissa, H. (2002) "Subjective Evaluation of Architectural Lighting via Computationally Rendered Images", *Journal of Illuminating Engineering Society* 32(2): 11–20.

Mahdavi, A., Liu, G. and Ilal, M.E. (1997) "CASCADE: A Novel Computational Design System for Architectural Acoustics", in *Proceedings of IBPSA (International Building Performance Simulation Association) Conference*, Prague, pp. 173–180.

Mahdavi, A., Becker, G. and Lechleitner, J. (2005) "An Empirical Examination of the Evaluative Utility of Design Auralization", in Beausoleil-Morrison, I. and Bernier, M. (eds) *Building Simulation 2005, Ninth International IBPSA Conference*, Montreal, Canada, pp. 677–684.

Mahdavi, A., Lechleitner, J. and Pak, J. (2008a) "Measurements and Predictions of Room Acoustics in Atria", *Journal of Building Performance Simulation* 1(2): 67–74.

Mahdavi, A., Kainrath, B., Orehounig, K. and Lechleitner, J. (2008b) "Measurement and Simulation of Room Acoustics Parameters in Traditional and Modern Bath Buildings", *Building Simulation* 1(3): 223–233.

Naylor, G.M. (1993) "ODEON – Another Hybrid Room Acoustical Model", *Applied Acoustics* 38: 131–143.

Niese, H. (1965) "Vorschlag für die Definition und Messung der Deutlichkeit nach subjektiven Grundlagen", *Hochfrequenztechnik und Elektroakustik* 65(4).

Peutz, V.M.A. (1971) "Articulation Loss of Consonants as a Criterion for Speech Transmission in a Room", *Journal of the Audio Engineering Society* 19(11): 915–919.

Rindel, J.H. (2000) "The Use of Computer Modeling in Room Acoustics", *Journal of Vibroengineering* 3(4): 41–72, in *Proceedings of the International Conference BALTIC-ACOUSTIC 2000*, pp. 219–224.

Sabine, W.C. (1927) *Collected Papers on Acoustics*, Cambridge, MA: Harvard University Press.

Steeneken, H.J.M. and Houtgast, T. (1980) "A Physical Method for Measuring Speech-transmission Quality", *The Journal of the Acoustical Society of America* 67(1): 318–326.

Stephenson, U. (1985) "Eine Schallteilchen-Computer-Simulation zur Berechnung der für die Hörsamkeit in Konzertsälen massgebenden Parameter", *Acustica* 59(1): 1–20.

Thiele, R. (1953) "Richtungsverteilung und Zeitfolge der Schallrückwürfe in Räumen", *Acustica* 3: 291.

Vian, J.P. and van Maercke, D. (1986) "Calculation of the Room Impulse Response Using a Ray-Tracing Method", in *Proceedings of ICA Symposium on Acoustics and Theatre Planning for the Performing Arts*, Vancouver, Canada, pp. 74–76.

Vorländer, M. (1989) "Simulation of the Transient and Steady-state Sound Propagation in Rooms Using a New Combined Ray-tracing/Image-source Algorithm", *The Journal of the Acoustical Society of America* 86(1): 172–178.

Vorländer, M. (1995) "International Round Robin on Room Acoustical Computer Simulations", in *Proceedings of the 15th International Congress on Acoustics*. Trondheim, Norway, pp. 689–692.

Recommended reading

Given the paucity of comprehensive textbooks dedicated solely to computational room acoustics, the reader should consult the technical references above. As to the introductory texts to room acoustics in general, many sources exist. Examples are: Barron 1993, Beranek 1962, Fasold and Veres 2003, Jordan 1980, and Kuttruff 2000.

Activities

A simulation expert in the area of the room acoustics must not only command the scientific foundations and computational experience, but also familiarity with the empirical measurement methods in architectural acoustics. A recurrent cross-checking between simulation and measurement results is essential for the development of expertise and a sense of professional judgment in this area.

Answers

There exist, for a number of performance indicators in the area of room acoustics (such as reverberation time), simple calculation formulas. The comparison of simulation results with those obtained via such simple formulas provides a number of useful learning opportunities.

9 Daylight performance predictions

Christoph Reinhart

Scope

This chapter provides an overview of the different steps required to use daylight simulations for building design. Following an brief introduction to the theory of daylight simulations including common simulation algorithms, sky models and material descriptions, the debate turns towards the question of what to calculate, i.e. what daylighting performance metrics to use, how to address building occupant behavior and expectations and how to interpret simulation results in order to help design teams to make more informed design decisions. This chapter does not provide a general introduction into the larger theory of daylighting, e.g. how a lightshelf functions or how to commission a lighting control system.

Learning objectives

This chapter is concerned with the assessment of the luminous environment in daylit spaces via simulation. The reader will learn about the basic algorithms that are commonly used to model daylight and be exposed to current and emerging metrics to evaluate the performance of daylit spaces.

Key words

Daylight simulations/daylighting metrics/energy savings from lighting controls/glare analysis

Introduction

Daylighting describes the use of natural light in buildings. The objectives for daylighting a building are manifold and range from questions of aesthetics, occupant health and comfort to energy savings for lighting and space conditioning. Daylight simulations can help design teams to address these multiple aspects by allowing them to (a) predict the amount of light available inside or outside of buildings under selected sky conditions or over the course of a whole year and (b) to interpret the results by converting them into meaningful performance metrics. Daylight simulations can be combined with simulations of electric light sources in order to make sure that both forms of space lighting complement each other. They can also be embedded within an integrated lighting-thermal simulation in order to assess the overall effect of a daylighting strategy on energy use for lighting, heating and cooling (Janak 1997; Winkelmann and Selkowitz 1985). Within recent years it has become increasingly evident that the success of a daylighting strategy in real buildings is closely related to how exactly the electric lighting *and* shading systems are being controlled, e.g. manually, automatically or automatically with manual override. As a consequence, a new building performance simulation branch is emerging that specifically focuses on developing behavior models that mimic occupant use of personal controls in buildings such as light switches, venetian blinds, thermostats and operable windows (Newsham 1994; Reinhart 2004;

Bourgeois, MacDonald and Reinhart 2006; Mahdavi *et al.* 2008; Hoes *et al.* 2009) (see also Chapter 4).

Before diving right into the theory of daylight simulations it is worthwhile to explore the larger context within which these simulations are typically being carried out. Daylighting is an established design component within any building that claims to be "sustainable" and/or "green" and that houses occupants. The reason for the strong association of daylighting with sustainable design practice is that its benefits are well aligned with the larger goals of the green building movement namely promoting resource efficiency of and improved occupant well-being in buildings. According to two recent surveys, daylight simulations are one amongst several "tools" that are routinely being used to design for daylight in buildings (Galasiu and Reinhart 2007; Reinhart and Fitz 2006). Other popular tools are "experience from previous work", "rules of thumb" and (to a lesser degree) "scale models". Figure 9.1 shows that in a recent survey of close to two hundred design practitioners with an interest in sustainable design both "designers" and "engineers" rated daylight simulations as their most important design tool during design development along with "experience from previous work" in the case of the designers. While this is an impressive statistic for daylight simulations, the reader should note that the target survey audience were practitioners with an explicit interest in sustainable design (Galasiu and Reinhart 2007). The market penetration of daylight simulations in the design community at large is probably significantly lower at this point. Nevertheless, the number of design practitioners in North America using the US Green Building Council's (USGBC) LEED rating system is continuously growing: Over 75,000 individuals have passed the "LEED Accredited Professional" exam between 2001 and 2009 (USGBC 2006). Given that LEED requires/promotes the use of building performance simulations for energy and daylighting, the use of simulations is destined to rise even further.

Apart from the above described "push" towards the use of daylight simulations through the green building movement, the ubiquitous proliferation of CAD software in building design practice and education is another factor that is helping to promote the use of daylight simulations due to the fortuitous fact that the kind of three dimensional building models

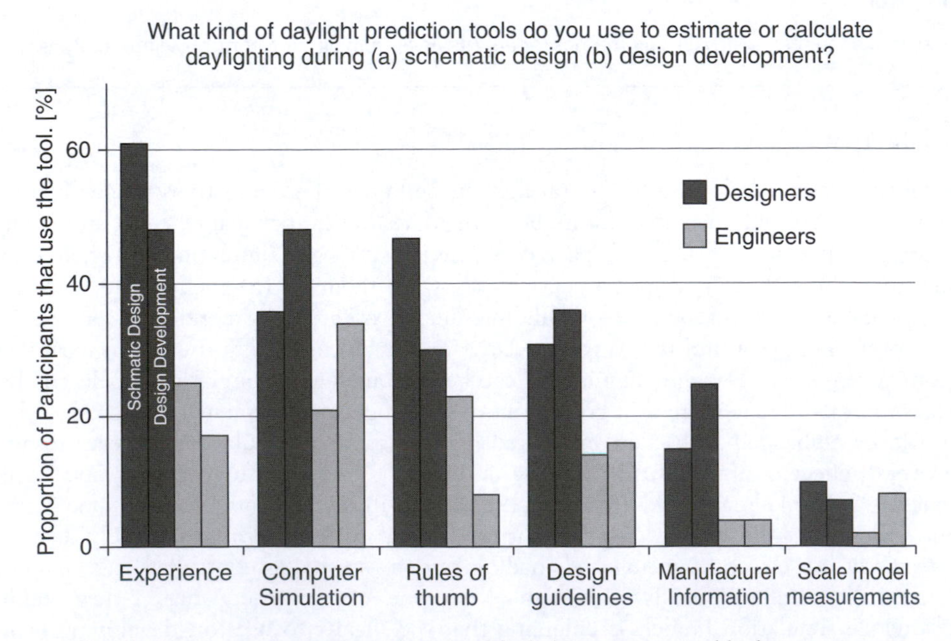

Figure 9.1 Results from a recent survey on the role of daylighting in sustainable design.

Source: Galasiu and Reinhart 2007.

that are commonly used for design presentation already contain a large amount of geometrical detail that can be reused for a computer-based daylighting analysis. These geometric models hence constitute a "free" resource that greatly reduces the additional effort and costs required to carry out a daylighting analysis. Having said this, the author would like to highlight that at the time of writing software interoperability issues were still often hindering or at least impeding the conversion of a geometric CAD model into a format suitable for further daylighting analysis. This caveat notwithstanding, the author is optimistic that the emerging practice of integrated building design, model sharing and building information modeling (BIM) will help the design industry to overcome these challenges within the coming years.

For the remainder of this chapter it is assumed that the reader has decided to use daylight simulations for building design. The chapter accordingly provides an overview of the different steps required to do so. Following a brief introduction to the theory of daylight simulations including common simulation algorithms, sky models and material descriptions, the debate turns towards the question of *what* to calculate, i.e. what daylighting performance metrics to use, how to address building occupant behavior and expectations and how to interpret simulation results in order to help design teams to make more informed design decisions. In the spirit of the overall book, this chapter is based on the premise that an *informed* design decision is a *better* decision.

This chapter does *not* provide a general introduction into the larger theory of daylighting, e.g. how a lightshelf functions or how to commission a lighting control system. It also avoids focusing on the limitations and capabilities of any one particular daylighting analysis tool. Many of the examples in this chapter are based on simulations using the Radiance backward raytracer (Ward and Shakespeare 1998), a rigorously validated and flexible open source tool that is used by practitioners and researchers world-wide.

Computational models

This section deals with the various elements needed to simulate daylighting levels within a scene. While the study scope may vary in scale from a façade detail to an urban neighborhood, the underlying simulation algorithms will often be identical. Figure 9.2 shows the basic elements required for a daylight performance simulation:

- The *scene* consists of a three-dimensional geometric model of the investigated daylit object(s) including optical material descriptions for all surfaces in the scene.
- The *sky model* quantifies the amount of direct sunlight and diffuse daylight coming from the different parts of the celestial hemisphere.
- *Areas of interest* in the scene can be selected viewpoints and/or discrete sensors such as a grid of upward facing illuminance sensors.
- *Space usage* information describe the type of space investigated (office, classroom, . . .), required lighting levels and occupancy schedules.
- The *daylight simulation engine* combines the sky model with the scene and calculates illuminances and/or luminances within the scene.
- The *results processor* translates the raw simulation results into a format that can directly inform design decisions. Example formats are scene visualizations and falsecolor maps of the daylight factor and/or other performance metrics including daylight autonomy and useful daylight illuminance. These metrics will be discussed in detail on pages 258–267.

While a substantial amount of time and resources in a daylighting analysis typically go towards building a scene and running the daylight simulations, the author would like to stress that the results processor is the most important, albeit frequently hastily performed step within a

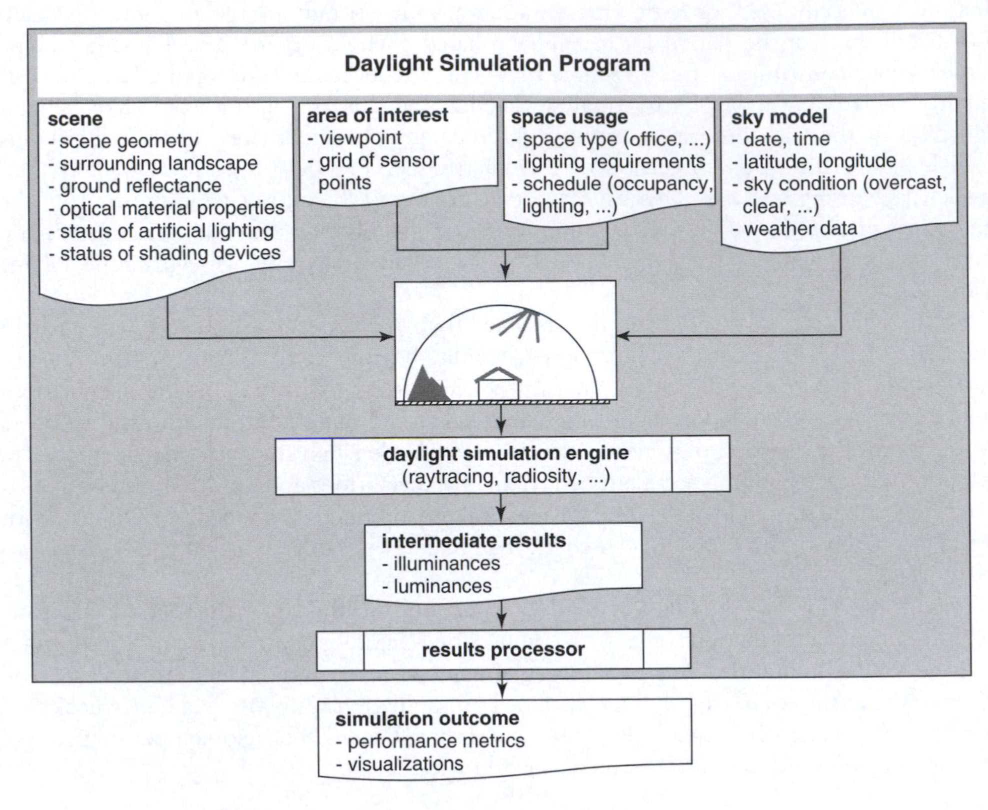

Figure 9.2 Elements needed for a daylight performance simulation.

daylight performance simulation. It is of course true that a user should pay attention to work with a validated simulation engine and carefully review the scene model and simulation parameters. But, getting the lighting simulation right is just a prerequisite for a successful daylighting analysis. What determines *whether* the analysis is going to influence the design outcome is *what* metrics are being calculated. Following Steven Covey's second habit, to "begin with the end in mind" (Covey 1989), is particularly good advice for daylight simulationists simply because the use of different daylighting metrics may lead to different, if not opposing design conclusions.

Building the scene

A three-dimensional model is an obvious requirement for any daylight simulation. But, for simulation newcomers it is often less clear how detailed a daylighting model has to be at the different design phases and how the optical surfaces properties should be set?

The required geometric model complexity strongly depends on what is being calculated. From the earliest design stages onwards it is advisable to model a site's immediate surroundings such as neighboring buildings and landscape that may affect the daylight availability on site. It is also good practice to explicitly model the ground plane and assign it a material that reflects the onsite conditions instead of relying on the simulation program to treat ground reflectances automatically (Reinhart and Walkenhorst 2001). As shown in Figure 9.3 it generally suffices to model neighboring buildings as simple blocks. While for direct shading studies the only required material distinction is whether a surface is transparent or opaque, it is generally good practice to make sure that all surfaces within a scene have been assigned "meaningful" optical properties for a daylighting analysis. At this point the reader should take note that even though a number of currently available daylighting analysis programs currently come with default material libraries, the materials in these libraries are not necessarily populated with meaningful values. The author therefore recommends that the reader carefully reviews the default materials used by his or her program of choice and – if necessary – defines a custom library with common materials. Table 9.1 provides a list of common material descriptions used by the author. These materials should suffice for early design explorations. As a design evolves, product specific databases such as the *International Glazing Database* from the US National Fenestration Rating Council provide optical surface properties of the most common glazing products in the US. The *Window 6.3 Research Version* further provides additional data on the optical properties of more complex glazing systems (CFS) such as window glazings combined with venetian blinds (LBNL last accessed in November 2009). A number of daylight simulation programs further allow the user to enter either measured or simulated bidirectional transmission and reflection distribution functions (BTDFs and BRDFs) for arbitrarily CFSs. BTDFs/BRDFs describe the amount or light coming from a certain direction (characterized by two incident angles) that is transmitted through or reflected off a CFS in a certain direction which is characterized by two outgoing angles. BTDFs/BRDFs are hence four dimensional functions. The optical device that is used to measure or simulate these functions is called a goniophotometer (Andersen and deBoer 2006).

It is worthwhile noting that a façade surface with a mostly diffuse reflectance corresponds to, e.g., a brick wall with a moderate window-to-wall ratio. A fully glazed neighboring building

Table 9.1 Optical properties of common material surfaces

Interior floor	20% diffuse reflectance[1]
Interior wall	50% diffuse reflectance[1]
Interior ceiling	80% diffuse reflectance[1]
Interior ceiling (high reflectance)	90% diffuse reflectance
Double glazing	72% direct visual transmittance
Single glazing	90% direct visual transmittance
Translucent glazing	20% diffuse-diffuse hemispherical transmittance[2]
Exterior building surfaces	44% diffuse reflectance[3]
Exterior ground	20% diffuse reflectance

Sources:
1 IESNA Handbook (IESNA 2000).
2 Reinhart and Andersen 2006.
3 Leder, Pereira and Moraes 2007.

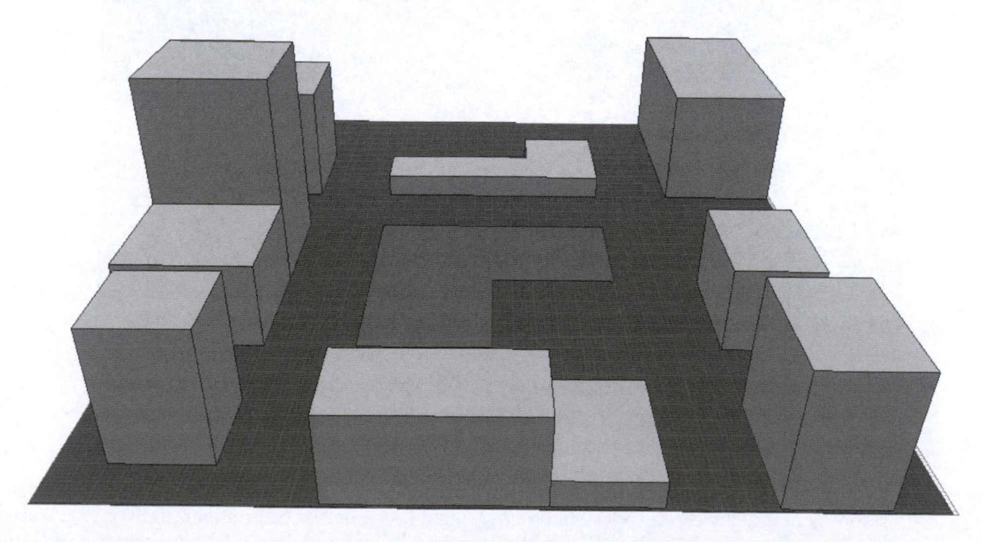

Figure 9.3 Basic geometric model of a new building site (L-shaped in the centre) with surrounding objects.

obviously has a considerable specular reflective component that may lead to glare from direct solar reflections. Depending on the type of daylighting analysis at hand it might therefore become necessary to explicitly model a specular neighboring façade that could have a decisive impact on the visual comfort conditions at the building site itself.[1] Another often overlooked fact is also that the surface property of the ground in immediate vicinity to a vertical window has a strong impact on the amount and quality of daylighting entering though the window especially for interior points that do not "see" the celestial hemisphere such as the back of a room and/or a ceiling mounted photocell-controlled dimming system.

Same as the design of a building evolves in various stages so should the daylighting model of a building. During the programming and massing stage simple block models suffice, e.g. for direct shading studies (shadow analysis onto the building envelope) and incident daily or seasonal irradiation patterns (Figure 9.4). At this stage exposure of different façade sections to

(a)

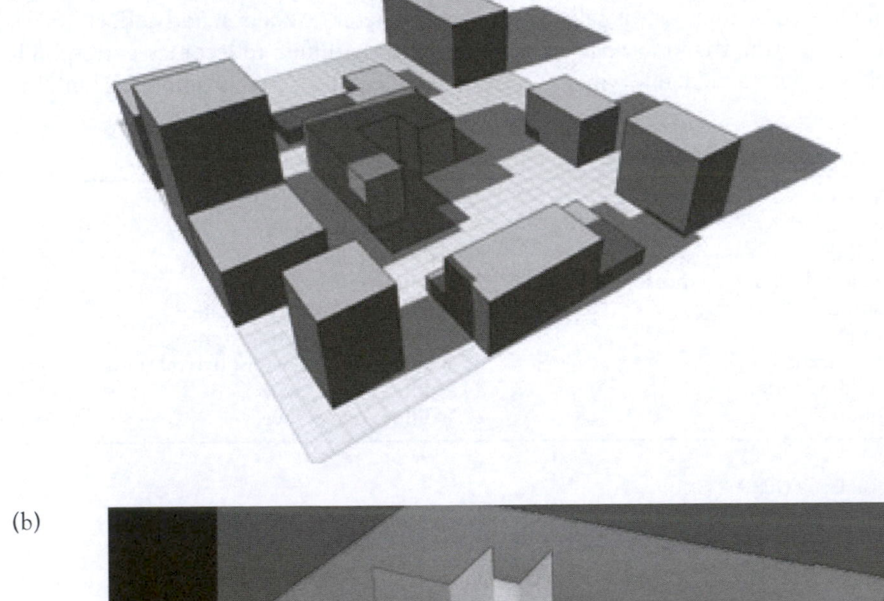

(b)

Figure 9.4 Two applications for a massing model of an urban site: (a) Direct shading study on Dec 21 at noon (generated with Autodesk Ecotect (Marsh, last accessed September 2009)); (b) Annual irradiation pattern (generated with Radiance and *GenCumulativeSky* (Robinson and Stone 2004)).

the celestial hemisphere can further be used to test the "daylight feasibility" for different parts of the massing model (Reinhart and LoVerso 2010).

As the design analysis moves "into" the building, façade openings, interior walls and large pieces of furniture, especially partitions, need to be added to the model to develop a feeling of how deep the daylight may penetrate into the building. At this point the thickness of exterior walls and roofs should be modeled along with specific material properties and dimensions of all elements around building envelope openings such as overhangs, light shelves, complex fenestration systems and/or other light redirecting devices. Getting wall thicknesses, positions of building envelope openings and interior dimensions right is crucial. Unfortunately, modeling window frames and mullions is time consuming and usually premature at the earlier design stages. In order to still yield meaningful simulation results of interior lighting conditions the simulationists may either assume that a fixed proportion of 20% of window openings are used up by frame and mullions, which thus effectively reduce the visual transmittance of the window by about 20%, or use a generic framing system from a CAD library with roughly the same opaque proportion as expected for the final design. Given this myriad of details required for a proper daylight simulation, it is worthwhile remembering the reader that – same as for other fields of building performance simulation – a relative comparison of different design options is typically easier to do and yields more reliable results than the use of absolute simulation results against benchmark levels and standards. That is, it is easier to estimate the relative effect of increasing a window by 20% than to report the absolute illuminance level that one might expect to encounter in a room under specific sky conditions.

As the façade design becomes increasingly detailed, modeling movable shading devices – such as venetian blinds – might become a necessity. As explained on pages 266–267, adding this level of detail only makes sense for some performance metrics. Similarly, modeling the luminance coming of a computer screen is usually not necessary for metrics based on upward facing illuminances but might become crucial for image-based glare analysis procedures.

As a side note, if a daylighting analysis is to be combined with a thermal simulation for a fully integrated energy performance assessment, the model detail and settings required for the thermal calculation generally diverge from the daylighting/lighting model: The daylighting model may benefit from a lot more geometric detail model than is typically provided or required for a thermal model. On the other hand, the division of a building into thermal zones whose boundaries may be defined as "air walls" is a concept foreign to (day)lighting calculations. These different modeling approaches are challenging to resolve from a model maintenance standpoint (BIM approaches might provide solutions one day) and potentially explain the author's informal observation that architectural students – trained in considering geometric detail – tend to initially experience difficulties building thermal building models (Wasilowski and Reinhart 2009).

Modeling the sky

Once the scene has been modeled, it has to be combined with one or several sky conditions. One of the challenges of evaluating the daylighting performance of a space, building or a neighborhood is that ambient sky conditions are in permanent flux and depend on prevailing weather conditions as well as the position of the sun in the sky. Sky models are mathematical constructs, used by computer simulation tools, to describe the *sky luminous distribution*, i.e. the amount of light coming from different parts of the celestial hemisphere. Sky models usually separate between *direct sunlight*, i.e. light that is incident from the geometric position of the sun with respect to the site, and *diffuse daylight*, i.e. sunlight that is scattered of clouds, aerosols, air molecules, and water vapor before it hits the Earth's surface (Figure 9.5). The ratio of direct sunlight to all daylight falling onto an unshaded horizontal surface over the course of a year depends on the local climate and may vary from over 70% for sunny climates such as Phoenix, Arizona, to around 40% for predominantly overcast climates such as London, England.

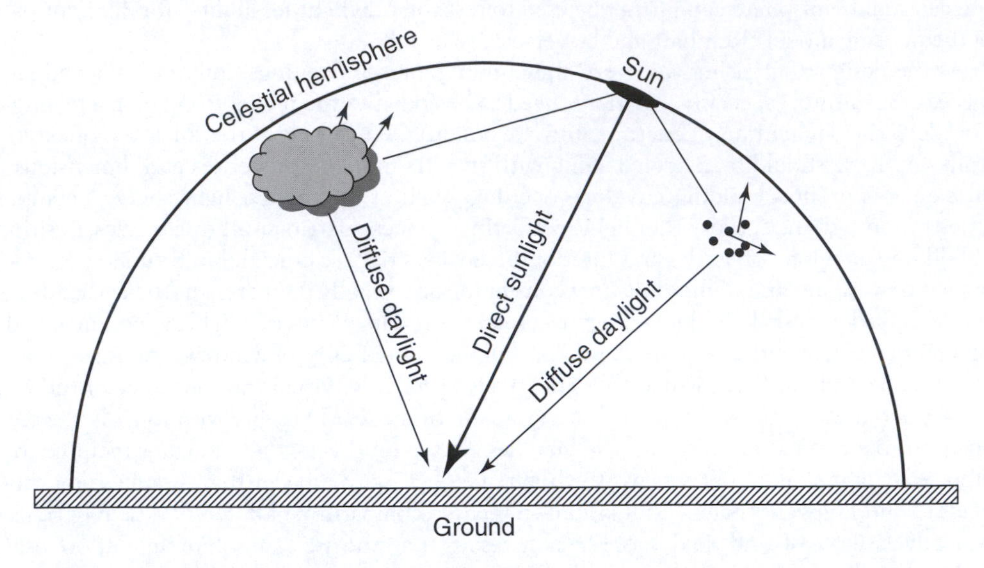

Figure 9.5 Daylight falling onto a site is usually divided into direct sunlight and diffuse daylight.

Different sky models require different types of inputs. The simplest sky models only consider the position of the sun with respect to a site and require time of day, date, latitude and longitude as input. The position of the sun in the sky is typically expressed by two angles, solar azimuth and solar altitude (Figure 9.6).

A detailed description of the trigonometric functions that are commonly used to calculate both angles for a particular site, date and time can be found in the Duffie and Beckmann book

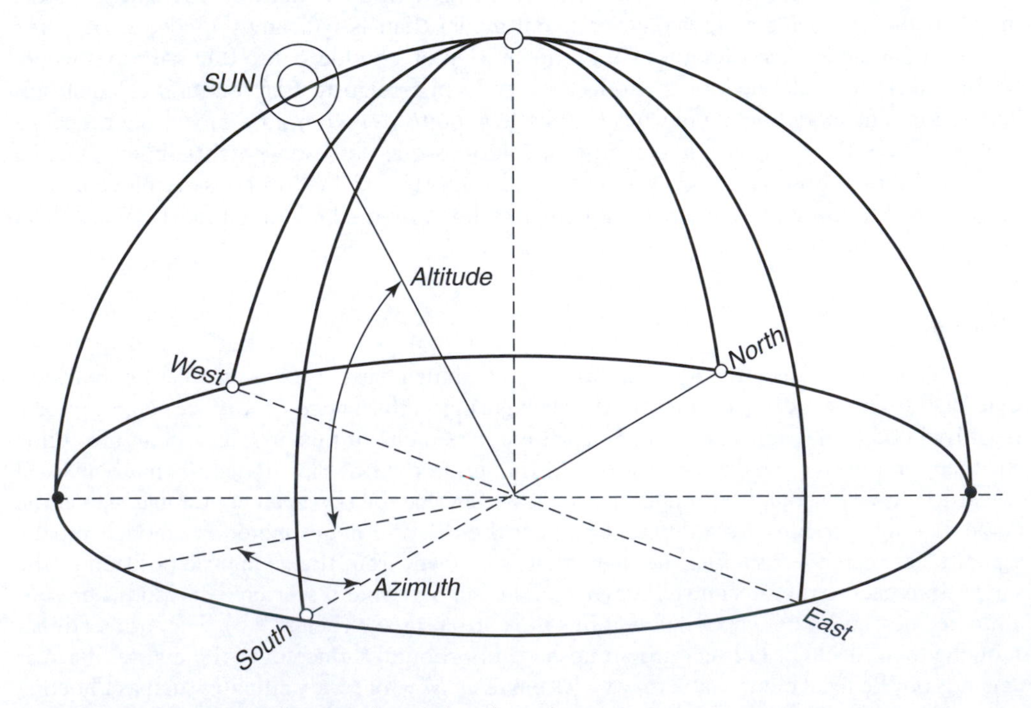

Figure 9.6 Definition of solar azimuth and altitude angles.

(Duffie and Beckman 1991). Direct sun models are used for shading studies (Figure 9.4(a)) or sun charts. Sun charts are two-dimensional representations of the position of the sun with respect to a site over the course of a year. There exist a number of different projection modes for sun charts. In *orthographic* sun charts the solar azimuth is plotted along the X axis with zero corresponding to due south and the solar altitude is plotted along the Y axis with ninety degrees corresponding to the zenith. Figure 9.7 shows an orthographic sun chart diagram for Boston, MA, USA (42°N 71°W), with a shading mask for the south-facing wall in Figure 9.4(a) marked with an "X". The figure reveals when direct sunlight is incident on the wall and can be used in combination with a local climate file to design for solar gain management and glare protection.

Historically, a series of increasingly refined standardized sky luminance distributions have been presented over the past one hundred years which consider diffuse daylight as well as direct sunlight. These distributions are mathematical functions of the *relative* luminous distribution of the celestial hemisphere with respect to the zenith luminance. Local radiation measurements combined with luminous efficacy models are required to translate a *relative* sky luminous distribution into *absolute* values.

The oldest, *uniform sky* model assumes an isotropic sky luminance distribution without any direct sunlight and initially served as the reference sky for the daylight factor metric which will be introduced below (Waldram 1909). The *CIE overcast sky* with a 3:1 gradation of the sky luminance from the zenith to the horizon later replaced the uniform sky as the reference sky for the daylight factor (Moon and Spencer 1942). In 1973, the Commission Internationale de l'Éclairage (CIE) published a standard sky luminous distribution for clear skies (CIE 1973), the *CIE clear sky*. The CIE overcast and clear skies served for several decades as two extreme sky

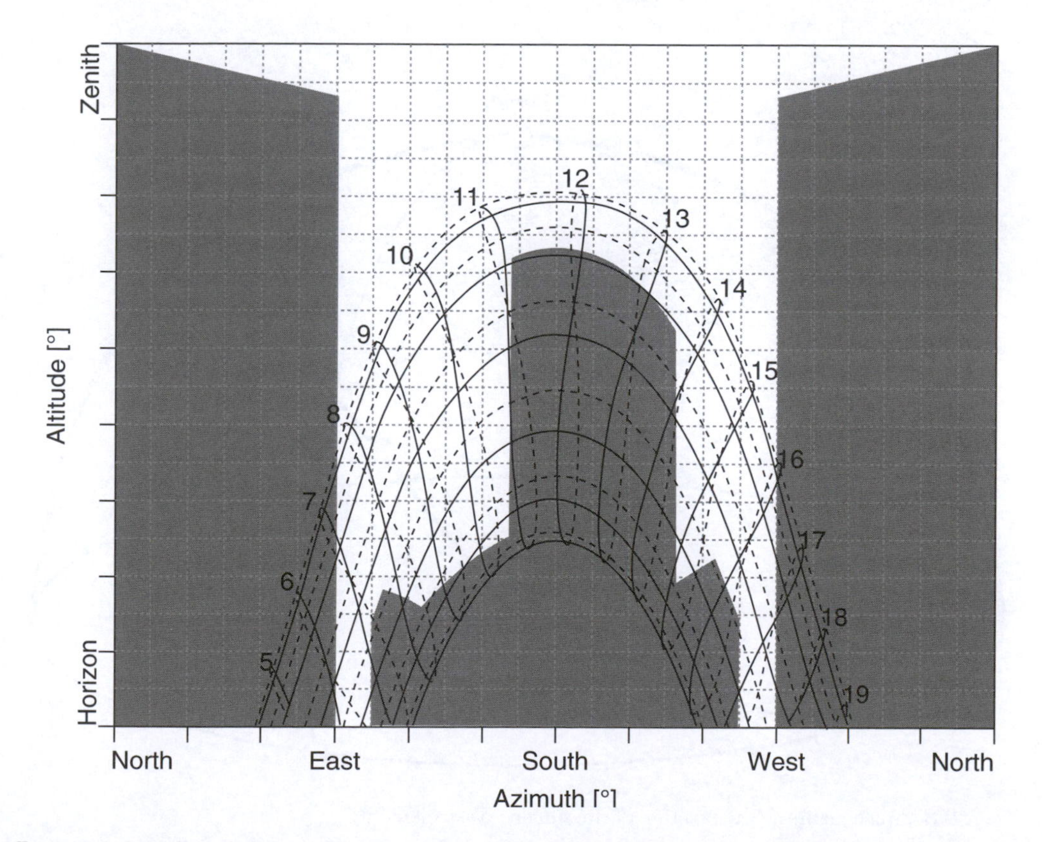

Figure 9.7 Sun chart diagram for Boston with a shading mask for a south-facing wall.

conditions to test building designs under. Then, during the mid 1990s, the *CIE standard general sky* (Kittler, Perez and Darula1997) as well as the *Perez sky* (Perez, Seals and Michalsky 1993) were introduced. Both approaches attempt to "cover the spectrum of intermediate and cloudy skies between the two already standardized clear and overcast sky distributions" (Kittler *et al.* 1997; Kittler, Perez and Darula 1998). The *CIE general sky* and Perez sky both rely on the same basic mathematical functions to describe the ratio of the luminance, L_a, of an arbitrary sky element to the zenith luminance, L_Z (Equation 9.1).

$$\frac{L_a}{L_Z} = \frac{f(\chi)\cdot\varphi(Z)}{f(Z_S)\cdot\varphi(0)} \quad \text{with } \varphi(Z) = 1 + a\cdot e^{\left(\frac{b}{\cos(Z)}\right)} \quad \text{and}$$
$$f(\chi) = 1 + c\cdot\left[e^{d\chi} - e^{d\frac{\pi}{2}}\right] + e\cdot\cos^2(\chi) \tag{9.1}$$

Parameters *a* to *e* are used to adopt the sky luminous distribution function to a range of clear, intermediate or overcast sky conditions. As shown in Figure 9.8, χ is the shortest angular distance between a given sky element and the sun, Z is the angular distance between the sky element and the zenith and Z_S is the angular distance between the sun and the zenith.

φ(Z) is called the *luminance gradation* function. It defines the changes of luminance from horizon to zenith. For a uniform sky this function corresponds to unity (*a*=0). f(χ) is called the *scattering indicatrix*. It relates the changes of luminance of a sky segment to its angular distance

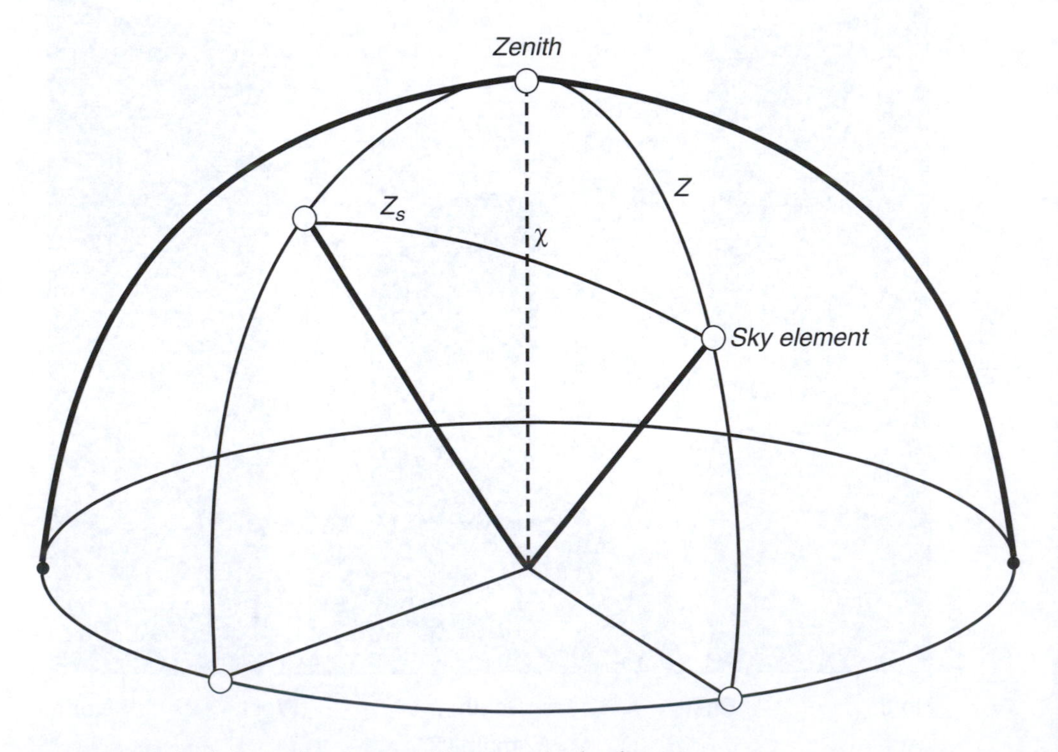

Figure 9.8 Angles defining the position of the sun and a sky element.

Source: CIE 2004.

from the sun. Its main purpose is to define the luminance in the *circumsolar region* surrounding the sun (CIE 2004).

The difference between the CIE general sky and the Perez sky is that the former defines fifteen discrete parameters sets for variables *a* to *e* whereas the Perez sky model continuously changes those parameters as a function of direct and diffuse irradiances/illuminances. Both models vary between and include the earlier mentioned "old" CIE overcast and clear skies. Same as Perez, the CIE general sky requires direct and diffuse radiation pairs palong with building site, date and time to identify the one of the fifteen standardized general skies that best matches a particular sky condition for a given set of date, time location and irradiation data. While the CIE general sky is being promoted through the CIE and ISO standards, the building energy simulation community tends to use the Perez sky model with leading programs such as ESP-r, EnergyPlus and TRNSYS all using it for modeling the celestial hemisphere for solar gains and daylighting predictions. CIE general sky and Perez have both been implemented into a number of commercial and open source daylight simulation programs making them accessible to the larger design community. The author has thus far used the Perez model to generate annual indoor and outdoor illuminance time series for a number of reasons: Perez can easily be used in combination with hourly annual climate files (see pages 42–43); using the same sky model for both the daylighting and thermal simulations makes the two more consistent and the author has yielded convincing results comparing outside illuminance predictions based on Perez with measurements taken in Freiburg, Germany, and Ottawa, Canada (Reinhart and Breton 2009; Reinhart and Walkenhorst 2001). Another, more fundamental reason why the author conceptually prefers the continuous Perez sky model to the fifteen discrete CIE general skies is that these skies seem to be viewed by its promoters to be reference skies under which a designer should test the daylighting performance of a particular building design. Rather than testing a design under a confusing number of intermediate skies, the author recommends to validate a design using visualizations under the CIE overcast sky and the CIE clear sky on solstice and equinox days to develop a general intuition of how the design performs under extreme skies. For a more representative performance analysis that takes the local climate into account, an annual daylight simulation is recommended (see Climate-based metrics on page 260).

Figure 9.9 visually compares the sky luminous distribution over Boston on April 2 at noon for the uniform, CIE overcast, CIE clear and Perez skies. According to the typical meteorological year (TMY2) for Boston Logan Airport, April 2 is a day with intermediate sky conditions. The Perez model therefore lies between the extremes of the rotationally invariant, overcast CIE sky and the CIE clear sky distribution.

In order to carry out an annual daylighting or energy analysis of a building, time series of climatic data are required (see Chapter 3). This usually comprises hourly data for temperature, relative humidity, wind speed and direction as well as direct and diffuse solar radiation and illuminances. An ever increasing number of climate files are available for download from the internet. The US Department of Energy offers complete annual typical meteoro-logical years for over 2100 sites worldwide free of charge in a standardized format that can be directly imported in a number of daylight simulation programs (Crawley, Hand and Lawrie 1999; US-DOE 2009). A caveat that a daylighting modeler should be aware of is that these data sets are frequently based on local airport weather stations that typically measure total solar radiation using a pyranometer. If this is the case for a particular site, the direct and diffuse radiation and illuminance components that are reported in the climate files have thus been theoretically derived from total solar radiation using, e.g., the Reindl (Reindl, Beckman and Duffie 1990) and Perez daylight luminous efficacy models (Perez *et al.* 1990).

In case a reader is interested in modeling the short time step behavior of daylight, e.g. for glare analysis or to predict energy savings from automated lighting controls (Clarke and Janak 1998), Olseth and Skartveit developed an autocorrelation model to generate sub-hourly time

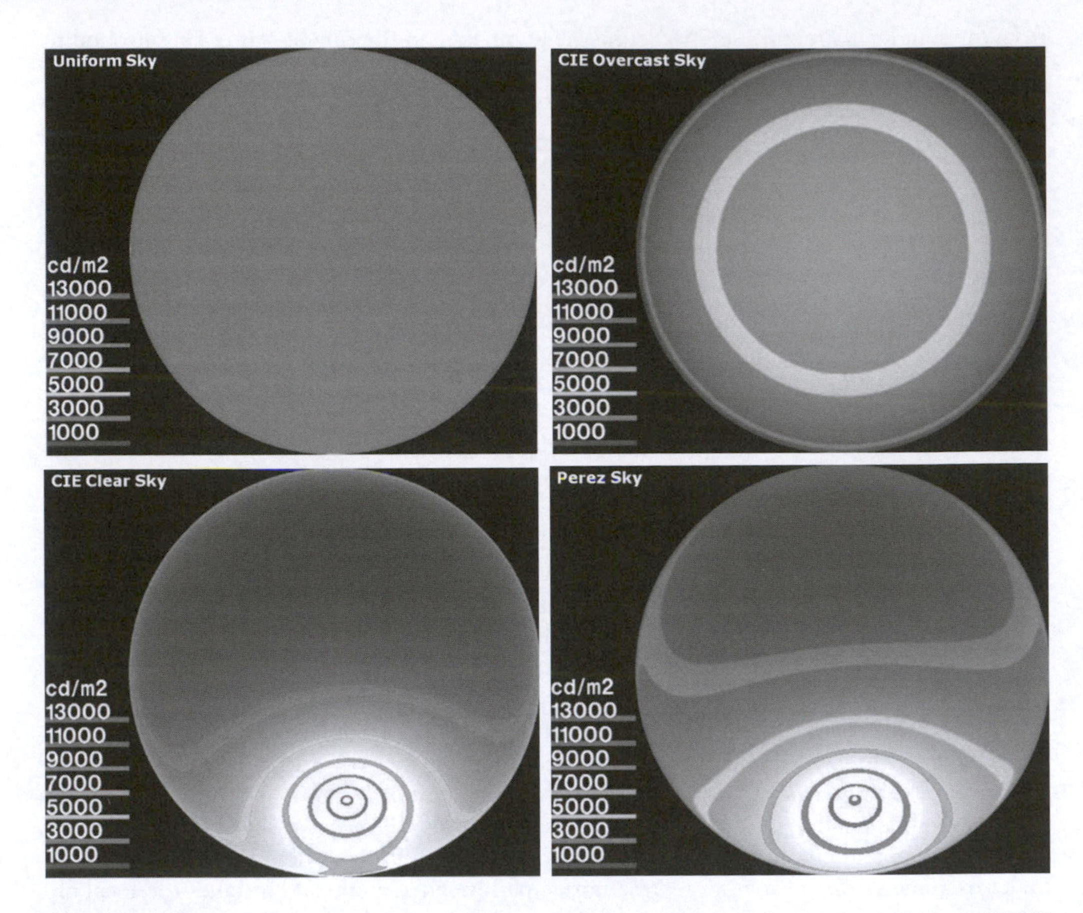

Figure 9.9 Sky luminous distribution over Boston on April 2 at noon according to the uniform, "old" CIE overcast and clear, as well as the Perez sky models.

series from hourly global solar radiation (Olseth and Skartveit 1989). The model has been validated and implemented into the Radiance-based Daysim simulation software (Walkenhorst, Luther, Reinhart and Timmer 2002).

For cumulative light exposure levels (e.g. annual and seasonal) a number of authors have presented the concept of a cumulative sky (Compagnon 2004; Mardaljevic and Rylatt 2003; Robinson and Stone 2004).The idea is to add series of different sky conditions on top of each other and then run a regular daylight simulation under this theoretical, very bright sky. The advantage of this method is that cumulated radiation and light exposure can be calculated in a single simulation run. An example annual radiation map generated with Radiance and "GenCumulativeSky" (Robinson and Stone 2004) is shown in Figure 9.4(b).

Before moving on to the next section, a few emerging topics related to modeling the celestial hemisphere for daylight simulations are presented in the following.

Sky scanners that measure the sky luminous distribution directly have been commercially available for several decades but – due to prohibitively high investment costs – their use has thus far been restricted to academic applications such as validation studies (Aizlewood 1993; Mardaljevic 1995) and the development of sky models. With the recent advances in digital photography, a high-end digital camera in combination with a fisheye lens and a luminance meter now allow the collection of high dynamic range (HDR) photographs of the sky luminous distribution making sky scanner data accessible to a larger number of researchers and design teams (Figure 9.10) (Inanici 2006). A few sites have started to collect hourly HDR sky images which can then be mapped onto the celestial hemisphere and act as a light source in daylight

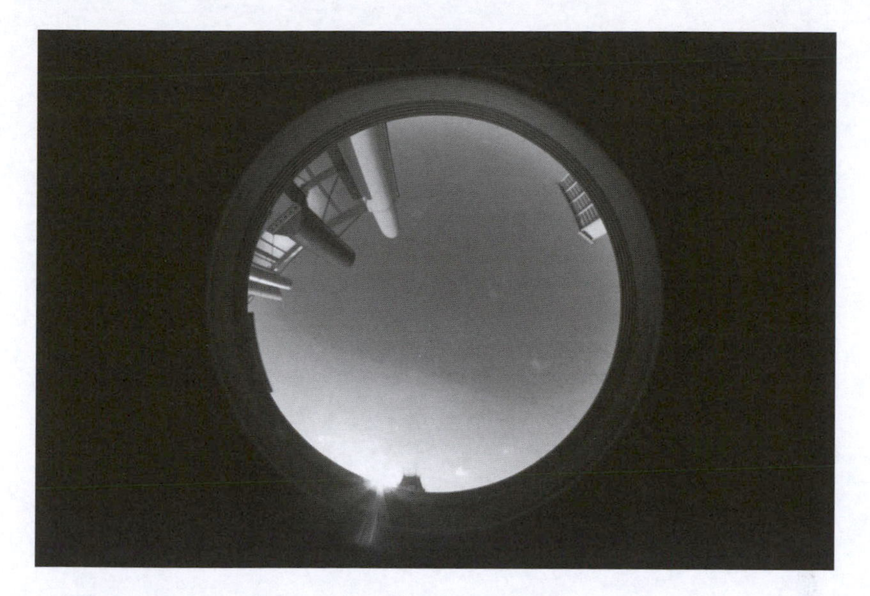

Figure 9.10 Calibrated high dynamic range (HDR) photograph generated with a digital camera, a luminance meter and Photosphere (Ward, last accessed in May 2009).

simulations (Inanici 2009). It is likely that time series of HDR sky images will soon become available for a variety of sites worldwide complementing the earlier mentioned climate files.

Another recent improvement to sky models is the addition of physically based color gradients, e.g. for modeling sunsets. A prominent example is the *Utah sky model* (Preetham, Shirley and Smits 1999). So far the resulting colored sky seems to have mainly been used to add realism to design visualizations but, as the spectral effects of light on human health and alertness are becoming better understood (Webb 2006), modeling the changing spectral composition of daylight throughout the day and year might become a basis for more quantitative, health-related design metrics (Pechacek, Andersen and Lockley 2008).

Areas of interest

Once a scene has been built and a sky condition has been selected, an "area of interest" has to be defined within the scene. Commonly used formats are selected viewpoints (Figure 9.11), two dimensional grids of illuminance sensors for falsecolor representations (Figure 9.12) and/or a line of upward facing illuminance sensor positioned along the central axis of a space (see Figure 9.25(b)). The choice of what area of interest a modeler chooses is obviously closely related to the purpose of the simulation and the performance metric used (see below). Generally speaking, visualizations and falsecolor distributions are more engaging than simple line plots revealing additional visual detail of a design. On the flipside, line plots allow the direct comparison of several design variants in a single diagram.

Depending on the daylight performance metric used, a varying number of information regarding the usage pattern of the various space types in a scene (offices, classrooms . . .) is required. Typical requirements are target work plane illuminances, occupancy schedules, required minimum daylight factors etc. Suitable levels for these numbers can, for example be found in the handbooks of the Illuminating Engineering Society of North America (IESNA 2000) as well as the American Society of Heating, Refrigerating and Air-Conditioning Engineers (ASHRAE 2007b).

Figure 9.11 Radiance visualization of an office work place. The surrounding landscape was mapped onto the sky dome to visualize how the building interior might "feel".

Daylight simulation algorithms

Once a complete daylighting model has been entered into a simulation program, the simulation engine needs to be invoked to carry out a global illumination calculation within a scene. A number of popular daylight simulation programs now support multiple simulation engines to carry out the same type of analysis. It is hence of paramount importance that the program user understands *which* simulation engine is used for a particular calculation and *whether* the simulation engine is capable of reliably modeling the task at hand. Figure 9.12 shows an example of the exact same model of a sidelit space modeled in Autodesk Ecotect with the daylight factor distribution calculated using either Ecotect's internal daylighting engine (a variation of the BRE split flux method (Dufton 1946)) or the Radiance backward raytracer (Ward and Shakespeare 1998). The choice of simulation engine used has a profound impact on the simulation results which vary in this example by *over 400 percent* between Radiance and BRE split flux (Ibarra and Reinhart 2009). The reason for this large discrepancy is that Radiance is capable of explicitly modeling light that is reflected multiple times whereas the BRE split flux method bases its calculation solely on objects which are in direct view of a sensor (see below). As a consequence, it underestimates the actual daylight factor distribution in the space. The objective of Figure 9.12 is to sensitize the reader to the fact that the choice of simulation engine used – even within the same simulation program – is crucial.

At the time of writing, the US Department of Energy's *Building Energy Software Tools Directory* (US-DOE) listed over twenty different programs that claimed to be able to "model daylight". Which one of them should a practitioner use? First of all, a reader should have some knowledge of the basic approaches commonly used to model the phenomena of (day)light in buildings. Three common methods are therefore briefly introduced in the following: BRE split flux, radiosity and raytracing.

The BRE split flux or protractor method, developed at a research institution that is now called the British Building Research Establishment (BRE), only models diffuse sky conditions, e.g. for daylight factor calculations, and splits internal daylighting levels into three components: A sky component (SC), an externally reflected component (ERC) and an internally reflected component (IRC) (Figure 9.13) (Dufton 1946). Originally the various components had to be geometrically derived using so-called protractor diagrams which were placed on building

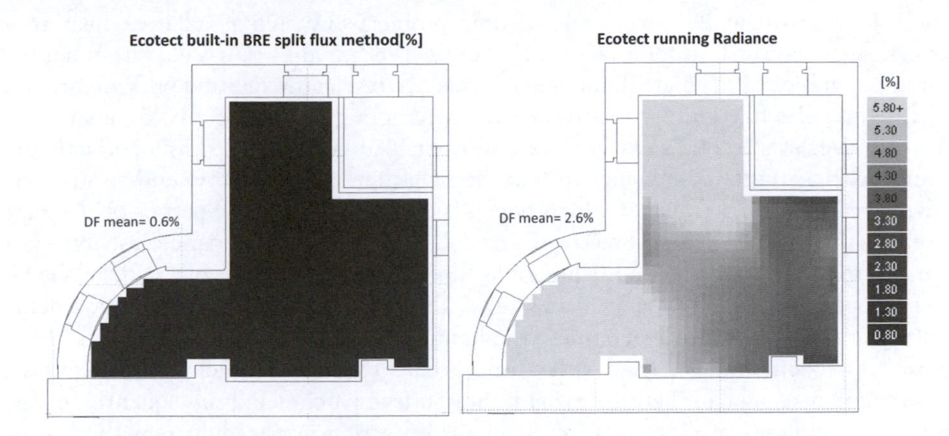

Figure 9.12 Daylight factor calculation in a sidelit space in Autodesk Ecotect v5.6 using (a) the built-in BRE split flux method and (b) Radiance v3.8.

Source: Ibarra and Reinhart 2009.

floor plans and sections. This manual process, tedious in its execution, can nowadays be dramatically accelerated and streamlined using a three-dimensional computer model and a set of rays traced backwards from a sensor point (Figure 9.13). The BRE split flux method tends to yield meaningful results for scene locations that "see" a lot of sky and for which the internally reflected component is small. For heavier obstructed spaces and points further away from any facade opening the method becomes increasingly unreliable since the internally reflected component is derived from static parameters that might have been derived in spaces that are significantly different from the space under investigation.

Two numerical approaches that are commonly used to simulate interior lighting conditions under varying sky conditions in arbitrarily shaped spaces are *radiosity* and *raytracing*.

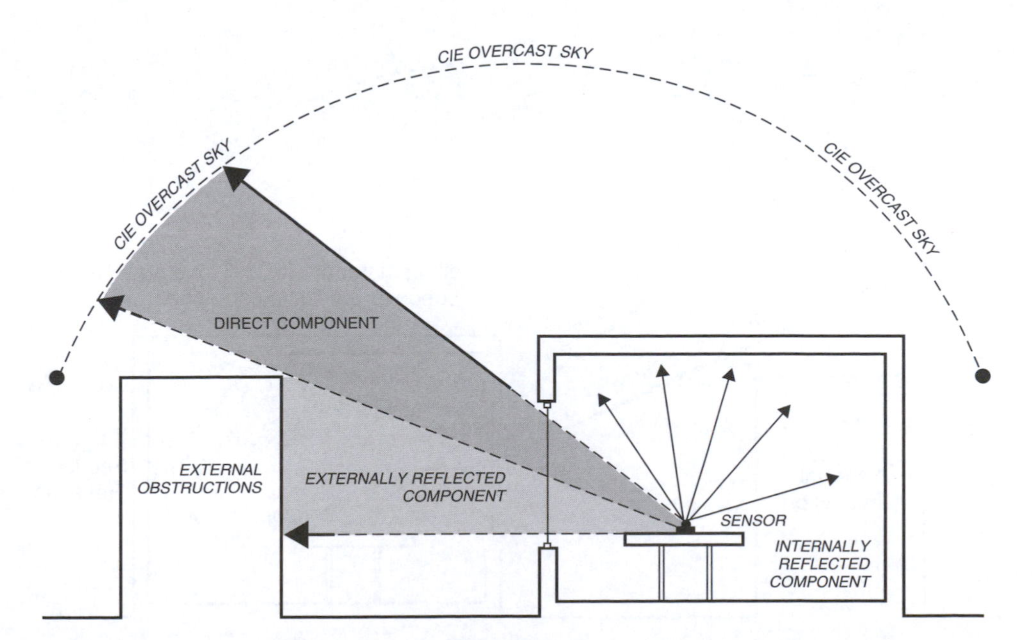

Figure 9.13 Principle of the BRE split flux method.

Radiosity was originally developed to solve problems involving radiative heat transfer between surfaces based on form factors. Since the 1980s, radiosity has also been applied in computer graphics to calculate illuminance levels due to electric lighting or daylight. A form factor defines the fraction of radiative energy leaving a given surface to the energy which directly arrives at a second surface. In radiosity, each surface is treated as a perfectly diffuse reflector with a constant luminance so that the radiation exchange between two surfaces can be described by a single number which depends on the reflective properties of the surfaces as well as the overall scene geometry. To calculate the indoor luminance distribution in a room due to daylight, the incoming luminous flux through all non opaque parts of the building envelope is set equal to the available flux within the building. This assumption defines a set of equations that uniquely determine the luminances of all considered surfaces. The basic radiosity approach can be coupled with a finite element approach which detects regions with a large luminance gradient between neighboring surface patches and subsequently subdivides the affected surfaces into sub-surfaces. A detailed description of radiosity methods can, for example, be found in Cohen and Wallace (1993).

The basic approach of *raytracing* is to simulate individual, representative light rays in a scene and to extrapolate the overall luminous distribution of the scene from the available lighting information. Forward raytracing (rays traveling *from* a light source) is generally used to analyze the interaction of light with individual components such as a luminaire or Venetian blinds. The simulation output can be, for example, IESNA photometric files (IESNA 2000) or bidirectional transmission and/or distribution functions (BTDF/BRDF). For more complex scenes, e.g. of a whole neighborhood, *backward raytracing* is generally used. In backward raytracing rays are emitted from the point of interest and traced backwards until they either hit a light source, such as the sun or celestial hemisphere, or another object. In the former case, the luminance distribution function of the light source determines the luminance contribution at the view point. If a ray hits a object other than a light source, the luminance of that object needs to be calculated by secondary rays which are emitting from the object (Figure 9.14). The angular distribution under which secondary rays are *spawned* depends on the optical properties of the surface. Raytracing allows for arbitrarily complex optical surface properties including purely specular surfaces such as a mirror, Lambertian surfaces such as conventional wall paint, transparent surfaces such as

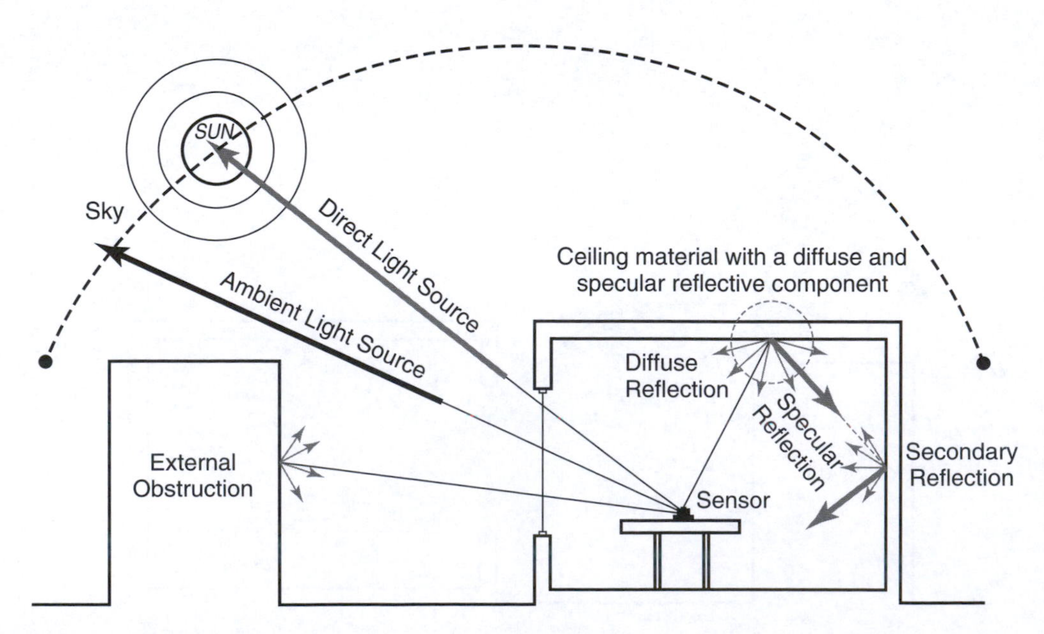

Figure 9.14 Principle of a backwards raytracer.

glazings as well as arbitrary mixtures of these basic surface types. A ray path is usually aborted once a certain number of reflective bounces has been reached or if the relative weight of a ray falls below a given threshold value[2] (Ward and Shakespeare 1998).

An advantage of following a light ray "backwards" from the point of measurement (view point or virtual sensor) to a light source is efficiency as the majority of rays that are emitted by the potentially many light sources within a scene never reach the particular view point or sensor of interest (Ward and Shakespeare 1998). To illustrate the point, the reader should consider the visualization of the workspace in Figure 9.11 which could be one of many similar workspaces in that building which in turn could be one of several buildings within the scene. For the visualization in Figure 9.11 the luminance distributions of objects in close vicinity to the view point are primary of interest and "going backwards" the simulation naturally concentrates on those objects.

In scenes which involve both, complex space geometries as well as complex fenestration systems in which the light is redirected multiple times (the reader might think of a light pipe), regular backward raytracing becomes inefficient and (potentially) unreliable. Approaches that have been explored to tackle such scenes include replacing the CFS with a single surface that has the optical properties of the CFS (Andersen and deBoer 2006), using a hybrid forward and backward raytracing approach in which light is first forward traced from the light source in the scene and then collected from the areas of interest using backward raytracing (mental-images 2007; Schregle 2004) and treating the CFS layer as a two-dimensional matrix with separate interior and exterior raytracing runs (Laouadi, Reinhart and Bourgeois 2008; Ward 2009).

Is radiosity or raytracing better suited to address a specific design question? The answer obviously depends on the design question. A general advantage of radiosity compared to raytracing is that it requires less calculation time for straightforward geometries which do not contain too many surface elements. This advantage diminishes with rising model complexity as the calculation time in radiosity increases with the square of the number of considered elements while in raytracing this relation is roughly linear (Tregenza 1983) or even sublinear (Ward 1994). A radiosity calculation yields the total luminance distribution in a room independent of the point of view of the spectator. Therefore, a scene walk-through can be realized a lot faster with radiosity than with raytracing as each new viewpoint requires a new raytracing run. Radiosity based walk throughs also tend to look smoother. An advantage of raytracing over radiosity is that it is able to simulate specular and partly specular materials (Ward and Shakespeare 1998). This quality can be important in daylit spaces especially for visual comfort analysis purposes since window frames, shading devices and furniture finishes often exhibit distinctly non-Lambertian surface properties.

Both radiosity and raytracing require the modeler to set several simulation parameters in accordance with the scene complexity and expected luminance gradient. A key simulation parameter for raytracing is the number of reflectances (bounces) considered before a ray is discarded. Setting the simulation parameters can be a complicated task – especially for simulation beginners – which is why graphical user interfaces are increasingly "hiding" the actual parameters and rather ask the user to provide more qualitative descriptions of a scene such as "scene detail" or "scene complexity". These advances notwithstanding, setting the simulation parameters remains a delicate issue as selecting too low values may lead to wrong results whereas "too high" values reduce the simulation uncertainty but also lead to unjustifiably long simulation runs.

A fundamental problem of raytracing is that it cannot provide a reliable scene-specific estimate of the remaining calculation errors. This can turn out to be a real impediment as simulation results may lie above *or* below the true illuminance levels. Simulation results are, for example, too low if a raytracer misses a small window or skylight in a room and therefore grossly underestimates the real illuminance. Results are too high if a raytracer interpolates between two bright luminances directions, e.g. from two neighboring windows, and ignores that a wall lies in the interpolated region.

Program validation

Once a modeler has developed a basic appreciation of the various daylight simulation algorithms that are commonly used, he or she can start searching for a simulation program that uses the desired simulation algorithm and calculates the daylight performance metrics the modeler is interested in (see the next section). Once a potential program has been identified, the modeler should also verify that the program has been validated. Validation may happen through three methods, (a) the comparison of simulation results to lighting scenarios that can also be solved analytically, (b) inter-program comparison and (c) through comparison to measurements. In 2006 the International Commission on Illumination, CIE, released a document titled "CIE 171 – Test cases to assess the accuracy of lighting computer programs" (CIE 2006). The report proposed a series of analytical (electric lighting and daylight) and experimental (electric lighting only) test cases and had the objective to help "program users and developers to assess the accuracy of lighting simulation programs and to identify areas of strength and weakness within different programs". Similarly to the BESTEST/ASHRAE Standard 140 (ASHRAE 2007a) cases for thermal simulation engines, the main focus of the CIE 171 test cases was to "test very isolated aspects of a simulation software such as the capability to model light sources including daylight, light transfer through openings and the ability to model inter-reflections" (CIE 2006). While the author agrees that the use of simple, idealized lighting scenarios as promoted by CIE 171 are very useful for basic software testing and "bug finding", the author also feels that a complementary set of more complex daylighting test cases that compare experimental measurements in "real" daylit spaces with simulations is necessary to help design practitioners to develop a feeling of how close a lighting simulation program, operated by a knowledgeable user, can model reality. This attitude seems to be shared by others given that a 2006 survey of one hundred and eighty-five daylighting modelers from twenty-seven countries yielded a strong preference for Radiance, a mixed stochastic deterministic backwards raytracer that had been previously validated based on illuminance measurements in full-scale spaces with a clear glazing and a lightshelf (Jarvis and Donn 1997; Mardaljevic 1995), venetian blinds (Reinhart and Walkenhorst 2001) and a translucent glazing (Reinhart and Andersen 2006). The survey participants named over forty different software packages that they frequently used but over 50% of all votes went to tools that are based on Radiance[3] (Reinhart and Fitz 2006).

For software developers interested in validating their tools against measurements at least two datasets are currently available: The British Building Research Establishment (BRE) offers a very rich data set of indoor illuminances in a full scale sidelit test room with a clear glazing and lightshelf including sky scanner data (Aizlewood 1993). More recently the National Research Council Canada (NRC) in Ottawa, Ontario, Canada collected a new data set that consists of measured indoor and outdoor illuminances as well as direct and diffuse outdoor irradiances for five daylighting test cases of varying complexity (Reinhart and Breton 2009). The NRC data set has thus far been used to validate the mental ray raytracer within Autodesk 3ds Max Design 2009 and the Radiance-based Daysim 3.0 program. Given the rising interest in physically accurate daylight simulations, the author expects that other simulation programs will soon go through comparable experimental validation exercises using the BRE, NRC or other experimental data sets. These validation studies should make it easier for beginning daylighting modelers to compare the validity and limitations of various simulation programs.

Dynamic simulations

The simulation methods discussed thus far are static in the sense that they were developed to simulate interior or exterior lighting levels under a particular sky condition, e.g. the CIE overcast sky. Since actual ambient daylighting conditions depend on the local climate and are constantly changing, dynamic or annual daylight simulations calculate time series of lighting levels due to daylight throughout the whole year. These time series are called annual illuminance profiles and typically have a time step varying from 1 hour down to a minute. Annual illuminance profiles can be used to predict electric lighting energy use in daylit spaces and/or combined with a thermal simulation program to predict overall energy use for heating, lighting and cooling. Another application is to translate them into climate-based daylighting metrics that consider the quantity and character of daily and seasonal variations of daylight for a given building site (see the next section).

In principle, it is possible to carry out an annual simulation by simply repeating a static simulation multiple times under varying sky conditions. But, this procedure typically leads to exorbitant simulation times (Reinhart and Herkel 2000). A series of different dynamic daylight simulation methods were introduced in the past but most researchers seem to nowadays use a daylight coefficient based approach (Janak 1997; Mardaljevic 2000; Reinhart and Walkenhorst 2001; Tsangrassoulis, Santamouris and Asimakopoulos 1996).

The concept of daylight coefficients, originally proposed by Tregenza and Waters (Tregenza and Waters 1983), is a numeric method to speed up an annual daylight simulation. The basic approach is to conceptually divide the celestial hemisphere into disjoint sky segments (Figure 9.15(a)), and to calculate the contribution of each sky segment to the total illuminance at various sensor points in a building (Figure 9.15(b)). These illuminance contributions, $E_\alpha(x)$, at a sensor point and orientation, x, for a sky segment, α, are then normalized with the luminance, L_α, and angular size, ΔS_α, of the sky segment (Equation 9.2). The resulting quantify is called a daylight coefficient, $DC_\alpha(x)$.

$$DC_\alpha(x) = \frac{E_\alpha(x)}{L_\alpha \Delta S_\alpha} \tag{9.2}$$

A complete set of daylight coefficients for all sky segments defines the relationship between a point within a scene and celestial hemisphere. It takes into account scene geometry and material properties but is independent of the local climate. Daylight coefficients can hence be coupled with an arbitrary sky luminance distribution, L_α, with $\alpha = 1 \ldots N$, by a simple linear superposition to calculate the total illuminance E(x) at x (Equation 9.3):

$$E(x) = \sum_{\alpha=1}^{N} DC_\alpha(x) L_\alpha \Delta S_\alpha \tag{9.3}$$

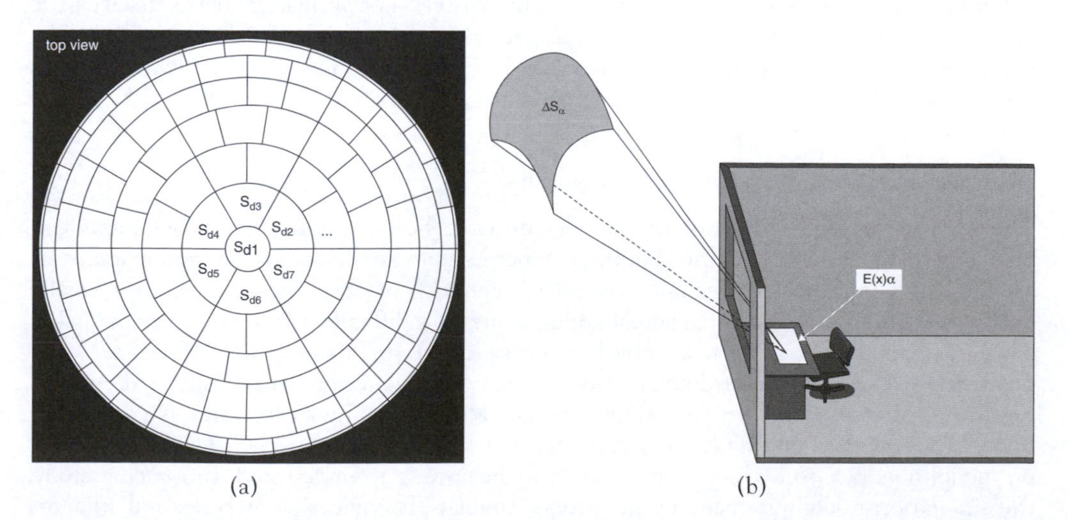

Figure 9.15 (a) Division of the celestial hemisphere into disjoint sky segment; (b) Visual definition of a daylight coefficient.

The time-consuming part of an annual daylight simulation based on daylight coefficients is the calculations of the coefficients, e.g. via raytracing. This step is hence carried out in an initial, pre-processing step before the annual simulation is invoked. Once a set of daylight coefficients is available, the illuminance calculation under any sky conditions only requires milliseconds. The daylight coefficient concept can also be extended to luminance time series. If applied to all pixels in an image, the concept even allows the generation of annual time series of scene visualizations (Wienold 2007). Daylight coefficient approaches mainly differ in how they differentiate between contributions from diffuse daylight and direct sunlight. For diffuse daylight most available methods split the celestial hemisphere into 145 sky segments according to the Tregenza division (Tregenza 1987) (Figure 9.15(a)) as well as one or several ground segments. Contributions from direct sunlight can be divided into those that are directly incident onto a sensor and those that have been reflected at least once within the scene before hitting a sensor (Bourgeois, Reinhart and Ward 2008; Mardaljevic 2000).

A number of simulation studies have yielded that dynamic daylight simulations using a daylight coefficient approach, backward raytracing and the Perez all weather and/or general sky models can simulate indoor illuminances under a range of overcast, intermediate and clear sky conditions with a relative error below 25% compared to measurements (Mardaljevic 2000; Reinhart and Andersen 2006; Reinhart and Breton 2009; Reinhart and Walkenhorst 2001). This error might initially sound high, but the reader should consider, that the human eye is a logarithmic sensor and that we can barely notice illuminance changes over time in that order of magnitude. In other words, the resolution of daylight simulation programs roughly corresponds to the sensitivity of the human eye. On the other hand, indoor illuminance levels in daylit spaces may vary from tens to ten thousands of lux, a range that can be captured by simulation programs. These facts suggest that the current generation of simulation programs provides sufficient accuracy to provide reliable design advice, at least for spaces of typical geometric complexity and with daylight components such as clear and translucent glazings, lightshelves and venetian blinds. As mentioned above, accurately modeling even more complex fenestration systems including mirror blinds and light wells can be tackled using photon mapping and/or matrix multiplications. But, this is still an area of active research.

From a practical point of view, the additional required effort for carrying out a dynamic as opposed to a static daylight simulation is small. As far as model preparation is concerned, the only difference is that a local climate file has to be specified which does often not require any serious preparation effort since climate data is nowadays available for many parts of the worlds in a standardized format (see Chapter 3 for more information). The annual simulation itself can be fully automated and, again, does not require any additional effort on behalf of the modeler. The main difference is that – in the case of raytracing – the simulation time is roughly five to eight times longer for a dynamic than for a static simulation (Reinhart and Walkenhorst 2001).

Simulation uncertainty, user errors and quality control

The earlier discussed validation studies demonstrate how close daylight simulations can approach reality if they are carried out by an expert user and if all simulation input parameters are known. The critical reader might wonder how much of these information are typically available during the design of an actual building and how difficult it is to build a high quality daylighting model of a space when one first starts using daylight simulations? These are valid concerns. A modeler is indeed commonly required to make a number of "educated guesses" regarding material properties and building usage patterns which can in turn have a decisive impact on how the building is going to perform. The effect of a modeler's level of expertise on model quality is something that we are only beginning to understand. In a recent study, the simulation results produced by sixty-nine "simulation novices" – architectural students that took a one semester lighting class – were compared to a "best practice" model (Ibarra and Reinhart 2009). The investigated space was the sidelit space from Figure 9.12 and the program used was Autodesk Ecotect version 5.6. Figure 9.16 shows the mean daylight factors for the sixty-nine models calculated using Radiance as the simulation engine. The data is separated into models that were generated in the fall 2005 and fall 2006 terms, respectively.

Figure 9.16 shows that the majority of student models yielded significantly higher results than the best practice model with the average mean daylight factors in 2005 and 2006 being 8.3% and 5.3%,[4] respectively, compared to 2.6% for the best practice model. A detailed analysis of the models showed that simulation novices made any number of modeling mistakes ranging from incomplete and/or inaccurate geometries to not adjusting material properties or encountering software interoperability errors when importing geometries that had been prepared in another CAD tool. It is worthwhile mentioning that the daylight factor calculation in Autodesk Ecotect was actually chosen for this study since it is a mature and visually intuitive workflow that the author had assumed could be easily followed by simulation novices. Chances are that modeling errors made using other simulation programs would have been comparable or worse. Figure 9.16 paints a rather bleak image of the quality of simulation results novices can expect to get. On a positive note, the overall model quality was somewhat better during 2006 compared to the 2005 class, i.e. in 2006 students avoided certain errors such as modeling walls with zero thickness and many considered shading due to neighboring buildings and trees. The author partly attributes this improvement to the "simulation tips" that were provided

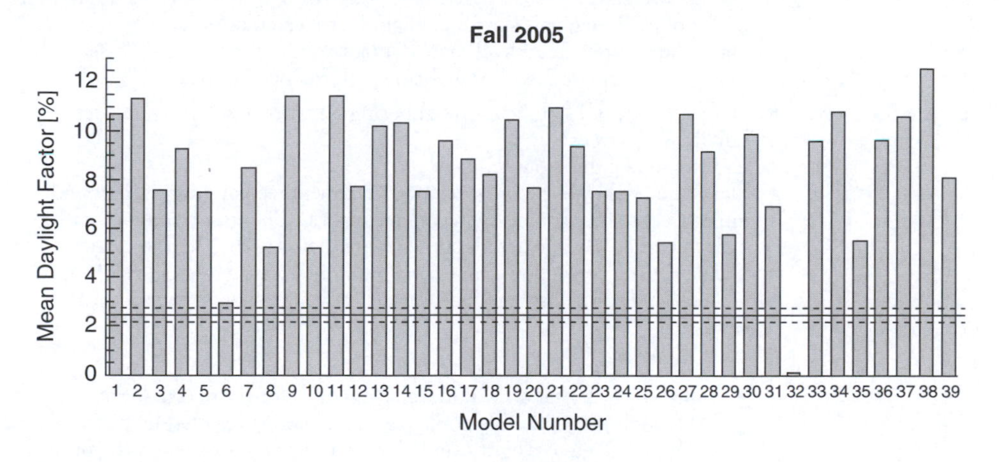

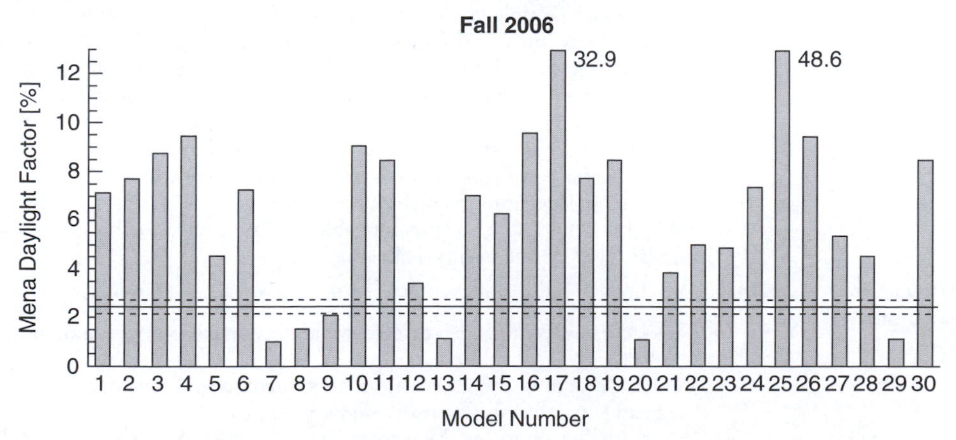

Figure 9.16 Calculated mean daylight factors of the sidelit space from Figure 9.12 based on sixty-nine simulation novices models. The solid line shows the mean daylight factor based on a "best practice" model of the space with a relative error margin (dashed lines) of ±10%.

Source: Ibarra and Reinhart 2009.

with the assignment in 2006 that prompted students to remember to model wall thicknesses, neighboring trees etc. The lesson learnt from this study is that simple workflow instructions are effective tools to help simulation beginners to build better models and to develop a simulation "mindset". Based on this finding the author recommends the use of the simulation checklist form Table 9.2 to simulation novices. Experienced modelers might also find it to be a useful reference list for model quality control.

Finally, Figure 9.17 compares predictions of the Lynes mean daylight factor rule of thumb (Equation 9.4) (Lynes 1979) to simulated mean daylight factors calculated in Radiance for

Table 9.2 Daylight simulation checklist

Before you start	• Did you decide which daylighting performance metrics to simulate and how to interpret the results? • Do you have a general idea of what the results should look like? E.g. a mean daylight factor in a standard sidelit space should lie between 2% and 5%; interior illuminance should lie between 100 lux and 3000 lux and daylight autonomies should range from 60% to 90% throughout the space. • Have you verified that the simulation program that you intend to use has been validated for the purpose that you intend to use it for, i.e. that the simulation engine produces reliable results and that the program supports the sky models related to your performance metric of choice? (An example would be the old CIE overcast sky for daylight factor calculations.) • Have you secured credible climate data for your building site? (This is only required for climate-based daylighting performance metrics.)
Preparing the scene	• Did you model all significant neighboring obstructions such as adjacent buildings and trees? • Did you model the ground plane? • Did you model wall thicknesses, interior partitions, hanging ceilings and larger pieces of furniture? Try to model all space dimensions at least within a 5cm tolerance. Façade details should be modeled with a 2cm tolerance. • Did you consider window frames and mullions by either modeling them geometrically or by using reduced visual transmittances for windows and skylights? Window glazings: • Did you check that all window glazings only consist of one surface? Several CAD tools model double/triple glazings as two/three closely spaced parallel surfaces whereas daylight simulation programs tend to assign the optical properties of multiple glazings to a single surface. • Did you check that all windows are "inserted" into the wall planes and not "overlaid" on the wall surfaces? Several CAD tools suggest that you can create and visualize a window in many different ways, one simply being the placement of a window surface on top of a wall surface which case end up with two coplanar surfaces. As a result the simulation program will either ignore the window or somehow "guess" which surface to consider. • Did you assign meaningful material properties to all scene components (see Table 9.1)? Did you model any movable shading devices such as venetian blinds? If yes, do the results make sense?
Setting up the simulation	Make sure that you set up your project files correctly. This may involve: • Checking that your project directory and file names do not contain any blanks (" "). • Verifying that all sensors have the correct orientation, i.e. work plane sensors are facing up and ceiling sensors are facing down. • Setting the resolution of the work plane to 0.5m × 0.5m or 1ft × 1ft and placing it around 0.85m above the floor. • Selecting simulation parameters that correspond to the "scene complexity". To do so you should consult the technical manual of your simulation program. • Selecting the correct sky model (CIE, Perez, etc.).

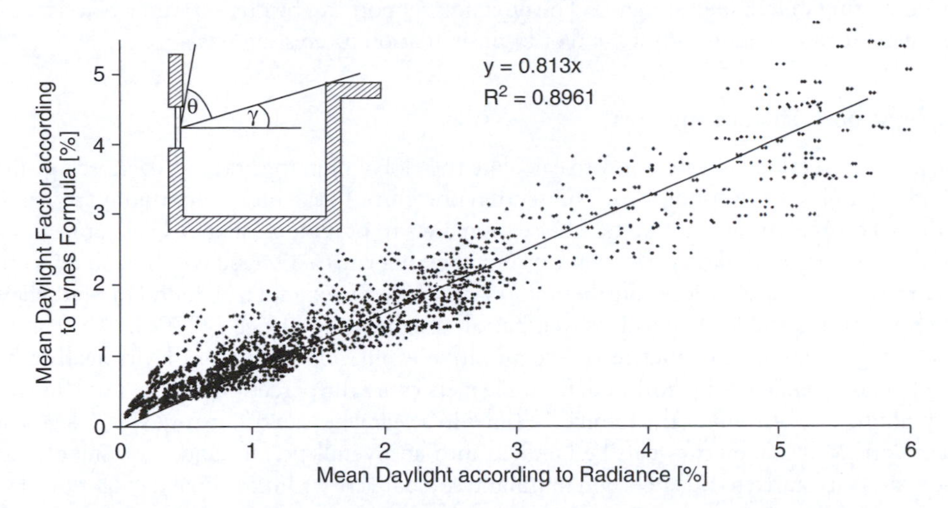

Figure 9.17 Mean daylight factor calculated with Radiance versus the mean daylight factor according to the Lynes formula (Eqn 9.4) for 2304 sidelit spaces with varying levels of external obstructions; inset: graphical definition of the sky and obstruction angles.

Source: Reinhart and LoVerso 2010.

over two thousand sidelit spaces. The figure shows a strong correlation (R^2=0.9) between the two prediction methods. While the daylight factor is not be the premiere daylight metric that the author would recommend modelers to use from a design standpoint, the figure shows that crosschecking mean daylight factor predictions of a model against the Lynes formula provides for a quick and easy "plausibility check". If a simulation model predicts a significantly different mean daylight factor than the Lynes rule, chances are that the modeler has made a mistake somewhere along the way. In that case the model should be carefully reviewed before the modeler tries to derive more advanced daylighting metrics from the model.

$$DF = \frac{A_{glazing}\tau_{vis}\theta}{A_{total}2(1-R_{mean})} \tag{9.4}$$

Where DF is the mean daylight throughout the space; R_{mean} is the area-weighted mean surface reflectance; $A_{glazing}$ is the net glazing area; A_{total} is the total area of interior surfaces (including windows); θ is the sky angle and τ_{vis} is the visible glazing transmittance.

In closing, Table 9.3 lists three practical questions that a novice modeler should consider when initially choosing a simulation program. The first "test" for a software to pass is whether it can actually calculate the set of metrics that the modeler is interested in (see section on daylight performance metrics). Next, the novice modeler should verify whether a potential software candidate has been validated and how (CIE test cases, measurements, . . .). This point become a key concern when the software is to be used in a consulting type environment as the modeler has to prove that due diligence was used when choosing a software tool. Finally, the modeler should find out whether there is a support network available for the software. Such a network can have many forms: Having a group of users of the same tool within a company provides by far the best learning opportunities. One can strategically build up such a group, e.g.

Table 9.3 Three questions to consider when choosing a daylight simulation tool

1. Which software can model the daylight performance metrics that I am interested in?
2. Has that software been validated (what are the supported sky model(s) and rendering algorithm(s))?
3. How easy is the software to learn? Is there any support network available?

through hiring well trained students. The next best support can be offered through active mailing lists, software vendor hotlines and/or local simulation interest groups.

Daylight performance metrics

The previous section discussed how to calculate the global illumination in a daylit scene under a single or multiple sky conditions. This section now turns to the most important part of a simulation: The analysis of the results. The section aims to provide a balanced account of established and emerging *daylight performance metrics* together with a discussion of how these metrics can influence design decisions. Further information can, for example, be found in Mardaljevic, Heschong and Lee (2009), as well as Reinhart, Mardaljevic and Rogers (2006).

A daylight performance metric is a quantitative evaluation of certain daylit qualities of a space, building and/or neighborhood. Example metrics are the percentage of the total floor area of a building that is daylit or the number of times in a year when an occupant experiences visual discomfort. Multiple metrics can be bundled into an overall performance assessment, e.g. a space can be required to meet certain minimum standards regarding annual daylight availability, visual comfort and energy usage. Performance metrics become especially useful design tools when they are combined with a target range so that a design iteration can be said to either "satisfy" a metric or not. Given that the earlier mentioned dynamic daylight simulation techniques have only been around for a short period of time, research into daylight performance metrics that can be derived from these simulations is still in its early stages. There is accordingly still a lot of "intellectual space" left for design practitioners and researchers to define their own daylighting metrics. The author actually recommends that design teams initially formulate their project-specific definition of "good daylight" along with a series of concrete metrics and target values. This exercise is useful as good daylighting is interpreted by some to be the "interaction of natural light and building form to achieve high quality workspaces" whereas other think of it as being primarily an energy-efficiency design measure (Galasiu and Reinhart 2007). Both definitions are valid but may at times favor different design choices.

The remainder of this section is organized as follows. Initially, illuminance-based daylighting metrics are being introduced followed by luminance-based metrics and metrics that consider occupant behavior.

Illuminance-based metrics

Illuminance based metrics are mostly derived from a grid of upward facing illuminance sensors at work plane height in order to describe the distribution of daylight within a space (see, e.g., Figure 9.12). The oldest and arguably still the most widely used daylighting metric today is the daylight factor.[5]

Daylight factor

The daylight factor is defined as the ratio of the internal illuminance at a point in a building to the unshaded, external horizontal illuminance under a CIE overcast sky (Figure 9.18) (Moon and Spencer 1942).

It is important to stress that the daylight factor is only defined under a CIE overcast sky and – since it is a ratio of inside to outside illuminances – it is independent of building site, time of day and year. Given that the CIE overcast sky is rotationally invariant and not site-specific (Figure 9.9(b)), the daylight factor is independent of building site, orientation, date and time. It might seem odd to have a daylighting metric that does not include such key design parameters. But, this quality was the original motive for working with a ratio rather than with an absolute value as a way of dealing with the "frequent and often severe fluctuations in the intensity of daylight" (Waldram 1950). In fact, using the daylight factor as a preliminary design metric

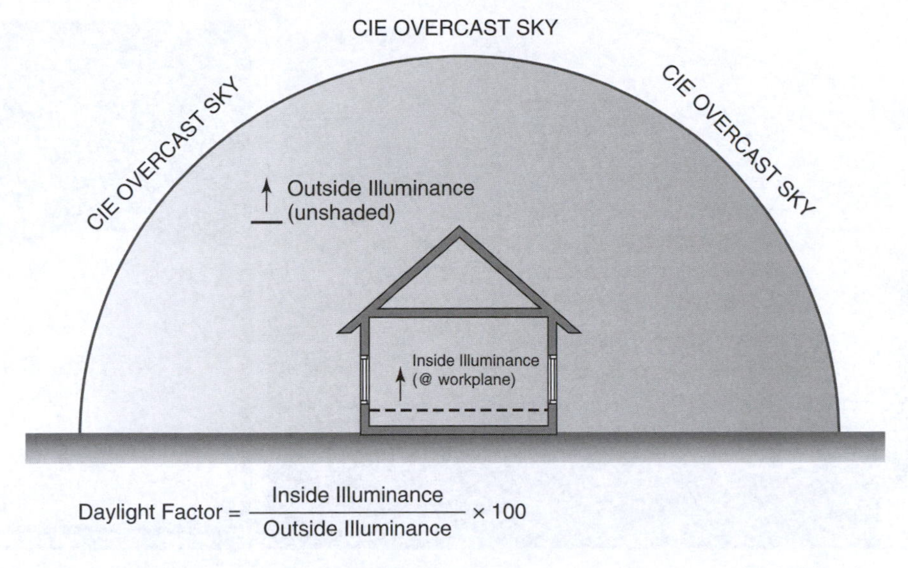

Figure 9.18 The daylight factor is defined as the ratio of an inside illuminance divided by the outside horizontal illuminance at an unshaded point under a CIE overcast sky.

for spaces that primarily receive diffuse daylight or in order to verify whether a certain amount of daylight can be maintained under diffuse sky conditions makes perfect sense to this day. The concept just becomes meaningless in the presence of direct sunlight. By itself the daylight factor is also of limited use to ensure that a visually stimulating, healthful, and productive interior environment is accomplished. It is merely a measure of "daylight availability".

One advantage of using the daylight factor for building design is that the designer can refer to various design guides and standards that recommend average daylight factor values for different space types. Most standards recommend a mean daylight factor of 2% for a "primarily daylit" space (British Standards Institution 1992; IESNA 2000). Figure 9.19(b) shows the daylight factor distribution in the example sidelit space (Figure 9.19(a)). Using a 2% daylight factor target level the space is divided into a daylit and a non-daylit area with the former extending approximately 4 m or 1.4 times the window-head-height away from the façade into the space.[6]

Due to its climate and orientation independence, the daylight factor for a south-facing office in Miami is the same as for a north-facing classroom in Anchorage – if both spaces have identical geometry and material properties. In order to make the daylight factor concept a bit more climate-specific, Szokolay introduced the concept of the design "sky illuminance" as the 15th percentile outdoor illuminance during daylit hours (9 a.m. to 5 p.m.) (Szokolay 2004). Once the design sky illuminance has been determined for a particular site, the product of the daylight factor and the design sky illuminance approximates the illuminance that should be exceed 85% of the working hours. The design sky illuminance for Boston, MA, USA (42°N 7°W), is about 5900 lux which implies that the average illuminance in the sidelit space from Figure 9.19 is above 5900 lux times 4.6% equals 271 lux for 85% of working hours between 9 a.m. and 5 p.m. in Boston.

As a design metrics the daylight factor promotes several design measures that are known to enhance the daylight in a space such as high window-head heights, high reflective ceiling and wall finishes, narrow floor plans and large facade and skylight openings with high transmittance glazings. A weakness of the daylight factor is that it promotes one-dimensional, "the more the better" approach to daylighting. Taking this to the extreme, the daylight factor optimized

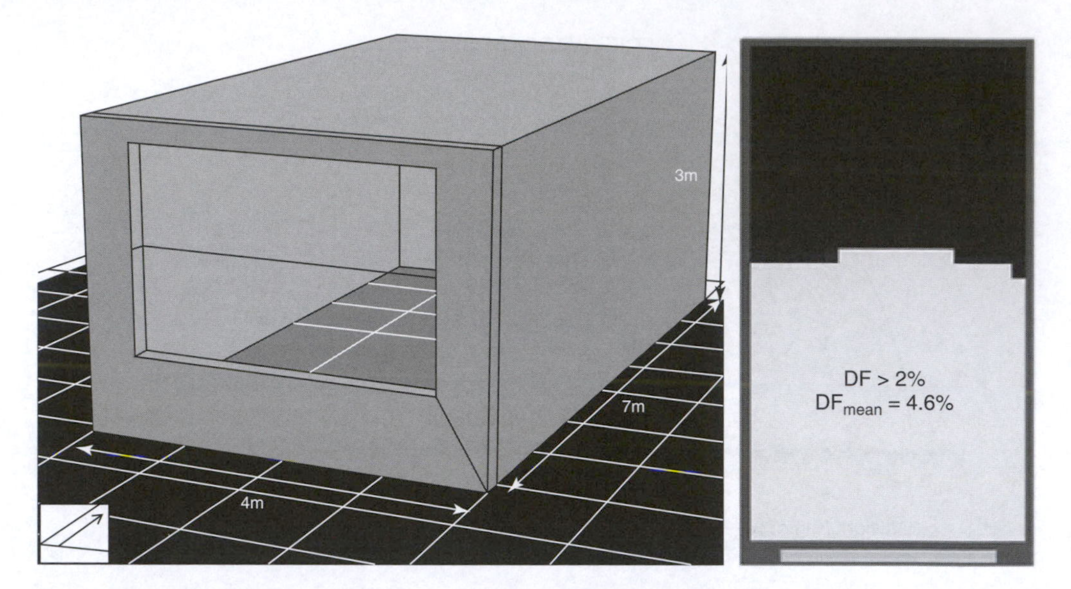

Figure 9.19(a) Visualization of an example sidelit space; (b) daylight factor distribution in the space.

building has a fully glazed building envelope. Therefore, the use of other daylighting metrics is required to keep the daylight factor "in check".

CIE clear sky studies

A popular way of exploring how a design performs under clear sky conditions is to model a space at 9 a.m., noon and 3 p.m. on solstice and equinox days. Figure 9.20 shows the results of the sidelit space from Figure 9.19 for Boston. Assuming that more than around 2000 to 5000 lux constitute an "oversupply" of daylight that may potentially lead to glare and/or overheating, the lighter areas in Figure 9.20 show where daylight in the space has to be further controlled. The June 21 figures show that summer solar gains could be largely controlled using an external overhang. For the March/September transition periods, with lower solar altitudes, a movable shading device would likely be more effective. This type of design analysis should be complemented with some type of thermal analysis using either a thermal simulation tool or simpler, balance-point-type spreadsheet methods. The use of the clear sky study from Figure 9.20 suggests that a space is mainly used during daytime such as an office or classroom. For other space types such as places of worship the designer should rather concentrate on daylighting levels on key holidays.

The use of the daylight factor combined with CIE clear sky simulations leads to a more balanced daylighting design as far as solar gains and glare are concerned since the two metrics somewhat work "against each other": Larger windows lead to higher daylight factors but also to more areas that are oversupplied under clear sky conditions. But, this approach has the drawback that specific climate data is not considered (what if it is virtually never sunny in December at the considered building site?). The approach also requires a larger number of individual simulations and a review of multiple figures. The next logical improvement is therefore to use a metric that explicitly considers annual climatic conditions.

Climate-based metrics

Climate-based metrics are derived from the earlier discussed annual illuminance profiles, i.e. hourly or even sub-hourly time series of interior illuminances due to daylight. In order to become usable for design, this massive amount of data has to be converted into an intuitive metric.

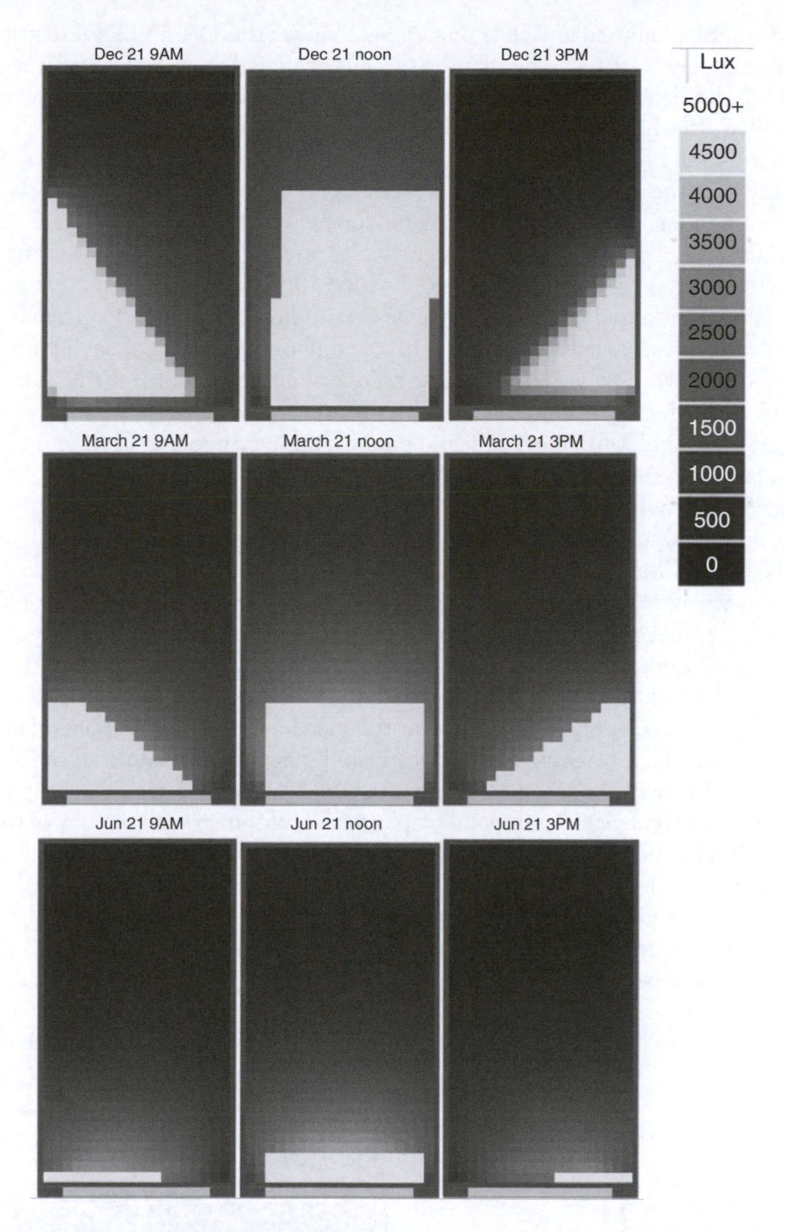

Figure 9.20 Indoor illuminance distribution in the sidelit space from Figure 9.19 under clear sky conditions on solar solstice and equinox days at 9 a.m., noon and 3 p.m.

The first step in this conversion process is to decide on which times of the year the analysis should be based. A common choice is to concentrate on the times when the investigated space will be occupied since daylight "needs witnesses" to have an effect. This approach corresponds to that taken by energy simulation programs that also assume typical usage patterns. An alternative argument of why one might instead concentrate on the daylit hours of the year is that these hours will not change over the lifetime of a building whereas usage patterns might change as warehouses get converted into apartments etc. This "loose fit" approach to building design can be especially attractive for developers who originally might not exactly know how a new building is going to be used.

The next step of the analysis is to decide what daylighting levels to consider "adequate". Here the two currently most commonly used approaches are daylight autonomy (DA) and useful daylight illuminance (UDI). Figure 9.21 shows the DA and UDI distributions for the space from Figure 9.19 assuming that the space is occupied on weekdays between 8 a.m. until 5 p.m. DA is defined as the "the percentage of the occupied hours of the year when a minimum illuminance threshold is met by daylit alone" (Reinhart and Walkenhorst 2001). The metric uses the target illuminance for a space according to IESNA requirements (IESNA 2000). Assuming that the investigated space is an office, a target level of 500 lux was chosen for the DA diagram in Figure 9.21. The DA diagram further distinguishes between three areas, a "daylit area" (white), a partially daylit area (grayscale) and an "over lit" area (red) near the window. The daylit area has a DA value above 46% which corresponds to *half of the maximum DA value* in the space (Reinhart 2005). The grayscale area indicates varying DAs from 0% to 46%. The "over lit" area presents a warning that is invoked if an oversupply of daylight is assumed for at least 5% of the working year. In this case an oversupply is assumed if the illuminance level is above 10 times the target illuminance. This criteria corresponds to the DA_{max} metric proposed by Rogers (Rogers 2006). The DA figure predicts an oversupply of daylight in the third of the space adjacent to the façade.

Figure 9.21 also shows the three related UDI metrics. UDI uses lower and upper thresholds of 100 lux and 2000 lux and accordingly divides the year into three bins (Nabil and Mardaljevic 2005; Nabil and Mardaljevic 2006). The upper bin ($UDI_{>2000lux}$) is meant to represent times when an oversupply of daylight might lead to visual and/or thermal discomfort, the lower bin ($UDI_{<100lux}$) represents times when there is "too little" daylight and the intermediate bin ($UDI_{100–2000lux}$) represents "useful" daylight. The $UDI_{100–2000lux}$ metrics suggests that there is useful daylight in the *back* two thirds of the space whereas – same as the DA_{max} metric – the $UDI_{>2000lux}$ metric flags an oversupply of daylight near the façade. These warnings are plausible given the size of the south-facing glazing of the investigated space (window-to-wall ratio = 50%).

Figure 9.22 shows a way to further investigate the area of daylight oversupply from Figure 9.21 using a *temporal map*. A temporal map is a falsecolor map with the day of the year mapped along the X axis and the time of day along the Y axis (Kleindienst, Bodart and Andersen 2008; Mardaljevic 2004). Figure 20.22 shows a temporal map of the point marked with the black triangle in the daylight autonomy diagram in Figure 9.21(a). The temporal map shows that the horizontal illuminance near the center of the triangle experiences levels above 3500 lux throughout the year from about 10 a.m. to 4 p.m., suggesting that the use of a shading device is required during those times. Since the oversupply happens all year around, the most meaningful design advice is to either reduce the window size or to introduce a movable shading device.

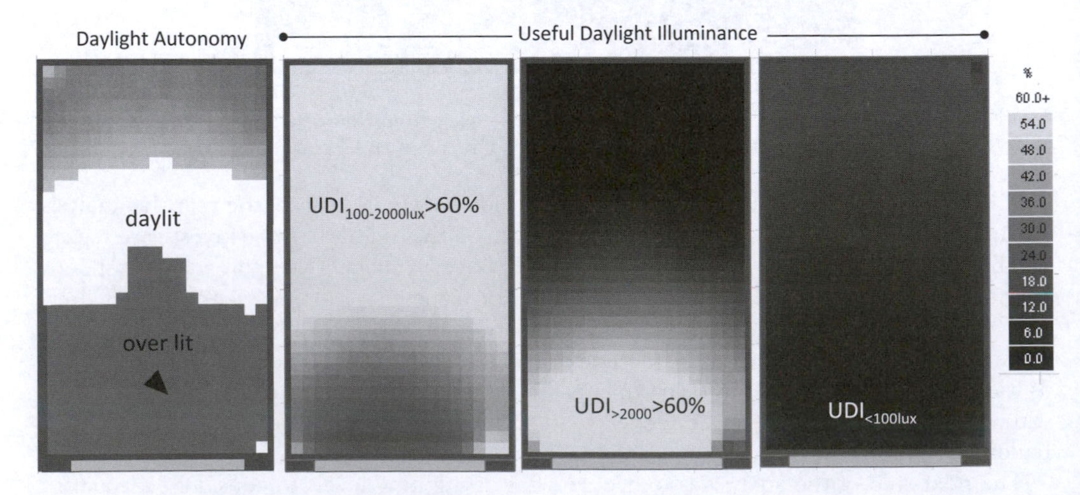

Figure 9.21 DA and three UDI for the sidelit space from Figure 9.19.

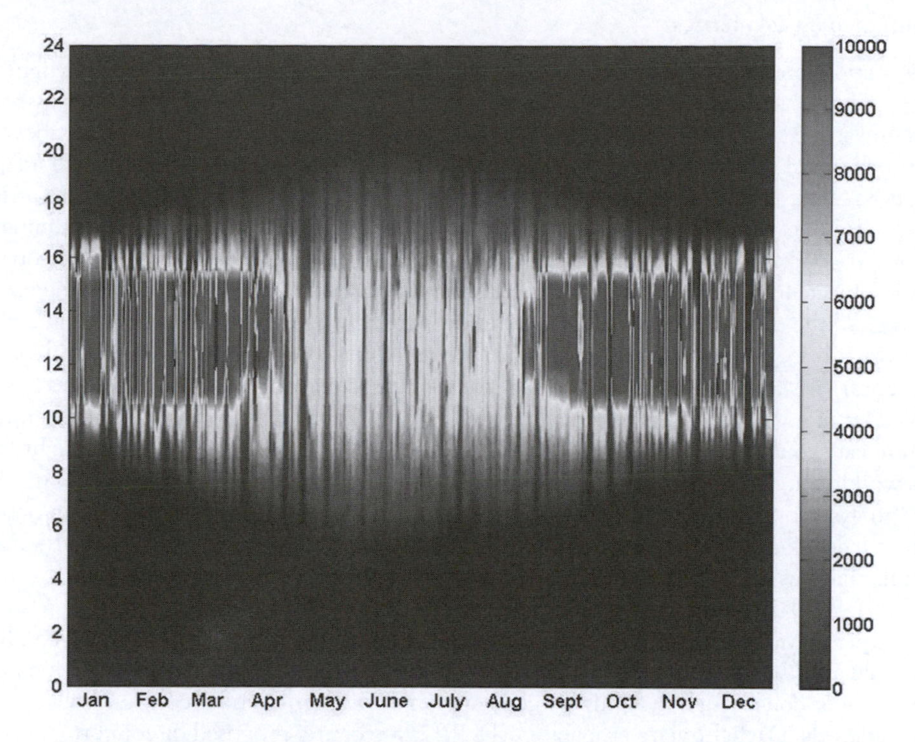

Figure 9.22 Temporal map of the point marked with a triangle in the DA diagram in Figure 9.21

Source: Simulation by S. Kleindienst.

As the name suggests, a temporal map presents performance over time concentrating on one point in a space or a mean of the whole space. In contrast, the earlier discussed DA and UDI metrics concentrate on the spatial patterns of daylight but summarize over the whole year. Temporal and spatial maps should ideally be used in combination. Figures 9.21 and 20.22 show in combination that there is too much daylight near the façade (special map) and that the times oversupply are evenly distributed from 10 a.m. to 4 p.m. throughout the year (temporal map).

Annual light exposure

Before moving on to luminance-based metrics, the concept of annual light exposure is briefly described. Designing spaces for people is challenging because individuals have different requirements and tolerances regarding their visual and thermal environment and due to the uncertainty associated with the use of personal controls (movable shading devices and light switches). Designing spaces for artwork and/or plants instead is in that sense "easier" because requirements and tolerances are better definable. Paintings and trees do also not "tinker" with their environment. As a consequence the use of annual daylight simulations for the placement of artwork and indoor plants is already more established (Franks 2005). Two frequently used design metrics for art galleries are annual light exposure, i.e. the cumulative amount of visible light incident on a point of interest over the course of a year, as well as a maximum illuminance level that may fall on the point at any given time. The International Commission on Illumination (CIE) Division 3 TC3.22 "Museum lighting and protection against radiation damage" recommends upper levels for both metrics for a variety of art forms (International Commission on Illumination (CIE) Division 3 TC-22, 2004). The earlier mentioned cumulative sky methods can be used for annual light exposure calculations whereas annual illuminance profiles are required to verify that the second, maximum illuminance criteria is consistently met.

Luminance-based metrics

This section presents a more refined approach to detect an "oversupply" of daylight in a space. What constitutes an "oversupply" of daylight? Too much daylight is either associated with unwanted solar gains or glare. Glare is a subjective human sensation that describes "light within the field of vision that is brighter than the brightness to which the eyes are adapted" (HarperCollins 2002). To establish whether solar gains are welcome or not at a particular point in time requires the use of a thermal balance analysis ideally using computer simulations (Chapter 5). Glare is typically divided into *disability glare*, which is the inability of a person to see certain objects in a scene due to glare, and *discomfort glare*, the premature tiring of the eyes due to glare. While disability glare in daylit interiors is relatively easy to identify, discomfort glare is a more subtle, subjective phenomena that is closely linked to a person's overall indoor environmental satisfaction. Predicting the appearance of discomfort glare in daylit spaces has been studied for many years. There is a widely shared notion that horizontal work plane illuminance – while easy to measure – is a poor predictor of discomfort glare as the amount of light falling on a working area has little in common with a person's visual experience of a space. Instead, a glare analysis should be based on the luminance distribution in the field of view of an observer. A *glare index* is a numerical evaluation of high dynamic range images using a mathematical formula that has been derived from human subject studies. Example indices include the unified glare rating (UGR) and the daylight glare index (DGI). All of these equations were derived from experiments with artificial glare sources – none of them under real daylight conditions (Wienold and Christoffersen 2006). The reason for this is that until recently it has been next to impossible to collect high dynamic range images of daylit scenes under continuously changing lighting levels. Daylight glare probability (DGP) is a recently proposed discomfort glare index that was derived by Wienold and Christoffersen from laboratory studies in daylit spaces using 72 test subjects in Denmark and Germany. In the experimental setup, two identical, side-by-side test rooms were used. In one of the rooms a CCD camera based luminance mapping technology was installed at the exact position and orientation as the head of the human subject in the other room (Wienold and Christoffersen 2006). DGP is defined by the following equation:

$$DGP = 5.87\times10^{-5}E_v + 9.18\times10^{-5}\log\left(1+\sum_i \frac{L_{s,i}^2\,\omega_{s,i}}{E_v^{1.87}P_i^2}\right) \tag{9.5}$$

where E_v is the vertical eye illuminance; $L_{s,i}$ is the luminance of source 'i'; $\omega_{s,i}$ is the solid angle of source l and P is the Guth position index, a weighing function that varies with the distance of a glare source form the field of vision. Figure 9.23 shows two visualizations and associated DGPs for the workplace and orientation marked by the triangle in the DA diagram in Figure 9.21 under a CIE clear and overcast sky on March 21 at 3 p.m. The associated DGPs for the overcast and clear skies are 22% and 41%, respectively. The values fall into subjective glare ratings of less than "imperceptible" and "disturbing" (Wienold 2009).

In order to evaluate the overall appearance of discomfort glare at a workspace over the course of a year, it becomes necessary to repeat the analysis from Figure 9.23 for all hours of the year using site-specific climate data. Wienold proposed a simplified method that uses daylight coefficient based vertical eye illuminance and simplified images in order to calculate annual DPG profiles more efficiently than running one full scene visualization at each time step (Wienold 2009). As an example, Figure 9.24 shows the cumulative annual DGP distribution for the workspace from Figure 9.23 assuming working hours on weekdays from 8 a.m. to 5 p.m. using Wienold's *enhanced simplified DGP* method. In order to guide the eye the figure shows the DGP ranges in which human subjects rated the glare within their field of view to be imperceptible, perceptible, disturbing and intolerable (Wienold 2009).

The figure reveals that the DGP value at the workplace is 1312 working hours (50% of the occupied time) in the "disturbing" or worse and 1015 working hours (39% of the occupied time)

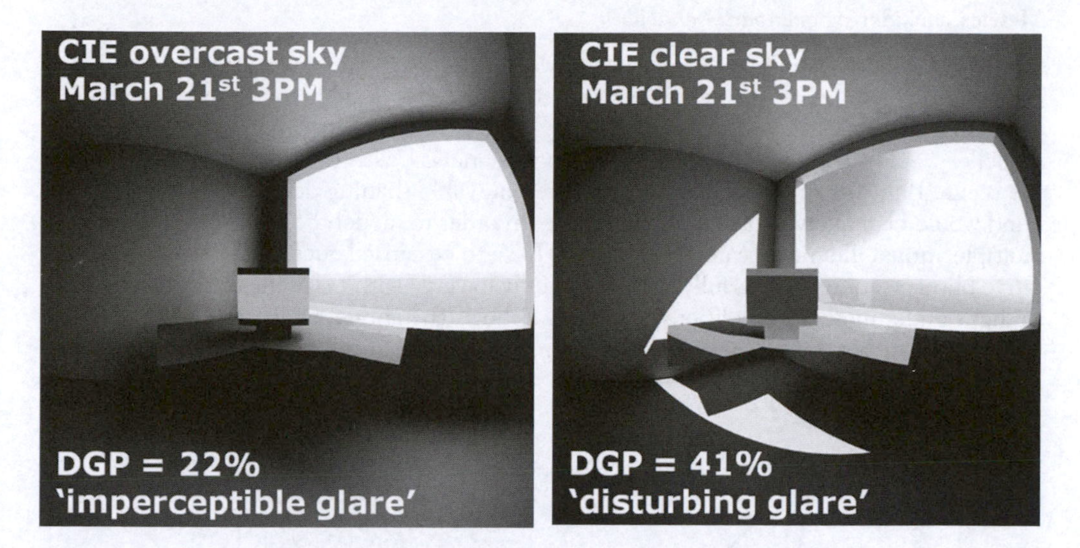

Figure 9.23 Radiance visualizations and associated DPGs for the workspace marked by the triangle on Figure 9.21(a) on March 21 at 3 p.m. under a CIE clear (a) and overcast (b) sky.

in the "intolerable" glare range suggesting that the workplace has unacceptable visual comfort conditions. The analysis shows that annual DGP profiles present a powerful way of evaluating the appearance of discomfort glare in a space. Form a simulation workflow standpoint this type of analysis can be fully automated and requires minimal additional simulation effort on behalf of the modeler since the only required additional input are one or several representative viewpoints throughout a scene. The simulation time required to calculate the annual DGP profile for Figure 9.24 was about an hour on a standard laptop. With venetian blinds (next section) the calculation ran for an additional one and a half hours.

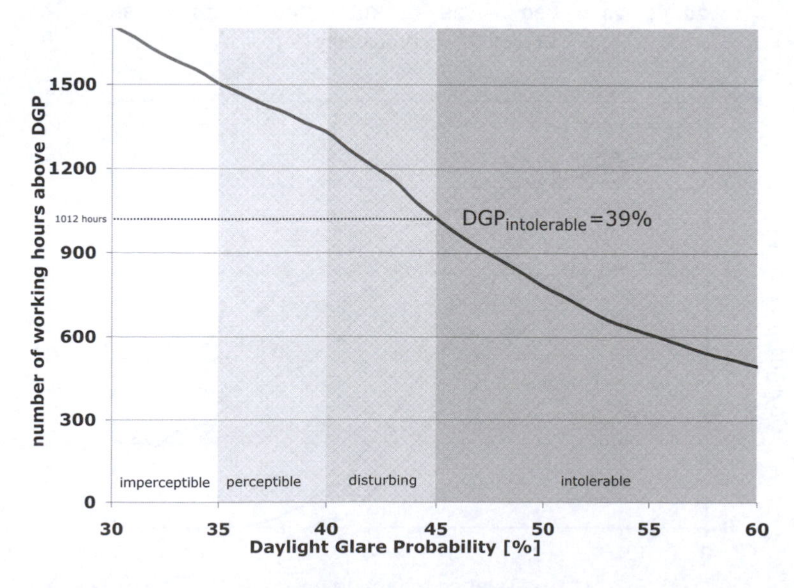

Figure 9.24 Annual cumulated DGP profile the workspace marked by the triangle on Figure 9.21(a) for the Boston climate assuming working hours on weekdays from 8 a.m. to 5 p.m.

Source: Simulation by J. Wienold.

Metrics considering occupant behavior

It has by now been established in multiple ways that the space from Figure 9.19 – if used as an office with the occupant sitting in vicinity to the window – would require the use of a glare protection device. If the use of a static shading device (or a smaller window) was desired, the modeller would simply have to repeat the previous analysis with the shading device added to the scene. But, in real life, a manually operated movable shading device such as a venetian blind would be a likely choice for such a space. In order to model the effect of such a system, multiple annual illuminance and DGP profiles have to be carried out for at least two representative blind positions such as fully opened and fully lowered with a slat angle that blocks direct sunlight under most sky conditions. Figure 9.25 shows the annual DGP profiles and daylight autonomy distributions on the center axis of the space for a variety of different blind usage pat-

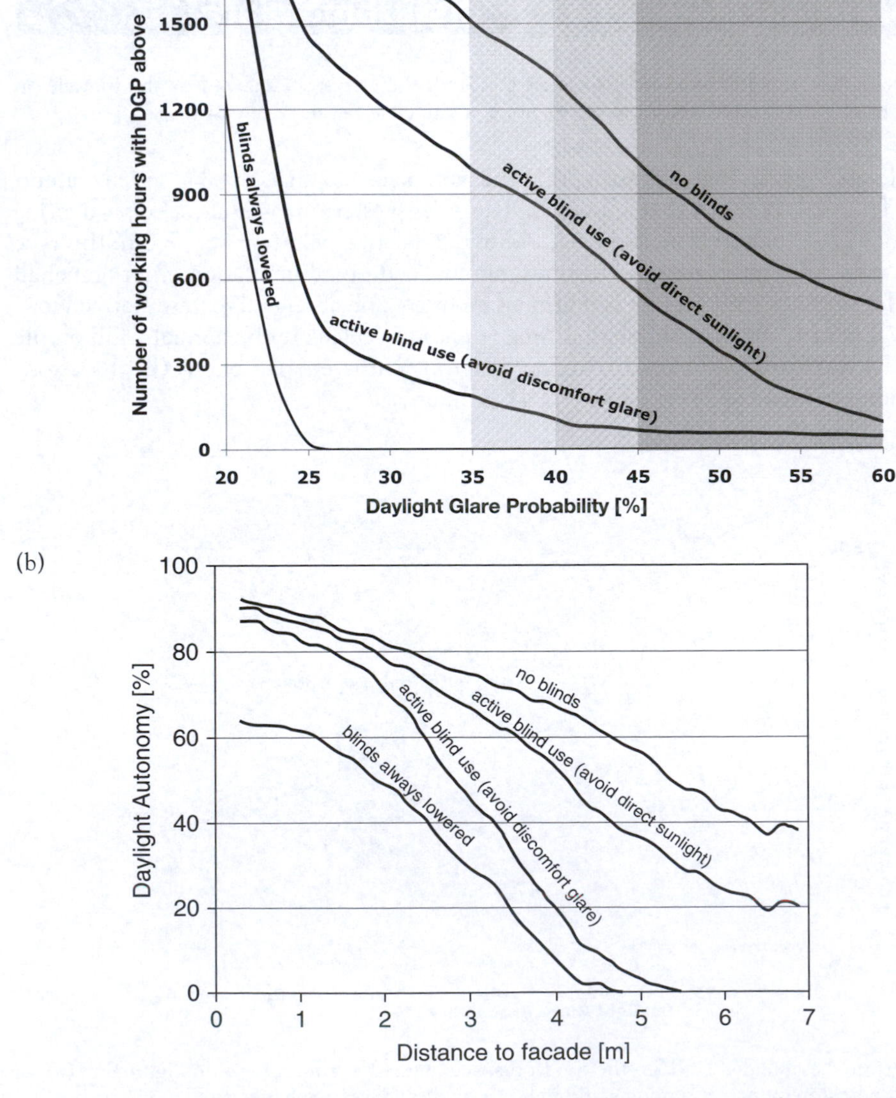

Figure 9.25 Annual DGP profiles (a) and DA distributions (b) for the space from Figure 9.19 assuming different blind control strategies.

terns. The "blinds always opened" and "passive user (blinds always lowered)" scenarios show the magnitude of the simulation uncertainty introduced by the occupant. The "true" occupant behavior is likely to lie somewhere in between these two extremes. In order to get a better estimate, an "occupant behavior model" such as Lightswitch 2002 can be used. The Lightswitch model was derived from field study data that monitored long term occupancy and use of light switches and shading devices in private and two-person offices (Reinhart 2002). Lightswitch accordingly mimics how occupants interact with manually and automatically operated lighting controls and shading devices and uses occupancy, work plane illuminance and the appearance of direct sunlight at the workspace as input data. A key assumption of the model is that occupant behavior can be described through two archetypical behavior patterns called "active" and "passive" users. For the active user an expanded version of the original model further distinguishes between a user who closes the blinds for the day once direct sunlight above 50Wm^{-2} is incident on the work space ("active blind use (avoid direct sunlight)") and a user who closes the blinds for the day once the DGP at the workplace becomes "disturbing" (>40%). Both active users raise their blinds once a day during arrival. In contrast, a passive user keeps the blinds lowered all year.

Figure 9.25(a) shows that the presence of a venetian blind reduces the number of hours with "intolerable" DGP to 553, 66 and no hours for the two active and the passive users, respectively, revealing that a venetian blind can alleviate visual discomfort in the space, if operated accordingly by the user. These hour values translate into occupancy rates of 21%, 3% and 0% of the year. Figure 9.25(b) reveals that the addition of venetian blinds places the daylit area (half of the maximum DA within the space) to about 3m to 4m away from the façade for all occupant behavior patterns.

In closing it is worthwhile noting that the presence of a manually operated shading device reduces the lighting levels throughout a space to the position where the occupant operating the device is usually located. This can lead to situations during which discomfort glare near a façade leads to the closing of venetian blinds which in turn results in an undersupply of daylight for another occupant further away from the façade. The resulting group dynamics between occupants related to the use of personal controls is not well understood, yet. There is also little known about how occupants operate multiple sets of blinds in a space or how many occupants in a building a designer can expect to be either active or passive users. Occupant behavior models – such as Lightswitch – are further based on data collected from a very limited number of human subjects and buildings and hence need further validation through more field work. Despite of these shortcomings, the author believes that not considering the use of blinds at all leads to misleading design interpretations: The reader might consider the UDI diagrams from Figure 9.21. There seems to be plenty of usable daylight in the back of the space paired with an oversupply of daylight near the façade. But, if used as an office the space would not be operated this way but instead a venetian blind system would be used that provides good daylight near the façade but insufficient daylight in the back (Figure 9.25). The design conclusions are exactly opposite.

Discussion

This chapter has presented a series of tips and insights into how to properly set up a daylight simulation and provided some examples of how the simulation outcomes can be transferred into effective daylight performance metrics. This last section discusses the relationship between daylight simulations and rules of thumb, physical models, performance benchmarks and electric lighting simulations.

Rules of thumb

The use of daylight simulations has increasingly gone main stream for a number of years and this chapter aims to further this trend by providing novice users with information of how to get started. Many of the metrics introduced in the previous section can only be calculated using computer simulations. But, while building design is getting increasingly digitized with electronic drawings becoming required construction documents in many jurisdictions, early design decisions are still commonly based on rules of thumb (Figure 9.1). Electronic documents are typically generated at later design stages. It is therefore important to better link rules of thumb and simulations so that designers know when to "naturally" switch from one analysis method to the other. One recent approach to bring rules of thumb and simulations closer together was to validate a series of interconnected rules of thumb based on simulations making the predictions of both methods more complementary (Reinhart and LoVerso 2010). The above noted use of the Lynes formula (Eqn. 9.4) as a plausibility check for simulations is also something for the daylight modeller to consider.

Physical models

Physical or scale models of buildings can be analyzed inside using either a heliodome (artificial sun) or an artificial sky (emulated overcast sky) or outside under real sky conditions. Typical techniques for analyzing physical models are through direct observation, photography and/or video animations and illuminance measurements. The use of physical models for daylighting design generally seems to be on a downward trend if findings from recent surveys can be trusted (Galasiu and Reinhart 2007; Reinhart and Fitz 2006). On the other hand, it is the author's experience that advanced daylighting consultants go back and forth between scale models and simulations using the former when material imperfections and visual experience become important.[7] When one compares measurements under real sky conditions in a scale model and the corresponding full size space, the errors introduced by the scale model tend to be comparable in magnitude to those introduced by a daylight simulation (Cannon-Brookes 1997). If one on the other hand uses a scale model under an under an artificial sky dome, a theoretical argument presented by Mardaljevic shows that "on the basis of a credible design goal for the sky simulator dome, high accuracy illuminance predictions (±10%) are practically unattainable" due to parallax errors (Mardaljevic 2002). Parallax errors are measurement inaccuracies that appear under artificial skies since the actual, physical light sources of the artificial sky structure are not infinitely far away from the physical model which leads to within-model overshadowing of parts of the artificial sky.

The author believes that there is a very valid place for the use of physical models in daylighting design, especially in schools of architecture which anyhow tend to be producers of physical models for other purposes. For many designers physical models are clearly more engaging than a two dimensional model on a computer screen. But, the author is cautioning the users of physical models not be lured into thinking that their measurements using these models are particularly accurate.

Need for performance benchmarks

A number of exciting new climate-based daylighting analysis methods have been introduced in this chapter. A key requirement to make these metrics more useful for building design is that performance benchmarks have to be established for these metrics, similar to the widely used 2% daylight factor target level. Such target levels are invaluable for both designers who work with these metrics as well as the developers of building standards who can reference them. A number of pilot studies have presented benchmark data for different climate-based metrics: Rogers presented an initial rating scheme for classrooms based on continuous DA (Rogers 2006); Wienold suggested glare classifications for shading devices based on annual DGP profiles (Wienold 2009); Mardaljevic presented UDI calculations in residential dwellings with and without skylights (Mardaljevic 2008) and Kwan and Reinhart proposed a rating system for daylit gymnasia based on IESNA requirements. These initial studies still need to be supplemented with more extensive work.

Electric lighting

Modeling electric light and daylight is largely the same process and a number of mature programs are available to simulate most commercially available luminaires. The light emitted by an electric light source is called the light source's luminous intensity distribution. There are several commonly available photometric file formats to describe this distribution with the IES format by the IESNA being the predominantly used file at least in North America (ANSI/IES_LM-63–02 2002). Most manufacturers provide IES files for all of their lighting products. The underlying data is mostly based on measurements but several luminaire design software packages also write out IES files. Compared to daylight, electric lighting is somewhat "easier" to model since lighting output is constant or scales linearly (for dimmed systems). Electric lighting and daylight can be easily combined in a simulation by defining both types of light sources in a scene. One remaining challenge is to accurately model the interaction between a photocell controlled dimming system and the light in a space which is usually a mixture of daylight and electric light with varying spectral compositions. Most daylighting and energy simulation programs currently predict dimming levels based on work plane illuminances essentially assuming a "perfectly commissioned" dimming system that ideally dims the electric lighting so that a desired target illuminance is maintained on the work plane. There are a number on inaccuracies associated with this idealization since real photocells tend to be located on a wall or ceiling and made out of inexpensive, not cosine-corrected sensors that cannot even accurately measure illuminances (Bierman 2007). Several researchers have presented methods of how to better model the true performance of a dimming system including modeling the actual special sensitivity of a photo sensor (Ehrlich *et al.* 2002) and using simulations to identify sensor positions that yield a optimized correlation between the electronic signal at the photosensor and the illuminance at the target level over the course of the year (Rogers, Scheib and Fitzpatrick 2009).

Outlook

During a time of increasingly faster computers and more integrated simulation programs the use of daylight simulations for the design of buildings and neighborhoods is likely to keep on increasing. We are already at a point at which the use of existing programs can add considerable value to the design process and requires only moderate effort on behalf of the modeler, especially if optimized design workflows are repeatedly used.

What type of practical barriers is the daylight simulation community still facing? One recurring problem is for modelers to find reliable material definitions for non-standard elements especially complex fenestration systems. Apart from a lack of optical data for CFS, reliably modeling techniques remain underdeveloped and requires expert simulation knowledge.

Occupant use of personal controls are known to have a decisive impact on the success of a daylighting concept but related occupant behavior models need to be expanded and based on more field data. The simulation uncertainty introduced by occupant behavior is particularly significant for daylight simulations but is also increasingly becoming a liability for energy modelers working on low-energy buildings (see Chapter 4).

There is still a need for better daylighting metrics that go beyond daylight availability and visual comfort and also quantify effects such a view to the outside, privacy, occupant well being, alertness etc. As the metrics are developed, suitable benchmark levels have to be identified as well.

Finally, the author is detecting a strong trend in current building design towards parametric design, i.e. the use of multiple, computer generated design variants in order to identify a solution that best suits multiple design needs. A new brand of building performance simulation called "animated building performance simulation" (ABPS) is emerging and likely going to influence BPS in the years to come (Lagios, Reinhart and Niemasz 2010).

In closing, it is worth pointing out that while designers now have access to advanced numerical analysis methods that predict annual daylight availability, energy use and the likelihood of discomfort, little progress has been made to describe more elusive lighting qualities of a space such as the "sparkle" that makes some spaces more desirably than others.

Synopsis

If carried out properly, validated daylight simulation programs are nowadays capable of correctly simulating annual daylighting levels in daylit spaces equipped with clear glazings, non-specular venetian blinds and lightshelves as well as translucent panels. The resulting information can be (a) combined with occupant behavior models that mimic occupant use of movable shading devices and lighting controls and (b) used to predict the availability of daylight throughout the year, the likeliness occupants experiencing discomfort glare and the amount energy savings due to automated controls. Further research is needed to adequately translate the results form daylight simulations into adequate design metrics and increase number of the complex fenestration systems that can be reliably modeled.

Notes

1 As a side note, while modeling a fully glazed façade explicitly, e.g. as a "mirror" material in Radiance may be required for a glare analysis, this added realism often comes at a price as far as the required simulation time and model handling are concerned and might not be required if work plane illuminances are calculated instead.
2 The weight assigned to a single ray usually depends on the optical properties of the material(s) from which it has been reflected before as well as the number of reflections since the primary ray has been emitted from a sensor or viewpoint.
3 A caveat of the study – as stated by the authors – is that invitations to participate in the survey (although they were disseminated as widely as possible) were somewhat skewed towards Radiance users due to the simple fact that the study authors are part of that community and therefore yielded a high feedback rate on the survey.
4 The two outliers in 2006 (models numbers 17 and 25) were ignored in the calculation.
5 The "glazing factor" promoted by the earlier mentioned LEED green building rating system calculates a quantity that is in essence identical to the mean daylight factor distribution in a space: Reinhart, C.F. and LoVerso, V.M. (2010) "A rules of thumb-based design sequence for diffuse daylight", *Lighting Research & Technology* 42: 17–32.
6 Note that a target daylight factor level of 2% anywhere in the space differs from a mean daylight factor criteria of 2% for a whole space.
7 Private communication C. Human (Loisos + Ubbelohde) and K. Yancey (Lam Partners).

References

Aizlewood, M.E. (1993) "Innovative Daylighting Systems: An Experimental Evaluation", *Lighting Research and Technology* 25(4): 141–152.

Andersen, M. and deBoer, J. (2006) "Goniophotometry and Assessment of Bidirectional Photometric Properties of Complex Fenestration Systems", *Energy and Buildings* 38(7): 836–848.

ANSI/IES_LM-63–02 (2002) "ANSI Approved Standard File Format for Electronic Transfer of Photometric Data and Related Information", *Lighting Measurement Standard (LM)*, New York: Illuminating Engineering Society of North America, pp. 1–26.

ASHRAE (2007a) "ANSI/ASHRAE Standard 140–2007 Standard Method of Test for the Evaluation of Building Energy Analysis Computer Programs", American Society of Heating, Refrigerating and Air-Conditioning Engineers.

ASHRAE (2007b) *ASHRAE Handbook of Fundamentals*, Atlanta, GA: American Society of Heating, Refrigerating and Air-Conditioning Engineers.

Bierman, A. (2007) "Photosensors – Dimming and Switching Systems for Daylight Harvesting", *Specifier Reports*, Troy, NY: National Lighting Product Information Program (NLPIP).

Bourgeois, D., MacDonald, I.A. and Reinhart, C.F. (2006) "Adding Advanced Behavioral Models in

Whole Building Energy Simulation: A Study on the Total Energy Impact of Manual and Automated Lighting Control", *Energy and Buildings* 38(7): 814–823.

Bourgeois, D., Reinhart, C.F. and Ward, G. (2008) "A Standard Daylight Coefficient Model for Dynamic Daylighting Simulations", *Building Research and Information* 36(1): 68–82.

British Standards Institution (1992) "Lighting for Buildings – Part 2: Code of Practice for Daylighting", *British Standard* BS8206, London: BSI.

Cannon-Brookes, S.W.A. (1997) "Simple Scale Models for Daylighting Design: Analysis of Error Illuminance Prediction", *Lighting Research and Technology* 29(3): 135–142.

CIE (1973) "Standardization of Luminance Distribution on Clear Skies", Vienna, Austria, Publication CIE No. 22 (TC-4.21).

CIE (2004) "ISO 15469/ CIE S 011/E Spatial Distribution of Daylight – CIE Standard General Sky", 2nd ed., Geneva, Switzerland: ISO/CIE.

CIE (2006) "CIE 171 – Test Cases to Assess the Accuracy of Lighting Computer Programs", Vienna, Austria: International Commission on Illumination (CIE).

Clarke, J. and Janak, M. (1998) "Simulating the Thermal Effects of Daylight-controlled Lighting", *Building Performance* (BEPAC journal) 1.

Cohen, M.F. and Wallace, J.R. (1993) *Radiosity and Realistic Image Synthesis*, 1st ed., San Diego, CA: Academic Press Professional.

Compagnon, R. (2004) "Solar and Daylight Availability in the Urban Fabric", *Energy and Buildings* 36(4): 321–328.

Covey, S.R. (1989) *The 7 Habits of Highly Effective People*, New York: Fireside.

Crawley, D.B., Hand, J.W. and Lawrie, L.K. (1999) "Improving the Weather Information Available to Simulation Programs", in *Proceedings of the Sixth International IBPSA Conference (BS '99)*, Kyoto, Japan, pp. 529–536.

Duffie, J.A. and Beckman, W.A. (1991) *Solar Engineering of Thermal Processes*, New York: John Wiley & Sons.

Dufton, A.F. (1946) "Protractors for the Computation of Daylight Factors", *D.S.I.R. Building Research Technical Paper* No. 28, London: Her Majesty's Stationery Office.

Ehrlich, C., Papamichael, K., Lai, J. and Revzan, K. (2002) "A Method for Simulating the Performance of Photosensor-based Lighting Controls", *Energy and Buildings* 34(9): 883–889.

Franks, M. (2005) "Daylighting in Museums", *International Radiance Workshops* (www.radiance-online.org), last accessed January 2009.

Galasiu, A. and Reinhart, C.F. (2007) "Current Daylighting Design Practice: A Survey", *Building Research and Information* 36(2): 159–174.

HarperCollins (2002) *Collins Thesaurus of the English Language – Complete and Unabridged*, 2nd ed., HarperCollins Publishers.

Hoes, P., Hensen, J.L.M., Loomans, M.G.L.C., de Vries, B and Bourgeois, D. (2009) "User Behavior in Whole Building Simulation", *Energy and Buildings* 41(3): 295–302.

Ibarra, D. and Reinhart, C.F. (2009) "Daylight Factor Simulations – How Close Do Simulation Beginners 'Really' Get?", in *Proceedings of Building Simulation 2009*, Glasgow, Scotland, pp. 196–203.

IESNA (2000) *IESNA Lighting Handbook*, 9th ed., New York: Illuminating Engineering Society of North America,

Inanici, M. (2006) "Evaluation of High Dynamic Range Photography as a Luminance Data Acquisition System", *Lighting Research and Technology* 38(2): 123–136.

Inanici, M. (2009) "Applications of Image Based Rendering in Lighting Simulation: Development and Evaluation of Image Based Sky Models", in *Proceedings of Building Simulation 2009*, Glasgow, Scotland, pp. 264–271.

Janak, M. (1997) "Coupling Building and Lighting Simulation", in *Proceedings of the 5th International IBPSA Conference*, Prague, Czech Republic, pp. 313–319.

Jarvis, D. and Donn, M. (1997) "Comparison of Computer and Model Simulations of a Daylit Interior with Reality", in *Proceedings of the 5th International IBPSA Conference*, Prague, Czech Republic.

Kittler, R., Perez, R. and Darula, S. (1997) "A New Generation of Sky Standards", *Proceedings of the Lux Europa Conference*, Amsterdam: Netherlands Institution of Illuminating Engineering.

Kittler, R., Perez, R. and Darula, S. (1998) "A Set of Standard Skies Characterizing Daylight Conditions for Computer and Energy Conscious Design", US SK 92 052 Final Report, ICA SAS Bratislava.

Kleindienst, S., Bodart, M. and Andersen, M. (2008) "Graphical Representation of Climate-Based Daylight Performance to Support Architectural Design", *Leukos* 5(1): 39–61.

Lagios, K., Reinhart, C.F. and Niemasz, J. (submitted) "Animated Building Performance Simulation (ABPS) – Linking Rhinoceros/Grasshopper with Radiance/Daysim", *Proceedings of BuildSim 2010*, New York City, August 2010.

Laouadi, A., Reinhart, C.F. and Bourgeois, D. (2008) "Efficient Calculation of Daylight Coefficients for Rooms with Dissimilar Complex Fenestration Systems", *Journal of Building Performance Simulation* 1(1): 3–15.

LBNL (last accessed in November 2009) "Windows and Daylighting >> Software Tools", Lawrence Berkeley National Laboratory, Website: http://windows.lbl.gov.

Leder, S.M., Pereira, F.O.R. and Moraes, L.N. (2007) "Characterization of a Medium Reflection Coefficient to Urban Vertical Surfaces" (in Portuguese), in *Encontro Nacional e Encontro Latino Americano de Conforto no Ambiente Construído Ouro* Preto, Brazil.

Lynes, J. (1979) "A Sequence for Daylighting Design", *Lighting Research and Technology* 11(2): 102–106.

Mahdavi, A., Mohammadi, A., Kabir, E. and Lambeva, L. (2008) "Shading and Lighting Operation in Office Buildings in Austria: A Study of User Control Behavior", *Building Simulation* 1(2): 111–117.

Mardaljevic, J. (1995) "Validation of a Lighting Simulation Program under Real Sky Conditions", *Lighting Research and Technology* 27(4): 181–188.

Mardaljevic, J. (2000) "Daylight Simulation: Validation, Sky Models and Daylight Coefficients", PhD Thesis, De Montfort University, Leicester, UK.

Mardaljevic, J. (2002) "Quantification of Parallax Errors in Sky Simulator Domes for Clear Sky Conditions", *Lighting Research & Technology* 34(4): 313–332.

Mardaljevic, J. (2004) "Spatio-temporal Dynamics of Solar Shading for a Parametrically Defined Roof System", *Energy and Buildings* 36(8): 815–823.

Mardaljevic, J. (2008) *Climate-Based Daylight Analysis for Residential Buildings*, Technical Report (www.thedaylightsite.com/), IESD, Leicester, De Montfort University.

Mardaljevic, J. and Rylatt, M. (2003) "Irradiation Mapping of Complex Urban Environments: An Image-based Approach", *Energy and Buildings* 35(1): 27–35.

Mardaljevic, J., Heschong, L. and Lee, E. (2009) "Daylight Metrics and Energy Savings", *Lighting Research and Technology* 41(3): 261–283.

Marsh, A. (last accessed September 2009) Autodesk Ecotect (version 5.6), Autodesk Inc.

mental-images (2007) Mental Ray(R) Functional Overview Version 1.5, <http://www.mentalimages.com/fileadmin/user_upload/PDF/mental_ray_Functional_Overview.pdf> Berlin.

Moon, P. and Spencer, D.E. (1942) "Illumination from a Non-uniform Sky", *The Illuminating Engineer* 37(10): 707–726.

Nabil, A. and Mardaljevic, J. (2005) "Useful Daylight Illuminance: A New Paradigm to Access Daylight in Buildings", *Lighting Research and Technology* 37(1): 41–59.

Nabil, A. and Mardaljevic, J. (2006) "Useful Daylight Illuminances: A Replacement for Daylight Factors", *Energy and Buildings* 38(7): 905–913.

Newsham, G.R. (1994) "Manual Control of Window Blinds and Electric Lighting: Implications for Comfort and Energy Consumption", *Indoor Environment* 3(3): 135–144.

Olseth, J.A. and Skartveit, A (1989) "Observed and Modeled Hourly Luminous Efficacies under Arbitrary Cloudiness", *Solar Energy* 42(3): 221–233.

Pechacek, C.S., Andersen, M. and Lockley, S.W. (2008) "Preliminary Method for Prospective Analysis of the Circadian Efficacy of (Day)Light with Applications to Healthcare Architecture", *Leukos* 5(1): 1–26.

Perez, R., Ineichen, P., Seals, R., Michalsky, J. and Stewart, R. (1990) "Modeling Daylight Availability and Irradiance Components from Direct and Global Irradiance", *Solar Energy* 44(5): 271–289.

Perez, R., Seals, R. and Michalsky, J. (1993) "All-Weather Model for Sky Luminance Distribution – Preliminary Configuration and Validation", *Solar Energy* 50(3): 235–245.

Preetham, A.J., Shirley, P. and Smits, B. (1999) "A Practical Analytic Model for Daylight", in *Proceedings of the International Conference on Computer Graphics and Interactive Techniques*, pp. 91–100.

Reindl, D.T., Beckman, W.A. and Duffie, J.A. (1990) "Diffuse Fraction Correlations", *Solar Energy* 45(1): 1–7.

Reinhart, C.F. (2004) "LIGHTSWITCH 2002: A Model for Manual and Automated Control of Electric Lighting and Blinds", *Solar Energy* 77(1): 15–28.

Reinhart, C.F. (2005) "A Simulation-based Review of the Ubiquitous Window-head-height to Daylit Zone Depth Rule of Thumb", in *Proceedings of Building Simulation 2005*, Montreal, Canada, pp. 1011–1018.

Reinhart, C.F. and Andersen, M. (2006) "Development and Validation of a Radiance Model for a Translucent Panel", *Energy and Buildings* 38(7): 890–904.

Reinhart, C.F. and Breton, P.-F. (2009) "Experimental Validation of Autodesk 3ds Max® Design 2009 and Daysim 3.0.", *Leukos* 6(1).

Reinhart, C.F. and Fitz, A. (2006) "Findings from a Survey on the Current Use of Daylight Simulations in Building Design", *Energy and Buildings* 38(7): 824–835.

Reinhart, C.F. and Herkel, S. (2000) "The Simulation of Annual Daylight Illuminance Distributions – A State of the Art Comparison of Six RADIANCE-based Methods", *Energy and Buildings* 32(2): 167–187.

Reinhart, C.F. and LoVerso, V.M. (2010) "A Rules of Thumb Based Design Sequence for Diffuse Daylight", *Lighting Research and Technology* 42(1): 7–32.

Reinhart, C.F. and Walkenhorst, O. (2001) "Dynamic RADIANCE-based Daylight Simulations for a Full-scale Test Office with Outer Venetian Blinds", *Energy and Buildings* 33(7): 683–697.

Reinhart, C.F., Mardaljevic, J. and Rogers, Z. (2006) "Dynamic Daylight Performance Metrics for Sustainable Building Design", *Leukos* 3(1): 1–20.

Robinson, D. and Stone, A. (2004) "Irradiation Modeling Made Simple: The Cumulative Sky Approach and its Applications", in *Proceedings of PLEA 2004 – The 21st Conference on Passive Low Energy Architecture*, Eindhoven, The Netherlands, pp. 1–5.

Rogers, Z. (2006) "Daylighting Metric Development Using Daylight Autonomy Calculations in the Sensor Placement Optimization Tool", Boulder, Colorado, USA: Architectural Energy Corporation, http://www.archenergy.com/SPOT/download.html.

Rogers, Z., Scheib, J. and H. Fitzpatrick (2009) SPOT (Sensor Placement and Optimization Tool) Users Manual ver. 4. http://www.archenergy.com/SPOT/support.html, Architectural Energy Corporation.

Schregle, R. (2004) "Daylight Simulation with Photon Maps", Dr. Ing. Thesis, Universität des Saarlandes, Saarbrücken, Germany.

Szokolay, S.V. (2004) *Introduction to Architectural Science*, Oxford, UK: Elsevier Ltd.

Tregenza, P.R. (1983) "The Monte Carlo Method in Lighting Calculations", *Lighting Research and Technology* 15(4): 163–170.

Tregenza, P.R. (1987) "Subdivision of the Sky Hemisphere for Luminance Measurements", *Lighting Research and Technology* 19(1): 13–14.

Tregenza, P.R. and Waters, I.M. (1983) "Daylight Coefficients", *Lighting Research and Technology* 15(2): 65–71.

Tsangrassoulis, A., Santamouris, M. and Asimakopoulos, D. (1996) "Theoretical and Experimental Analysis of Daylighting Performance for Various Shading Systems", *Energy and Buildings* 24(3): 223 –230.

US-DOE "Building Energy Software Tools Directory", retrieved 2008, from www.eere.energy.gov/buildings/tools_directory/.

US-DOE (2009) "EnergyPlus Weather Data", last accessed September 2009, from http://apps1.eere.energy.gov/buildings/energyplus/cfm/Weather_data.cfm.

USGBC (2006) "LEED-NC (Leadership in Energy and Environmental Design) Version 2.2.", from www.usgbc.org/LEED/.

Waldram, P. J. (1950) *A Measuring Diagram for Daylight Illumination*, London: B.T. Batsford Ltd.

Walkenhorst, O., Luther, J., Reinhart, C.F. and Timmer, J. (2002) "Dynamic Annual Daylight Simulations Based on One-hour and One-minute Means of Irradiance Data", *Solar Energy* 72(5): 385–395.

Ward, G. (2009) "Complex Fenestration and Annual Simulation", 8th International Radiance Workshop (www.gsd.harvard.edu/research/gsdsquare/RadianceWorkshop2009.html). Cambridge, MA, USA.

Ward, G. (last accessed in May 2009) "Photosphere Software Version 1.8 by Software Anyhere".

Ward, G. and Shakespeare, R. (1998) *Rendering with RADIANCE – The Art and Science of Lighting Visualization*, San Francisco, CA: Morgan Kaufmann Publishers.

Ward, G.J. (1994) "The RADIANCE Lighting Simulation and Rendering System: Computer Graphics", in *Proceedings of the 1994 SIGGRAPH Conference*, pp. 459–472.

Wasilowski, H.A. and Reinhart, C.F. (2009) "Modeling an Existing Building in DesignBuilder/EnergyPlus: Custom versus Default Inputs", in *Proceedings of Building Simulation 2009*, Glasgow, Scotland, pp. 1252–1259.

Webb, A.R. (2006) "Considerations for Lighting in the Built Environment: Non-visual Effects of Light", *Energy and Buildings* 38(7): 721–727.

Wienold, J. (2007) "Dynamic Simulation of Blind Control Strategies for Visual Comfort and Energy Balance Analysis", in *Proceedings of Building Simulation 2007*, Beijing, China, pp. 1197–1204.

Wienold, J. (2009) "Dynamic Daylight Glare Evaluation", in *Proceedings of Building Simulation 2009*, Glasgow, Scotland, pp. 944–951.

Wienold, J. and Christoffersen, J. (2006) "Evaluation Methods and Development of a New Glare Prediction Method for Daylight Environments with the Use of CCD Cameras", *Energy and Buildings* 38(7): 743–757.

Winkelmann, F.C. and Selkowitz, S. (1985) "Daylight Simulation in the DOE-2 Building Energy Analysis Program", Energy and Buildings 8(4): 271–286.

Recommended reading

Lam, W.M.C. (1986) *Sunlighting as Formgiver for Architecture*, New York: Van Nostrand Reinhold.

O'Connor, J., Lee, E., Rubinstein, F. and Selkowitz, S. (1997) "Tips for Daylighting with Windows", Lawrence Berkeley National Laboratory, Report # LBNL-39945.

Ward, G. and Shakespeare, R. (1998) *Rendering with RADIANCE – The Art and Science of Lighting Visualization*, San Francisco, CA: Morgan Kaufmann Publishers.

Activities

The purpose of the following assignments is to practice the use of manual and computer-based calculation methods to estimate the mean daylight factor and annual daylight autonomy distribution in a rectangular, sidelit space (Figure 9.26). The space is 5m wide, 6m deep and 3m high (interior measures). The space is daylit by a 4m wide and 1.5 m high window in the south wall with a visual transmittance of 72%. The wall thickness is 0.3m and interior ceiling, walls and floor and diffuse reflectances of 80%, 50% and 20%, respectively. The space is obstructed by an 11.5m high building that lies 25m south of the space. All exterior walls have a 44% reflectance. The ground has a reflectance of 20%. The space has a target illuminance of 500 lux and is occupied on weekdays from 8 a.m. to 5 p.m. It is located in Boston, Massachusetts.

Assignment 1:

Calculate the mean daylight factor of the space using the Lynes formula (Equation (9.4)).

Assignment 2:

Calculate the daylight factor distribution and mean daylight factor of the space using a validated simulation program of your choice. Rotate the scene by 180° and repeat the calculation.

Assignment 3:

Calculate the daylight autonomy distribution of the space using a validated simulation program of your choice. Rotate the scene by 180° and repeat the calculation.

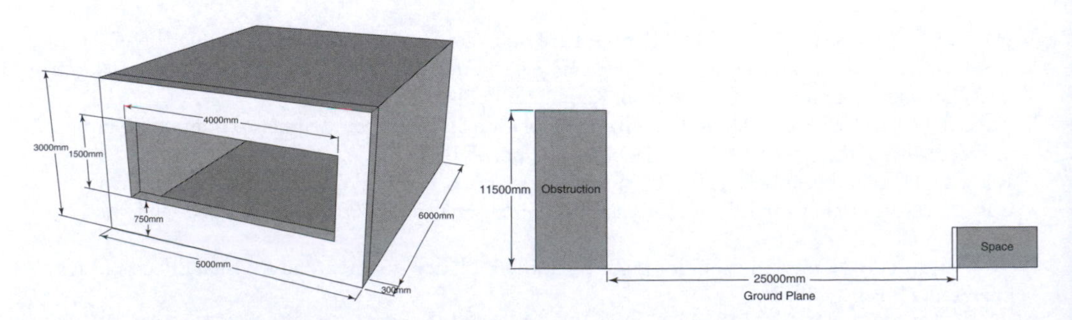

Figure 9.26 Illustration of the office room (a) and surrounding area (b).

Answers

Assignment 1:

Calculate the mean daylight factor of the space using the Lynes formula (Equation (9.4)).

$A_{glazing}$ = net glazing area = 4m × 1.5 m = 6m^2

R_{mean} = mean interior surface area

= (5·6·0.2+5·6·0.8+2·(3·6+5·3)·0.5/126=0.5

τ_{vis}= visual glazing transmittance = 72%

A_{total} = total interior surface area = 2· (5·6+3·6+5·3) m^2 = 126 m^2

$$\theta = sky\ angle = 90° - tan^{-1}\left(\frac{11500mm - 1500mm}{25000mm}\right) = 68°$$

$$DF = \frac{A_{glazing}\tau_{vis}\theta}{A_{total}2(1-R_{mean})} = \frac{6m^2 \cdot 0.72 \cdot 68}{126m^2 \cdot 2 \cdot (1-0.5)} = 2.3\%$$

The mean daylight factor in the space is 2.3%.

Assignment 2:

Calculate the daylight factor distribution and mean daylight factor of the space using a validated simulation program of your choice. Rotate the scene by 180° and repeat the calculation.

Figure 9.28 shows the results for the two spaces using Radiance. The mean daylight factor for both spaces is identical since the reference CIE overcast sky is rotationally invariant (Figure 9.9). The mean daylight factor is around 2.6% depending on the exact resolution and position of the sensor point grid. In this case the manual method and the simulation lie within a 15% error band with respect to each other revealing that the manual method can serve as a good quality control test for the simulation.

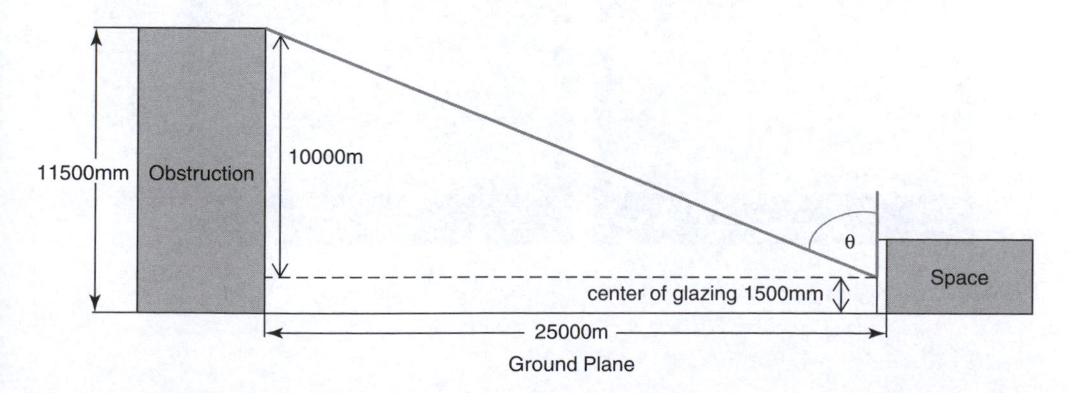

Figure 9.27 Definition of the sky angle, θ, for the investigated space.

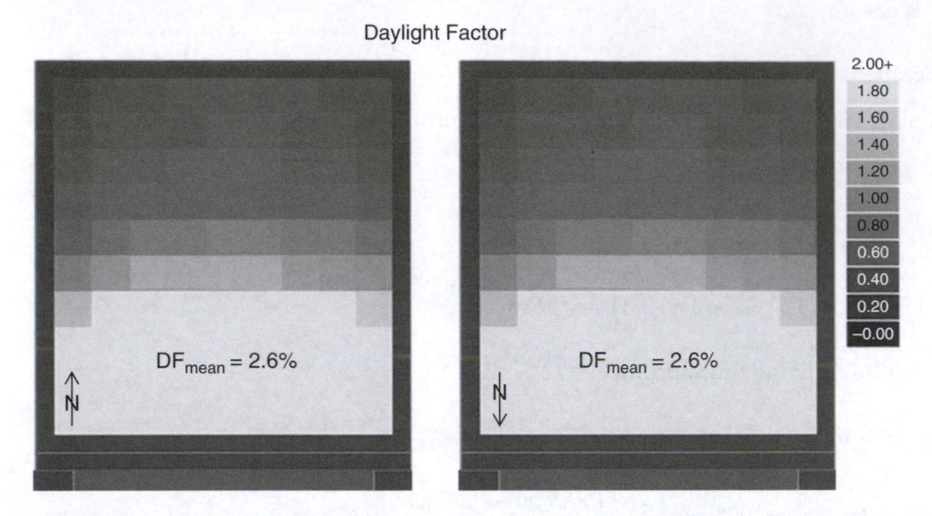

Figure 9.28 Daylight factor distribution for the space facing south and north.

Assignment 3:

Calculate the daylight autonomy distribution of the space using a validated simulation program of your choice. Rotate the scene by 180° and repeat the calculation.

Figure 9.29 shows the results for the two spaces using Radiance/Daysim. For this metric the mean daylight autonomy is around 40% for the south-facing and 35% for the north-facing space. Visually, the difference between the two distributions is not that pronounced since the window is large enough so that there is plenty of daylight near the façade for all façade orientations. During the winter the obstructing building is blocking the sun for the south-facing space, further reducing the difference between both façade orientations.

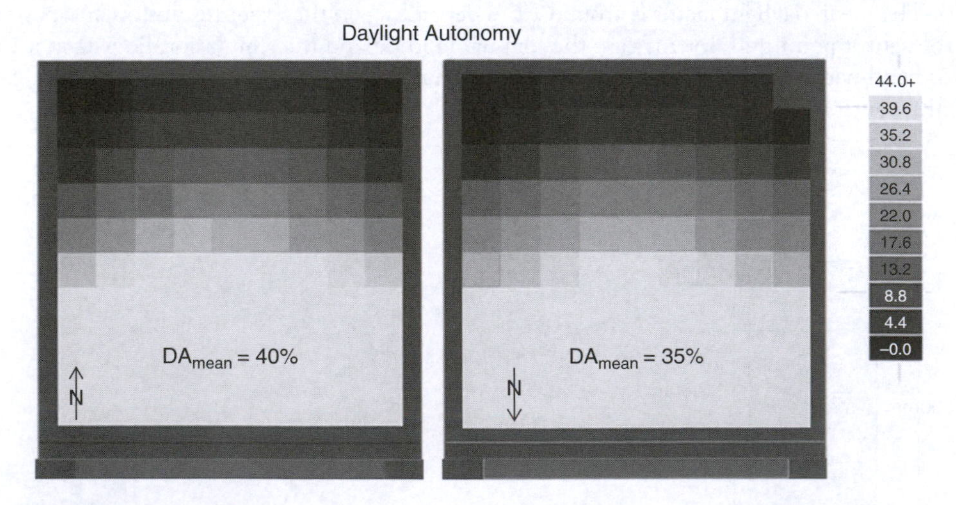

Figure 9.29 Daylight autonomy distribution for the space facing south and north.

10 Moisture phenomena in whole building performance prediction

Jan Carmeliet, Bert Blocken, Thijs Defraeye and Dominique Derome

Scope

This chapter focuses on modeling aspects of some important moisture phenomena in whole building simulations. The term 'whole building' refers to the outside environment of buildings (microclimate), the building envelope and the indoor environment of buildings. In particular, this chapter treats computational modeling of moisture buffering in rooms, computational modeling of wind-driven rain and computational modeling of convective heat and moisture transfer at exterior building surfaces. The chapter is primarily oriented to the deterioration of building materials due to high variations of relative humidity or wind-driven rain. However, the methodology and computational models can be extended to analyze the effect of moisture on other issues like energy consumption in buildings, hygrothermal comfort, indoor air quality and the reduction of active cooling in urban heat islands by means of evaporative cooling.

Learning objectives

The objectives are to present the state of the art on computational modeling in these three topics and to provide physical insight in these phenomena.

Key words

Heat-air-moisture transfer/moisture buffering/moisture damage risks/reliability analysis/ wind-driven rain/driving rain/computational fluid dynamics/surface convection coefficient

Introduction

Moisture can play an important role in the building performance and should be taken into account in the simulation of the behavior of buildings and building components. In this chapter, recent advances in modeling moisture phenomena, relevant for whole building simulation, are presented. These advances deal with the modeling of boundary conditions, i.e. the interaction of the building envelope with the indoor and outdoor environments.

In a first part, we discuss the modeling and analysis of moisture buffering in a room and propose a design method for the moisture buffering capacity using a reliability approach. This analytical approach links the moisture conditions of the indoor environment and of the interior finishing. The second part presents wind-driven rain (WDR) modeling as an important outside climatic boundary condition for heat-air-moisture (HAM) analysis of building envelopes. This method combines CFD and Lagrangian particle tracking to determine the spatial and temporal distribution of the rain load over the building facades. The third part proposes a method to predict the convective heat/moisture transfer coefficient (CHTC/CMTC) for a building immersed in an atmospheric boundary layer. These transfer coefficients play an important role in the correct modeling of the microclimate around the building. In this case, CFD is coupled to a HAM model to determine the CHTC/CMTC spatially, for different wind conditions. Examples are presented in each part.

By demonstrating how to simulate the varying moisture interactions of the building with the indoor and outdoor environments, this chapter aims at illustrating the different approaches that can be used, but more significantly it indicates the importance of integrating moisture aspects in whole building simulation.

Scientific foundation and computational methods

Moisture buffering in rooms

Problem statement

The control of the indoor relative humidity is important, since it influences the indoor thermal comfort (Berglund and Cunningham 1986; Fang *et al.* 1998; Toftum *et al.* 1998), it has an impact on the perceived indoor air quality and occupant health (Arundel *et al.* 1986), it determines moisture damage risks (Merril and ten Wolde 1989; Oreszczyn *et al.* 2006) and it can have an effect on energy consumption (Osanyintola and Simonson 2006). ASHRAE (2004) specifies an upper humidity limit of 0.012 kg/kg to meet thermal comfort conditions. For museums, guidelines to limit moisture damage specify a lower and upper limit of respectively 40 and 60 per cent relative humidity (RH), and a maximal RH of 80 per cent on cold surfaces. Too high relative humidities may cause moisture damage, due to condensation, mould growth, deterioration of surface finishing (paint, wallpaper) and wood rot. Too high fluctuations in RH result in shrinkage/swelling of materials, which can lead to damage, e.g. to wooden cultural heritage artefacts. An overview of moisture damage and its assessment is given in Carmeliet *et al.* (2009). Moisture buffering by hygroscopic materials in rooms can (1) reduce RH fluctuations and moisture damage risk; (2) increase the perceived indoor air quality; (3) lead to a reduction of energy consumption. Therefore, it is important to include moisture buffering in the design and retrofit of buildings, aiming at high energy efficiency, low cycle costs and low moisture damage risks.

A simplified model for moisture buffering in a room

In this part we address moisture buffering by hygroscopic materials in rooms in order to reduce fluctuations in relative humidity and moisture damage risk. We present a simple model for evaluating moisture buffering in a room. The model is based on the formulation of the moisture balance in a room. The model only applies for natural ventilation. The interaction with the walls is modelled using the effective penetration depth model. In the section on pages 293–297, as an application, this simplified model will be used to evaluate the moisture damage risk in a room based on a stochastic formulation of the reliability problem.

The inside relative humidity (RH_i) in a room (without HVAC control) depends on the ventilation rate G_v (or supply of air from outside having a different RH and temperature), the moisture production rate G_p and the moisture uptake/release rate by the moisture buffering materials. The moisture balance can be written as:

$$\frac{d\rho_i V}{dt} = -G_v V(\rho_i - \rho_e) - \sum_{j=1}^{n} A_j g_j + G_p V \tag{10.1}$$

with t the time (s), ρ the vapour concentration (kg/m³) (subscript i = interior, e = exterior), V the volume (m³), G_v ventilation rate (l/s), A_j the surface of the materials (m²), g_j the vapour density flow rate (kg/m²s), G_p the water vapour production rate (kg/m³s). The term on the left hand side describes the change of vapour concentration in the room with time. The first term on the right hand side describes the exchange of inside and outside air by ventilation, the second term describes the moisture uptake/release by the *n* materials of the walls and the third term describes the water vapour production in the room. The moisture buffering by air and furniture is not considered, but can be treated as an additional moisture buffering term.

Using the ideal gas law, which relates the vapour concentration to vapour pressure $p = \rho R_v T$, with R_v the ideal gas constant (462 J/kg K) and T the temperature (K), we get:

$$\frac{dp_i}{dt} = -G_v\left(p_i - \frac{T_i}{T_e}p_e\right) - \sum_{j=1}^{n} \alpha_j R_v T_i g_j + R_v T_i G_p \tag{10.2}$$

The surface to volume ratio of a wall is given by $\alpha_j = A_j/V$ (1/m) with A_j the surface (m²) and V the volume (m³) of the room. The water vapour uptake/release g_j (kg/m²s) by the surface materials is modelled using the effective penetration depth model (EPDM) (Cunningham 1992), where an equivalent surface coefficient $\beta_{eq,j}$ (s/m) for the wall j is introduced to model the vapour transport from the room to the (painted) wall:

$$g_j = -\frac{p_{w,j} - p_i}{\dfrac{1}{\beta_j} + \dfrac{\mu d_{pa,j}}{\delta_a} + \dfrac{d_{p,j}}{2\delta_j}} = -\beta_{eq,j}\left(p_{w,j} - p_i\right) \tag{10.3}$$

with $p_{w,j}$ the vapour pressure in the wall (Pa), β_j the water vapour surface coefficient (s), $\mu d_{pa,j}$ (m) the equivalent water vapour resistance thickness for the paint layer, δ_a the vapour permeability of dry air (s), $d_{p,j}$ the penetration depth (m) (subscript p referring to penetration) and δ_j the water vapour permeability of the material (s). The penetration depth is defined as the thickness of the surface layer where hygric interaction with the indoor air occurs, or:

$$d_{p,j} = \sqrt{\frac{\delta_j p_{sat} t_p}{\xi_j \pi}} \tag{10.4}$$

with p_{sat} the water vapour saturation pressure (Pa) and t_p the period of time (one day). In the EPDM, it is assumed that the hygroscopic moisture capacity ξ (kg/m³) and water vapour permeability δ_j (s) of the material are constants. The differential equation for the moisture buffering of the jth wall then is:

$$\frac{dp_{w,j}}{dt} = -\gamma_j \beta_{eq,j}\left(p_{w,j} - p_i\right), \gamma_j = \frac{p_{sat}}{d_{p,j}\xi_j} \tag{10.5}$$

with γ_j a measure for the capacity of the effective material layer. Equations (10.2) and (10.5) define a set of $(n+1)$ ordinary differential equations, which are solved numerically.

We now discuss in more detail the different input variables for this model. The mass surface coefficient β and all the material properties are considered to be deterministic. The inside temperature is considered to be constant and equal to 20 °C. The exterior conditions, i.e. vapour pressure $p_e(t)$ and temperature $T_e(t)$, are represented by measured climatic data. Figure 10.1(a) gives as an example the time variation of the exterior vapour pressure during 10 days. The ventilation rate G_v depends on the airtightness of the building, on the air pressure difference between the inside and the outside due to wind and stack effects and on human behavior (opening of windows). For simplicity, we consider the ventilation rate as a stochastic process characterized by a random variation around daily and hourly values. The daily variation accounts for windy versus non-windy days, while the hourly variation accounts for the short-time stochastics of wind pressure and human behavior. The minimum ventilation rate is limited to G_{vbas} = 0.2 air changes an hour (1/h). The average ventilation rate is 0.6 1/h. A typical daily variation of $G_v(t)$ is given in Figure 10.1(b). We consider the vapour production rate G_p to consist of two components: a basic production rate accounting for permanent vapour production, for example, due to the presence of people and plants, and two peak production rates during the morning and late afternoon. The morning period, between 6 a.m. and 8 a.m., accounts for human activities such as showers and washing. The late afternoon period from 4 p.m. until 10 p.m. simulates certain activities like cooking, cleaning and laundry. Since the peak water vapour production rates depend on human activity, they are considered to be stochastic. Parts c and d of Figure 10.1 give a typical variation of the vapour production rate during one day and ten days respectively.

Figure 10.2 shows the relative humidity in the room calculated by the simple model. The input data of Figure 10.1 are used for these calculations. The inside relative humidity is determined for two cases: a room without moisture buffering capacity, putting $\alpha_j = 0$, and a case with moderate moisture buffering capacity. We observe that the inside relative humidity follows strongly the trend of the outside vapour pressure: a higher inside RH observed during the first period is clearly due to a higher outside vapour pressure, while in the second period the inside RH is globally lower due to the lower vapour pressure outside. The two daily peaks in the inside RH are explained by the presence of the two peaks in moisture production rate. We also observe a reduction of the daily maximum and of the daily variation of the RH inside due to

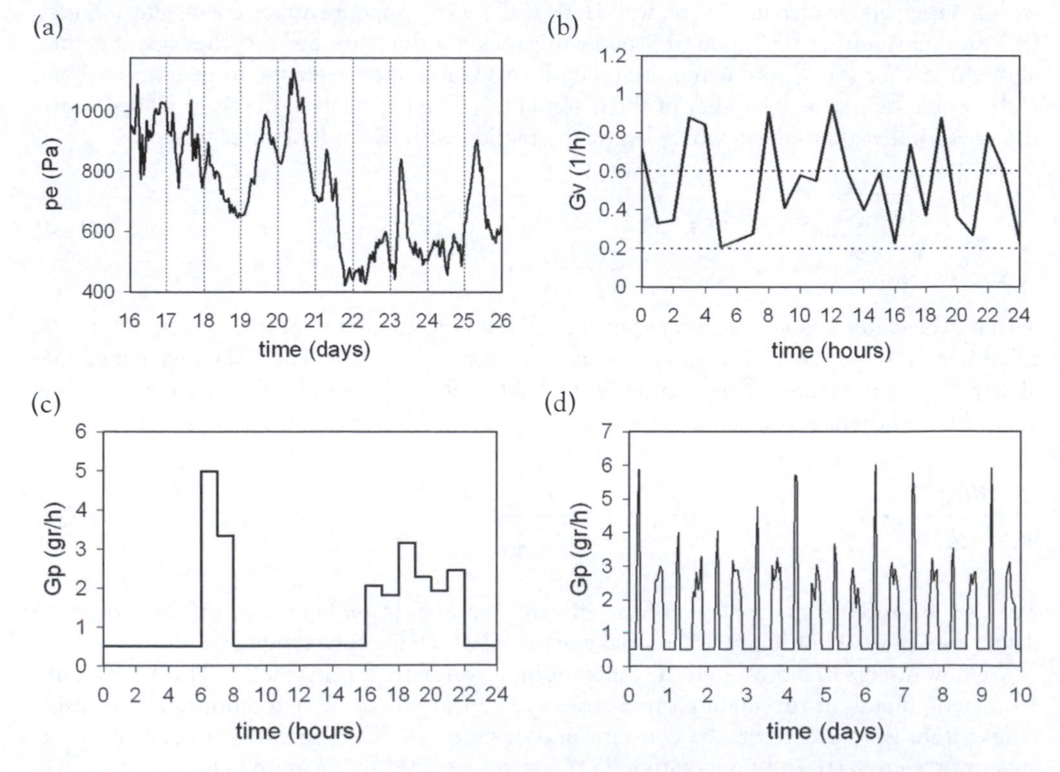

Figure 10.1 Input data: (a) 'climatic': outside vapour pressure variation during 10 days in April for Belgium; (b) random variation of ventilation rate during 1 day; (c) random variation of water vapour production rate during 1 day; (d) random variation of water vapour production rate during 10 days.

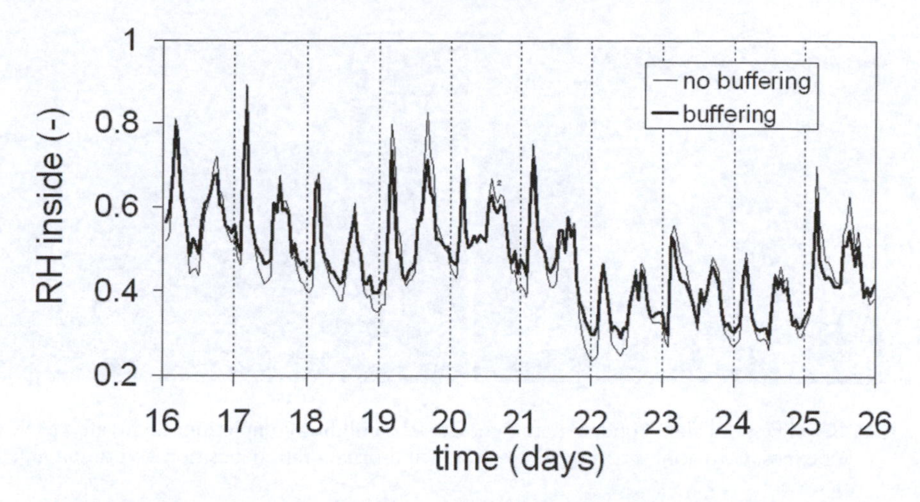

Figure 10.2 The variation of the relative humidity inside the room during 10 days in April for Belgium. Two cases are considered: a room without buffering capacity by putting $\alpha_j = 0$ and a case with moisture buffering. The input data of Figure 10.1 are used for the simulation.

moisture uptake and release by the walls. However the mean daily RH inside the room shows not to depend strongly on the moisture buffering capacity of the room. This indicates that moisture buffering by the walls can be used to control the variations in RH in the room but to a lesser extent for the control of the mean RH in the room.

On pages 293–297, this model will be applied to determine the necessary moisture buffering capacity in order to prevent moisture damage due to too high daily variations in RH, which is important for cultural heritage preservation in buildings like museums, churches, etc. In the Activities section, an exercise is given to determine the statistical distribution of inside relative humidity variations in a room.

Computational modeling of wind-driven rain

Problem statement

Wind-driven rain (WDR) or driving rain is rain that is given a horizontal velocity component by the wind. It is one of the most important moisture sources affecting the hygrothermal performance and durability of building facades. The negative consequences of WDR can take many forms. Moisture accumulation in porous materials can lead to water penetration (e.g. Day *et al.* 1955; Eldridge 1976; Garden 1963; Lacy 1977; Marsh 1977), frost damage (Eldridge 1976; Franke *et al.* 1998; Maurenbrecher and Suter 1993; Price 1975; Stupart 1989), moisture-induced salt migration (Charola and Lazzarini 1986; Franke *et al.* 1998; Price 1975), discoloration by efflorescence (Eldridge 1976; Franke *et al.* 1998), structural cracking due to thermal and moisture gradients (Franke *et al.* 1998), to mention just a few. WDR impact and runoff is also responsible for the appearance of surface soiling patterns on facades (Camuffo *et al.* 1982; Davidson *et al.* 2000; Eldridge 1976; Etyemezian *et al.* 2000; Robinson and Baker 1975; White 1967). As an example, Figure 10.3 illustrates the Royal Festival Hall in London, right after the building completion (left) and after a few years of atmospheric exposure and wind-driven rain deposition and runoff along the facade (right). The combined effect of atmospheric deposition of pollutants and WDR on the facade leads to 'white washing' and 'dirt washing'. At the top of the facade, the WDR with high intensities rinses away the deposited dirt ('white washing') and moves it to the lower facade parts, where it is redeposited ('dirt washing').

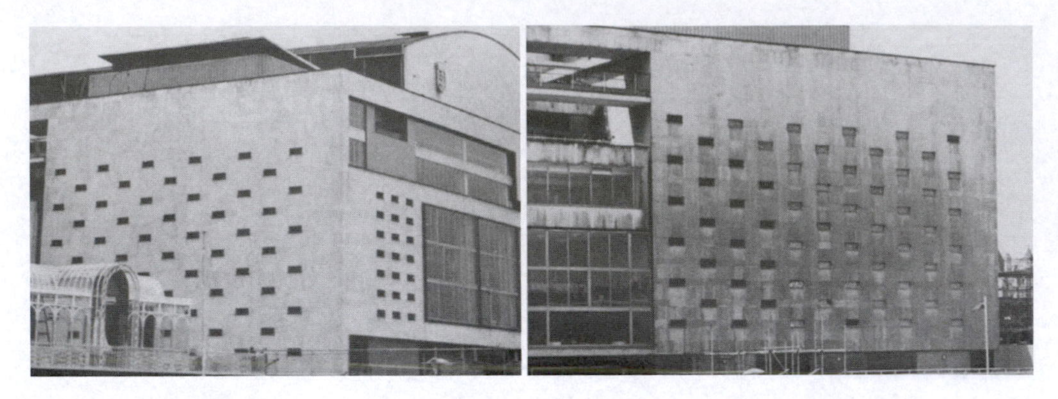

Figure 10.3 Royal Festival Hall, London. Left: right after the building completion; Right: after a few years of exposure to atmospheric pollution and wind-driven rain deposition and runoff along the facade.

Source: © White 1967, reproduced with permission.

Locally, rain water runoff from window glass and sills can also cause 'white washing' and 'dirt washing'.

In the past 80 years, many efforts have been made to increase the understanding of the interaction between wind, rain and buildings. WDR research in building engineering can be divided into two parts (Figure 10.4): (1) assessment of the impinging WDR intensity and (2) assessment of the response of the building facade to the impinging WDR. The impinging WDR intensity is governed by a diversity of parameters: building geometry, environment topography, position on the building facade, wind speed, wind direction, rainfall intensity and raindrop-size distribution. The impinging WDR intensity is the total amount of rainwater that comes into contact with the building surface. What happens at and after impact/impingement is the focus of the second part of WDR research. It comprises the study of contact and surface phenomena such as splashing, bouncing, adhesion, runoff, evaporation, absorption and the distribution of the moisture in the facade (rain penetration and wetting-drying). It also includes the wide variety of rainwater penetration mechanisms including hydrostatic pressure, wind pressure, surface tension, gravity, etc. The response of the wall to WDR is also determined by a diversity of parameters: the WDR intensity and raindrop-size distribution, the raindrop diameter, impact angle and impact speed, material-surface characteristics such as wall roughness, material characteristics such as moisture permeability and moisture retention, construction details, cracks

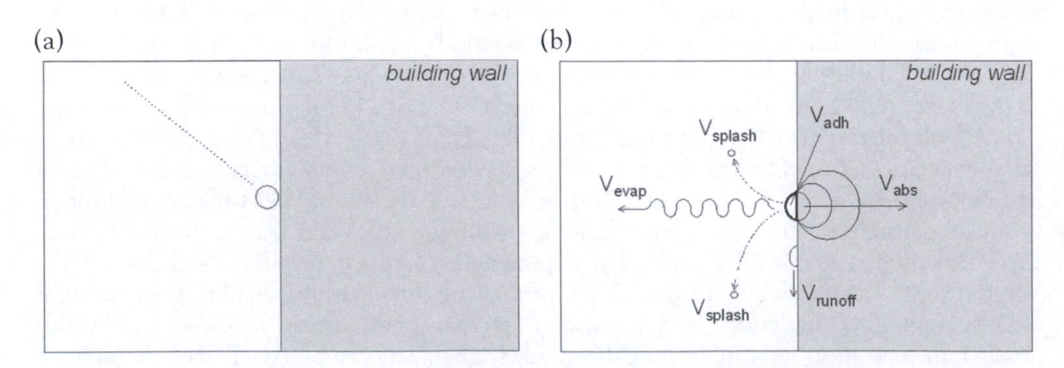

Figure 10.4 Schematic representation of the two parts in wind-driven rain research: (a) assessment of the impinging wind-driven rain intensity (before rain impact) and (b) assessment of the response of the wall (at and after rain impact).

and joints, etc. Given the complexity of WDR and its wide range of influencing parameters, it is not surprising that despite research efforts spanning more than 80 years, WDR is still an active research topic in building engineering, and a lot of work remains to be done.

Most computational research on WDR has focused on the first part in WDR research: assessment of the impinging WDR intensity. Three categories of methods exist to assess the intensity of WDR impinging on building facades: experimental, semi-empirical and numerical methods. An extensive review of these methods was provided by Blocken and Carmeliet (2004). Measurements have always been the primary tool in WDR research, but are nowadays only rarely conducted. The most important reasons are the fact that WDR measurements are time-consuming and expensive, and that measurements at a particular building site have very limited applicability to other sites. Additionally, it has been shown that WDR measurements can easily suffer from large errors (Blocken and Carmeliet 2005, 2006a; Högberg *et al.* 1999; van Mook 2002). Recently, guidelines have been proposed that should be followed for selecting accurate and reliable WDR data from experimental WDR datasets (Blocken and Carmeliet 2005). The strict character of these guidelines, however, implies that only very few rain events in a WDR dataset are accurate and reliable and hence suitable for WDR studies. Semi-empirical methods are an alternative to WDR measurements (ISO 2009; Sanders 1996; Straube 1998; Straube and Burnett 2000). The main advantage of semi-empirical methods is their ease of use; their main disadvantage is that generally only rough estimates of the WDR exposure can be obtained (Blocken and Carmeliet 2004). Given the limitations of measurements and semi-empirical methods, and as research efforts employing these methods continued to reveal the inherent complexity of WDR, researchers realized that further achievements were to be found through computational modeling.

State of the art in computational modeling of wind-driven rain

Numerical modeling of WDR with Computational Fluid Dynamics (CFD) has provided the capability to meticulously study the complex interaction between WDR and buildings. In such simulations, the wind-flow pattern around the building is first calculated with a CFD code, generally using the steady Reynolds-averaged Navier-Stokes (RANS) equations, after which raindrops are injected in the flow pattern and their equations of motion are solved. This process is referred to as Lagrangian particle tracking. As an example, Figure 10.5 illustrates trajectories of raindrops of 1 and 5 mm diameter in the wind-flow pattern around the VLIET test building in Flanders, Belgium. The CFD simulation technique for WDR was developed by Choi (1991, 1993, 1994) and extended in the time domain by Blocken and Carmeliet (2002, 2007a). The steady-state simulation technique by Choi allows determining the spatial distribution of WDR on buildings under steady-state conditions of wind and rain, i.e. for fixed, static values of wind speed, wind direction and horizontal rainfall intensity (i.e. the rainfall intensity falling through a horizontal plane). This technique has been adopted by a large number of researchers for their WDR analyses. Validation studies of the steady-state simulation technique were performed by, for example, Hangan (1999) and van Mook (2002). Hangan (1999) used the innovative wind tunnel studies of WDR that were performed by Inculet and Surry (1994) and Inculet (2001) at the Boundary Layer Wind Tunnel Laboratory of the University of Western Ontario, London, Canada. Van Mook used full-scale measurements of WDR on the west facade of the main building of Eindhoven University of Technology in the Netherlands. The extension of this technique in the time domain by Blocken and Carmeliet (2002, 2007a) allowed the numerical determination of both the spatial and temporal distribution of WDR on buildings. Validation studies for the low-rise VLIET test building and for different rain events have indicated that this extended numerical method can provide quite accurate predictions of the WDR amount and the WDR deposition pattern on the building facade (Blocken and Carmeliet 2002, 2006b, 2007b). Figure 10.6 shows a comparison between CFD and experimental WDR results on the south-west facade of the VLIET test building, for wind direction perpendicular to the facade. A

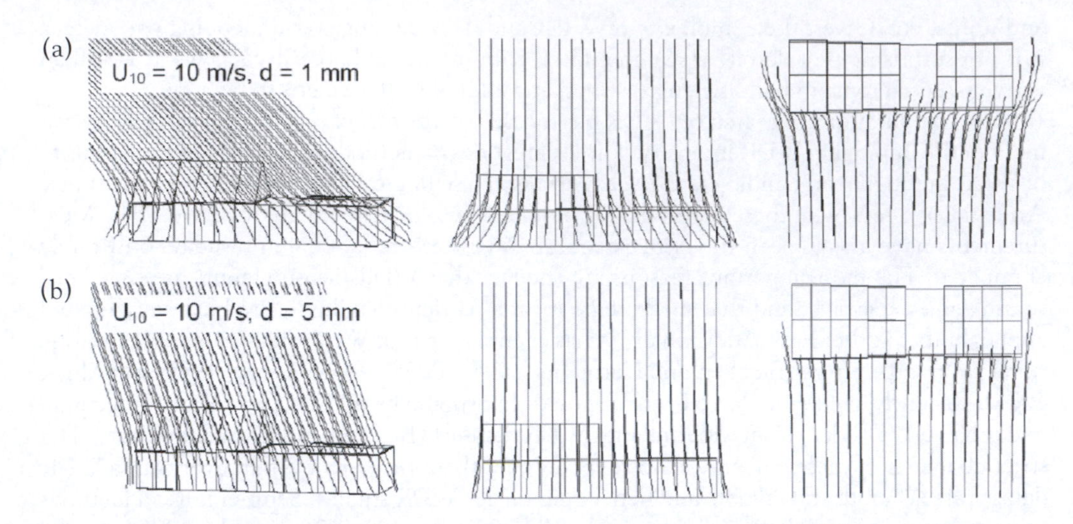

Figure 10.5 Perspective, front and top view of raindrop trajectories in the U_{10} = 10 m/s wind-flow pattern around the VLIET test building. U_{10} is the reference wind speed at 10 m height. The wind direction is perpendicular to the wide facade. (a) Raindrop diameter d = 1 mm; (b) d = 5 mm.

good agreement between simulations and measurements is observed, except at the left building edge. The discrepancies at this position are due to the fact that the trees upstream of this edge were not included in the model. Recently, additional validation studies, based on the extended numerical method, were made for two rather complex high-rise buildings and for a simple rectangular test building (Abuku *et al.* 2009; Briggen *et al.* 2009; Tang and Davidson 2004). Abuku

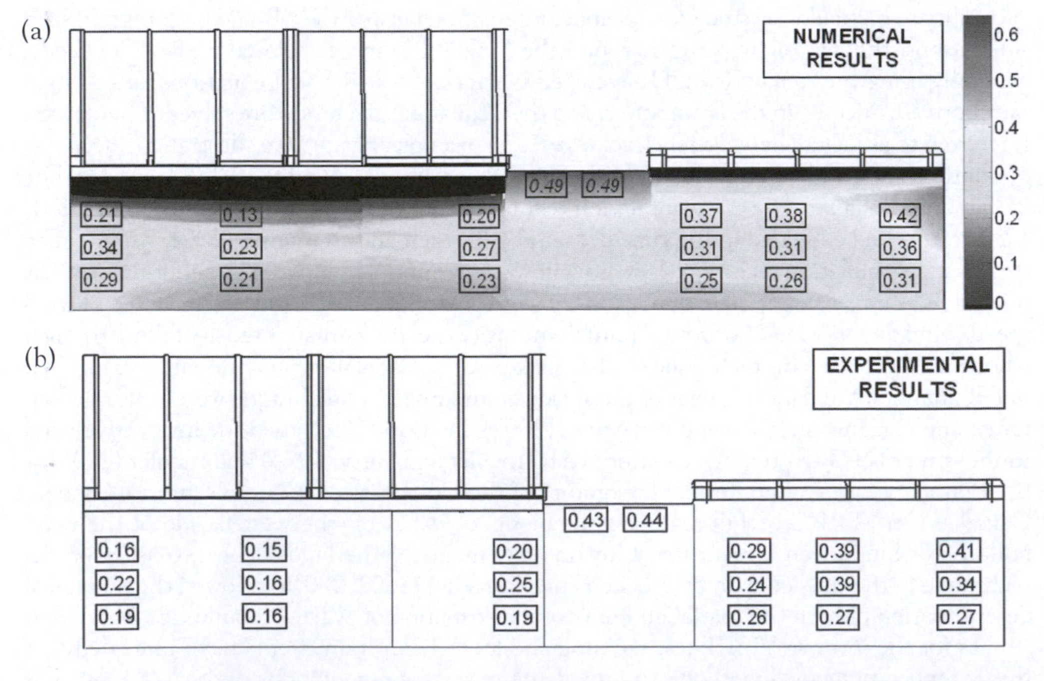

Figure 10.6 Spatial distribution of the ratio S_{wdr}/S_h (total wind-driven rain amount to total horizontal rainfall amount) for a given rain event with total horizontal rainfall amount S_h = 15.7 mm. (a) Numerical results, with indication of the numerical values at the location of the wind-driven rain measurements. (b) Measurement results.

et al. (2009) specifically focused on validation for oblique wind directions. Briggen *et al.* (2009) found good results for the top part of a high-rise tower building, but a significant underestimation of the measured WDR intensity at the bottom part. This could be attributed to not modeling the turbulent dispersion of raindrops (Blocken *et al.* 2009a; Briggen *et al.* 2009). Previous studies (e.g. Choi 1997; Lakehal *et al.* 1995; and van Mook 2002) disagree on the importance of turbulent dispersion modeling. Future work in this area should address this issue, preferably by enhanced turbulence modeling using Large Eddy Simulation (LES) or hybrid LES/Unsteady RANS (URANS) methods, instead of the commonly used steady RANS approach.

It is important to note that computational modeling of WDR is receiving strong support from several international initiatives that specifically focus on the establishment of guidelines for CFD in wind engineering (Franke *et al.* 2004; Franke *et al.* 2007; Tominaga *et al.* 2008a; Yoshie *et al.* 2007). In addition, other guidelines, for more general CFD applications, are available (e.g. Casey and Wintergerste 2000). Other research efforts have focused on specific aspects for improving the quality of CFD simulations, such as those concerning the simulation of equilibrium atmospheric boundary layers in computational domains (e.g. Blocken *et al.* 2007a, 2007b; Franke *et al.* 2007; Gorlé *et al.* 2009; Hargreaves and Wright 2007; Richards and Hoxey 1993; Yang *et al.* 2009). Among others, the existing CFD guidelines focus on the importance of grid quality, in terms of cell shape, cell size gradients and overall grid resolution, and on the importance of grid-sensitivity analysis and the assessment of discretisation errors (Franke and Frank 2008). In addition, an efficient and systematic procedure to generate high-quality and high-resolution grids for complex urban environments has been presented (van Hooff and Blocken 2010), which can be used in future WDR computational modeling studies.

Most computational studies of WDR on buildings so far were performed for isolated buildings, located on flat terrain. Two exceptions to this are the work of Karagiozis *et al.* (1997) and Blocken *et al.* (2009a), in which a two-building configuration was considered and in which the mutual influence of these buildings on their WDR exposure was examined. These studies showed that complexity of the WDR distribution pattern significantly increases by the interaction effects between the buildings, and that computational modeling with CFD is required to understand these effects.

Another example in which CFD has provided new insights in WDR on building facades, is the influence of the wind-blocking effect by a building on its WDR exposure.

Wind-driven rain and the wind-blocking effect

The influence of the wind-blocking effect by a building on its WDR exposure was identified by Blocken and Carmeliet (2006b) and demonstrated by performing CFD simulations for different isolated building configurations. Four of these configurations are shown in Figure 10.7: a low-rise cubic building, a medium-rise wide building, a high-rise building slab and a tower building. The wind-blocking effect refers to the disturbance of the wind-flow pattern by the presence of the building. In particular for WDR, the decrease of the streamwise horizontal wind-velocity component upstream of the building (wind-speed slow-down) is important, as this component carries the rain to the facade. The higher and the wider the building, the larger the wind-blocking effect, and the larger the region of wind-speed slow-down will be. The raindrops travelling through this region will therefore receive less momentum by the wind, resulting in lower WDR intensities on the facade.

Figure 10.8 illustrates the wind-blocking effect by comparing contours of the dimensionless streamwise velocity component U/U_{10} in the centerplane of the four building configurations in Figure 10.7. The wind direction is perpendicular to the windward facade. The upward shift of the isolines indicates the extent of the wind-blocking effect. It is most pronounced for the high-rise building slab (Figure 10.8(c)), due to its large width and height, and least pronounced for the low-rise cubic building (Figure 10.8(a)). Figure 10.9 shows the contours of the catch ratio (ratio of WDR intensity to horizontal rainfall intensity) on the windward facade of each

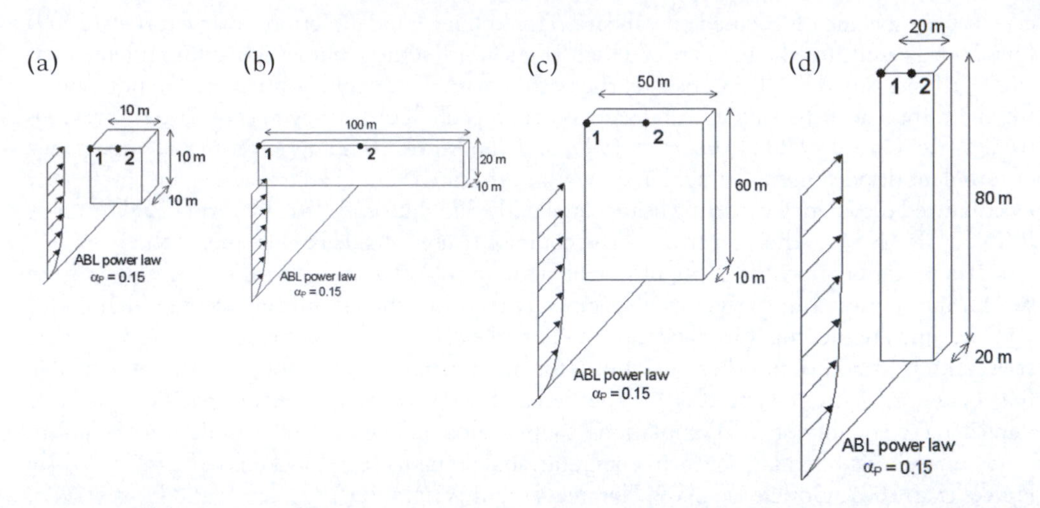

Figure 10.7 Four of the building configurations for which the wind-blocking effect was investigated: (a) low-rise cubic building; (b) medium-rise wide building; (c) high-rise building slab; (d) tower building.

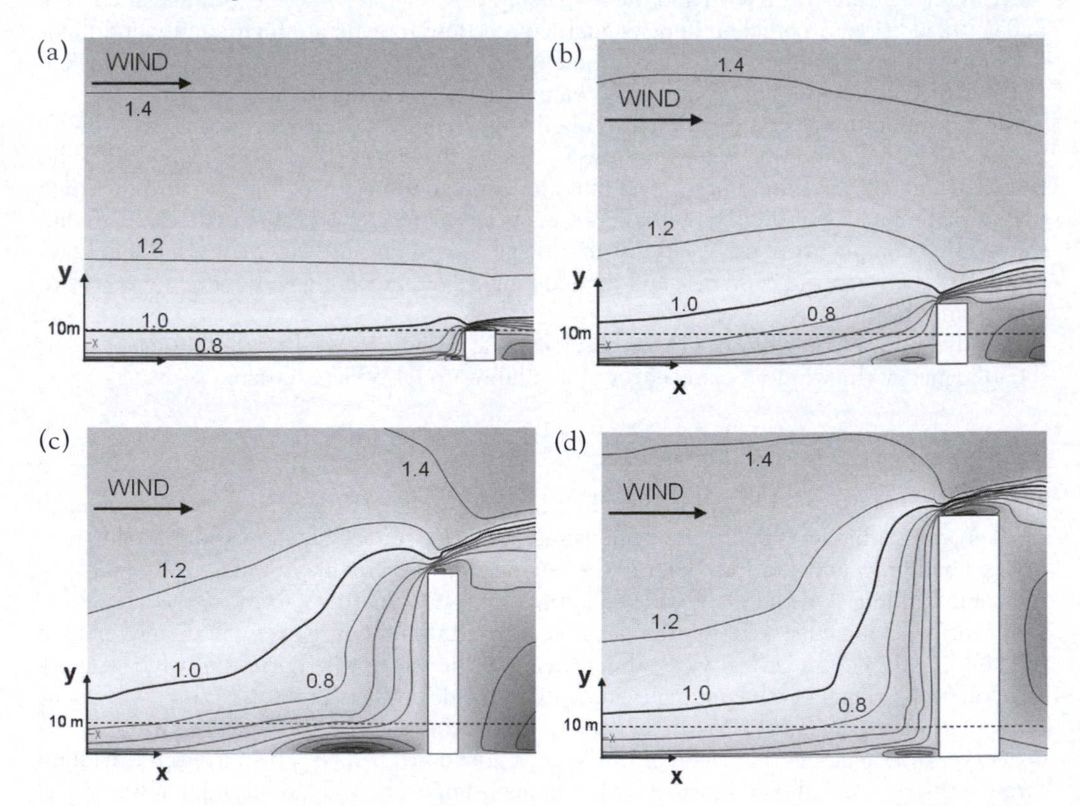

Figure 10.8 Contours of the streamwise horizontal wind-velocity component (dimensionless: U/U_{10}) in a vertical plane through the center of the four buildings in Figure 10.7. The wind direction is perpendicular to the windward facade.

of the four buildings. The maximum value occurs at the top corner of each building. It is highest for the low-rise cubic building (Figure 10.9(a)), for which the wind-blocking effect is least pronounced, and lowest of the high-rise building slab (Figure 10.9(c)), for which the wind-blocking effect is most pronounced.

The higher and the wider the building, the larger the wind-blocking effect will be. The dimensions (height and width) of the windward facade can be combined into the parameter 'building scaling length', *BSL* (m), which was defined by Wilson (1989) for estimating recirculation lengths on buildings:

$$BSL = \left(B_L B_S^2\right)^{\frac{1}{3}} \tag{10.6}$$

where B_L (m) is the largest and B_S (m) the smallest dimension of the windward facade. Figure 10.10 shows that the WDR exposure in terms of the maximum catch ratio is indeed strongly related to the BSL.

The wind-blocking effect is responsible for the counter-intuitive observation that the WDR exposure of a low-rise, cubic building (h = 10 m) can be larger than the WDR exposure of a high-rise building slab (h = 60 m) or even a high tower building (h = 80 m). These observations appear to be in contrast to the general notion that the WDR exposure at the top of a building increases with the building height. This notion stems from the fact that the free-field wind speed increases with height and that higher wind-speed values yield higher WDR amounts. However, from the CFD simulations, it is clear that a wider and higher building represents a larger obstruction to the wind-flow pattern (wind blocking) which in turn can cause a lower

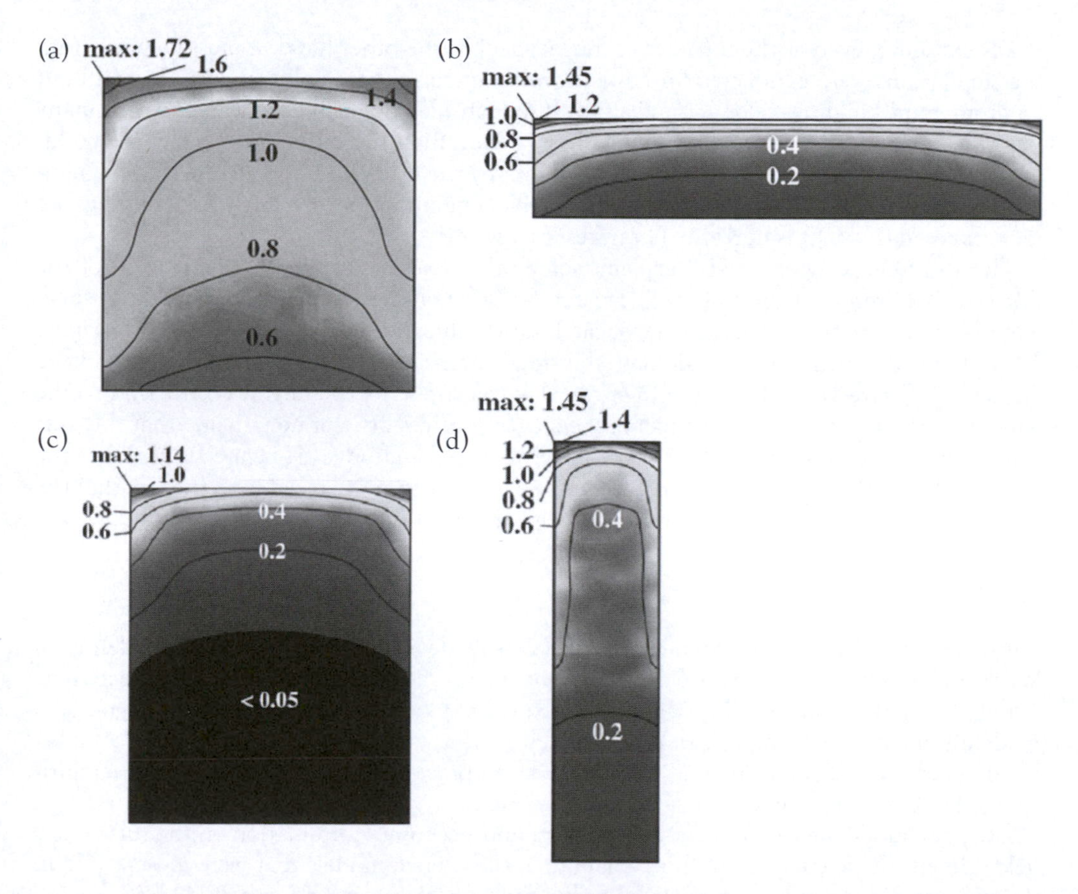

Figure 10.9 Contours of the wind-driven rain catch ratio on the windward facade of each of the four buildings, for the U_{10} = 10 m/s wind-flow pattern and horizontal rainfall intensity R_h = 1 mm/h. (a) low-rise cubic building; (b) medium-rise wide building; (c) high-rise building slab; (d) tower building.

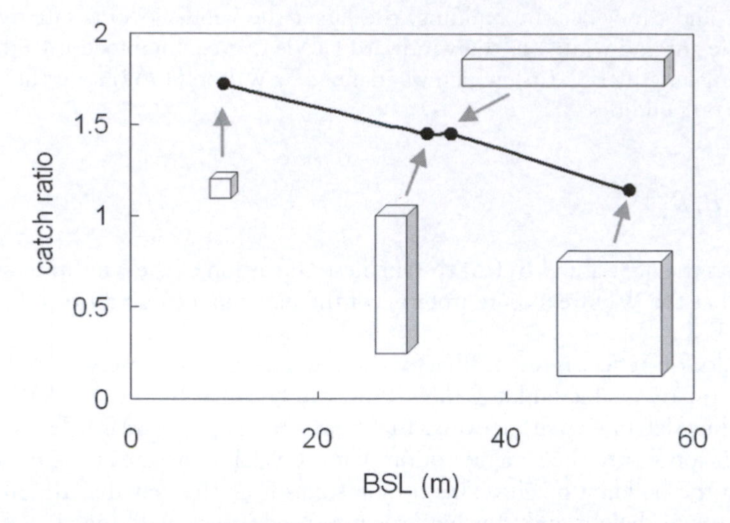

Figure 10.10 Catch ratio at the top corner of the windward facade versus building scaling length, for the four buildings, for the U_{10} = 10 m/s wind-flow pattern and for horizontal rainfall intensity R_h = 1 mm/h.

WDR exposure, even at the top edge of the facade. On the other hand, it must be noted that the simulations were conducted for buildings without surrounding obstructions. In real built environments, buildings seldom stand alone. If we consider a typical European city with many low-rise and medium-rise buildings and with only a few high-rise buildings, the low-rise buildings will generally be sheltered from wind and rain by the other buildings, while the few high-rise buildings will not be sheltered. Therefore, the general notion that high-rise buildings are most exposed to WDR is in reality not necessarily wrong.

The wind-blocking effect also explains some other features of the wetting pattern of the different building configurations: (1) Figure 10.9(a): For the low-rise cubic building: small wind-blocking effect and therefore large catch ratio values across the entire facade. (2) Figure 10.9(b): For the medium-rise wide slab: the small catch ratio values at the lower half of the facade. (3) Figure 10.9(c): For the high-rise slab that presents the largest obstruction to the wind flow: the fact that the lower half of the facade receives little or no rain and that the wetting gradients are only pronounced at the upper part of the facade. (4) Figure 10.9(d): For the tower building: the small catch ratio values at the lower part of the facade and the fact that the wetting gradients are only pronounced at the upper part of the facade.

Future perspectives

Although considerable advancements have been made in the computational modeling of WDR on buildings in the past two decades, some important future tasks remain. These include (1) modeling the turbulent dispersion of WDR; (2) exploring Eulerian instead of Lagrangian modeling for the rain phase; (3) WDR modeling and validation studies for more complex building/urban configurations; and (4) further development and improvement of semi-empirical models based on computational modeling results.

Accurate modeling of turbulent dispersion requires accurate information on the turbulence field. Given the deficiencies of the k-ε family of turbulence models that were mostly used in the past, second-moment closure modeling but probably transient LES or hybrid LES/URANS will be required. The problem is that computational WDR modeling based on Lagrangian particle tracking with the 'steady-stream tube' approach (Blocken and Carmeliet 2002, 2007b; Choi 1993, 1994) is already a very time-consuming procedure, because particle tracking has to be performed for the wide range of raindrop diameters, and this should be repeated for every reference wind speed and every reference wind direction. Adding turbulent dispersion and maybe resorting to transient wind-flow pattern calculations with LES or hybrid LES/URANS will drastically increase computation times. The Lagrangian approach is the natural approach for dealing with particle motion. As an alternative to Lagrangian particle tracking, however, Eulerian particle dispersion could be applied (Loth 2000; Shirolkar *et al.* 1996). The Eulerian description applied to the rain phase assumes that the characteristics of the raindrops can be described as a continuum. This assumption allows the rain phase to be treated with the same discretisation and numerical techniques as used for the continuous (air) phase. While this can significantly reduce the computational cost, the description of turbulent diffusion is, however, more straightforward with the Lagrangian approach (Loth 2000).

Up to now, the vast majority of WDR modeling efforts have focused on isolated buildings or – exceptionally – on very simple two-building configurations. For these configurations, steady RANS can be suitable, as shown previously (Abuku *et al.* 2009; Blocken and Carmeliet 2002, 2006b, 2007b; Blocken *et al.* 2009a; Briggen *et al.* 2009; Tang and Davidson 2004). Complex urban configurations require the use of more advanced turbulence modeling that allows for a correct representation of – among others – the separated and recirculating flow in the wake behind buildings (Murakami 1993; Tominaga *et al.* 2008b). This confirms the need to use LES or hybrid LES/URANS methods for further advancements in WDR research.

For practical WDR assessment, computational modeling is too time-consuming and expensive. As opposed to computational modeling, semi-empirical methods, such as the method in the new ISO Standard on WDR (ISO 2009) and the method by Straube (1998) and Straube and Burnett (2000), are fast and easy to use. However, they only allow a rough estimate of the WDR intensity to be obtained. Current work by the authors is focusing on improving these semi-empirical methods, based on detailed insights obtained by computational modeling, such as the wind-blocking effect.

Computational modeling of convective heat and moisture transfer at exterior building surfaces

Problem statement

Quantifying convective heat and moisture transfer (H&MT) at exterior building surfaces is of interest for thermal performance analysis of building components and hygrothermal analysis of buildings, one example could be the drying of facades wetted by wind-driven rain (WDR). Usually, convective heat and moisture transfer coefficients (CHTCs and CMTCs) are used in modeling, which relate the convective heat and moisture flux normal to the wall, $q_{c,h,w}$ (W/m²) and $q_{c,m,w}$ (kg/sm²) respectively, to the difference between the surface temperature T_w (°C) or vapour pressure $p_{v,w}$ (Pa) at the wall and the outside reference air temperature T_e (°C) or vapour pressure $p_{v,e}$ (Pa):

$$CHTC = \frac{q_{c,h,w}}{T_w - T_e} \quad ; \quad CMTC = \frac{q_{c,m,w}}{p_{v,w} - p_{v,e}} \tag{10.7}$$

Previous research on convective transfer coefficients (CTCs) for buildings mainly focussed on the CHTC. CHTCs were determined by wind-tunnel tests for flat plates or bluff bodies, by full-scale experiments on various types of buildings and recently also by CFD simulations (e.g. Blocken *et al.* 2009b; Defraeye *et al.* 2010). CMTCs are often determined out of CHTC data, using the Chilton-Colburn analogy. However, the assumptions on which this analogy is based limit its applicability, for example for drying processes (e.g. Chen *et al.* 2002; Derome *et al.* 1999).

This CTC information is used in HAM (heat-air-moisture) and BES (Building Energy Simulation) modeling. Apart from the specific conditions under which these CTCs have been determined, there are still some limitations to the use of these CTCs in such models: (1) Often limited or even no spatial variation of the CTCs over a building component or building facade is accounted for; (2) A temporal variation is not considered, which inherently assumes a constant relation between wall flux and wall temperature or vapour pressure difference (see Equation (10.7)); (3) Convective heat and moisture transfer are considered to be uncoupled, even for phenomena where they are in fact strongly coupled (e.g. drying).

From the above, it is clear that the use of such CTCs simplifies coupled 2D or 3D convective H&MT problems. A more accurate quantification of the convective transport requires that both the coupled H&M transport in the porous material and in the air are taken into account. From a numerical point of view, this can be done by solving the coupled H&MT in the porous materials of the building envelope with a HAM model and by using CFD to provide information on the convective transfer at the exterior wall surface. This approach was already proposed by Erriguible *et al.* (2006), and also Neale *et al.* (2007) and Defraeye *et al.* (2009), and allows for spatially and temporally varying CTCs due to, among others, the specific features of the air-flow pattern around the porous material. Such a coupled CFD-HAM model (referred to as CHAM) is described below and is demonstrated by application to the drying of a building facade after a WDR shower. The results are compared with the conventional approach, namely using a HAM model with constant (spatial and temporal) and uncoupled CTCs.

CFD-HAM prediction of CHTC/CMTC

Methodology

The HAM model is coupled with CFD, where HAM is used to model the porous material and CFD is used to model the air flow in the outside environment. This coupling is done by exchanging boundary condition information at the interface, namely at the exterior wall surface. This coupling between CFD and HAM programs is done in an explicit way by executing the CFD program for one time step, after which the boundary condition information, i.e. the heat and moisture fluxes, is transferred to the HAM program which is subsequently executed for the same time step. At the end of the time step, boundary condition information (temperatures and vapour pressures) is transferred to CFD for the calculation of the next time step. This explicit coupling is justified if sufficiently small time steps (e.g. 0.1 s) are used so the fluxes do not change significantly over the time step.

The CFD simulations in this study are performed with the commercial CFD code Fluent 6, using the steady Reynolds-averaged Navier-Stokes equations supplemented with the realizable k-ε turbulence model (Shih *et al.* 1995) for closure in the high-Reynolds number region and low-Reynolds number modeling with the Wolfshtein model for the near-wall region (Wolfshtein 1969). Only forced convection is taken into account in the simulations. The HAM simulations are performed with an in-house HAM code. It solves for heat, air and moisture (vapour and liquid) transport in porous materials. Air transport in the material is not taken into account in this study. Modeling details can be found in Janssen *et al.* (2007).

Case study

A 2D (infinitely long) square-shaped building in an atmospheric boundary layer (ABL) is considered (see Figure 10.11(a)) where only the H&MT in the windward vertical wall is included in the analysis. The building has a height (H) of 10 m but the actual height of the windward wall is 9.8 m, accounting for the thickness of the roof. The wall is a cavity wall, consisting of an outer and inner leaf and an insulated cavity in between. The numerical simulations are performed after a considerable WDR event since they aim to predict the drying behavior of the outer leaf. A period of 12 hours after the rain shower is considered. The boundary conditions of the numerical model are specified in Figure 10.11.

The computational domain to model the wind environment (with CFD) is shown in Figure 10.11(a). At the inlet of the domain, a logarithmic ABL velocity profile is imposed with a reference wind speed at building height (U_{10}) of 2.5 m/s. For the porous material modeling (with HAM), only the outer leaf, consisting of ceramic brick, is considered in the numerical model. Since the outer leaf is well insulated on its inside with an air and moisture tight material (XPS; Extruded Polystyrene), the interior surface of the outer leaf is assumed to be adiabatic and impermeable for moisture. The initial temperature and moisture distributions in the outer leaf, namely after the rain event, are determined from a 1D HAM simulation of the outer leaf, including WDR modeling. For the HAM simulations with constant CTCs, the CHTC (4.7 W/m²K) is determined according to Sharples (1984), based on U_{10}, and the CMTC (3.4×10^{-8} s/m) is determined according to the Chilton-Colburn analogy.

Results and discussion

Drying of outer leaf

In Figure 10.12(a), the drying rates are compared for HAM (with constant CTCs) and CHAM for different locations on the wall. Also the average drying rate for CHAM is given and does not differ significantly from the drying rate of HAM. A notable variation over the wall is however found with CHAM, showing high drying rates near the roof top which are caused by the high wind speeds at that location.

The significant decrease in drying rates (for CHAM at 4h, 6h and 9h for y = 9.8, 7.5 and 5 m respectively and for HAM at 10h) can be explained in combination with Figure 10.12(b), where the temperature and RH at the exterior wall surface (interface) are shown. As long as the surface is almost saturated (RH ≈ 100 per cent), a relatively high but steadily decreasing drying rate is found together with a temperature decrease (below T_e), caused by the extraction of heat required for evaporation at the surface. When the RH at the surface drops below 100 per cent, the dry layer at the exterior of the material results in an additional vapour resistance for drying, causing a sudden decrease in drying rate. This decrease in drying rate results in an increase in temperature since less heat is required for the evaporation of water. Note that a relatively low wind speed is used in the simulations (U_{10} = 2.5 m/s) and that the drying time will decrease at higher wind speeds.

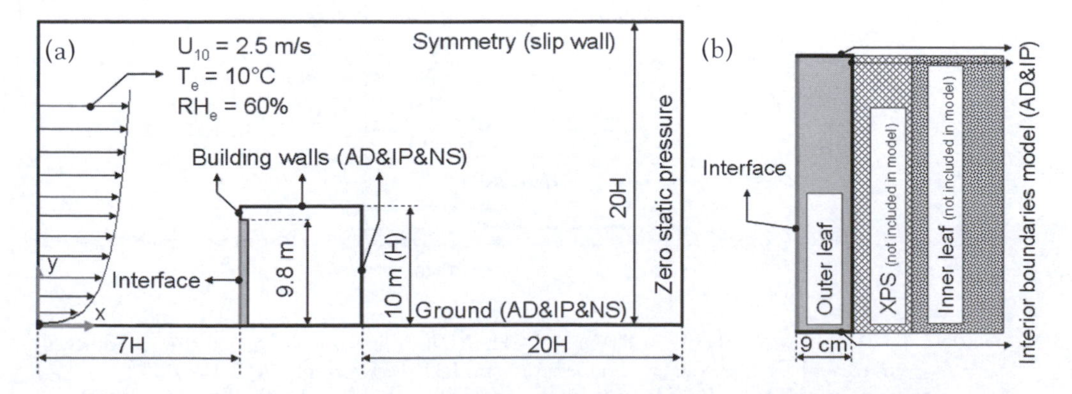

Figure 10.11 Model for numerical analysis with indication of boundary conditions: (a) CFD computational domain; (b) Wall composition (AD = adiabatic, IP = impermeable for moisture, NS = no-slip wall with zero roughness) (Note that the figure is not to scale).

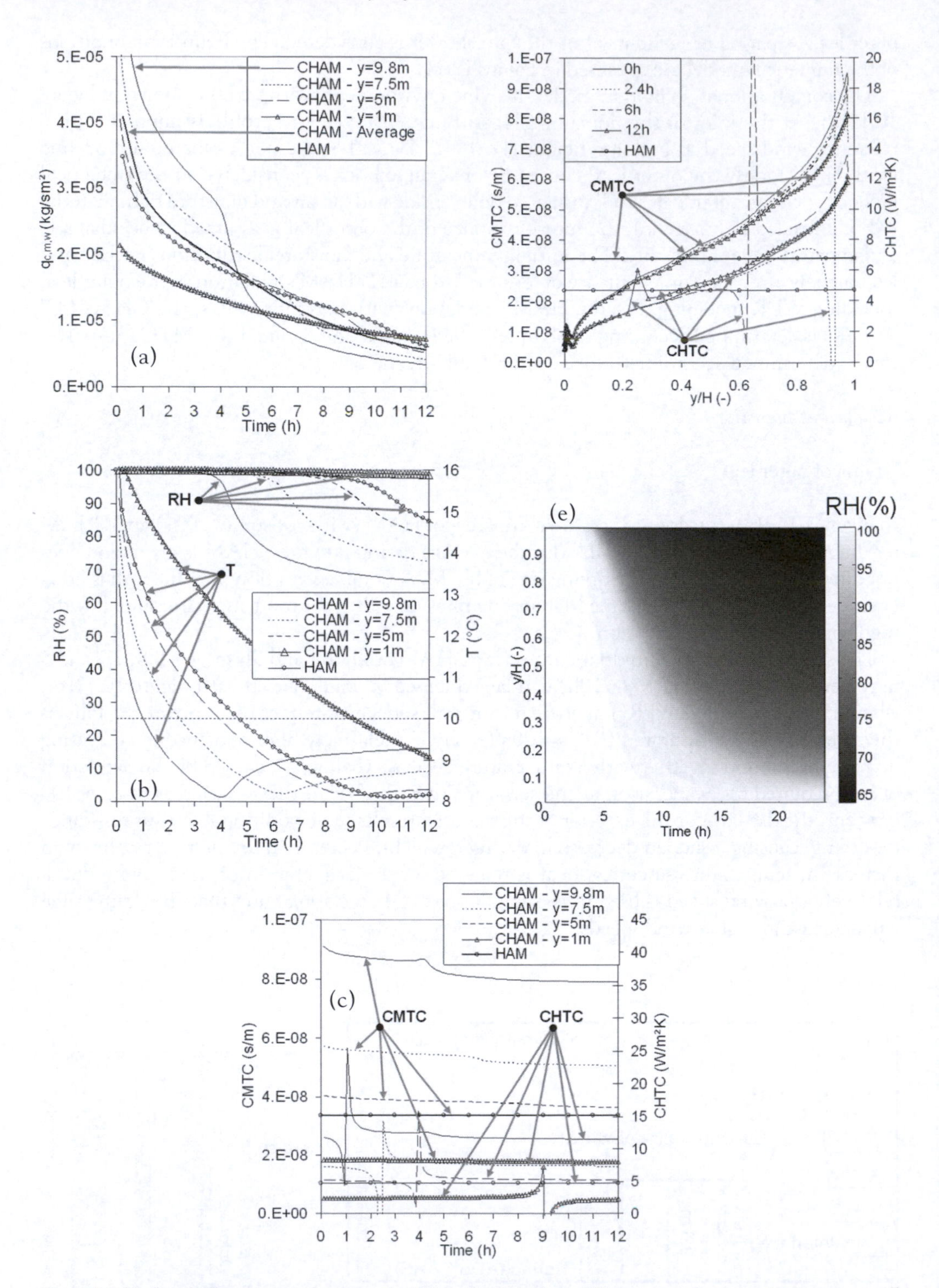

Figure 10.12 (a–c) Comparison between HAM and CHAM at different positions on the exterior wall surface: (a) Drying rate; (b) Temperature and RH (RH = thick lines, T = thin lines); (c) CHTC and CMTC (CMTC = thick lines, CHTC = thin lines); (d) Distribution of CHTC and CMTC over exterior wall surface at different points in time (CMTC = thick lines, CHTC = thin lines) for HAM and CHAM; (e) RH (%) distribution (with CHAM) over exterior wall surface as a function of time and location on surface.

CTCs

The spatial variation of the CHTC and CMTC over the outer leaf is shown in Figure 10.12(d), at different points in time. As expected, the increased drying rate at near the roof top is also reflected in the CTC distribution. In Figure 10.12(e), the RH distribution on the exterior wall surface is shown, as a function of time (for a period of 24h) and location along the height of the surface, clearly showing the importance of the spatial variation of the CTCs in the analysis of drying of the outer leaf, which could be successfully investigated with the CHAM approach.

A relatively small temporal variation of the CTCs can also be noticed in Figure 10.12(d), more pronounced for the CMTC. In Figure 10.12(c), the variation of the CTCs in time is shown for different locations on the exterior wall surface. In both figures, the peaks found for the CHTC are due to the fact that the temperature drops below the reference temperature (see Figure 10.12(b)) due to the evaporation at the surface, resulting in local peaks. These anomalies are intrinsically resulting from the way the CHTC is defined, namely based on the outside air temperature (T_e).

Conclusions

The influence of the use of spatially and temporally constant CTCs, i.e. the conventional HAM modeling approach, on the convective drying of an exterior wall was compared with a more detailed approach, which considered coupling between HAM and CFD, in order to obtain more accurate information on the convective transfer at the exterior surface. An observable spatial variation in the drying behavior and of the CTCs was noticed (with CHAM) although the temporal variation was quite limited. The use of the coupled approach, using CFD, allowed a detailed analysis of 2D H&MT and showed an added value compared to the use of constant CTCs.

Conclusion

Long since recognized as a main agent of building envelope failure, moisture is nowadays more and more taken into account when simulating the whole building. This chapter has presented different modeling approaches for the determination of the moisture loads and the prediction of the effects of the presence of moisture. The role of moisture in terms of the performance of the building can thus be assessed. The next challenge in the field of building simulation will be to predict the durability of the building. For such prediction, the adequate determination of the exposure to moisture will be a key element.

Application examples

Moisture buffering in rooms

In this part we apply the model as formulated in the section on pages 281–283 to evaluate the necessary moisture buffering capacity in a room in order to prevent moisture damage due to too high daily variations in RH. This problem is important for cultural heritage preservation in buildings like museums, churches, etc. The method is based on a statistical approach to the reliability problem. The method results in a fast evaluation of the necessary moisture buffering capacity by an 'easy-to-use' chart, which is valid for a given room in a building exposed to known climatic conditions and with known (or estimated) moisture production and natural ventilation schemes.

Defining the reliability problem

Due to the stochastic characteristics of the input variables, the fluctuation of the relative humidity in a room is also a stochastic process, denoted $Y(t)$. An example of a time train for

$Y(t)$ is given in Figure 10.13(a). We consider as measure for the moisture damage risk the maximum daily variation in RH, ΔY, or the difference in maximum and minimum relative humidity during a day:

$$\Delta Y(n) = \max Y_n(t) - \min Y_n(t) \tag{10.8}$$

where n is the index for the day. For instance, in museums, a significant daily RH variation may be a moisture damage indicator for precious artefacts sensitive to shrinkage/swelling. Other damage conditions, such as maximum relative humidity, can be formulated. Since $Y(t)$ is a stochastic process, the damage indicator $\Delta Y(n)$ is also random and can be described by a cumulative distribution function F_Y. Figure 10.13(b) gives an example of the cumulative distribution of the maximum daily variation in relative humidity. We observe that, for this particular case, the daily maximum variation in RH is higher than or equal to 30 per cent for 20 per cent of the time during a year. We also plotted the cumulative distribution function F_{YA} for the same room without moisture buffering, which is obtained by putting for all walls $\alpha_j = 0$. We observe that variations in RH higher than 30 per cent now occur during 66 per cent of the time.

In order to identify the moisture buffering effect of surface materials in a room, we plot the maximum daily variation in relative humidity for the room with moisture buffering $\Delta Y(n)$ versus the results for the room without buffering $\Delta Y_A(n)$ (Figure 10.13(c)). The figure suggests a linear relation between $\Delta Y(n)$ and $\Delta Y_A(n)$ with a as the regression coefficient. Using the regression coefficient a, we rescale the distribution F_Y to F_{Ym} by the linear transformation $\Delta Y_m = a\Delta Y$, where the subscript m refers to 'modelled'. Figure 10.13(d) shows that the rescaled distribution F_{Ym} coincides with the distribution F_{YA}, without moisture buffering. We observe that all distributions rescale into one curve. This means that the regression coefficient a can be interpreted as a measure to describe the moisture buffering of a room. We define the moisture buffering capacity MBC = 1 – a. The MBC value indicates how much the maximum daily variation in relative humidity is reduced on average due to moisture buffering compared to the reference case without moisture buffering. The MBC value is independent of the outside climatic conditions. The MBC value is also found to be independent of the stochastic characteristics, as long the mean value of the stochastic processes is preserved.

Having defined the daily maximum variation in RH as a damage indicator, we can now introduce a damage criterion, defining the limit between safe and unsafe state. As damage criterion, we consider a maximum allowed change in relative humidity during a day, denoted as ΔY_{lim}, where the subscript *lim* refers to 'limit' state. The condition $\Delta Y_{lim} \leq \Delta Y(n)$ denotes the damage or unsafe state. The failure probability P_f is defined as the probability that the damage criterion is violated, or $P_f = P\,[\Delta Y_{lim} \leq \Delta Y(n)]$, with P the probability operator. The failure probability P_f can be determined using $P_f = 1 - F_Y(\Delta Y_{lim})$ with F_Y the cumulative distribution function. To close the reliability problem, we define the maximum allowed failure probability, denoted by P_{max} (as shown on Figure 10.13(b)). For example, we could assume a limit value ΔY_{lim} of 30 per cent with a maximum failure probability P_{max} of 20 per cent, meaning daily variations in relative humidity higher than 30 per cent are allowed for 20 per cent of the time. Having defined the limit value for the daily maximum variation in relative humidity ΔY_{lim} and the maximum failure probability P_{max}, the design problem reduces to finding the cumulative distribution F_Y for which the failure probability equals the maximum failure probability, or $F_Y(\Delta Y_{lim}) = 1 - P_{max}$. For a given room geometry, described by α-values of the different walls, and for given stochastic processes of ventilation rate G_v, moisture production G_p and external climatic conditions T_e and p_e, the distribution function F_Y depends now only on the moisture buffering capacity of the walls, represented by the equivalent surface coefficient β_{eq} (s/m). The design problem now reduces to identifying the equivalent surface coefficients β_{eq} given the constraint $F_Y(\Delta Y_{lim}) = 1 - P_{max}$. Using the rescaling property of the cumulative RH distribution onto the distribution of

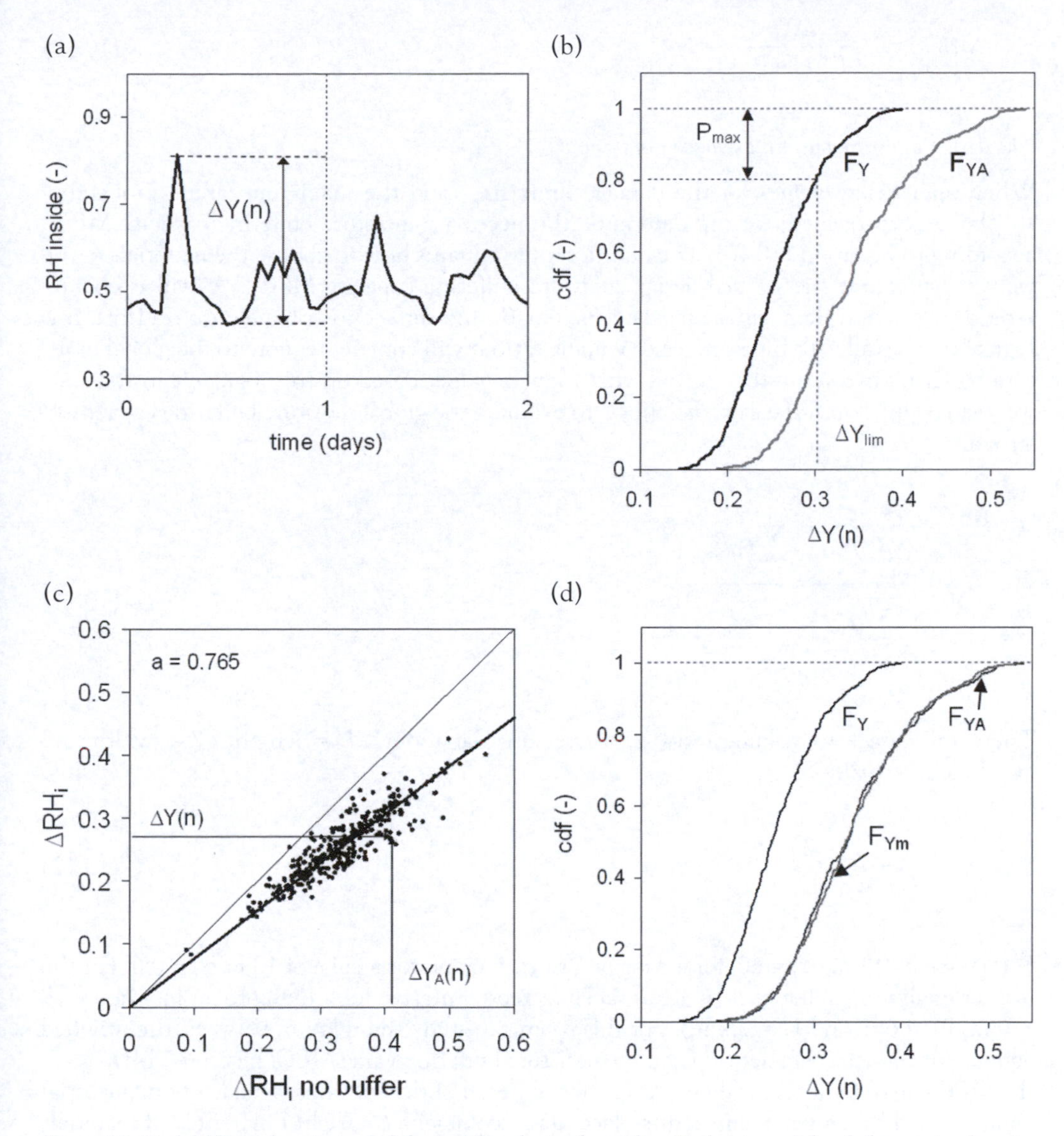

Figure 10.13 (a) Typical variation of the inside relative humidity. Definition of the daily maximal variation of the inside relative humidity $\Delta Y(n)$ as damage risk indicator. (b) Cumulative distribution function (cdf) of the maximum daily variation of inside relative humidity F_Y and F_{YA} respectively for the room with and without moisture buffering. (c) Maximum daily variation of inside relative humidity for the room with moisture buffering versus maximum daily variation of inside relative humidity for the room without buffering. The slope of the linear approximation is defined as 'a'. (d) Comparison of scaled cumulative distribution function F_{Ym} with probability function of the room without buffering F_{YA}.

the room without moisture buffering F_{YA}, the reliability problem can be defined as finding the moisture buffering capacity, for which the following equation holds:

$$F_Y(\Delta Y_{\text{lim}}) = F_{YA}\left(\frac{1}{a}\Delta Y_{\text{lim}}\right) = F_{YA}\left(\frac{1}{1-MBC}\Delta Y_{\text{lim}}\right) = 1 - P_{\text{max}} \tag{10.9}$$

Inverting Equation (10.9), we find the required *MBC* value, referred to as the design moisture buffering capacity (subscript *D*):

$$MBC_D = 1 - \frac{\Delta Y_{\text{lim}}}{F_{YA}^{-1}\left(1 - P_{\text{max}}\right)} \tag{10.10}$$

The design moisture buffering capacity of a room

When appropriate values for the damage limit ΔY_{lim} and the maximum failure probability P_{max} have been defined, we can determine the necessary moisture buffering capacity MBC_D according to Equation (10.10). The question now remains how to choose the materials of the walls to guarantee the needed design moisture buffering capacity MBC_D. A wall is characterised by its equivalent surface mass coefficient β_{eq} and surface to volume ratio α (1/m). It is logical that a wall with high surface to volume ratio α will contribute more to the global moisture buffering in a room than a wall with a low α value. Based on this logic, we introduce a weighed equivalent surface coefficient $\bar{\beta}_{eq}$ to evaluate the global moisture buffering capacity of a room:

$$\bar{\beta}_{eq} = \frac{\sum_{j=1}^{n} \alpha_j \beta_{eq,j}}{\sum_{j=1}^{n} \alpha_j} = \frac{\sum_{j=1}^{n} \alpha_j \beta_{eq,j}}{\alpha_T} \tag{10.11}$$

The total surface to volume ratio α_T for rectangular rooms (L = length, W = width and H = height) equals:

$$\alpha_T = \frac{2}{L} + \frac{2}{W} + \frac{2}{H} \tag{10.12}$$

For a cube of 0.1 m, $\alpha_T = 60$, for a cube of 1 m $\alpha_T = 6$ and for a cube of 10 m $\alpha_T = 0.6$. For the further analysis, a value of $\alpha_T = 1.5$ is taken as representative for a living room in a house (L = 6 m, W = 6 m and H = 2.4 m). For this room, we study the relation between the weighed equivalent surface coefficient $\bar{\beta}_{eq}$ and the moisture buffering value *MBC*. Figure 10.14(a) gives the *MBC*-curve for a room where the surface material of the six walls is made of one material characterised by a given equivalent surface mass coefficient β_{eq}. Eight different wall materials are considered. The solid line gives the obtained relation. In a second step we consider that the surface of the six walls of the room can be composed of different materials. Fifty simulations are performed for random combinations of wall materials chosen out of a set of eight materials with different equivalent surface mass coefficients β_{eq}. We observe that all simulations coincide into a single curve, called the *MBC*-curve. In Figure 10.14(b), the *MBC*-curve is determined for different values of α_T. Small rooms with high α_T value only need a low moisture buffering capacity (low $\bar{\beta}_{eq}$ value), while large rooms with low α_T value need a higher moisture buffering capacity of the walls.

In summary, the moisture buffering design method consists of the following steps: (1) select proper stochastic processes for the ventilation and moisture production rate; (2) determine the cumulative distribution function of the maximum daily variation of the inside relative humidity for the room without moisture buffering; (3) choose a maximum value for the daily variation in RH as damage criterion and a maximum probability of failure; (4) determine the design moisture buffering capacity (MBC_D) value using Equation (10.10); (5) determine the necessary moisture buffering capacity of the surface materials using the moisture buffering capacity curve, as shown in Figure 10.14(b).

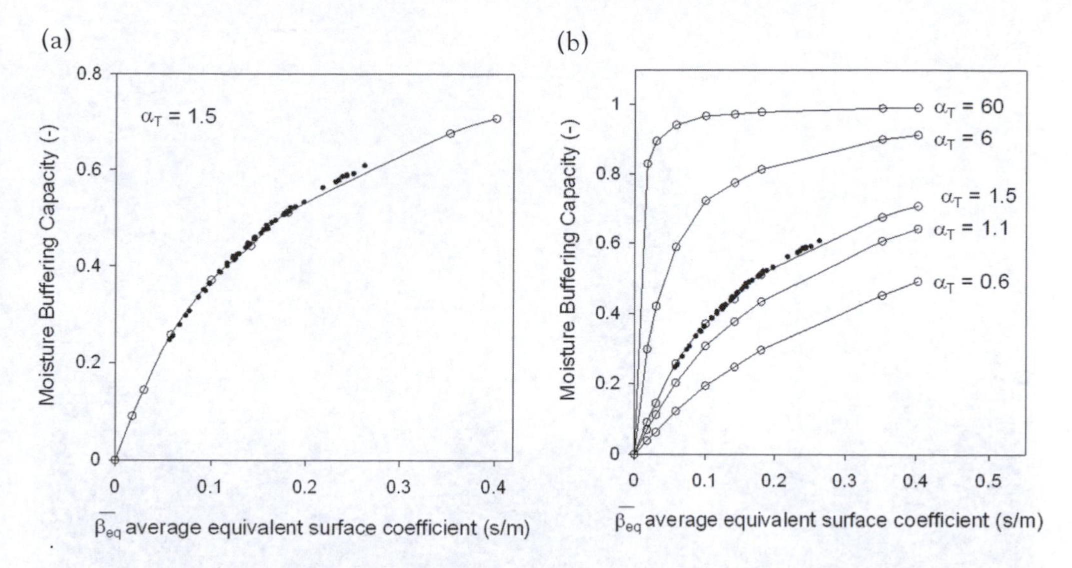

Figure 10.14 Moisture buffering design (MBD) curves representing the moisture buffering capacity versus the average equivalent surface mass coefficient. (a) Comparison of MBC curves for a room with one and the same wall material (solid line) and with random wall materials (solid dots); (b) Moisture buffering design curves for rooms with different total surface to volume ratios.

Modeling of wind-driven rain

Computational modeling of WDR has many applications. It is one of the most important boundary conditions for heat-air-moisture (HAM) transfer analysis of building components exposed to the outside climate (Blocken *et al.* 2007c; Janssen *et al.* 2007). In addition, it can also reduce the effectiveness of rain sheltering devices that are often designed with only vertical rainfall in mind, such as the roofs that cover the stands of football stadia (Persoon *et al.* 2008; van Hooff *et al.* 2011). In this section, a case study of WDR on building facades is presented.

The monumental building St. Hubertus (Figure 10.15(a)) is situated in the National Park 'De Hoge Veluwe' in the Netherlands. Especially the south-west facade of the building shows severe deterioration caused by WDR and subsequent phenomena such as rain penetration, mold growth, frost damage, salt crystallization and efflorescence, and cracking due to hygrothermal gradients (Figure 10.15(b–e)). As part of a larger project to analyse the causes and to provide remedial measures, CFD simulations of WDR on the south-west facade of the tower were performed. The CFD WDR simulations are important to obtain accurate spatial and temporal distribution records of WDR, to be used as input for numerical HAM transfer simulations. Validation of the CFD simulations was performed by WDR measurements at a few selected locations at the south-west facade. Note that only performing WDR measurements would have been insufficient, because these measurements at the few discrete positions do not give enough information to obtain a complete picture of the spatial distribution of WDR on the south-west facade of the tower. Therefore, they were supplemented with numerical simulations. Figure 10.16 displays the computational grid on the surfaces of the building. The numerical method for simulation of WDR on buildings was applied, yielding the spatial distribution of WDR across the south-west facade. Figure 10.17 shows a comparison of the CFD simulations with the corresponding measurements, indicating a good agreement on the top part of the facade, but a poor agreement at the bottom part, which was attributed to the absence of modeling the turbulent dispersion effect on the raindrops. These errors were compensated based on the measurement results, which demonstrates the importance of high-quality measurements for CFD calibration and validation. The complete study, including sensitivity analysis, is reported in (Briggen *et al.* 2009).

Figure 10.15 (a) Hunting Lodge St. Hubertus and (b–e) moisture damage at the tower due to wind-driven rain: (b) salt efflorescence; (c) cracking/blistering due to salt crystallisation; (d) rain penetration and discoloration; (e) cracking at inside surface.

Source: From Briggen *et al.* 2009.

Discussion

Modeling of moisture buffering

Modeling of moisture buffering includes the modeling of different transport processes. First, we have to consider the heat-air-moisture (HAM) transport in the room itself including the entrance/exit of air by natural ventilation or HVAC systems. This zone is called the fluid zone. Second, we have the HAM transport in the walls surrounding the room, which can be inside and outside walls, which we refer to as the porous zone. Third, we have the hygrothermal interaction between the fluid and porous zone, occurring in a thin boundary layer. In this chapter, we do not consider air transport in the porous zone, e.g. due to air leakage through the wall.

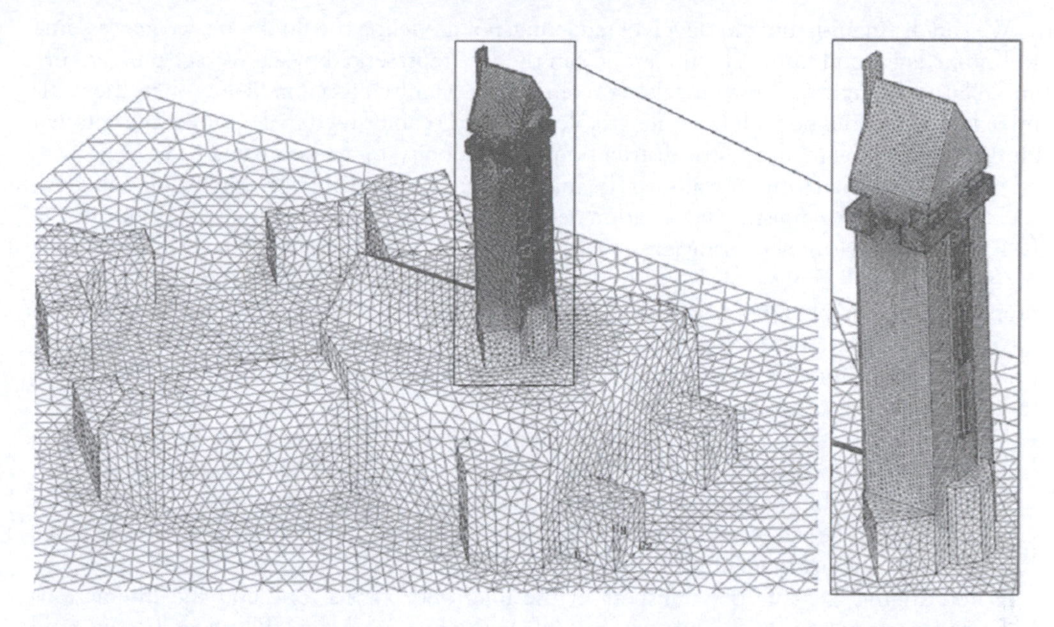

Figure 10.16 Part of the building models and surface grids used for calculation of the wind-flow pattern around the building and wind-driven rain on the south-west building facade. The grid contains 2,110,012 cells.

Source: From Briggen *et al.* 2009.

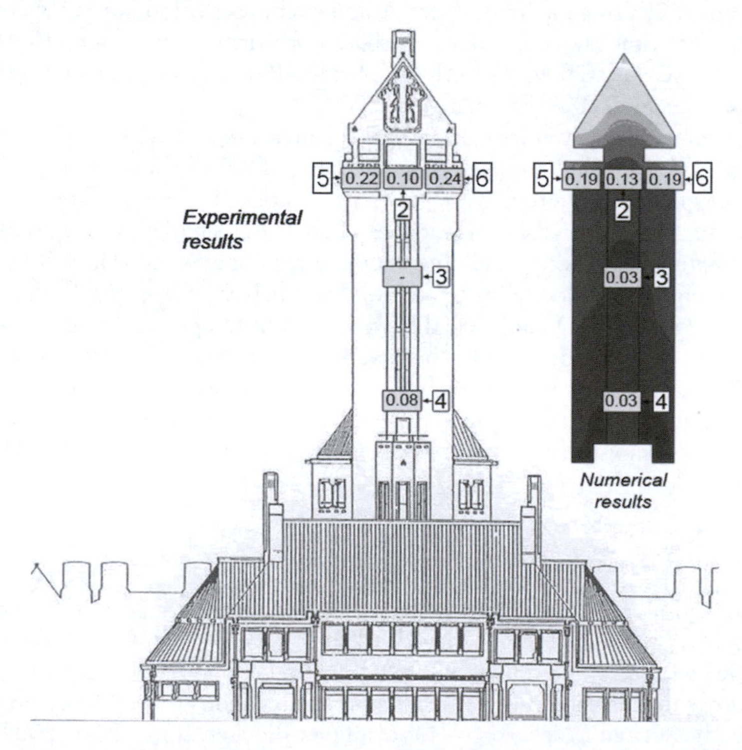

Figure 10.17 Spatial distribution of the catch ratio (ratio of wind-driven rain to horizontal rainfall) at the end of a given rain spell. The experimental results at the locations of the wind-driven rain gauges are shown on the left, the numerical results are shown on the right.

Source: From Briggen *et al.* 2009.

We can distinguish different levels of modeling. For modeling the fluid zone, we may assume well-mixed air conditions. The fluid zone can then be represented by an average temperature and relative humidity. These models are referred to as (multi-) zonal models. When the well-mixed assumption is not valid (e.g. for an inlet jet), CFD can be used to determine the detailed air flow field in the room and the distribution of air velocity, temperature and RH.

The moisture buffering in walls can be modelled in a simplified way using simple models, such as the effective moisture penetration depth (EMPD) model (Cunningham 1992). In the EMPD model, moisture buffering is assumed to occur only in a thin material layer at the surface of the wall. The thickness of this effective penetration layer depends on the moisture transport properties and the period of the relative humidity variation. More accurate transport models for the porous zone solve the coupled heat (air) and moisture transport equations based on finite difference, finite element or control volume techniques, where also interactions with the indoor climate are taken into account.

The transport in the boundary layer is commonly described by convective heat and moisture transfer (surface) coefficients (CHTC/CMTC). Standardized coefficients are commonly used in combination with zonal models. Better predictions of the surface-averaged or local CHTCs and CMTCs can be determined using CFD, resolving the boundary layer using for instance low-Reynolds number modeling (Steeman *et al.* 2009).

When aiming to solve the transport in the fluid and porous zone and the interactions between these zones in the boundary layer, CFD has to be used. Most CFD codes, however, do not include moisture transfer in porous materials as they do not include hygroscopic or capillary effects. For such complete models, CFD for the fluid zone and the boundary layer has to be combined with a HAM model for the porous zone. The coupling between CFD and HAM can be solved externally or internally within the CFD code. External coupling solves the fluid and porous domain by using different programs and exchanges information between the both domains at discrete time steps. Examples of these approaches can be found in Neale *et al.* (2007) and Defraeye *et al.* (2009). Directly coupled models solve the fluid and porous domain in one program (Steeman 2009; Steeman *et al.* 2009).

In the literature, different combinations of the presented modeling approaches for the two zones can be found. Building energy models, like TRNSYS and EnergyPlus, are zonal models and take moisture buffering by the walls into account by simplified models such as the EPDM model using standardized CHTCs and CMTCs. HAM simulation models for walls have recently been extended to whole building simulation programs, allowing the simultaneous evaluation of energy use, indoor climate and moisture behavior (e.g. WUFI Plus, CHAMPS BES). Steeman *et al.* (2009) found that the prediction of temperature and RH in the room by these building energy models can be improved by using average transfer coefficients, predicted by CFD. In the same work, zonal models in combination with local transfer coefficients were found to be insufficiently accurate to predict local moisture behavior in the wall. In these cases (e.g. in cases of moisture damage of wooden artefacts), fully coupled CFD-HAM modeling is required.

Modeling of wind-driven rain

Generally, problems in computational modeling of WDR are associated with correctly simulating the atmospheric boundary layer (ABL) (Blocken *et al.* 2007a; Franke *et al.* 2007), reproducing the complex wind-flow pattern around buildings and appropriately calculating the movement of raindrops in this wind-flow pattern. Potential solutions to wall function problems in ABL modeling have been addressed (in, for example, Blocken *et al.* 2007a, 2007b; Franke *et al.* 2007; Gorlé *et al.* 2009; Hargreaves and Wright 2007; Richards and Hoxey 1993; Yang *et al.* 2009). Some general problems in calculating the wind-flow pattern and the raindrops and potential ways to solve them were addressed on pages 288–289. More information can be found in best practice documents (Casey and Wintergerste 2000; Franke *et al.* 2007; Tominaga *et al.*

2008a). For raindrop modeling, additional guidelines can be found in Choi (1993, 1994) and Blocken and Carmeliet (2002, 2004).

It is important to mention that CFD solution verification and validation are essential and that every CFD user should perform these studies. Solution verification refers to assessing and limiting the numerical errors, such as the discretisation error and the iterative convergence error. Performing a grid-sensitivity analysis is a minimum verification requirement for every type of CFD study. CFD validation refers to analysing the physical modeling errors, by comparing the CFD results with high-quality on-site or wind-tunnel measurements. For WDR studies, two validation steps can be distinguished: validation of the wind-flow pattern around the building and validation of the WDR intensities on the building facade. For the first step, some experimental data of mean wind velocities and turbulence intensities exist in the literature, although mostly for relatively simple building geometries. For the second step, unfortunately, only a limited amount of high-quality and complete experimental datasets exist. Three extensive datasets are known to the authors, two for low-rise buildings and one for a high-rise building. They are reported in (van Mook 2002; Blocken and Carmeliet 2005; Nore *et al.* 2007). There is an important need for additional, extensive and complete experimental wind, rain and WDR data.

Modeling of convective heat and moisture transfer at exterior building surfaces

Computational modeling of convective heat and moisture transfer by coupling CFD with HAM models is relatively new and therefore still under development (Defraeye *et al.* 2009; Erriguible *et al.* 2006; Neale *et al.* 2007; Steeman 2009). The modeling problems are related to: (1) CFD modeling of the wind-flow field around buildings; (2) HAM modeling of heat and moisture transport in porous building materials; and (3) the coupling of both models.

The CFD modeling issues have already been discussed above (pages 300–301). Several HAM models (Delphin; Janssen *et al.* 2007) have been developed in the past and were extensively validated against benchmark experiments, e.g. HAMSTAD (Hagentoft 2002). Note that a critical component in the accuracy of these models is however related to an accurate determination of the heat and moisture transport properties of the porous material, such as permeabilities and capacities, which is usually done experimentally. For the coupling of both models, different strategies can be used which are discussed in Zhai *et al.* (2002). Two are of specific interest, namely quasi-dynamic coupling (Defraeye *et al.* 2009; Erriguible *et al.* 2006; Neale *et al.* 2007), where information is only exchanged once every time step, and fully-dynamic coupling, where iterations between both models are performed each time step (Steeman 2009). Quasi-dynamic coupling requires very small time steps, which can lead to long simulation times. It is however often more straightforward to implement that a fully-dynamic coupling strategy.

Verification of the coupled CFD-HAM code is imperative, which includes verification of both CFD and HAM models separately (see also pages 300–301) but most of all the verification of the coupling, which means, amongst others, that heat and mass are conserved. Apart from spatial (grid) sensitivity analysis for both models, temporal sensitivity (time step) analysis has to be performed for the coupled model. For validation, laboratory experiments are preferred over full-scale experiments on buildings due to the higher spatial resolution and a better control of boundary conditions. Few experiments are however available for validation of these coupled models. A detailed benchmark experiment of hygroscopic loading of gypsum board can be found in Roels (2008).

Synopsis

This chapter has focused on modeling aspects of some important moisture phenomena in whole building simulations, in particular modeling of moisture buffering in rooms, modeling of wind-driven rain and of convective heat and moisture transfer at exterior building surfaces.

It has provided the background and state-of-the-art in these areas, as well as some case studies illustrating the importance for building performance simulation.

References

Abuku, M., Blocken, B., Nore, K., Thue, J.V., Carmeliet, J. and Roels, S. (2009) 'On the Validity of Numerical Wind-driven Rain Simulation on a Rectangular Low-rise Building under Various Oblique Winds', *Building and Environment* 44(3): 621–632.

Arundel, A.V., Sterling, E.M., Biggin, J.H. and Sterling, T.D., (1986) 'Indirect Health Effects of Relative Humidity in Indoor Environments', *Environmental Health Perspectives* 65: 351–356.

ASHRAE (2004) Thermal Environmental Conditions for Human Occupancy. *ANSI/ASHRAE Standard 55–2004*, Atlanta: ASHRAE.

Berglund, L.G. and Cunningham, D.J. (1986) 'Parameters of Human Discomfort in Warm Environments', *ASHRAE Transactions* 92(2): 732–746.

Blocken, B. and Carmeliet, J. (2002) 'Spatial and Temporal Distribution of Driving Rain on a Low-rise Building', *Wind and Structures* 5(5): 441–462.

Blocken, B. and Carmeliet, J. (2004) 'A Review on Wind-driven Rain Research in Building Science' *Journal of Wind Engineering and Industrial Aerodynamics* 92(13): 1079–1130.

Blocken, B. and Carmeliet, J. (2005) 'High-resolution Wind-driven Rain Measurements on a Low-rise Building – Experimental Data for Model Development and Model Validation', *Journal of Wind Engineering and Industrial Aerodynamics* 93(12): 905–928.

Blocken, B. and Carmeliet, J. (2006a) 'On the Accuracy of Wind-driven Rain Measurements on Buildings', *Building and Environment* 41(12): 1798–1810.

Blocken, B. and Carmeliet, J. (2006b) 'The Influence of the Wind-blocking Effect by a Building on Its Wind-driven Rain Exposure', *Journal of Wind Engineering and Industrial Aerodynamics* 94(2): 101–127.

Blocken, B. and Carmeliet J. (2007a) 'On the Errors Associated with the Use of Hourly Data in Wind-driven Rain Calculations on Building Façades', *Atmospheric Environment* 41(11): 2335–2343.

Blocken, B. and Carmeliet, J. (2007b) 'Validation of CFD Simulations of Wind-driven Rain on a Low-rise Building Façade', *Building and Environment* 42(7): 2530–2548.

Blocken, B., Stathopoulos, T. and Carmeliet, J. (2007a) 'CFD Simulation of the Atmospheric Boundary Layer: Wall Function Problems', *Atmospheric Environment* 41(2): 238–252.

Blocken, B., Carmeliet, J. and Stathopoulos, T. (2007b) 'CFD Evaluation of Wind Speed Conditions in Passages between Parallel Buildings – Effect of Wall-function Roughness Modifications for the Atmospheric Boundary Layer Flow', *Journal of Wind Engineering and Industrial Aerodynamics* 95(9–11): 941–962.

Blocken, B., Roels, S. and Carmeliet, J. (2007c) 'A Combined CFD-HAM Approach for Wind-driven Rain on Building Façades', *Journal of Wind Engineering and Industrial Aerodynamics* 95(7): 585–607.

Blocken, B., Dezsö, G., van Beeck, J. and Carmeliet, J. (2009a) 'The Mutual Influence of Two Buildings on their Wind-driven Rain Exposure and Comments on the Obstruction Factor', *Journal of Wind Engineering and Industrial Aerodynamics* 97(5–6): 180–196.

Blocken, B., Defraeye, T., Derome, D. and Carmeliet, J. (2009b) 'High-resolution CFD Simulations of Forced Convective Heat Transfer Coefficients at the Facade of a Low-rise Building', *Building and Environment* 44(12): 2396–2412.

Briggen, P.M., Blocken, B. and Schellen, H.L. (2009) 'Wind-driven Rain on the Facade of a Monumental Tower: Numerical Simulation, Full-scale Validation and Sensitivity Analysis', *Building and Environment* 44(8): 1675–1690.

Camuffo, D., Del Monte, M., Sabbioni, C. and Vittori, O. (1982) 'Wetting, Deterioration and Visual Features of Stone Surfaces in an Urban Area', *Atmospheric Environment* 16(9): 2253–2259.

Carmeliet, J., Roels, S. and Bomberg, M. (2009) 'Chapter 29: Towards Development of Methods for Assessment of Moisture-originated Damage', *ASTM Moisture Manual*, in press.

Casey, M. and Wintergerste, T. (2000) *Best Practice Guidelines*. ERCOFTAC Special Interest Group on Quality and Trust in Industrial CFD, Brussels: ERCOFTAC.

CHAMPS-BES: Coupled Heat, Air, Moisture and Pollutant Simulation in Building Envelope Systems, TU Dresden.

Charola, A.E. and Lazzarini, L. (1986) 'Deterioration of Brick Masonry Caused by Acid Rain', *ACS Symposium Series* 318, pp. 250–258.

Chen, X.D., Lin, S.X.Q. and Chen, G. (2002) 'On the Ratio of Heat to Mass Transfer Coefficient for Water Evaporation and Its Impact upon Drying Modelling', *International Journal of Heat and Mass Transfer* 45(21): 4369–4372.

Choi, E.C.C. (1991) 'Numerical Simulation of Wind-driven Rain Falling onto a 2-D Building', in *Proceedings of the Asia Pacific Conference on Computational Mechanics*, Hong Kong, pp. 1721–1728.

Choi, E.C.C. (1993) 'Simulation of Wind-driven Rain around a Building', *Journal of Wind Engineering and Industrial Aerodynamics* 46–47: 721–729.

Choi, E.C.C. (1994) 'Determination of Wind-driven Rain Intensity on Building Faces', *Journal of Wind Engineering and Industrial Aerodynamics* 51: 55–69.

Choi, E.C.C. (1997) 'Numerical Modeling of Gust Effect on Wind-driven Rain', *Journal of Wind Engineering and Industrial Aerodynamics* 72: 107–116.

Cunningham, M.J. (1992) 'Effective Penetration Depth and Effective Resistance in Moisture Transfer', *Building and Environment* 27(3): 379–386.

Davidson, C.I., Tang, W., Finger, S., Etyemezian, V., Striegel, M.F. and Sherwood, S.I. (2000) 'Soiling Patterns on a Tall Limestone Building: Changes over 60 Years', *Environmental Science & Technology* 34(4): 560–565.

Day, A.G., Lacy, R.E. and Skeen, J.W. (1955) 'Rain Penetration through Walls, A Summary of the Investigations Made at the UK Building Research Station from 1925 to 1955', *Building Research Station Note*. No. C.364, unpublished.

Defraeye, T., Blocken, B. and Carmeliet, J. (2009) 'Computational Modelling of Convective Heat and Moisture Transfer at Exterior Building Surfaces', in *Proceedings of the 7thInternational Conference on Urban Climate*, Yokohama, Japan.

Defraeye, T., Blocken, B. and Carmeliet, J. (2010) 'CFD Analysis of Convective Heat Transfer at the Surfaces of a Cube Immersed in a Turbulent Boundary Layer', *International Journal of Heat and Mass Transfer* 53(1–3): 297–308.

Delphin: Simulation Program for the Calculation of Coupled Heat, Moisture, Air, Pollutant, and Salt Transport. Available from: www.bauklimatik-dresden.de/delphin.

Derome, D. (1999) 'Moisture Occurrence in Roof Assemblies Containing Moisture Storing Insulation and Its Impact on the Durability of the Building Envelope', PhD Thesis, Concordia University, Montreal, Canada.

Eldridge, H.J. (1976) *Common Defects in Buildings*, London: Her Majesty's Stationery Office, pp. 486.

Erriguible, A., Bernada, P., Couture, F. and Roques, M. (2006) 'Simulation of Convective Drying of a Porous Medium with Boundary Conditions Provided by CFD', *Chemical Engineering Research and Design* 84(2): 113–123.

Etyemezian, V., Davidson, C.I., Zufall, M., Dai, W., Finger, S. and Striegel, M. (2000) 'Impingement of Rain Drops on a Tall Building', *Atmospheric Environment* 34(15): 2399–2412.

Fang, L., Clausen, G. and Fanger, P. O. (1998) 'Impact of Temperature and Humidity on the Perception of Indoor Air Quality', *Indoor Air* 8(2): 80–90.

Fluent Inc. (2006) *Fluent 6.3 User's Guide*, Lebanon, New Hampshire.

Franke, J. and Frank, W. (2008) 'Application of Generalized Richardson Extrapolation to the Computation of the Flow across an Asymmetric Street Intersection', *Journal of Wind Engineering and Industrial Aerodynamics* 96(10–11): 1616–1628.

Franke, L., Schumann, I., van Hees, R., van der Klugt, L., Naldini, S., Binda, L., Baronio, G., Van Balen, K. and Mateus, J. (1998) 'Damage Atlas: Classification and Analyses of Damage Patterns Found in Brick Masonry', *European Commission Research Report*, No. 8, Vol. 2, Fraunhofer IRB Verlag.

Franke, J., Hirsch, C., Jensen, A.G., Krüs, H.W., Schatzmann, M., Westbury, P.S., Miles, S.D., Wisse, J.A. and Wright, N.G. (2004) 'Recommendations on the Use of CFD in Wind Engineering', in van Beeck, J. (ed.), *Proceedings of the International Conference on Urban Wind Engineering and Building Aerodynamics. COST Action C14, Impact of Wind and Storm on City Life Built Environment*, Sint-Genesius-Rode, Belgium: Von Karman Institute.

Franke, J., Hellsten, A., Schlünzen, H. and Carissimo, B. (2007) *Best Practice Guideline for the CFD Simulation of Flows in the Urban Environment*, Brussels: COST Office.

Garden, G.K. (1963) 'Rain Penetration and Its Control', *Canadian Building Digest*, Division of Building Research, National Research Council, CBD40, 401–404.

Gorlé, C., van Beeck, J., Rambaud, P. and Van Tendeloo, G. (2009) 'CFD Modelling of Small Particle

Dispersion: The Influence of the Turbulence Kinetic Energy in the Atmospheric Boundary Layer', *Atmospheric Environment*, 43(3): 673–681.

Hagentoft, C.-E. (2002) *HAMSTAD – Final Report: Methodology of HAM-Modeling*. Report R-02:8, Department of Building Physics, Chalmers University of Technology, Gothenburg.

Hangan, H. (1999) 'Wind-driven Rain Studies. A C-FD-E Approach'. *Journal of Wind Engineering and Industrial Aerodynamics* 81: 323–331.

Hargreaves, D.M. and Wright, N.G. (2007) 'On the Use of the k-ε Model in Commercial CFD Software to Model the Neutral Atmospheric Boundary Layer', *Journal of Wind Engineering and Industrial Aerodynamics* 95(5): 355–369.

Högberg, A.B., Kragh, M.K. and van Mook, F.J.R. (1999) 'A Comparison of Driving Rain Measurements with Different Gauges', in *Proceedings of the 5th Symposium of Building Physics in the Nordic Countries*, Gothenborg, Sweden, pp. 361–368.

Inculet, D. and Surry, D. (1994) 'Simulation of Wind-driven Rain and Wetting Patterns on Buildings', BLWTL-SS30-1994, Final Report.

Inculet, D.R. (2001) 'The Design of Cladding against Wind-driven Rain', PhD Thesis, The University of Western Ontario, London.

ISO (2009) *Hygrothermal Performance of Buildings – Calculation and Presentation of Climatic Data – Part 3: Calculation of a Driving Rain Index for Vertical Surfaces from Hourly Wind and Rain Data*, ISO 15927–3:2009 International Organization for Standardization.

Janssen, H., Blocken, B. and Carmeliet, J. (2007) 'Conservative Modelling of the Moisture and Heat Transfer in Building Components under Atmospheric Excitation', *International Journal of Heat and Mass Transfer* 50(5–6): 1128–1140.

Karagiozis, A., Hadjisophocleous, G. and Cao, S. (1997) 'Wind-driven Rain Distributions on Two Buildings', *Journal of Wind Engineering and Industrial Aerodynamics* 67–68: 559–572.

Lacy, R.E. (1977) *Climate and Building in Britain*, London: Her Majesty's Stationery Office.

Lakehal, D., Mestayer, P.G., Edson, J.B., Anquetin, S. and Sini, J.-F. (1995) 'Eulero-Lagrangian Simulation of Raindrop Trajectories and Impacts within the Urban Canopy', *Atmospheric Environment* 29(23): 3501–3517.

Lienhard, J.H.IV, and Lienhard, J.H.V (2006) *A Heat Transfer Textbook*, 3rd ed., Cambridge, MA: Phlogiston Press.

Loth, E. (2000) 'Numerical Approaches for Motion of Dispersed Particles, Droplets, and Bubbles', *Progress in Energy and Combustion Science* 26: 161–223.

Marsh, P. (1977) *Air and Rain Penetration of Buildings*, Lancaster, England: The Construction Press Ltd, p. 174.

Maurenbrecher, A.H.P. and Suter, G.T. (1993) 'Frost Damage to Clay Brick in a Loadbearing Masonry Building', *Canadian Journal of Civil Engineering* 20: 247–253.

Merril, J.L. and Ten Wolde, A. (1989) 'Overview of Moisture Related Damage in One Group of Wisconsin Manufactured Homes', *ASHRAE Transactions* 95(1): 405–414.

Murakami, S. (1993) 'Comparison of Various Turbulence Models Applied to a Bluff Body', *Journal of Wind Engineering and Industrial Aerodynamics* 46–47: 21–36.

Neale, A., Derome, D., Blocken, B. and Carmeliet, J. (2007) 'Coupled Simulation of Vapor Flow between Air and a Porous Material', *Performance of Exterior Envelopes of Whole Building X Conference*, Atlanta: ASHRAE.

Nore, K., Blocken, B., Jelle, B.P., Thue, J.V. and Carmeliet, J. (2007) 'A Dataset of Wind-driven Rain Measurements on a Low-rise Test Building in Norway', *Building and Environment* 42(5): 2150–2165.

Oreszczyn, T., Ridley, I., Hong, S.H. and Wilkinson, P. (2006) 'Mould and Winter Indoor Relative Humidity in Low Income Households in England', *Indoor and Built Environment* 15(2): 125–135.

Osanyintola, O.F. and Simonson, C.J. (2006) 'Moisture Buffering Capacity of Hygroscopic Materials: Experimental Facilities and Energy Impact', *Energy and Buildings* 38(10): 1270–1282.

Persoon, J., van Hooff, T., Blocken, B., Carmeliet, J. and de Wit, M.H. (2008) On the Impact of Roof Geometry on Rain Shelter in Football Stadia, *Journal of Wind Engineering and Industrial Aerodynamics* 96(8–9): 1274–1293.

Price, C.A. (1975) 'The Decay and Preservation of Natural Building Stone', *Chemistry in Britain* 11(10): 350–353.

Richards, P.J. and Hoxey, R.P. (1993) 'Appropriate Boundary Conditions for Computational Wind Engineering Models Using the k-ε Turbulence Model', *Journal of Wind Engineering and Industrial Aerodynamics* 46–47: 145–153.

Robinson, G. and Baker, M.C. (1975) *Wind-driven Rain and Buildings*, Technical Paper No. 445, Division of Building Research, National Research Council, Ottawa, Canada.

Roels, S. (2008) *IEA Annex 41: Whole Building Heat, Air, Moisture Response. Subtask 2: Experimental Analysis of Moisture Buffering*, Technical Report, International Energy Agency.

Sanders C. (1996) *IEA Annex 24: Heat, Air and Moisture Transfer in Insulated Envelope Parts: Environmental Conditions*, Technical Report, International Energy Agency.

Sharples, S. (1984) 'Full-scale Measurements of Convective Energy Losses from Exterior Building Surfaces', *Building and Environment* 19(1): 31–39.

Shih, T.H., Liou, W.W., Shabbir, A. and Zhu, J. (1995) 'A New k-ε Eddy-viscosity Model for High Reynolds Number Turbulent Flows – Model Development and Validation', *Computers and Fluids* 24(3): 227–238.

Shirolkar, J.S., Coimbra, C.F.M. and McQuay, M.Q. (1996) 'Fundamental Aspects of Modeling Turbulent Particle Dispersion in Dilute Flows', *Progress in Energy and Combustion Science* 22: 363–399.

Steeman, H.-J. (2009) 'Modelling Local Hygrothermal Interaction between Airflow and Porous Materials for Building Applications', PhD Thesis, Ghent University, Ghent, Belgium.

Steeman, H.-J., Janssens, A., Carmeliet, J. and De Paepe, M. (2009) 'Modelling Indoor Air and Hygrothermal Wall Interaction in Building Simulation: Comparison between CFD and a Well-mixed Zonal Model', *Building and Environment* 44(3): 572–583.

Straube, J.F. (1998) 'Moisture Control and Enclosure Wall Systems', PhD Thesis, University of Waterloo, Ontario, Canada.

Straube, J.F. and Burnett, E.F.P. (2000) 'Simplified Prediction of Driving Rain on Buildings', in *Proceedings of the International Building Physics Conference*, Eindhoven, The Netherlands, pp. 375–382.

Stupart, A.W. (1989) 'A Survey of Literature Relating to Frost Damage in Bricks', *Masonry International* 3(2): 42–50.

Tang, W. and Davidson, C.I. (2004) 'Erosion of Limestone Building Surfaces Caused by Wind-driven Rain: 2. Numerical Modelling', *Atmospheric Environment* 38(33): 5601–5609.

Toftum, J., Jorgensen, A.S. and Fanger, P.O. (1998) 'Upper Limits of Air Humidity for Preventing Warm Respiratory Discomfort', *Energy and Buildings* 28(1): 15–23.

Tominaga, Y., Mochida, A., Yoshie, R., Kataoka, H., Nozu, T., Yoshikawa, M. and Shirasawa, T. (2008a) 'AIJ Guidelines for Practical Applications of CFD to Pedestrian Wind Environment around Buildings', *Journal of Wind Engineering and Industrial Aerodynamics* 96(10–11): 1749–1761.

Tominaga, Y., Mochida, A., Murakami, S. and Sawaki, S. (2008b) 'Comparison of Various Revised k–ε Models and LES Applied to Flow around a High-rise Building Model with 1:1:2 Shape Placed within the Surface Boundary Layer', *Journal of Wind Engineering and Industrial Aerodynamics* 96(4): 389–411.

van Hooff, T. and Blocken, B. (2010) 'Coupled Urban Wind Flow and Indoor Natural Ventilation Modelling on a High-resolution Grid: A Case Study for the Amsterdam Arena Stadium', *Environmental Modelling & Software* 25(1): 51–65.

van Hooff, T., Blocken, B. and van Harten, M. (2011) '3D CFD Simulations of Wind Flow and Wind-Driven Rain Shelter in Sports Stadia: Influence of Stadium Geometry', *Building and Environment* 46(1): 22–37.

van Mook, F.J.R. (2002) 'Driving Rain on Building Envelopes', PhD Thesis, Eindhoven University of Technology, Eindhoven, The Netherlands.

White, R.B. (1967) *The Changing Appearance of Buildings*, London: Her Majesty's Stationery Office, p. 64.

Wilson, D.J. (1989) 'Airflow around Buildings', *ASHRAE Handbook of Fundamentals*, Atlanta: ASHRAE, pp. 14.1–14.18.

Wolfshtein, M. (1969) 'The Velocity and Temperature Distribution of One-dimensional Flow with Turbulence Augmentation and Pressure Gradient', *International Journal of Heat and Mass Transfer* 12(3): 301–318.

WUFI®, WUFI Plus: PC-Program for Calculating the Coupled Heat and Moisture Transfer in Building Components, Fraunhofer Institut für Bauphysik, Holzkirchen.

Yang, Y., Gu, M., Chen, S. and Jin, X. (2009) 'New Inflow Boundary Conditions for Modelling the Neutral Equilibrium Atmospheric Boundary Layer in Computational Wind Engineering', *Journal of Wind Engineering and Industrial Aerodynamics* 97(2): 88–95.

Yoshie, R., Mochida, A., Tominaga, Y., Kataoka, H., Harimoto, K., Nozu, T. and Shirasawa, T. (2007) 'Cooperative Project for CFD Prediction of Pedestrian Wind Environment in the Architectural Institute of Japan', *Journal of Wind Engineering and Industrial Aerodynamics* 95(9–11): 1551–1578.

Zhai, Z., Chen, Q., Haves, P. and Klems, J.H. (2002) 'On Approaches to Couple Energy Simulation and Computational Fluid Dynamics Programs', *Building and Environment* 37(8–9): 857–864.

Recommended reading

ASHRAE (2009) 'Airflow around Buildings', Chapter 24 in *Handbook of Fundamentals*, Atlanta: ASHRAE.
Ferziger, J.H. and Peric, M. (1996) *Computational Methods for Fluid Dynamics*, Berlin: Springer.
White, F.M. (1991) *Viscous Fluid Flow*, 2nd ed., New York: McGraw-Hill.
Wilcox, D.C. (2004) *Turbulence Modeling for CFD*, 2nd ed., DCW Industries Inc.

Activities

Exercise 1: Moisture buffering in a room. Determine the maximal daily RH variation in a room with an occurrence probability of less than 20 per cent of the time

In this exercise we determine the variation of the inside relative humidity as a function of time for a room without moisture buffering capacity. The exercise can be extended to a room with moisture buffering following the same procedure. In a second step we determine the cumulative distribution function of the maximal daily variation in relative humidity. Based on this distribution we then determine the maximal daily RH variation for an occurrence probability of less than 20 per cent of the time.

1. Solution method

The inside vapour pressure of a room without moisture buffering by the walls is, according to Equation (10.13) given by:

$$\frac{dp_i}{dt} = -G_v\left(p_i - \frac{T_i}{T_e}p_e\right) + R_v T_i G_p \tag{10.13}$$

This equation can be solved numerically using an explicit time discretisation scheme:

$$\frac{p_{i,t+\Delta t} - p_{i,t}}{\Delta t} = -G_{v,t+\Delta t}\left(p_{i,t} - p_{e,t+\Delta t}\frac{T_{i,t+\Delta t}}{T_{e,t+\Delta t}}\right) + R_v T_{i,t+\Delta t} G_{p,t+\Delta t} \tag{10.14}$$

where the subscript denotes the time when the variable is evaluated, t or $t+\Delta t$, and Δt denotes the time step. In this exercise we take the time step equal to one hour. Solving for the unknown vapour pressure $p_{i,t+\Delta t}$ at time $t+\Delta t$, we get:

$$p_{i,t+\Delta t} = p_{i,t} + \Delta t\left(-G_{v,t+\Delta t}\left(p_{i,t} - p_{e,t+\Delta t}\frac{T_{i,t+\Delta t}}{T_{e,t+\Delta t}}\right) + R_v T_{i,t+\Delta t} G_{p,t+\Delta t}\right) \tag{10.15}$$

This formula allows to determine the unknown vapour pressure $p_{i,t+\Delta t}$ from the vapour pressure $p_{i,t}$ at time t and the given temperatures, ventilation rate and moisture production rate at time $t+\Delta t$. From the vapour pressure, the relative humidity can then be determined using:

$$RH = \frac{p_i}{p_{sat}(T_i)};\ p_{sat}(T) = \exp\left(65.8094 - \frac{7066.27}{T} - 5.976\ln(T)\right) \tag{10.16}$$

2. Generating the input data

The outside temperature is considered to be constant and equal to 8°C. The inside temperature also is considered to be constant due to heating control and is equal to 20°C. The moisture production rate is considered to follow a daily pattern: a constant production of 0.5 grams/hour and peak production rates of 4 grams/hour between 6 and 8 a.m. and 2 grams/hour between 4 and 10 p.m (Figure 10.18).

The hourly variation of the outside vapour pressure is described by a sinusoidal variation according to the equation:

$$p_e(t) = \bar{p}_e + \hat{p}_e \sin\left(\frac{2\pi}{t_p} t + \phi\right) \qquad (10.17)$$

with $\bar{p}_e$ (Pa) the average value in vapour pressure, $\hat{p}_e$ (Pa) the daily amplitude and ϕ (°) the phase angle which is taken equal to zero as a reference. We consider a daily cycle with hourly variation so t_p equals 24 h. The daily amplitude is assumed to vary randomly according to $\hat{p}_e = \hat{p}_{e,0} + \alpha\,\hat{p}_{e,amp}$ with $\hat{p}_{e,0}$ a mean value, the $\hat{p}_{e,amp}$ daily amplitude and α a random variable between –0.5 and 0.5. An hourly variation of the outside vapour pressure can be generated by the following procedure: generate a random number between –0.5 and 0.5 and calculate the daily amplitude $\hat{p}_e$, determine an hourly value of the vapour pressure using Equation (10.17). The input data for the random outside vapour pressure variation are: $\bar{p}_e$ = 800 *Pa*, = 100 *Pa*, $\hat{p}_{e,amp}$ = 50 *Pa*.

The ventilation rate follows a Weibull distribution. The cumulative distribution of the ventilation rate is given by:

$$F_{G_v}(x) = 1 - \exp\left(-\left(\frac{x-\varepsilon}{u-\varepsilon}\right)^k\right)_e \qquad (10.18)$$

with x the independent variable and u, k and ε distribution parameters. A random time train of the ventilation rate can be generated using the Monte Carlo technique. Taking the inverse of Equation (10.18) we get:

$$G_v = \left(-\ln(1-y)\right)^{\frac{1}{k}}(u-\varepsilon)+\varepsilon \qquad (10.19)$$

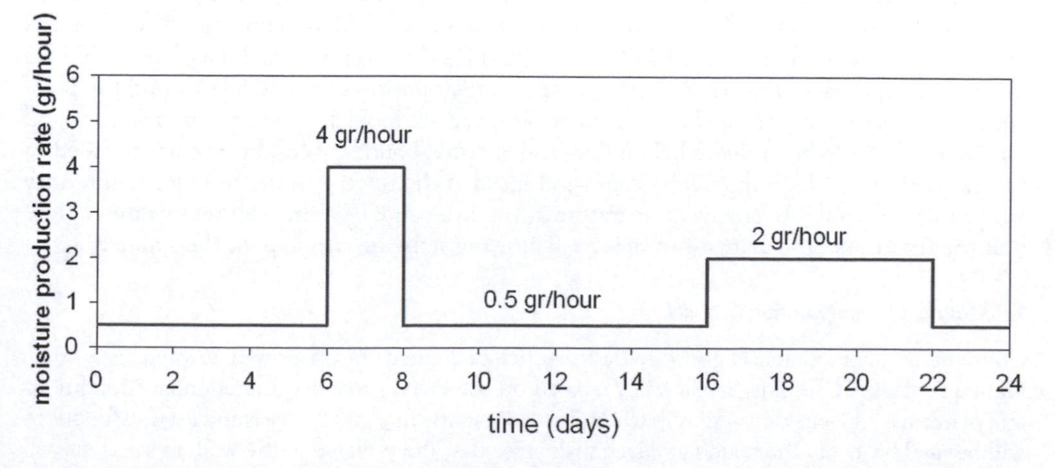

Figure 10.18 The moisture production rate during a typical day.

where y is the dependent variable. Using Equation (10.19), we can now generate an hourly value of the ventilation rate following the Weibull distribution by introducing a randomly generated value between 0 and 1 for y in Equation (10.19). The parameters used are $u = 0.6$, $k = 2$ and $\varepsilon = 0.2$.

3. *Solving procedure*

In a first step a time train for a period of 30 × 24 hours (n = 30 days) is generated using the given input variables: outside vapour pressure (Equation (10.17)), moisture production rate and ventilation rate (Equation (10.19)).

In a second step, the time variation of the inside vapour pressure is determined using Equation (10.15). The initial inside vapour pressure at time zero is determined assuming stationary conditions with mean values of the variables:

$$p_i(t=0) = \bar{p}_e \frac{\bar{T}_i}{\bar{T}_e} + \frac{R_v T_i \bar{G}_p}{\bar{G}_v} \tag{10.20}$$

where the average values are given or calculated from the generated time series. In a third step, the inside vapour pressure is then recalculated into an inside relative humidity using Equation (10.16). For every day, the maximal daily variation in relative humidity is determined.

In a fourth step, a cumulative distribution of the maximal daily variation is determined. The cumulative distribution is obtained by sorting the n values of the maximal daily RH variation, plotting these values against $F_{RHi,}(j)$ determined by:

$$F_{RHi}(j) = \frac{j-0.5}{n} \tag{10.21}$$

where j is the ranking number of the sorted values. Repeat finally steps 1 to 4 for three different random realisations and average the three cumulative distributions for these realisations.

In a fifth step, determine the maximal daily RH variation for an occurrence probability of less than 20 per cent of the time. This maximal daily RH equals to the value found for an F_{RHi} equal to 0.8. This value of maximal daily RH can be found by linear interpolation.

Exercise 2: Calculation of convective heat transfer coefficient for turbulent flow over a heated flat plate with CFD

Often convective mass transfer coefficients for building surfaces are estimated from convective heat transfer coefficients (CHTCs) by using the heat and mass transfer analogy. Therefore this exercise aims at determining the CHTC (Equation (10.7)) over a heated flat plate with CFD for uniform approach flow where both the spatial distribution of the CHTC along the plate and the correlation of the CHTC with the wind speed are looked at. Note that different wind speeds will have to be evaluated. The flow and thermal boundary conditions are specified in Figure 10.19. Here, U_∞ is the free-stream wind speed at the inlet, I_T is the turbulence intensity at the inlet, T_{ref} is the reference temperature at the inlet and T_w is the wall temperature. Note that the first 1 m of the plate is not heated. Following steps are required for the simulations.

1. *Making the computational model*

Compose an appropriate 2D grid (see best practice documents of Casey and Wintergerste 2000; Franke *et al.* 2007; Tominaga *et al.* 2007), based on the given geometry and boundary conditions and perform a grid sensitivity analysis to obtain a sufficiently fine grid. Since standard wall functions will be used to model the boundary layer, make sure that the y^+ value of the wall-adjacent cell on the surface of the flat plate lies within the interval 30–500 for the velocity range that is specified.

2. Set the simulation parameters

Use steady RANS and the standard k-ε model, with standard wall functions to model the boundary layer. Use second-order discretisation schemes. Estimate the turbulence inlet conditions for k and ε using following equations:

$$k = \frac{3}{2}(U_\infty I_T)^2 \;\; ; \; \varepsilon = C_\mu^{3/4} \frac{k^{3/2}}{0.14L} \tag{10.22}$$

where C_μ is 0.09 and L is the height of the computational domain (1 m). For air, use a constant density of 1.225 kg/m³, a dynamic viscosity of 1.7894 × 10^{-5} kg/ms and a thermal conductivity of 0.0242 W/mK. Do not include buoyancy effects in the simulation, which is justified since forced convection is dominant at these high Reynolds numbers.

3. Iterative solution of the problem

Perform simulations for different wind speeds (at least three) in the specified range (5–10 m/s). Iterate until the solution is fully converged. Check the convergence by monitoring flow variables, for example the heat flux at the flat plate.

4. Analyse the results

Plot the convective heat transfer coefficient (Equation (10.7)) on the heated part of the flat plate as a function of the distance along the plate (x) for a wind speed of 5 m/s. Compare with following empirical correlation (from Lienhard and Lienhard 2006):

$$CHTC_x = 0.032 \frac{\lambda}{x} \mathrm{Re}_x^{0.8} \, \mathrm{Pr}^{0.43} \tag{10.23}$$

where $CHTC_x$ is the CHTC at each location x along the plate, Pr is 0.744, Re_x is the Reynolds number, based on the distance along the plate x and U_∞, and λ is the thermal conductivity of air. Note that this empirical correlation is derived assuming heating of the entire plate, in contrast to the numerical model, which has an unheated starting length of 1 m.

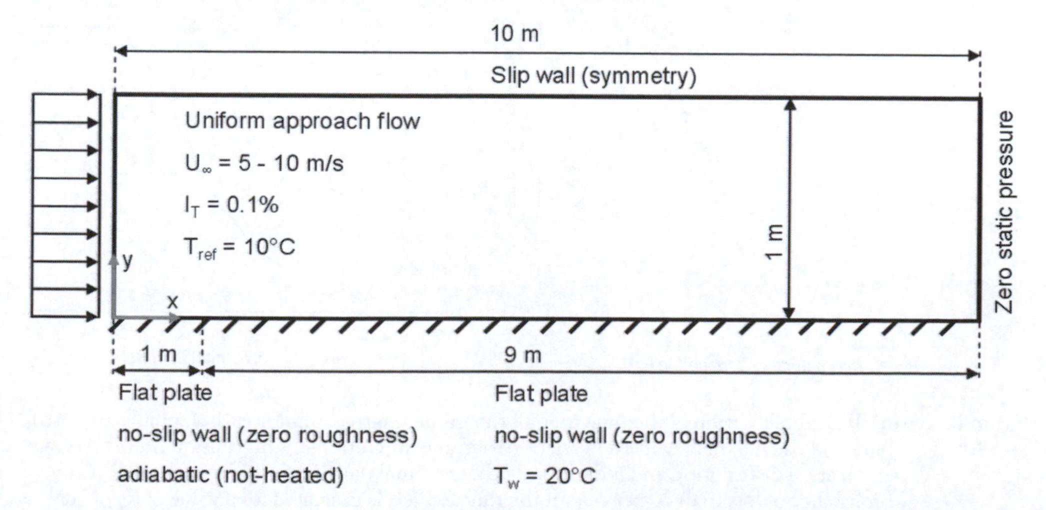

Figure 10.19 Model for numerical analysis with indication of geometry and boundary conditions.

Calculate the (surface) average convective heat transfer coefficient on the heated part of the flat plate for every wind speed. Plot these averaged values ($CHTC_{AVG}$) together with the wind speed (U_∞) in a $CHTC_{AVG} - U_\infty$ graph. Approximate the simulation data with a power-law function and determine the coefficient A and exponent B:

$$CHTC_{AVG} = AU_\infty^B \tag{10.24}$$

Model answers

Exercise 1

A typical variation of the inside relative humidity according to exercise 1 in a room is given in Figure 10.20(a). Three different random realisations of the cumulative distribution of the maximal daily variation of the inside RH are given in Figure 10.20(b). Figure 10.20(c) gives the average cumulative distribution for these three distributions. The maximal daily RH varia-

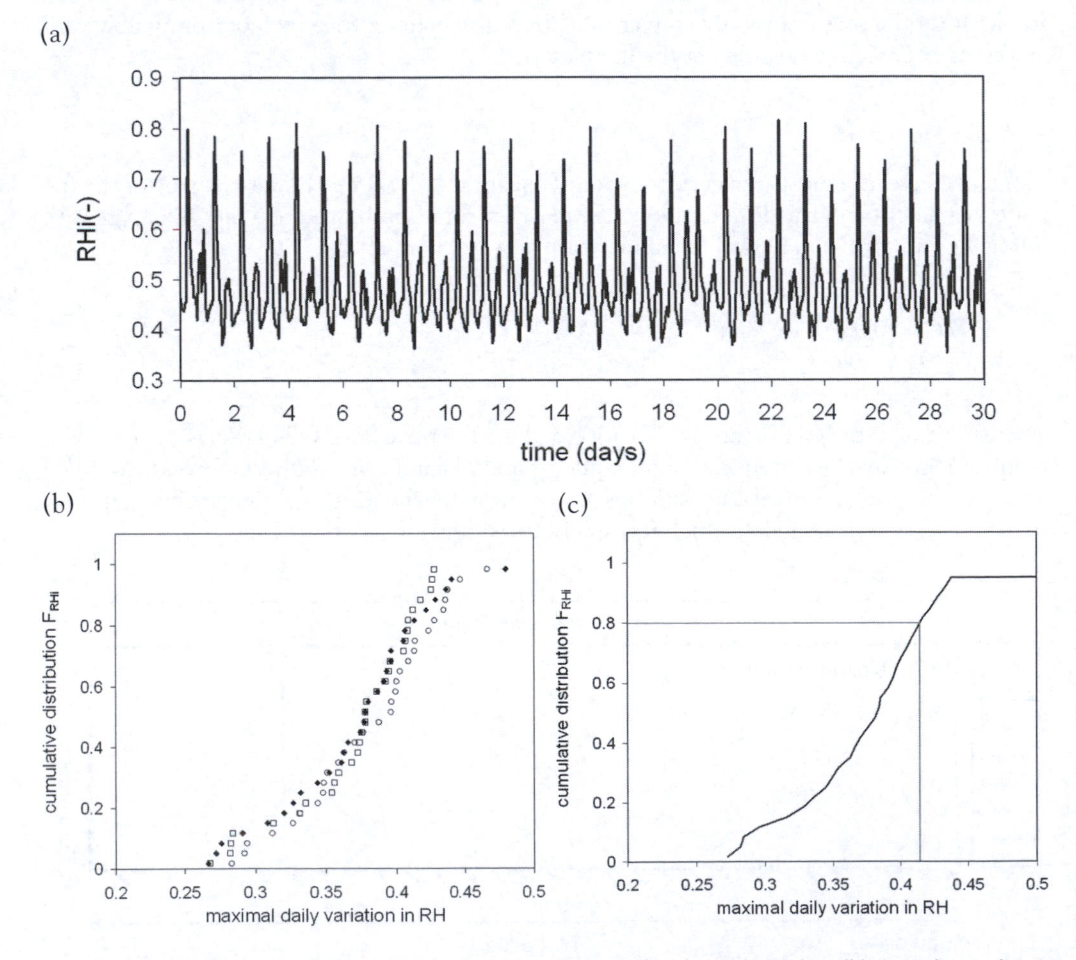

Figure 10.20 (a) Typical realisation of the time train of the inside relative humidity; (b) Cumulative distribution of the maximal daily RH variation for three different random realisations; (c) Average cumulative distribution and indication of the maximal daily RH variation for an occurrence probability of less than 20 per cent of the time, which is indicated for a value of F_{RHi} equal to 0.8. The corresponding maximal daily RH variation in the room equals 0.41 or 41 per cent.

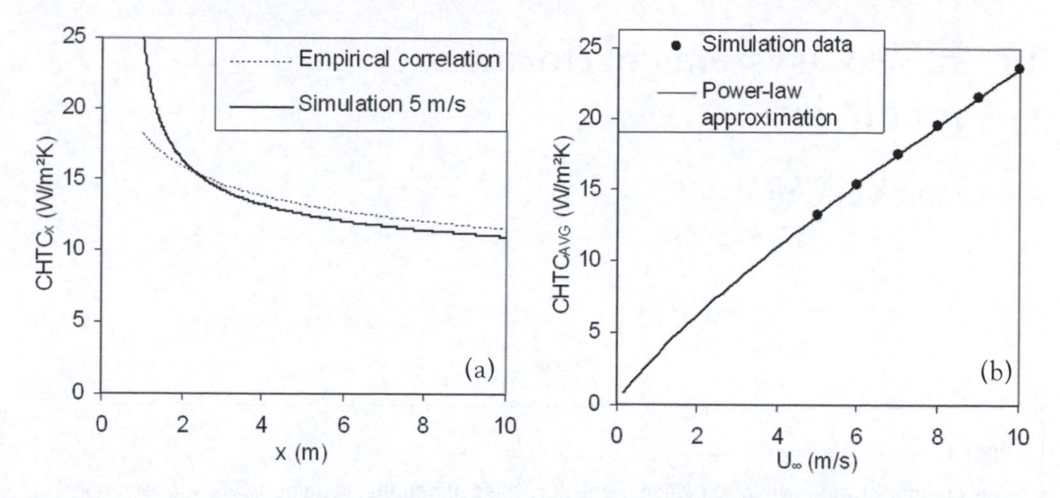

Figure 10.21 (a) $CHTC_x$ as a function of the distance along the plate: simulation data and empirical correlation; (b) $CHTC_{AVG}$ as a function of U_∞: simulation data and power-law approximation.

tion for an occurrence probability of less than 20 per cent of the time is indicated in the figure for a value of F_{RHi} equal to 0.8. This value is found to be equal to 0.41 or a maximal daily variation in the room equal to 41 per cent.

Exercise 2

The $CHTC_x$ as a function of the distance along the plate is shown in Figure 10.21(a) for both numerical simulation and empirical correlation. Note that the discrepancy at the beginning of the plate (x = 1–2 m) is related to the unheated starting length in the numerical simulation. For the remainder of the plate, both $CHTC_x$s agree within about 5 per cent. The $CHTC_{AVG}$ as a function of U_∞ is shown in Figure 21(b), together with the power-law approximation. The coefficient A and exponent B are respectively 3.47 and 0.83.

11 HVAC systems performance prediction

Jonathan Wright

Scope

This chapter provides an introduction to the principles of heating, ventilating and air-conditioning (HVAC) system performance modeling, including a description of the alternative forms of steady-state model and the computational methods for the solution of the model equations. The application of HVAC performance models to HVAC system fault-detection and optimization is also described.

Learning objectives

To develop an understanding of the principles of HVAC system modeling, in terms of the level of detail required for a particular application and the alternative modeling approaches available.

Key words

HVAC system performance/steady-state component models/system modeling/solution of simultaneous equations.

Introduction

HVAC system performance prediction has a number of applications; in evaluating the impact of a particular HVAC system configuration on building energy use and the extent to which the system is able to maintain the desired indoor-air conditions; in HVAC system condition monitoring; and the study of supervisory and local-loop control strategies.

Where the impact of a particular system type on building energy use is to be investigated, the HVAC system performance simulation is integral with the simulation of the building thermal performance. However, for some applications, such as HVAC system condition monitoring, the performance of the system is simulated independently of the building thermal performance. The integrated simulation of building and HVAC system performance is not described here, this chapter being focused on the development of HVAC component and system models.

There are a number of alternative types of HVAC system available, but each of these is formed from a similar set of HVAC components. This observation led to the development of two approaches to HVAC system modeling. In some simulation environments, system models are selected from a list of pre-configured system types, while in other environments, the system is defined component-by-component. Pre-configured system models have the advantage that the user is spared the effort and difficulty of defining the system model component-by-component. The use of pre-configured system models also has the advantage that it may be possible to tailor the solution algorithms to the particular set of model equations and therefore improve the speed and robustness of their solution. However, restricting the user to selecting from a choice of pre-configured system types has the major disadvantage that often the system to be modeled is not available in the list of pre-configured systems. The probability of this being the case increases with the development of innovative design solutions, and so for this reason, contemporary simulation approaches are based on the definition of the HVAC system by linking the input-output variables of individual system components to form a model of the complete HVAC system. Therefore, this chapter is focused on the development of HVAC component models and the connection of their input-output variables in forming a HVAC system model. The development of a component orientated simulation environments is also described in Chapter 17.

The particular application of the HVAC system model can influence the extent to which the non-linear and dynamic characteristic of the system are modeled. Where the aim of the simulation is to evaluate the building energy use, it is often assumed that the HVAC systems and their controls have an idealized linear characteristic and operate under perfect control. In this case, a simple steady-state linear model can be used to represent the relationship between the system input and output variables, although a non-linear efficiency relationship may be used to model the impact of part-load operation on the systems energy use. Conversely, the application of HVAC system models to condition monitoring and the study of control system behavior demands the use of models which encapsulate the full non-linear performance characteristics of the system; the system dynamics must also be modeled in the analysis of HVAC control system performance, particularly when local loop control strategies are being investigated. This chapter describes the formulation of models that include the non-linear static characteristic of HVAC systems.

HVAC component models consist of; the variables to be simulated at the input and output of the component; one or more generically applicable equations defining the relationship between the input-output variables; a set of equation parameters (constants) that define a specific HVAC component (size). The component model equations are formulated in one of two ways, either in a way that fixes the input and output variables, or in a way that provides some flexibility in the choice of input-output variables. An "input-output free" model has the flexibility of being used in a variety of applications. For example, an input-output free pump model could be used to simulate the pumps output pressure during system design, and say, the pump speed during commissioning. The model equations can be; derived directly from first principles; they can be a simple curve-fit of the measured performance of a specific product; or they can be a combination of both. This chapter describes the different types of model equation, and their use in formulating both fixed input-output, and input-output free models. The use of polynomial curve-fitting of a HVAC component manufacturer's data is also described.

The HVAC system component models are linked through their input-output variables to form a model of the HVAC system (or sub-system). The choice of linking variables must be done in such a way that the set of system equations resulting from the linking of the components can be solved for the selected set of simulation variables. This is particularly the case for the linking of input-output free models where there is a greater choice of potential linking variables than fixed input-output variable models. Although greater care must be taken in selecting the linking variables when using input-output free models, the solution of the input-output free model equations is potentially more robust than the solution of the fixed input-output model equations (Sowell and Haves 2001). The linking of the components to form a system, methods for checking that the resulting set of system equations can be solved for the simulation variables, and the methods for the solutions of the system equations are described in this chapter.

Scientific foundation

The time constant of an HVAC system is generally in the order of minutes,[1] whereas the time constant of the building fabric is in the order of hours and as such, in comparison to the thermal response of the building fabric, the HVAC system can be considered to be in a quasi-steady state condition for the majority of its operation (the exception being during the system start and stop operation). Since, the weather and other simulation boundary conditions are normally specified at hourly intervals, steady-state models are typically used to represent the HVAC system performance in building thermal simulations (particularly where the focus of the simulation is on the building energy use). However, where the focus of the simulation is on the control system performance, dynamic models of the HVAC systems should be used, (with the boundary conditions being either specified or interpreted to be in sub-hourly intervals). This chapter describes the principles of developing steady-state HVAC component models, together with the formation of HVAC system models and the solution of the model equations.

HVAC component models are defined in terms of the simulation variables, the model equations, and the model parameters (Figure 11.1). The simulation variables in HVAC component models are normally those that represent the thermo-fluid properties of mass flow-rate ($\dot{m}$), temperature (T), humidity ratio (w), and pressure (P); variables related to the control of the component, such as a control signal (u), or motor speed (n), are also included in the set of variables for some component models. Note that separate sets of input-output variables are specified for each fluid stream entering (i, j), or leaving (o, k), the component. For a given type of component, the model equations ($F = \{f_1(\cdot), f_2(\cdot), \ldots, f_1(.)\}$), represent the physical relationship between the simulation variables at input and output of the component and are normally of a generically applicable form. The equations are of an algebraic form in steady-state models, but include differential terms in dynamic models. The model parameters ($A = \{\alpha_1, \alpha_2, \ldots, \alpha_m\}$), define a specific product or "size" of component. The form of model parameters depends on the form of model equations; in some instances they can have a physical meaning (such as a heat transfer coefficient), while in other models, they can be a dimensionless coefficients of, for example, a polynomial curve-fit. In addition to the model equations that define the relationship between the variables at input and output, it is likely that the model will require the calculation of thermodynamic property relationships ($g(\cdot)$), and other intermediate calculations which can be iterative (↻). This section focuses on defining the main categories of model equations (and associated model parameters).

The steady-state model equations can be categorized as being derived from first-principles, from empirical data, or from a mix of first-principles and empirical data.

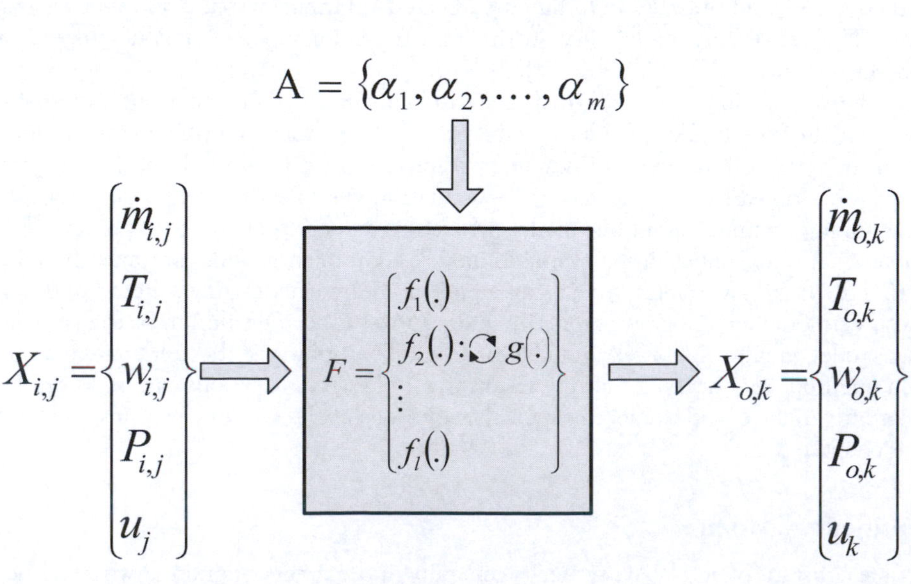

$$A = \{\alpha_1, \alpha_2, \ldots, \alpha_m\}$$

$$X_{i,j} = \begin{Bmatrix} \dot{m}_{i,j} \\ T_{i,j} \\ w_{i,j} \\ P_{i,j} \\ u_j \end{Bmatrix} \Rightarrow F = \begin{Bmatrix} f_1(.) \\ f_2(.) : \circlearrowright\, g(.) \\ \vdots \\ f_l(.) \end{Bmatrix} \Rightarrow X_{o,k} = \begin{Bmatrix} \dot{m}_{o,k} \\ T_{o,k} \\ w_{o,k} \\ P_{o,k} \\ u_k \end{Bmatrix}$$

Figure 11.1 HVAC component model structure.

First principle models

First principle models are derived from fundamental physical relationships and do not have any empirically derived element. An example is a model of an air mixing section (Figure 11.2), where three model equations can be derived from a mass and energy balance on the fluid streams (Equations (11.1) to (11.3)).

$$\dot{m}_o = \dot{m}_{i,1} + \dot{m}_{i,2} \tag{11.1}$$

$$\dot{m}_o w_o = \dot{m}_{i,1} w_{i,1} + \dot{m}_{i,2} w_{i,2} \tag{11.2}$$

$$\dot{m}_o h_o(T_o, w_o) = \dot{m}_{i,1} h_{i,1}(T_{i,1}, w_{i,1}) + \dot{m}_{i,2} h_{i,2}(T_{i,2}, w_{i,2}) \tag{11.3}$$

where, $h_i(\cdot)$ is enthalpy, this being a function of the fluid temperature and humidity ratio. Note that for clarity, the model equations are given in the form of the mass and energy balance. These equations would normally be transformed to be in either a form that allowed the direct calculation of the component output variables (generally taken to be $\dot{m}_o, T_o$ and w_o), or in an input-output free form. Note also that there are no model parameters in this case.

Unfortunately, there are very few instances where a model can be developed entirely from first-principles. In fact, a complete model of a ducted air-mixing section would include the air pressure loss across the duct section, the pressure loss coefficient being derived from experimental (empirical) measurement.

Empirical models

Empirical models are used where the HVAC components internal processes are not well understood, or where the effort required to generate a detailed first-principle model of the

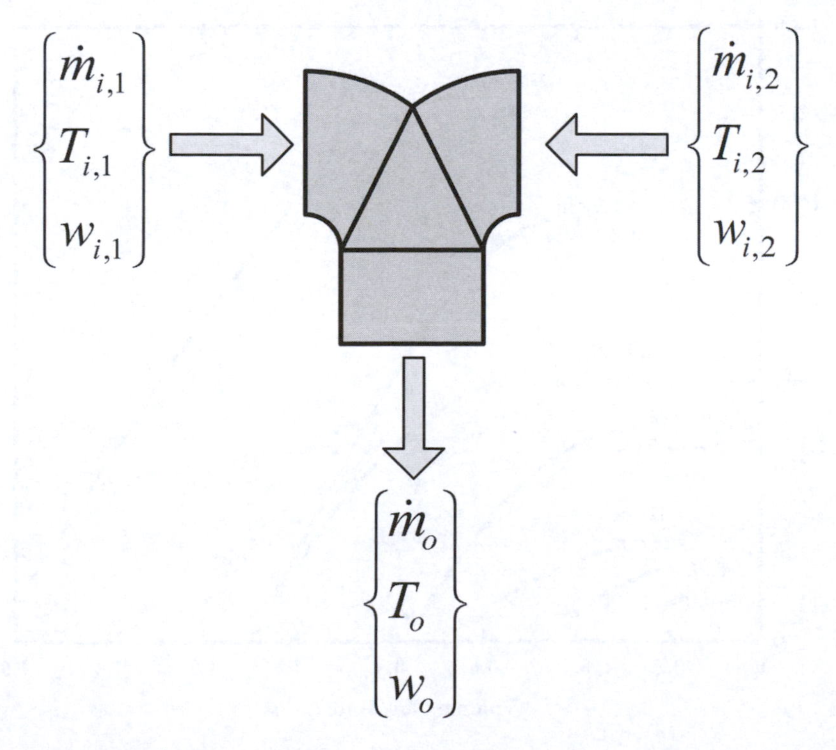

Figure 11.2 Air-mixing duct section.

component is not justified. As such, this type of model is frequently used to model the performance of turbo-machines such as compressors, pumps, and fans. Although it is possible to develop a detailed model of a pump or fan that is based on a geometric specification of the components impeller, such detail is not readily available and is not of importance in building design. The approach can also been used to model the part-load performance of refrigeration sub-systems, where a detailed simulation of the refrigeration components and cycle is not warranted (Braun 1988).

Many empirical models take the form of a polynomial curve-fit of the component manufacturers test data, with one or more polynomial expressions forming the model equations, and the model parameters being the polynomial coefficients (Wright 1986). Although not described here, it is also possible to use other "black-box" models such as those based on fuzzy-logic (Angelov *et al.* 2000).

In the case of pumps and particularly fans, two elements of the component's performance are modeled, fluid pressure rise and fluid temperature rise, both as a function of flow-rate. Both fan and pump performance data can be provided for a range of impeller speeds and several impeller diameters. Figure 11.3 illustrates the pressure-flow performance curves for three impeller speeds (n), and two diameters (d), of a centrifugal fan (a similar set of performance characteristics is normally given for the fan power or efficiency as a function of air flow-rate).

Rather than generate a separate curve-fit for each impeller diameter and speed, the number of curve-fits can be reduced in one of two ways; by generating a curve-fit for one impeller speed and diameter, and predicting the performance at other diameters and speeds by using fan and pump scaling rules; or by normalizing the performance data, the component performance for a number of impeller speeds and diameters often then being represented by a single normalized performance curve. Both the scaling rules and expressions for normalization have been derived from dimensional analysis. Normalizing the data generally results in the complete range of speeds and diameters being represented by a single non-dimensional curve (the exception

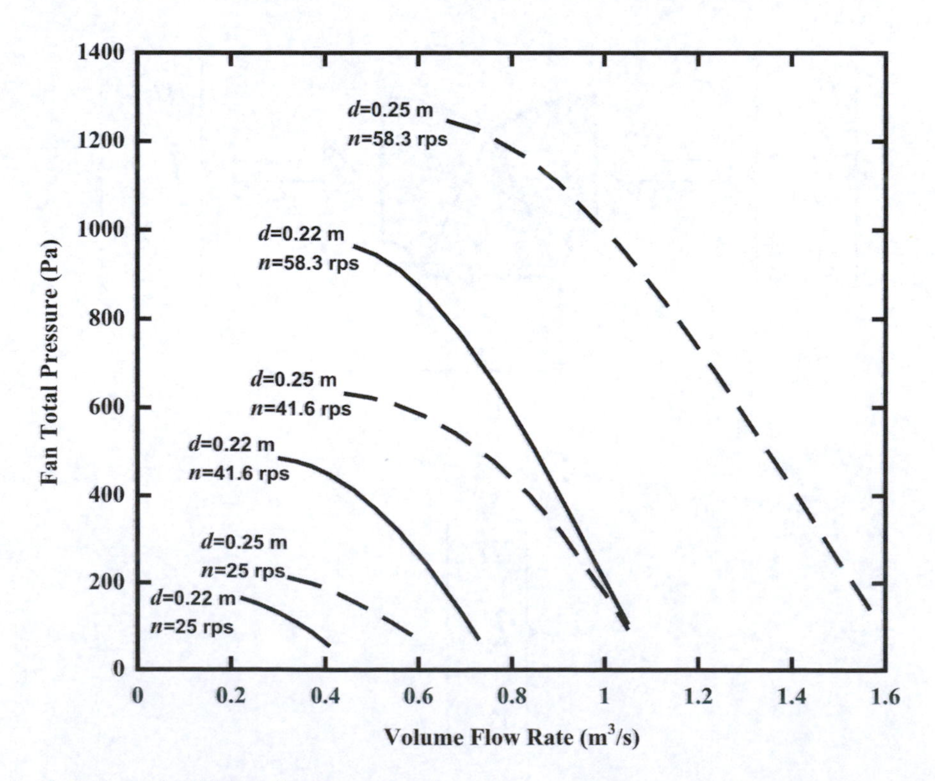

Figure 11.3 Centrifugal fan performance curves.

being for axial-flow fans where the performance data is given as a function of the impeller blade angle as well as the impeller diameter and speed, this requiring the use of a two-dimensional curve-fit as it is not possible to scale or normalize for the blade angle (Wright 1991)).

The applicability of using scaling rules depends on the complete pump or fan characteristic being available at a given speed; this may not be available when the component performance data is given in a tabular form (since such data would normally be presented with the rows and columns of the table being formed from the air-flow rate and pressure rise, with the speed data being fragmented as a result). Since the complete characteristic is not required in order to generate a normalized performance curve, performance modeling using data normalization is the more widely applicable approach. The non-dimensional volume-flow rate (ϕ), total pressure rise (ψ), and absorbed power (λ) are given by:

$$\phi = \frac{m}{\rho_i \pi^2 d^2\, n/4} \tag{11.4}$$

$$\psi = \frac{2P}{\rho_i \left(\pi d n\right)^2} \tag{11.5}$$

$$\lambda = \frac{8W}{\rho_i \pi^4 d^5 n^2} \tag{11.6}$$

where, $\dot{m}$ is the fluid mass flow rate through the component, P is total pressure rise of the fluid, W, the absorbed power, ρ_i the fluid density at the inlet to the component, d the impeller diameter, and n the impeller speed.[2]

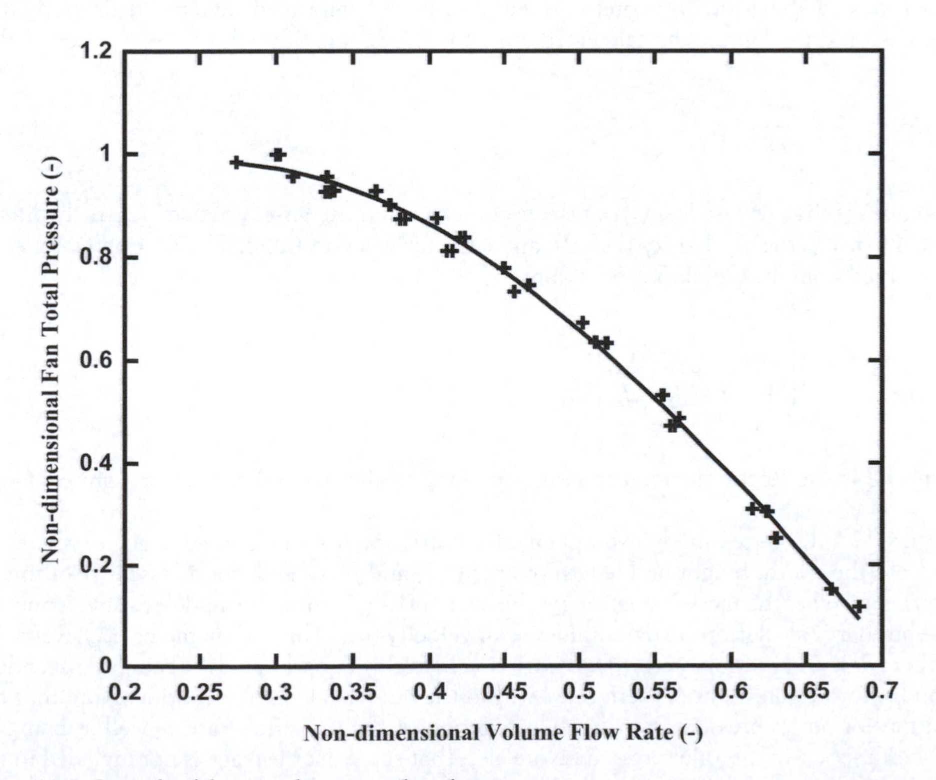

Figure 11.4 Centrifugal fan non-dimensional performance curves.

Figure 11.4, shows the non-dimensional fan flow-pressure rise curve for the fan performance data illustrated in Figure 11.3. The six performance curves have been reduced to one curve which can be modeled, in this case, using a second order polynomial having only three parameters, $\{\alpha_1, \alpha_2, \alpha_3\}$; $\psi = \alpha_1\phi^2 + \alpha_2\phi + \alpha_3$.

The scatter in the data about the curves is a result of the digitizing processes used to extract numerical values for flow rate and pressure rise from the fan performance curves. Errors (scatter) can also occur due to a lack of geometric similarity among the different component diameters (Wright 1991), or the effects of the working fluid being compressible (although this effect is limited for the range of fluid pressure rise occurring in most HVAC systems). Errors can also occur due to the poor or unreliable performance measurement by the component manufacturer. Normalization of the data has the benefit that it makes such errors apparent and tends to average their effect.

Quasi-first principle models

The majority of HVAC component models can be classified as being quasi-first principle models, particularly models of components that promote heat exchange (heating and cooling coils, evaporators, condensers, and boilers). The equations of quasi-first principle models are derived from fundamental principles, but require some empirical input (generally in relation to the model parameters). They are distinguished from purely empirical models in that the equations describing the performance characteristic of the components are derived from fundamental principles, whereas in a purely empirical model, the performance characteristic is a result of a curve-fit of measured performance data. Since the model equations are derived from fundamental principles, this class of model is also sometimes referred to as a "first-principle" model (for instance, by Haves *et al.* 1996).

A model of duct friction loss can be classified as a quasi-first principle model, it being developed from a combination of theoretical relationships and empirically determined performance data. Duct friction loss can be calculated using the Darcy equation:

$$\Delta P = \frac{fL}{D_h} \times \frac{\rho v^2}{2} \tag{11.7}$$

where, ΔP, is the pressure loss (Pa), f the friction factor, L the length of duct, D_h the hydraulic diameter (m), ρ the air density (kg/m^3), and v the air velocity (m/s). The friction factor f can be obtained from the Colebrook equation:

$$\frac{1}{\sqrt{f}} = -2\log\left(\frac{\varepsilon}{3.7Dh} + \frac{2.51}{Re\sqrt{f}}\right) \tag{11.8}$$

where, Re is the Reynolds number, and ε the empirically derived absolute roughness factor (m).

Figure 11.5 illustrates an input-output model for friction loss in rectangular ducts (where, W, H, L, are the width, height and length of duct (m), and μ is the dynamic viscosity of the air (Pa.s)). As well as the model equation for the duct outlet pressure, the model requires a number of preliminary calculations to determine the air velocity (v), hydraulic diameter (D_h), Reynolds number (Re), and friction factor (f), the friction factor in Equation (11.8) being transcendental and solved by an iteration performed within the model calculation routine. Note that the equation for outlet pressure is an implicit function of the mass flow rate as well as being an explicit function of the inlet pressure. Note also, that since duct leakage is not included in the model, the air mass-flow rate is treated as a "through" variable ($\dot{m}$), its value being the same

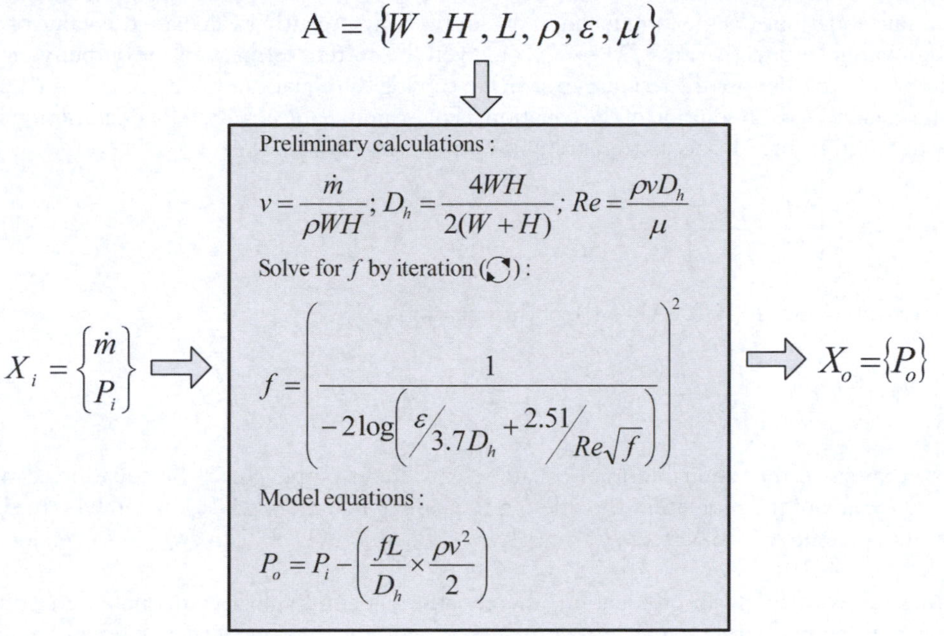

$$A = \{W, H, L, \rho, \varepsilon, \mu\}$$

Preliminary calculations:

$$v = \frac{\dot{m}}{\rho W H};\ D_h = \frac{4WH}{2(W+H)};\ Re = \frac{\rho v D_h}{\mu}$$

Solve for f by iteration:

$$f = \left(\frac{1}{-2\log\left(\varepsilon / 3.7 D_h + 2.51 / Re\sqrt{f} \right)} \right)^2$$

Model equations:

$$P_o = P_i - \left(\frac{fL}{D_h} \times \frac{\rho v^2}{2} \right)$$

$$X_i = \begin{Bmatrix} \dot{m} \\ P_i \end{Bmatrix} \Rightarrow \quad \Rightarrow X_o = \{P_o\}$$

Figure 11.5 Rectangular duct friction loss model.

at the duct outlet as the inlet; "through" variables are often illustrated as being a model input variable, as in the case of the air mass-flow rate in Figure 11.5.

Computational methods

The computational tools required for use in HVAC system performance prediction are concerned with curve-fitting empirical data, and the solution of the model equations.

Polynomial curve-fitting

Methods for identifying the coefficients of a polynomial curve-fit are widely available and are not described here. However, for model developers who are faced with polynomial curve-fitting for a large number of different components, it may be useful to develop a dedicated curve-fit software package that is able to search for an optimum order of polynomial, this being particularly the case when a curve-fit is required for a function of more than one variable (Wright 1986).

Model developers who employ a polynomial curve-fit should also be aware that a curve-fit is only reliable within the range of the curve-fit data; extrapolation beyond the range of data can result in the prediction of component operation that is infeasible in practice. This is particularly the case when higher order polynomials are used to curve-fit the data, or where the curve-fit is a function of more than one variable (Wright 1986). When this is the case, it is wise to either add extra data points to the curve-fit in an attempt to ensure the extrapolated component performance curves represent practicable operation (Wright 1986), or to form functions that can be used to check the simulated performance lies within the original measured performance (Wright 1991).

Formation of system models and solution of model equations

HVAC component models consist of a set of algebraic and differential equations; in many instances, the dynamic performance of the component is not modeled so only the algebraic equations exist. For a given set of input variables (X_i) and model parameters (A), the model algebraic equations (F) are formulated in one of two ways; to either calculate the value of the output variables directly (F:(X_i,A) $\rightarrow$ X_o); or, given the current estimate of the output variable values (X_o), to calculate the residual error (R) in solving the equations (F:(X_i, X_o,A) $\rightarrow$ R), this being equal to zero on solution of the equation). For example, the equation for duct friction loss illustrated in Figure 11.5, is written in input-output form (F: (X_i,A) $\rightarrow$ X_o):

$$P_o = P_i - \left(\frac{fL}{D_h} \times \frac{\rho v^2}{2} \right) \tag{11.9}$$

The residual form (F:(X_i,X_o,A) $\rightarrow$ R), of the equation is:

$$R = (P_i - P_o) - \left(\frac{fL}{D_h} \times \frac{\rho v^2}{2} \right) \tag{11.10}$$

Note that since the residual form of equations are "input-output free", the equations can be solved for any of the simulation variables; in the case of Equation (11.10), it can be solved for the inlet pressure or the outlet pressure (whereas, Equation (11.9), can only be solved for the outlet pressure).

Regardless of the form of equation, the equations for individual components are linked through the input-output variables to form a set of simultaneous algebraic equations. The type of algorithm used to solve the equations depends on the form of equation, directed input-output equations generally being solved using approaches based on successive substitution, and input-output free equations using methods based on a Newton-Raphson iteration.

The linking of the component models to form a set of simultaneous equations, the choice of system boundary conditions, and the form of solution algorithm, are illustrated here using the components of a heating coil sub-system.

Example heating coil sub-system component models

The example heating coil sub-system model (Figure 11.6), is formed here from three component models; a model of the heating coil, a model of the proportional controller, and a model

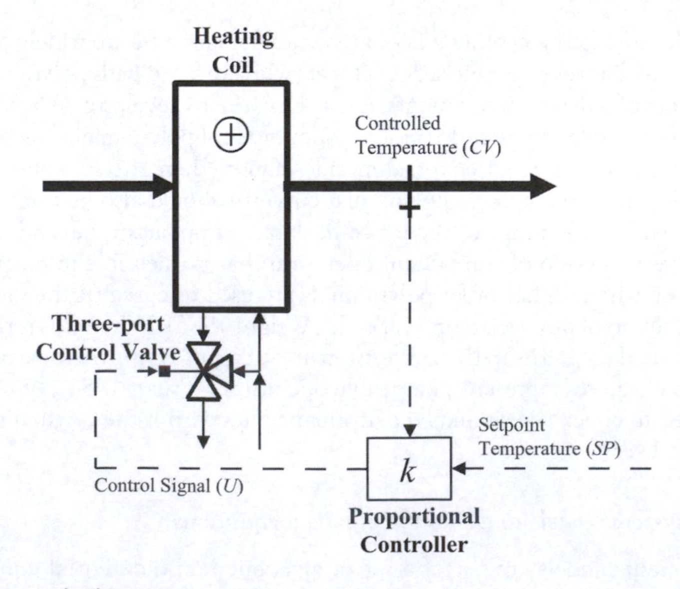

Figure 11.6 Heating coil sub-system.

of the three-port valve. The temperature sensor has not been modeled, it being assumed that the sensor bias error is zero; the valve actuator has also been omitted from the model on the assumption that the actuator hysteresis is zero. The model equations are written in input-output form ($F:(X_i,A) \rightarrow X_o$), with the input-output variables relating directly to the physical processes and the conditions at inlet and outlet of the components.

The **heating coil model** is illustrated in Figure 11.7. The model is based on the "NTU-effectiveness" method, this form of heat exchanger model having been widely used in HVAC system performance modeling (Holmes 1982; Zhou and Braun 2007). The model input variables are: the air and water mass-flow rates ($\dot{m}_a$, $\dot{m}_w$), although these are strictly "through" variables as the fluid-flow also appears at the component outlet; and the air and water inlet temperatures (T_{ai}, T_{wi}). The model output variables are the air and water outlet temperatures (T_{ao}, T_{wo}). The model parameters are: the width (W) and height (H) of the coil that are exposed to air-flow; the number of water-tube rows (N_r), number of water-tube circuits (N_c), and the water-tube internal diameter (D); and the heat transfer resistance coefficients, r_a relating to the air-side thermal resistance, r_m the thermal resistance of the water-tubes and fins and r_w the water-side thermal resistance (typical values for the thermal resistances are given in Table 11.1, Holmes 1982).

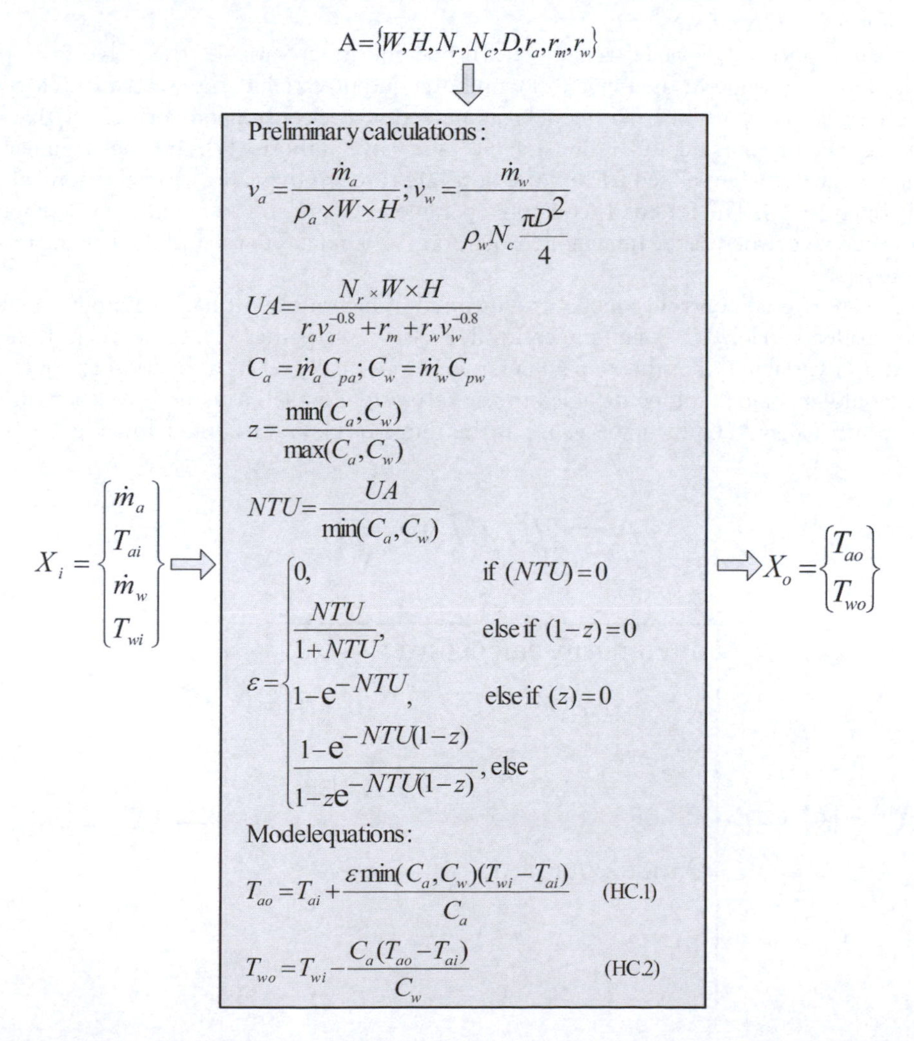

Figure 11.7 Heating coil model.

Table 11.1 Heating coil model thermal resistance parameters

Coil Type	*Fin Density*	*Water Turbulator*	r_a $\left(\frac{k.\ s^{-0.8}}{kg^{-0.8}m^2kW.row}\right)$	r_m $\left(\frac{k}{m^2kW.row}\right)$	r_w $\left(\frac{k.\ s^{-0.8}}{kg^{-0.8}\ m^2kW.row}\right)$
Low "efficiency"	Low	×	1.32	0.68	1.1
Nominal coil	Low	✓	0.44	0.2	0.2
High "efficiency"	High	✓	0.49	0.38	0.38

Source: Holmes 1982.

The overall heat transfer conductance (UA), is a function of the coil dimensions and fluid velocities (v_a, v_w); note that the fluid densities (ρ_a,ρ_w) are internally computed variables, as are the fluid specific heat capacities (C_{pa}, C_{pw}).

The model (Figure 11.7) has two final equations, the first (HC.1), is derived from an energy balance between the conditioned air and the rate of heat transfer through the metal surfaces of the coil (this being a function of the coils effectiveness, ε). The second equation (HC.2), is derived from an energy balance between the air and water. Both equations are written in input-output form ($F:(X_i, A) \rightarrow X_o$).

The **three-port valve model** (Figure 11.8), has one input variable, the valve stem position (S, having a value in the range 0–1); and one output variable, the water mass-flow rate through the control port ($\dot{m}_w$). The model parameters are: the control port curvature (β), with a value of 0.0 being equivalent to a "linear valve"; the valve authority (γ); the fractional leakage through the valve when closed (l); and the flow rate through the valve when the control port is fully open ($\dot{m}_{max}$). The internally computed parameter C_{ih}, relates to the inherent characteristic of the valve, whereas the final model equation (V.1), relates to the installed characteristic (Salsbury 1996).

The **proportional controller model** is illustrated in Figure 11.9. It has two input variables, the controlled variable (CV), and the controlled variable setpoint (SP); the setpoint has been defined as a variable rather than a parameter as it is normally set by a supervisory controller, which could at some point, be included in the sub-system model. The model output variable is the control signal (U, having a value in the range 0–1), and the model parameter is the

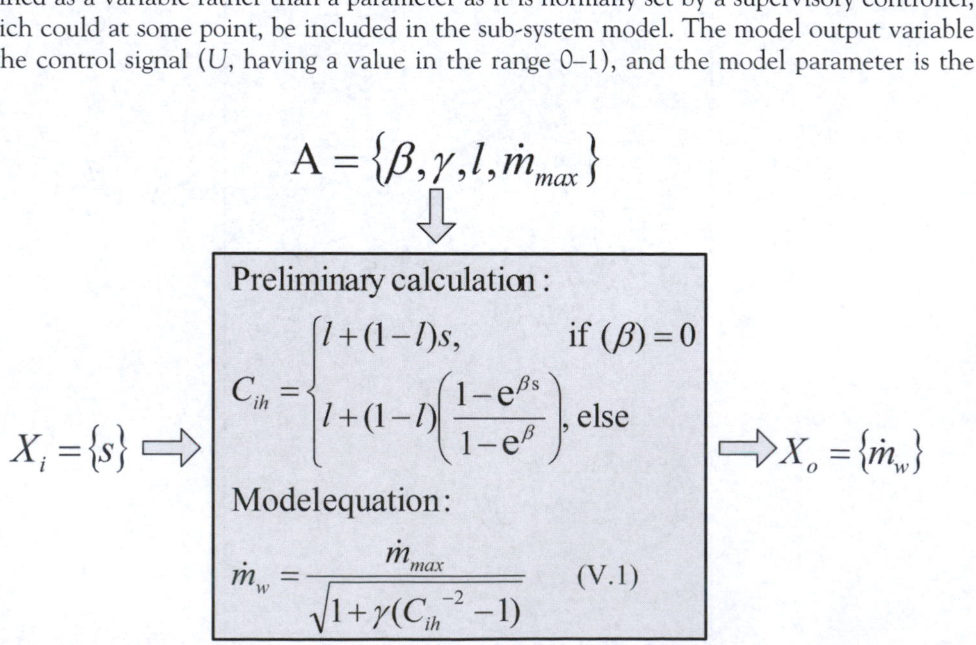

$$A = \{\beta, \gamma, l, \dot{m}_{max}\}$$

$$X_i = \{s\} \Rightarrow$$

Preliminary calculation:

$$C_{ih} = \begin{cases} l + (1-l)s, & \text{if } (\beta) = 0 \\ l + (1-l)\left(\dfrac{1-e^{\beta s}}{1-e^{\beta}}\right), & \text{else} \end{cases}$$

Model equation:

$$\dot{m}_w = \frac{\dot{m}_{max}}{\sqrt{1+\gamma(C_{ih}^{-2}-1)}} \qquad \text{(V.1)}$$

$$\Rightarrow X_o = \{\dot{m}_w\}$$

Figure 11.8 Three-port valve model.

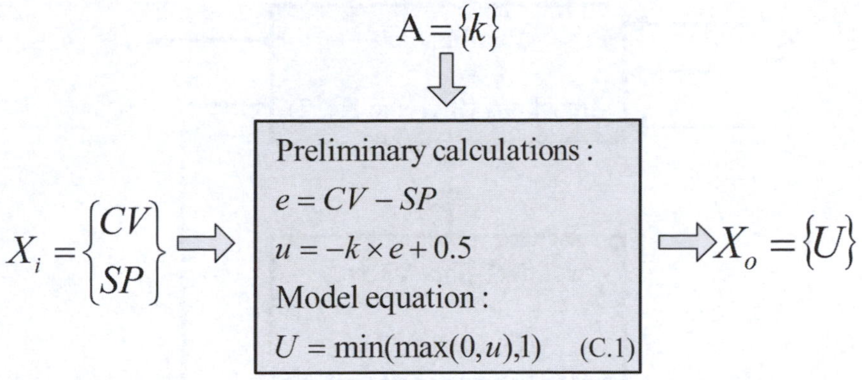

$$A = \{k\}$$

$$X_i = \begin{Bmatrix} CV \\ SP \end{Bmatrix} \Rightarrow$$

Preliminary calculations:

$$e = CV - SP$$

$$u = -k \times e + 0.5$$

Model equation:

$$U = \min(\max(0, u), 1) \quad \text{(C.1)}$$

$$\Rightarrow X_o = \{U\}$$

Figure 11.9 Proportional controller model.

proportional gain (k). The model has one final equation (C.1), which in this case ensures that the output signal is in the range 0 to 1.

Formation of system models

There are two tasks in forming a system model: the first is to identify the variables used to link the component models (and in effect, the model equations); the second task is to identify the set of boundary variables (or alternatively, the simulation variables).

The components are linked to form a system model through the input-output variables, the matching of pairs of variables mirroring the physical processes. With reference to the sub-system schematic diagram (Figure 11.6), and the component model descriptions (Figures 11.7 to 11.9), the air temperature at the outlet of the heating coil is the controlled variable; hence, in the absence of a sensor model, the air temperature at the outlet of the heating coil model (T_{ao}), is linked to the controlled variable (CV) at the "inlet" to the controller model (Figure 11.10). The linking results in T_{ao} and CV being mapped to a single variable ($\{T_{ao}, CV\} \rightarrow T_{ao}$), in this case having the notation, T_{ao}.

Similarly, in the absence of an actuator model, the output signal (U), from the proportional controller is linked directly to the valve stem position (S), the two variables being mapped ($\{U,S\} \rightarrow U$), to a single variable (U). Finally, the water mass-flow rate leaving the three-port valve ($\dot{m}_w$), is linked to the water mass-flow rate through the heating coil ($\dot{m}_w$), the two flow rates being mapped ($\{\dot{m}_w, \dot{m}_w\} \rightarrow \dot{m}_w$), to a single variable ($\dot{m}_w$). The connected system model is illustrated by the information flow diagram in Figure 11.11.

A set of n algebraic equations can only be solved for n variables. Figure 11.11 indicates that there are 8 unique variables ($\dot{m}_a$, T_{ai}, T_{wi}, T_{wo}, T_{ao}, SP, U, $\dot{m}_w$), and 4 algebraic equations (HC.1, HC.2, V.1, C.1), and therefore, 4 of the variables must be defined to be boundary variables. A poor choice of boundary variables and can result in an insoluble set of equations. Although, in most simulation studies, the system inlet conditions naturally form the boundary variables and result in a solvable set of equations, the viability of the simulation should be checked. This can be achieved through examination of an adjacency matrix (Table 11.2), which describes the

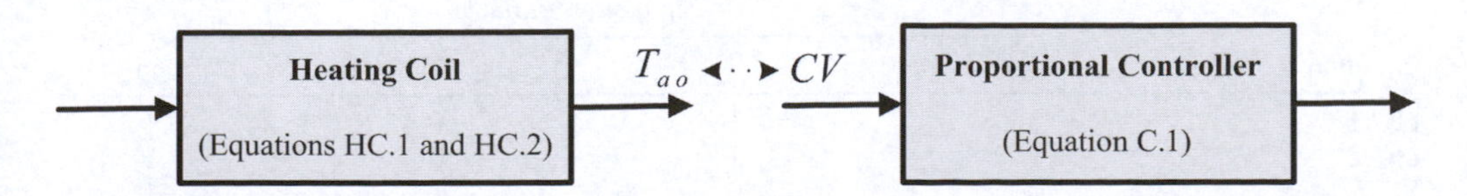

Figure 11.10 Heating coil and controller linking variables.

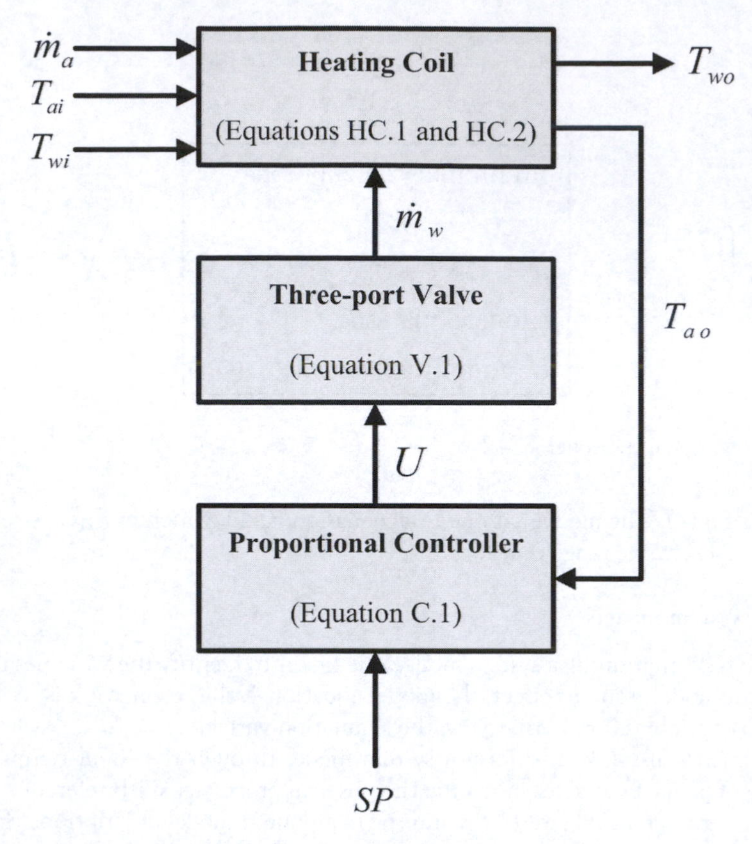

Figure 11.11 Heating coil sub-system information flow diagram.

relationship between the simulated variables and the model equations. The matrix indicates the presence of a variable in an equation by a non-zero entry in the matrix (a zero entry indicating that there is no relationship between the variable and equation).

For example, if the system inlet conditions and controller setpoint ($\dot{m}_a$, T_{ai}, T_{wi}, SP), of the heating coil sub-system model are taken as the system boundary conditions, then the simulated variables become the coil air and water outlet temperatures, together with the control signal and the water mass flow rate (T_{wo}, T_{ao}, U, $\dot{m}_w$). The resulting adjacency matrix is illustrated in Table 11.2. The key indication that the simulation model and the choice of boundary conditions have led to a well-formed simulation model, is that every row of the adjacency matrix has at least one non-zero entry (this being the case in Table 11.2). The fact that every equation has more than one non-zero entry also suggests that the set of equations must be solved simultaneously, although graph-theoretic methods can be used to identify the sub-set of the equations that must be solved simultaneously, with the remaining equations being solved sequentially (Sowell and Haves 2001).

Table 11.2 System adjacency matrix (well-formed model)

	Simulation Variable			
Equation	T_{wo}	T_{ao}	U	$\dot{m}_w$
HC.1	0	1	0	1
HC.2	1	1	0	1
V.1	0	0	1	1
C.1	0	1	1	0

To illustrate a choice of boundary conditions that result in an insoluble set of equations, consider the case, perhaps during system commissioning, were the control setpoint and control signal are known together with measurements of the air temperature at inlet and outlet of the coil; but the air and water flow rates and water temperatures are unknown. If the known conditions (T_{ai}, T_{ao}, U, SP), are set as the boundary conditions, and the unknown ($\dot{m}_a$, $\dot{m}_w$, T_{wi}, T_{wo}), as the simulated variables, the adjacency matrix (Table 11.3), indicates that the equations are insoluble as none of the simulation variables are linked to the controller equation (C.1). In effect, this reduces the number of equations to 3, with the result that the equation set can only be solved for 3 unknowns rather than 4. Inspection of the information flow diagram (Figure 11.11), indicates that the assertion that the controller equation is redundant is correct, as all of the controllers input and output variables (T_{ao}, U, SP), are known and have been set as boundary condition. Intuition would also suggest that the resulting set of simultaneous equations is insoluble as it is physically impossible to identify both the water inlet and outlet temperatures (the best that could be achieved is to model the change in temperature across the heating coil).

The linking of the component models to form a system model is generally independent of the form of model equations, as the linking variables are normally those related to the physical processes (outlet temperature links to inlet temperature; outlet humidity ratio links to inlet humidity ratio; and so on). However, the form of algorithm that can be used to solve the set of simultaneous algebraic equations is dependent on the form of equations.

Solution of directed input-output algebraic equations

System models consisting of directed input-output equations (of the form, $F{:}(X_i, A) \rightarrow X_o$) are generally solved using methods based on the successive substitution algorithm. The solution procedure begins by identifying one or more iteration variables and an associated information flow diagram. For instance, consider the example heating coil sub-system (Figure 11.6), and associated component models (Figures 11.7 to 11.9). If the boundary conditions are taken to be the air mass-flow rate ($\dot{m}_a$), the air and water inlet temperatures (T_{ai}, T_{wi}), and the controller setpoint (SP), and the water mass-flow rate ($\dot{m}_w$), is selected as the iteration variable, then the resulting information flow diagram is illustrated in Figure 11.12. Figure 11.12 makes clear that the coil outlet air temperature (T_{ao}), and the control signal (U), are also candidates for the iteration variable. However, the coil water outlet temperature (T_{wo}), is not a viable iteration variable as it is not used to link any components (and therefore equations). Inspection of the adjacency matrix for the system (Table 11.2), indicates that the water outlet temperature is only linked to heat exchanger equation HC.2, and as such can be calculated directly from HC.2 provided that values for the water mass flow rate ($\dot{m}_w$), and outlet air temperature (T_{ao}), are known. This makes it clear that only three equations need to be solved simultaneously using successive substitution (HC.1, V.1, and C.1).

An implementation of the successive substitution algorithm for the heating coil sub-system is given in Figure 11.13. The algorithm begins by specifying an initial guess for the iteration variable, together with a convergence tolerance on the iteration, given here as a fractional change (ζ), in the iteration variable. The algorithm continues by sequentially calculating the model output variables from the model equations, the sequential iteration continuing until the

Table 11.3 System adjacency matrix (insoluble model)

	Simulation Variable			
Equation	T_{wo}	T_{wi}	$\dot{m}_a$	$\dot{m}_w$
HC.1	0	1	1	1
HC.2	1	1	1	1
V.1	0	0	0	1
C.1	0	0	0	0

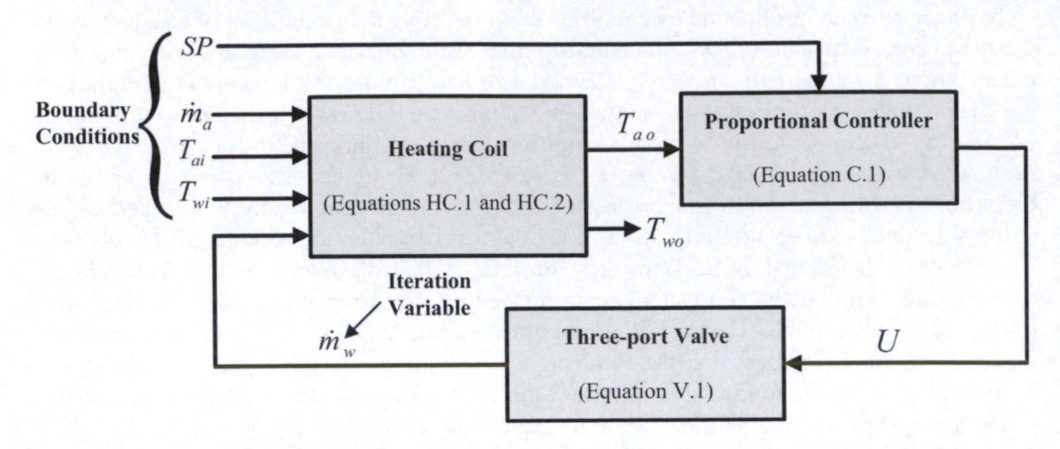

Figure 11.12 Heating coil sub-system information flow diagram for successive substitution.

//Set boundary conditions for current time period, t

set $SP^t, \dot{m}_a^t, T_{ai}^t, T_{wi}^t$;

//Initialize iteration

guess $\dot{m}_w$; //Iteration variable

set ζ; //Convergence tolerance

set $\dot{m}_w^p = \dot{m}_w / (1-\zeta)$; //Previous value of iteration variable

//Iteration

while $\left(\dfrac{|\dot{m}_w^p - \dot{m}_w|}{\dot{m}_w^p} > \zeta \right)$

set $\dot{m}_w^p = \dot{m}_w$; //Reset previous value of iteration variable

calculate T_{ao}; //From the coil model, Equation HC.1

calculate U; //From the controller model, Equation C.1

calculate $\dot{m}_w$; //From the valve model, Equation V.1

endwhile

calculate T_{wo};//From the coil model, Equation HC.2

//Solution = $T_{ao}, U, \dot{m}_w, T_{wo}$;

Figure 11.13 Heating coil sub-system successive substitution pseudo-code.

change in iteration variable is less than or equal to the convergence tolerance.[3] In the case of the heating coil sub-system, the water outlet temperature is calculated on completion of the iteration.

The successive substitution algorithm is seemingly simple to implement, although it can fail to converge due to divergence or oscillation of the iteration variable value. The robustness of the algorithm is dependent, in part, on the choice of iteration variable and the equation iteration sequence, the choice of a robust iteration sequence increasing in difficulty with the size and complexity of the system being modeled. The difficulty in forming a robust iteration sequence has led to an increase in the use of residual model equations and more robust solvers based on the Newton-Raphson iteration.

Solution of input-output free algebraic equations

Input-output free models ($F:(X_i, X_o, A) \rightarrow R$) can be solved using methods based on the Newton-Raphson algorithm. The Newton-Raphson algorithm is derived from a first-order Taylor-series expansion of the model equations about the solution. For the case of four residual equations (and four simulation variables), the Taylor-series expansion would result in equation set (11.11a):

$$\begin{bmatrix} \frac{\partial R_1}{\partial x_1} & \frac{\partial R_1}{\partial x_2} & \frac{\partial R_1}{\partial x_3} & \frac{\partial R_1}{\partial x_4} \\ \frac{\partial R_2}{\partial x_1} & \frac{\partial R_2}{\partial x_2} & \frac{\partial R_2}{\partial x_3} & \frac{\partial R_2}{\partial x_4} \\ \frac{\partial R_3}{\partial x_1} & \frac{\partial R_3}{\partial x_2} & \frac{\partial R_3}{\partial x_3} & \frac{\partial R_3}{\partial x_4} \\ \frac{\partial R_4}{\partial x_1} & \frac{\partial R_4}{\partial x_2} & \frac{\partial R_4}{\partial x_3} & \frac{\partial R_4}{\partial x_4} \end{bmatrix} \begin{bmatrix} x_1 - x_1^s \\ x_2 - x_2^s \\ x_3 - x_3^s \\ x_4 - x_4^s \end{bmatrix} = \begin{bmatrix} R_1 \\ R_2 \\ R_3 \\ R_4 \end{bmatrix} \quad (11.11a)$$

where, $x_1^s, x_2^s, x_2^s, x_4^s$, represent the values of the simulation variables at the solution (these being unknown); R_1, R_2, R_3, R_4 are the residual equation values which are computed for the current estimate of the solution (x_1, x_2, x_3, x_4); similarly, the gradients of the residual equations ($\partial R_i / \partial x_j$), with each simulation variable are calculated about the current estimate of the solution.

Equation (11.11b) gives equations (11.11a) in matrix form where; J is the Jacobian matrix of partial derivatives; D the vector of differences between the current estimate of the solution and the value at the solution; and $\vec{R}$ the vector of residual equation values.

$$J \times D = \vec{R} \quad (11.11b)$$

Having computed the partial derivatives and residual equation values, Equation (11.11b) can be solved for the vector of differences D, between the variables current values and the values at the solution. The differences can then be used to correct the current value of the simulation variables:

$$(X - D) \rightarrow X \quad (11.12)$$

where, $X = \begin{bmatrix} x_1 \\ x_2 \\ x_3 \\ x_4 \end{bmatrix}$.

Figure 11.14 illustrates the Newton-Raphson procedure, which begins by assigning values to the model parameters (A), the convergence tolerance (ζ), and the differencing intervals (∂X); in some algorithms, the differencing interval is also adapted within the do-while iteration loop. The algorithm continues with the boundary conditions (X_t^b), being set for the current time-step (t). An initial guess of the value of the simulation variables (X), must also be given for each time-step, the closer the guess to the solution the more likely it being that a solution will be

```
//Begin
set A;                                   //Vector of model parameters

set ζ;                                   //Convergence tolerance

set ∂X = {∂x_1, ∂x_2, ..., ∂x_n};        //Variable differencing intervals

//Initialize iteration for current time period,t

set X_t^b;                               //Vector of boundary variables

guess X = {x_1, x_2, ..., x_n};          //Simulation variables

//Iteration
do
   calculate R^T = [R_1(A, X_t^b, X), R_2(A, X_t^b, X), ... R_n(A, X_t^b, X)];
                                                     //Residual equations
   calculate J = f(A, X_t^b, X, ∂X);        //Jacobian matrix

   solve     J×D = R;                       //Solve for D

   set       X = X − D;                     //Update simulation variable values

while( Σ_{i=1}^{n} |R_i| > ζ )

//Solution = X = {x_1, x_2, ..., x_n}
```

Figure 11.14 Newton-Raphson pseudo-code.

found. Once a solution has been found for the initial time-step ($t=0$), the variable values found in the previous time-step (t–1), can be used to initialize the guess of the solution in the current time-step (t), the rational being, that in many instances, the system operation does not change significantly from one time-step to the other.

The iteration sequence begins by calculating the value of the residual equations ($\vec{R}$), for the current estimate of the solution (X), the value of the residuals also being a function of the model parameters (A), and the boundary variables (X_t^b). The iteration continues with the calculation of the Jacobian matrix of partial differentials (J), this also being a function of the model parameters, boundary variables, current estimate of the solution, as well as the differencing intervals (∂X). The set of simultaneous equations (Equation 11.11b), can then be solved for the estimated distances of the variable values from the solution (D), these then being used to update the current estimate of the solution. The iteration continues until the absolute sum of the residual equation values lies within the convergence tolerance (ζ).

The residual equations for the heating coil sub-system (Figure 11.11), can be formed by subtracting the left-hand side of the model equations from the right-hand side of the equations:

$$R_1 = T_{ai} + \frac{\varepsilon \min(w_a, w_w)(T_{wi} - T_{ai})}{w_a} - T_{ao} \tag{HC.1R}$$

$$R_2 = T_{wi} - \frac{w_a(T_{ao} - T_{ai})}{w_w} - T_{wo} \tag{HC.2R}$$

$$R_3 = \frac{\dot{m}_{max}}{\sqrt{1+\gamma(C_{ih}{}^{-2}-1)}} - \dot{m}_w \qquad \text{(V.1R)}$$

$$R_4 = \min(\max(0,u),1) - U \qquad \text{(C.1R)}$$

If the components are connected in the same manner as illustrated in Figure 11.11, and we choose the boundary conditions to be equal to the inlet conditions and the control setpoint ($\dot{m}_a$, $\dot{m}_w$, T_{ai}, T_{wi}, SP), the adjacency matrix for the sub-system (Table 11.4) is equivalent to that for the solution of the directed input-output equations (Table 11.2).

Since the residual equations are not a function of all of the simulation variables, the Jacobian matrix will be sparse as it can contain many zero entries (the zero entries in the Jacobian matrix corresponding to the zero entries in adjacency matrix). The sparse nature of the Jacobian indicates that simultaneous equations (Equation 11.11b), are best solved using sparse matrix methods (Sowell and Haves 2001).

Application examples

The use of HVAC component models is illustrated with two applications, HVAC system condition monitoring,[4] and HVAC system optimization. The factors having most impact on the success of the modeling differ between the two applications.

HVAC system condition monitoring often requires models that are easily calibrated to a correctly operating system and as such they can have some model parameters that have no real physical meaning. Conversely, models used in HVAC system optimization must have parameters that represent the manufactured components.

Since the control signals are often recorded during condition monitoring, there is no need to model the local-loop controls. However, since the controller outputs are not available as boundary conditions during the design stage, the local-loop controllers must be modeled in system optimization studies. The availability of the control signals during condition monitoring can allow the system equations to be solved sequentially rather than simultaneously, but simultaneous solution being required when the control loops are modeled in optimization studies.

Since the modeled outputs are used to assess the state of a real system in condition monitoring schemes, the uncertainty arising from measurements and the modeling procedure are critical to the success of the scheme. Although the uncertainty should be considered during the system optimization process, it is more critical to develop optimization constraints that ensure the optimization search remains within the component performance envelopes (the performance envelopes being defined by the manufactures measured data, and/or the valid region of the component model).

Reference models for HVAC system commissioning and fault detection

HVAC component models have been used in the commissioning of HVAC systems, as well as the detection of HVAC system faults. In both cases a model is used as a reference for correct operation of the system (Figure 11.15).

Table 11.4 System adjacency matrix

	Simulation Variable			
Residual Equation	T_{wo}	T_{ao}	U	$\dot{m}_w$
HC.1R	0	1	0	1
HC.2R	1	1	0	1
V.1R	0	0	1	1
C.1R	0	1	1	0

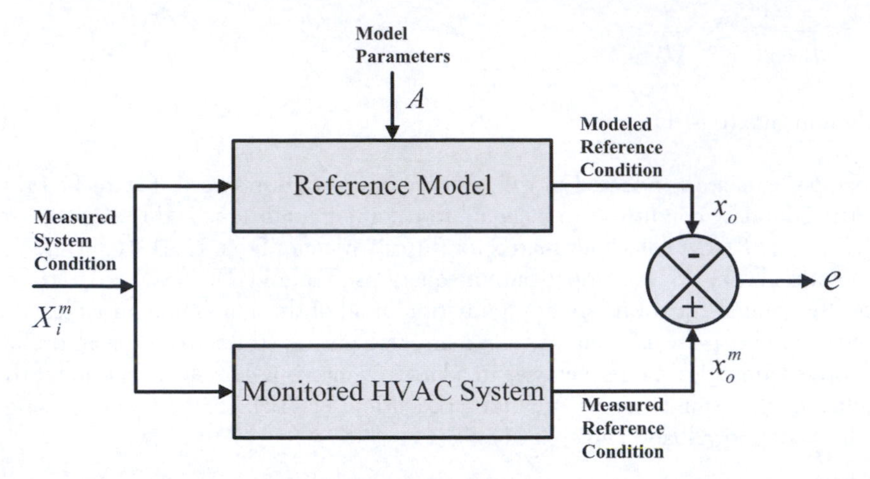

Figure 11.15 HVAC system condition monitoring scheme.

Figure 11.15 illustrates the approach, in which, given a set of measured input conditions[5] (X_i^m), and a set of model parameters (A), the model is used to predict the value of a single quantity at the "outlet" of the component or system (x_o). This is compared to the measured value of the same quantity (x_o^m), the difference in the measured and modeled values (e), being used to infer the existence of a fault (or in the case of system commissioning, that the system is poorly commissioned, or that incorrect equipment has been installed).

The form of reference model and associated parameters depends on the application. In the case of system commissioning, the model must encapsulate the design intent, and therefore it is important for the model to have parameters that relate to the design specification of the installed system components (for instance; the diameter of a fan, or the number of tube rows in a heating coil). However, following the completion of the system commissioning, models for use in fault detection are often calibrated to a correctly operating system to produce near-zero prediction error (e), the rational being that an increase in the model error following calibration would indicate the presence of a new fault condition. In this case, the model equations and parameters may be simplified and should be of a form that enables their calibration to the system; for instance, a heating coil model may include a heat transfer correction parameter that can be tuned to account for the difference in predicted and installed performance.

The form of model equations also depends on which system conditions are being measured, these often being restricted to measurements used in the direct control of the system. This can result in the simplification of the model equations, or required model inputs being estimated from other measurements; for instance, a flow temperature or flow rate being estimated through an energy balance on different flow streams.

Given that the measured conditions can include the control signals (U), at the output of the local-loop controllers, the model equations can often be solved sequentially. For example, consider the heating coil sub-system in Figure 11.6, and the associated information flow diagram in Figure 11.11; if the outlet air temperature (T_{ao}), is used as a modeled reference condition and the control signal (U), is measured together with the air mass-flow rate ($\dot{m}_a$), and the air and water inlet temperatures (T_{ai}, T_{wi}), then the iterative loop associated with the outlet air temperature is broken and the proportional controller model is redundant. This enables the equations for the three-port valve and heat exchanger to be solved in sequentially (Figure 11.16).

Sources of uncertainty have a significant impact on the effectiveness of HVAC system condition monitoring schemes. Uncertainty exists, not only in the measured values, but also in the model parameters and the form of model equations. Uncertainty in the model parameters

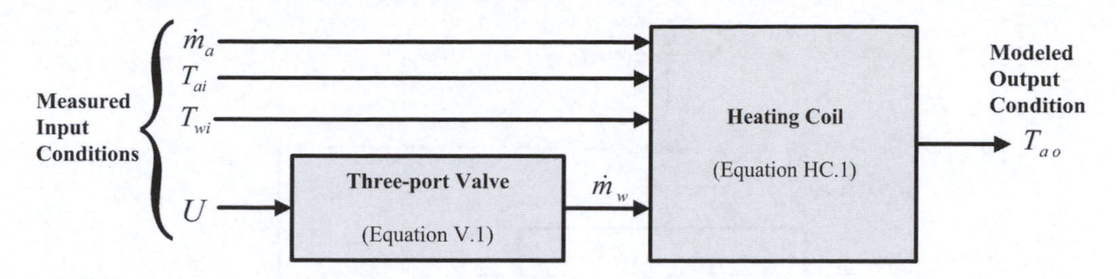

Figure 11.16 Information flow diagram for sequential equation solution.

can occur in the manufacturer's measurement of component performance, or through the procedure used to calibrate the model parameters to a set of measured data.

The uncertainty in the model equations is a result of the assumptions made during the model development. For example, the equations for heat exchanger effectiveness (ε), used in the heating coil model (Figure 11.7), are for a pure counter-flow heat exchanger, whereas the flow regime in an installed heating coil is more likely to be a combination of cross-flow and counter-flow. Where the model equations are solved simultaneously, "uncertainty" also arises due to the convergence tolerance used by the solution algorithm (the equations rarely being solved exactly using numerical solution procedures).

If the condition monitoring scheme is based on the use of steady-state models, then there is also some uncertainty about the extent to which the system was in steady-state when the performance measurements were taken. The combined impact of several sources of uncertainty can mask faults that have a magnitude equivalent to 14% of the peak capacity of the coil in a cooling coil sub-system (Buswell and Wright 2004).

Component models for HVAC system optimization

The design and operation of HVAC systems can be optimized by using an optimization algorithm to search for the value of the model parameters and/or model input variables that minimize a specified criterion such as energy use, capital cost, or life-cycle cost (Wright 1996). The optimization algorithm successively generates new trial solution in a manner that minimizes the objective function ($f(X_o,X_i,A)$), with each new trail solution being evaluated using the HVAC system model (Figure 11.17). Where the optimization is concerned with optimizing the size of the HVAC components, the optimization variables (X^{opt}), are mapped onto the model parameters (since the physical size of the components are normally defined by the model parameters). Similarly, since the control setpoints are normally specified as a sub-set of the model input variables, optimization of the systems supervisory control strategy is achieved by mapping the optimization variables onto the model input variables. In some instances, the size and operation of the HVAC systems is optimized simultaneously with the form of building fabric construction (Wright *et al.* 2002).

Since the optimization is often concerned with sizing the components, the component models must be formulated to represent a range of possible sizes of component. Quasi-first principle models (such as of coils and other heat exchangers), tend to be generic and therefore valid for range of components. However, empirical models (typically used for pumps and fans), require some form of data normalization, or use of scaling rules in order to be able to represent a range of components (Wright 1991).

Optimization of HVAC system size and operation is normally performed subject to one or more constraints ($g(X_o,X_i,A)$), that ensure the design solutions do not exceed practicable performance limits (such as those set on fluid velocities), and are within the measured and/or valid performance envelope of the component. Constraint functions that represent the limits of the measured component performance envelope are particularly important for fully empirical

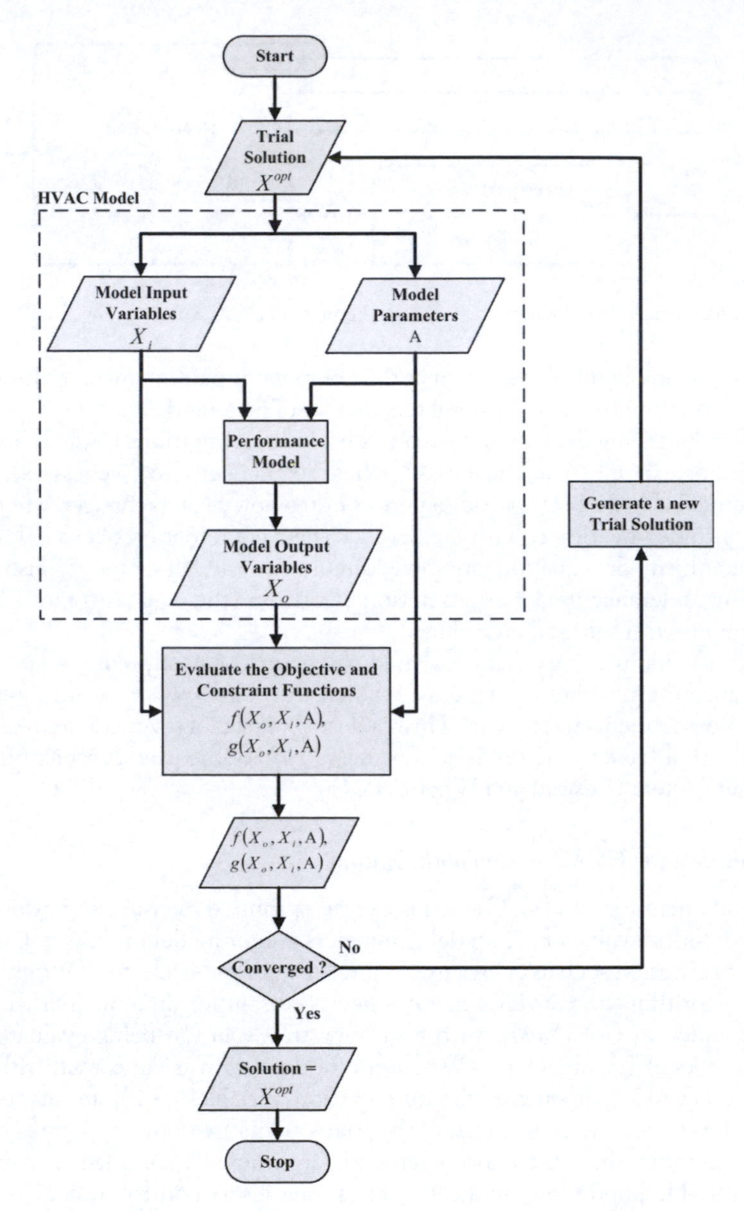

Figure 11.17 Optimization process.

models (Wright 1991), as extrapolation beyond the measured range of performance is likely to be erroneous or impracticable (erroneous due to unpredictable behavior of the model beyond the range of curve-fit data; and impracticable, as the optimized size of component may not be manufactured and available in practice). It may also be necessary to place constraints on the solutions obtained from quasi-first principle models. For example, the Colebrook equation for duct friction factor (Equation 11.8), is only valid for fully turbulent flow, and in this respect, a duct size optimization based on the model in Figure 11.5, should include a constraint on the Reynolds number to ensure that the optimization solution is for fully turbulent flow.

In summary, HVAC component models for use in HVAC system optimization must be able to model the performance of a range of component sizes (and products from different manufacturers). Separate post-processed models must also be developed for calculation of derived quantities used in formulation of the optimization objective and constraint functions, these being a function of the model input variables and model parameters, as well as the model output variables.

Discussion

This chapter describes; the fundamental forms of HVAC component models; the connection of the component models to form a HVAC system model; and two approaches to solving the system algebraic equations. These principles can be applied to the modeling of a wide range of HVAC components and systems. However, there is an increasing tendency for the use of HVAC systems in which the air-paths are integral with the building fabric. This results in HVAC component or sub-system models that must include elements of building heat transfer and thermal storage. Further, it is important to consider the sources of available component and system performance data before developing a component model, as the form of available data will inevitably dictate the form of the component model parameters.

Modeling of integrated ventilation systems

There are three technologies in which the ventilation passes through the building fabric; hollow-core ventilated floor slabs; under floor air distribution systems; ventilated facades.

Passing the ventilation air through the hollow-core of concrete floor slabs enhances its thermal coupling to the heat energy stored in the building fabric. As such, a model of the ventilated floor slab must encapsulate the dynamic heat storage and the heat transfer between the ventilation air and the concrete slab. Since the slabs form the floors and ceilings of the zones, the model must also include the heat exchange between the slab and the adjacent zone air and surfaces. Models which have these characteristics are described by Ren and Wright (1998) and Barton *et al.* (2002).

The ventilation air can also be supplied to the room using an under-floor distribution system. These systems also result in a close thermal coupling between the ventilation air and the floor slab. As well as modeling the thermal storage in the floor slab, the modeling of these systems must also include the heat transfer through the light-weight floor surface and through the concrete slab from the adjacent zone. The air distribution in the ventilated zoned is also normally included in under-floor ventilation system models. The under-floor air distribution heat transfer paths and modeling are described by Bauman *et al.* (2006), and Hui *et al.* (2006).

The transmission of solar heat energy into the building can be useful when the building is being heated and detrimental when being cooled. Some control of the solar heat flux can be achieved by ventilating a cavity between layers of glazing (or between an opaque inner layer, and glazed outer layer). Models of such façade systems need to account for not only the heat transfer, but the flow regime within the cavity. A model of a mechanically ventilated façade is described by Infield *et al.* (2006); models for buoyancy driven flow have also been developed, but these façades are not always integral with the HVAC system.

Performance data and model development

Given that there are very few HVAC components that can be modeled using purely first principles, the majority of models rely on some measured performance data. The development of HVAC component models therefore depends on the availability and form of measured performance data. The test procedure used to obtain the data can also have an impact on the development of the model. For instance, one goal of a performance test of a fin-tube heating or cooling coil is to identify the air-side heat transfer coefficient. However, the air-side heat transfer coefficient could be defined in several ways. For use in design practice, the air-side heat transfer coefficient might be given as a function of the coil face-area and associated air-velocity (with different functions being given for each fin-density). However, a more detailed performance evaluation might give the air-side heat transfer coefficient as a function of the air velocity in the passage between the fins. In both cases, it is also necessary to make assumptions about the form of heat transfer coefficient inside the water-tubes and for the fin efficiencies. The form of the coil model will therefore be restricted by the form of the air-side heat transfer coefficients, and must also incorporate the same assumptions made about the water-side heat transfer coefficient and fin-efficiency (as the measured heat transfer coefficients are an implicit function of these assumptions). Similar observations can be made regarding the performance testing of fans as the measured performance is a function of the particular test-duct, the implication being that the model should attempt to include the "installation effect".

To date, the format of HVAC performance data has varied from one manufacturer to another, this making it difficult to develop a generic model structure. This deficiency is being addressed through initiatives in "building information modeling" (BIM), and in particular, the development of HVAC industry foundation classes (Bazjanac 2004). Further discussion on data modeling and the development of BIM, can be found in Chapter 17. Chapter 17 also describes a modeling paradigm that extends the component based concepts described here to the development of a modeling and simulation environment in which whole building models are implemented by "formulating equality and conservation equations that link shared interface variables on an equation-level".

Synopsis

HVAC systems are constructed from components. In the same manner, HVAC system models can be constructed from models of the individual system components. This chapter describes the principles used in the development of steady-state component models together with the procedures for linking components to form a system model. Two forms of model equation (directed input-output equations and input-output free residual equations), are described together with the common algorithms used to solve the set of simultaneous system-model equations. The principles described in the chapter can be applied to a wide range of HVAC equipment and systems. However, their limitation in respect to the modeling of the systems dynamic response required for control system performance evaluation is discussed, as is the need to model the dynamic storage effects for HVAC systems that have a close thermal coupling with the building fabric. The relationship between the format of the HVAC equipment manufacturers measured performance data and the development of HVAC component models is also described. Two applications of HVAC system modeling are described in the chapter: the use of HVAC system models in system condition monitoring; and the application of model-based optimization methods in the optimum sizing of HVAC systems.

Notes

1 Some HVAC systems are strongly coupled to thermal storage elements (such as described on pages 333–334), which significantly increases the time-constant of this part of the system.
2 Note that an impeller speed is in revolutions per second.
3 The check for convergence can also be extended to identify divergence or oscillation of the iteration.
4 The calibration of simulation models and their use in building performance monitoring and performance optimization is also discussed in Chapter 13.
5 A "condition" is taken to be a "measured" quantity, including, temperature, relative humidity, flow rate, pressure, and control signal.

References

Angelov, P.P., Hanby, V.I., Buswell, R. A. and Wright, J. A. (2000) "A Methodology for Modelling HVAC Components using Evolving Fuzzy Rules", in *Proceedings of IEEE International Conference on Industrial Engineering, Control and Instrumentation, IECON-2000*, Nagoya, Japan, pp. 247–252.

Barton, P., Beggs, C.P. and Sleigh, P.A. (2002) "A Theoretical Study of the Thermal Performance of the TermoDeck Hollow Core Slab System", *Applied Thermal Engineering* 22(13): 1485–1499.

Bauman, F.S., Hui, J. and Webster, T. (2006) "Heat Transfer Pathways in Underfloor Air Distribution (UFAD) Systems", *ASHRAE Transactions* 112(2): 567–580.

Bazjanac, V. (2004) "Building Energy Performance Simulation as Part of Interoperable Software Environments", *Building and Environment* 39(8): 879–883.

Bourdouxhe, J.P., Grodent, M. and Lebrun, J. (1998) *Reference Guide for Dynamic Models of HVAC Equipment*, Atlanta, Georgia: ASHRAE.

Brandemeuhl, M. (1993) *HVAC2 Toolkit: Algorithms and Subroutines for Secondary HVAC Systems Energy Calculations*, Atlanta, Georgia: ASHRAE.

Braun, J.E. (1988) "Methodologies for the Design and Control of Central Cooling Plants", PhD Thesis, University of Wisconsin-Madison, USA.

Buswell, R.A. and Wright, J.A. (2004) "Uncertainty in Model Based Condition Monitoring", *Building Services Engineering Research and Technology* 25(1): 65–75.

Haves, P., Salsbury, T.I. and Wright, J.A. (1996) "Condition Monitoring in HVAC Subsystems Using First Principles Models", *ASHRAE Transactions* 102(1): 519–527.

Holmes, M.J. (1982) "The Simulation of Heating and Cooling Coils for Performance Analysis", in *Proceedings of the 1st International Conference on System Simulation in Buildings*, Liège, Belgium, pp. 245–280.

Hui, J., Bauman, F.S. and Webster, T. (2006) "Testing and Modeling of Underfloor Air Supply Plenums", *ASHRAE Transactions* 112(2): 581–591.

Infield, D., Eicker, U., Fux, V., Mei, L. and Schumacher, J. (2006) "A Simplified Approach to Thermal Performance Calculation for Building Integrated Mechanically Ventilated PV Façades", *Building and Environment* 41(7): 893–901.

Lebrun, J., Bourdouxhe, J.P. and Grodent, M. (1999) *HVAC 1 Toolkit: A Toolkit for Primary HVAC System Energy Calculation*, Atlanta, Georgia: ASHRAE.

Ren, M.J. and Wright, J.A. (1998) "A Ventilated Slab System Thermal Storage Model", *Building and Environment* 33(1): 43–52.

Salsbury, T.I. (1996) "Fault Detection and Diagnosis in HVAC Systems Using Analytical Models", PhD Thesis, Loughborough University, Loughborough, Leicestershire, UK.

Sowell, E.F. and Haves, P. (2001) "Efficient Solution Strategies for Building Energy System Simulation", *Energy and Buildings*, 33(4): 309–317.

Underwood, C.P. (1999) "HVAC Control Systems: Modelling, Analysis and Design", London: E&FN Spon Press.

Wright, J.A. (1986) "The Application of the Least Squares Curve Fitting Method to the Modelling of HVAC Component Performance", in *Proceedings of the Second International Conference on System Simulation in Buildings*, Liège, Belgium, pp. 27–45.

Wright, J.A. (1991) "HVAC Optimisation Studies: Steady-state Fan Model", *Building Services Engineering Research and Technology* 12(4): 129–135.

Wright, J.A. (1993) "HVAC Simulation Studies: Solution by Successive Approximation", *Building Services Engineering Research and Technology* 14(4):179–182.

Wright, J.A. (1996) "HVAC Optimisation Studies: Sizing by Genetic Algorithm", *Building Services Engineering Research and Technology* 17(1): 7–14.

Wright, J.A., Loosemore, H.A. and Farmani, R. (2002) "Optimization of Building Thermal Design and Control by Multi-Criterion Genetic Algorithm", *Energy and Buildings* 34(9): 959–972.

Zhou, X. and Braun, J.E. (2007) "A Simplified Dynamic Model for Chilled-Water Cooling and Dehumidifying Coils-Part 1: Development", *HVAC&R Research* 13(5): 785–824.

Recommended reading

Numerous academic journal papers can be found that describe the development and application HVAC component performance models. However, in particular, descriptions of several steady-state HVAC system component models can found in two reference "toolkits" developed by the American Society of Heating, Refrigerating, and Air-Conditioning Engineers (ASHRAE); there being separate toolkits for primary systems (Lebrun *et al.* 1999), and secondary systems (Brandemeuhl 1993). Both toolkits give the theoretical basis for each component model as well as each model coded as a FORTRAN routine. The models described in these toolkits have been adopted for use in several established building performance simulation programs.

The principles of developing dynamic HVAC component models are described by Bourdouxhe *et al.* (1998). The models described in this reference are a result of an extensive survey of dynamic model development, with models being described for ducts and pipes, heat exchangers, boilers, heat pumps and chillers. This publication also gives models for control system components such as sensors and actuators; control system component models are also described in the textbook by Underwood (1999).

Mathematical methods for the solution of algebraic equations can be found in any mathematics text book. The extent to which the computational performance of these methods can be improved through the application of graph-theoretical methods in reducing the set of simultaneous equations is described by Sowell and Haves (2001).

Activities

Exercise 1

Using the centrifugal fan performance data given in Table 11.5, develop an input-output model based on a curve-fit of the non-dimensional fan performance. Use the model to predict the total pressure and air temperature at the fan outlet for a 0.25m diameter fan running at a speed of 3000 rpm when the air mass flow rate is 0.9kg/s, the inlet air temperature is 20.0 °C, and the inlet air pressure is 0.0Pa. Assume a fixed inlet air density of 1.2 kg/m^3 and an air specific heat capacity of 1.005 kJ/kg K.

Exercise 2

Implement the rectangular duct friction loss model illustrated in Figure 11.5, and use it to calculate the outlet pressure of a 0.5m by 0.5m square duct when the inlet mass flow rate is 0.9 kg/s and the inlet pressure is 100 Pa. Take the air density and dynamic viscosity to be 1.2 kg/m^3 and 1.85×10^5 Pa.s respectively. Assume that the duct is constructed from galvanized sheet steel having an absolute roughness of 0.09mm.

Answers

Exercise 1

Figure 11.18, illustrates an input-output centrifugal fan model based on a polynomial curve-fit of the non-dimensional performance data. The model parameters are the fan diameter,

Table 11.5 Centrifugal fan performance data

D *(m)*	*n* *(rpm)*	$\dot{m}$ *(kg/s)*	P *(Pa)*	W *(W)*
0.22	1500	0.26	166	52
0.22	1500	0.32	145	51
0.22	2500	0.55	417	266
0.22	2500	0.36	490	216
0.22	3500	0.83	759	728
0.22	3500	0.55	974	649
0.25	1500	0.72	71	95
0.25	1500	0.65	109	102
0.25	2500	1.28	97	326
0.25	2500	1.09	313	483
0.25	3500	1.85	146	752
0.25	3500	1.5	670	1352

$$A = \{d, n, \alpha_1, \alpha_2, \alpha_3, \alpha_4, \alpha_5, \alpha_6, \alpha_7, \rho_i, C_p\}$$

$$X_i = \begin{Bmatrix} \dot{m} \\ T_i \\ P_i \end{Bmatrix} \Rightarrow$$

$$\phi = \frac{\dot{m}}{\rho_i \pi^2 d^3 \, n/(60 \times 4)}$$

$$\psi = \alpha_1 \phi^2 + \alpha_2 \phi + \alpha_3$$

$$\lambda = \alpha_4 \phi^3 + \alpha_5 \phi^2 + \alpha_6 \phi + \alpha_7$$

$$P_o = P_i + \frac{\psi \rho_i \left(\pi d \, n/60\right)^2}{2}$$

$$T_o = T_i + \frac{\lambda \rho_i \pi^4 d^5 \left(n/60\right)^3}{8 \dot{m} C_p}$$

$$\Rightarrow X_o = \begin{Bmatrix} T_o \\ P_o \end{Bmatrix}$$

Figure 11.18 Input-output centrifugal fan model.

polynomial coefficients, inlet air density and specific heat capacity. The fan speed has also been treated as a parameter, although it could equally be a model input variable which would allow the model to be used in variable-volume system simulations. The air density and specific heat capacity would also normally be calculated within the model as functions of the inlet air temperature. Figure 11.19 illustrates the normalised fan performance curves; a '+' indicates a

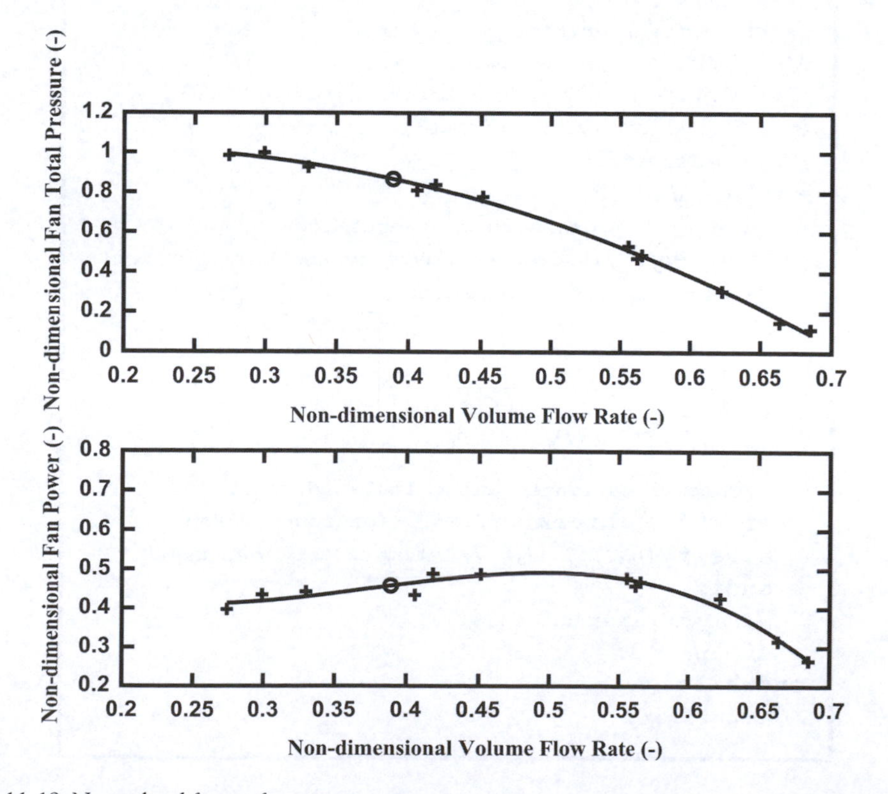

Figure 11.19 Normalised fan performance curves.

normalised data point from Table 11.5, and a 'o' the normalised data point for the example operating conditions. The normalised fan curves can be modelled using a 2nd order polynomial for the relationship between normalised flow and pressure rise, and a 3rd order fit for the relationship between normalised flow and power. The model predicts that a 0.25 m diameter fan running at 3000 rpm produces a fan total pressure of 802 Pa and a power output of 816W; this translates to an outlet temperature of 20.9 °C and pressure of 802Pa.

Exercise 2

The air outlet pressure is 99.38 Pa, resulting from an air friction loss of 0.62 Pa (with a friction factor f of 0.019; an air velocity v of 3.0m/s; a hydraulic diameter D_h of 0.5m; and a Reynolds number Re of 9.73×10^4). These values were obtained using the calculation sequence illustrated in Figure 11.5. Full implementation of the model requires the development of an iteration procedure to solve the transcendental friction factor equation (Equation 11.8). Solution of this equation was achieved here using a using a successive approximation algorithm (Wright 1993). The algorithm is robust and easily implemented, and can be used to solve a range of iteration problems, including those associated with calculating the psychrometric properties of air. However, the method relies on an understanding of the relationship between the iteration variable (termed the "estimate" in Figure 11.20), and the value calculated from the equation being solved. Figure 11.21, illustrates that for this problem, when the estimated value is lower than the solution, the calculated friction factor is higher than the estimate. Given this relationship, the algorithm has been initialized with an initial guess of the solution for the friction factor being equal to the maximum likely friction factor (Figure 11.20). The algorithm proceeds by continuously subtracting a specified increment (Δf), from the iteration

```
//Initialize iteration parameters
Δf = 0.05;      //Increment size
f  = 0.1;       //Maximum friction factor
ζ = 0.00025; //Convergence tolerance
//Begin iteration
while (Δf > ζ)
    fo = f;     //Store current estimate
    f = f-Δf; //Reduce estimate by current increment
    //Calculate friction factor
```

$$f_t = \left(\frac{1}{-2\log\left(\varepsilon / 3.7D_h + 2.51 / Re\sqrt{f} \right)} \right)^2$$

```
    //Compare estimated and calculated values
    if (ft ==f) break;     //Solution found, stop
    elseif (ft >f) f=fo; //Estimate too low, reset
    endif
    //Halve increment size
    Δf=Δf/2.0;
endwhile
//Solution = f
```

Figure 11.20 Pseudo-code for the friction factor successive approximation.

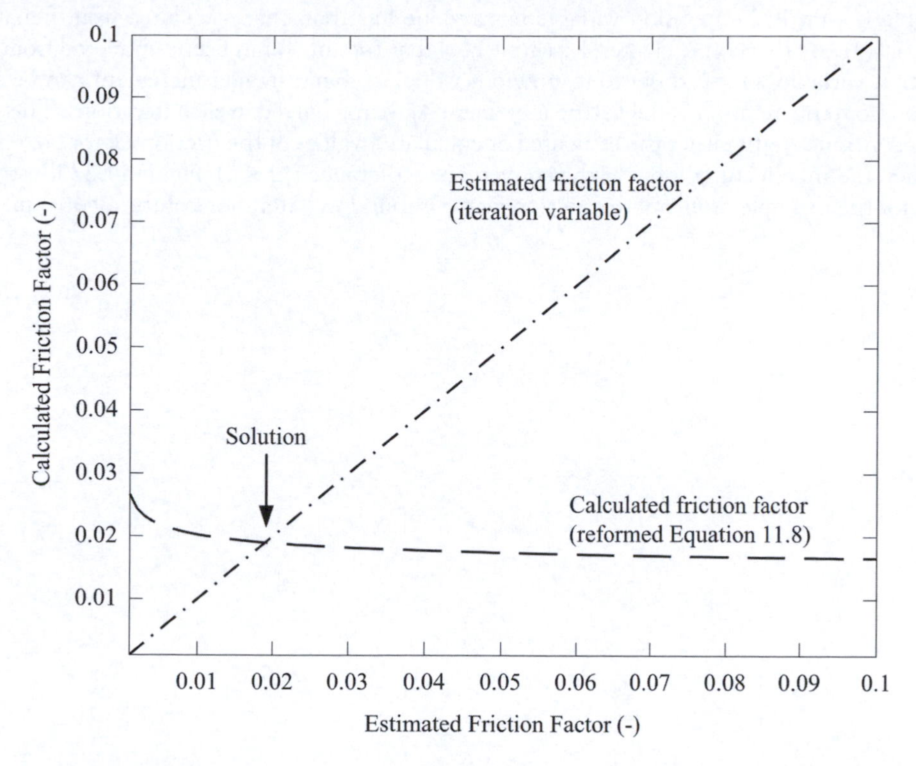

Figure 11.21 Comparison of estimated (iterated) and calculated friction factor.

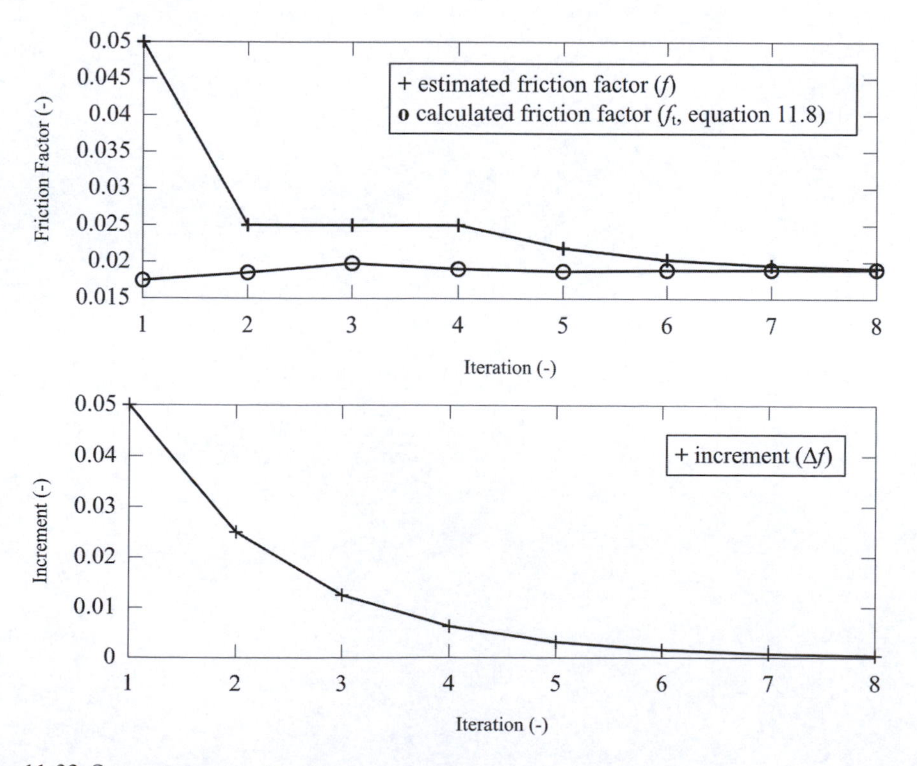

Figure 11.22 Successive approximation convergence.

variable (f), until the iteration variable has a value less than that calculated from Equation 11.8 ($f_t > f$). At this point, Figure 11.21 indicates that too much has been subtracted from the iteration variable, so it is re-set to its previous value, so that a smaller increment can be subtracted from the iteration variable (the increment Δf, being halved in each iteration). The iteration continues until either the estimated or calculated values of the friction factor are equal, or when the increment falls below the convergence tolerance ($\Delta f \leq \zeta$). Figure 11.22 illustrates that, for the example problem, the friction factor is found in 8 iterations of the algorithm.

12 Micro-cogeneration system performance prediction

Ian Beausoleil-Morrison

Scope

Fuel-cell and combustion-based cogeneration devices with electrical outputs under 15 kW.

Learning objectives

Develop an understanding of small-scale cogeneration (micro-cogeneration) technologies and their potential applications for heating, cooling, and powering buildings. Develop an understanding of modelling the performance of these devices concurrently with building thermal and electrical systems.

Key words

Combined heat and power/cogeneration/micro-cogeneration/fuel cell/Stirling engine/Internal combustion engine

Introduction

Micro-cogeneration, residential cogeneration, and small-scale combined heat and power are synonyms for technologies that concurrently produce electricity and heat from a single fuel source. By convention these terms apply to systems with electrical outputs less than 15 kW. These emerging technologies offer a number of potential benefits: reduced primary energy consumption, reduced greenhouse gas emissions, reduced electrical transmission and distribution losses, and alleviation of electrical grid congestion during peak periods. Numerous companies are actively developing fuel cell (FC), internal combustion engine (ICE), and Stirling cycle engine (SE) micro-cogeneration devices and introducing these to the market (Knight and Ugursal 2005; Onovwiona and Ugursal 2006). Although these technologies are technically immature and their potential benefits remain unproven, promising indications of their performance have led to financial incentives and favourable electricity tariff structures in many countries to encourage their adoption.

Micro-cogeneration devices have only modest fuel-to-electrical conversion efficiencies. The existing prototypes have been found to have efficiencies in the range of 5 to 30% in terms of the net alternating current (AC) electrical output relative to the source fuel's lower heating value, or LHV (Beausoleil-Morrison 2008; Carbon Trust 2007; Entchev *et al.* 2004). These electrical efficiencies are relatively low compared to state-of-the-art natural-gas-fired central power generation: combined-cycle central power plants can achieve efficiencies in the order of 55% (Colpier and Cornland 2002; DeMoss 1996). As such, it is critical that the thermal portion of the micro-cogeneration device's output be well utilized, otherwise micro-cogeneration will compare poorly to the best-available central power generation technologies. Potential uses of this thermal output include space heating, domestic hot water (DHW) heating, and space cooling. The latter could be accomplished by using the micro-cogeneration device's thermal output to activate a thermal refrigeration cycle (the candidate cycles are outlined in Henning (2007), for example).

Residential buildings have a strong diversity in both electrical and thermal loads and there is no inherent coincidence between these loads. This fact is illustrated in Figure 12.1, which plots space heating and electrical demands for a house over a typical spring day. (The electrical demand was measured while the space heating demand was simulated.) The space heating demand is modest on this day due to relatively warm weather. As can be seen, this thermal demand is greatest from late evening to early morning and low during the day when passive solar gains are available. In contrast, the electrical demand peaks during the morning and before and after the dinner hour, this due to the operation of appliances for cooking and cleaning. Each combination of house, occupants, and climate will produce different patterns and these will vary day to day. However, what will be consistent is the diversity of these loads and their non-coincidence.

This reality necessitates some kind of thermal (and perhaps electrical) storage in any practical micro-cogeneration system. The volume and thermal characteristics of the thermal store, occupant electrical usage patterns, the house's thermal characteristics, prevailing weather, and dispatch strategies all influence whether the micro-cogeneration device's thermal output will be exploited or wasted. Accurately assessing this highly integrated system can only be achieved within the context of whole-building performance simulation and only when the micro-cogeneration device is simulated concurrently with the building's thermal and electrical demands and with supporting HVAC components.

Scientific foundation

Stirling engines and internal combustion engines are heat engines. They first convert the internal energy associated with the bonds between atoms in the reacting fuel to thermal energy

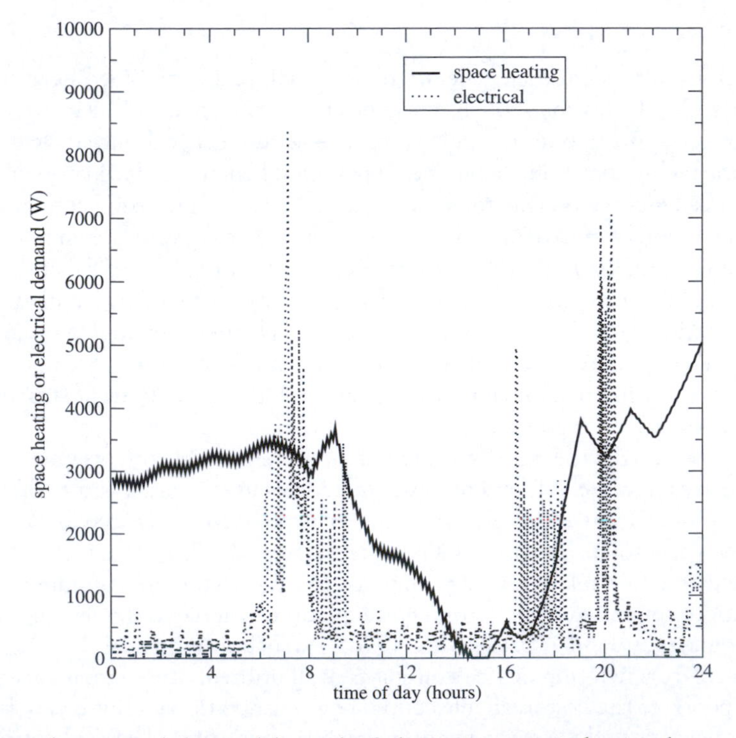

Figure 12.1 Space heating and electrical demands of a house over a typical spring day.

through chemical reactions (combustion). The heat released by the combustion is then converted to mechanical energy in a thermodynamic cycle. The well-known Otto and Diesel cycles are employed in internal combustion engines whereas Stirling engines are external combustion devices operating with closed cycles. The interested reader is referred to Finkelstein and Organ (2001) for a detailed treatment of the Stirling cycle and the design and operation of Stirling engines. The mechanical energy produced by both ICE and SE devices is then converted to electrical energy in a generator. Thermal energy can be recovered from exhaust gases and cooling water circuits to provide the device's useful thermal output. Both ICE and SE devices can be fueled from a variety of sources, such as gasoline, petrodiesel, natural gas, bioethanol, and biodiesel.

In contrast to the heat engines, fuel cells are energy conversion devices that directly convert the internal energy associated with the bonds between atoms in the reacting fuel into electrical energy. This is accomplished through the electrochemical oxidation of a fuel and the electrochemical reduction of oxygen. These electrochemical reactions occur at electrodes which are continuously fed with fuel and oxygen and which are separated by an electrolyte layer. A number of fuel cell varieties exist and these are typically classified according to the electrolyte material employed. Two types are pertinent for micro-cogeneration: solid oxide fuel cells (SOFCs) and proton exchange membrane fuel cells (PEMFCs), also referred to as polymer electrolyte membrane fuel cells.

SOFCs use a solid metal oxide as the electrolyte. Due to their high operating temperature (800 to 1000°C) they are able to operate directly on natural gas and other light hydrocarbon fuels. These temperatures are sufficient to internally reform the gas's constituent hydrocarbon molecules (methane, ethane, propane, etc.) to hydrogen and carbon monoxide which are then supplied to the anode where they are partially oxidized with the oxygen ions crossing the electrolyte.

In contrast, PEMFCs employ a polymer as the electrolyte to conduct hydrogen ions. These are low-temperature fuel cells that operate at temperatures below 100°C. Because of these low operating temperatures PEMFCs must be fueled by hydrogen. This hydrogen can either be delivered and/or stored at the building site or it can be produced from a hydrocarbon fuel (e.g. natural gas) by an external reformer.

The electrochemical reactions occurring in both types of fuel cells produce a direct current (DC), which must then be inverted to AC in a power conditioning unit to produce the micro-cogeneration device's electrical output. Due to thermodynamic irreversibilities, some of the source fuel's chemical energy is converted to thermal energy. Additionally, with SOFCs fuel that is unreacted at the fuel cell electrodes is often combusted in an afterburner resulting in more heat production. SOFC micro-cogeneration devices are typically equipped with heat exchangers to extract much of this thermal energy from the fuel cell's product gases prior to their exhaust. In the case of PEMFCs, the stack is typically cooled with an external water circuit to remove the thermal energy resulting from the irreversibilities and this warmed water stream can form the device's useful thermal output.

The interested reader is referred to Mench (2008) and Larminie and Dicks (2003) for a thorough review of fuel cell electrochemistry and fuel processing.

It is important to note that for SOFC and PEMFC systems the fuel cell stack itself is only a single component within a micro-cogeneration device. Similarly, the thermodynamic cycle (Otto, Diesel, Stirling) is only one component of combustion-based systems. This reality is illustrated in Figure 12.2 which shows one possible system configuration for a SOFC micro-cogeneration device.

Besides the fuel cell stack (shown in grey), the system might include: an afterburner to combust unreacted fuel; an air filter and pre-heater; a fuel desulfurizer, pre-heater, pre-reformer, and reformer; and water preparation. A compressor may be required to supply pressurized fuel while a blower will likely be present to supply air to provide oxygen to support the electrochemical and combustion reactions. A pump may also be required to supply liquid water for steam

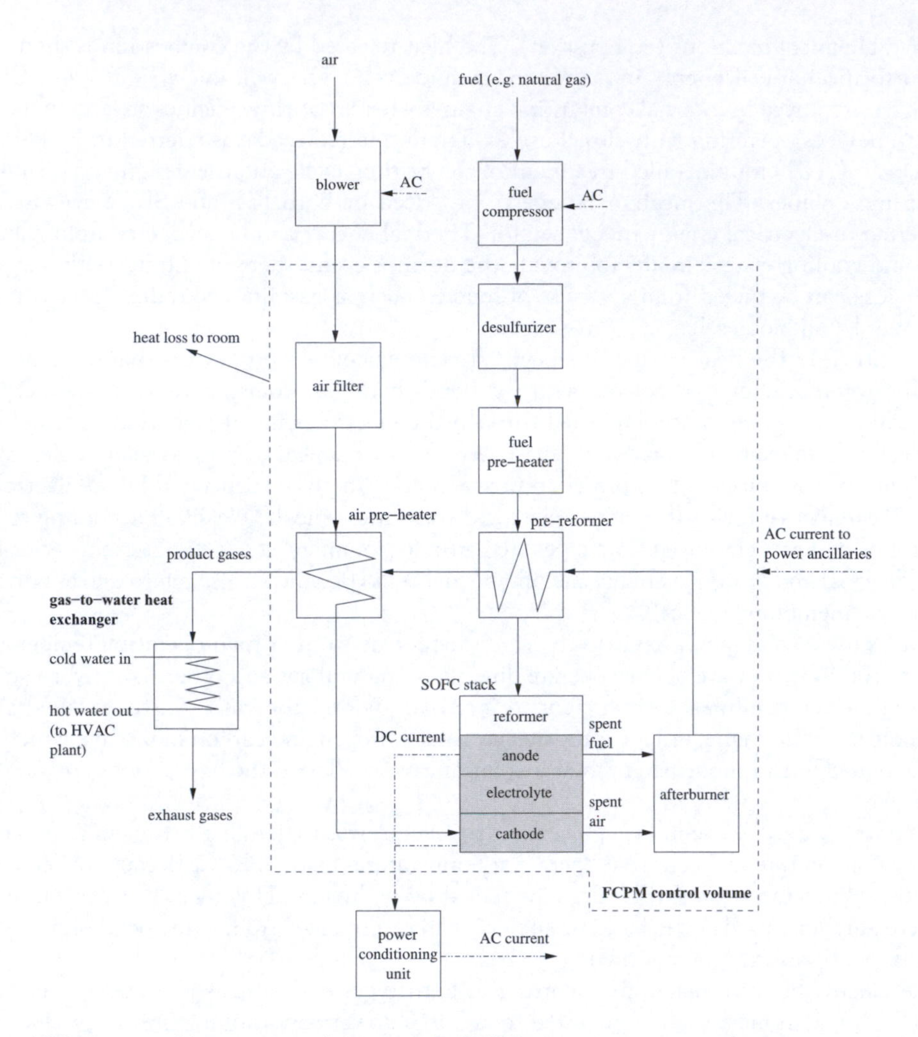

Figure 12.2 One possible system configuration of a SOFC micro-cogeneration device.

reformation purposes. A battery could be used for buffering the fuel cell stack's DC electrical production and for meeting load transients and the system will include a power conditioning unit to convert the electrical output to AC. All SOFC micro-cogeneration systems will include a heat recovery device that transfers the heat of the hot product gases to the building's HVAC system. Some systems may include an integrated auxiliary burner that is activated when the fuel cell cannot satisfy the building's thermal loads. The complexities of these systems must be considered in the modelling of micro-cogeneration devices.

The computational methods that are recommended for treating micro-cogeneration devices within the context of building performance simulation are elaborated in the next section.

Computational methods

The pragmatic modelling of micro-cogeneration devices within the context of building-performance simulation requires models at an appropriate level of resolution.

In the past it has been common to model the performance of micro-cogeneration devices using performance-map methods wherein the device's electrical and thermal efficiencies are treated as constant or are a parametric function of the device's loading (e.g. Alanne *et al.* 2006; Boait *et al.* 2006; Cockroft and Kelly 2006; DePaepe and Mertens 2007; Dorer *et al.*

2005; Hawkes *et al.* 2007; Hawkes and Leach 2005, 2007; Maalla and Kunsch 2008; Peacock and Newborough 2006; Pearce *et al.* 2001; Possidente *et al.* 2006; Sibilio *et al.* 2007; Sicre *et al.* 2005). This approach essentially decouples the modelling of the micro-cogeneration device from that of the house and other aspects of the thermodynamic system and as such precludes an accurate treatment of the thermal coupling to the building and its HVAC system, as demonstrated by Ribberink *et al.* (2009).

A research project (known as Annex 42) undertaken within the International Energy Agency's Energy Conservation in Buildings and Community Systems Programme (IEA/ECBCS) addressed this shortcoming by developing two new models for micro-cogeneration devices. One of these models is suitable for the combustion-based ICE and SE micro-cogeneration devices while the other treats SOFC and PEMFC devices. These are recommended for use in simulating the performance of micro-cogeneration devices while concurrently modelling building thermal and electrical demands within the context of whole-building simulation. Both models are available in the publicly released versions of ESP-r and EnergyPlus while TRNSYS TYPEs based upon the ESP-r source code as well as an IDA-ICE version of the fuel cell model can be acquired upon request.

The following subsections describe the methodologies employed by the IEA/ECBCS Annex 42 models by focusing on a few key aspects. Additionally, the methods recommended for calibrating these models to represent actual micro-cogeneration devices are treated. The interested reader is referred to IEA/ECBCS Annex 42 final reports and associated journal publications for a more complete description of these models, their validation, and their calibration (Beausoleil-Morrison 2007, 2008; Beausoleil-Morrison and Ferguson 2007; Beausoleil-Morrison and Lombardi 2009; Beausoleil-Morrison *et al.* 2006; Ferguson *et al.* 2009; Kelly and Beausoleil-Morrison 2007).

Modelling fuel-cell-based devices

Topology of fuel cell micro-cogeneration model

The model topology of the IEA/ECBCS Annex 42 FC micro-cogeneration model is illustrated in Figure 12.3.

The model discretizes the FC micro-cogeneration system into groupings of components that comprise major sub-systems, such as those that produce electrical power, supply air, provide auxiliary heat, capture heat from hot product gases, convert DC to AC, etc. In this manner, once the model is calibrated for a specific FC micro-cogeneration device analyses can be conducted to explore the benefits of improving the performance of individual sub-systems. For example, the impact of improving the DC-AC power conditioner upon overall system performance can be simulated without recalibrating the portions of the model that represent the fuel cell power module, gas-to-water heat exchanger, and other sub-systems. Additionally, such a structure facilitates the future development of more detailed modelling methods for specific sub-systems.

The model includes the 12 control volumes shown in Figure 12.3, a subset of which will be required to represent any given FC micro-cogeneration device. For example, the treatment of an internally reforming SOFC micro-cogeneration system that operates on standard natural gas delivery pressures would not require the inclusion of the control volumes that represent the fuel compressor and pump. Likewise, as an SOFC does not require stack cooling, the control volumes representing the pump, air cooler, and water-water heat exchanger on the stack cooler would be neglected. The control volumes representing the auxiliary burner, dilution air system, and battery may or may not be required depending upon the system configuration.

Each of the 12 control volumes is modelled in as rigorous a fashion as possible given the constraints of computational efficiency and the need to calibrate model inputs based upon the testing of coherent systems. It is worth noting that the equations described in this section will

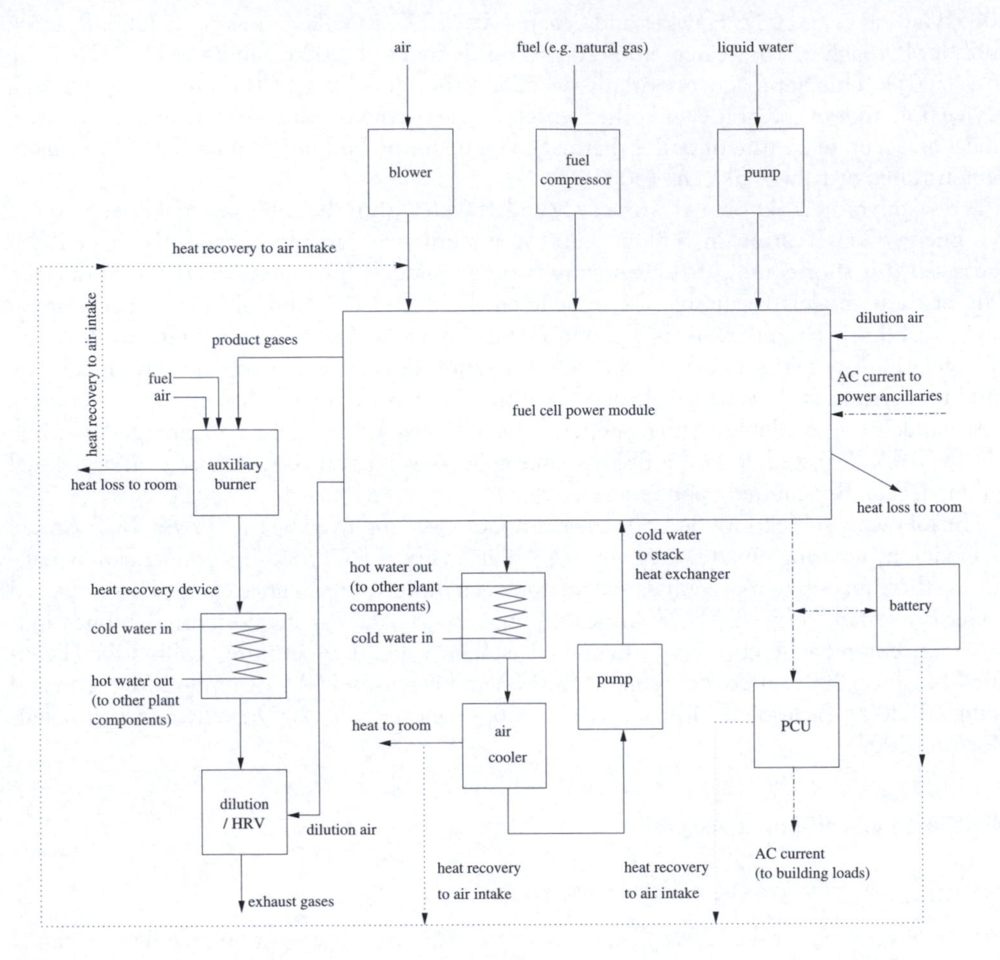

Figure 12.3 Topology of FC micro-cogeneration model.

be recalculated hundreds of thousands of times to perform a single annual simulation at two minute time-steps.

The following sub-sections illustrate the modelling methods employed by focusing on two control volumes. One represents the exhaust-gas-to-water heat exchanger that produces the useful thermal output. The other control volume represents the fuel cell power module (FCPM) which includes the stack, after-burner, and the other components enclosed by the dashed line in Figure 12.2.

Exhaust-gas-to-water heat exchanger

The gas-to-water heat exchanger that transfers energy from the hot gases exiting the FCPM and/or the auxiliary burner is shown on the left side of Figure 12.3. A schematic representation of the control volume encapsulating this heat exchanger is shown in Figure 12.4. The state point labels shown in the figure are used in the development that follows.

The sensible component of the heat transfer from the hot gases to the water is characterized with the log mean temperature difference (LMTD) method for counterflow heat exchangers,

$$q_{sensible} = (UA)_{eff} \cdot \frac{(T_{g-in} - T_{w-out}) - (T_{g-out} - T_{w-in})}{ln\left[(T_{g-in} - T_{w-out}) / (T_{g-out} - T_{w-in})\right]} \tag{12.1}$$

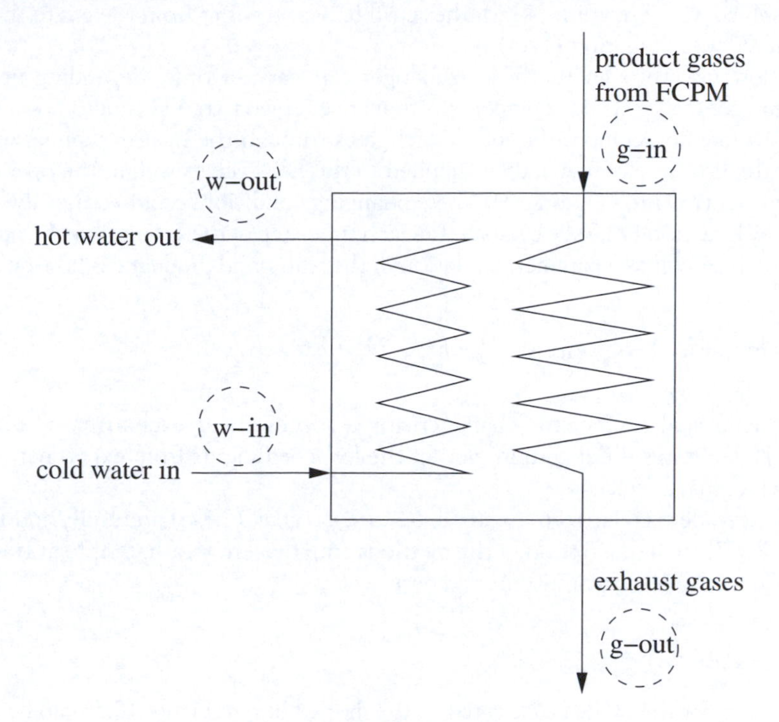

Figure 12.4 Gas-to-water heat exchanger in FC micro-cogeneration model.

where T_{g-in} and T_{g-out} are the temperatures of the gases entering and exiting the heat exchanger and T_{w-in} and T_{w-out} are the entering and exiting water temperatures. $(UA)_{eff}$ is the effective product of the heat exchanger's heat transfer coefficient and area (W/K).

If it is assumed that heat loss from the heat exchanger to the ambient is negligible and that the heat capacity of each fluid stream remains constant through the heat exchanger, then the following energy balance can be written for the heat transfer between the fluid streams,

$$q_{sensible} = (\dot{N}\hat{c}_P)_g \cdot [T_{g-in} - T_{g-out}] = (\dot{N}\hat{c}_P)_w \cdot [T_{w-out} - T_{w-in}] \tag{12.2}$$

where $\hat{c}_p$ is the fluid's molar heat capacity (J/kmolK) and $\dot{N}$ is its molar flow rate (kmol/s).

The rearrangement of Equations (12.1) and (12.2) allows the solution of the temperature of the gas exiting the heat exchanger based upon the entering gas and water temperatures,

$$\begin{aligned} T_{g-out} &= \left\{ \frac{1 - \dfrac{(\dot{N}\hat{c}_P)_g}{(\dot{N}\hat{c}_P)_w}}{exp\left[(UA)_{eff} \cdot \left(\dfrac{1}{(\dot{N}\hat{c}_P)_g} - \dfrac{1}{(\dot{N}\hat{c}_P)_w}\right)\right] - \dfrac{(\dot{N}\hat{c}_P)_g}{(\dot{N}\hat{c}_P)_w}} \right\} \cdot T_{g-in} \\ &+ \left\{ \frac{exp\left[(UA)_{eff} \cdot \left(\dfrac{1}{(\dot{N}\hat{c}_P)_g} - \dfrac{1}{(\dot{N}\hat{c}_P)_w}\right)\right] - 1}{exp\left[(UA)_{eff} \cdot \left(\dfrac{1}{(\dot{N}\hat{c}_P)_g} - \dfrac{1}{(\dot{N}\hat{c}_P)_w}\right)\right] - \dfrac{(\dot{N}\hat{c}_P)_g}{(\dot{N}\hat{c}_P)_w}} \right\} \cdot T_{w-in} \end{aligned} \tag{12.3}$$

Once T_{g-out} is solved with Equation (12.3), the sensible heat transfer from the gas to the water can be determined with Equation (12.2).

The water flow rate through the heat exchanger can vary in time, depending upon the arrangement and control of HVAC components that are coupled to the FC micro-cogeneration device. The flow rate and composition of product gases through the heat exchanger will vary in response to the flow rate of fuel and air supplied to the FC stack as well as the operation of the auxiliary burner (refer to Figure 12.3). These changing conditions required that the $(UA)_{eff}$ value required by Equation (12.3) be evaluated at each time-step of the simulation. An approach is employed which casts it as a parametric relation of the water and product gas flow rates,

$$(UA)_{eff} = hx_{s,0} + hx_{s,1} \cdot \dot{N}_w + hx_{s,2} \cdot \dot{N}_w^2 + hx_{s,3} \cdot \dot{N}_g + hx_{s,4} \cdot \dot{N}_g^2 \tag{12.4}$$

where $hx_{s,i}$ are empirical coefficients characterizing a specific heat exchanger. The form of Equation (12.4) facilitates the determination of the $hx_{s,i}$ coefficients from experimental data, as will be shown on page 352.

The interested reader is referred to Beausoleil-Morrison *et al.* (2006) and Kelly and Beausoleil-Morrison (2007) for a description of the methods employed to treat latent heat transfer as well as alternate methods to the one outlined here.

Fuel cell power module

The energy balance for the FCPM (indicated by the dashed line in Figure 12.3) can be written in the following form,

$$\begin{aligned} &\dot{H}_{fuel} + \dot{H}_{air} + \dot{H}_{liq-water} + P_{el-ancillaries-AC} \\ &= P_{el} + \dot{H}_{FCPM-gas-out} + q_{stack-cooling} + q_{skin-loss} + q_{dilution-air-heat-loss} \end{aligned} \tag{12.5}$$

where $\dot{H}_{fuel}$ and $\dot{H}_{air}$ represent the enthalpy carried into the control volume by the fuel and air required to support electrochemical and combustion reactions (including excess air). Similarly, $\dot{H}_{liq-water}$ represents the enthalpy of the liquid water that might be supplied for fuel reformation purposes. The last term on the left side of the equation represents the power draw of ancillaries that are included within the control volume and that are powered by AC electricity that is supplied to the FC micro-cogeneration device. All terms in Equation (12.5) are in terms of Watts.

P_{el} is the net DC power production, that is the stack power less ohmic losses in cabling and the power draw of DC-powered ancillaries included within the control volume. $\dot{H}_{FCPM-gas-out}$ represents the enthalpy carried out of the control volume by the exiting gas stream. These are the hot products of the electrochemical and combustion reactions as well as the excess air and the inert constituents of the fuel.

$q_{stack-cooling}$ represents the heat transfer from the FCPM to the stack cooler, as would be present in a PEMFC. The final two terms in Equation (12.5) represent thermal losses: $q_{skin-loss}$ is the radiant and convective heat transfer from the skin of the FC micro-cogeneration device to the containing room while $q_{FCPM-to-dilution}$ represents the heat transfer from the FCPM to the air stream which might be drawn through the micro-cogeneration device's cabinet to limit skin temperatures and for gas venting purposes.

The enthalpy flow rates in Equation (12.5) represent summations of the enthalpies of constituent gases, e.g.

$$\dot{H}_{air} = \sum_i (\dot{N}_i \hat{h}_i)_{air} \tag{12.6}$$

where $\hat{h}_i$ is the molar enthalpy (J/kmol) and $\dot{N}_i$ is the molar flow rate (kmol/s) of gas constituent i (e.g. N_2, O_2).

Standardized enthalpies are used since chemical reactions are occurring within the FCPM. This ensures that the enthalpy of each reactant or product is properly related to the enthalpies of other elements and compounds by using a standard reference state. By convention, the standard state is taken to be 25 °C and 1 atmosphere pressure and the enthalpies of all elemental substances (e.g. O_2, H_2) are taken to be zero at the standard state.

The enthalpy of each reactant or product gas can be expressed as a sum of its enthalpy at the standard state (i.e. the standard enthalpy of formation) and the deviation between its enthalpy and that at the standard state. Furthermore, the definition of the specific heat can be introduced to provide the following relationship,

$$\begin{aligned} \hat{h}_i &= \Delta_f \hat{h}_i^o + \left[\hat{h}_i - \Delta_f \hat{h}_i^o \right] \\ &= \Delta_f \hat{h}_i^o + \int_{T^o}^{T} \hat{c}_P dT \end{aligned} \tag{12.7}$$

where $\Delta_f \hat{h}_i^o$ is the standard molar enthalpy of formation (J/kmol) and $\hat{c}_P$ is the molar specific heat (J/kmolK) for gas *i*.

The fuel's lower heating value (LHV) can be conveniently introduced into the energy balance represented by Equation (12.5) if it is assumed that the reactions of the fuel constituents are complete, as given by the following stoichiometric reaction,

$$C_x H_y + \left(x + \frac{y}{4} \right) . O_2 \rightarrow x . CO_2 + \frac{y}{2} . H_2O \tag{12.8}$$

The LHV (J/kmol) of a fuel is expressed using the standard enthalpies of formation of the reactants and products,

$$\begin{aligned} LHV_{fuel} &= \frac{\Delta_f H^o_{fuel} + \Delta_f H^o_{O_2} - \Delta_f H^o_{CO_2} - \Delta_f H^o_{H_2O}}{\dot{N}_{fuel}} \\ &= \frac{\Delta_f H^o_{fuel} - \Delta_f H^o_{CO_2} - \Delta_f H^o_{H_2O}}{\dot{N}_{fuel}} \end{aligned} \tag{12.9}$$

where $\Delta_f H^o_{fuel}$ is the total flow rate of the standard enthalpy of formation of the fuel entering the FCPM control volume and $\Delta_f H^o_{CO_2}$ and $\Delta_f H^o_{H_2O}$ are the total flow rates of the enthalpies of formation of the product gases created by the complete reaction of the fuel (all in W). $\dot{N}_{fuel}$ is the molar flow rate of the fuel (kmol/s).

The FCPM's efficiency at converting the energy released through the chemical reaction of the fuel to electrical energy is expressed as follows,

$$\varepsilon_{el} = \frac{P_{el}}{\dot{N}_{fuel} \times LHV_{fuel}} \tag{12.10}$$

where ε_{el} is the electrical conversion efficiency (–).

The electrical conversion efficiency can be introduced into the energy balance by substituting Equations (12.6), (12.7), (12.9), and (12.10) into Equation (12.5),

$$\sum_i\left(\dot{N}_i\int_{T^o}^{T}\hat{c}_P dT\right)_{fuel}+\sum_i\left(\dot{N}_i\int_{T^o}^{T}\hat{c}_P dT\right)_{air}+\sum_i\left(\dot{N}\int_{T^o}^{T}\hat{c}_P dT\right)_{liq-water}$$
$$-\dot{N}_{liq-water}\cdot\hat{h}_{fg}^{o}+P_{el-ancillaries-AC}+(1-\varepsilon_{el})\cdot\dot{N}_{fuel}\cdot LHV_{fuel} \tag{12.11}$$
$$=\sum_i\left(\dot{N}_i\int_{T^o}^{T}\hat{c}_P dT\right)_{FCPM-gas-out}+q_{stack-cooling}+q_{skin-loss}+q_{dilution-air-heat-loss}$$

where $\hat{h}_{fg}^{o}$ is the latent heat of vapourization of water at the standard state (J/kmol). This term represents the energy required to vapourize the liquid water which is introduced for fuel reformation.

Once the flow rates and temperatures of the fuel, air, and liquid water streams that are introduced into the control volume are defined the first three terms of Equation (12.11) can be calculated. These temperatures are established through the solution of energy balances that represent the fan, fuel compressor, and water pump that are represented in Figure 12.3.

Empirical relations are used to establish a number of terms in Equation (12.11). For example, the air supply rate is cast as a parametric function of the electrical production and the entering air temperature,

$$\dot{N}_{air}=\left(a_0+a_1P_{el}+a_2P^2{}_{el}\right)\left(1+a_3T_{air}\right) \tag{12.12}$$

where T_{air} is the temperature of the air (°C) supplied to the FCPM.

The form of Equation (12.12) facilitates the determination of the a_i coefficients from experimental data, as will be shown on page 352.

Another example of an empirical relation that is used to establish the terms of Equation (12.11) is that for the electrical conversion efficiency,

$$\varepsilon_{el}=\left[\varepsilon_0+\varepsilon_1\cdot P_{el}+\varepsilon_2\cdot P_{el}^{\;2}\right]\cdot\left[1-N_{stops}\cdot D\right]$$
$$\cdot\left[1-L\cdot MAX\left|\int_0 dt-t_{threshold},0\right|\right] \tag{12.13}$$

The $[1-N_{stops}\,D]$ term in Equation (12.13) represents the degradation of the efficiency as a result of start-stop cycling while the $\left[1-L\,MAX\left|\int_0 dt-t_{threshold},0\right|\right]$ term represents operational degradation.

The ε_{el} terms of Equation (12.13) along with D and L are empirical constants that characterize the performance of a given FC micro-cogeneration device. The methods that can be used to establish these values are treated on pages 352–353.

Techniques similar to those elaborated above for the air supply and electrical conversion efficiency are employed to establish the individual terms in Equation (12.11) each time-step of the simulation. This equation is then rearranged and solved each time-step of the simulation to predict $T_{FCPM-gas-out}$, the temperature of the hot gases exiting the FCPM and thus entering the subsequent control volume (see Figure 12.3). Similar methods are used to treat the other control volumes represented in the figure.

Modelling combustion-based devices

The FC micro-cogeneration model described on pages 345–350 employed a topology wherein the system was discretized into sub-systems (see Figure 12.3). As indicated, with this approach state points at the couplings between the sub-systems are solved. As will be shown on pages 352–353, calibrating this type of model necessitates data that can only be gathered through invasive measurements.

A more idealized representation is recommended for combustion-based micro-cogeneration devices for the pragmatic reason that invasive measurements are more difficult to effect in these technologies. Additionally, unlike their fuel cell counterparts, combustion-based micro-cogeneration devices tend to be configured to thermal load follow. In this mode of operation the devices will spend significant amounts of time in warming up and cooling down modes. Consequently, a representation that considers the thermal mass of the device is required in order to accurately treat these transient situations.

The IEA/ECBCS Annex 42 combustion micro-cogeneration model is based upon three abstract control volumes. One control volume represents the conversion of internal energy to electrical and thermal energy, while the other two control volumes represent the thermal mass of the engine and its cooling water.

The micro-cogeneration device's electrical conversion efficiency is given by,

$$\varepsilon_{el} = \frac{P_{el-net-AC}}{N_{fuel} \cdot LHV_{fuel}} \tag{12.14}$$

where $P_{el-net-AC}$ is the net AC electrical production from the micro-cogeneration device (W). As with the FC micro-cogeneration model, the denominator represents the product of the fuel flow rate (kmol/s) and the fuel's LHV (J/kmol).

When in steady-state operation (i.e. following completion of the warm-up cycle), the electrical conversion efficiency is determined using a parametric relation of the following form (higher-order terms are not shown here for the sake of clarity),

$$\begin{aligned} \varepsilon_{el} &= \varepsilon_0 + \varepsilon_1 \cdot (\dot{N}_{fuel} \cdot LHV_{fuel}) + \varepsilon_2 \cdot (\dot{N}_{fuel} \cdot LHV_{fuel})^2 \\ &+ \varepsilon_3 \cdot (\dot{m}_w) + \varepsilon_4 \cdot (\dot{m}_w)^2 + \varepsilon_5 \cdot (T_{w-in}) + \varepsilon_6 \cdot (T_{w-in})^2 \end{aligned} \tag{12.15}$$

where $\dot{m}_w$ is the mass flow rate (kg/s) of the water stream transporting the device's useful thermal output and T_{w-in} is the temperature of that water at the inlet to the device.

An adjustment to the efficiency calculated with Equation (12.15) is made during the warm-up period to account for the variation of fuel flow and power production as the engine temperature rises to its steady-state value. During the cool-down period it is assumed that there is no electrical production but rather that electricity is consumed for controls and auxiliary systems.

The useful thermal output is calculated as follows,

$$\begin{aligned} q_{thermal} &= (\dot{m}c_P)_w \cdot (T_{w-out} - T_{w-in}) \\ &= \eta_{th} \cdot \dot{N}_{fuel} \cdot LHV_{fuel} - (UA)_{losses} \cdot (T_{engine} - T_{room}) \\ &- (Mc_P)_{engine} \frac{dT_{engine}}{dt} - (Mc_P)_{HX} \frac{dT_{w-out}}{dt} \end{aligned} \tag{12.16}$$

where $q_{thermal}$ is the useful thermal output from the device (W). $(\dot{m}c_P)_w$ represents the product of the flow rate (kg/s) and heat capacity (J/kgK) of the water stream transporting the device's thermal output and T_{w-in} and T_{w-out} represent the temperatures of this fluid at the inlet and exit of the device.

The first term in the second line of Equation (12.16) represents the gross conversion of internal energy to thermal energy in the device. As with the electrical conversion efficiency, η_{th} is expressed as a parametric equation (again, the higher order terms are not shown here for the sake of clarity),

$$\begin{aligned} \eta_{th} &= \eta_0 + \eta_1 \cdot (\dot{N}_{fuel} \cdot LHV_{fuel}) + \eta_2 \cdot (\dot{N}_{fuel} \cdot LHV_{fuel})^2 \\ &+ \eta_3 \cdot (\dot{m}_w) + \eta_4 \cdot (\dot{m}_w)^2 + \eta_5 \cdot (T_{w-in}) + \eta_6 \cdot (T_{w-in})^2 \end{aligned} \tag{12.17}$$

Some of this gross thermal energy liberation is lost due to radiant and convective heat transfer from the skin of the micro-cogeneration device to the containing room. This is represented by the second term in the second line of Equation (12.16).

The final two terms of Equation (12.16) represent the transient energy storage of the micro-cogeneration device. It was found that two separate thermal storage terms were required to correctly represent the thermal storage of both the engine and the cooling water contained within the engine's heat exchanger.

The form of Equation (12.16) is such that it can calculate the thermal output during steady-state operation as well as during the warm-up and cool-down phases.

The following section will treat the methods that can be employed to calibrate the model by establishing the values of the coefficients in Equations (12.15), (12.16), and (12.17). The interested reader is referred to Ferguson *et al.* (2009) and Beausoleil-Morrison and Ferguson (2007) for a complete description of the IEA/ECBCS Annex 42 combustion micro-cogeneration model.

Model calibration

The sections on pages 345–350 and 350–352 described the recommended methodologies for simulating the performance of fuel-cell and combustion-based micro-cogeneration devices within the context of building performance simulation. As was described, these models rely upon empirical coefficients that are unique to specific micro-cogeneration devices (see Equations (12.4), (12.12), (12.13), (12.15), and (12.17)). The process of establishing these coefficients (i.e. inputs) for a particular device is known as model calibration.

Ideally a series of experiments are conducted on a particular micro-cogeneration device under controlled conditions in order to gather the necessary data to calibrate the models. Beausoleil-Morrison and Lombardi (2009) describe the experimental programme that was designed and conducted to calibrate the FC model to represent the performance of a prototype 2.8 kW_{AC} SOFC micro-cogeneration device. In these experiments, for example, the values of $\dot{N}_w$ and $\dot{N}_g$ in Equation (12.4) were varied and measurements were made on the performance of the gas-to-water heat exchanger. The measured temperatures and flow rates were used to assemble a set of derived $(UA)_{eff}$ values using Equations (12.1) and (12.2). A non-linear regression was then performed to determine the values of the $hx_{s,i}$ coefficients that produced the best fit between the controlled $\dot{N}_w$ and $\dot{N}_g$ and the derived $(UA)_{eff}$ values.

Figure 12.5 compares the $(UA)_{eff}$ values determined with Equation (12.4) and the calibrated coefficients with the $(UA)_{eff}$ values derived from the measurements. The error bars represent the uncertainty of the values derived from the measurements at the 95% confidence level. The left side of the figure contrasts the calibration with variations in $\dot{N}_w$ while the right side contrasts the calibration with variations in $\dot{N}_g$.

Similar experiments must be conducted to gather the necessary data to calibrate each aspect of the FC model, such as Equations (12.12) and (12.13). Although powerful and accurate, this calibration approach necessitates invasive measurements. For example, the above-mentioned calibration of the gas-to-water heat exchanger required measuring the temperatures and flow rates of the water and gas streams at the heat exchanger inlets and outlets.

In contrast, the design of the combustion-based model averts the need for invasive measurements. Rather, measurements taken external to the micro-cogeneration device – the electrical output, fuel supply rate and the temperature and flow rate of water entering and exiting the heat exchanger – are sufficient to calibrate Equations (12.15) and (12.17). Ideally a series of experiments are conducted on a particular micro-cogeneration device under controlled conditions in order to gather the necessary calibration data. However, as demonstrated by Ferguson *et al.* (2009) it is possible to calibrate the model using less complete data with a dynamic parameter estimation method.

The interested reader is referred to Beausoleil-Morrison (2007) which provides the full set of calibration parameters for prototype SOFC, SE, and ICE micro-cogeneration devices.

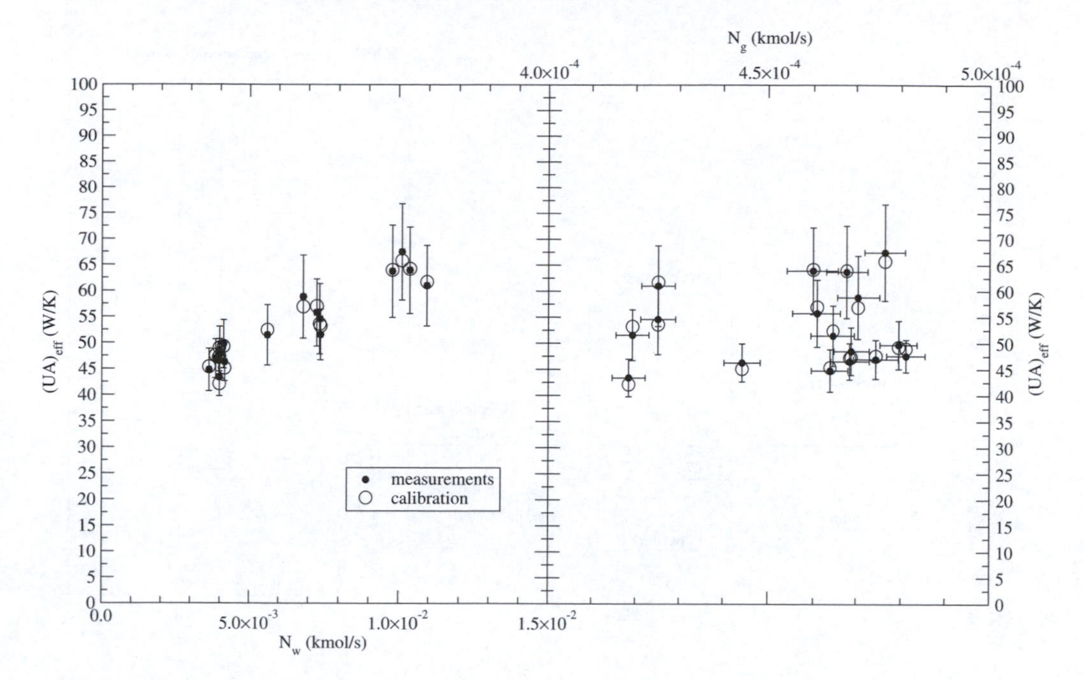

Figure 12.5 Calibration of Equation (12.4) using experimental data.

Application example

The introduction to this chapter provided the motivation for modelling the behaviour of micro-cogeneration devices within the context of building performance simulation. This section demonstrates how the models described in this chapter can be applied to concurrently simulate the performance of micro-cogeneration devices, supporting HVAC components, and the building's thermal and electrical demands. This application focuses upon a representative detached house served by a hypothetical SOFC micro-cogeneration system. ESP-r is used for the purposes of this demonstration, but as explained earlier the same modelling capabilities are available in other simulation programs.

Figure 12.6 illustrates how the hypothetical SOFC micro-cogeneration device is coupled to the house's convective heating system in this example. The thermal output of the SOFC micro-cogeneration device is transferred to a water storage tank. The pump that circulates water between the tank and the SOFC runs continuously and draws 20 W of electrical power. When the house's temperature drops below 20.5 °C, the second pump cycles on to circulate hot water from the tank through a water-to-air heat exchanger. Likewise, the fan cycles on to circulate return air through the heat exchanger and to deliver this warmed air to heat the house. When operating, this pump and fan draw 20 W and 50 W of electrical power, respectively.

The tank's make-up burner is assumed to have an efficiency of 95% (higher heating value, HHV, basis) and cycles on when then tank temperature drops below 50 °C. The tank has a volume of 1000 L and is well insulated. To protect the tank from excessive thermal input, energy is extracted from the tank and rejected to the environment when its temperature rises to 92 °C. This fictitious representation of a heat dump accurately accounts for the need to periodically reject excessive thermal production to the environment, but neglects the electrical draw of ancillary devices that would be required to provide this functionality.

Rather than conducting this simulation using the calibrated parameters for the prototype 2.8 kW_{AC} SOFC micro-cogeneration device treated in Beausoleil-Morrison and Lombardi (2009), for this example the model was configured to represent a hypothetical device. This hypothetical device has the same electrical conversion efficiency as the tested prototype. Similarly, the

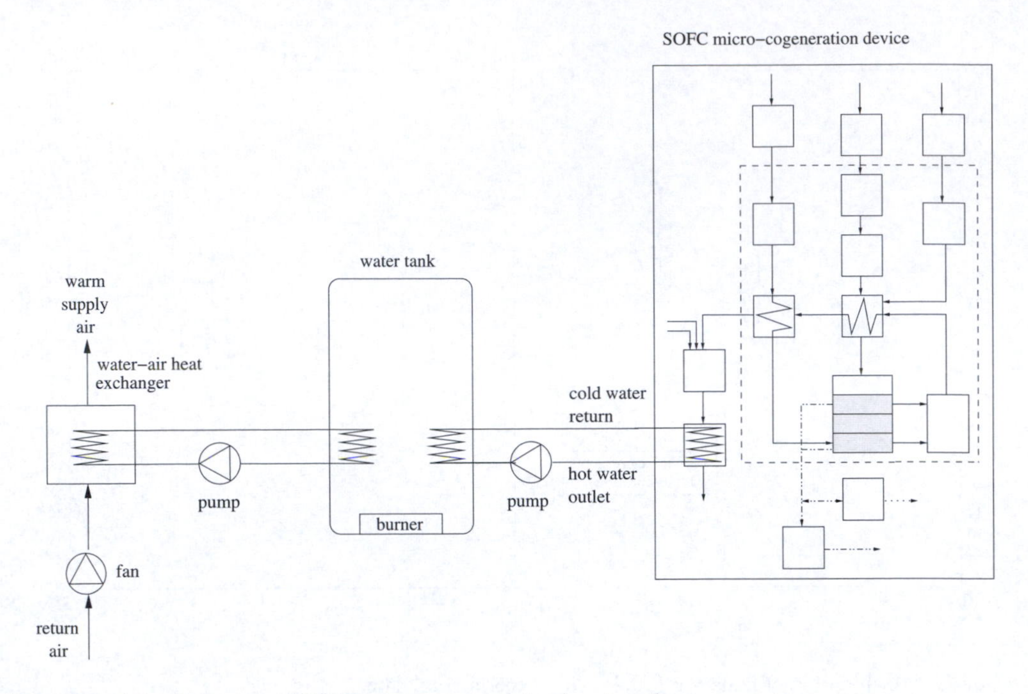

Figure 12.6 Configuration of SOFC micro-cogeneration heating plant.

prototype's measured data were used to calibrate many of the model's terms that were previously described, such as the air supply rate, heat exchanger, and transient response. However, unlike the tested prototype, this hypothetical device is given the capability of modulating its output between 500 W and 2.5 kW (net DC production). As well, it is assumed to have a constant DC-AC power conditioning efficiency of 95% and to have only 100 W of convective and radiative heat losses.

Initially, the SOFC micro-cogeneration device is assumed to operate continuously throughout the year producing 1 kW of AC power. It is assumed that the house can export power to the central electrical grid when there is a surplus of power, and import from the grid when there is a deficit. A simulation was conducted for a typical house using Ottawa (Canada) weather data. The non-HVAC electrical demands imposed by the occupants were assumed to follow one of the demand patterns recommended by Armstrong *et al.* (2009).

The monthly integrated electrical balance for this initial simulation is illustrated in Figure 12.7. Over the year the SOFC micro-cogeneration device generates 31.5 GJ of electrical energy (net AC) while the occupants and HVAC equipment consume 30.5 GJ. Considerable grid interaction is required due to temporal mismatches between production and demand: over the year 11.7 GJ of electrical energy are exported to the grid while 10.6 GJ are imported. As shown in the figure, there is little seasonal variation in these observations.

The monthly integrated thermal energy balance on the water storage tank (see Figure 12.6) for this initial simulation is illustrated in Figure 12.8. As can be seen, during the winter months the SOFC micro-cogeneration device's thermal output is insufficient to meet the house's space heating demand. Consequently, extensive use of the tank's make-up burner is required. The make-up burner is also required in the spring and fall. The impact of temporal mismatches between the SOFC micro-cogeneration device's thermal production and the building's space heating demand can be seen in April and October. During each of these two months, the space heating demand is greater than the SOFC micro-cogeneration device's thermal production. Notwithstanding, some of this thermal production must be dumped due to overheating of the tank during some periods. And during the summer months the vast majority of the

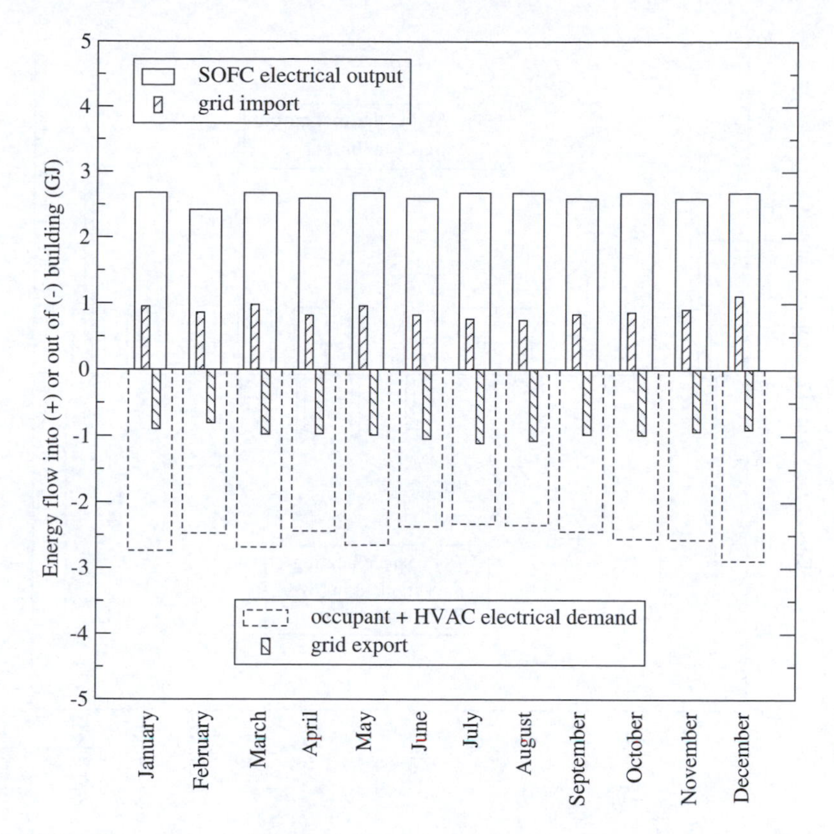

Figure 12.7 Monthly integrated electrical balance when SOFC produces a constant 1 kW_{AC} output.

micro-cogeneration device's thermal output is wasted through the heat-dump facility.

This initial simulation predicts that over the year, 65.8% of the fuel's LHV is converted to useful electrical and thermal output, as defined by,

$$\eta_{overall} = \frac{E_{net\ AC} + E_{gross\ thermal} - E_{tank\ skin\ losses} - E_{dump}}{N_{fuel} \cdot LHV} \tag{12.18}$$

where $\eta_{overall}$ is the overall efficiency (LHV basis), N_{fuel} is the fuel consumption over the year (kmol), and LHV is the fuel's lower heating value. $E_{net\ AC}$ is the SOFC micro-cogeneration device's net AC electrical production and $E_{gross\ thermal}$ is its gross thermal output, both in GJ. $E_{tank\ skin\ losses}$ represents the thermal losses from the tank to the containing room while E_{dump} is the amount of energy that must be dumped from the tank due to overheating, both in GJ.

To demonstrate how the modelling methods outlined in this chapter can be used to study the impact of dispatch strategy upon system operation, two additional simulations were performed with this example. In one case, rather than producing a constant 1 kW of electrical output, the SOFC micro-cogeneration device was made to follow the electrical demand of the occupants and HVAC equipment. However, its load following capabilities were constrained to the hypothetical device's modulating range of 500 W to 2.5 kW (net DC). As well, measured data from the tested prototype were used to impose limits on how quickly the device could modulate its electrical output in response to load variations. In the second case the SOFC micro-cogeneration device was made to follow the house's demand for space heating.

Figure 12.9 contrasts the annually integrated performance of these three dispatch strategies.

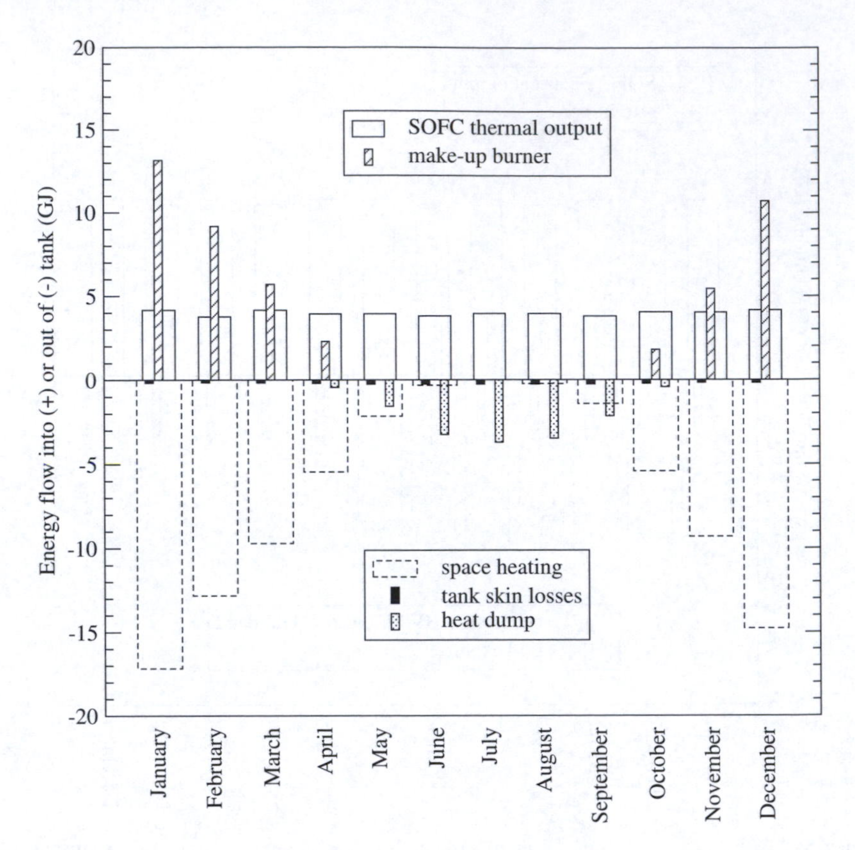

Figure 12.8 Monthly integrated thermal balance on tank when SOFC produces a constant 1 kW_{AC} output.

In this figure, the useful thermal output is taken to be the gross output less tank skin losses and the amount of energy that must be dumped due to tank overheating,

$$E_{useful\ thermal\ output} = E_{gross\ thermal} - E_{tank\ skin\ losses} - E_{dump} \tag{12.19}$$

As can be seen, both alternate dispatch strategies result in greater electrical and useful thermal production from the SOFC micro-cogeneration device. The thermal load following scenario requires the greatest reliance upon the electrical grid, whereas grid interaction is minimized with electrical load following, as expected. The thermal load following strategy results is a substantially higher overall system efficiency (see Equation (12.18)): 80% versus 65 to 66%.

This section demonstrates only a single application of the micro-cogeneration device modelling capabilities treated in this chapter. These methods can be used to examine various system configurations and dispatch strategies and to assess the performance of specific or hypothetical devices in various climates, buildings, and with various occupancy scenarios.

Discussion

The accurate treatment of micro-cogeneration devices requires that the user configure building simulation programs to treat building thermal and electrical demands concurrently with electrical and thermal energy conversion systems. This demands the user to operate across multiple modelling domains and that the interactions between the modelling domains be considered.

The greatest challenge in modelling the electrical and thermal performance of these technologies is in acquiring sufficient data to calibrate the recommended models. These

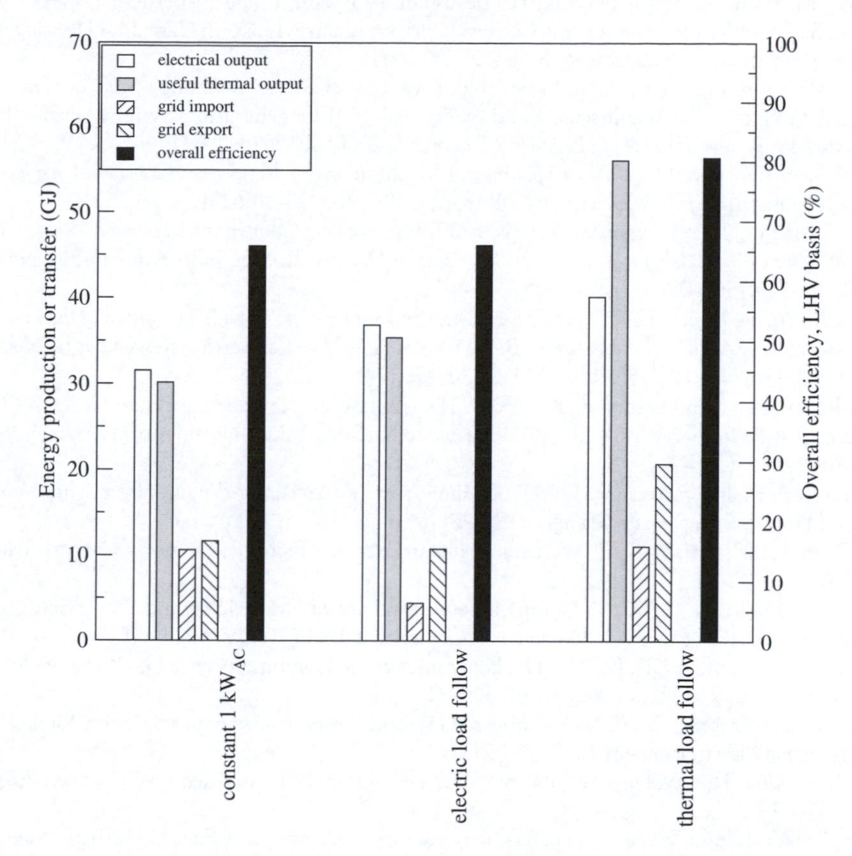

Figure 12.9 Comparison of annual results for three dispatch strategies.

technologies are only emerging onto the market so few detailed performance data exist. Notwithstanding, this chapter has enumerated the existing data sets and has treated the methods that are recommended for model calibration.

The assignments that follow allow the reader to develop the necessary knowledge and skills to treat these nascent technologies within building performance assessments.

Synopsis

This chapter has provided an introduction to micro-cogeneration technologies based upon fuel cells, internal combustion engines, and Stirling engines for the combined production of heat and electricity within buildings. These emerging technologies have the potential to deliver energy services more efficiently while reducing environmental impacts. However, due to the complex performance characteristics, these technologies can only be accurately assessed by building performance simulation programs that treat building thermal performance concurrently with electrical and HVAC systems.

This chapter has outlined the recommended methods for modelling micro-cogeneration devices as well as techniques for calibrating these models to represent the performance of specific devices using measured data.

References

Alanne, K., Saari, A., Ugursal, V. and Good, J. (2006) 'The Financial Viability of an SOFC Cogeneration System in Single-family Dwellings', *Journal of Power Sources* 158(1): 403–416.

Armstrong, M., Swinton, M., Ribberink, H., Beausoleil-Morrison, I. and Millette, J. (2009) 'Synthetically Derived Profiles for Representing Occupant-driven Electric Loads in Canadian Housing', *Journal of Building Performance Simulation* 2(1): 15–30.

Beausoleil-Morrison, I., Schatz, A. and Maréchal, F. (2006) 'A Model for Simulating the Thermal and Electrical Production of Small-scale Solid-oxide Fuel Cell Cogeneration Systems within Building Simulation Programs', *Journal of HVAC&R Research Special Issue* 12(3a): 641–667.

Beausoleil-Morrison, I. (ed.) (2007) *Experimental Investigations of Residential Cogeneration Devices and Model Calibration*, IEA/ECBCS Annex 42 Report, ISBN No. 978-0-662-47523-1.

Beausoleil-Morrison, I. and Ferguson, A. (eds) (2007) *Inter-model Comparative Testing and Empirical Validation of Annex 42 Models for Residential Cogeneration Devices*, IEA/ECBCS Annex 42 Report, ISBN No. 978-0-662-47562-0.

Beausoleil-Morrison, I. (ed.) (2008) *An Experimental and Simulation-Based Investigation of the Performance of Small-Scale Fuel Cell and Combustion-Based Cogeneration Devices Serving Residential Buildings*, IEA/ECBCS Annex 42 Report, ISBN No. 978-0-662-47923-9.

Beausoleil-Morrison, I. and Lombardi, K. (2009) 'The Calibration of a Model for Simulating the Thermal and Electrical Performance of a 2.8 AC Solid-oxide Fuel-cell Micro-cogeneration Device', *Journal of Power Sources* 186(1): 67–79.

Boait, P., Rylatt, R. and Stokes, M. (2006) 'Optimisation of Consumer Benefits from Micro-combined Heat and Power', *Energy and Buildings* 38(8): 981–987.

Carbon Trust (2007) *Micro-CHP Accelerator: Interim Report*, Technical Report, Carbon Trust, UK, CTC726.

Cockroft, J. and Kelly, N. (2006) 'A Comparative Assessment of Future Heat and Power Sources for the UK Domestic Sector', *Energy Conversion and Management* 47(15–16): 2349–2360.

Colpier, U.C. and Cornland, D. (2002) 'The Economics of the Combined Cycle Gas Turbine–An Experience Curve Analysis', *Energy Policy* 30(4): 309–316.

De Paepe, M. and Mertens, D. (2007) 'Combined Heat and Power in a Liberalised Energy Market', *Energy Conversion and Management* 48(9): 2542–2555.

DeMoss, T. (1996) 'They're He-e-re (almost): The 60% Efficient Combined Cycle', *Power Engineering* 100(7): 17–21.

Dorer, V., Weber, R. and Weber, A. (2005) 'Performance Assessment of Fuel Cell Micro-cogeneration Systems for Residential Buildings', *Energy and Buildings* 37(11): 1132–1146.

Entchev, E., Gusdorf, J., Swinton, M., Bell, M., Szadkowski, F., Kalbfleisch, W. and Marchand, R. (2004) 'Micro-generation Technology Assessment for Housing Technology', *Energy and Buildings* 36(9): 925–931.

Ferguson, A., Kelly, N., Weber, A. and Griffith, B. (2009) Modelling Residential-scale Combustion-based Cogeneration in Building Simulation, *Journal of Building Performance Simulation* 2(1): 1–14.

Finkelstein, T. and Organ, A. (2001) *Air Engines: The History, Science, and Reality of the Perfect Engine*, New York: ASME Press.

Hawkes, A., Aguiar, P., Croxford, B., Leach, M., Adjiman, C. and Brandon, N. (2007) 'Solid Oxide Fuel Cell Micro Combined Heat and Power System Operating Strategy: Options for Provision of Residential Space and Water Heating', *Journal of Power Sources* 164(1): 260–271.

Hawkes, A. and Leach, M. (2005) 'Solid Oxide Fuel Cell Systems for Residential Micro-combined Heat and Power in the UK: Key Economic Drivers', *Journal of Power Sources* 149(1): 72–83.

Hawkes, A. and Leach, M. (2007) 'Cost-effective Operating Strategy for Residential Micro-combined Heat and Power', *Energy* 32(5):711–723.

Henning, H.-M. (2007) 'Solar Assisted Air Conditioning of Buildings – An Overview', *Applied Thermal Engineering* 27(10): 1734–1749.

Kelly, N. and Beausoleil-Morrison, I. (eds) (2007) *Specifications for Modelling Fuel Cell and Combustion-Based Residential Cogeneration Devices within Whole-Building Simulation Programs*, IEA/ECBCS Annex 42 Report, ISBN No. 978-0-662-47116-5.

Knight, I. and Ugursal, V. (eds) (2005) *Residential Cogeneration Systems: A Review of the Current Technologies*, IEA/ECBCS Annex 42 Report, ISBN No. 0-662-40482-3.

Larminie, J. and Dicks, A. (2003) *Fuel Cells Explained*, 2nd ed., New York: John Wiley & Sons.

Maalla, E.M.B. and Kunsch, P.L. (2008) 'Simulation of Micro-chp Diffusion by Means of System Dynamics', *Energy Policy* 36(7): 2308–2319.

Mench, M. (2008) *Fuel Cell Engines*, New York: John Wiley & Sons.

Onovwiona, H. and Ugursal, V. (2006) 'Residential Cogeneration Systems: Review of the Current Technology', *Renewable and Sustainable Energy Reviews* 10(5): 389–431.

Peacock, A. and Newborough, M. (2006) 'Impact of Micro-combined Heat-and-power Systems on Energy Flows in the UK Electricity Supply Industry', *Energy* 31(12): 1804–1818.

Pearce, J., Zahawi, B.A. and Shuttleworth, R. (2001) 'Electricity Generation in the Home: Modelling of Single-house Domestic Combined Heat and Power', in *Proceedings of Science, Measurement and Technology* 148, pp. 197–203, Institute of Engineering and Technology.

Possidente, R., Roselli, C., Sasso, M. and Sibilio, S. (2006) 'Experimental Analysis of Micro-cogeneration Units Based on Reciprocating Internal Combustion Engine', *Energy and Buildings* 38(12): 1417–1422.

Ribberink, H., Bourgeois, D. and Beausoleil-Morrison, I. (2009) 'A Plausible Forecast of the Energy and Emissions Performance of Mature-technology Stirling Engine Residential Cogeneration Systems in Canada', *Journal of Building Performance Simulation* 2(1): 47–61.

Sibilio, S., Sasso, M., Possidente, R. and Roselli, C. (2007) 'Assessment of Micro-cogeneration Potential for Domestic Trigeneration', *International Journal of Environmental Techology and Management* 7(1–2): 147–164.

Sicre, B., Bühring, A. and Platzer, B. (2005) 'Energy and Financial Performance of Micro-chp in Connection with High-performance Buildings', in *Proceedings of Building Simulation 2005*, Montréal, Canada, pp. 1131–1138.

Recommended reading

Beausoleil-Morrison, I. (ed.) (2008), *An Experimental and Simulation-Based Investigation of the Performance of Small-Scale Fuel Cell and Combustion-Based Cogeneration Devices Serving Residential Buildings*, IEA/ECBCS Annex 42 Report, ISBN No. 978-0-662-47923-9.

Activities

Assignment 1

In this assignment you will develop familiarity with modelling micro-cogeneration devices by incorporating a FC micro-cogeneration device into a building and HVAC system model. A simple idealized arrangement has been selected to allow efforts to be focused upon adding the micro-cogeneration component to a model and configuring the model using supplied data.

Use any of the simulation programs that have implemented the IEA/ECBCS Annex 42 FC micro-cogeneration model (see pages 344–356). In this assignment the building is inconsequential and a simple HVAC system is utilized in which a pump circulates a fixed flow rate of constant-temperature water through the micro-cogeneration device. The model's stack heat exchanger and gas-to-water heat exchanger should be configured in a cascading arrangement, as illustrated in Figure 12.10.

Configure the micro-cogeneration model as follows:

- The device is able to modulate its electrical output at the rate of 100 W/s such that the FCPM can produce from 500 W to 2.5 kW of DC power. However, in this simulation a constant 1 kW of net AC power is produced by the micro-cogeneration device.
- The electrical conversion efficiency (see Equations (12.10) and (12.13)) is a constant 35%.
- There is no battery and the efficiency of the DC to AC power conditioner is 100%.
- The blower, fuel compressor, and water pump do not draw any power.
- The fuel is 100% methane and the air is 79% nitrogen and 21% oxygen.
- Air is supplied to the FCPM at a constant rate of $1\ 10^{-4}$ kmol/s (see Equation (12.12)).
- Air and fuel are supplied to the micro-cogeneration device from the building, which is maintained at a constant 10 °C.
- The gas-to-water heat exchanger has a constant $(UA)_{eff}$ of 9 W/K.

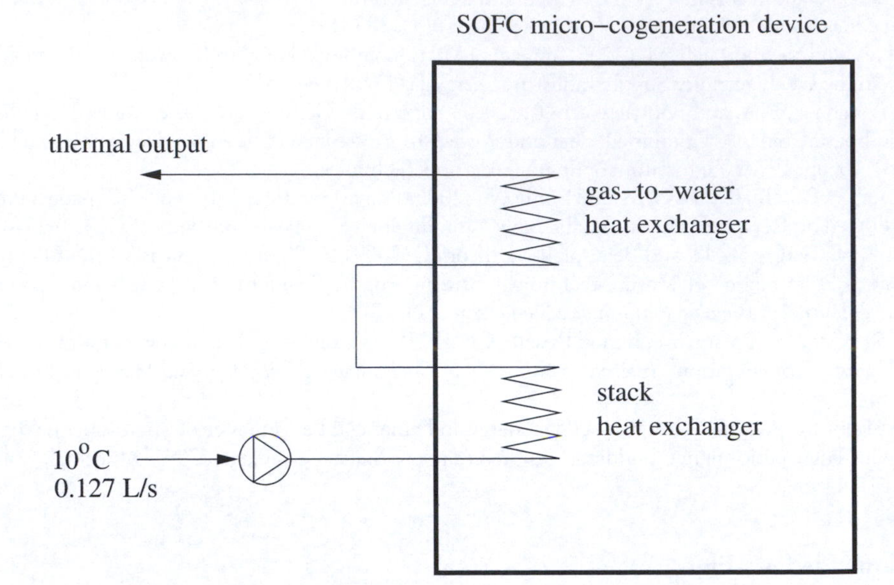

Figure 12.10 HVAC configuration for Assignment 1.

- There are no skin losses from the FCPM, no liquid water supply, ancillary devices consume no electrical power, and there is no dilution air system or stack cooling system (refer to Equation (12.5)).

Conduct a simulation for a single day using a time-step between 1 and 10 minutes. Examine the results file generated by the simulation program and plot the net electrical output and thermal output from the micro-cogeneration device as a function of time. Determine the fuel consumption rate, the rate of heat transfer from the micro-cogeneration device to the HVAC system's water stream, and the fraction of the fuel's energy (LHV basis) that is converted to useful electrical and thermal output using Equation (12.18).

Assignment 2

In this assignment you will examine how the micro-cogeneration model can be calibrated to represent the performance of a specific (hypothetical) device. This assignment is built upon Assignment 1 but drops the assumption of a constant electrical conversion efficiency. Rather, the data given in Table 12.1 are to be used to calibrate this model input using the methods

Table 12.1 Electrical efficiency data for Assignment 2

P_{el} (W)	ε_{el}
500	0.3025
700	0.3150
900	0.3443
1100	0.3641
1300	0.3797
1500	0.3900
1700	0.3832
1900	0.3629
2100	0.3440
2300	0.3212
2500	0.2974

outlined on pages 352–353. Use these data to calibrate the SOFC's electrical conversion efficiency using Equation (12.13). For the purposes of this assignment, ignore the degradation terms in this equation.

Configure the micro-cogeneration model using these calibrated parameters. All other inputs should be as in Assignment 1. Then conduct a simulation for a single day using a time-step between 1 and 10 minutes. Determine the fuel consumption rate and the rate of heat transfer from the micro-cogeneration device to the HVAC system's water stream.

Assignment 3

In the previous assignments the micro-cogeneration device operated with a constant electrical and thermal output. In this assignment you will explore how controls can be employed in your chosen simulation platform to allow the micro-cogeneration device to respond to building loads.

Configure the micro-cogeneration model as in Assignment 2. However, rather than producing a constant 1 kW net AC output, have the micro-cogeneration device follow the building's electrical demand profile as illustrated in Figure 12.11.

All other inputs should be as in Assignment 2. Conduct a simulation for a single day using a time-step between 1 and 10 minutes. Plot the FCPM's electrical efficiency and the rate of heat transfer from the micro-cogeneration device to the HVAC system's water stream as a function of time over the day. Calculate the total net AC electrical production and thermal production over the day.

Assignment 4

The previous assignments have considered hypothetical micro-cogeneration devices. In this assignment you will configure the model to represent the performance of an actual prototype SOFC micro-cogeneration device.

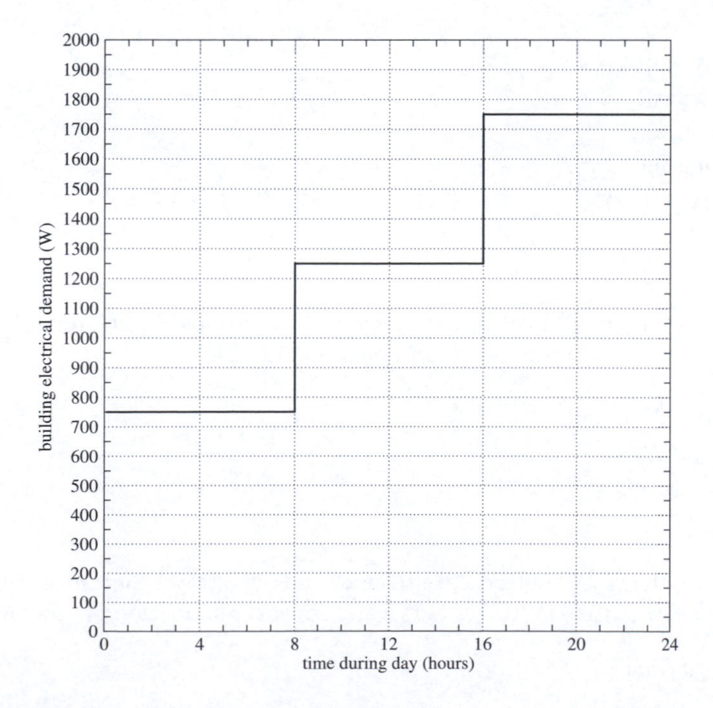

Figure 12.11 Building electrical demand for Assignment 3.

Using the calibration parameters provided in Beausoleil-Morrison and Lombardi (2009) reconfigure the model of Assignment 2 to represent the performance of the tested 2.8 kW_{AC} SOFC micro-cogeneration device. Configure the model so that the micro-cogeneration device produces a constant 2.8 kW of net AC power. In the previous assignments, the temperature of the water supplied to the micro-cogeneration device's heat exchanger was 10 °C. The temperature of this water is raised to 40 °C in this assignment. All other inputs should be as in Assignment 2.

Conduct a simulation for a single day using a time-step between 1 and 10 minutes. Calculate the total fuel consumption and the thermal production and skin losses over the day.

Assignment 5

Configure a model to represent the application example given in on page 354 and conduct simulations to represent the three control scenarios demonstrated.

Assignment 6

The previous assignments have focused upon FC micro-cogeneration systems. In this assignment you will develop familiarity with modelling a SE micro-cogeneration device.

Replace the FC in Assignment 1 with a SE. Ferguson *et al.* (2009) provide the calibrated model inputs for a natural-gas-fired SE micro-cogeneration that produces 699 W of net AC output under steady conditions. Use the calibrated parameters provided in this paper to configure the model to represent this device. Furthermore, configure the device to operate at a constant electrical output of 699 W. All other inputs should be as in Assignment 1.

Conduct a simulation for a single day using a time-step between 1 and 10 minutes. Calculate the rate of useful thermal output and the rate of heat loss.

Answers

Assignment 1

- $\dot{N}_{fuel} = 3.56 \cdot 10^{-6}$ kmol/s
- $q_{sensible} = 1\,822$ W
- $\dfrac{P_{net-AC} + q_{sensible}}{\dot{N}_{fuel} \cdot LHV_{fuel}} = 98.8\%$

Assignment 2

The calibration of the FCPM's electrical efficiency is illustrated in Figure 12.12. The figure's legend gives the calibration parameters for Equation (12.13).

- $\dot{N}_{fuel} = 3.48 \cdot 10^{-6}$ kmol/s
- $q_{sensible} = 1754$ W

Assignment 3

Figure 12.13 plots the FCPM's electrical efficiency and the rate of heat transfer from the micro-cogeneration device to the HVAC system's water stream as a function of time over the day.

- $E_{net\ AC} = 108.0$ MJ
- $E_{gross\ thermal} = 184.3$ MJ

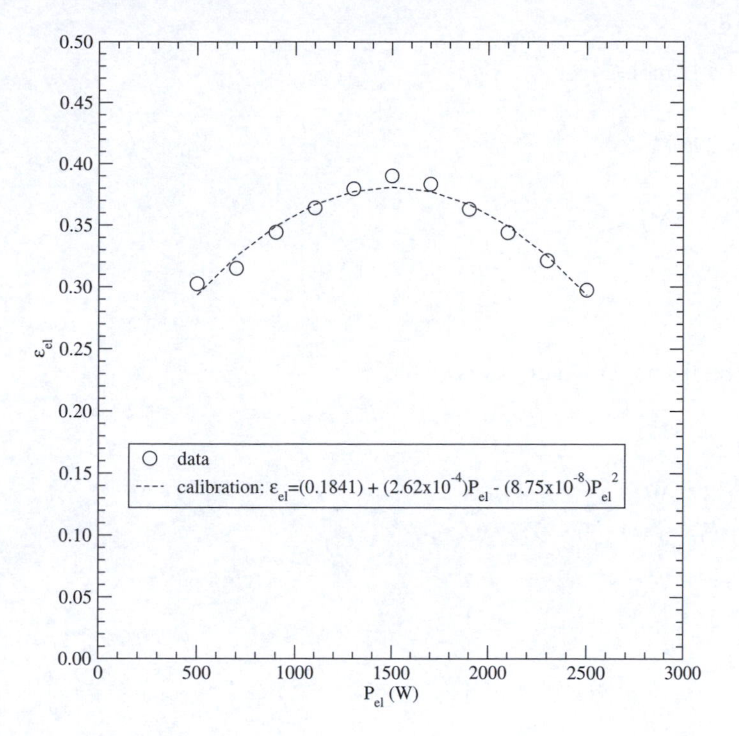

Figure 12.12 ε_{el} calibration for Assignment 2.

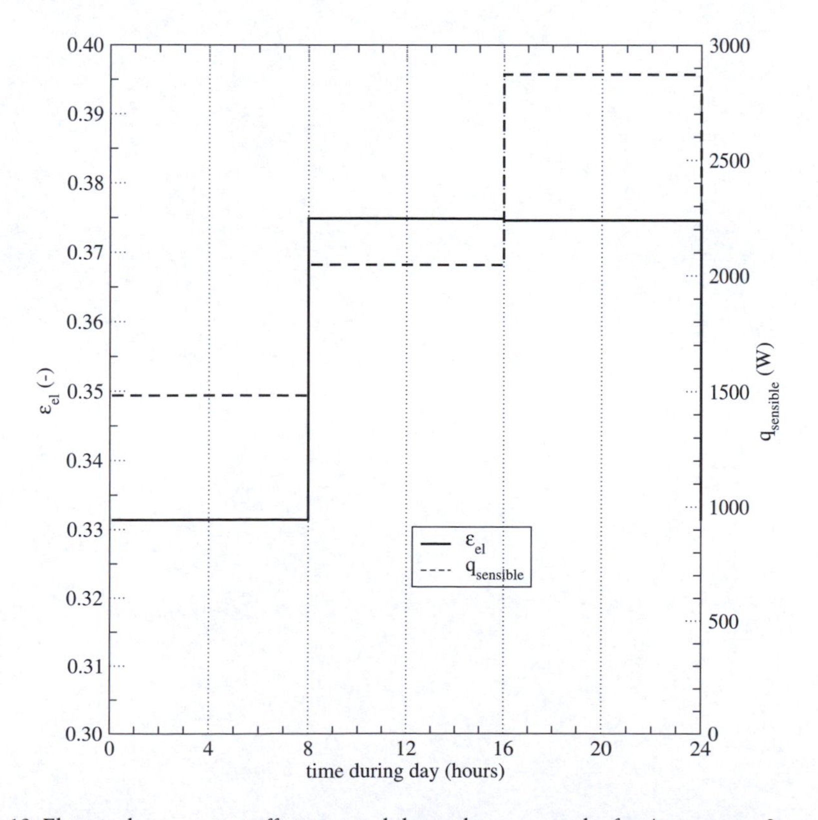

Figure 12.13 Electrical conversion efficiency and thermal output results for Assignment 3.

Assignment 4

- $\int_{day} \dot{N}_{fuel} = 1.14$ kmol
- $\int_{day} q_{sensible} = 296$ MJ
- $\int_{day} q_{skin-loss} = 63$ MJ

Assignment 5

Refer to the results presented on pages 353–356.

Assignment 6

- $q_{thermal} = 6\,321$ W.
- $(UA)_{losses}\,(T_{engine} - T_{room}) = 977$ W

13 Building simulation for practical operational optimization

David E. Claridge

Scope

Whole building simulation is increasingly used in the design of building heating, ventilating and cooling (HVAC) systems but is seldom used subsequently to optimize operation of the building. However, there are a significant number of cases where simulation has been/is being used in some way to optimize operation of the building. This chapter provides an overview of these applications with specific examples. These uses of simulation typically require that the inputs of the simulation program be adjusted so the simulated values of heating and cooling closely match the measured consumption values, a process typically called "calibrating" the simulation to the measured performance.

When calibrated simulations are used for practical operational optimization of building operation, this normally occurs as part of the Existing Building Commissioning (EBCx) Process. This process will be new to many readers, so it is described in some detail. A systematic and relatively simple procedure for rapidly calibrating simulations is presented and illustrated with two detailed examples. The chapter concludes with two case studies illustrating the use of calibrated simulation in the EBCx process for practical operational optimization of building operation.

Learning objectives

To systematically and quickly calibrate a whole building simulation to measured heating and cooling consumption data. How a calibrated simulation may be used to optimize building operation and/or to maintain optimized operation of a building.

Key words

Operational optimization/calibrated simulation/existing building commissioning

Introduction

Whole building simulation is increasingly used in the design of building heating, ventilating and cooling (HVAC) systems but is seldom used subsequently to optimize operation of the building. The continuous use of simulation throughout the life cycle of a building has been envisioned for more than a decade (e.g. Selkowitz *et al.* 1992), but the author is unaware of an instance where simulation was used in the building design and subsequently incorporated to maintain and optimize the routine ongoing operation of the building.

There are a significant number of reported cases where simulation has/is being used in some way subsequent to the design of a building. Most of these post-design uses of simulation occur as part of one of the forms of what is called building commissioning.

Building commissioning and calibrated simulation

The term "commissioning" has been used to refer to the process of making a navy ship completely ready for active duty for generations. Over the last two decades, the term has come to refer to the process that makes a building or some of its systems completely ready for use. In the context of this chapter, commissioning will refer to making the HVAC systems of a building fully functional.

It is reasonable to expect that any new building will operate properly when it is finished, but the manager of a large medical center once told the author "I don't want any more new buildings. I want three-year-old buildings where the problems have been fixed." The objective of the HVAC commissioning process that is now increasingly being used in new buildings is to make sure that when a building is delivered to the owner, the problems have been fixed and the systems operate as they were designed to operate. When the term "new building commissioning" is used in this chapter, it will refer to this process.

There is also an increasing recognition that most existing buildings do not operate as well or as efficiently as they could and should. A recent study at Texas A&M University found that simply applying a "commissioning" process to over 60 existing buildings had produced average reductions in energy use of 14% (Bynum *et al.* 2008). These savings were not achieved by installing more efficient equipment in the buildings; they were achieved simply by fixing or replacing broken components and changing control settings to optimize the operation of the HVAC systems for the way the building was being used while maintaining, and in many cases improving, the comfort conditions in the buildings. The process of identifying HVAC system problems, diagnosing the problem solutions, determining optimum control algorithms, and then implementing these changes in existing buildings is more complex than it may appear, and often becomes an iterative process. For example, many constant volume air distribution systems operate with more air flow than actually required, using more fan power and more heating and cooling energy than necessary. Reducing this air flow can make one zone not cool properly. This necessitates going through the complete problem identification-diagnosis-optimization-implementation sequence again. A common reason for this particular problem is imbalance in the air distribution system. This commissioning process goes by multiple names, including retro-commissioning, re-commissioning, Continuous Commissioning® (CC®),[1] and existing building commissioning. The American Society of Heating, Refrigerating and Air Conditioning Engineers (ASHRAE) Guideline Committee 0.2 on Commissioning of Existing Buildings has adopted "existing building commissioning," so this term will be used throughout this chapter.

Once a building is operating properly, various HVAC failures, control changes and building usage changes that do not cause comfort problems will generally reduce the operating efficiency over time unless there is an active program that identifies and responds to these changes. For example, in a volley ball arena on the Texas A&M campus, control failures in two of the twelve air supply units in the arena caused them to go into a continuous heating mode. This caused no comfort problems since the other 10 units readily compensated by providing additional cooling – but the operating cost went up by U.S. $75,000/year (Claridge *et al.* 2004). Different procedures are used to identify and diagnose such problems – they will be called "on-going commissioning" in this chapter.

While there are post-design applications of standard design simulations, many post-design applications of simulation fine-tune the simulation inputs so the heating and cooling consumption values predicted by the simulation closely match the measured consumption of the building. This type of simulation is customarily called "calibrated simulation."

Post-design applications of simulation

Post-design applications of simulation may be considered in five different categories. These categories include the use of:

- Design simulation in post-construction commissioning of new buildings
- Design simulation or calibrated simulation for on-going commissioning
- Calibrated simulation for existing building commissioning
- Simulation to evaluate new control code
- Calibrated simulation as a baseline for savings determination.

The first four applications of simulation listed above are typically related to operational optimization in some manner, and will be discussed in more detail. The use of simulation for diagnostic purposes is frequently part of optimization, but will be discussed under these four headings. The use of calibrated simulation as part of a savings determination process (e.g. Option D of the *International Performance Measurement and Verification Protocol* (EVO 2007) and *ASHRAE Guideline 14* (ASHRAE 2002)) is not typically part of operational optimization and will not be discussed further.

Important factors in whole building simulation for operational optimization

Use of whole building simulation for optimization is conceptually most practical in cases where a simulation has already been performed as part of the design process or as part of a retrofit evaluation. The input values for the existing simulation may then be used to predict building performance and explore opportunities for optimization; likewise, if the operation has been optimized, deviations will indicate the need for investigation to bring the building back to optimum operation. However, the specific comparisons that can be performed will be dictated by the capabilities of the simulation model and by the performance data available for comparison with the simulation. If energy management control system (EMCS) data will be used for comparison with the simulation, appropriate energy consumption sensors must be available on the EMCS.

The simulation will generally be used to evaluate what may be termed "passive testing" or "active testing" (Claridge 2004). The term "passive testing" refers to use of data collected during normal operation of the building, without any intervention to extend the range of operating variables implemented during any particular time interval. In contrast active testing entails use of specified control sequences to determine response to an extended range of operating variables, or a particular dynamic sequence of operating variables.

Passive testing

Passive testing for optimization has generally involved examination of simulated data and of measured consumption data. These comparisons are sometimes performed with time series data, but often with plots of heating/cooling consumption plotted as functions of outside air temperature. If the simulation is an idealized simulation, then significant differences between the simulated and measured consumption are taken as indications that the system is not performing optimally or as designed.

An illustrative sequence of passive tests is described below. This sequence is not exhaustive. Many more tests are used and many others will undoubtedly be developed and used.

a. *Check room temperatures and humidity levels*. Trend logs of the temperature and humidity in individual zones can be tracked over one or more days. As long as these values stay within the set points (and any control undershoot, overshoot or throttling range), temperature and humidity control are deemed acceptable. If not, diagnostics (that may or may not involve simulation) are needed to diagnose reasons for the excursions observed. If this test is passed,
b. *compare energy use with predictions* over a period of at least a few days. If measured consumption is within an acceptable range of predicted consumption, this test is passed. However,

it is non-trivial to develop practical "passing" criteria for this type of test. If "pass" criteria are narrowly missed, this may indicate a need to change inputs and

c. *extrapolate performance from limited trend data to design conditions*. It is certainly necessary to determine whether equipment capacity is adequate.

Depending on the capability of the simulation and the sensors available on the EMCS, it is desirable to verify a wide range of operating parameters such as airflows and supply temperatures to individual zones, waterside system parameters, and primary system performance.

Active testing

Active testing involves specific active tests implemented for diagnostic purposes; these may be triggered in response to failure of one or more passive tests. Active testing may also include functional tests devised to explore the comfort control capabilities of the system and its dynamic response over a wide range of operating conditions. Active tests normally provide some empirically determined input variables to be used in the simulation program being used.

Active testing will normally control an input variable with a major impact on building comfort and energy performance through a specified sequence of values. It may include the cycling of a known lighting load or other major load on/off on a specific cycle devised to test response of the HVAC system and test system performance. Likewise, it may involve variation of space temperature set points in a way that will test system capacity and control.

Multiple tests are needed to test system response to a range of loads in the spaces and to determine the efficiency with which the primary and secondary systems are working with the control system to meet the space loads.

Different applications of building-level simulation for operational optimization

Four different applications of simulation for operational optimization will be described in some detail with subsequent examples.

1. *Use of design simulation in post-construction commissioning of new buildings*. A design simulation of the building may be used to predict heating and cooling performance and the predictions may be compared with measured use – significant deviations then serve as clues to identify problems in the building. The design simulation should have the occupancy schedules changed if necessary to reflect the actual occupancy of the building. Simulation may also be used at this stage to refine and optimise the control strategy. Relatively complex simulations are used for this purpose.
2. *Use of a design simulation or calibrated simulation for on-going commissioning*. The same simulation developed in the design process may then be run at specified intervals, e.g. weekly, monthly, etc. and the model predictions compared with the measured energy consumption. This simulation must be modified to reflect any differences between the design assumptions and actual use of the building. Deviations then serve to trigger an alarm when building performance degrades. The nature of the deviations observed may also be used to help diagnose reasons for performance degradation. Alternatively, if no design simulation was performed, subsequent to existing building commissioning to ensure that the building is operating optimally, a simulation may be calibrated to the operation of the building and used to trigger an alarm when building performance degrades. These simulations may be run on-line or off-line. The response time of maintenance staff to issues affecting only energy use and not comfort is normally measured in days or weeks, so daily or weekly use of such simulations is appropriate.
3. *Use of calibrated simulation for existing building commissioning*. A rapidly calibrated simulation may be used as a diagnostic aid and to predict the savings that will be achieved from implementing proposed existing building commissioning measures.

4. *Use of simulation to evaluate new control code*. Either the design simulation or a calibrated simulation may be used to test the energy impact of proposed changes in control code before implementation. This will generally be done off-line.

Examples using building-level simulation for operational optimization

Each of the four different applications of building-level simulation for operational optimization is illustrated in this section in the context of one or more specific applications.

Use of design simulation in post-construction commissioning of new buildings

Keranen and Kalema (2004a, 2004b) have utilized simulations of the 9500 m^2 IT-Dynamo Building in the commissioning process. This building, which houses the Department of Information Technology at Jyvaskyla Polytechnic in Jyvaskyla, Finland, was completed in May 2003 and placed in service in August 2003. The building was simulated using the programs IDA-ICE and RIUSKA based on DOE-2. The heating and electricity use of the building are monitored and the electricity used for HVAC is separately monitored. Indoor conditions including numerous temperatures and CO_2 levels are monitored and stored for two days on the building automation system. During commissioning, these variables were transferred to memory for longer terms.

Comparison of the simulated and measured heating consumption showed measured heating consumption almost twice the simulated consumption during the first three months of operation. Investigation revealed that heating and cooling thermostat dead bands were too narrow in about 20% of the building area, resulting in continuous operation of either the heating or the cooling at maximum values in these areas. Correction of this problem reduced the difference between measured and simulated heating consumption to less than 10% in each of the next three months.

Carling *et al.* (2004a, 2004b) have tested the use of whole building simulation in the commissioning of an office building in Katsan, Sweden. This building utilizes multiple innovative HVAC systems. They used the IDA simulation environment to assess these innovative systems during design and for component sizing. During the initial commissioning they evaluated the performance using extensive measurements from the EMCS. Five-minute values for about 200 signals were collected during a full year following the initial occupancy of the tenants. A detailed whole building simulation model was calibrated with adjusted internal loads based on measured electrical power and measured weather inputs. To determine whether the HVAC-systems performed as intended, the results of the calibrated model were compared with measurements of the whole-building energy use as well as with some important temperatures and control signals.

Comparison between the extensive measurements and the calibrated model were made and use of the different residuals for evaluation of the HVAC-system performance and on-going commissioning led to detection and correction of five problems with control set point and/or operation.

Carling *et al.* found the cost of the procedure to be too high for routine application for several reasons: generating the detailed model took several weeks; model run times were long; additional sensors (at additional cost) would have improved model contribution to commissioning; and dealing with poor data was time consuming. They concluded that the approach has a large potential to support better design and commissioning of buildings *provided* that the costs can be decreased to an acceptable level.

Use of calibrated simulation for on-going commissioning

A simulation calibrated to a building after commissioning is performed may be used to check the measured consumption on an on-going basis. Comparison once a month or once a quarter is probably adequate. Significant increases can then serve as an alarm to indicate when additional commissioning follow-up is justified. Building operators generally don't get very interested in following up on an alarm that does not directly impact comfort and create occupant complaints unless it has a rather substantial cost impact. Hence, there is little need for this information on an hourly or even a daily basis when tracking whole building consumption. This type of tracking may also be performed with simple regression models of consumption, as well as with more detailed physical models. If the simulation is coupled to a diagnostic system that can indicate probable causes of deviations, this will increase its value.

The need for this type of tracking has been explored by Turner *et al.* (2001) and Claridge *et al.* (2004), and findings in two specific buildings have been investigated and reported by Chen *et al.* (2002) and Liu *et al.* (2002b).

On-line simulation as a commissioning follow-up tool

A design simulation or a calibrated simulation may be embedded in the EMCS. It can serve as an alarm any time consumption deviates beyond an alarm limit. It may also be used to evaluate the impact of any control changes implemented – comparison of measured performance with simulation results would show whether performance has improved or degraded as a result of the changes.

Liu *et al.* (2002a) presents the results of a study of the potential for using simulation programs for on-line fault detection, problem diagnosis, and operational schedule optimization for large commercial buildings with built-up HVAC systems. He specifically examines the potential for the use of calibrated whole building simulation for existing building-commissioning and for on-going commissioning.

This study reviewed over a dozen simulation programs and determined that AirModel and EnergyPlus (LBL 2001) were most suitable for initial use in the on-line simulation applications that were the focus of the study. These programs cover both ends of a spectrum from a relatively simple program that can be used quickly and embedded in an EMCS for on-line simulation to one of the more detailed and flexible simulation programs available.

Tsubota and Kawashima (2004) developed a detailed load and HVAC system simulation program for a 28,481 m^2 building that consists of an 11-story office wing and a three-floor conference wing. The building uses double bundle heat pump chillers with heat recovery, an ice storage tank and hot water storage tanks. The simulation was carefully calibrated to measured consumption from the building. The error in cooling load for one week shown was 2.2%, with an expected annual discrepancy of 5.7%. The discrepancy between measured and simulated electric consumption was only 1.4% on an annual basis. The program has been operated online using data from the building so the operators can view hourly comparisons between target (or simulated) consumption and measured consumption. Operators can also view 8-day "weekly" plots of the hourly data comparisons that permit examination of a week's performance and with the same day of the previous week. Similar "weekly" plots of daily total target and measured consumption are also available.

Others performing investigations of on-line simulation include Claridge and Liu (2004).

Use of calibrated simulation for existing building commissioning

Simulation at the building level may be used as a tool in conjunction with data on the demands and needs of the building to determine the potential for energy savings in the building. This application is particularly apt when commissioning an older building. For older buildings, utility billing history is generally available. The increasing use of interval metering means that hourly or 15-minute data is more frequently available at the whole building level. The decreasing cost of metering and recording such data means that it will also increasingly be available on the building EMCS for additional end uses such as heating and cooling, though this is not yet common.

Such data can be used to calibrate a simulation program to the measured consumption data from the building. When this is done, the simulation can readily be used to accurately explore the impact of a wide range of building changes, ranging from operational changes that may be implemented as part of a commissioning program to evaluation of thorough energy efficiency retrofit measures, and demand reduction measures. The simulation can also be used to investigate the comfort impact of certain measures before they are implemented.

Wei *et al.* (1998) as modified by Claridge *et al.* (2003) have developed an approach to calibration of a cooling and heating energy simulation for a building to measured heating and cooling consumption data that addresses some of the time/cost constraints reported by Carling *et al.* (2004a, 2004b). They present a methodology for the rapid calibration of cooling and heating energy consumption simulations for commercial buildings based on the comparison of "calibration signatures" that characterize the differences between measured and simulated performance with "characteristic signatures" that characterize the influence of changes in key parameters on the heating and cooling consumption. The method is described in Claridge *et al.* (2003) and its use is demonstrated in two illustrative examples and two case studies. The report contains characteristic signatures suitable for use in calibrating energy simulations of large buildings with four different system types: single-duct variable-volume, single-duct constant-volume, dual-duct variable-volume and dual-duct constant-volume. Separate sets of calibration signatures are presented for each system type for climates typified by Pasadena, Sacramento and Oakland, California.

Liu *et al.* (2002a) emphasizes the use of simulation for on-going commissioning, but also contains two case studies in which the AirModel simulation was used to identify and diagnose system problems at the whole building level. These case studies illustrate the value of calibrated whole building simulation for existing building commissioning. The first case study conducted a calibrated simulation of a 28,000 m^2 hospital in Galveston, Texas as part of an existing building commissioning project. The calibration process lead to identification of 2–4 °C differences between pre-cooling, cold deck and hot deck temperatures, and their respective set points. The simulation was subsequently used to develop optimum schedules for these quantities. In the second case study, simulation was performed on an 11,400 m^2 medical laboratory building in Houston, Texas. The calibration process indicated a probable error in the chilled water metering and serious lack of control in the chilled water valves. Subsequent field inspections revealed that the chilled water meter was reading only 50% of the correct consumption due to an open bypass valve and found that leaks in the pneumatic control lines had caused the chilled water valves to operate in the full open position much of the time.

The simulation effectively identified HVAC component problems and was used to develop optimized HVAC operation and control schedules in the hospital. Likewise, it successfully identified metering and valve leakage problems in the laboratory building. Re-heat valve leakage problems and excessive airflow problems were identified after fixing the leaking chilled water valve. The simulation indicated that building thermal energy consumption would be reduced by 23% or $191,200/yr by using the optimized operating schedules in the hospital. The measured energy savings were consistent with the simulated savings.

Ginestet and Marchio (2004) developed a simulation tool with several features that are useful for existing building commissioning and on-going commissioning. The simulation utilizes typical weather and building thermal parameters, but also incorporates a number of flow parameters and control parameters that are not used in many building simulation tools. These parameters enable the program to be used to evaluate indoor air quality in individual zones and evaluate different ventilation strategies. It is also able to simulate a number of air handler faults since it models duct pressure drops, fan operation, and sensor problems related to location, bad connections, bad set points, etc.

Andre *et al.* (2003, 2004) describe the existing building commissioning of a relatively new building. Additional investigators who have performed relevant calibrated simulation investigations include Adam *et al.* (2004) and Carling *et al.* (2004a, 2004b).

Use of simulation to evaluate new control code

New control code can be tested by simulation before actually putting it into the system and activating the new or modified mode of control. Baumann used this approach to optimize control of the water supply temperature for heating and cooling in a new school. Wang and Xu (2003) have developed a new simulation tool particularly intended to permit rapid evaluation of the dynamic performance of a building at short time steps so it is suitable for evaluation of control code.

Baumann (2003) used a TRNSYS simulation in conjunction with the GenOpt program to develop an optimal control strategy for the heating and cooling water supply temperatures in a 10,000 m^2 vocational school in Biberach, Germany. This school was completed in the summer of 2004 and incorporates an embedded hydronic heating and cooling (EHHC) system consisting of flexible tubes embedded in the massive concrete ceilings. Water heated by a heat pump or cool ground water is supplied to the EHHC system, depending on whether heating or cooling is required in the building.

The massive ceiling has a very long time constant, so controlling the temperature of the water supplied to the system is critical to minimize the number of occupied hours when the space temperatures are either too hot or too cold. The control scheme adopted uses the median temperature values for the last three days (with the most recent day double-weighted) to define the "median" outdoor temperature that determines the temperature of the supply water. The supply water temperature varies between a maximum value of 28 °C and a neutral value of 21 °C as the "median" outdoor temperature varies between values of $T_{o,heat,max}$ and $T_{o,heat,mln}$. The supply temperature is maintained at 21 °C until the "median" outside temperature increases to $T_{o,cool,min}$ where the supply temperature starts to decrease linearly toward 18 °C at a temperature of $T_{o,cool,max}$. The supply temperature is held constant at 18 °C for temperatures above $T_{o,cool,max}$ and is held constant at 28 °C for temperatures below $T_{o,heat,max}$.

The simulation was used to optimize the values of $T_{o,heat,max}$, $T_{o,heat,min}$, $T_{o,cool,min}$, and $T_{o,cool,max}$ to minimize heating and cooling energy subject to the constraints that the room temperature never go below 21 °C and go above 26 °C for only a small number of hours. It was found that the optimum heating consumption was only half the maximum value that produced equivalent comfort.

Wang and Xu (2003) developed a hybrid model suitable for rapid short time-step simulation of the dynamic performance of buildings – hence suitable for control simulation and optimization. This hybrid model uses a 3-resistor, 2-capacitor (3R2C) model of the building envelope, with the sum of the three resistances constrained by the total resistance of the envelope and the total capacitance constrained by the total thermal capacitance of the envelope. Wang and Xu use a genetic algorithm to choose optimum values for the resistances and capacitances so the frequency response and phase lag of the 3R2C model closely approximates that of the theoretical model for the walls at all but the highest frequencies. The same technique is used to develop optimal parameters for a 2R2C model of the interior mass in the building by searching in the time domain.

This model was developed in part to determine optimal control strategies for night ventilation and off-peak pre-cooling to take advantage of a Time of Use Rate that was implemented in Hong Kong in 2001.

The following applications of building-level simulation for operational optimization have been described and illustrated:

- Design simulation in post-construction commissioning of new buildings
- Design simulation or calibrated simulation for on-going commissioning
- Calibrated simulation for existing building commissioning
- Simulation to evaluate new control code.

Examples of each of these applications are provided. The only application cited that has found use outside research projects to date is the use of calibrated simulation for existing building commissioning. The other examples have been applied in a research setting, and costs must be lowered for routine application, but there appears to be potential for significant application of simulation in the operational optimization process.

Using building simulation for practical optimization of operational performance

The remainder of this chapter will focus on use of building simulation in the existing building commissioning process since this is the application that has found the most use to date, and is arguably, the easiest to implement. An overview of the Existing Building Commissioning (EBCx) process is presented for those not familiar with this process followed by description of a method for rapid calibration of simulations. The use of simulation in the EBCx process for fault detection is then described followed by two case study examples.

Overview of the existing building commissioning (EBCx) process

Commissioning of existing buildings has become a key energy management activity in North America over the last decade, often resulting in energy savings of 10%, 20% or sometimes 30% without significant capital investment. It generally provides an energy payback of less than three years. In addition, building comfort is improved, systems operate better and maintenance cost is reduced. Commissioning measures typically require no capital investment, though the process often identifies maintenance that is required before the commissioning can be completed. Potential capital upgrades or retrofits are often identified during the commissioning activities, and knowledge gained during the process permits more accurate quantification of benefits than is possible with a typical audit. Involvement of facilities personnel in the process can also lead to improved staff technical skills.

Existing building-commissioning is typically defined as the first-time commissioning of an existing building. The focus of EBCx is usually on energy-using equipment such as mechanical equipment and related controls. It may or may not bring the building back to its original design intent, since the usage may have changed or the original design documentation may no longer exist. The version of EBCx practiced by the author is Continuous Commissioning (CC®). CC is an ongoing process to resolve operating problems, improve comfort, optimize energy use, and identify retrofits for existing commercial and institutional buildings and central plant facilities. CC focuses on improving overall system control and operations for the building, as it is currently utilized, and on meeting existing facility needs. It is not intended to ensure that a building's systems function as originally designed, but it ensures that the building and its systems operate optimally to meet the current uses of the building. Near-optimal operational parameters and schedules are developed based on actual building conditions and current occupancy requirements. Once operation has been optimized, optimization is maintained by a continuous ongoing process that monitors building performance and corrects problems that arise. In the remainder of this chapter, EBCx will refer to the CC process.

The EBCx process includes the following steps:

- Assessment
- Develop EBCx plan and form project team
- Develop performance baselines
- Conduct system measurements and develop EBCx measures
- Implement EBCx measures
- Document comfort improvements and energy savings
- On-going commissioning to maintain optimum operation.

Simulation may be utilized in three steps of the process. The assessment of the building operation identifies major potential EBCx measures, the potential savings from implementing these measures, and the cost of implementation. Potential savings may be evaluated using simulation, but are often estimated using simpler methods since the assessment budget seldom includes enough time for careful simulation. The detailed development of the EBCx measures and expected performance may be materially aided by the use of simulation. Simulation has also been found to be an effective way to track continuing performance and identify the initiation of future changes or faults that degrade building performance.

Developing the EBCx plan, forming the project team, and developing performance baselines does not involve simulation, unless IPMVP Option D (EVO 2007) is used for the performance baselines. Likewise, implementation and documentation of savings does not involve simulation.

Experience shows that detailed dynamic envelope simulation is not needed for the operational optimization implemented in the EBCx process (Liu and Claridge 1998). The approach that has generally used has been the ASHRAE Simplified Energy Analysis Procedure (Knebel 1983) that basically combines simple envelope simulation with relatively detailed simulation of the secondary systems and plants.

An essential element of the successful use of simulation in EBCx projects as practiced to date has been calibration of the simulation to the measured performance of the building.

The calibration processes used to achieve agreement between simulation and measured consumption have generally been quite time-consuming and required a great deal of specialized expertise. The approach presented here is based on the work of Wei *et al.* (1998) and Claridge *et al.* (2003) and has been simplified so calibration of a building simulation has been incorporated within the author's classes at Texas A&M University for several years.

A simple process for calibrating simulations

Building energy simulations have become increasingly important both in new building design and in evaluating potential energy efficiency measures in existing buildings. Simulations can be valuable tools in many applications, such as calculating energy savings from proposed or implemented retrofits, existing building and new building commissioning, fault detection and diagnosis, and program evaluation. Today a number of simulation tools exist, ranging from complex, whole building modeling programs such as DOE-2 and BLAST, to simplified air-side modeling programs like AirModel. Each of these programs requires a number of input parameters from the user to model the building, or specific systems within the building. The accuracy in using these and all simulation programs depends on the ability of the user to input parameters that result in a good model of actual building energy use. The initial parameters used are generally obtained from building design data, which can result in output errors of 50% or more when compared with actual building performance. An example of simulated energy use versus actual use is shown below in Figures 13.1 and 13.2. Figure 13.1 shows the daily simulated and measured heating usage in Harrington Tower on the Texas A&M University campus for the year 2003, while Figure 13.2 is a plot of the same data as a function of average daily dry bulb temperature.

As evidenced in these figures, a simulation may not accurately reflect the performance of the building; in this case, the simulated consumption reasonably approximates the building consumption at temperatures below 15 °C, but is about 2/3 the actual consumption at 30 °C. Any conclusions made about cooling conservation measures and their savings based on data from this simulation have a high probability of significant error. The simulated data must closely match the measured data to allow good analyses to be made, and the closer the data match, the better the expected results of analyses. Unfortunately, in many cases the process of calibrating a simulation can be extremely tedious and time consuming, especially with simulation programs that require a large number of input parameters. Because changing two or more different parameters in the input could result in similar changes to the output, the problem is over defined, and it is difficult to know which parameters to change in order to achieve the desired

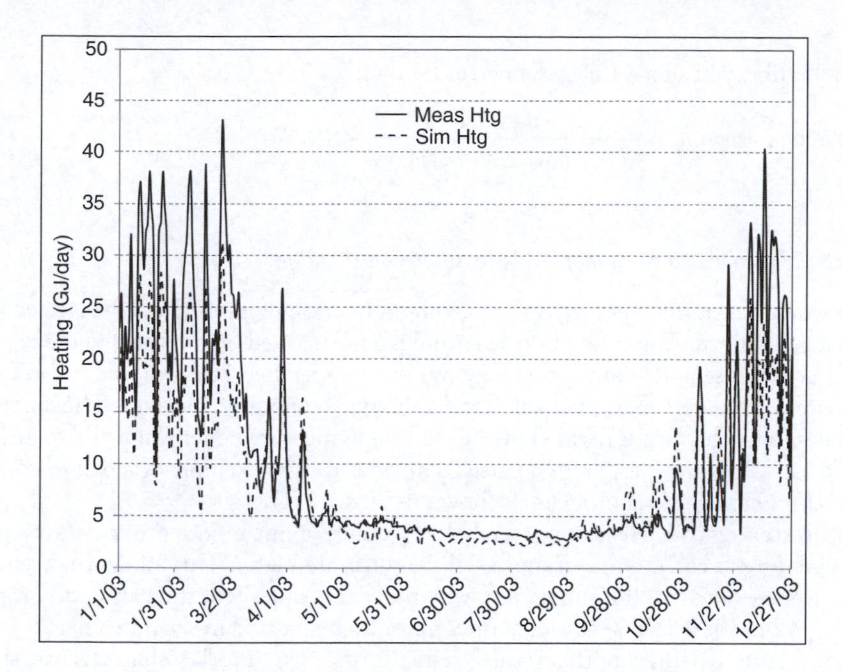

Figure 13.1 Simulated and measured heating usage in Harrington Tower for 2003.

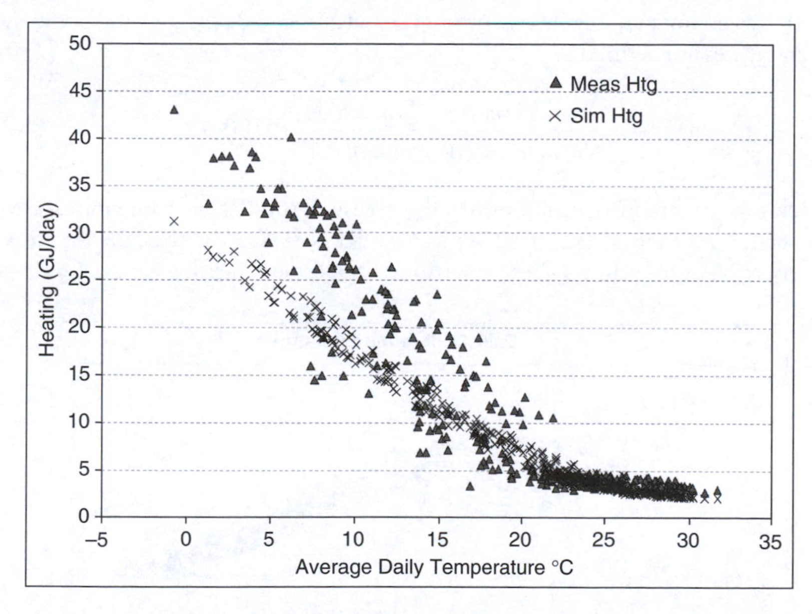

Figure 13.2 Simulated and measured daily heating usage for Harrington Tower versus daily average dry bulb temperature.

results, particularly for inexperienced users. Without a straightforward method of calibrating, the process can be too expensive and time-consuming to be deemed worthwhile.

A method has been developed in recent years to calibrate simulations with relative ease based on a concept called "calibration signatures." Experienced users of this method have shown the ability to calibrate a two-zone simulation in a few hours, significantly less time than previously needed with a "trial and error" approach.

Definitions

The term "calibration signature" is defined as follows:

$$Calibration\ signature = \frac{-Residual}{Maximum\ measured\ energy} \times 100\% \tag{13.1}$$

where

$$Residual = Simulated\ consumption - Measured\ consumption \tag{13.2}$$

The "maximum measured energy" is the maximum heating or cooling energy use recorded in the data set examined. The calibration signature is a normalized plot of the difference between measured energy use and simulated energy use over a specified temperature range. For each temperature, a measured energy use value and a simulated energy value exist. The difference in these values for each point is divided by the maximum measured energy use and multiplied by 100%. These values are then plotted versus temperature. An example of a pair of calibration signatures for heating and cooling use is shown in Figure 13.3.

Due to the scatter that exists, it is often helpful to draw a line representing the average value of each signature at each temperature, as will be illustrated later. This allows the trend of the signature to be evaluated. If a simulation were perfectly calibrated, the calibration signatures would be flat lines at 0%. The two signatures may also be plotted in separate charts.

It is now useful to define another term, "characteristic signature." A characteristic signature is identical to a calibration signature, except that instead of comparing simulated and measured values, it compares two simulations. One simulation is taken to be the baseline or "measured" value. Then, by varying parameters one by one, signatures can be plotted and compared. Characteristic signatures are defined as:

$$Characteristic\ signature = \frac{Change\ in\ energy\ consumption}{Maximum\ energy\ consumption} \times 100\% \tag{13.3}$$

where the change in energy consumption is the changed simulation value minus the baseline simulation value. As mentioned, the baseline model is treated as the "measured" case, and maximum energy consumption comes from this model. Characteristic signatures are generated

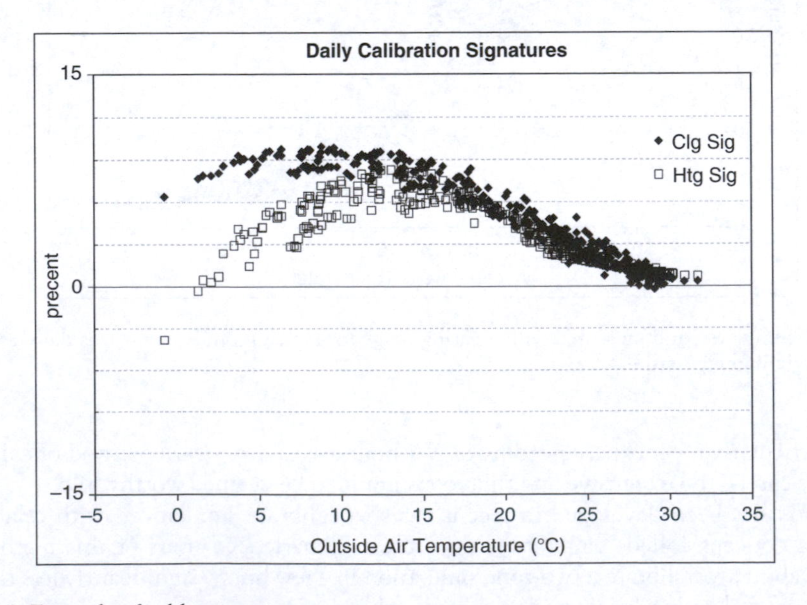

Figure 13.3 Example of calibration signatures.

for a range of parameters for the building being calibrated. An example of a pair of characteristic signatures is shown in Figure 13.4. This signature was generated for a dual duct VAV system in Harrington Tower by changing the simulation minimum flow rate from the baseline value of 1.75 L/(s-m^2) to 1.25 L/(s-m^2).

Notice that for a change in a parameter, characteristic signatures should be generated for both heating and cooling energy use. Ideally, characteristic signatures should be generated for the building to which the simulation is being calibrated. However, it is still helpful to use characteristic signatures for a similar building in a similar climate if it is not practical to generate characteristic signatures for the building to be calibrated. Characteristic signatures for the dual-duct variable-air-volume (DDVAV) system in Harrington Tower are given in the Appendix. The parameters of major importance for which characteristic signatures should be generated include cold deck temperature, hot deck temperature (DD systems), supply air flow rate (CV systems), minimum air flow rate (VAV systems), floor area, preheat temperature, internal gains, outside air flow rate, room temperature, envelope U-value, and economizer. Characteristic signatures from a climate with different correlations between relative humidity and dry-bulb temperature are unlikely to work well.

Two indices used for evaluating the accuracy of a simulation are the "Root Mean Square Error" and the "Mean Bias Error." The Root Mean Square Error (RMSE) is defined as:

$$RMSE = \sqrt{\frac{\sum_{t-1}^{n} Re\,sidual_t^2}{n-2}} \tag{13.4}$$

where n is the total number of data points. The RMSE is a good measure of the overall magnitude of the errors, but does not give any reflection of bias, since no indication is made as to whether the errors are positive or negative. A good simulation will minimize the RMSE, though it is generally difficult to reduce this below 10% of the mean of the larger of the heating and cooling consumption. The Mean Bias Error (MBE) is defined as:

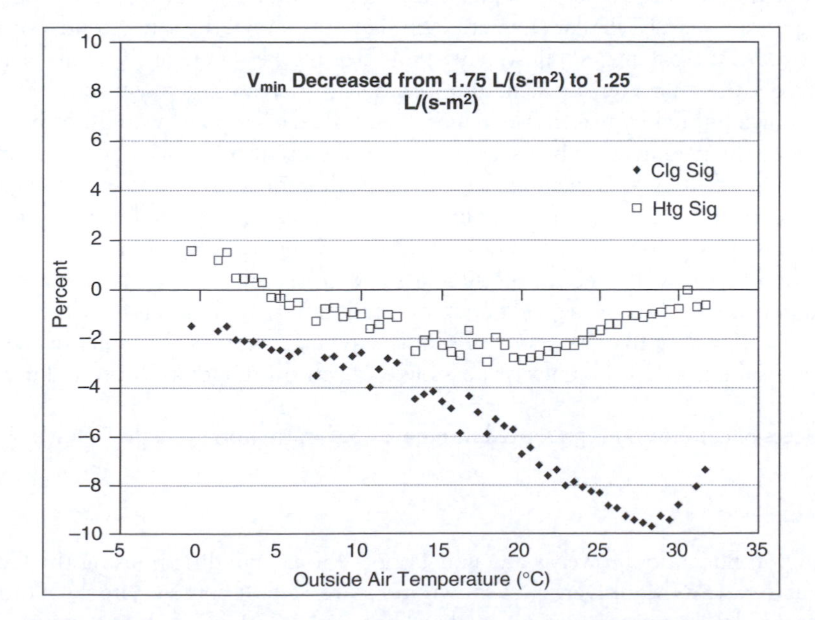

Figure 13.4 Characteristic signatures for a dual duct VAV system in Harrington Tower with a change in minimum flow rate.

$$MBE = \frac{\sum_{t-1}^{n} Re\,sidual_t}{n} \tag{13.5}$$

where n is the number of data points. The MBE is an overall measure of the data bias, since positive and negative errors cancel each other. The relative importance of MBE and RMSE varies depending on the purpose of the calibrated model. Both should be considered if the model will be used to evaluate energy efficiency measures, while MBE becomes more important if the model will be used as the baseline for evaluating energy savings after efficiency measures have been implemented.

The approach used in the remainder of this chapter will minimize a combined error measure defined as:

$$\begin{aligned} ERROR_{TOT} &= \left[RMSE_{TOT}^2 + MBE_{TOT}^2 \right]^{1/2} \\ &= \left[\left(RMSE_{CLG}^2 + RMSE_{HTG}^2 \right) + \left(MBE_{CLG} + MBE_{HTG} \right)^2 \right]^{1/2} \end{aligned} \tag{13.6}$$

This error measure arbitrarily combines the root mean square heating and cooling errors in quadrature and the heating and cooling mean bias errors algebraically before combining both in quadrature. The algebraic combination of mean bias errors is a combination that reduces the mean bias in the total cost of heating and cooling energy.

Implementing the calibration process

Calibration signatures combined with characteristic signatures are used to speedily calibrate a simulation. The calibration signatures for heating and cooling generated for the simulation are compared with the characteristic signatures from the corresponding system and climate type, to see which change of parameter or parameters most closely resemble the calibration signature. Normally one parameter is changed at a time in the correct direction and according to the magnitude needed. For example, if the calibration signature is in the range of 20% for low temperatures, and a similar characteristic signature shows the same trend, but is in the range of only 5%, the parameter adjustment would need to be significantly greater than what was done to get the characteristic signature in order to increase the magnitude. The adjustment is of course limited by reasonable values – a cold deck set point would not be 3 °C, for example. Once the parameter to be varied has been chosen, it is changed and the simulation is run again. The $ERROR_{TOT}$ is calculated again, and calibration signatures are again generated and compared with the characteristic signatures. This process is repeated until an acceptable low value of $ERROR_{TOT}$ is achieved, and the calibration signatures approximate a normal distribution around zero. At this point, the simulation can be considered to be calibrated to the measured data. As noted above, a good target for a calibrated simulation is RMSE values of the heating and cooling data that are about 10% of the average value of the larger of the heating or cooling consumption. It is not always possible to reach this target as illustrated in Example 2 below.

This process of calibration is presented in Figure 13.5 in the form of a flow chart.

Calibration Example 1

The Harrington Education Tower is an eight-floor 9,430 m^2 building housing the College of Education at Texas A&M University as shown in Figure 13.6. It consists largely of offices, but there are some classrooms and computer labs. There are 1,770 m^2 of glazing, with 50% of it being on the first, second, and eighth floors, as they are predominantly glass sided. Harrington

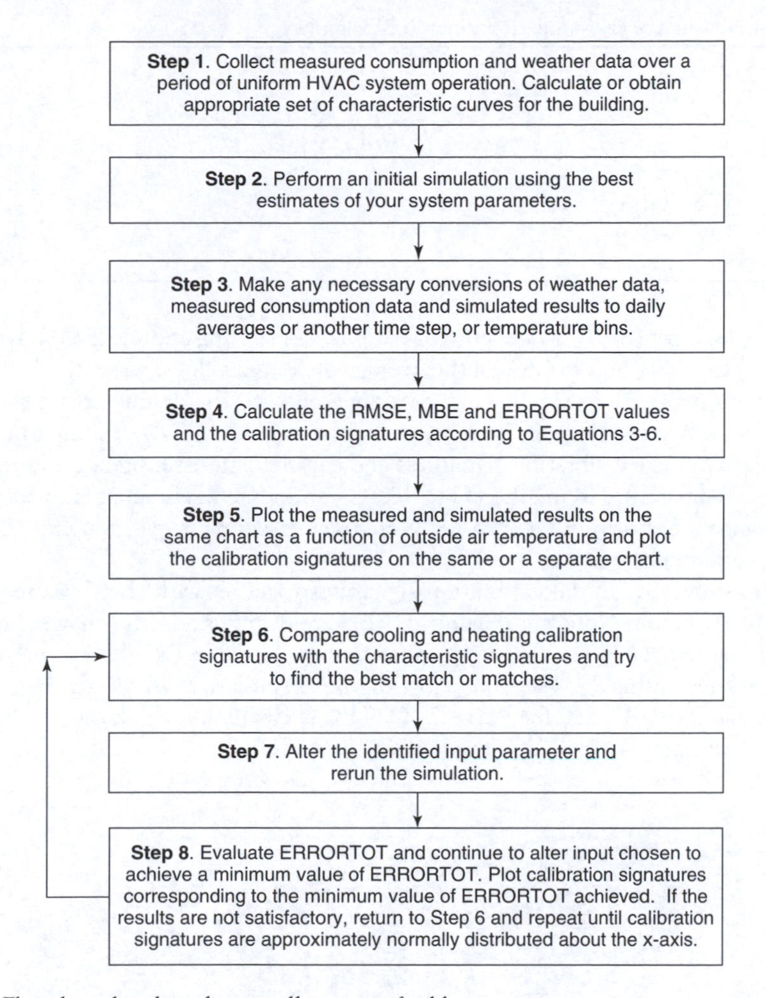

Figure 13.5 Flowchart detailing the overall process of calibration.

Tower receives hot water and chilled water from the Texas A&M Physical Plant for its HVAC systems and provides heating and cooling to about 90% of the building with a single dual-duct variable air volume (DDVAV) air handler powered by a 150 kW fan providing up to 65,000 L/s of air. To illustrate the calibration procedure, the building is simulated using an implementation of the ASHRAE Simplified Energy Analysis Procedure (Knebel 1983). To provide a cleaner example of the procedure, one or more inputs to the simulation were changed and the output of that simulation is used as the "measured" consumption data for this example. Example 2 will show the calibration of this building to measured data instead of simulated data. Table 13.1 shows initial simulation inputs for Example 1. The entry "Linearize" means that the hot deck temperature varies according to $T_{HD} = 43.3 - 1.33(T_{oa} - 4.4)$ when $4.4 \leq T_{oa} \leq 21.1$ °C. Figure 13.7 compares the "measured" and initially simulated heating and cooling consumption as functions of daily average temperature, while Figure 13.3 shows the initial set of calibration signatures for this example.

Figure 13.7 shows that the simulated values of both heating and cooling consumption appear to be systematically low at temperatures below about 20 °C. The calibration signatures in Figure 13.3 show that both heating and cooling values are 5–10% (of the maximum measured value) below the measured values over much of this range. $ERROR_{TOT}$=3.72 GJ/day for this initial simulation with values of 1.76 GJ/day for cooling and 1.05 GJ/day for heating. The

Table 13.1 Initial values of key simulation inputs for Example 1

A_{floor}	9430	m^2	U-values		Q_{int}	18.2	W/m^2
A_{walls}	3457	m^2	0.85	$W/(m^2 \cdot K)$	T_{room}	22.2	°C
$A_{windows}$	1767	m^2	6.25	$W/(m^2 \cdot K)$	T_{CL}	12.8	°C
A_{roof}	1520	m^2	0.28	$W/(m^2 \cdot K)$	T_{ph}	1.7	°C
V_{oa}	15	%			T_{HD}	43.3	$T_{oa} < 4.4$
V_{min}	1.78	$L/(s \cdot m^2)$				Linearize	$4.4 \le T_{oa} \le 21.1$
V_{max}	5.6	$L/(s \cdot m^2)$				23.9	$T_{oa} > 21.1$
$A_{IntZone}$	58	%					

initial Mean Bias Error (MBE) values are –1.76 GJ/day for cooling and –1.05 GJ/day for heating corresponding to –8.8% and –14.8% of the average measured values, respectively.

The calibration signatures should now be compared with the characteristic signatures for the DDVAV system in Harrington Tower shown in the Appendix. When comparing the signatures, a similarity in the calibration signatures and the characteristic signatures for decreasing the cold deck temperature from 12.8 °C to 10 °C can be seen. The characteristic signatures from this change are shown in Figure 13.8. Note the decreasing slopes that exist for both the chilled water and the hot water.

Figure 13.8 shows the initial calibration signatures together with the three most relevant characteristic signatures. Note that Figure 13.8(c) is relevant as a mirror image, i.e. it implies that an increase may be needed in V_{min} rather than a decrease. Examining the characteristics signatures, decreasing T_{CL} seems to show the most similarity. We decrease T_{CL} and get a minimum value of $ERROR_{TOT}$ for T_{CL} ~ 7 °C. This is clearly a non-physical value since we know the chilled water is supplied to the building at ~6 °C.

Figure 13.6 Harrington Tower on the Texas A&M University campus.

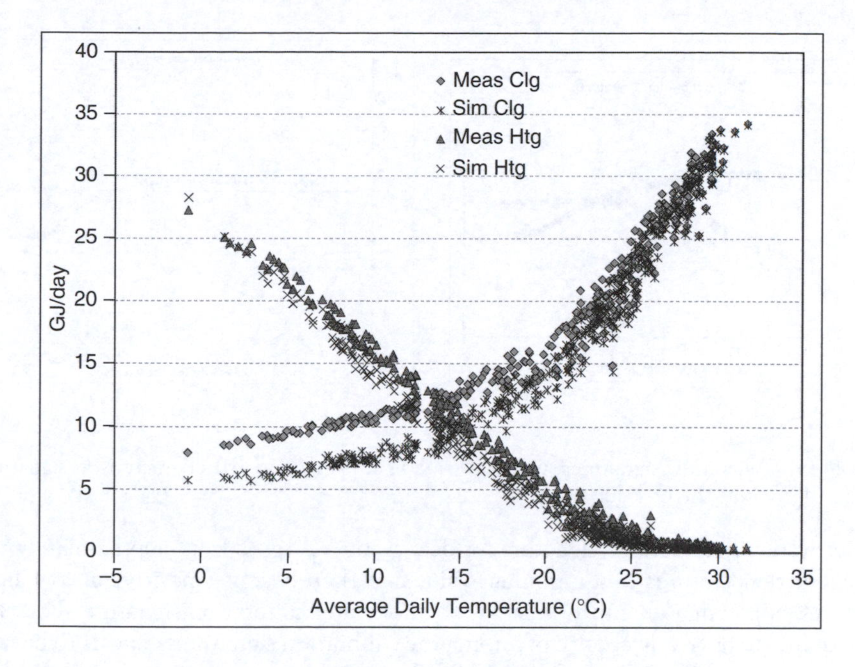

Figure 13.7 Comparison of initial simulated and "measured" heating and cooling consumption values as functions of daily average temperature for Example 1.

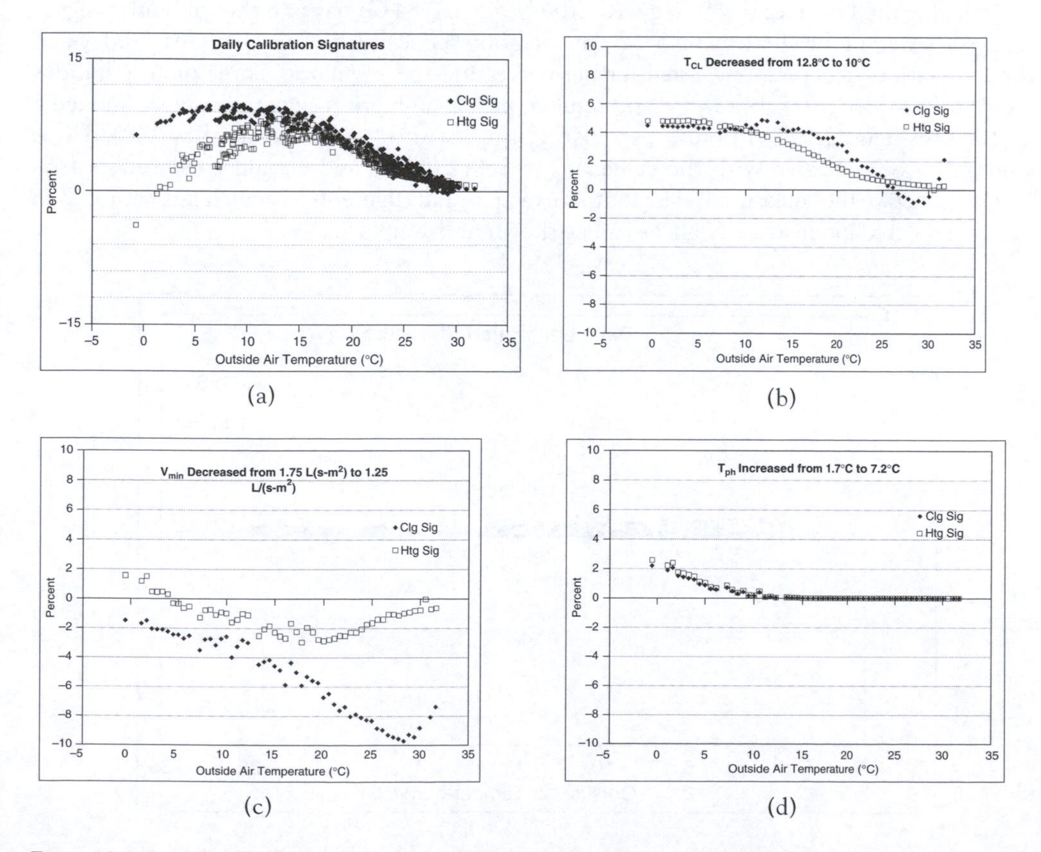

Figure 13.8 Initial calibration signatures and relevant characteristic signatures.

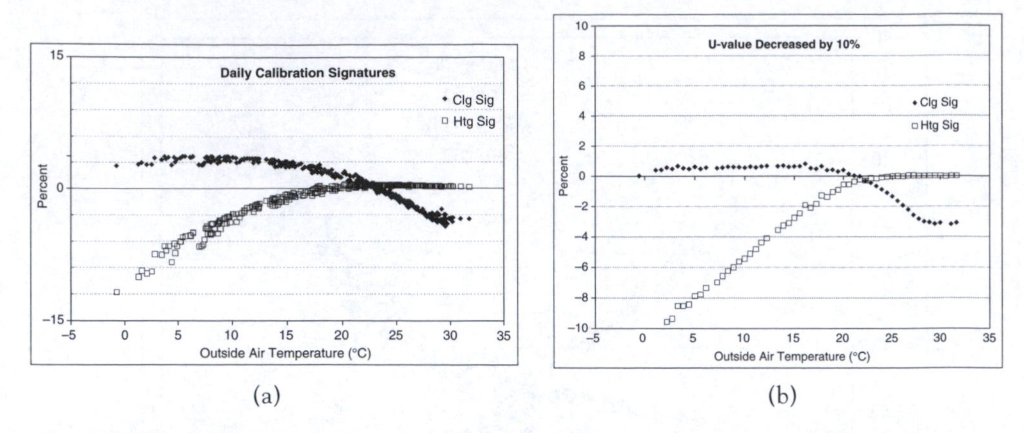

Figure 13.9 (a) Calibration signatures with V_{min} = 2.34 L/(s·m²) and (b) characteristic signatures for decreasing the U-value.

The hot water characteristic signature for decreasing V_{min} goes from approximately zero at high ambient temperatures to a minimum value and then becomes positive at low ambient temperatures, approximately mirror imaging the hot water calibration signature. The preheat change required to have any chance of zeroing the calibration signatures appears to be large so we then try increasing V_{min}. The value of $ERROR_{TOT}$ reaches a minimum value of 1.13 GJ/day when V_{min} is increased to 2.34 L/(s·m²). The calibration signatures for V_{min} = 2.34 L/(s·m²) are shown in Figure 13.9(a).

Reducing the U-values by 8% reduced $ERROR_{TOT}$ to 0.84 GJ/day and the calibration signatures suggested further increasing V_{min}. This iteration reduced $ERROR_{TOT}$ to 0.61 GJ/day and the calibration signatures suggested further decreasing the U-values. Three more iterations reduced $ERROR_{TOT}$ to 0.14 GJ/day and resulted in the calibration signatures shown in Figure 13.10. This figure resulted from V_{min} = 2.46 L/(s·m²) with U-values reduced by 16%. These values may be compared with the values V_{min} = 2.49 L/(s·m²) and U-values reduced by 14% used to generate the "measured" data for this example. This figure shows much less scatter than can be expected for measured data as will be shown in Example 2.

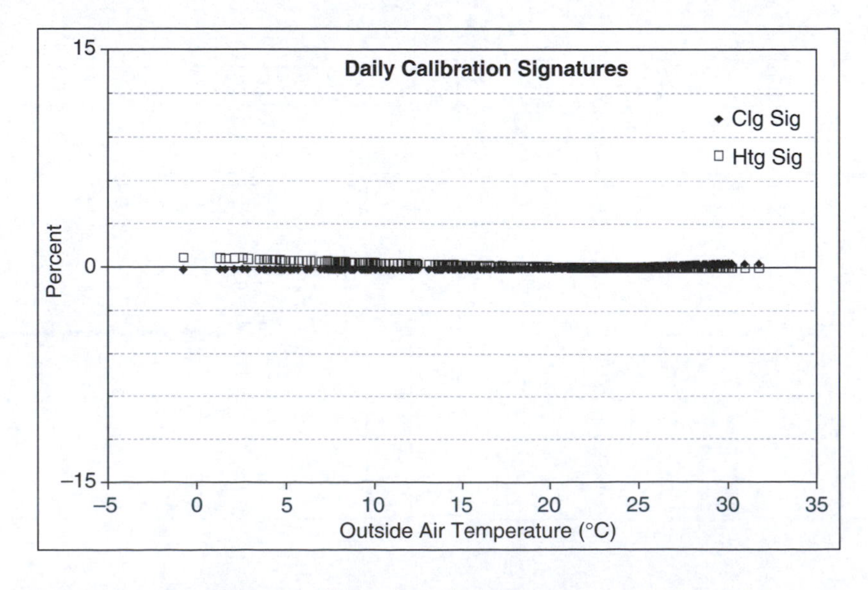

Figure 13.10 Calibration signatures for Example 1 after calibration.

Calibration Example 2

The same building simulated in Example 1 is simulated again, but this time, it will be calibrated to the actual measured heating and cooling data. As noted earlier, one large DDVAV air handler serves about 90% of the building. The remaining 10% is served by a small single duct constant volume air handler with terminal reheat. Table 13.2 shows the key initial inputs used for AHU1, the DDVAV AHU. The major changes from Example 1 are use of the measured non-HVAC electric use for Q_{int} and a reduction in the various areas to account for the area served by AHU2. Table 13.3 shows the inputs used for AHU2, the SDCV AHU.

Using this set of inputs, Harrington Tower was simulated and Figure 13.11 plots the simulated and measured heating and cooling consumption values as functions of ambient temperature. The calibration signatures for this simulation are shown in Figure 13.12 and the error measures for the initial simulation vs. the measured data are given in Table 13.4.

Table 13.2 Initial inputs for DDVAV AHU

A_{floor}	8780	m^2	U-values		Q_{int}	26.9	W/m^2
A_{walls}	3320	m^2	0.85	$W/(m^2{\cdot}K)$	T_{room}	22.2	°C
$A_{windows}$	1360	m^2	6.25	$W/(m^2{\cdot}K)$	T_{CL}	12.8	°C
A_{roof}	1520	m^2	0.28	$W/(m^2{\cdot}K)$	T_{ph}	1.7	°C
V_{oa}	4950	L/s			T_{HD}	43.3	$T_{oa}<4.4$
V_{min}	1.78	$L/(s{\cdot}m^2)$				Linearize	$4.4 \le T_{oa} \le 21.1$
V_{max}	5.6	$L/(s{\cdot}m^2)$				23.9	$T_{oa}>21.1$
$A_{IntZone}$	58	%					

Table 13.3 Inputs used for the SDCV AHU

A_{floor}	650	m^2	U-values		Q_{int}	26.9	W/m^2
A_{walls}	50	m^2	0.85	$W/(m^2{\cdot}K)$	T_{room}	22.2	°C
$A_{windows}$	400	m^2	6.25	$W/(m^2{\cdot}K)$	T_{CL}	12.8	°C
A_{roof}	0	m^2	0.28	$W/(m^2{\cdot}K)$	T_{ph}	1.7	°C
V_{oa}	660	L/s			$A_{IntZone}$	50	%
V_{tot}	7.1	$L/(s{\cdot}m^2)$					

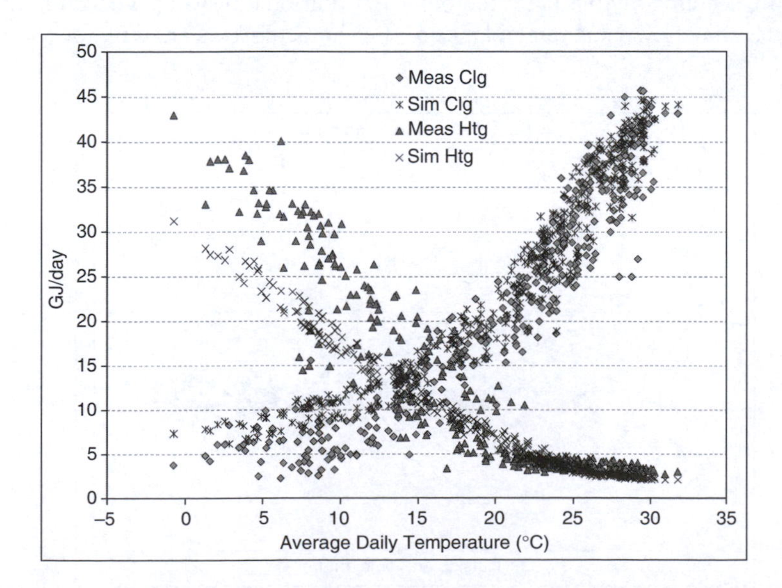

Figure 13.11 Measured and initial simulated heating and cooling consumption as functions of average daily temperature for Example 2.

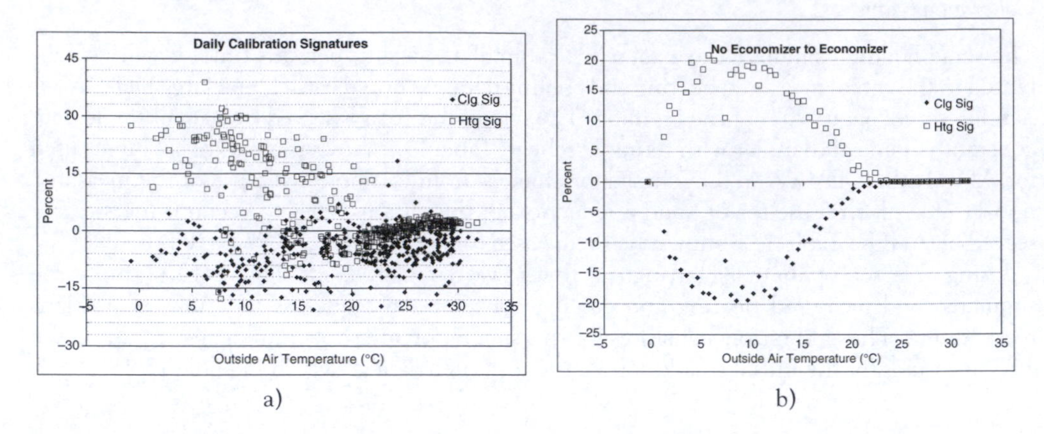

Figure 13.12 Initial calibration signatures for Example 2 and characteristic signatures for adding an economizer.

Table 13.4 Initial error measures for Example 2

	GJ/day		*GJ/day*
$RMSE_{Htg}$	5.02	$RMSE_{TOT}$	6.14
$RMSE_{Clg}$	3.54	MBE_{TOT}	0.18
MBE_{Htg}	–2.21	$ERROR_{TOT}$	6.14
MBE_{Clg}	2.40		

Examining the initial calibration signatures for Example 2 and comparing with the characteristic signatures in the Appendix, the heating calibration signature matches the characteristic signature for going from no economizer to an economizer quite well. The cooling signatures do not match as well, but after implementing an economizer, $ERROR_{TOT}$ is reduced somewhat to 5.31 GJ/day. It is known that the building has an economizer, but it has frequently been turned off since cooling savings are often offset by additional heating. The calibration signatures (Figure 13.13) show some resemblance to the characteristic signatures for increasing V_{oa}.

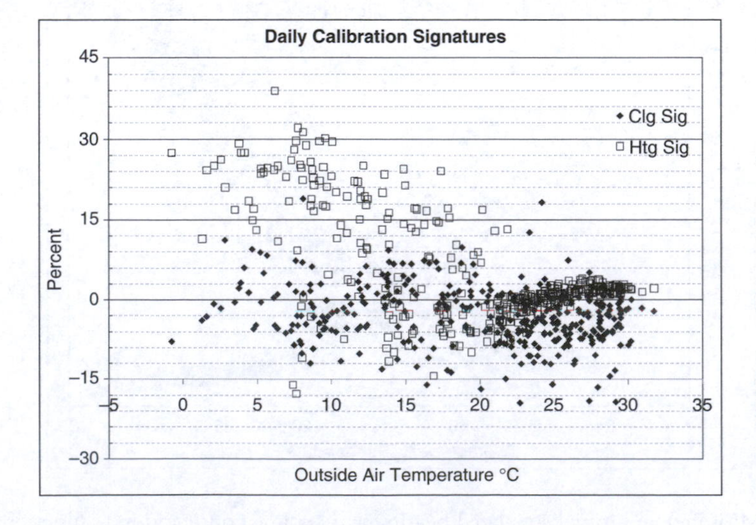

Figure 13.13 Calibration signatures after economizer was added.

V_{oa} is increased from 1.78 to 2.08 L/(s·m²), reducing $ERROR_{TOT}$ slightly to 5.13 GJ/day. Subsequently, increasing interior zone area to 73% and increasing the room temperature to 22.5 °C reduces $ERROR_{TOT}$ to 4.94 GJ/day as shown in Table 13.5.

The final calibration signatures are shown in Figure 13.14 and the simulated heating and cooling values for the calibrated simulation are compared with the measured values in Figure 13.15. The average cooling consumption for this building was 23.25 GJ/day. So the heating RMSE is a relatively high 16.3% of the average cooling consumption. Since it is known that the economizer is not always operated on this building, it is possible that the relatively high scatter in the final calibrated model is due to this irregular operation of the building. It can be seen that the measured heating points in Figure 13.15 for temperatures from approximately 10 °C to 20 °C appear to fall in two groups that may correspond to the economizer operating or not operating.

Summary – a simple process for calibrating simulations

Calibrated simulations can be powerful tools for analyzing building performance and predicting future behavior. Using the process of calibration signatures and characteristic signatures that has been explained, the calibration of a simulation can be greatly simplified. The best way to be successful with this approach is to practice it until familiarity with the procedures and most important parameter variations is achieved. It is also helpful to generate characteristic signatures for the specific building to be calibrated, to better match the climate and system type of the building. In all simulation, the more that is known about actual building conditions being simulated, the more accurately the building can be simulated. But even without extensive knowledge of the building, this process can generally provide acceptable results in a

Table 13.5 Error measures for the calibrated simulation

	GJ/day		*GJ/day*
$RMSE_{Htg}$	3.81	$RMSE_{TOT}$	4.90
$RMSE_{Clg}$	3.08	MBE_{TOT}	0.61
MBE_{Htg}	0.49	$ERROR_{TOT}$	4.94
MBE_{Clg}	0.11		

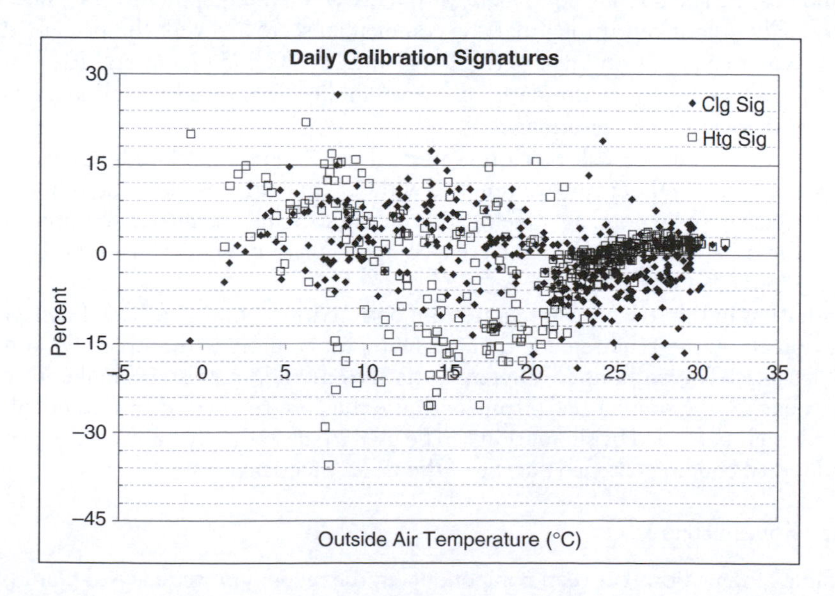

Figure 13.14 Final calibration signatures for Example 2.

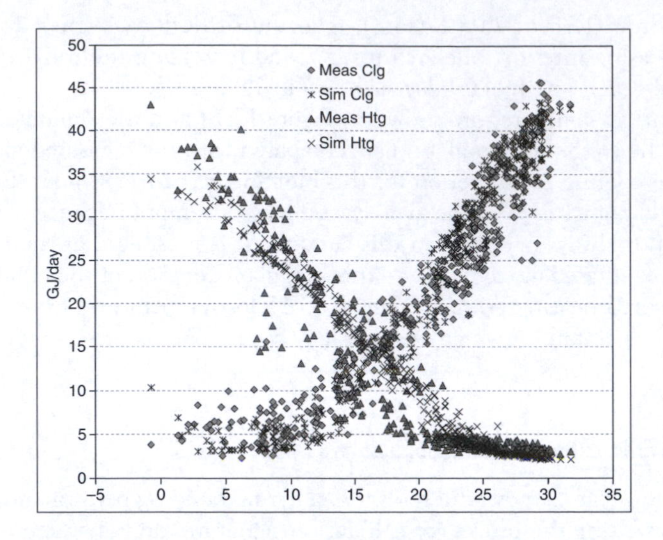

Figure 13.15 Measured and calibrated simulated heating and cooling consumption as functions of average daily temperature.

reasonable amount of time when done correctly. It is therefore useful to become acquainted with the procedures and practiced in the application of the process.

Using calibrated simulation for practical optimization of operational performance

Optimizing operational performance of a building can be defined in a number of different ways. The fundamental approach of the simulator is to minimize the loads in the building, minimize waste in the secondary system, and then optimize plant performance. Operational measures in the residential sector tend to emphasize moving interior temperatures to the edges of the comfort zone and turning off lights and appliances when not needed. Retrofits tend to emphasize improving envelope performance by adding insulation and replacing inefficient lighting, appliances, and heating and cooling equipment with more efficient equipment. The same approach has historically been taken throughout the commercial sector as well; the primary difference has been greater emphasis on improved lighting and HVAC equipment and less emphasis on envelope retrofits since the performance of these buildings tends to be dominated by internal gains rather than envelope performance.

Within this chapter, we consider only internal gain dominated buildings. In many such buildings, energy waste that can be eliminated without changing the equipment in the building primarily occurs in the secondary systems and the plant. Some occurs due to improper control settings while other waste is due to defective components that compromise efficiency without causing obvious comfort problems.

Optimization in the strict sense requires the optimization of an objective function. Energy use constrained by the requirement that we maintain comfort is the objective function we wish to optimize. The millions of different variations in building systems make it impractical (at least today) to use simulation to truly optimize the performance of an individual building. However, insights afforded by simulation can be used to substantially improve the performance of an individual building – hence our use of the term "practical optimization."

Key system parameters

The complex interactions between components in the secondary systems and plant of a building make it difficult to understand the impact of changing a single operating set-point or control schedule within a system. Hence it is important to identify the key parameters that significantly impact system energy use and efficiency. Fortunately, the same system parameters that are most valuable when calibrating a simulation are likewise useful in the practical optimization of system operation. For example, optimizing the minimum flow settings in one laboratory office building was the principle measure implemented to reduce the HVAC consumption in the building by over 50%.

Just as we need building specific characteristic signatures to best calibrate, these signatures can be used to identify the most important variables that should be examined when improving operation.

Detection of faulty building operation

The simplest detection of faulty operation might be comparison of a calibrated simulation with a design simulation. This might immediately show significant differences between the design system settings and apparent operation. After verifying the differences with a field visit, the settings may be changed and operation improved.

It is critical to note that I have said "apparent operation." A calibrated simulation should never be taken as truth without verifying differences in the field. A simulation model never models every facet of a building's operation so there are always deviations between predicted behavior and actual behavior, even with the best calibrated simulation. There are often multiple sets of input variables that will give an equally close approximation to a building's performance – sometimes with a large difference(s) in one or more parameters.

More typically, the calibrated simulation, after field verification of major parameters may be used to predict the savings that are expected from different changes in control settings and assist the commissioning professional or building staff in selecting the changes they wish to make.

Case study: off-line simulation applied to existing building commissioning

The following case study illustrates the use of off-line simulation in the process of commissioning an existing building. However, it also serves to illustrate the diagnostic capability of simulation, whether used off-line or on-line. It is taken from Claridge and Liu (2004).

The Basic Research Building (BRB) at M. D. Anderson (MDA) Cancer Center is a seven-story building with 11,400 m^2 gross floor area, which includes 8,600 m^2 for the laboratory and office section, 1900 m^2 for a library, and 900 m^2 for mechanical rooms and other purposes. The HVAC systems operate 24 hours per day.

Four single duct constant volume air handling units (AHUs) provide cooling and heating to the laboratory and office section. The design airflow rate is 71,000 L/s with 100% outside air. Figure 13.16 presents the schematic diagram of a typical AHU. The pre-heat deck set point is

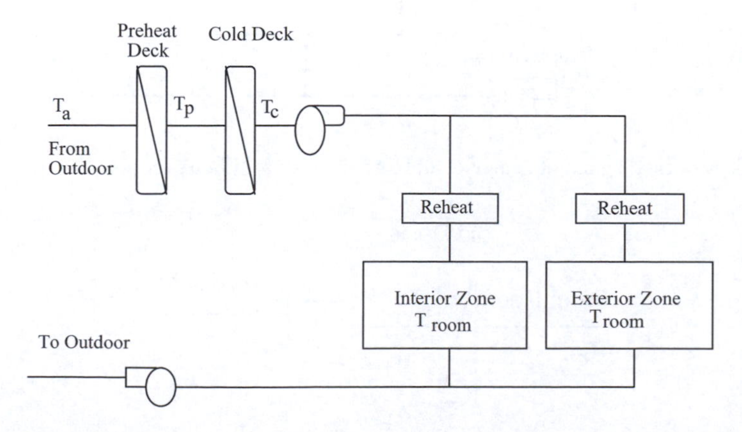

Figure 13.16 Schematic diagram of single duct air handling units.

12.8 °C. If the outside air temperature is below 12.8 °C, the pre-heat coil warms the air temperature to 12.8 °C. If the outside air temperature is higher than 12.8 °C, the pre-heat valve is closed. The cold deck temperature is set at 12.8 °C. The room temperature is controlled using reheat. If the room temperature is below the set point, which varies from 22.2 °C to 23.9 °C from room to room, the reheat coil is turned on to maintain the room temperature.

In addition to the single duct system that serves most of the building, there is one dual duct constant volume air handling unit, which provides cooling and heating to the library section. The design airflow is 12,700 L/s with 50% outside air intake. Figure 13.17 presents the schematic diagram of the dual duct air handling unit for the library section. The cold deck set point is 12.8 °C. The hot deck set point varies from 29.4 °C to 43.3 °C as the outside air temperature decreases from 29.4 °C to 4.4 °C.

Three single duct air handling units provide heating and cooling to mechanical rooms and other spaces. The design airflow is 6,600 L/s with 100% return air. Figure 13.18 presents the schematic diagram of these systems. The cold deck set point is 12.8 °C. If the room temperature is satisfied, the AHUs will be turned off.

Baseline development

AirModel was used to simulate the building heating and cooling energy consumption using simplified building and system models. The building was divided into two parts: the laboratory section, which uses 100% outside air and the library section, which uses 50% outside air. Each part was simplified to two zones: interior and exterior. The design operational schedules were used in the simulation.

Figure 13.20 compares the measured heating and cooling with baseline heating and cooling energy consumption. The baseline energy consumption was simulated using actual Houston

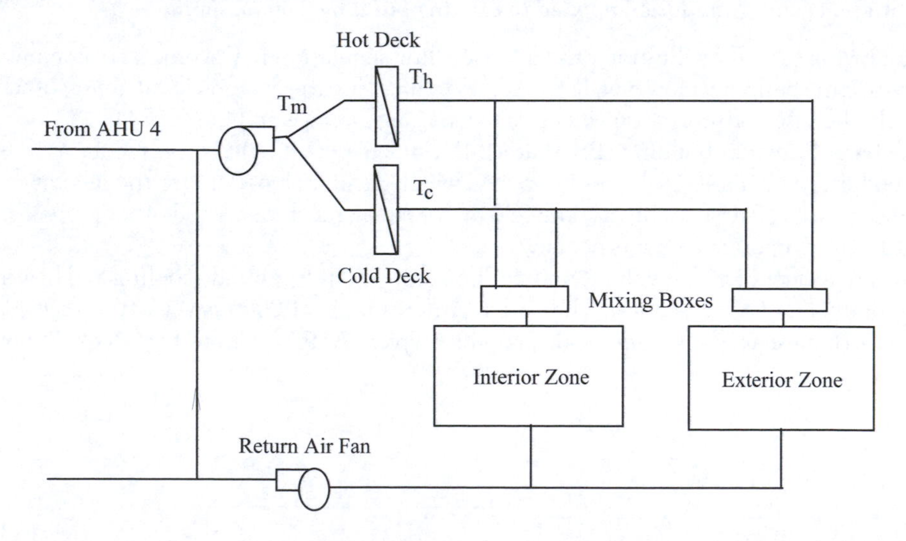

Figure 13.17 Schematic diagram of dual duct air handling unit for library section.

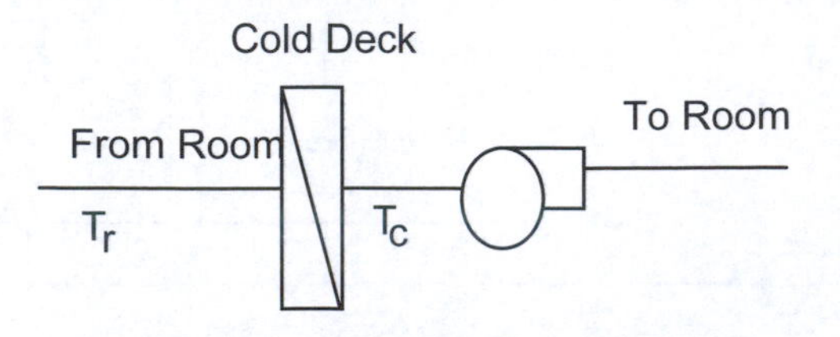

Figure 13.18 Schematic diagram of cooling only single duct air handling unit.

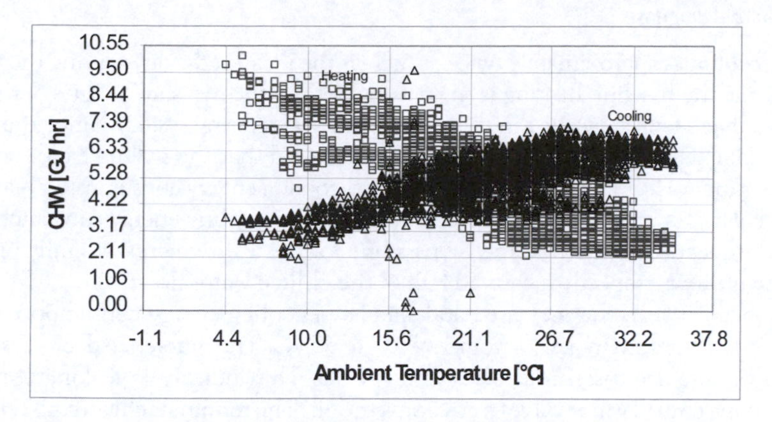

Figure 13.19 Measured hourly heating and cooling energy consumption versus the ambient temperature.

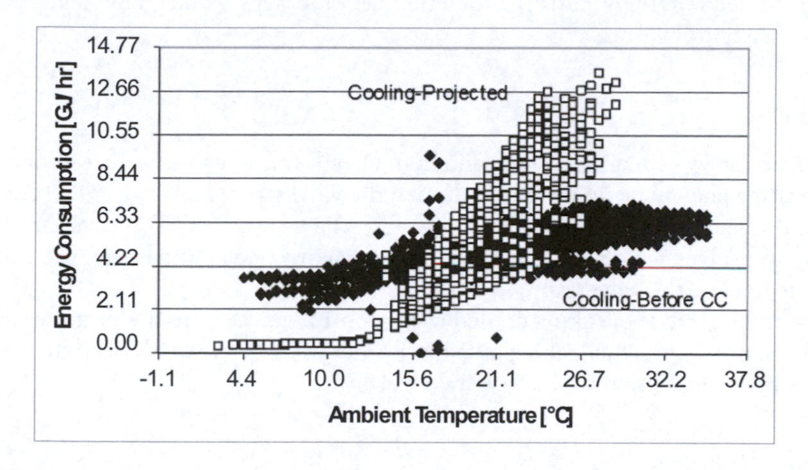

Figure 13.20a Comparison of baseline and measured cooling energy consumption.

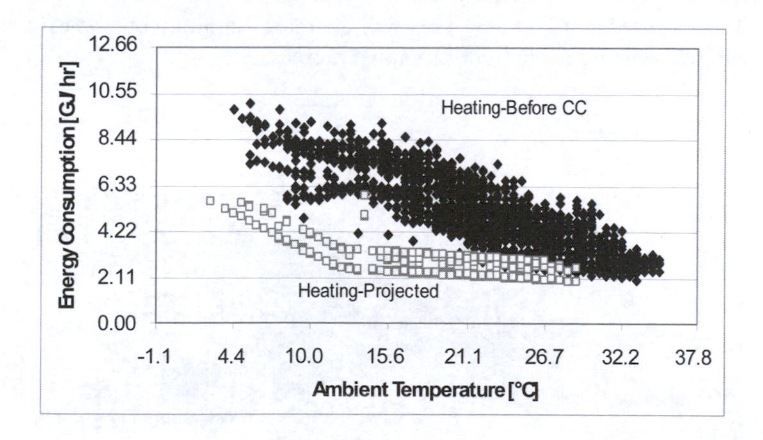

Figure 13.20b Comparison of baseline and measured heating energy consumption.

weather but not the weather data corresponding to the measured energy consumption, since the measured dew point temperature was missing for the measured energy consumption period. The baseline heating is significantly less than the measured heating while the baseline cooling is significantly higher than the measured cooling.

Fault detection and diagnosis

The baseline cooling is approximately twice as high as the measured values during the peak summer period while the baseline heating is slightly lower than the measured values during winter. This indicates that a fault may exist in the cooling energy metering system. Since the measured value is approximately 50% of the baseline, it was suggested that the scaling factor or the engineering conversion was set incorrectly. The measured cooling energy consumption is adjusted by a factor of 2. Figure 13.21 presents the corrected heating and cooling energy consumption.

Later, field inspection found that the by-pass line for the utility meter was fully open. Consequently, the utility meter only measured half of the chilled water flow.

The difference between the measured and simulated cooling energy consumption decreases as the ambient temperature increases from 12.8 °C to 35 °C. The difference decreases when the ambient temperature decreases from 12.8 °C to –1 °C. This indicates a leaking chilled water valve. The leaking chilled water valve over-cooled the air. The terminal reheat-coils reheated the air to maintain room temperature, causing significant waste of heating and cooling energy.

Leaking chilled water flow can arise for a number of reasons. A recommendation was given to inspect the control valves.

Field inspection

A field inspection was conducted and found that: (1) all control valves were less than 3 months old; (2) existing pneumatic lines were used when the valves were replaced; (3) all chilled water valves are normally open with a range of 20 to 55 kPa gauge; and (4) the maximum control pressure to the valves was 35 kPa gauge due to old, leaking pneumatic lines. As a result, it was not possible to close the valves fully.

This confirmed that the leaking chilled water control valves were the primary cause of the poor performance. Fixing the leaking pneumatic lines was expected to reduce the heating and cooling energy consumption to the baseline level.

Operational optimization

It was suggested to reset the supply air temperature from 13.9 °C to 15 °C as the outside air temperature decreases 38 °C to 15 °C. This will decrease simultaneous heating and cooling significantly with a moderate room humidity level increase.

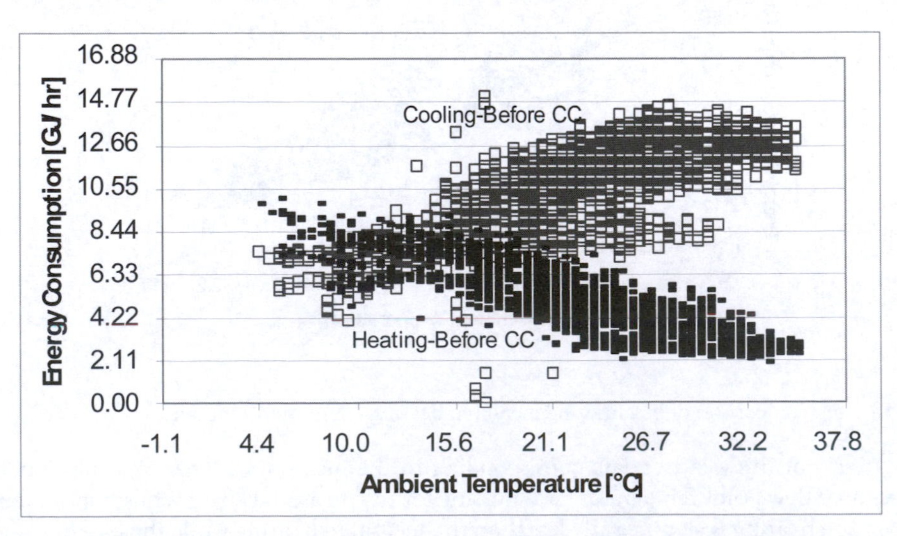

Figure 13.21 Measured heating and cooling energy consumption after meter correction.

Implementation

The implementation included replacing the pneumatic lines and programming the reset schedule into the building automation system. These changes were made at the same time.

Figure 13.22 compares the measured chilled water consumption before fault detection and diagnosis, the chilled water consumption after fixing the pneumatic lines and implementing the optimized schedule, and the simulated optimal consumption. Figure 13.28 provides the same comparisons for heating water.

The measured annual cooling energy savings are 30,500 GJ/yr, and heating energy savings are 17,050 GJ/yr. The total annual cost savings are $369,000/yr, which includes heating savings of $129,000 and cooling savings of $240,000.

When the ambient temperature is lower than 10 °C, the measured energy consumption agrees with the simulated energy consumption. When the ambient temperature is higher than 10 °C, the measured energy consumption is somewhat higher than the simulated energy consumption. It appears that the building has other problems such as leaking reheat valves and excessive airflow.

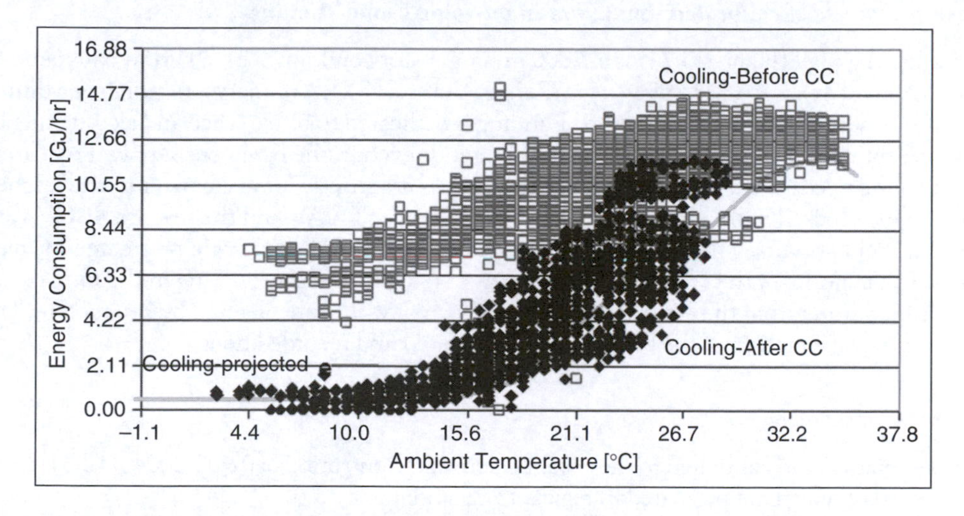

Figure 13.22 Comparison of measured cooling energy consumption before and after repair of leaky pneumatic lines and implementation of optimal reset schedule.

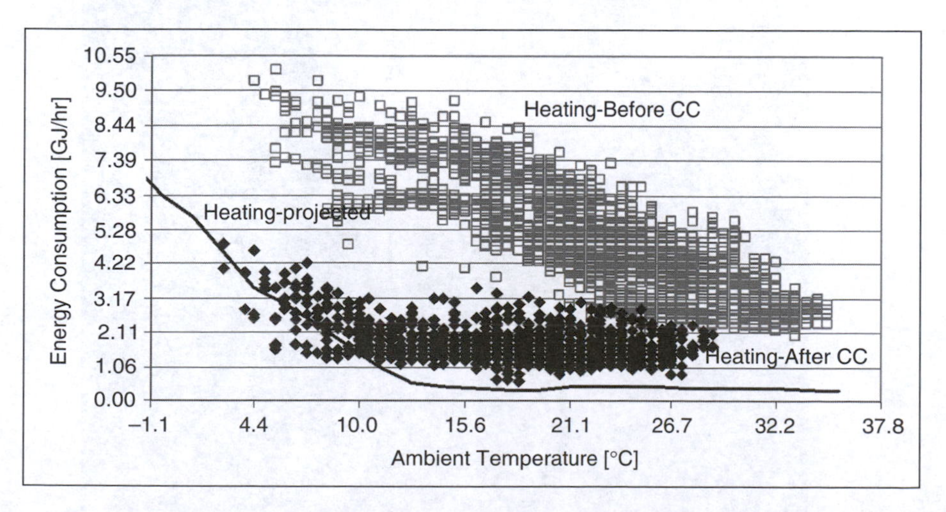

Figure 13.23 Comparison of measured heating energy consumption before and after repair of leaky pneumatic lines and implementation of optimal reset schedule.

Off-line simulation case study summary

The simulation effectively identified the metering and valve leakage problems successfully in this case. Re-heat valve leakage problems and excessive airflow problems were identified after fixing the leaking chilled water valve. This suggests that on-line fault detection should be an on-going process.

The simulation identified HVAC component problems and was used to develop optimized HVAC operation and control schedules in this case study. It further indicated that building thermal energy consumption would be reduced by 23% or $191,200/yr by using the optimized operating schedules in this building. The measured energy savings after implementing the optimized schedules were consistent with the simulated savings.

These results, coupled with similar experience in other buildings strongly support the value of simulation in the commissioning process, both for off-line and on-line applications as a diagnostic and optimization tool for building operation.

Case study: use of calibrated simulation in on-going commissioning

Sbisa Dining Hall (Figure 13.24) is a 7,600 m^2 single-story building with a partial basement on the campus of Texas A&M University in College Station, TX. Its primary function is as a dining facility, although a small fraction of the building is dedicated to office space and a convenience store. A total of 19 AHUs, 12 on the main floor and seven in the basement supply the heating and cooling needs of the building. Four of these units are strictly for makeup air in the kitchen and are interlocked with individual fume hoods, two are SDVAV and the rest are SDCV with terminal reheat units. Three constant volume dedicated OAHUs provide pretreated makeup air for the majority of the AHUs. Thermal energy is supplied to the building in the form of hot and chilled water from the central utility plant, both of which are metered by Btu meters. The electricity consumption of the building is also metered and recorded hourly.

Case-study calibration

The simulation was calibrated to the baseline consumption period of 02/02/2004 – 12/31/2004, the results of which are presented in Figure 13.25 and Table 13.6.

Figure 13.24 Sbisa Dining Hall.

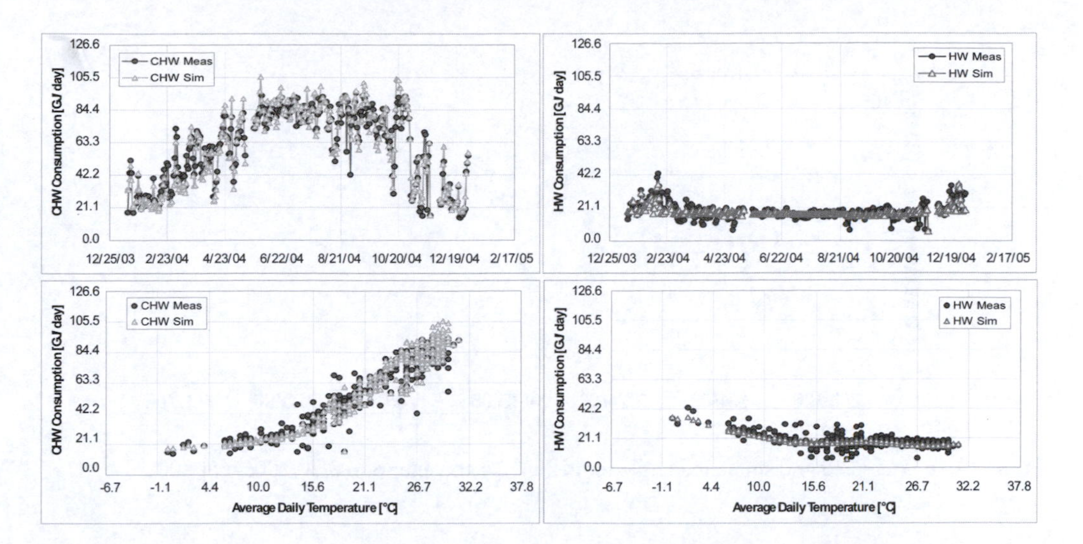

Figure 13.25 Time series and OA temperature dependent plots of the measured and simulated cooling and heating consumption for the period used to calibrate the Sbisa Dining Hall simulation.

Table 13.6 Calibration statistics of the Sbisa Dining Hall

	RMSE	*MBE*	*Max*	*Average*	*CV-RSME*	
CHW:	7.34	0.63	95.46	59.00	12.45%	GJ/day
HW:	3.54	–0.30	43.04	18.74	18.92%	GJ/day

Initial Findings

During the early spring of 2006 an unexpected increase in cooling energy consumption was detected, at an additional cost to the campus of approximately $950/week. The fault was determined to be the result of excessively low discharge air temperatures in two of the three dedicated OAHUs in the building. Several figures are presented below to explain how, through use of the calibrated simulation, the conclusion was drawn that this fault existed. Also described is how the simulation, along with the addition of some key trended control system data, assisted in narrowing down the likely cause of the fault. Note that all of the plotted data are presented as daily averages.

Fault Detection

Figure 13.26 displays the measured and simulated chilled water energy consumption for the period of 2/20/2006–6/4/2006, which spans the period during which this cooling energy consumption fault appeared.

The significance of the daily deviation seen in Figure 13.26 may be difficult to gage unless it is also accumulated and compared with past performance. The cumulative effect of this increased consumption can be seen in Figure 13.27, where the daily difference between the measured and simulated consumption is multiplied by an estimated cooling energy cost of $12.74/GJ, and added to the previous day's total. This figure shows that costs had increased by nearly $9,500 over those expected for the 10 week period of time (3/27/2006–6/4/2006).

Figure 13.28 is a plot of the difference (Meas – Sim), as a percent of the maximum consumption during the baseline period (2004), which was 95.5 GJ/day. This means that a 10% difference in Figure 13.28 corresponds to a measured cooling energy consumption of 10% of 95.5 or

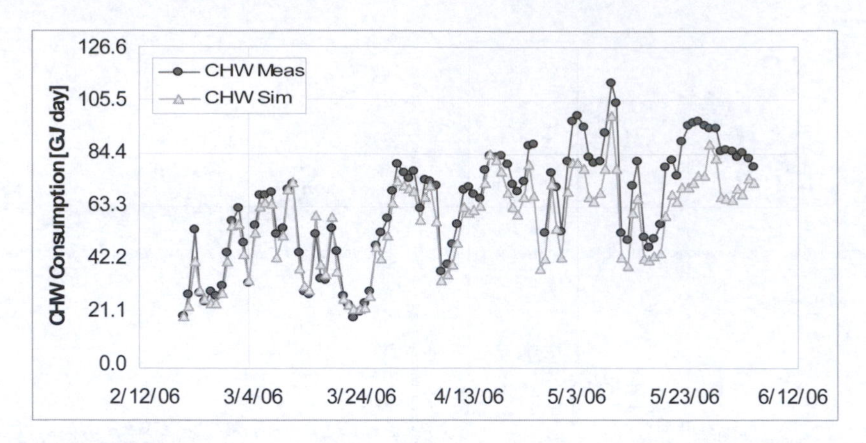

Figure 13.26 Measured and simulated chilled water energy consumption 2/20/2006–6/4/2006.

9.55 GJ/day greater than that which was simulated. This error plot shows that the seven day moving average of the percent difference had been within the +/–10% range for nearly the entire period of the plot (more than 2 years) except for the period starting around April 25th, 2006, and continuing through the beginning of June.

Diagnosis

To assist in identifying the cause of the fault, Figure 13.29 compares the WB_{Elec} consumption for this period of increased cooling energy consumption, alongside the WB_{Elec} from the same dates in 2005. The similarity of the two series in Figure 13.29 indicates that the cause of this increase in cooling energy is not likely related to an increased electric load in the building. Additionally, in referring back to Figure 13.27, it can be seen that the measured HW heating energy consumption did not vary from the simulation nearly as much as the CHW consumption, and actually was consistently slightly less than the simulated consumption since the beginning of 2006. This revealed that the increased cooling energy was not caused by an increase in heating energy consumption.

Further investigation into trended control data points has led to the discovery of exceptionally low discharge air temperatures (extended periods at < 10 °C) in two of the three OAHUs

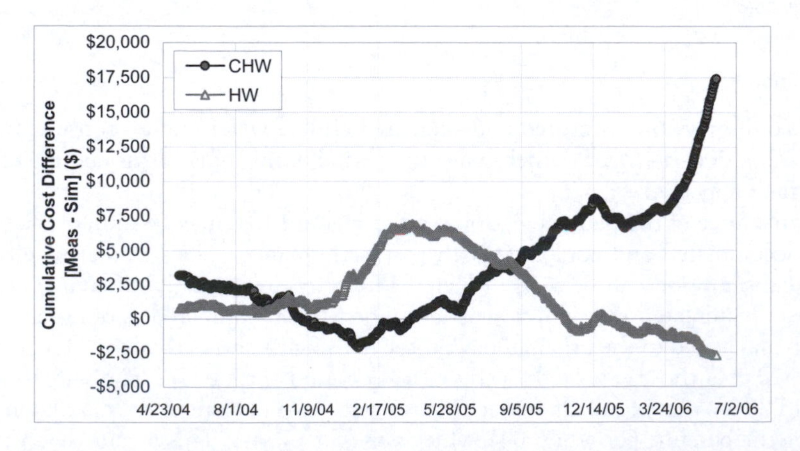

Figure 13.27 Cumulative cost difference (assuming $12.74 and $16.73/GJ for CHW and HW respectively).

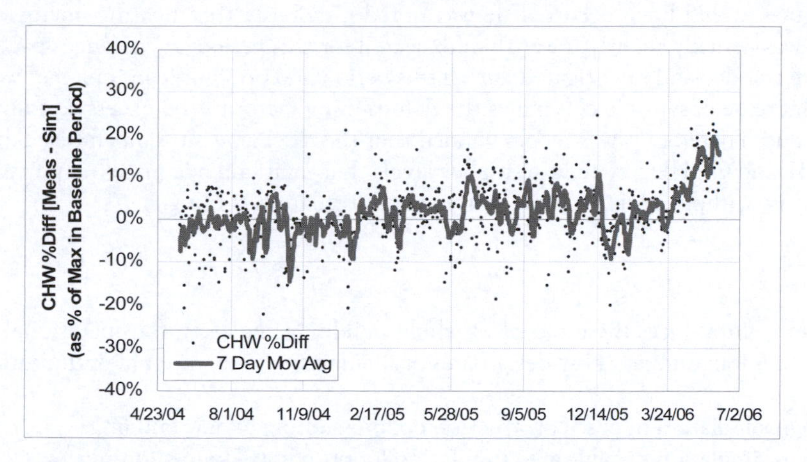

Figure 13.28 CHW % difference plot (10% error → (Meas – Sim)) of 9.55 GJ/day.

(OAHUB1 and OAHU2) in the building. From analyses of the trend data it can be determined that these same operating conditions have been in place for an extended period of time (at least since November 2005), and therefore the change in discharge air temperature of these units does not coincide directly with the identified period of increased cooling consumption. But this does not mean that the two are unrelated. The average ambient dew point temperature in College Station does not tend to consistently rise above 10 °C until late March/early April (Figure 13.30). This means that the low discharge air temperatures for the outside air handling units would not result in a significant increase in latent cooling loads until this spring time period had arrived.

Case study conclusion

This investigation surmised that the OAHU's discharge air temperature control set points were lowered in late September/early October 2005 during a time when the outside dew point had already fallen well below its summer highs. Although the trended data is not available from that time period to prove this conjecture, there are other reasons to believe this is a latent cooling issue. First, if it was an increase in sensible cooling, either hot water reheat energy would have increased to compensate for the increased cooling, or a noticeable drop in space

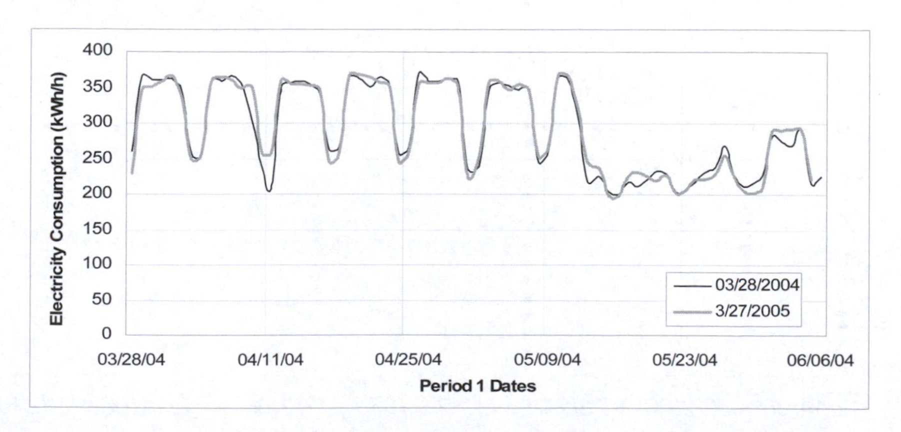

Figure 13.29 Electric consumption for two 10-week periods, one year apart; start dates of periods are 3/28/2005 and 3/27/2006.

temperatures would have occurred. It was noted previously that heating energy use in the building was actually slightly lower than expected for this period. Also, since space temperatures were trended and investigated during this period and no significant changes were identified, an increase of sensible cooling as the determining factor in the excess consumption can be ruled out. Therefore it was recommended that the discharge air temperature schedules for OAHUB1 and OAHU2 be reset to higher levels that maintain adequate humidity control in the building, without adding a penalty from excessive latent cooling.

Synopsis

This chapter introduces the concept of whole building simulation for operational operation and describes four different post-design uses of simulation for operational optimization:

- Design simulation in post-construction commissioning of new buildings
- Design simulation or calibrated simulation for on-going commissioning
- Calibrated simulation for existing building commissioning
- Simulation to evaluate new control code.

The majority of the chapter is devoted to the use of calibrated simulation in the process of commissioning existing buildings and the use of calibrated simulation in the follow-up on-going commissioning process. The existing building commissioning (EBCx) process is introduced and a detailed description of a process for calibrating simulations that is suitable for use in the EBCx process follows. The calibration process is further clarified with two case study examples.

A case study in which off-line simulation was used in the EBCx process to aid in diagnosis of operating problems and predict the savings potential of these problems is presented followed by a case study in which on-line calibrated simulation was used to track the performance of a building and detect the appearance of an operational problem and diagnose a change in the outside air pre-treatment temperature as the cause of the problem is presented. Both of these uses are beginning to find application in the field today and are likely to become increasingly important in the next decade.

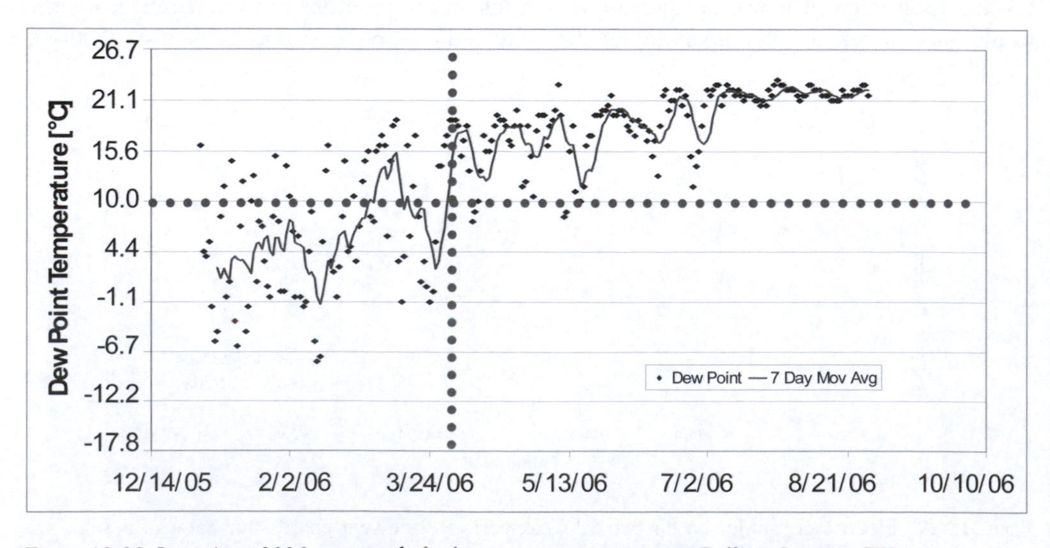

Figure 13.30 Jan.–Aug. 2006 average daily dew point temperature in College Station, TX.

Acknowledgements

Portions of this chapter are based on the work of Subtask D2 of the International Energy Agency ECBCS Annex 40 "Commissioning of Buildings and HVAC Systems for Improved Energy Performance." Subtask D2 is titled Use of Whole Building Simulation in Commissioning. Regular participants of the subtask meetings included Oliver Baumann (Germany), David Claridge, Chair (USA), Jorgen Eriksson (Sweden), Nestor Fonseca, Johnny Holst (Norway), Per Isaakson (Sweden), Timo Kalema (Finland), Hannu Keranen (Finland), Jean Lebrun (Belgium), Noriyasu Sagara (Japan), Makato Tsubaki (Japan), and Sheng Wei Wang (Hong Kong). Eighteen others attended at least one meeting. While this chapter has been written by the author, it reflects contributions from all of these individuals as well as the contributions specifically referenced.

Note

1 Continuous Commissioning and CC are registered trademarks of the Texas Engineering Experiment Station.

References

Adam, C., Andre, P., Aparecida Silva, C., Hannay, J. and Lebrun, J., (2004) "Commissioning Oriented Building Loads Calculations. Application to the CA-MET Building in Namur (Belgium)," in *Proceedings of the International Conference for Enhanced Building Operations*, Paris, France.

Andre, P., Cuevas, C., Lacote, P. and Lebrun, J. (2003) "Re-Commissioning of the CAMET HVAC System: A Successful Case Study," in *Proceedings of the International Conference for Enhanced Building Operations*, Berkeley, CA.

Andre, P., Fonseca, N., Hannay, J., Hannay-Silva, C., Lacote, P. and Lebrun, J. (2004) "Re-Commissioning of a VAV Air-Distribution System," IEA Annex 40 Working Paper A40-B2-M6-B-FUL-ULG-01. Online. Available at: http://www.commissioning-hvac.org/files/doc/A40-B2-M6-B-FUL-ULG-01.pdf.

ASHRAE (2002) *Guideline 14–2002 Measurement of Energy and Demand Savings*, Atlanta, GA: American Society of Heating, Refrigerating and Air-Conditioning Engineers.

Baumann, O. (2003) "Design and Optimization of Control Strategies and Parameters by Building and System Simulation," in *Proceedings of the International Conference for Enhanced Building Operations*, Berkeley, CA.

Bynum, J., Claridge, D.E., Turner, W.D., Deng, S. and Wei, G. (2008) "The Cost-Effectiveness of Continuous Commissioning® Over the Past Ten Years," *Proc. 8th Int. Conf. for Enhanced Building Operation*, Berlin, Germany, Oct. 22–24, Session 6.3, CD.

Carling, P., Isakson, P., Blomberg, P. and Eriksson, J. (2004a) "Simulation-Aided Commissioning of the Katsan Building," IEA Annex 40 Working Paper A40-D-M5-SWE-AF/KTH. Online. Available at: http://www.commissioning-hvac.org/files/doc/A40-D-M5-SWE-ÅFKTH.pdf.

Carling, P., Isakson, P., Blomberg, P. and Eriksson, J. (2004b) "Experiences from Evaluation of the HVAC Performance in the Katsan Building Based on Calibrated Whole building simulation and Extensive Trending," IEA Annex 40 Working Paper A40-D-M6-SWE-AF/KTH-2. Online. Available at: http://www.commissioning-hvac.org/files/doc/A40-D-M6-SWE-ÅF_KTH-2.pdf.

Chen, H., Deng, S., Bruner, H.L., Claridge, D.E. and Turner, W.D. (2002), "Continuous Commissioning^{SM} Results Verification and Follow-up for an Institutional Building – A Case Study," in *Proceedings of the 13th Symposium on Improving Building Systems in Hot and Humid Climates*, Houston, TX, pp. 87–95.

Claridge, D.E. (ed.) (2004) "Using Models in Commissioning – Using Simulation Models at the Building Level," in Visier, J.C., (ed.) "Commissioning Tools for Improved Energy Performance," Final Report of IEA ECBCS Annex 40, International Energy Agency, http://www.commissioning-hvac.org/.

Claridge, D.E. and Liu, M. (2004) "Using Whole Building Simulation Models in Commissioning," IEA Annex 40 Working Paper A40-D2-M4-US-TAMU-02. Online. Available at: http://www.commissioning-hvac.org/files/doc/A40-D2-M4-US-TAMU-02.pdf.

Claridge, D.E., Bensouda, N., Lee, S.U., Wei, G., Heinemeier, K. and Liu, M. (2003) Manual of Procedures for Calibrating Simulations of Building Systems, submitted to the California Energy Commission Public Interest Energy Research Program, Report HPCBS#E5P23T2b. Online. Available at: http://buildings.lbl.gov/hpcbs/pubs/E5P23T2b.pdf.

Claridge, D.E., Turner, W.D., Liu, M., Deng, S., Wei, G., Culp, C., Chen, H. and Cho, S.Y. (2004) "Is Commissioning Once Enough?." *Energy Engineering* 101(4): 7–19.

EVO (2007) *International Performance Measurement and Verification Protocol: Concepts and Options for Determining Energy and Water Savings*, Vol. 1, Efficiency Valuation Organization (EVO) 10000–1.2007, San Francisco, CA, U.S.A. 94121–0518.

Ginestet, S. and Marchio, D. (2004) "IAQ-Op (Indoor Air-Quality in Operation): A Simulation Tool for Retro-Commissioning, On-Going-Commissioning and Fault Simulation," IEA Annex 40 Working Paper A40-D2-M5-F-ELYO-03. Online. Available at: http://www.commissioning-hvac.org/files/doc/A40-D2-M5-F-ELYO-03.pdf.

Keranen, H. and Kalema, T. (2004a) "Commissioning of the IT-Dynamo Building," IEA Annex 40 Working Paper A40-D2-M6-FI-TUT-01. Online. Available at: http://www.commissioning-hvac.org/files/doc/A40-D2-M6-FI-TUT-01.pdf.

Keranen, H. and Kalema, T. (2004b) "Commissioning Educational Building IT-Dynamo," IEA Annex 40 Working Paper A40-D2-M7-FI-TUT-01. Online. Available at: http://www.commissioning-hvac.org/files/doc/A40-D2-M7-FI-TUT-01.pdf.

Knebel, D. (1983) *Simplified Energy Analysis Using the Modified Bin Method*, Atlanta, GA: American Society of Heating, Refrigerating, and Air-Conditioning Engineers.

LBL (2001) *The Reference to EnergyPlus Calculations*, Energy Plus Engineering Document, Lawrence Berkeley National Laboratory.

Liu, M. and Claridge, D.E. (1998) "Use of Calibrated HVAC System Models to Optimize System Operation," *Journal of Solar Energy Engineering* 120(2): 131–138.

Liu, M., Song, L., and Claridge, D.E. (2002a) "Potential of On-Line Simulation for Fault Detection and Diagnosis in Large Commercial Buildings with Built-Up HVAC Systems," IEA Annex 40 Working Paper A40-D2-M4-US-TAMU-01. Online. Available at: http://www.commissioning-hvac.org/files/doc/A40-D2-M4-US-TAMU-01.pdf.

Liu, C., Turner, W.D., Claridge, D., Deng, S. and Bruner, H.L. (2002b) "Results of CC Follow-Up in the G. Rollie White Building," in *Proceedings of the 13th Symposium on Improving Building Systems in Hot and Humid Climates*, Houston, TX, pp. 96–102.

Selkowitz, S.E., Haberl, J.S. and Claridge, D.E. (1992) "Future Directions: Building Technologies and Design Tools," in *Proceedings of the ACEEE 1992 Summer Study on Energy Efficiency in Buildings*, Vol. 3, Washington, D.C., pp. 269–290.

Tsubota, Y. and Kawashima, M. (2004) "Online Building Energy Evaluation System," IEA Annex 40 Working Paper A40-D-M5-J-TEPCO-2-rev1. Online. Available at: http://www.commissioning-hvac.org/files/doc/A40-D-M5-J-TEPCO-2-rev1.pdf.

Turner, W.D., Claridge, D.E., Deng, S., Cho, S., Liu, M., Hagge, T., Darnell, C. and Bruner, H. (2001) "Persistence of Savings Obtained from Continuous Commissioning," in *Proceedings of 9th National Conference on Building Commissioning*, Cherry Hill, NJ.

Wang, S. and Xu, X. (2003) "Hybrid Model for Building Performance Diagnosis and Optimal Control," in *Proceedings of the International Conference for Enhanced Building Operations*, Berkeley, CA.

Wei, G., Liu, M. and Claridge, D.E. (1998) "Signatures of Heating and Cooling Energy Consumption for Typical AHUs," in *Proceedings of the Eleventh Symposium on Improving Building Systems in Hot and Humid Climates*, Fort Worth, Texas, pp. 387–402.

Activities

Consider the four floor building shown in Figure 13.31, with the SW wall in the left portion and the SE wall in the right portion of the picture. The NW and SE faces are 108 m long and consist of four floors, each 4m high, while the NE and SW facing walls are 69 m long. The ground floor is level with the sidewalk in the foreground.

The building is supplied with hot water and chilled water from a central plant and distributes heating and cooling throughout the building using 12 dual-duct variable-volume air handlers with design flows of 19,000 L/s powered by 30 kW fans on variable speed drives. **Calibrate a**

Figure 13.31 Building on the Texas A & M campus that is to be simulated in the 'Activity'.

simulation to the measured daily heating, cooling, and electric consumption data available in the file at http://tmpwebesl.tamu.edu/pub/bldg_data/problem_data.xlsx

This file also contains measured daily average temperature and humidity data corresponding to the measured energy data. In addition, the measured daily pump and fan power for the building is provided, along with a week of typical hourly electrical consumption data.

For your initial simulation you may use the following assumptions:

- The building is located at 30° N latitude
- 14% of the wall area is composed of clear single glazed windows approximately 2m wide, recessed by 1.25 m and fitted with venetian blinds
- The wall U-value is 0.018 W/m^2-K
- The roof U-Value is 0.009 W/m^2-K
- Minimum flow setting is 9,000 L/s for each AHU
- Fixed cooling coil leaving air temperature is 13 °C
- Hot deck set-point is 50 °C
- Exterior zone is 6m wide around the exterior of the building
- No economizer
- Mixed air is preheated to 7 °C if it falls below 7 °C at the preheat coils
- 15% outside air
- Interior space temperature setting is 23 °C.

Use the characteristic signatures for the Harrington Building to assist in your calibration.

Answers

Your calibrated simulation should have an $\text{ERROR}_{\text{TOT}}$ value of 17 GJ/day or less.

Appendix

Characteristic Signatures for DDVAV System in the Harrington Building in College Station, Texas USA.

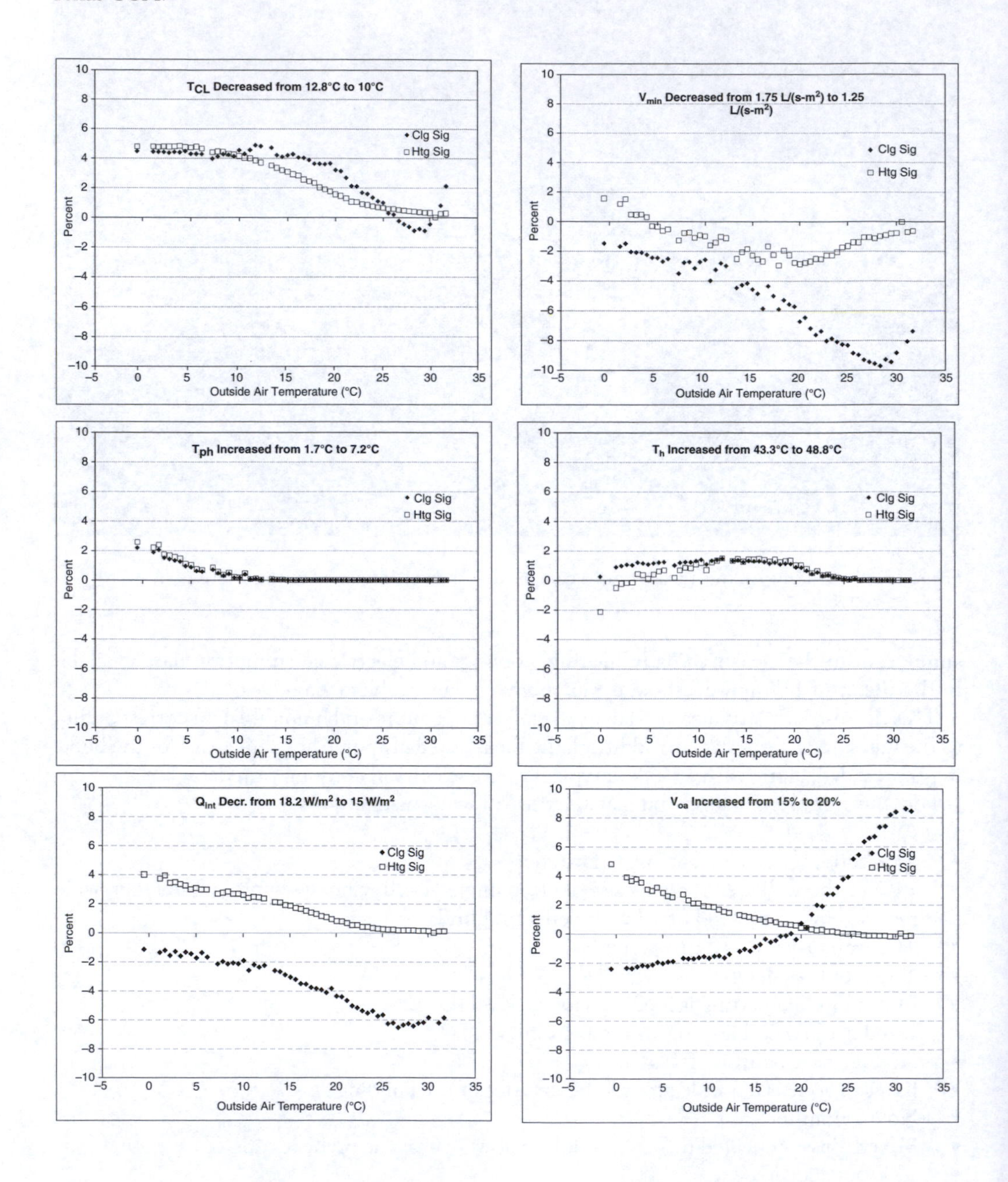

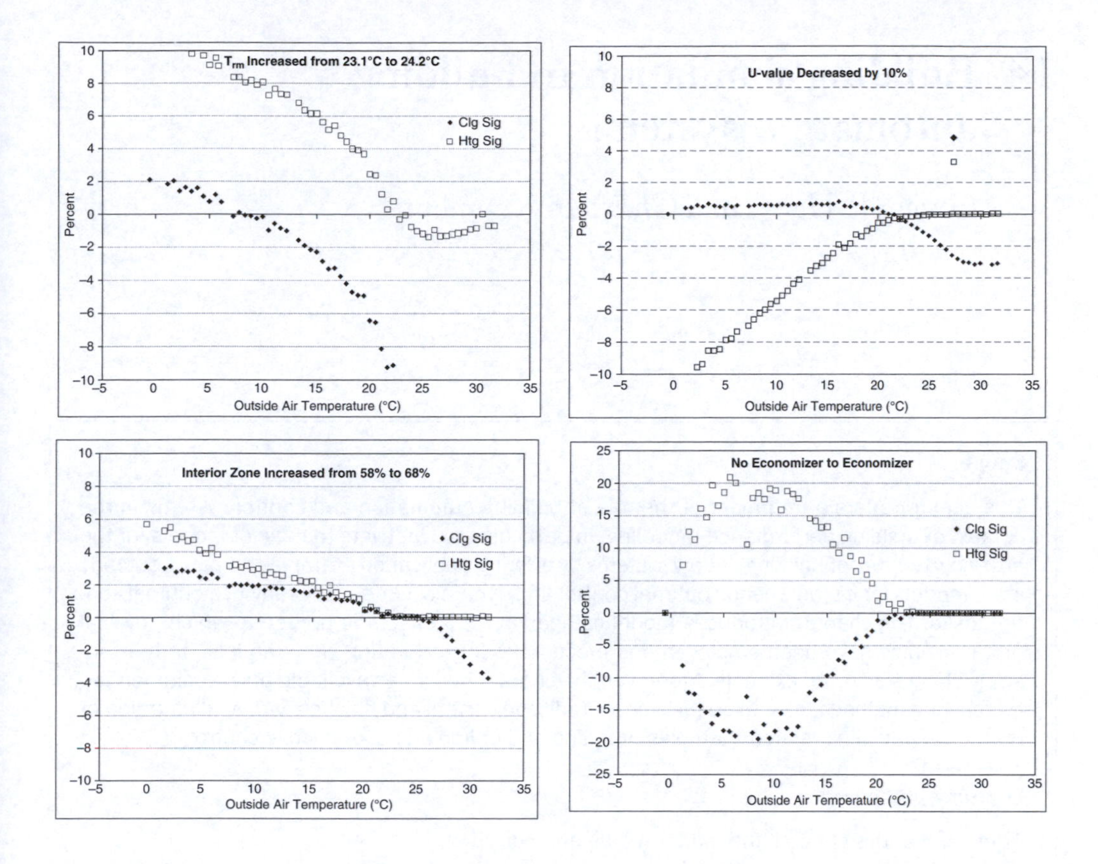

T_rm Increased from 23.1°C to 24.2°C
Clg Sig
Htg Sig
Percent
Outside Air Temperature (°C)
U-value Decreased by 10%
Percent
Outside Air Temperature (°C)
Interior Zone Increased from 58% to 68%
Clg Sig
Htg Sig
Percent
Outside Air Temperature (°C)
No Economizer to Economizer
Clg Sig
Htg Sig
Percent
Outside Air Temperature (°C)

14 Building simulation in building automation systems

Gregor P. Henze and Christian Neumann

Scope

This chapter offers a treatment of the role of building automation and control systems in the context of building performance modeling. In particular, we motivate the use of models for the advanced control of building energy systems by offering application examples employing steady state models for instantaneous optimal control and dynamic models for predictive optimal control tasks. The chapter introduces model categories that reflect the degree to which physical first principles have been employed and procedures for calibrating such models, both extensively illustrated by an example. Moreover, the chapter describes available and recommended forecasting methods to be used for model predictive control and finishes with an illustration of two current building related examples of offline and online model predictive control.

Learning objectives

At the end of this chapter, the reader will be able to:

(i) Differentiate between white box, gray box, and black box models, (ii) Assess the relative model characteristics for any application scenario and select the proper approach, (iii) Apply the model development, calibration, and reduction process to any linear time invariant building energy problem using lumped parameter analysis and state space representation, (iv) Select a range of appropriate forecasting algorithms from the wide spectrum of available approaches, (v) Extrapolate to other model predictive control opportunities based on the examples provided.

Key words

Building automation/optimal control/state space/forecasting/model predictive control

Introduction

Commercial building performance simulation requires not only a description of building geometry, construction materials, energy systems, and equipment but also a characterization of the building utilization through occupancy, lighting, and equipment schedules. In addition, a specification of the building operation is necessary, commonly through a definition of setpoint schedules (such as for temperature and/or relative humidity), HVAC system availability, sequencing of multiple pieces of HVAC equipment (chiller staging), control sequences of several devices for a single control loop (e.g. outside air damper, followed by heating coil, and then by cooling coil for supply air temperature control) and conventional PID control loop parameters such as proportional gain and integral time. While high-end building performance simulation programs such as EnergyPlus and TRNSYS allow for the realistic representation of common control approaches found in commercial buildings today, modeling sophisticated energy management strategies as found in well-run modern facilities (e.g. occupancy-based override of minimum airflow settings on terminal boxes, demand controlled ventilation, or complex reset schedules that relate control setpoints with environmental conditions such as outside air temperature) may be difficult to impossible to model.

The emphasis of this chapter is model-based control, i.e., building control problems where a model of the process under control is required and utilized to guide the operation of either an entire building or selected subsystems. The intriguing facet is that such a model would be used *in* the existing building automation system (BAS) but relevant control aspects *of* the existing BAS must be embedded in the model itself. Thus, the title of this chapter could have been either "Modeling and Simulation *in* Building Automation Systems" or "Modeling and Simulation *of* Building Automation Systems". In relationship to the control of an entire building, model-based control implies a model to be used in the BAS for real-time control as well as a representation of the BAS in the model. In trying to move the reader to the edge of the seat, simply consider how the performance of a building under model-based control could be simulated? The modeled building would include a BAS, which for its control would use another model of the same building! But do not worry, we will not take matters to this extreme, instead, we will focus on relevant applications of model-based control for commercial buildings.

Applications for modeling in building automation systems

The strongest justification for modeling in general and model-based control in particular is that we can exploit such models to guide us toward a performance optimum, which could be lowest energy use, lowest energy cost, carbon emissions, or other performance metrics, while maintaining acceptable indoor environmental conditions. Forty percent of the source energy and 75% of the electricity are consumed in buildings in the U.S. (22% for residences and 18% for commercial buildings[1]) and similar figures can be found for European buildings. Hence, the existing building stock is a bounty for improved operation through optimal control. Moreover, with new construction becoming more and more complex, model-based control in real time using a detailed building energy model holds the promise of serving as a guidepost for continuous fault detection and diagnostics (FDD), maintaining the building operation close to the design intent.

Examples of model-based building control include (a) instantaneous optimal building control without storage effects such as the global setpoint optimization of a combination of chiller, cooling tower and associated air handling units (Brandemuehl and Bradford 1998; Braun *et al.* 1989a, 1989b), operation of building power plants (Van Schijndel 2002), or central chilled water plant control (Ma and Wang 2009), as well as (b) predictive optimal building control with storage effects including building thermal mass control (Braun 2003), ice and chilled water storage control (Henze 2003; Kintner-Meyer and Emery 1995; Massie 2002), and mixed-mode building control (Spindler 2004; Spindler and Norford 2008a, 2008b) as shown in Figure 14.1. The absence of storage allows for the use of time-independent equipment models such as polynomial curve fits or performance maps, while the presence of storage requires the use of time-dependent dynamic models that account for the memory associated with the massive building construction and its thermal response. A comprehensive review of supervisory and optimal control applications has been recently provided by Wang and Ma (2008) and is recommended for further case studies.

Optimal control requires a functional and controllable building energy system after any fundamental faults have been removed. Through optimization, the energy efficiency of the system and/or the indoor climate is improved given the current boundary conditions such as weather, comfort requirements or the presence of occupants. Thus, optimization can be described as the targeted improvements of the building operation subject to the currently imposed boundary conditions. Typical approaches for optimizing building operation include the adjustment of HVAC system operating times, setpoints, and energy sources.

Such model-based control can be accomplished either as:

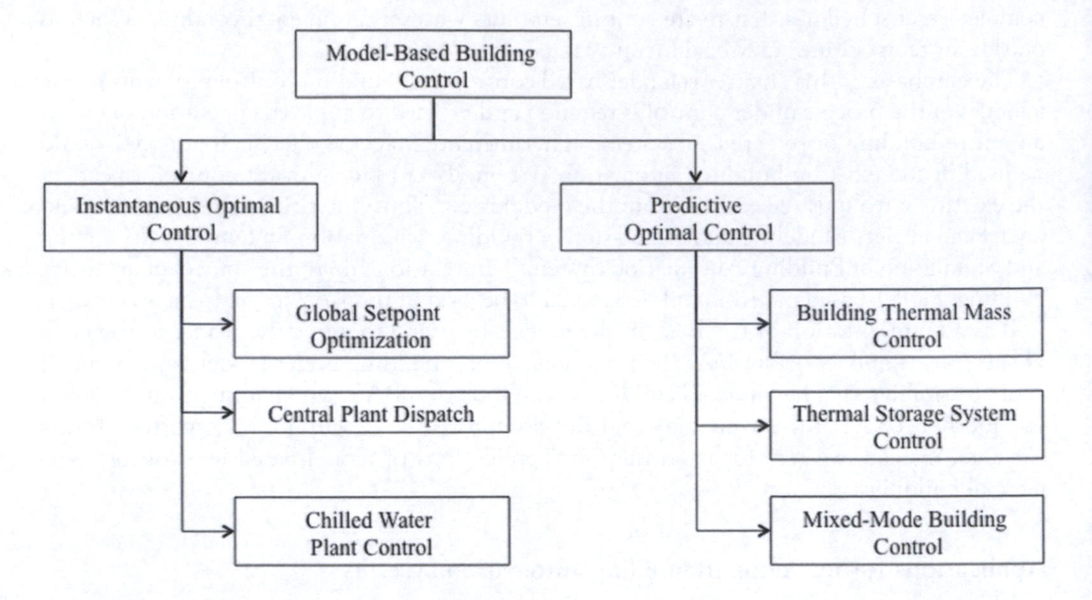

Figure 14.1 Examples of model-based building control.

- Offline optimization: In this case, the building operation is optimized externally, i.e., the building is simulated offline, where certain user profiles and a suitable weather data set are assumed. The optimized control parameters are then passed back (possibly manually or through control rules) and used for the building operation. The aim here is to determine general control strategies for the systems in the building in order to minimize annual energy consumption or costs.
- Online optimization: Model-based predictive and learning controllers aim to optimize the system by accounting for current and future building operation boundary conditions, e.g. in terms of weather, indoor climate or presence of occupants. From this information, the actual operation can be optimized, theoretically every hour. Yet, this type of optimization requires significant effort, and a close coupling with the building automation system.

In the scientific community, numerous approaches have been developed and tested for optimizing the operation of HVAC systems (Wetter 2004). In practice, however, the system operation is often optimized manually. Even during manual optimization, computer models are sometimes called upon for assistance in a trial and error approach (Claridge 2004).

Fault detection and diagnosis (FDD) is a necessary prerequisite for achieving a fault-free operation and is therefore a prerequisite to optimization. As a result, a two-step process prevails, which is schematically shown in Figure 14.2.

The process begins with data collection, which can be acquired through the building automation system (BAS) in the ideal case or with short-term metering equipment. Information about the building in the form of measurement data, setpoints and status information as well as stock data are gathered by means of an extended building energy audit.

Fault detection and diagnosis is part of the first step. The fault detection can take place either manually (e.g. by the operating personnel) or automatically. When a fault is detected during operation, a fault diagnosis is carried out whereby the cause for the fault is investigated. Whether or not a fault is judged to be critical or not depends on the given boundary conditions with respect to energy, costs and indoor climate. These factors are weighted according to their importance from the viewpoint of the building operator or owner.

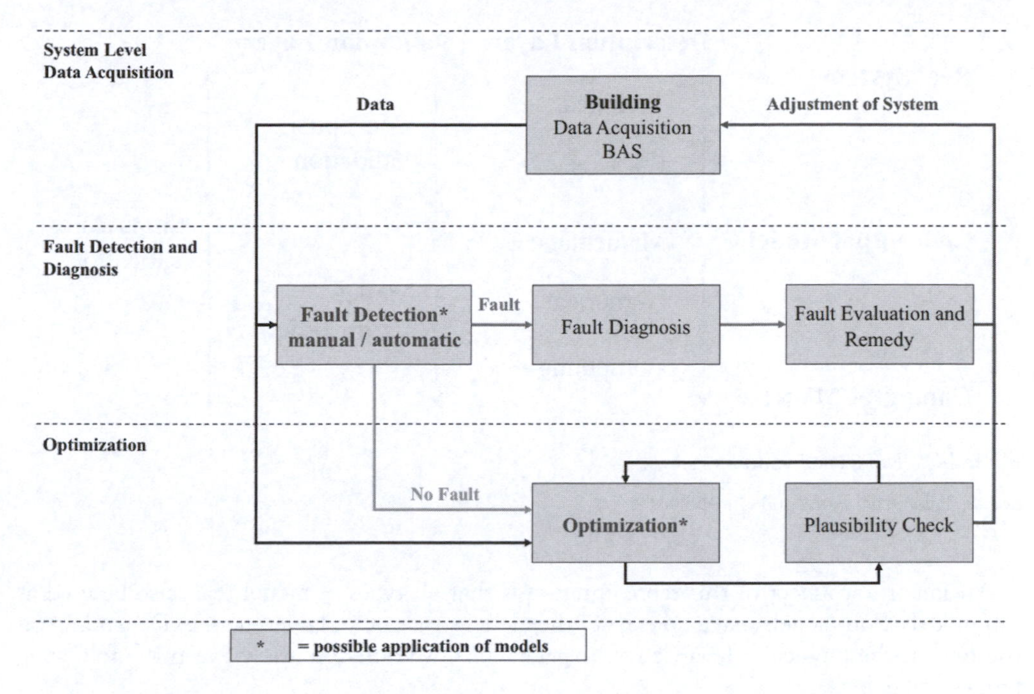

Figure 14.2 Process for fault detection and optimization in building operation.

Once all critical faults are addressed and remedied, the optimization can commence to generate specifications for the setpoints of the system operation. Before these optimized setpoints are fed into the building automation system, a plausibility test should be carried out that determines if all requirements have been considered.

To ensure the energy efficient operation over the entire building lifetime, it is pivotal that this process be performed either continuously or at regular intervals in order to be able to react promptly to any changes. These changes can be intentional (e.g. building renovation, reconstruction of part of the building) as well as unintentional (e.g. degradation of an energy system due to a fault).

Model types, development tools and their implementation

Models are representations of systems, which when properly developed, allow conclusions to be drawn about the system performance under specified boundary conditions. In a suitable model it is attempted to reduce the complexity as much as possible without losing the details and features, which are necessary and important for the system description. Model complexity should be reduced to simplify the model development and its validation as well as to achieve fast calculation time. Models can be considered on different levels: The actual systems are described in terms of mathematical formulas as physical or as conceptual models as shown in Figure 14.3. In most cases, these mathematical formulas must be implemented using numerical methods.

The central elements of a model are the mathematical calculations for connecting inputs and outputs as well as parameters that are constituent parts of the model structure. The representation of the inputs and outputs and the parameters at the boundary of the model shown in Figure 14.4 indicates that external forces influence these. Input variables and parameters are specified by analysts or taken from other models or processes. In the same way, the output data can be used for other models or for evaluation by the analysts. The generic structure, on the other hand, remains unchanged.

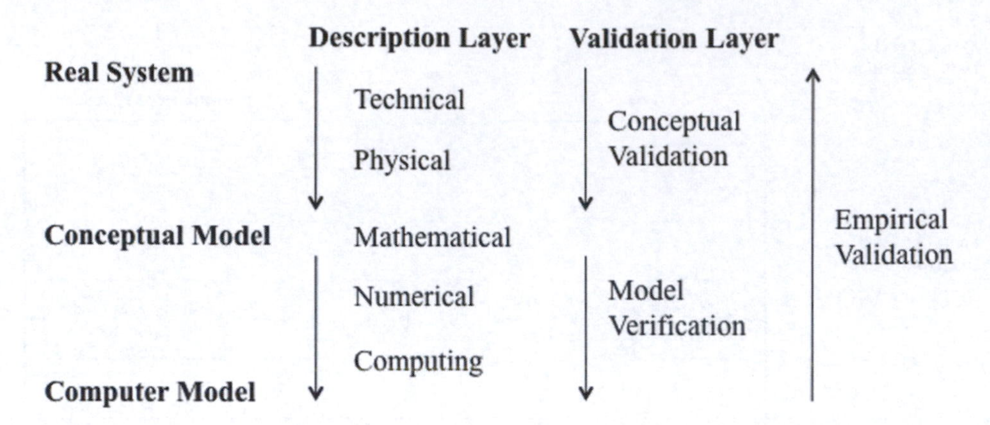

Figure 14.3 Model and validation levels.

Source: Palomo del Barrio and Guyon 2002.

An important aspect of this representation is that all types of models can also be used as sub-models. Sub-models make up parts of the model structure in larger models. Depending on the type or the physical relevance of the parameters, we differentiate between the following types of models:

- White Box Model: Physical model with exclusively physically meaningful parameters (internal structure of the process is modeled);

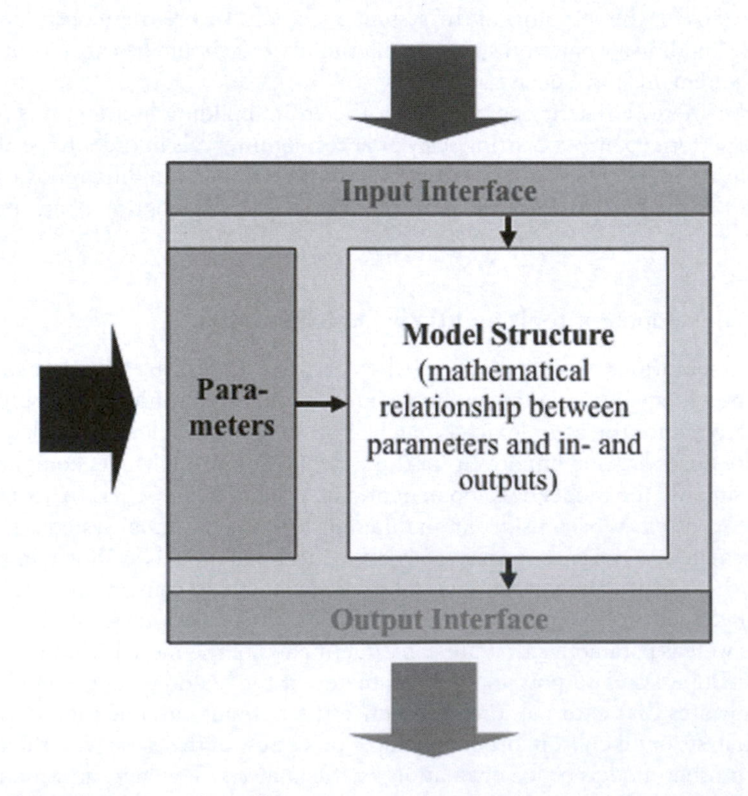

Figure 14.4 General structure of a model.

- Black Box Model: Non-physical model (e.g. statistical/stochastic) without physically relevant parameters (internal structure is not modeled);
- Gray Box Model: Model form combining both the white and black box models.

Before the individual types of models are described in more detail, the calibration of models is presented first in the following section.

Calibration of models

Good building energy models can provide parameter values for the expected or optimized operation of an overall building or a specific building energy system; these calculated values are used for comparison with the currently implemented ones and are thus of special interest for fault detection and optimization.

In general, it is necessary that the parameters of the model be fitted to the actual physical system. This process is called parameter estimation or model calibration. In creating the model, it can be differentiated between two different types of parameter specification (ASHRAE 2009):

- *Forward modeling:* In forward modeling, it is assumed that all of the necessary model parameters are adequately known so that a realistic model of the actual system can be created. For most of the models, this assumption is valid for only some of the parameters. Therefore, this type of modeling is limited, for the most part, for use in the design phase of new systems. The term "forward" here shall imply that there is no coupling to measurements. Most of the models that fall under the forward modeling category are based on fundamental principles and therefore require parameters like material characteristics, which tend to be known with relatively high confidence.
- *Inverse modeling:* In contrast to forward modeling, some or all of the parameters in inverse modeling are estimated using measurements of the input and output quantities of the system. These estimates can be performed manually or automatically. Here the aim is to fit the calculated parameters to the measured values of the system in the best possible way, a process called model calibration. As a general rule, only simple models are suitable for determining the parameters using the inverse modeling procedure. This is especially true for models where the parameters are estimated in an automated procedure. Here so-called black box models (that do not contain any physical parameters) are used.

One often finds models calibrated with a combination of the above-named parameter estimation methods. Many models contain parameters that can be determined with sufficient accuracy based on the knowhow of analysts (forward) as well as parameters that are determined using measurement data (inverse). Typical examples for parameters that are suitable for forward modeling are the thermal conductivity of building components such as the external wall or the roof. An example for a parameter that is suitable for inverse modeling is the effective building mass.

In order to make reliable prognoses about the behavior of real systems, it is imperative to adequately calibrate models that are to be used for control or FDD. As a result, inverse modeling is always chosen (at least for parts of the modeling) for models that are used in the area of fault detection and optimization. Certainly, the degree of the automation in the estimation of the parameters can vary greatly depending on the model and the available measurement data.

Figure 14.5 shows a schematic of the calibration of a model. For the calibration, a training data set is used, which contains the measured input and output quantities of the physical system over a given time period. The model parameters must now be adjusted so that the model can replicate the relationship between the input and output quantities of the training data set with sufficient accuracy.

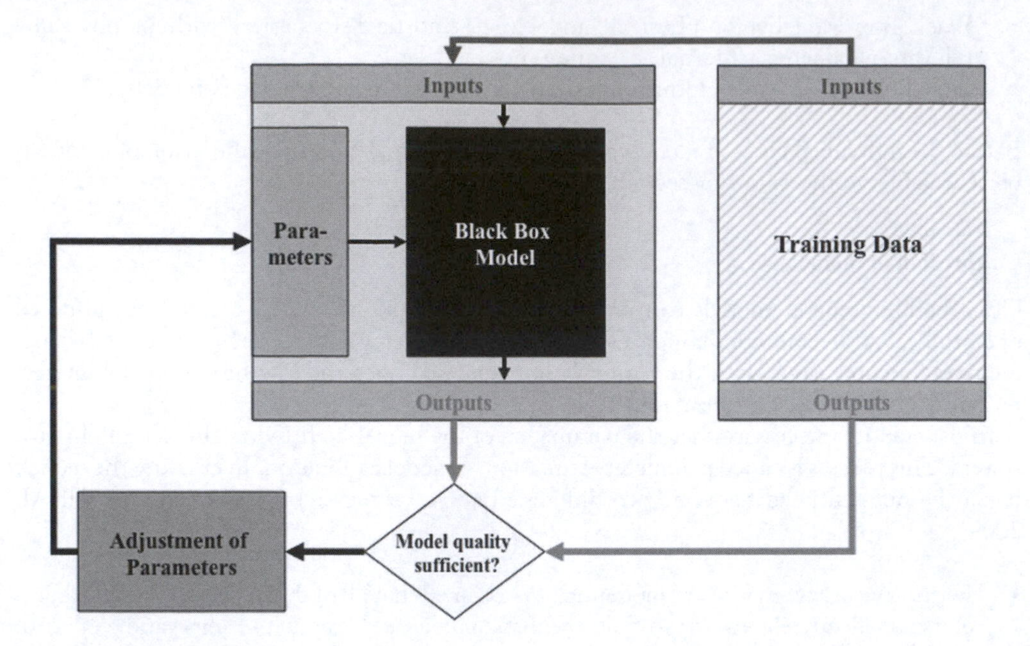

Figure 14.5 Schematic of model calibration.

As a rule, calibration is an iterative process by which the output variables of the system can be calculated with the assistance of the model structure, the measured inputs and the known, or estimated, parameters. The calculated outputs of the model are compared to those quantities in the training data set. If significant variations occur, the parameters must be fitted appropriately and the system output then newly calculated. If the discrepancy is small enough and thus the model output acceptable, then this process can be aborted. In some cases, the model structure may have to be adapted, a step which must be performed manually in most cases. Often an additional validation data set is used to see how the calibrated performs on previously unseen data. At the end of this process, the calibrated model is available for use in model-based control. An example of this model calibration process is described in detail below.

White box models

Background

White box models are based on the physical description of the system; their parameters have a physical meaning. Such models are also known as "first principle models". In developing these models, often the conservation laws (e.g. energy and mass balance) of classical engineering are applied.

The amount of detail contained in white box models can vary quite a bit, in spite of the use of the physical laws. If all phenomena typical for a building are modeled, the number of required parameters for an adequate description can quickly increase to several hundred. With increasing resolution in the models, for example starting from the simple one-zone model up to models that simulate indoor air flow and daylight distribution in individual rooms, the models can contain from several equations up to several thousand equations. In the same way, the effort for modeling and solving the equations as well as the application and usefulness spans a wide range.

White box models can be categorized according to the equations, on which they are based, summarized in Table 14.1:

Table 14.1 Examples for different models or types of equations

Type	*Example*	*Application*
Steady-state linear equations	$\dot{q}_{21} = U(T_2 - T_1)$ Transmission	Steady-state heat loss
Steady-state nonlinear equations	$\dot{q}_{21} = \varepsilon c f(T_2^4 - T_1^4)$ Radiation	Building simulation (radiative heat flow between walls)
Dynamic linear equations Ordinary differential equation ODE	$C\frac{dT}{dt} = U(T - T_a)$ Heat storage	Passive/active heat storage
Dynamic linear equations Partial differential equation PDE	$\frac{\partial}{\partial t}u(x,t) = a\frac{\partial^2}{\partial x^2}u(x,t)$ Heat diffusion equation	Thermal bridge calculations
Dynamic nonlinear equations Partial nonlinear PDE	$\rho\frac{\partial}{\partial t}u + \rho(u \cdot \nabla)u = -\nabla p + \eta\Delta u + (\lambda + \eta)\nabla(\nabla \cdot u) + f$ Navier-Stokes equation	Computational fluid dynamics (CFD)
Discontinuous relations	switches/if-clauses,	Building control system simulation

- Steady-state versus dynamic models,
- Linear versus nonlinear models,
- Differentiable, continuous, and discontinuous models.

In steady-state models, the output does not depend on the dynamic course of events over time. On the other hand, the output of dynamic models depends on time since storage effects need to be considered, which in turn requires the use of differential equations. Depending on the type of differential equation, a further differentiation can be made: ordinary, partial, linear, and nonlinear.

Depending on the system of equations that the white box model employs, different numerical models are used to generate the model output. In part, different numerical or analytical methods are used for different submodels in order to increase the performance of the model.

The system of equations that are used to describe the dynamic building performance can usually not be solved with analytical methods. Furthermore, it must be noted that the choice of the solution method, and the numerical parameters (e.g. tolerance, step size, maximum number of iteration, accuracy) impact the results, e.g. for instable or chaotic systems. For such models, the dependence on solver features can dominate the results.

Since evaluating the target function in calculations for building optimization requires, in part, great computational effort, it can be shown that convergence can be achieved faster when the accuracy is variable (Wetter 2004). For this purpose, the optimization algorithm must control the accuracy of the entire model and it is very important that the target function is continuous and differentiable, which avoids chaotic behavior. Many building simulation programs do not fulfill these requirements (controllable accuracy, continuous, differentiable target function); therefore, a building simulation program was developed that meets these specifications. Certain control features (such as an on and off switching, switching functions with hysteresis) prevent that these requirements (continuous differentiable target function) are met even in real buildings.

A greater amount of effort is not necessarily needed for modeling white box models as opposed to other model types. Above all, the effort depends on the amount of detail in the model. Very simple white box models can be developed that, in spite of their simplicity, accurately model the behavior of the actual system. Yet, a prerequisite for a white box model is that the underlying technical, practical and mathematical descriptions must be known and that a numerical solution is attainable.

For white box models, it is known a priori what the boundaries are for extrapolating results. Even classical methods for uncertainty analysis can be applied to white box models in order to determine the accuracy of the results. For a calibrated white box model, therefore, the domain of validity and thus the extrapolation behavior within the desired range of accuracy is known.

The values of white box model parameters are available over a wide range of confidence. In part, exact measurement values are available (e.g. thermal conductivity of certain materials) and, in part, only rough estimates are available for the values, since exact measurement values require a great deal of effort (e.g. window openings and air exchange rates due to natural ventilation). For the calibration of white box models, it makes sense to fit the parameters that are known less exactly. However, plausibility checks for these parameters must be determined (correct order-of-magnitude). If the model cannot be calibrated with this method, then the module structure must be reconsidered. Moreover, the literature points out that too many parameters represent a large potential source of error (Déqué *et al.* 2000; Lebrun 2001) as parameter uncertainties may compound significantly to give erroneous results.

Example development of a white box model for model-based control

An example of white box modeling is provided here (Henze *et al.* 2009), which applies conservation principles to attain a dynamic linear model of a tankless water heater formulated as a state space model and used in the context of model-based control. Tankless water heaters (TWH), also called "instantaneous" or "demand" water heaters, have both advantages and disadvantages when compared to storage water heaters. Some of the advantages are that they are smaller, they have a longer life, they can provide a continuous stream of heated water, and they typically use less energy than their storage counterparts. Two main disadvantages are that they require a large power input and that the outlet temperature is difficult to control. A high-fidelity model of the physical process has the benefit that the TWH configuration (e.g. heater chambers in series versus parallel), open-loop thermal response, and closed-loop overall system response can be investigated in relatively short time periods. This allows for rapid prototyping of both the thermal components of the TWH and the control strategy to be adopted.

THERMAL MODEL OF HEATER CHAMBER WATER CONTENT

For problems involving control system design, a quasi steady-state approach fails to capture the dynamics that describe the heat transfer problem so the TWH system needs to be described dynamically. The simplest approach is the stirred tank methodology, which is, in effect, a lumped parameter method for the TWH system. Here, each of the three heater chambers of the sample TWH is modeled as a series of interconnected zones such that the thermal properties of each zone can be assumed to be entirely uniform (Figure 14.6). Thus, the outlet water temperature is equal to the zone temperature. This is an acceptable approximation so long as

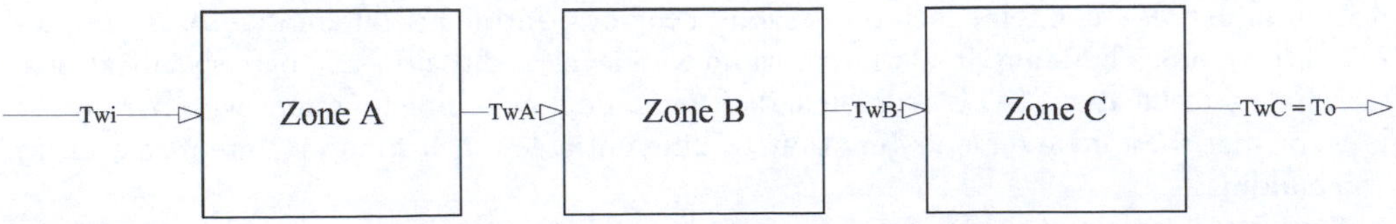

Twi → Zone A → TwA → Zone B → TwB → Zone C → TwC = To →

Figure 14.6 Zoning of a TWH heater chamber.

a sufficient number of zones are used. A third-order approach provides acceptable accuracy for most heat exchanger components. On this basis, we split each of the three heating chambers into three equally-sized series-connected subsystems (with index 'I' referring to the inlet condition and index 'O' to the outlet condition:

We further assume the water to flow through an annulus bounded by the cylindrical heating chamber shell on the exterior, and the sheath of the tubular heater on the interior. Although there are typically several tubular heaters or one that is repeatedly bent into hairpin-like structures, for the purpose of modeling the heat transfer in this problem, all of the heater elements were assumed to be concentrically located around the centerline of the heater chamber, as illustrated in Figure 14.7. This assumption allows us to sketch a modified zoning schematic for one chamber of the TWH as shown in Figure 14.8.

Given the stirred tank philosophy, we replace the mean water temperature with the zone water temperature. This leads to three energy balances for each chamber. The change of internal energy of each zone is due to (a) water flow into and out of the zone, (b) convective heat transfer from the surface of the sheath to the water and (c) convective heat transfer from the water to the outer heater chamber body. Conduction resistance in the chamber body (the *outer shell* in Figure 14.7) is considered negligible, and the chamber body is considered to be adiabatic. The three energy balances are expressed in the following equations:

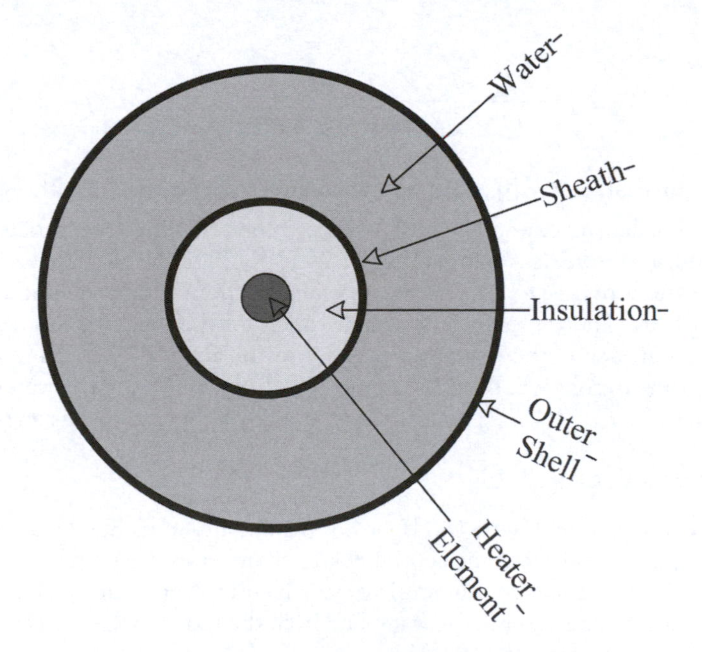

Figure 14.7 Section through an individual heater chamber (not to scale).

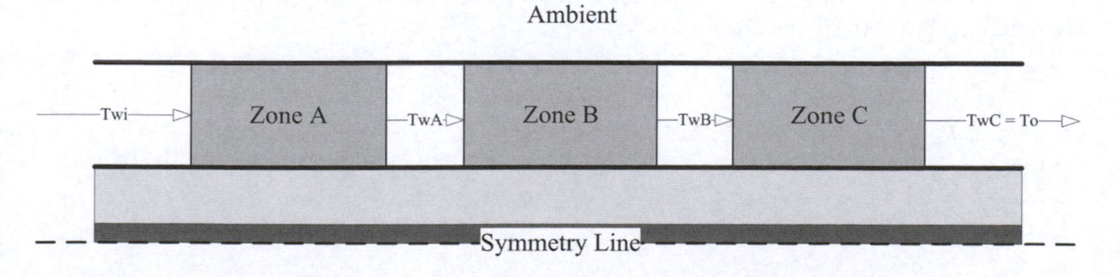

Figure 14.8 Zoning of heater chamber relative to heater element and ambient.

$$\frac{dU_A}{dt} = \rho V_{ch} c_{pw} \frac{dT_{wA}}{dt} = 3\rho \dot{V}_{ch} c_{pw} (T_{wi} - T_{wA}) + \frac{T_{sA} - T_{wA}}{R_{vs}} + \frac{T_{bA} - T_{wA}}{R_{vb}} \tag{14.1}$$

$$\frac{dU_B}{dt} = \rho V_{ch} c_{pw} \frac{dT_{wB}}{dt} = 3\rho \dot{V}_{ch} c_{pw} (T_{wA} - T_{wB}) + \frac{T_{sB} - T_{wB}}{R_{vs}} + \frac{T_{bB} - T_{wB}}{R_{vb}} \tag{14.2}$$

$$\frac{dU_C}{dt} = \rho V_{ch} c_{pw} \frac{dT_{wC}}{dt} = 3\rho \dot{V}_{ch} c_{pw} (T_{wB} - T_{wC}) + \frac{T_{sC} - T_{wC}}{R_{vs}} + \frac{T_{bC} - T_{wC}}{R_{vb}} \tag{14.3}$$

or

$$\frac{dT_{wA}}{dt} = 3\frac{\dot{V}_{ch}}{V_{ch}} (T_{wi} - T_{wA}) + \frac{T_{sA} - T_{wA}}{\rho V_{ch} c_{pw} R_{vs}} + \frac{T_{bA} - T_{wA}}{\rho V_{ch} c_{pw} R_{vb}} \tag{14.4}$$

$$\frac{dT_{wB}}{dt} = 3\frac{\dot{V}_{ch}}{V_{ch}} (T_{wA} - T_{wB}) + \frac{T_{sB} - T_{wB}}{\rho V_{ch} c_{pw} R_{vs}} + \frac{T_{bB} - T_{wB}}{\rho V_{ch} c_{pw} R_{vb}} \tag{14.5}$$

$$\frac{dT_{wC}}{dt} = 3\frac{\dot{V}_{ch}}{V_{ch}} (T_{wB} - T_{wC}) + \frac{T_{sC} - T_{wC}}{\rho V_{ch} c_{pw} R_{vs}} + \frac{T_{bC} - T_{wC}}{\rho V_{ch} c_{pw} R_{vb}} \tag{14.6}$$

Here, U is the internal energy [J], V_{ch} the total chamber volume [m^3], c_{pw} the specific heat of water [J/kgK], ρ the density of water [kg/m^3], $\dot{V}$ is the volumetric flow rate through the chamber [m^3/s], R_{vs} is the convective heat transfer resistance in [K/W] from the surface of the sheath to the water, R_{vb} is the convective heat transfer resistance in [K/W] from the water to the surface of the chamber body. The appearance of overall values for these parameters accounts for the scaling value of 3 in each of the water energy terms in the above. T_{sA}, T_{sB}, T_{sC}, are the sheath surface temperatures in zones A, B, and C, while T_{wA}, T_{wB}, and T_{wC} are the water temperatures in zones A, B, and C. T_{bA}, T_{bB}, T_{bC}, are the chamber body temperatures in zones A, B, and C.

THERMAL MODEL OF HEATER ELEMENTS

Within each heating chamber zone (A, B, or C), we simplify modeling by assuming that the heat transfer is symmetrical about the central heater chamber axis, which leaves a one-dimensional problem (in cylindrical coordinates) to be solved in each heating chamber zone, i.e., heat flow from the tubular heater to the water and from the water to the ambient in a direction perpendicular to the direction of water flow.

One can visualize the problem of transient heat transfer through an opaque multilayer construction element as an equivalent electric *RC* circuit using a lumped parameter modeling approach as shown in Figure 14.9:

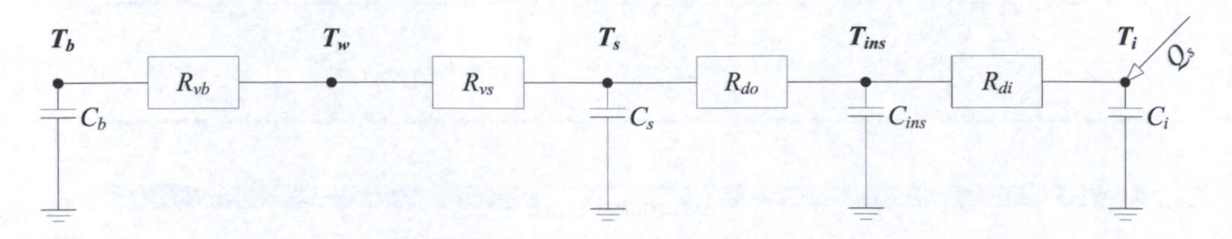

Figure 14.9 RC network for construction element.

This will represent a heater chamber subcomponent of multiple layers using four capacitors, C_i,C_{ins},C_s,C_b, four resistors, R_{di},R_{do},R_{vs},R_{vb}, and a heat source $\dot{Q}_s$ [W] at the core of the tubular heater. The heat source term is multiplied with a heat input efficiency η to account for the fact that not all of the heat delivered to the TWH ends up in the water stream, i.e., some it is lost to the environment. C_i [J/K] is the thermal capacitance of the heater element (wire), R_{di} [K/W] is the thermal resistance through the annular inner shell of insulation, C_{ins} [J/K] is the thermal capacity of the insulation, R_{do} [K/W] the thermal resistance through the annular outer shell of the insulation and the sheath, C_s [J/K] is the thermal capacity of the sheath, R_{vs} [K/W] is the convective resistance between the sheath and the water, R_{vb} [K/W] is the convective resistance between the water and the heating chamber body. This model is a fourth-order linear time invariant system with eight parameters (4R4C). As can be seen by the units, all relevant areas must be accounted for.

Based on the total sheath area A_s [m^2] and the convective heat transfer coefficient h_s [W/m^{2}K], we can determine the convective heat transfer resistance R_{vs} [K/W]

$$R_{vs} = \frac{1}{h_s A_s} \tag{14.7}$$

The conduction heat transfer resistance R_{di} or R_{do} [K/W] through the insulation and sheath may be determined from

$$R_d = \frac{1}{2\pi L\lambda}\ln\frac{r_o}{r_i} \tag{14.8}$$

where L [m] is the heating chamber length, λ [W/mK] the thermal conductivity and r_o [m] and r_i [m] the outer and inner radii of layer j.

STATE SPACE FORMULATION

Now, noting that $\dot{T} \equiv \frac{dT}{dt}$, this thermal network can be described by the following energy balances about the surface and inside temperature nodes when substitutions for C_i,C_{ins},C_s,C_b, are made:

$$C_b\dot{T}_b = (T_w - T_b)/R_{vb} \tag{14.9}$$

$$C_s\dot{T}_s = (T_w - T_s)/R_{vs} + (T_{ins} - T_s)/R_{do} \tag{14.10}$$

$$C_{ins}\dot{T}_{ins} = (T_s - T_{ins})/R_{do} + (T_i - T_{ins})/R_{di} \tag{14.11}$$

$$C_i\dot{T}_i = (T_{ins} - T_i)/R_{di} + \eta\dot{Q}_s \tag{14.12}$$

Rearranging:

$$\dot{T}_b = \frac{T_w - T_b}{C_b R_{vb}} \tag{14.13}$$

$$\dot{T}_s = \frac{T_w - T_s}{C_s R_{vs}} + \frac{T_{ins} - T_s}{C_s R_{do}} \tag{14.14}$$

$$\dot{T}_{\text{ins}} = \frac{T_{\text{s}} - T_{\text{ins}}}{C_{\text{ins}} R_{\text{do}}} + \frac{T_{\text{i}} - T_{\text{ins}}}{C_{\text{ins}} R_{\text{di}}} \quad (14.15)$$

$$\dot{T}_{\text{i}} = \frac{T_{\text{ins}} - T_{\text{i}}}{C_{\text{i}} R_{\text{di}}} + \frac{\eta \dot{Q}_{\text{s}}}{C_{\text{i}}} \quad (14.16)$$

The outputs, T_b, T_s, T_{ins}, T_i, are defined as the state variables; $\dot{T}_{\text{b}}$, $\dot{T}_{\text{s}}$, $\dot{T}_{\text{ins}}$, $\dot{T}_{\text{i}}$ are the time rate changes of the state variables; T_W, $\dot{Q}_{\text{s}}$ are the input variables and $R_{vs}, R_{vb}, R_{do}, R_{di}, C_b, C_s, C_{ins}, C_i, \eta$ are parameters. We can express these equations in matrix-variable or *state space* notation:

$$\begin{aligned} \dot{\mathbf{x}} &= \mathbf{Ax} + \mathbf{Bu} \\ \mathbf{y} &= \mathbf{Cx} + \mathbf{Du} \end{aligned} \quad (14.17)$$

where $\dot{\mathbf{x}}$ is a vector of state derivatives; **x** and **u** are vectors of state and input variables; **y** is a vector of output variables and **A, B, C, D** are matrices of parameters. Specifically:

$$\begin{bmatrix} \dot{T}_b \\ \dot{T}_s \\ \dot{T}_{ins} \\ \dot{T}_i \end{bmatrix} = \begin{bmatrix} -\frac{1}{C_b R_{vb}} & 0 & 0 & 0 \\ 0 & -\left(\frac{1}{C_s R_{vs}} + \frac{1}{C_s R_{do}}\right) & \frac{1}{C_s R_{do}} & 0 \\ 0 & \frac{1}{C_{ins} R_{do}} & -\left(\frac{1}{C_{ins} R_{do}} + \frac{1}{C_{ins} R_{di}}\right) & \frac{1}{C_{ins} R_{di}} \\ 0 & 0 & \frac{1}{C_i R_{di}} & -\frac{1}{C_i R_{di}} \end{bmatrix} \begin{bmatrix} T_b \\ T_s \\ T_{ins} \\ T_i \end{bmatrix} + \begin{bmatrix} \frac{1}{C_b R_{vb}} & 0 \\ \frac{1}{C_s R_{vs}} & 0 \\ 0 & 0 \\ 0 & \frac{\eta}{C_i} \end{bmatrix} \begin{bmatrix} T_w \\ \dot{Q}_s \end{bmatrix} \quad (14.18)$$

$$\begin{bmatrix} y_1 \\ y_2 \\ y_3 \\ y_4 \end{bmatrix} = \begin{bmatrix} 1 & 0 & 0 & 0 \\ 0 & 1 & 0 & 0 \\ 0 & 0 & 1 & 0 \\ 0 & 0 & 0 & 1 \end{bmatrix} \times \begin{bmatrix} T_b \\ T_s \\ T_{ins} \\ T_i \end{bmatrix} + \begin{bmatrix} 0 & 0 \\ 0 & 0 \\ 0 & 0 \\ 0 & 0 \end{bmatrix} \times \begin{bmatrix} T_w \\ \dot{Q}_s \end{bmatrix} \quad (14.19)$$

Next, we may want to formulate the coupled differential equations for the water content of the heater chamber in state space notation as well. However, the volumetric flow rate that will appear in state matrix **A** is variable, and thus we cannot adopt the state space notation. Part of the problem is nonlinear because two variables (flow and a dependent temperature) appear as a product. Therefore we will numerically solve the overall TWH model, using an assembly of state space submodels and the numerical integration of differential equations.

Complete model of the tankless water heater assembly

The modeling of the prototype TWH structure is now complete and shown in Figure 14.10. Input variables are the water flow rate $\dot{V}$ water inlet temperature T_{wi}, and total TWH input power $\dot{Q}_{s,TWH}$. The series connection of three chambers and the uniform electric power into the three chambers can be seen from Figure 14.11. The next step will be the validation of the TWH structure model using experimental data gathered from the prototype TWH.

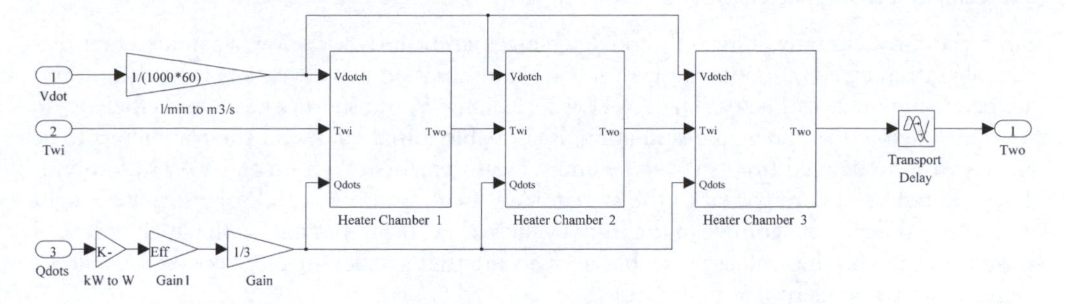

Figure 14.10 Block model of the complete prototype TWH assembly.

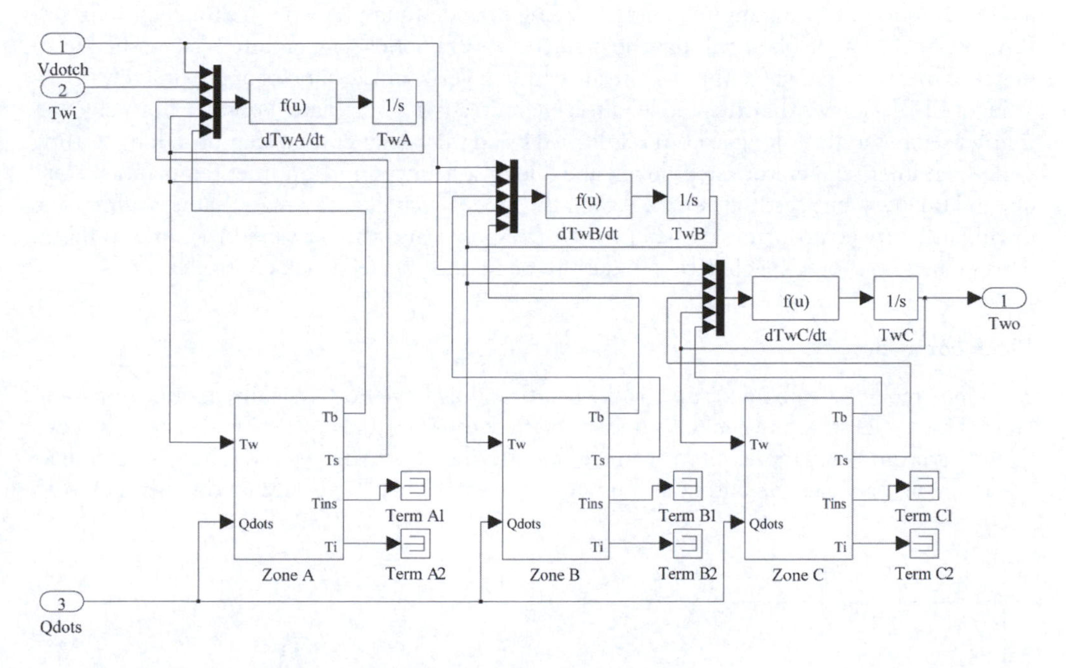

Figure 14.11 Block model of an individual TWH chamber.

Validation and parameter estimation of example white box model

In order to validate the established model, a test environment was set up in which the behavior of the new dynamic model can be simulated and compared with measured data from an actual prototype tankless water heater. The model validation is conducted in a three-step process consisting of parameter initialization, coarse parameter estimation of highly uncertain parameters, and finally fine parameter estimation of all model parameters. The parameter estimation is conducted within the technical computing environment by calling the block-based TWH model repeatedly, and examining outputs using a nonlinear least squares regression algorithm, until the predicted model response adequately mirrors the measured data.

DETERMINATION OF INITIAL PARAMETER VALUES

Initially, the thermal capacitances C_i, C_{ins}, C_s, and C_b and the conduction resistances R_{di} and R_{do} are calculated based on the available measured and manufacturer's data.

LOOSELY BOUNDED PARAMETER ESTIMATION OF CONVECTION RESISTANCES AND VOLUME

Since the convection resistances R_{vs} and R_{vb} change significantly with flow, a parameter estimation algorithm based on nonlinear least squares minimization was developed to find appropriate parameter value estimates. The TWH water volume V, transport delay t_d, and efficiency η were included in the parameter estimation. Reasonable initial values for the parameters to be estimated were derived from shorter 1-minute measurements taken on the TWH prototype. These initial values are passed to the system identification routine and solutions are sought within two orders of magnitude of the initial values. Long measurements with a high degree of variance in the driving values were chosen to ensure that a wide range of flow rates would be seen in the system identification process.

Figure 14.12 shows experimental data from the test bed compared to simulated data. The TWH is being controlled by a PID controller and is subjected to flow rate changes. A long lead-in is followed by a ramp-up, a long settling period, a ramp down, and finally a long settling period. It can be observed that the predicted TWH outlet temperature (gray dashed line) matches the measured value (black) already within a very narrow band of approximately 0.1 K.

Figure 14.13 shows that this model also generalizes (or extrapolates) quite well to changes in inlet temperature. A long lead-in is followed by an inlet temperature step-up, a long settling period, an inlet temperature step-down, and a long settling period, an inlet temperature step-up, and finally a long settling period. Again, the model accuracy is so high that it is difficult to distinguish between the gray dashed predicted response and the measured response in black. This figure also shows the classical oscillations of an unstable feedback controller.

Black box models

Black box models are characterized by the fact that they have no physically significant parameters. The model structure generally does not reflect the structure of the actual system; it is of a general type and only represents a "vehicle" for arriving at a model, which is able to reproduce the mapping between the output and input variables of the actual system correctly, i.e., with

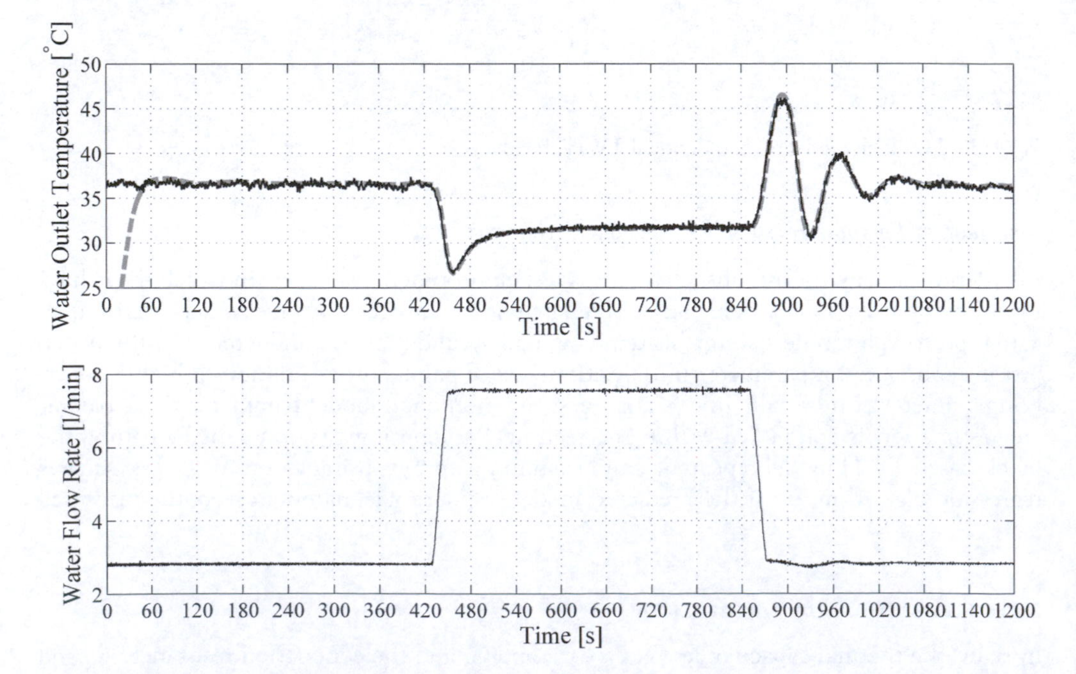

Figure 14.12 Initial parameter estimation result for training data set.

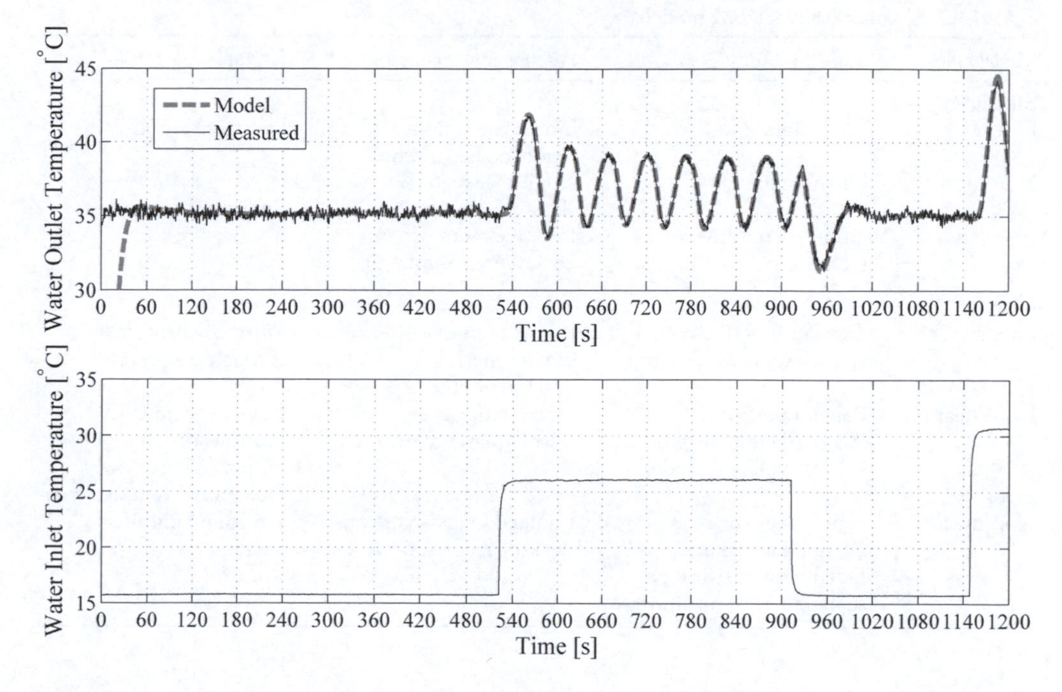

Figure 14.13 Generalization result of initial estimated model to unseen testing data

sufficient accuracy. The parameters are usually automatically estimated for this purpose (Ljung 1999).

This automatic parameter estimation represents a significant advantage over white box models. The setup cost and required computational effort is small, by comparison, for black box models. The downside, however, is the fact that the models allow little if any insight into the "inner structure" of the actual system. As a result, the analytical capability of black box models is limited.

Black box models are used commonly for the purpose of fault detection and not for optimization because their structure does not allow for the necessary parameter variation. Furthermore, the ability to extrapolate or generalize beyond the area of training data is often not available or only to a limited extent. The advantage of black box models lies in their fast and automated identification of patterns in the input-output relationship.

Regarding the structure of black box models, similar to the white box counterparts, a distinction can be made between steady-state and dynamic as well as between linear and nonlinear models. Depending on the model structure, various methods are available for estimating the parameter values. Table 14.2 provides a summary of the various approaches.

STEADY-STATE, LINEAR MODELS

Models of steady-state processes are time independent, i.e., they describe the mapping between input and output for stationary conditions. Linear black box models for steady-state processes therefore represent simple linear equations:

$$y = a_0 + a_1 u_1 + a_2 u_2 + \ldots + a_n u_n \tag{14.20}$$

where:

y = outputs
u_1–u_n = inputs
a_0–a_n = parameter

Table 14.2 Summary of black box models

Model type	*Model structure*	*Parameter estimation*	*Example*
Steady-state			
Linear	Linear function	Linear regression (least squares method)	Energy signatures
Nonlinear	Polynomial	Nonlinear regression (least squares method)	Pump and fan curves
	Arbitrary nonlinear Functions	Iterative methods, *e.g.* Levenberg-Marquardt	
Dynamic			
Linear	Transfer function model (ARMA, ARMAX, etc.)	Linear regression (least squares method), iterative methods	Unsteady heat flow through plane wall
Nonlinear	Polynomials (e.g. Wiener/Hammerstein Model, Volterra Model)	Linear regression (least squares method)	Linear system with nonlinearity in input or output (e.g. control element with saturation)
Nonlinear	Neural Network (sigmoid, wavelet, radial basis networks, support vector machines)	Damped Gauss-Newton, Back-propagation	Arbitrary nonlinear Systems

The parameters for these models can be calculated easily and quickly with the aid of linear multiple regression (linear least squares method LS).

An example of this type of model is, for example, the energy signature of a building. This represents a linear connection between the energy consumption and various determining factors such as outside temperature, radiation and operating times. See, for example, Kissock *et al.* (2003).

STEADY-STATE, NONLINEAR MODELS

Nonlinear, steady-state black box models produce nonlinear mappings between inputs and outputs. Polynomials are often used as the model structure. A simple example with only one input and output is, for example:

$$y = a_0 + a_1 u + a_2 u^2 + \ldots + a_n u^n \tag{14.21}$$

Not only this particular model but many other nonlinear models are linear in their parameters or can be transformed correspondingly (e.g., by logarithmic transformation). As with linear models, therefore, the parameter calculation can be carried out in this case using LS methods.

If the model cannot be transformed so that it is linear in its parameters, iterative processes such as the Gauss-Newton or Levenberg-Marquart procedures can be used for calculating the parameters. These are, however, numerically more complex and, in some circumstances, convergence is not guaranteed. An example of a nonlinear, steady-state model are performance maps such as pump and fan curves.

If the parameters of steady-state models are calculated using data from actual systems, it is necessary to ensure that the data represents steady-state conditions. For this reason, prior to the actual parameter calculation, a steady-state filter is often used in order to extract steady-state conditions from the data.

DYNAMIC, LINEAR MODELS

It is characteristic of dynamic systems that the mapping from inputs to outputs is time dependent. A plethora of black box models exists for dynamic linear systems, which can be described as transfer function models and for which there is an all-embracing theory (Ljung 1999). It is characteristic of the structure of these models that the current output variable is calculated using linear functions of previous input and output variables.

The easiest model of this type is the finite impulse response model (FIR), which produces the mapping between the current output and the current and previous inputs:

$$\begin{aligned} y(t) &= B(q)\, u(t) + e(t) \\ &= b_1 u(t-1) + \ldots + b_n u(t-n) + e(t) \end{aligned} \tag{14.21}$$

where:

$u(t)$ = measured inputs
$y(t)$ = measured outputs
$e(t)$ = disturbance variable (normal distribution)
$b_1 - b_n$ = parameters
q = so-called back shift operator:
$q^{-1}u(t) = u(t-1);\ B(q) = 1 + b_1 q^{-1} + \ldots + b_n q^{-n}$

The black box models of linear, dynamic systems used in practice are all modifications of this model. For the general case, the following connection arises:

$$A(q)y(t) = \frac{B(q)}{F(q)}u(t) + \frac{C(q)}{D(q)}e(t) \tag{14.22}$$

The five matrices or polynomials A, B, C, D, and F represent the transfer functions. In practical applications, one or more of these functions is unity, which gives rise to various model structures. The most common are listed in Table 14.3. Figure 14.14 shows the structure of an ARX model. The term "autoregressive" describes the inclusion of previously measured data in the model for the outputs, while "moving average" relates to the inclusion of external disturbance variables.

Generally, simple and tested methods are used for calculating parameters for models of this type (linear regression least squares) and it is assumed that the disturbance variable is a normally distributed variable with zero mean value.

These models are, however, limited to linear, time-invariant systems. Since actual systems – at least in the building sector – are mainly time-variant and nonlinear, the usefulness of these models is limited. One typical application is for modeling the time-dependent transfer of heat through flat walls.

Table 14.3 Typical linear black box models for dynamic systems

Polynomials in use	*Short form of the model structure*	*Remarks*
B	FIR	Finite impulse response
AB	ARX	Autoregressive with external input
ABC	ARMAX	Autoregressive moving average and external input
AC	ARMA	Autoregressive moving average
BF	OE	Output error
BFCD	BJ	Box-Jenkins

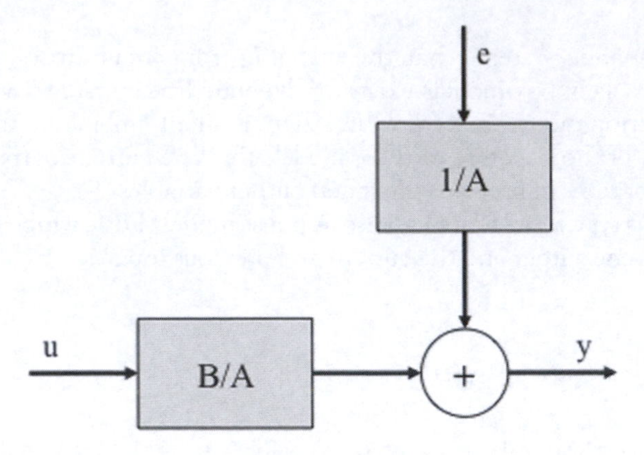

Figure 14.14 Structure of an ARX model.

To a certain extent, temporal variance (e.g. with recursive models or weighting functions for "older" inputs) or nonlinearity (e.g. inclusion of steady-state nonlinearities of the inputs or outputs in otherwise linear models, Wiener or Hammerstein models) can be handled by reducing the problem back to a linear model (Ljung 1999).

DYNAMIC, NONLINEAR MODELS

Nonlinear models clearly offer more freedom with regard to the structure of the model, which on the one hand makes them suitable for describing real systems, but on the other hand is numerically expensive. Apart from the choice of variables (measured data) and regressors for the model, the most important step is the selection of a model structure that can reflect any desired nonlinear feature of the variable space.

Classical methods for identifying nonlinear dynamic systems are largely based on polynomial models. These have the advantage of having linear parameters, which means that the estimation of the parameters can simply be carried out by means of linear regression.

Nevertheless, other forms of nonlinear models, such as neural nets (in particular the nonlinear autoregressive type with exogenous input NARX) and support vector machines, are now being employed extensively. Judging by the sheer number of publications dealing with the creation and application of neural network, support vector machine, and wavelet models, there is a strong interest apparent in these approaches.

The advantages of nonlinear black box models such as neural nets are obvious: in principle, they can be used to represent any real nonlinear, dynamic system with a manageable numerical effort.

A few distinct disadvantages must be mentioned though, which are closely associated with the black box character, i.e., these models require large amounts of data to estimate the parameters, while at the same time an extrapolation beyond the domain of training data is potentially hazardous. As the model parameters are not physical, they are not suitable for optimization of building parameters (Katipamula and Brambley 2005a, 2005b).

Gray box models

Gray box models are a hybrid form of white box and black box models and they are distinguished by the fact that they have physically significant as well as physically insignificant parameters (Braun and Chaturvedi 2002; Lebrun 2001).

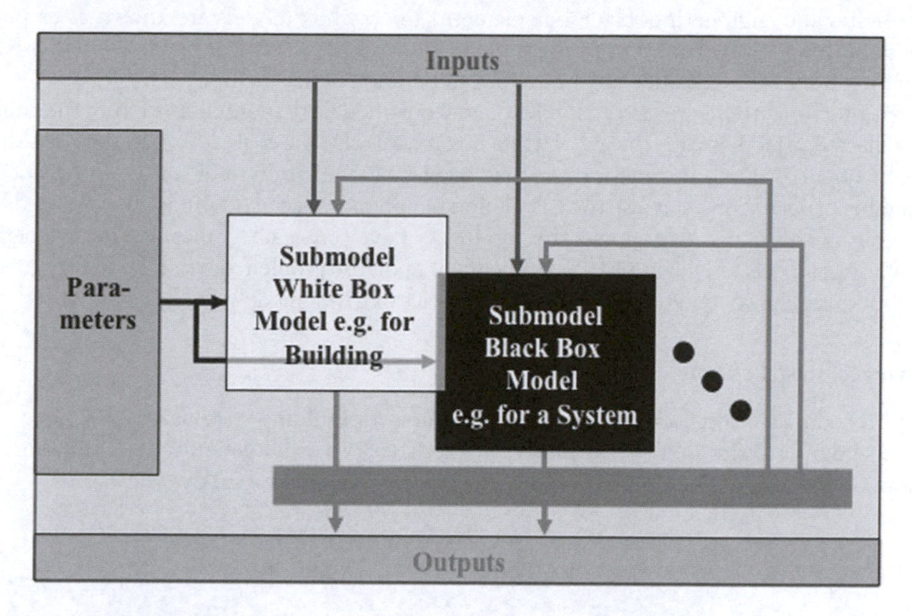

Figure 14.15 Gray box models that subsume models with mixed characteristics are combined into one overall model of different parts.

Gray box models can be developed for individual components or they are derived from the combination of white box and black box models for a large complete system (Figure 14.15). In addition, black box models can complement white box models, in order to incorporate higher-order features that are not captured by the physical model (Vilim *et al.* 2001) (Figure 14.16).

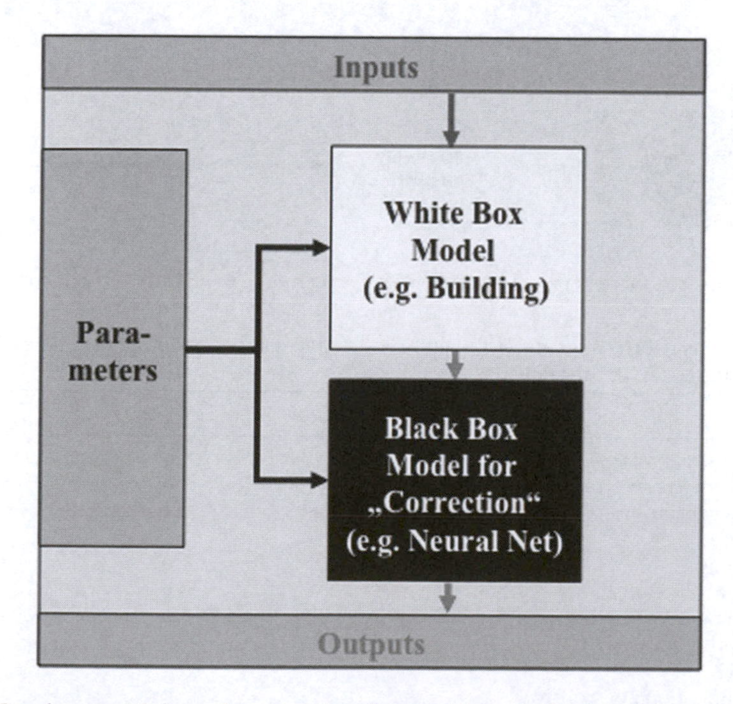

Figure 14.16 Gray box model that enhances a white box model using a black model for higher-order error correction.

In physics and engineering, black box elements in gray box models are known as empirical models and have a long tradition in these fields. In the past, statistical methods were used for this, which have been enhanced by newer developments such as neural nets.

In the literature, there are also examples of gray box models that are derived from the calibration of black box models with the aid of white box models (Déqué *et al.* 2000). In these cases, linear, polynomial or harmonic functions are matched to the simulation results. In this way, not only the number of parameters, but also the calculation requirements can be substantially reduced. The parameters of the matched functions thus no longer have physical significance, but are derived from physical models (see Figure 14.17). Dependent upon the problem set and the selection of the basic functions, the good extrapolation characteristics of white models can be retained.

Overview of model characteristics

In the preceding sections, a range of model structures for building systems was described. This section summarizes the most important characteristics of model generation, calibration and application for fault detection and optimization. See Table 14.4 for an overview.

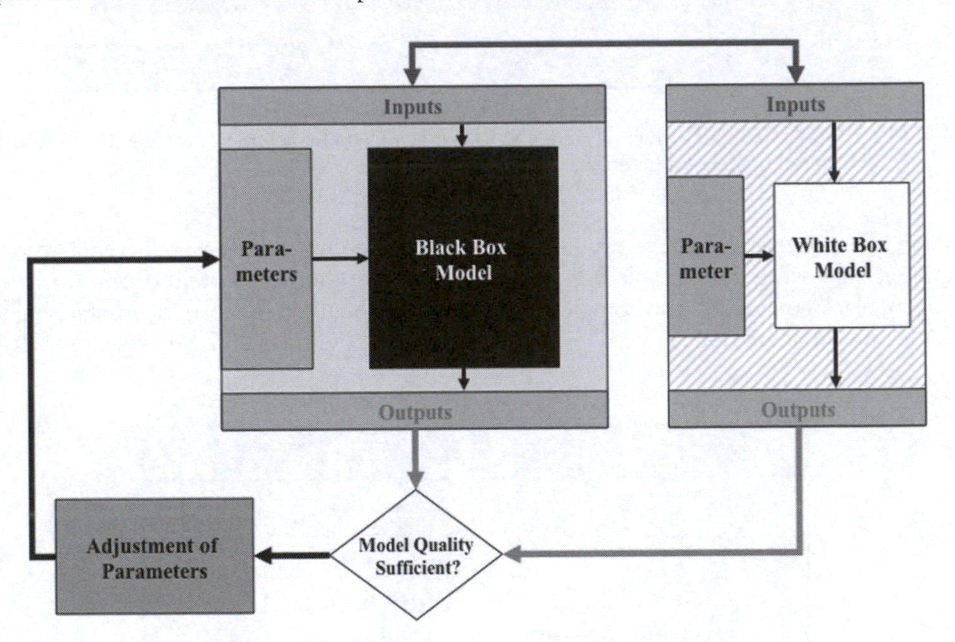

Figure 14.17 Gray box model created by using a white box model to produce a black box model.

Table 14.4 Summary of the advantages (+) and disadvantages (–) of various model types. (o) implies an intermediate satisfaction of the feature

Feature	*White Box*	*Gray Box*	*Black Box*
Insight into physical processes	+	o	–
Number of parameters → error source	–	o	+
Effort in model generation	+ –	+ o	+
Calculation speed	+ o	+ o	+
Need for training data	+	o –	–
Complexity of calibration	+ –	+ –	+
Generalization or extrapolation	+	o	–
Portability to other systems	+	+ –	– *
Usable for FDD	+	+	+
Usable for optimization and control	+	o	–

* only with new training data.

Essentially, white and black box models are opposites: while white box models have very good analytical capabilities, they are often more costly to produce and calibrate; the reverse is true for black box models. It is not necessarily true, however, that white box models (when of simple structure and resultant reduced prediction accuracy) are more difficult to produce than black box models.

To gain an understanding of the system, white box models are initially a good choice. It is hence advisable to search for a way to allow a white box model to be generated with simplified assumptions, which can be enhanced later, either with greater detail or with a matching black box model (Lebrun 2001).

On the other hand, black box models are most suitable for rapid and easy identification of patterns (e.g. usage profiles for buildings) and the recognition of outliers from the expected pattern. The patterns identified can additionally be used as input values for other, more detailed models. In particular, for complex processes of which the internal structure is unknown or not known in detail, black box models can prove to be helpful.

Whole building model development

Apart from the white, gray, and black box models described above, whole building models for the purpose of model predictive control can also be based on specialized simulation programs. A comprehensive overview of calculation tools specifically for commercial buildings is provided on the internet site of the United States Departments of Energy (DOE, Building Technologies Program);[2] where over 200 software tools are documented ranging from simulation environments that are specifically designed for building simulation to general simulation tools.

Specialized simulation environments

Typical examples of these traditional building simulation programs are: EnergyPlus,[3] DOE-2,[4] ESP-r,[5] or TAS.[6] Many of these simulation programs operate with models having a medium to high level of detail, i.e., they are multiple zone models with algorithms for heat transfer in multilayer building components, ventilation, solar and internal gain, building energy systems in buildings and control. It must be appreciated, however, that building energy system models often do not have the same level of detail and scope as building models.

What is required in many cases is the ability to simulate an entire year with a time step of an hour or less. This results in a considerable calculation cost, which has led to a situation where, at the beginning of the development, numerical aspects played a significant role and the simulation programs (models) were specifically optimized for high calculation speeds. A high level of modeling flexibility or a clear and comprehensible implementation of the physical principles was not of concern. This has the downside that the implementation of changes and enhancements at a later stage is both difficult and costly.

Due to the increased numerical performance of computers, the costs of the calculation requirements of many projects/cases is now very much less than the cost of the generation of the simulation models and the integration of new models. As a result, a high level of modeling flexibility or a particularly clear and comprehensible implementation of the physical principles is of much higher importance. A modeling approach for this is the so-called "equation-based approach", where the physical equations are also included in the simulation program as a component of the simulation model. This enables them to be matched to the current problem in terms of the level of detail and in relation to the interconnections (Sahlin 2000). The disadvantage of equation-based structures is the high calculation expense.

At scientific conferences in relation to buildings (building physics, IAQ, HVAC, air conditioning, building simulation), a large number of new mathematical/physical models (in the form of equations) have been presented (Sahlin 2000; Sahlin *et al.* 2004). The advantage of equation-based simulation structures is that such new models can be directly integrated into simulations. This means that the community that can participate in the further development of the simulation tools can be substantially expanded. For the integration of new physical models, no special numerical knowledge is necessary.

With TRNSYS,[7] developed in 1973 (Klein 1973) for the simulation of solar installations, certain elements of equation-based models were put into practice. It allows the arbitrary interconnection of subordinate models and the integration of algebraic equations. Modeling with equations was not seen as the primary interest here, but merely as an enhancement opportunity. The only example of an equation-based simulation tool produced specially for buildings and installations simulation is IDA-ICE.[8]

General simulation tools

In principle these tools allow any system to be modeled. Typical examples of this group are: Matlab/Simulink,[9] Dymola/Modelica,[10] Mathmodelica,[11] IDA-SE or EES.[12] Most of these tools belong to the "equation-based" type and were used successfully for buildings and in particular in the field of building control (Adam *et al.* 2006). While Matlab/Simulink is a de facto standard (in particular in the domain of controls analysis), Dymola/Modelica has become more significant in the last few years.

Implementation of model-based control

We differentiate between two types of model-based control implementations: (a) *Offline optimization* simulates and optimizes the control parameters independently of the current building operation (Figure 14.18); (b) by contrast, *online optimization* is a continuous process coupling both building and controllers. Current building state and possibly predictions (*e.g.* daily power or climate predictions) are considered for this optimization approach (Figure 14.19).

Simple feedforward control only works adequately when the system can be accurately represented with an instantaneous algebraic equation. However, when an accurate dynamic model is to be used for control, model-based predictive control (MPC) is the approach of choice. With MPC, the dynamic model will be used to find a control strategy over a sliding planning horizon that minimizes a performance criterion of choice, such as the cumulative root mean

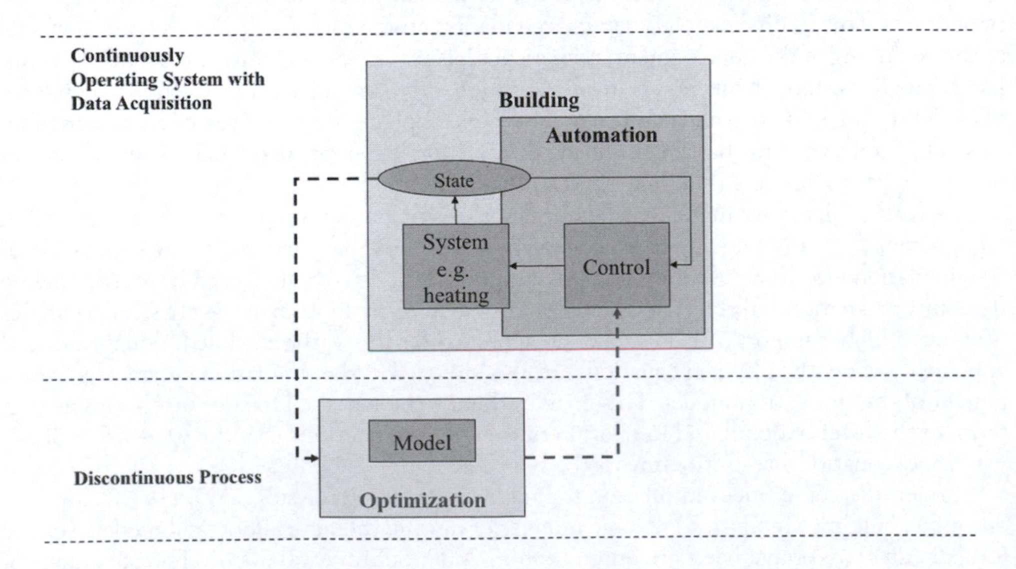

Figure 14.18 Process structure for offline building optimization.

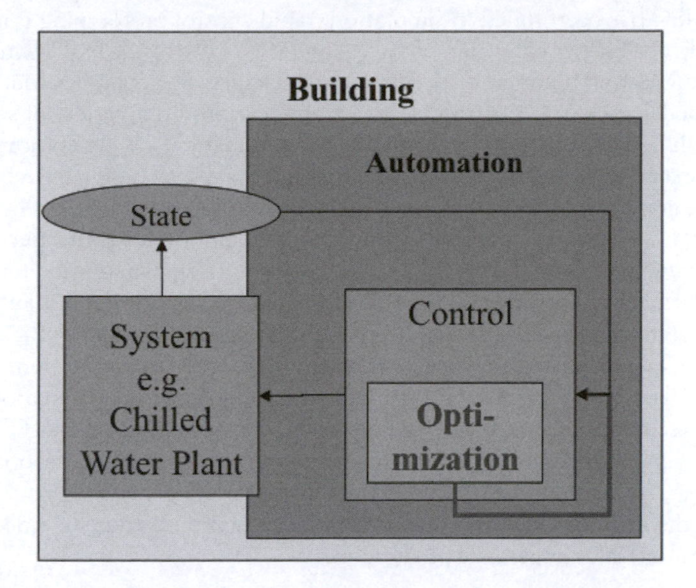

Figure 14.19 Process structure for online optimization.

square RMS error of the outlet water temperature deviation from the setpoint. Once the optimal strategy is found, the first action of the strategy at the current time is selected as the current control action. After the next time increment, the process is repeated.

As its name suggests, model predictive control automates a target system (the "plant") by combining a prediction and a control strategy. An approximate, linear plant model, as described above, provides the prediction. The control strategy compares predicted plant states to a set of objectives, then adjusts available actuators to achieve the objectives while respecting the plant's constraints. Such constraints can include the actuators' physical limits, boundaries of safe operation, and lower limits for indoor environmental quality. This constraint tolerance of model predictive control differentiates it from other optimal control strategies, such as the linear quadratic Gaussian (LQG) approach. The interested reader will find a short list of books on MPC at the end of this chapter.

Implementation of linear time invariant system MPC

For the implementation of an MPC for the discussed tankless water heater application (Henze *et al.* 2009), the overall model predictive controller design is supported by a commercial tool chain, which enables rapid prototyping of the desired accurate controller. The development environment consists of the following elements employed sequentially:

a. The TWH model development discussed above is conducted in the block-based dynamic systems analysis front end (Simulink 2007), while the parameter estimation and subsequent validation is achieved with the help of a popular technical computing enviroment and associated optimization routines (Matlab 2007).
b. The block-based dynamic systems analysis framework is extended by a toolbox for the design of model predictive controllers (MPC Toolbox 2007). The MPC design entails creating linearized plant models at one or several water flow rates. Because of the nonlinearity caused by the variation in TWH water flow rate, a high-flow and a low-flow plant model are created. The measured flow rate triggers which model the MPC controller uses, in what is called a bumpless transfer design.

c. Once the MPC has been tuned in simulation, rapid control prototyping is accomplished with the help of a further extension of the capabilities of the block-based dynamic systems analysis frontend that automatically generates, packages, and compiles source code from the block models to create real-time software applications on a variety of systems (Real-Time Workshop 2007). This extension provides automatic code generation tailored to a variety of target platforms, a rapid and direct path from system design to implementation, and a simple graphical user interface with an open architecture.
d. For the TWH, we selected a popular hardware target platform for controller evaluation in the laboratory setting (xPC Target 2007). The target platform is a solution for prototyping, testing, and deploying real-time systems using standard PC hardware. It is an environment that uses a target PC, separate from a host PC, for running real-time applications. In this environment, one uses a desktop computer as a host PC with the above mentioned tools to create a model using blocks. After creating the model, one can run simulations in nonreal time. The executable code is downloaded from the host PC to the target PC running the hardware target real-time kernel. After downloading the executable code, one can run and test the target control application in real time in the laboratory setting.
e. Eventually, the same tool chain is used to develop production code for embedded micro-controllers selected for large-scale production.

Implementation of whole-building model MPC

An example of the implementation of a MPC for combined active and passive building thermal storage control utilizing whole-building simulation programs is discussed in Henze *et al.* (2005). Here, a sequential approach to model-based control is employed: (1) short-term forecasting, (2) optimization, and (3) post-processing and control command implementation as shown in Figure 14.20. A real-time weather station provides the current weather data to the short-term weather predictor. If an on-site weather station is not available, the nearest available weather station operated by the National Weather Service can be used. This predictor provides an improved forecast for the next planning horizon to the optimal controller, which iteratively adjusts the control variables in the model until convergence is reached. The optimal solution is passed to a post-processor that interprets the optimal results and turns them into commands understood by the building automation system of the facility under control. The building is modeled in the transient systems simulation program TRNSYS (Klein 1973), while the general purpose technical computing environment Matlab (2007) including the

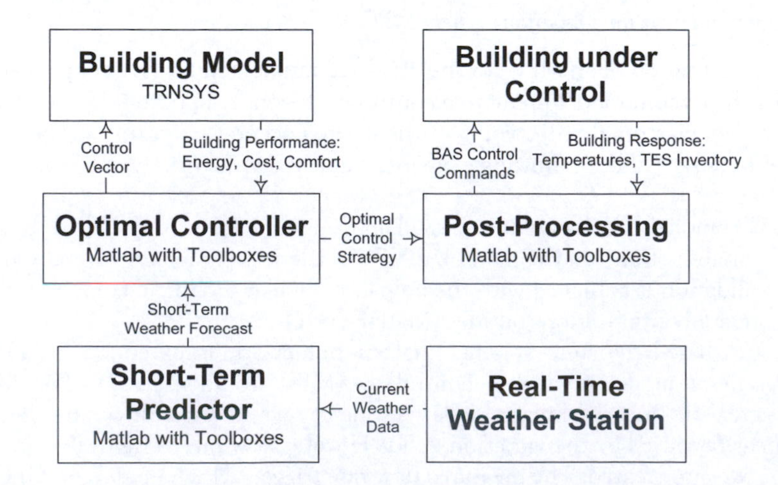

Figure 14.20 Real-time predictive optimal control schematic.

optimization toolbox was used to interface with the building simulation program. In this investigation, the building model is static and no system identification is employed to estimate the changes in model parameters during operation.

Forecasting models

Background

Significant motivation for time series prediction of local weather and other driving forces exists for model-based predictive control (MPC) applications, for example, building thermal mass control to respond to utility pricing signals, increased free cooling through superior economizer control and thermal load prediction for uniform chiller loading and improved part-load performance. The desire is to optimize operation based on a short-term prediction of weather and building utilization to yield energy and cost savings through the minimization of an objective function over the prediction horizon. Identifying the forecasting model complexity required will support MPC applications ranging from embedded controllers to high-level supervisory control within the BAS; the availability of accurate dynamic building energy simulation programs further promotes future application of MPC in commercial buildings.

Review of forecasting models

Many studies of short-term prediction come from the very active research domain of electrical power forecasting used to optimally dispatch central power generation equipment with weather forecasting typically influencing the unit commitments. Despite overlaps in forecasting strategies and algorithms, the present concerns limit the literature review to studies involving commercial buildings with weather forecasting and/or prediction features. In both domains a significant fraction of researchers are convinced that nonlinear forecasting models based on neural networks (NN) provide superior forecasting performance, nonetheless, many have found success using traditional time series analysis. For a detailed presentation of time series analysis and neural networks the reader is referred to Box *et al.* (1994) and Bishop (1995), respectively.

Time Series Analysis Models: In forecasting hourly cooling loads, MacArthur *et al.* (1989) used an autoregressive moving average (ARMA) model. A recursive least squares (RLS) algorithm was used for online model parameter adjustment with exponential forgetting factor to ensure a maximum covariance between the predicted and actual load. The quality of the load forecast was tied to ambient weather prediction because the method used historical data and external forecasts of maximum and minimum values for rescaling the profile.

Seem and Braun (1991) developed a model for electrical load forecasting by combining an autoregressive (AR) model for the stochastic part of the prediction task and an adaptive look-up table for the deterministic part. Using an exponentially weighted moving average (EWMA), hourly electrical power demand was determined using previous forecasts and measured values. The deterministic look-up table is often referred to as a cerebellar model articulation controller (CMAC). Its performance showed improvement by introducing electrical load profile temperature dependence, with daily peak values modified according to the maximal daily temperature forecast as provided by the National Weather Service (NWS).

Chen and Athienitis (1996) discuss an optimal predictive control methodology for building heating systems in a real-time dynamic operation. The system consisted of three loops: a conventional feedback control loop with predictive regulator, a parameter estimator using an RLS technique, and a setpoint optimizer. The weather prediction makes use of both historical records and local weather forecasts. Simple rules are devised to modify historic shape factors based on external weather service forecasts.

Henze *et al.* (2004) investigated the impact of forecasting accuracy on predictive optimal control of active and passive building thermal storage inventory. The short-term weather forecasting models including bin, unbiased random walk, and seasonal integrated ARMA predictors. It was shown that model complexity does not imply accuracy.

Neural Network Models: Ferrano and Wong (1990) used a feed-forward NN software package to predict the next day's cumulative cooling load by mapping hourly (24 input units) ambient dry-bulb temperatures, with results subsequently used in a real-time expert system. In order to allow for adequate generalization and avoid memorization, it was determined that training sets should contain a difference of at least 3% in the temperature patterns. This effectively reduced training time and reduced the prediction error from a maximum of 14% to only 4% for the validation sets.

Kreider and Wang (1992) used a feed-forward neural network with nine input units in the prediction of hourly electrical consumption integrated from 15-minute power values. The network was capable of responding to unusual weather phenomena; for example, during a non-cooling month (December) the weather was unusually hot for two days and the cooling plant had to come on during that time period. The network picked up this unusual condition and predicted the required power consumption for this period. Conventional methods of regression missed this event. It was shown that the quality of the forecast improves when the network is trained on data of the same season for which the prediction is made, i.e. predicting energy consumption for a non-cooling month such as December should be done with a network that is trained on November data rather than data from a typical cooling month such as July.

Gibson and Kraft (1993) used a recurrent NN for electric load prediction in conjunction with a thermal energy storage system. The network was trained using electric and cooling load data of an office building monitored over a cooling season. The ability to generalize was noted by the network's ability to achieve similar prediction errors for atypical days as compared to normal days when supplied with detailed occupancy load information. Using a recurrent network architecture allowed the network to not only pick up the steady-state physical processes, but also the temporal relationships between system states, i.e. dynamic aspects.

Dodier and Henze (2004) used neural networks as a general nonlinear regression tool in predicting building energy use data as part of the Energy Prediction Shootout II contest. By combining a NN with a version of Wald's test taken from conventional statistical analysis, the relevance of free parameter inputs were determined. Time-lagged input variables were determined by inspecting the autocovariance function. The strategy was found to be the most accurate predictor in the competition.

Comparative Studies in Prediction: The "1993 Great Energy Predictor Shootout", detailed in Kreider and Haberl (1994), was a competition in the prediction of hourly building energy data available to world-wide data analysts and competitors alike. The results showed that connectionist approaches were used in some form by all winners, i.e., NN approaches using different architectures and learning algorithms proved to achieve superior accuracy when compared with traditional statistical methods. Kawashima *et al.* (1995) compared hourly thermal load prediction for the coming day by means of linear regression, ARIMA, EWMA, and NN models. The NN models were confirmed to have excellent prediction accuracy and considered by the authors to be superior methods for utilization in HVAC system control, thermal storage optimal control, and load/demand management. The "Great Energy Predictor Shootout II" extended the understanding of prediction methods for building applications, evaluating the prediction of hourly whole-building energy baselines after energy conservation retrofits. The effectiveness of prediction models was compared for the top entries in Haberl and Thamilseran (1996). NN models were shown to be the most accurate in nearly all cases, but unique statistical approaches were shown to compete, in terms of accuracy, with the NN models.

In a recent investigation, Florita and Henze (2009) identified the complexity required for short-term weather forecasting in the context of a model predictive control environment. Moving average (MA) models with various enhancements and neural network models were used to predict weather variables seasonally in numerous geographic locations. Their performance was statistically assessed using coefficient-of-variation (CV) and mean bias error (MBE) values. When used in a cyclical two-stage model predictive control process of policy planning followed by execution, the results showed that even the most complicated nonlinear autoregressive neural network with exogenous input (NARX) does not appear to warrant the additional efforts in forecasting model development and training in comparison to the simple prior and exponentially weighted moving average (MA) models.

Examples of model-based control applications

Offline control optimization: mixed mode building HVAC control

Mixed mode (MM) building design refers to commercial building designs that combine mechanical and natural ventilation (also known as hybrid ventilation) and typically feature a range of passive cooling strategies such as building thermal mass control and low-exergy conditioning concepts including low-temperature heating and high-temperature cooling systems.

Spindler and Norford first examined model-based optimal control of an entire MM building through Spindler's PhD dissertation at MIT (Spindler 2004) later publishing the work as a two-part paper (Spindler and Norford 2008a, 2008b). The control of a MM office building was examined in two parts. First, an inverse modeling system identification framework was developed to assist in assembling an accurate black-box model of the building that provided lower prediction error than those published for many other naturally ventilated buildings. A Principal Hessian Direction Regression Tree (PHDRT), neural network, and Kernel Recursive Least Squares (KRLS) model were each developed, trained, and compared against measured building data. Despite the inherently non-linear nature of the processes of interest (particularly with regards to airflow), the linear PHDRT model provided the best overall predictive capability and about half the RMS error of a corresponding physical model of the building in question (0.42 K, compared to 0.74 K for the physical model).

In the second phase, a combination of optimization techniques, including dynamic programming, integer programming, genetic algorithms, and simulated annealing were investigated to optimally control the developed linear, inverse model with the goals of achieving basic thermal comfort and minimizing fan operation. Automatic control for a fan and certain operable windows were investigated; however, occupant control and override of these control strategies were not considered.

Due to computing constraints, general pattern search algorithms such as genetic algorithms and simulated annealing proved to be impractical for optimization purposes. To decompose the problem and reduce computing time, a number of different decision variables were discretized into lookup tables. For example, the number of different operating modes of the building was fixed and enumerated in an output table, representing a discrete set of combinations of fan operation and operable window opening areas. A binary search was then performed on the resulting simplified model, reducing compute time. The optimal control of night cooling was shown to maintain indoor temperatures over 4 K below ambient throughout the hottest parts of the day.

Spindler and Norford (2008) applied the same modeling and optimal control concepts to a second building, but were unable to obtain significant data on window openings to be able to either validate optimal control algorithms or implement a supervisory, model predictive control strategy for the building. Thus, it remains to be seen whether a similar strategy (linear, inverse building models coupled with a simplified, binary search optimal control algorithm) could be effectively extended and generalized to a larger number of buildings. Furthermore, the use of black box regressive building models makes the approach somewhat limited in evaluating buildings that are in the design phase; as such models must be trained empirically after construction and occupation.

In a research project sponsored by the U.S. Green Building Council, an international research team led by the author (G.P. Henze) is currently taking a white box model approach to MM building control, with the hopes of deriving near-optimal control guidelines from the offline optimization case studies. A meta-heuristic optimization of a simple, mixed-mode building was pursued in a proprietary Matlab/EnergyPlus (ME+) optimization environment to determine the optimal window opening schedule for one day. A variety of decision variables were investigated within the limitations of the window opening controls that EnergyPlus provides. A particle swarm optimization (PSO) algorithm was used to identify the near-optimal control pattern for the building. A simplified objective function of DX (direct expansion) cooling electrical energy was minimized. Penalties for thermal comfort were not taken into account.

Using a modified EnergyPlus build, the optimizer was able to explore multiple combinations of window opening positions for the space until a minimum DX cooling energy point was found. The building's HVAC systems were allowed to run at their standard cooling setpoints even with the windows open, meaning that window opening during hot periods of the day would be penalized. For the day in question, June 1, the optimal window opening schedule was to simply leave all windows open throughout the day except for a warm, late afternoon period during which openings were set back to 40%. For a hotter day, the 1% cooling design day for the Chicago climate, the optimal window opening schedule involved fixed openings at about 100% until 9 a.m., followed by a period of closed windows during the hottest part of the day, followed again by 100% openings in the late evening hours.

Table 14.5 provides the energy savings achieved by the various control schemes compared to the base case with closed windows and the default hybrid ventilation controller as implemented in EnergyPlus. For the part-load day, various optimization control sequences were able to achieve a 50% savings over the base case and a 30% savings over even the hybrid ventilation controller: "Constant" optimizes the hourly ventilation availability and commands the windows to open to their predefined opening factor when open, "temperature" refers to the optimization of temperature setpoints for a proportional controller, while "variable opening" optimizes both hourly availability and window opening factors. For the design day, savings were measurably smaller. For the many part-load days in a typical summer cooling period, significant cumulative energy savings can be expected and the derivation of near-optimal control rules and guidelines appears very attractive.

Results for both days are presented in Figure 14.21 for the default model (with a simple rule-based hybrid ventilation controller to manage the hand-off between natural and mechanical cooling), the constant window opening, temperature setpoint, and the variable opening optimization. In both the part-load and design day cases, the optimization that allowed fully variable window opening control was able to discover incremental savings over the previous

Table 14.5 Summary of cooling energy savings from optimal MM control

	Part-Load Day (June 1)			*Design Day*		
	DX Cooling Energy (kWh)	*Savings Over Base Case*	*Savings Over Hybrid Ventilation Controller*	*DX Cooling Energy (kWh)*	*Savings Over Base Case*	*Savings Over Hybrid Ventilation Controller*
Base case	36.6	–	–	42.3	–	–
Hybrid ventilation control	27.4	25%	–	39.3	7%	–
Optimized – constant	18.5	49%	33%	38.7	9%	2%
Optimized – temperature	17.7	52%	36%	37.1	14%	6%
Optimized – variable opening	16.2	56%	41%	38.7	9%	2%

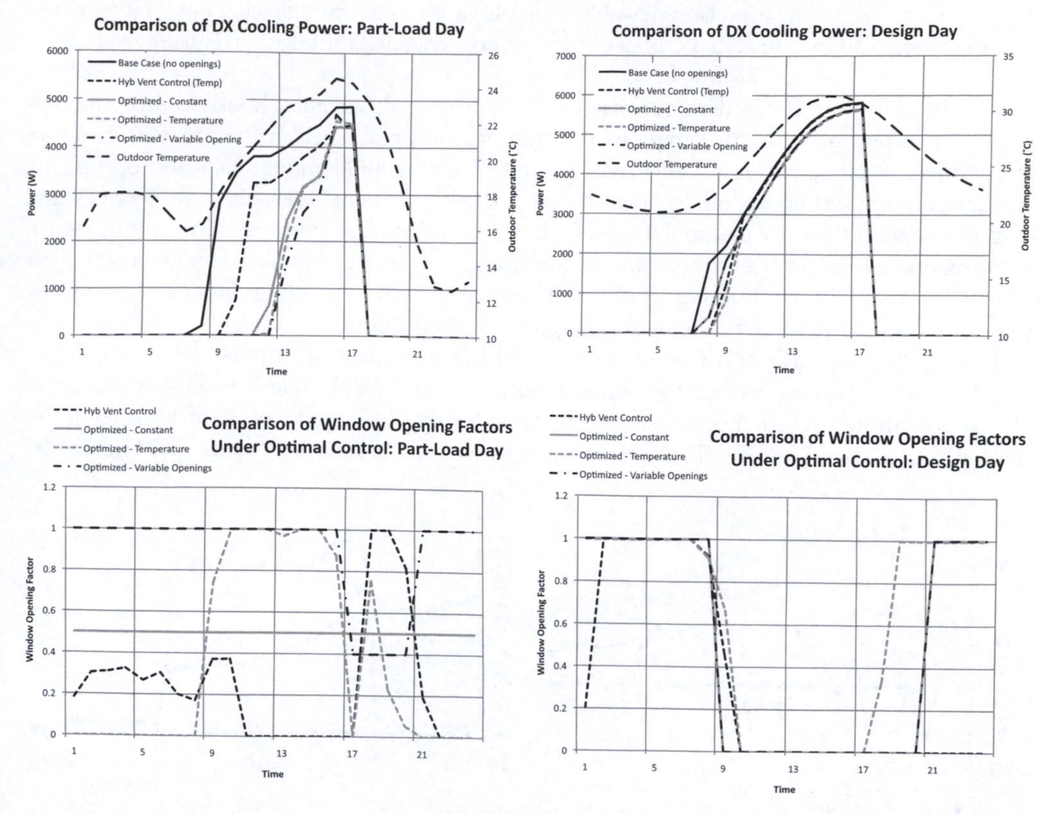

Figure 14.21 Results of natural ventilation optimizations, showing cooling DX power (above) and window opening positions (below) for various control schemes and optimal solutions. June 1 (left). 1% Chicago, IL design day (right).

optimization cases. It is apparent that the optimal control schemes have the greatest success for milder days such as June 1 when cooler evening and early morning air can be utilized. Extremely hot days offer small temperature differentials between indoor and ambient conditions, and so it is no surprise that all three optimizations explored provide only minor improvement over the built-in hybrid ventilation controller (May-Ostendorp and Henze 2009).

Online control optimization: price responsive thermal mass control

While optimization of building thermal mass control has been extensively investigated in scientific publications, e.g., see Braun (2003), Cheng *et al.* (2008) and Henze *et al.* (2008), online model-based control has only very recently been attempted in an effort to achieve the integration of electrical utility and commercial building operation.

Building thermal mass can be viewed as a thermal flywheel, providing energy storage opportunities. Such a thermal flywheel, on the multi-megawatt scale possible in metropolitan areas, creates the opportunity for:

- Continuous demand elasticity – competing and dispatching against electric generation to moderate market prices and market price volatility.
- Untapped source and end-use efficiency and reduced carbon – optimally controlling the thermal flywheel to more efficiently dispatch generators; to displace inefficient marginal generators in grid ancillary services markets; and to better use high efficiency systems within buildings.

- Accelerated penetration of renewables – deploying the thermal flywheel as a large, existing substitute for the added grid operating reserves needed for intermittent renewables.

Currently underway is the demonstration of an automated, scalable, hourly online, model-based optimization system that exploits the thermal mass embodied in the physical structure of commercial office buildings, effectively shaping heating and cooling loads and optimizing building energy system operation in response to highly dynamic electric market prices and carbon emission rates. By manipulating the global temperature setpoint within comfort limits, building energy system operations can optimally respond to price signals, as shown on the left side of Figure 14.22. Such opportunity, when captured, will lead to building demand profile that optimally responds to the prevailing supply side constraints.

The implementation of the real-time model-based control system is shown in Figure 14.23. Here, a real-time electricity market pricing data service provides the day-ahead cost of electricity while a weather forecast service provides the 24-hour forecast of weather information for the building site. These external variables are communicated as XML input files

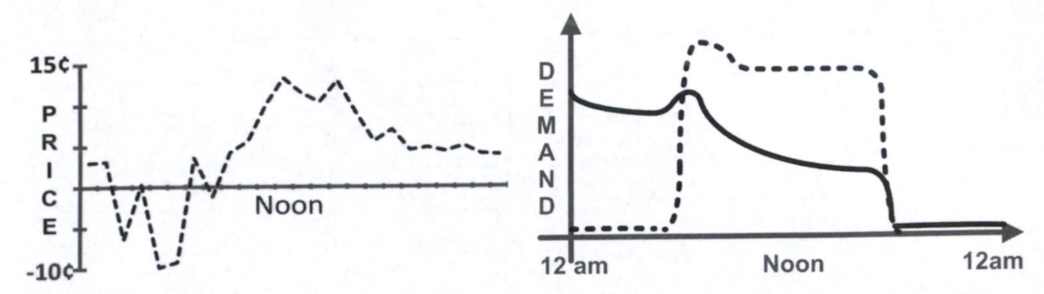

Figure 14.22 Wholesale hourly electric price [¢/kWh] for Chicago for June 16, 2008 (left); demand profile in response to volatile price signal (right).

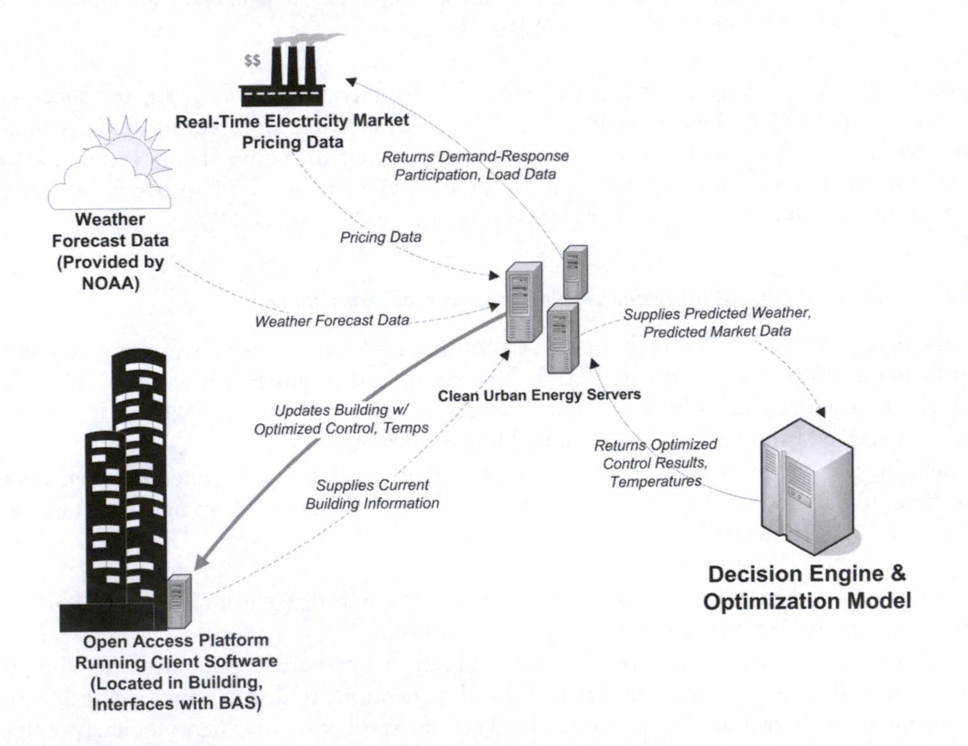

Figure 14.23 Model-based control system context diagram.

through an application server to the model-based optimization environment that calculates the cost minimal temperature setpoint trajectory for the building under control. The optimization server is hosted by a cloud computing service that scales the required computing hardware with the number of simultaneous optimizations. The EnergyPlus building energy model used in the online optimal control scenario has been generated through the help of an initial building audit, followed by manual and automated model calibration steps. The optimal control setpoints are communicated back through the application server as XML output files and then to the open-access BAS and implemented in the building.

Closed-loop optimization of model-based control is employed in this application,[13] i.e., the predictive optimal controller carries out an optimization over a predefined planning horizon L and of the generated optimal strategy only the first action is executed. At the next time step the process is repeated. The final control strategy of this near-optimal controller over a total simulation horizon of K steps is thus composed of K initial control actions of K optimal strategies of horizon L, where $L < K$. Figure 14.24 illustrates the procedure involved in determining the predictive optimal control policy. By moving the time window of L time steps forward and updating the control strategy after each time step, a new forecast is introduced at each time step and yields a policy (thick dotted line), which is different from the policy found without taking new forecasts into account (thin dashed line).

Notes

1 Source: http://www.eia.doe.gov/emeu/aer/pdf/pages/sec2_6.pdf
2 Source: http://www.eere.energy.gov/buildings/tools_directory.
3 Source: http://www.energyplus.gov.
4 Source: http://simulationresearch.lbl.gov.
5 Source: http://www.esru.strath.ac.uk.
6 Source: http://ourworld.compuserve.com/homepages/edsl.
7 Source: http://sel.me.wisc.edu/trnsys.
8 Source: IDA-ICE 3.0, www.equa.se, 2009.
9 Source: http://www.mathworks.de.
10 Source: http://www.dynasim.se bzw. http://www.modelica.org.

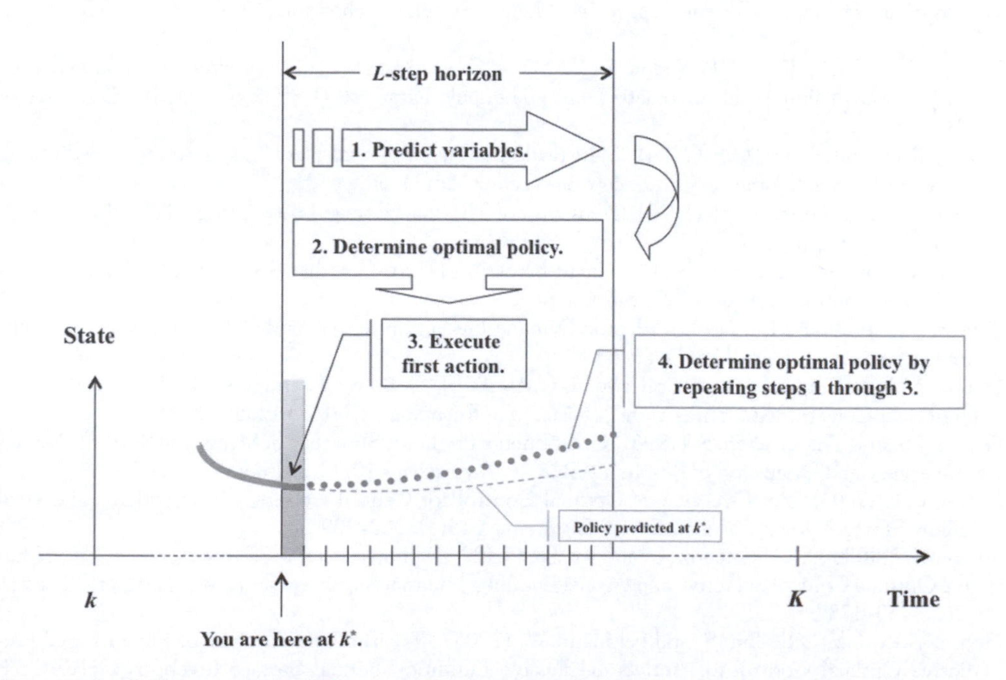

Figure 14.24 Schematic of closed-loop model-based optimal control.

11 Source: http://www.mathcore.com.
12 Source: http://www.fchart.com.
13 Source: http://www.cleanurbanenergy.com.

References

Adam, C., André, P., Hannay, C., Hannay, J., Lebrun, J., Lemort, V. and Teodorese, V. (2006) "A Contribution to the Audit of an Air-conditioning System: Modeling, Simulation and Benchmarking", *7th International Conference on System Simulation in Buildings*, Liège, Belgium.

ASHRAE (2009) *ASHRAE Handbook Fundamentals 2009*, Atlanta: American Society of Heating, Refrigerating and Air-Conditioning Engineers.

Bishop, C.M. (1995) *Neural Networks for Pattern Recognition*, Oxford: Oxford University Press.

Box, G.E.P., Jenkins, G.M. and Reinsel, G.C. (1994) *Time Series Analysis: Forecasting & Control*, Upper Saddle River, NJ: Prentice-Hall.

Brandemuehl, M.J. and Bradford, J. (1998) "Implementation of Online Optimal Supervisory Control of Cooling Plants without Storage", *Technical Report*, Joint Center for Energy Management, University of Colorado, Colorado.

Braun, J.E., Klein, S.A., Mitchell, J.W. and Beckman, W.A. (1989a) "Applications of Optimal Control to Chilled Water Systems without Storage", *ASHRAE Transactions* 95(1).

Braun, J.E., Klein, S.A., Mitchell, J.W. and Beckman, W.A. (1989b) "Methodologies for Optimal Control to Chilled Water Systems without Storage", *ASHRAE Transactions* 95(1).

Braun, J.E. (2003) "Load Control Using Thermal Mass", *Journal of Solar Energy Engineering* 125(3): 292–301.

Braun, J.E and Chaturvedi, P.D.N. (2002) "An Inverse Gray-Box Model for Transient Building Load Prediction", *HVAC&R Research* 8(1): 73–100.

Chen, T.Y. and Athienitis, A.K. (1996) "Ambient Temperature and Solar Radiation Prediction for Predictive Control of HVAC Systems and a Methodology for Optimal Building Heating Dynamic Operation", *ASHRAE Transactions* 102(1): 26–35.

Cheng, H., Brandemuehl, M.J., Henze, G.P., Florita, A.R. and Felsmann, C. (2008) "Evaluation of the Primary Factors Impacting the Optimal Control of Passive Thermal Storage", *ASHRAE Transactions, Technical Paper (1313-RP)* 114(2): 57–64.

Claridge, D.E. (2004) "Using Simulation Models for Building Commissioning", *International Conference on Enhanced Building Operation*, Energy Systems Laboratory, Texas A&M University, Texas.

Déqué, F., Ollivier, F. and Poblador, A. (2000), "Grey Boxes Used to Represent Buildings with a Minimum Number of Geometric and Thermal Parameters", *Energy and Buildings* 31(1): 29–35.

Dodier, R.H. and Henze, G.P. (2004) "Statistical Analysis of Neural Networks as Applied to Building Energy Prediction"*Journal of Solar Energy Engineering* 126(1): 592–600.

Ferrano, F.J. and Wong, K.F.V. (1990) "Prediction of Thermal Storage Loads Using a Neural Network", *ASHRAE Transactions* 96(2): 723–726.

Florita, A.R. and Henze, G.P. (2009) "Comparison of Short-Term Weather Forecasting Models for Model Predictive Control", *HVAC&R Research*, in press.

Gibson, F.J. and Kraft, T.T. (1993) "Electric Demand Prediction Using Artificial Neural Network Technology", *ASHRAE Journal* 35(3): 60–68.

Gouda, M.M., Danaher, S. and Underwood, C.P. (2002) "Building Thermal Model Reduction Using Nonlinear Constrained Optimization", *Building and Environment* 37(12):1255–1265.

Haberl, J.S. and Thamilseran, S. (1996) "Great Energy Predictor Shootout ii: Measuring Retrofit Savings – Overview and Discussion of Results", *ASHRAE Transactions* 102(2): 419–435.

Henze, G.P. (2003) "An Overview of Optimal Control for Central Cooling Plants with Ice Thermal Energy Storage", *Journal of Solar Energy Engineering* 125(3): 302–309.

Henze, G.P., Kalz, D.E., Felsmann, C. and Knabe, G. (2004) "Impact of Forecasting Accuracy on Predictive Optimal Control of Active and Passive Building Thermal Storage inventory", *HVAC&R Research* 10(2): 153–178.

Henze, G.P., Kalz, D.E., Liu, S. and Felsmann, C. (2005) "Experimental Analysis of Model-Based Predictive Optimal Control for Active and Passive Building Thermal Storage Inventory", *HVAC&R Research* 11(2): 189–214.

Henze, G.P., Felsmann, C., Florita, A.R., Brandemuehl, M.J., Cheng, H. and Waters, C.E. (2008) "Optimization of Building Thermal Mass Control in the Presence of Energy and Demand Charges", *ASHRAE Transactions, Technical Paper (1313-RP)* 114(2): 75–84.

Henze, G.P., Yuill, D.P. and Coward, A. (2009) "Development of a Model Predictive Controller for Tankless Water Heaters", *HVAC&R Research* 15(1): 3–25.

Katipamula, S.P. and Brambley, M.R.P. (2005a) "Methods for Fault Detection, Diagnostics, and Prognostics for Building Systems – A Review, Part I", *HVAC&R Research* 11(1): 3–25.

Katipamula, S.P. and Brambley, M.R.P. (2005b) "Methods for Fault Detection, Diagnostics, and Prognostics for Building Systems – A Review, Part II", *HVAC&R Research* 11(2): 169–187.

Kawashima, M., Dorgan, C.E. and Mitchell, J.W. (1995) "Hourly Thermal Load Prediction for the Next 24 Hours by arima, lr, and an Artificial Neural Network", *ASHRAE Transactions* 101(1): 186–200.

Kintner-Meyer, M. and Emery, A.F. (1995) "Optimal Control of an HVAC System Using Cold Storage and Building Thermal Capacitance", *Energy and Buildings* 23(1): 19–31.

Kissock, J.K., Haberl, J. and Claridge, D.E. (2003) "Inverse Modeling Toolkit: Numerical Algorithms.", *ASHRAE Transactions* 109(2): 425–434.

Klein, S.A. (1973) "TRNSYS – A Transient Simulation Program", Solar Energy Laboratory, University of Wisconsin-Madison, Wisconsin-Madison.

Kreider, J.F. and Haberl, J.S. (1994) "Predicting Hourly Building Energy Usage", *ASHRAE Journal* 36(6): 72–81.

Kreider, J.F. and Wang, X.A. (1992) "Improved Artificial Neural Networks for Commercial Building Energy Use Prediction", *Solar Engineering: Proceedings of Annual ASME Solar Energy Conference*, Maui, HI.

Lebrun, J. (2001) "Simulation of a HVAC System with the Help of an Engineering Equation Solver", *Building Simulation, 7th International IBPSA Conference*, Rio de Janeiro, Brazil.

Ljung, L. (1999) *System Identification – Theory for the User*, Upper Saddle River, NJ : Prentice-Hall.

Ma, Z. and Wang, S.W. (2009) "An Optimal Control Strategy for Complex Building Central Chilled Water Systems for Practical and Real-time Applications", *Building and Environment* 44(6): 1188–1198.

MacAthur, J.W., Mathur, A. and Chao, J. (1989) "On-line Recursive Estimation for Load Profile Prediction", *ASHRAE Transactions* 95(1): 621–628.

Massie, D.D. (2002) "Optimization of a Building's Cooling Plant for Operating Cost and Energy Use", *International Journal of Thermal Sciences* 41(12): 1121–1129.

Matlab (2007) "Matlab User's Guide v7.4", The Mathworks, Inc.

May-Ostendorp, P. and G.P. Henze (2009) "Preliminary MM Optimization Results: Addendum for Submittal with USGBC Q2'09 Progress Report", United States Green Building Council, Washington, DC.

MPC Toolbox (2007) "Model Predictive Control Toolbox User's Guide v2.24", The Mathworks, Inc.

Palomo del Barrio, E. and Guyon, G. (2002) *Using Parameters Space Analysis Techniques for Diagnostic Purposes in the Framework of Empirical Model Validation – Theory, Applications and Computer Implementation*, Report IEA Annex 22, Subtask A.

Real-Time Workshop (2007) "Real-Time Workshop User's Guide v6.6.1", The Mathworks, Inc.

Sahlin, P.P. (2000) "The Methods of 2020 for Building Envelope and HVAC Systems Simulation – Will the Present Tools Survive?", *CIBSE/ASHRAE Conference*, Dublin.

Sahlin, P., Eriksson, L., Grozman, P., Johnsson, H., Shapovalov, A. and Vuolle, M. (2004) "Whole-building Simulation with Symbolic DAE Equations and General Purpose Solvers", *Building and Environment* 39(8): 949–958.

Schijndel, A.W.M., van (2002) "Optimal Operation of a Hospital Power Plant", *Energy and Buildings* 34(10): 1055–1065.

Seem, J.E. and Braun, J.E. (1991) "Adaptive Methods for Real-time Forecasting of Building Electrical Demand", *ASHRAE Transactions* 97(1): 710–721.

Simulink (2007) "Simulink User's Guide v6.61", The Mathworks, Inc.

Spindler, H. (2004) "System identification and Optimal Control for Mixed-mode Cooling", PhD Thesis, Massachusetts Institute of Technology, USA.

Spindler, H. and Norford, L. (2008a) "Naturally Ventilated and Mixed-mode Buildings – Part I: Thermal Modeling", *Building and Environment* 44(4): 736–749.

Spindler, H. and Norford, L. (2008b) "Naturally Ventilated and Mixed-mode Buildings – Part II: Optimal Control", *Building and Environment* 44(4): 750–761.

Vilim, R.B., Garcia, H.E. and Chen, F.W. (2001) "An Identification Scheme Combining First Principle Knowledge, Neural Networks, and the Likelihood Function", *IEEE Transaction on Control System Technology* 9(1): 186–199.

Wang, S.W. and Ma, Z. (2008) "Supervisory and Optimal Control of Building HVAC Systems: A Review", *International Journal of HVAC&R Research* 14(1): 3–32.

Wetter, M. (2004) "Simulation-Based Building Energy Optimization", PhD Thesis, Department of Mechanical Engineering, University of California, Berkeley.

xPC Target (2007) "xPC Target User's Guide v3.2", The Mathworks, Inc.

Recommended reading

Allgower, F. and Zheng, A. (2000) *Nonlinear Model Predictive Control*, Berlin: Springer Verlag.

Camacho, E.F. and Bordons, C. (2007) *Model Predictive Control*, Berlin: Springer Verlag.

Kouvaritakis, B. and Cannon, M. (2001) *Non-Linear Predictive Control: Theory & Practice*, London: IEE Publishing.

Prett, D. and Garcia, C. (1988) *Fundamental Process Control*, Boston, MA: Butterworth-Heinemann.

Rossiter, J.A. (2003) *Model-Based Predictive Control: A Practical Approach*, Boca Raton, FL: CRC Press.

Activities

First, using lumped parameter analysis, you are to develop a state space model of a multilayer opaque construction element, something that experts refer to as a wall. Assume that the overall transmission heat transfer resistance R_T from an interior temperature T_i to an exterior temperature T_o can be broken into 10% interior convection resistance, 40% conduction resistance, and 50% exterior convection resistance. Further, assume that the total wall capacitance C_T can be broken into 15% capacitance of the internal mass node (inside surface) and 85% on the exterior mass node (outside surface). The result will be lumped parameter model with three resistors and two capacitors. To simplify the solar radiation treatment, assume that a radiative flux q_r acts on the inside surface only. Begin with an energy balance on both inside and outside surface temperature nodes and develop a state space model for use in a modeling environment such as R, Matlab, or Simulink.

Next, develop a model to compare the thermal response of a "high thermal capacity" element and a "low thermal capacity" element using the information given below. Excite the three input variables (internal air temperature; radiant flux; external air temperature) in turn and compare the responses of the two construction types using graphical outputs of the time responses with both results shown on the same graph. Comment on the practical significance of this by giving examples of applications in which the two construction types would be most suited.

"High" thermal capacity:
20mm light plaster
100mm heavy concrete block
70mm mineral wool insulation
25mm air gap
100mm outer brick

"Low" thermal capacity:
12mm slag
10mm air gap
100mm light concrete block
70mm mineral wool insulation
25mm air gap
2mm metal clad

Find the parameters (R_T, C_T) for the above using the standard data listed in Table 11 on page 28.19 of the 1997 ASHRAE Handbook of Fundamentals (SI units).

Answers

The solution to this activity borrows from an article by Gouda, Danaher, and Underwood (2002) and their contribution is gratefully acknowledged. We visualize the problem of transient heat transfer through an opaque multilayer construction element as an equivalent electric *R*-C circuit:

This can represent a wall, floor or roof of *n* layers using two capacitors, C_i, C_o, and three resistors, R_i, R_m, R_o From the problem statement

$$R_i = 0.1R_T \ ; R_m = 0.4R_T \ ; R_o = 0.5R_T$$

and

$$C_i = 0.15C_T \ ; C_o = 0.85C_T$$

(in which R_T, C_T are the overall resistance and capacitance for the construction element). Note that q_r is a radiant source term (e.g. solar radiation).

With $\dot{T} \equiv \frac{dT}{dt}$ the circuit can be described by the following energy balances about the internal temperature nodes when substitutions for C_i, C_o and R_i, R_m, R_o are made:

$$0.15C_T\dot{T}_s = q_r + (T_i - T_s)/0.1R_T - (T_s - T_s')/0.4R_T$$

$$0.85C_T\dot{T}_s' = (T_s - T_s')/0.4R_T - (T_s' - T_o)/0.5R_T$$

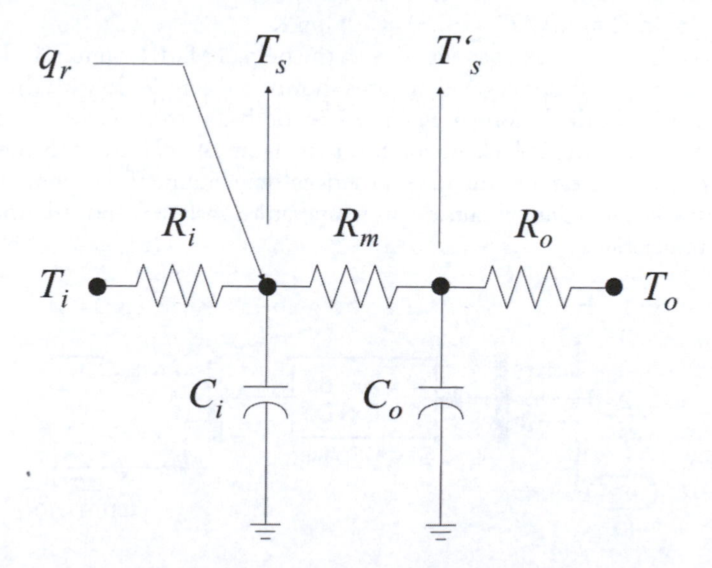

Figure 14.25 Equivalent thermal network for multilayer construction element.

Rearranging:

$$\dot{T}_s = \frac{q_r}{0.15C_T} + \frac{T_i}{0.015R_TC_T} - \frac{0.075T_s}{0.06R_TC_T} + \frac{T_s'}{0.06R_TC_T} \tag{14.23}$$

$$\dot{T}_s' = \frac{T_s}{0.34R_TC_T} - \frac{0.765T_s'}{0.1445R_TC_T} + \frac{T_o}{0.425R_TC_T} \tag{14.24}$$

(note that R_T, C_T are the overall thermal resistance and capacitance of the construction element *per unit surface area*, respectively). The outputs, T_s, T_s', are the state variables; $\dot{T}_s$,$\dot{T}_s'$ are state variable derivatives; T_i, q_r T_o are input variables and R_T, C_T are parameters.

We can express Eqs. (14.23) and(14.24) in state space notation:

$$\dot{\mathbf{x}} = \mathbf{Ax} + \mathbf{Bu}$$
$$\mathbf{y} = \mathbf{Cx} + \mathbf{Du}$$

As a result, we obtain

$$\begin{bmatrix} \dot{T}_s \\ \dot{T}_s' \end{bmatrix} = \begin{bmatrix} \frac{-0.075}{0.0009R_TC_T} & \frac{1}{0.06R_TC_T} \\ \frac{1}{0.34R_TC_T} & \frac{-0.765}{0.1445R_TC_T} \end{bmatrix} \times \begin{bmatrix} T_s \\ T_s' \end{bmatrix} + \begin{bmatrix} \frac{1}{0.015R_TC_T} & \frac{1}{0.15C_T} & 0 \\ 0 & 0 & \frac{1}{0.425R_TC_T} \end{bmatrix} \times \begin{bmatrix} T_i \\ q_r \\ T_o \end{bmatrix} \tag{14.25}$$

$$\begin{bmatrix} y_1 \\ y_2 \end{bmatrix} = \begin{bmatrix} 1 & 0 \\ 0 & 1 \end{bmatrix} \times \begin{bmatrix} T_s \\ T_s' \end{bmatrix} + \begin{bmatrix} 0 & 0 & 0 \\ 0 & 0 & 0 \end{bmatrix} \times \begin{bmatrix} T_i \\ q_r \\ T_o \end{bmatrix} \tag{14.26}$$

Using the information suggested in the problem statement, the total resistance and total capacitance for the two construction types can be found to be $R_{T,hi}$ = 2.327 m²K/W, $C_{T,hi}$ = 3.02 × 10⁵ J/m²K and $R_{T,lo}$ = 2.552 m²K/W, $C_{T,lo}$ =8.81 × 10⁴ J/m²K.

Results shown here have been generated with the help of Matlab/Simulink. The state space model of the multilayer construction element is shown in Figure 14.26 with the model parameters being shown in the dialog form in Figure 14.27. Both construction elements can be simultaneously investigated using a block model shown in Figure 14.28. Here, only the inside surface temperature T_s is of interest and the outside surface temperature T_s' is ignored. Figure 14.29 shows sample results for a diurnal outside air temperature variation and a diurnal step change in inside air temperature.

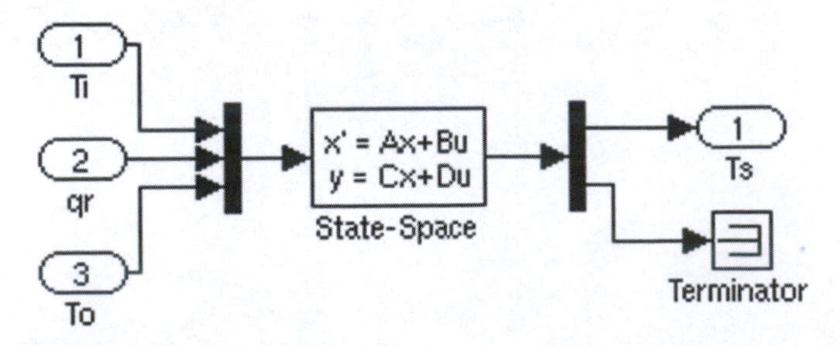

Figure 14.26 State space model simulation block.

Table 14.6 Material properties of individual layers and composite wall

High Thermal Mass

Code Number	Description	L	c_p	R	m	C	
E0	Inside surface resistance	0	0.00	0.121	0.00	0.00	
E1	20mm plaster	20	0.84	0.026	30.74	25.82	
C3	100mm high density concrete block	100	0.84	0.125	99.06	83.21	
B24	70mm insulation	70	0.84	1.76	6.34	5.33	
B1	Air space resistance	0	0.00	0.16	0.00	0.00	
A2	100mm face brick	100	0.92	0.076	203.50	187.22	
A0	Outside surface resistance	0	0.00	0.059	0.00	0.00	
Total				2.327	m^2K/W	301.58	kJ/m^2K
						3.02E+05	J/m^2K

Low Thermal Mass

Code Number	Description	L	c_p	R	m	C	
E0	Inside surface resistance	0	0.00	0.121	0.00	0.00	
E2	12mm slag	20	0.84	0.026	30.74	25.82	
B1	Air space resistance	0	0.00	0.16	0.00	0.00	
C2	100mm low density concrete block	100	0.84	0.266	61.98	52.06	
B24	70mm insulation	70	0.84	1.76	6.34	5.33	
B1	Air space resistance	0	0.00	0.16	0.00	0.00	
A3	2mm steel siding	2	0.42	0	11.71	4.92	
A0	Outside surface resistance	0	0.00	0.059	0.00	0.00	
Total				2.552	m^2K/W	88.13	kJ/m^2K
						8.81E+04	J/m^2K

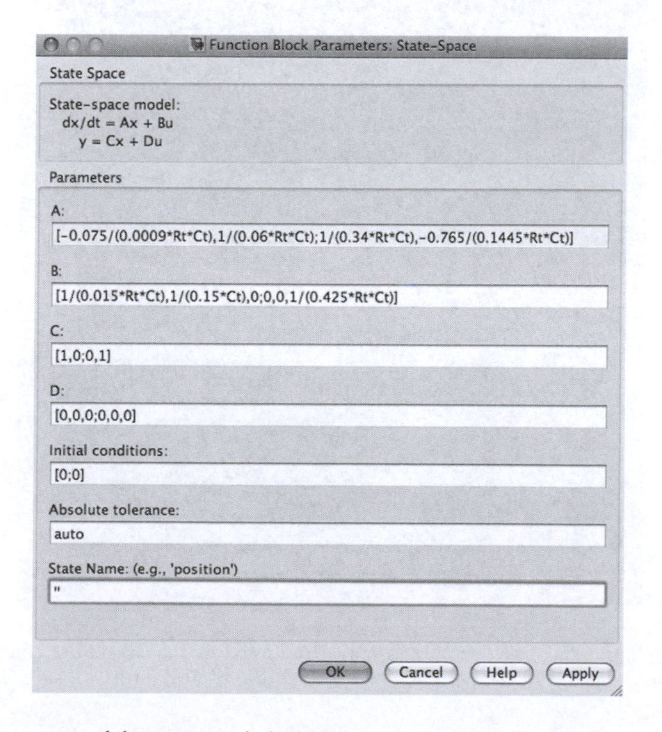

Figure 14.27 State space model parameter dialog form.

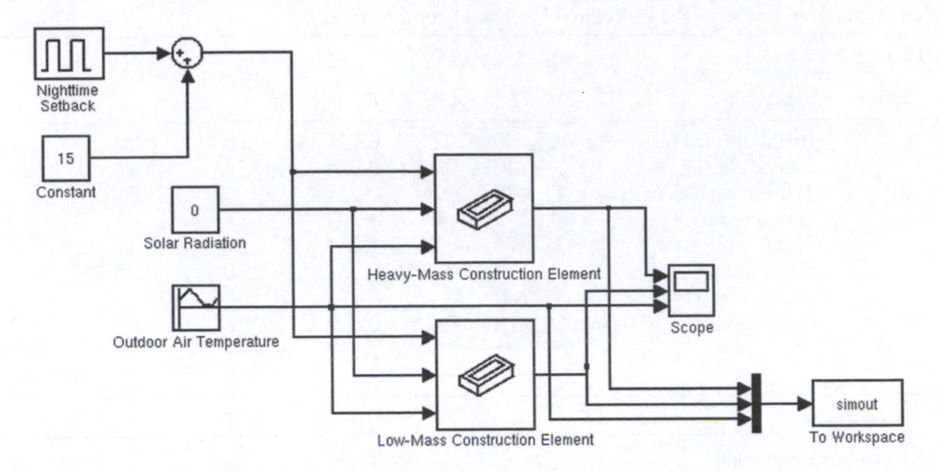

Figure 14.28 Simulation program to evaluate the thermal response of high-mass and low-mass construction elements subject to inside and outside air temperature variations.

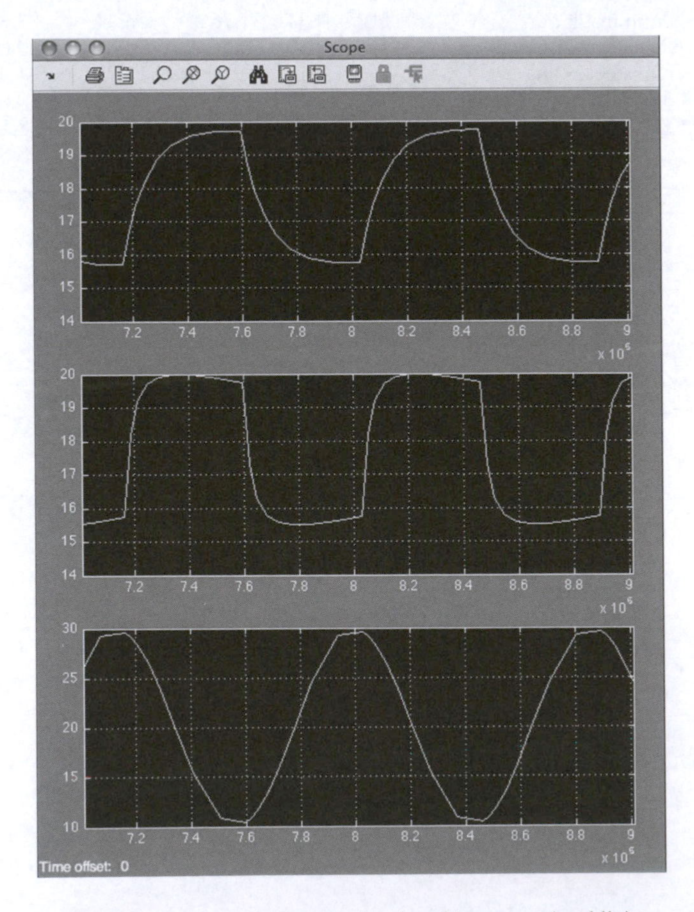

Figure 14.29 Sample thermal response of high-mass (top) and low-mass (middle) construction subject to diurnal outside air temperature variation (bottom) and diurnal inside temperature setback (not graphed).

15 Integrated resource flow modelling of the urban built environment

Darren Robinson

Scope

The vast majority of energy and matter to sustain human life is consumed in urban settlements, with profound environmental consequences. There is thus a growing interest in developing techniques to better understand this urban metabolism and ways of minimising it. With this in mind, this chapter reviews progress that has been made to predict the availability of resources for urban energy conversion as well as in the simulation of urban resource flows. From this it is clear that considerably more progress is required to support the design of new and development of existing settlements along a more sustainable trajectory. In particular, the need is discussed for progress in: more comprehensive resource flow modelling (including exchanges between buildings); urban microclimate modelling; dynamics of urban growth and land use change; transportation of good and services; heuristic optimisation algorithms. Clearly this is an interdisciplinary challenge.

Learning objectives

Although very much in its infancy, the modelling of resource flows at the urban scale is already a vast topic. Work thus far has led to improvements in the ways in which we simulate the presence and behaviour of building occupants and in the interactions between buildings and the urban climate (both radiant and thermal). Other innovations relate to the achievement of a good balance between predictive accuracy and computational efficiency for thermal and plant models as well as in describing urban scenes in an efficient way. There is also much work to do to achieve truly comprehensive models of urban resource flows, to be able to optimise these flows and how they may evolve with time.

The aims of this chapter are to give an overview of the achievements that have thus far been made in this new modelling domain as well as the many challenges that await us. The objectives here are both to provide an insight into the importance of considering buildings in their urban context as well as to stimulate further research in this domain.

Key words

Urban modelling/irradiation modelling/climate modelling/pedestrian comfort/behavioural modelling/ optimisation/land use and transport interaction modelling/CitySim.

Introduction

Non-ambient resources are imported into urban settlements to sustain them to: accommodate humans in a comfortable way, perform the commercial and industrial processes which employ most of them, transport them and to furnish the goods that they consume. In contrast to the circular flows in natural ecosystems, resource flows in urban settlements are essentially linear. This consumption causes a depletion of raw materials and the transformation of chemical substances tends to lead to a concentration of hazardous residues which pollute air, water and soil. Air pollution in particular is increasingly evident in climatic disorders and health hazards linked with climate change.

On a small scale, this may not be significant. But with around three quarters of the European population and half of the global population (or three billion) currently urbanised the scale of this resource consumption is profound. Indeed, some 75% of global resource consumption takes place within our increasingly interconnected urban settlements, which cover only 2% of the Earth's surface (Girardet 1999). Furthermore, it is projected that globally almost two thirds of humans will live in cities by 2030 and three-quarters by 2050 (UN 2004). Urban resource consumption is therefore set to further intensify unless natural and applied resources are managed more efficiently, perhaps inspired by the highly circular flows of resources found in natural ecosystems (Erkman 1998).

The more efficient management of urban resources implies neither a financial burden nor a reduction in quality of life; in fact, quite the contrary. For example, decentralised resource management benefits from economies of scale: costs of energy conversion systems do not scale linearly. In the case of energy conversion systems, the cost per kW of installed capacity tends to reduce with size, to the extent that in Europe a growing number of energy supply companies are being established who self finance the installation of district heat and co-generation systems in return for medium term energy purchasing contracts. Moreover, in Germany local urban cooperatives have been created to fund and manage these investments; the by-product being a stronger sense of local community (Barton 2000).

But these grass roots initiatives to more efficiently manage urban resource flows remain the exception, despite growing political support. Further progress would be facilitated with the availability of appropriate decision making support tools. The purpose of this chapter is to review the progress that has been made thus far to develop such tools as well to identify related research needs for their further development.

But the subject of urban modelling is both broad and diverse. Nevertheless, modelling approaches may be broadly classified into those which tackle the availability (supply side) of resources and those which simulate their flows (demand and supply side). To this end, models of resource availability are first discussed before attending to the more complete modelling of resource flows. Concerning the latter, models focussing on the energy demands of domestic and non-domestic are buildings are first discussed, before going on to describe more comprehensive tools which are designed for simulating energy flows in a considerably more comprehensive way as well as for simulating water and waste flows. Outstanding urban modelling challenges are then introduced; firstly with respect to urban climate modelling, which influences resource demands as well as pedestrian comfort, so that this is also discussed. Research needs for the more comprehensive modelling of resource flows and for the optimisation of these flows are then discussed before introducing the particularly challenging task of modelling urban spatial dynamics and the interactions between land use and transport.

Urban resource availability

Research on simulating the availability of ambient resources has thus far focused on evaluating the potential for using solar energy conversion systems in urban settings. In particular, this has involved the development of techniques for simulating annual solar irradiation I^T (MWh.m^{-2}) incident on urban surfaces to evaluate the potential for solar thermal and photovoltaic collectors. Two alternative approaches were developed in parallel – one based on statistical sky reduction, the other on the use of a cumulative sky – both using the backwards Monte Carlo ray tracing program RADIANCE (Ward and Shakespeare 1997).

Inspired by the use of daylight coefficients in the simulation of interior daylighting (see Chapter 10) Mardaljevic (2000, 2003) introduced a technique of post-processing simulation images from RADIANCE produced using unit (ir)radiance light sources representing the sun (s), anisotropic sky (an) and overcast sky (ov) respectively. Firstly, the sky is discretised into bins of equal range in azimuth and altitude (8° by default). From calculations of sun position throughout the year, these bins are then filtered into those that are active (the sun is within this bin for one or more time steps). For Leicester in the UK this results in b = 179 active bins for a year, based on a 15min time resolution. A RADIANCE simulation is then performed for a *unit radiance* light source located at the mean position within each of these bins. A similar process is repeated for the anisotropic sky calculation, but since the region of circumsolar brightening is large relative to the size of the sun, a cruder discretisation is appropriate (e.g. 16°). A RADIANCE simulation is then performed using the CIE (Conseil Internationale d'Eclairage) intermittent sky model, based on a *unit diffuse horizontal irradiance*. One final simulation is performed to account for overcast sky conditions, also with respect to unit diffuse horizontal irradiance. The irradiation I^T (Wh) for the period of interest is then determined, for each pixel in the resultant image, from combining results corresponding to each relevant sun position with the associated (ir)radiance to scale these results, according to the expression: $I^T = \Delta t \sum_{h=1}^{n} [(I_s R) + (I_{an} f I_{dh}) + I_{ov} (1 - f) I_{dh}]_h$. *In this* R is the radiance of the sun, I the incident irradiance, f is a clearness factor which governs a blending between intermittent and overcast skies, I_{dh} is the diffuse horizontal irradiance and Δt is the integration interval (e.g. 0.25h) throughout the n hours.

However, if our purpose is simply to predict the total incident irradiation for a given period rather than its temporal distribution, then an alternative is simply to perform a single simulation using a sky which represents the total annual radiance distribution ($Wh.m^{-2}.Sr^{-1}$): a pre-process as opposed to a post-process. This was the approach adopted by Compagnon (2000, 2004).

Compagnon's technique involved distributing a set of n round light sources throughout the sky vault, with the cumulative radiance of each light source being calculated relative to its centre, using the Perez luminance distribution model[1] (Perez *et al.* 1993) – so that for the ith light source and nth hourly interval of time $R_i^T = \sum_{h=1}^{n} R_{i,h}$, where $R_{i,h}$ is given by the Perez model for the diffuse component and the quotient of beam solar irradiance and light source solid angle (I_b / Φ_s) for the solar component. This approach is beneficial in the sense that only RADIANCE's direct calculation is used, so that it is efficient. A downside is that the means for subdividing partially obscured light sources in RADIANCE is a little crude, with corresponding implications for accuracy. The use of discrete light sources may also introduce parallax errors (Mardaljevic 1999).

An alternative is to divide the sky vault itself into discrete regions (or patches), for which the uniform radiance is calculated at the patch centroid (Robinson and Stone 2004a), also using a luminance/radiance distribution model due to Perez *et al.* (1993) for the diffuse component and the solar component may either be effectively smeared over the patch as with Compagnon's approach or represented as multiple discrete light sources of appropriate position and solid angle. Although this approach uses RADIANCE's ambient calculation, which is computationally less efficient (though not dramatically so), it does not suffer from the same loss of accuracy as with Compagnon's method.

Applications

The most straightforward application of the solar irradiation modelling techniques discussed above is to visualise the irradiation distribution throughout a given urban scene (Figure 15.1(a)) using the RADIANCE program *r-pict*. To further aid with results interpretation we may export simulation results from RADIANCE into a format readable by external software (e.g. MATLAB), using the command *p-value*. Brightness values for each pixel may then be processed, for example to display only those pixels for which we believe photovoltaic cells to be economically viable (e.g. for which $I^T > 800kWh.m^{-2}$) – Figure 15.1(b)).

An alternative to the rendering and processing of images is to calculate irradiation results for discrete points of a given position (x,y,z) and view vector (x,y,z) within our urban scene using the RADIANCE program *r-trace*.

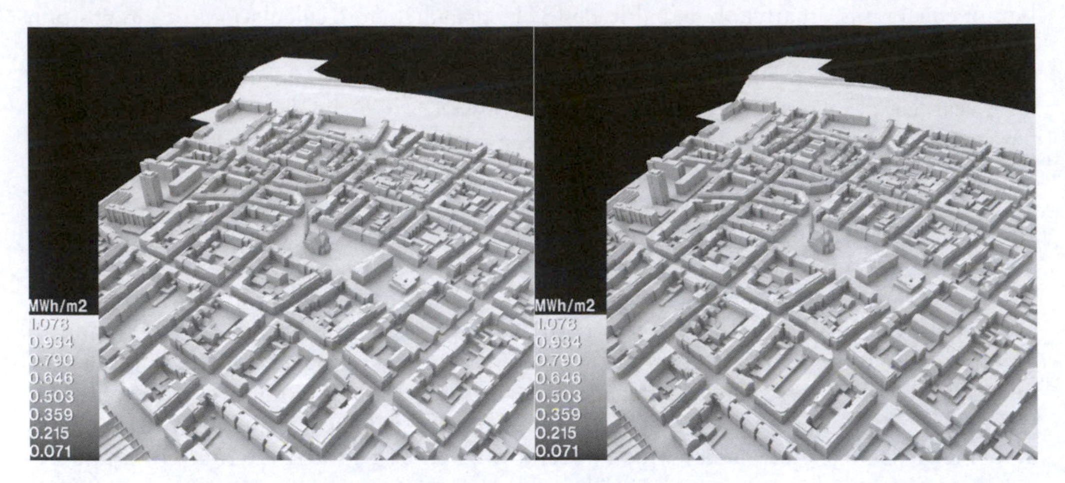

Figure 15.1 (a) Irradiation distribution, and (b) thresholded at 0.8MWh/m² to represent surfaces that are viable for the installation of solar cells.

With this approach Compagnon (2004) was able to quantify, to a reasonable degree of accuracy, the proportion of urban surfaces for which solar collectors (thermal and photovoltaic) are viable as well as for which high performance glazing should be considered for installation in façades to reduce buildings' lighting and thermal energy demands. Indeed taking this a step further Robinson *et al.* (2005) calculated the solar energy utilisation potential of three urban districts in Switzerland, each of up to 10E+6m² in size. Presented in Figure 15.2 for example are cumulative solar irradiation distributions for both horizontal and vertical surfaces for these three districts. From this we see that assuming a threshold of 1000 kWh/m² for economic viability 49%, 65% and 95% of roof surfaces are viable for the integration of PV cells in the decreasingly dense districts of Matthaüs (Basel), Bellevaux (Lausanne) and Meyrin (Geneva) respectively.

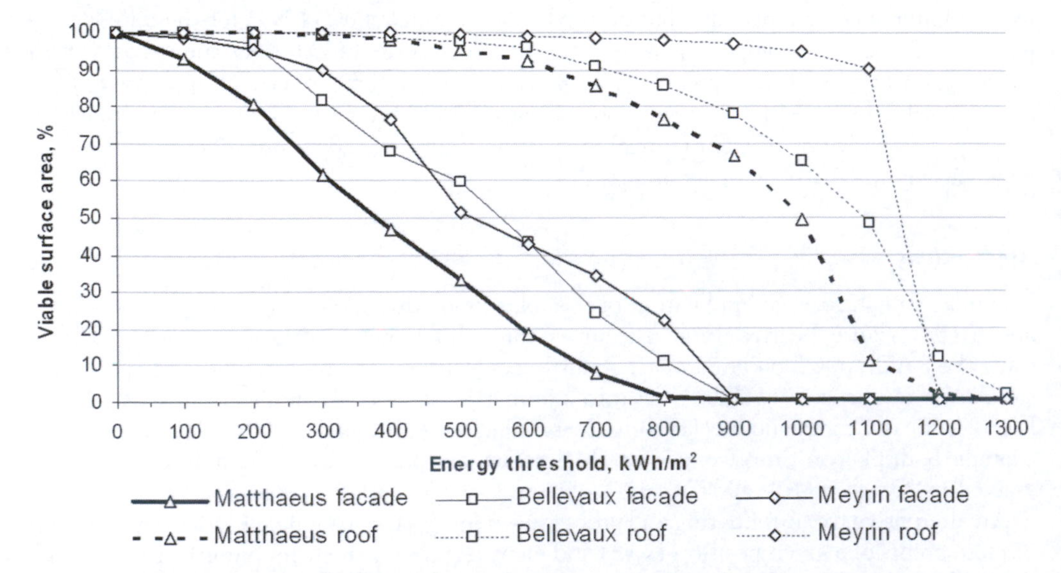

Figure 15.2 Cumulative annual solar irradiation distributions for three Swiss city districts.

Urban resource flows

As discussed in the introduction to this chapter, the key motivation for the development of urban resource flow modelling software is to support urban decision makers to improve the environmental sustainability of urban settlements. Depending upon the scale of a particular settlement, these decision makers may be politicians, regional planners, town and city planners, architects or specialist consultant engineers and physicists; with corresponding implications for the software design. With some notable exceptions, much of the stakeholder-oriented software developed thus far has targeted the broader planning scale; whilst other advances have been made from a more academic point of view, for example to better understand relationships between urban form and energy use. These developments are discussed below.

Energy modelling

Domestic energy modelling

In common with models developed for testing scenarios to improve the performance of the national stock of domestic buildings (Johnston *et al.* 2005; Natarjan and Levermore 2007; Shorrock and Dunster 1997), Jones *et al.*'s (1998, 1999, 2007) Energy and Environmental Prediction model (EEP) estimates the energy consumption of a municipality's stock of buildings based on extrapolation of results from representative building typologies. For the Welsh district of Neath Port Talbot they have constructed 100 such models which represent the average characteristics of buildings within a given category, based on building geometry and a combination of national and local field survey attribute data. In each case a stationary model, the UK Standard Assessment Procedure (SAP), is used to predict annual energy use and CO_2 emissions and to determine the costs and benefits of measures to upgrade the performance of this stock. Geographical Information System (GIS) software is used to test these scenarios and to visualise the results at a user-defined aggregation level (e.g. postcode and ward).

A similar tool has been developed to model the domestic building stock of Osaka in Japan (Shimoda *et al.* 2003), based on extrapolation of results from 460 combinations of building type, building size and household composition. A key difference in emphasis however, is in the representation of occupants' behaviour using detailed survey statistics which relate activities (particularly occupants' presence and use of appliances) to household composition. The underlying thermal model is also dynamic. Initially applied to test hypotheses regarding improvements to building fabric and space conditioning plant efficiency, this tool has also been employed (Taniguchi *et al.* 2007) to test the potential influence of behavioural changes on energy consumption.

In contrast with the above, is the modelling of individual buildings by the Solar Energy Planning tool (Rylatt *et al.* 2001; Gadsden *et al.* 2003); the purpose of which is to determine the potential for installing solar energy conversion systems in a municipality's district. After initial screening for technical and socio-economic feasibility, the remaining buildings' energy demands are predicted using a stationary monthly energy balance model. The potential for meeting this energy demand using solar thermal collectors and photovoltaic panels is then determined, to assist with their investment appraisal. Buildings with good potential are then highlighted for the municipality to target investments, also based on a GIS interface.

Non-domestic energy modelling

For the non-domestic building category there are thus far two examples. The first, LT-Urban (Ratti *et al.* 2000, 2005), was developed with a more academic objective of examining relationships between urban texture (or geometry) and energy consumption. For this a digital elevation model of an urban scene is first created. This is a two dimensional image in which each pixel has x,y coordinates and height (z) may be represented by the pixel value: 0 to 255 for a greyscale image, with 0 representing unbuilt space. Using the MATLAB™ image processing toolbox key geometric parameters are derived, relating to whether a given pixel is near to a façade or not (as a basis for judging whether or not the zone is "passive": daylight, solar gains and natural ventilation are available); what the orientation of the façade is and to what extent views from this façade are obstructed; likewise how obstructed views from opposing buildings are. This geometric information, together with a file containing default characteristics regarding the building (normally assumed to be an office) and its occupancy, is then parsed to a stationary energy model (Robinson and Baker 2000) to predict each building's energy use, accounting approximately for the effects of adjacent buildings on transmitted daylight and solar radiation. LT-Urban was employed to compare (assuming identical building characteristics in each case) the energy effectiveness of the different types of urban form found in London, Berlin and Toulouse. In each case passive zones for each 3m floor slice were also associated with a glazing ratio which minimised total energy consumption and this optimal ratio (and associated energy demand) was compared with a "typical" glazing ratio, assumed to be 30%. Such simulations, intended for comparison purposes rather than to estimate actual energy consumption, are attractive in that they take only a few seconds to complete.

In common with SEP the second example (Yamaguchi *et al.* 2003) was conceived to model the energy performance of individual buildings, based on building-specific descriptions of their constructional and occupancy characteristics. A basic stochastic model simulates occupants' weekday presence (arrival, lunch and then departure) and, whilst present, whether they are using one or two PCs or none at all; this depending upon their type of job. Other electrical loads (including lighting) are deterministically predicted on the basis of installed capacity and schedule of use. Heat gains from these electrical loads are then input to a space conditioning demand model, using a "weighting factor" method. Given a system coefficient of performance, this load is then translated into an energy demand. Results from all buildings within a given case study zone are then aggregated and input to a district energy supply model, which simulates district heating, cooling and co/tri-generation systems as well as thermal distribution losses. This model was initially applied to test scenarios for reducing the net energy consumption of two buildings, based on demand and decentralised supply options; then (Yamaguchi *et al.* 2007a) in a similar fashion to model the energy demands of and supply to fifty five large commercial buildings in a district of Osaka, Japan.

Following from this initial work Yamaguchi *et al.* (2007b) increased their scale of analysis to that of the city, again based on extrapolation from a sample of building types (in common with models of the national non-domestic building stock (Pout 2000)): six hundred and twelve for Osaka. This model has also been run in tandem with Shimoda *et al.*'s (2003) domestic model to predict the performance of the entire building stock of Osaka (Shimoda *et al.* 2007), to investigate the implications of telecommuting on building energy demand.

Summary

Owing to a few ambitious research projects, some good progress has been made in the development of software to assist urban planners to improve the energy performance of their districts and cities. Nevertheless, these software are still in their infancy. None of the above attempts to model the effects of the urban thermal microclimate on building performance and the few attempts to model the radiant environment have been highly simplistic. With the partial exception of progress made in Japan, the presence and behaviour of occupants also tends to be handled in a highly simplistic way. In no case has a rigorous attempt been made to develop a complete representation of the stochastic nature of occupants' presence and those aspects of behaviour which impact upon buildings' energy performance. This is important because such random variability between individuals' behaviours may have significant implications for both the magnitude and temporal variation of aggregate resource flow profiles. Finally, only limited progress has been made with respect to the modelling of supply from energy conversion systems, whether building-embedded or locally centralised.

Finally stakeholders have not been personally involved in applications of this urban energy modelling software. Rather it has remained in the hands of the developers.

Integrated resource flow modelling

Sustainable Urban Neighbourhood modelling tool (SUNtool)

At present the only software that fit clearly within this category are SUNtool (the Sustainable Urban Neighbourhood modelling tool) and its recent successor CitySim. In contrast with the above, SUNtool was conceived as a decision support system for *designers* to optimise the environmental sustainability of masterplanning proposals (Robinson *et al.* 2003, 2007), based on integrated resource (energy, water and waste) flow modelling.

A masterplan will typically be composed of many buildings, up to say a couple of hundred, and these may support a variety of uses (residential, commercial, schooling, healthcare, etc); not to mention adjacent obstructions, external landscaping and services (such as street lighting). For manageable predictions, we therefore need a rapid way of describing the geometric form of our masterplan and for attributing the described objects according to characteristics which influence the flows of resources. The solution of the flows of resources should also be arrived at quickly, so that the user is swiftly guided towards a well performing solution. Furthermore, the set of resource flows should respect their sensitivity to the urban microclimate, to human behaviour and to possible couplings between buildings and between buildings and systems, which may be embedded or centralised (Robinson 2005). These were the guiding principles for the development of SUNtool – the conceptual structure of which is outlined in Figure 15.3.

To model a masterplan with SUNtool the user first chooses/enters the geographical location, to set the site coordinates and associate the site with climate data. The user is then invited to select a relevant "iDefault" dataset; that is, a database of default attributes (occupancy, constructional, plant systems etc) for the range of building types. The software's basic sketching tool may then be used to develop a three-dimensional description of the site. Building objects are assigned a default use (residential), but this may be overridden, so that the appropriate characteristics are assigned to the building. Likewise any building or façade-specific characteristics may be easily overridden, by clicking on the corresponding object. Solar (thermal or electric) collectors may be associated with building surfaces in a similar fashion. Specific buildings may also be selected for association with a district energy centre, for locally centralised resource management.

When a site description is complete this is parsed, in XML file format, to the SUNtool solver. This solver has a reduced dynamic *thermal model* (Déqué *et al.* 2000) at its core, which simulates the energy required to maintain a defined temperature difference between each building and its outside environment. This model is also directly influenced by a *radiation model* which calculates, in an integrated way, the solar and thermal radiant exchanges at the building envelope and the daylight received within the building, and how these are influenced by other buildings (Robinson and Stone 2004b, 2005a, 2005b, 2006). The thermal model is also connected with deterministic profiles[2] to account for *occupants' presence and their interactions* with appliances (e.g. computers or cookers) and passive (e.g. windows or blinds) or active (e.g. lights or heating system) environmental controls for work, sustenance, leisure or comfort purposes. The deterministic or stochastic representation of artificial lighting use is itself also connected with the daylight model. Finally, a family of *plant and energy conversion system* models calculates the energy demands for space conditioning equipment and the supply of these thermal demands as well as the electrical demands of lights and appliances by renewable and/or non-renewable energy conversion systems. If there is insufficient capacity to maintain the desired environmental condition, then the initial predicted room state (assuming infinite plant capacity) is corrected.

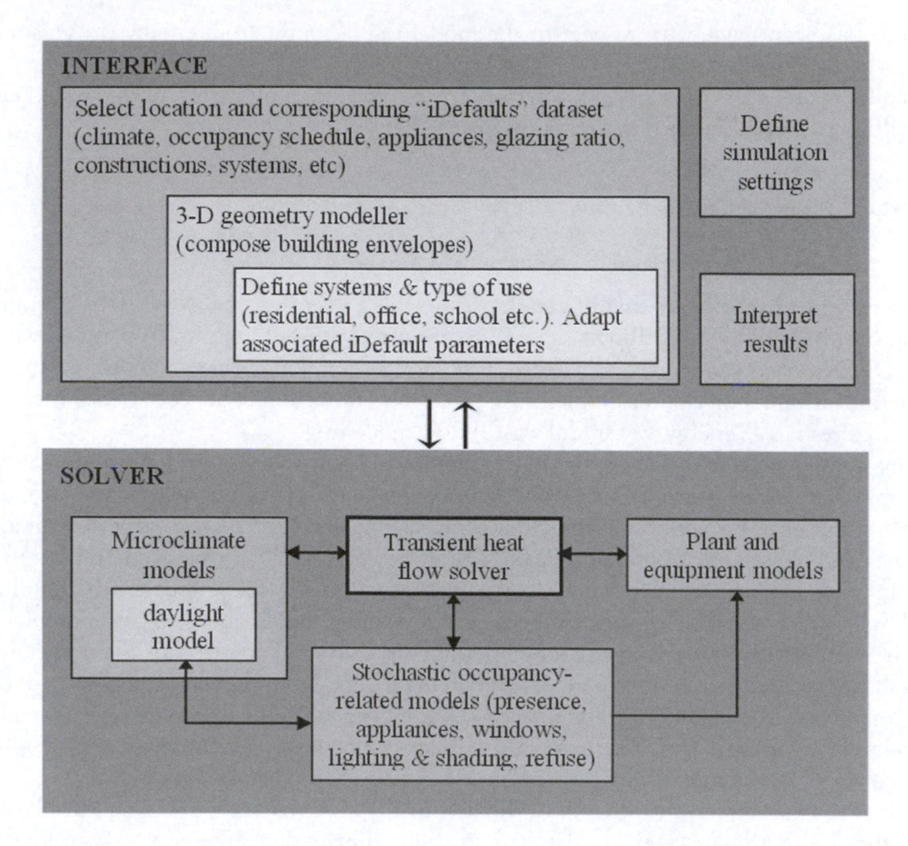

Figure 15.3 Conceptual structure of SUNtool.

Batches of simulations may be prepared, relating to a selected group of buildings and for a chosen period of time. Alternatively, parametric studies may be performed. For this latter certain parameters may be varied (one at a time) between user-defined lower and upper limits according to some chosen increment. The associated results are streamed back to the interface from the solver via XML files. When complete line graphs and summaries of results may be interrogated and buildings may be false-coloured according to the magnitude of certain variables. Animations of the evolution of these falsecoloured plots may also be displayed.

On the whole SUNtool responds well to the weaknesses identified above in relation to urban energy models. It accounts for the radiant environment, human presence and behaviour and a range of embedded and locally centralised energy conversion systems. It also explicitly models the energy needs of heating, ventilating and air conditioning plant. But SUNtool has some weaknesses of its own. The thermal model lacks generality, only basic prototypes of stochastic models of occupant's presence and their use of lighting and blinds have been integrated, not all key energy conversion systems are modelled, there is no model of energy storage associated with building or district energy centres and modelling of water and waste flows is limited.

CitySim

Partially in response to some of the above weaknesses a successor to SUNtool is under development at the Solar Energy and Building Physics Laboratory (LESO) of the Ecole Polytechnique Fédérale de Lausanne (EPFL). Called CitySim this is designed for simulating and optimising the sustainability of urban developments of various scales; from a small neighbourhood to a district and work is under way to model an entire city (Robinson 2009).

In common with its predecessor SUNtool, the use of CitySim's Java-based GUI to simulate and optimise building-related resource flows proceeds according to four key steps (Robinson *et al.* 2009):

1. Definition of site location and associated climate data.
2. Choice and adjustment of default datasets for the types and age categories of buildings to be studied.
3. Definition of 3D form of buildings and rendering of them using Java OpenGL; definition of energy supply and storage systems to be modelled; refinement of building and systems attributes; choice of simulation parameters.
4. Parsing of data in XML format from the GUI to the C++ solver for simulation of hourly resource flows; analysis of results parsed back to the GUI.

In the following we describe briefly the principles of each category of model in CitySim's solver.

THERMAL MODEL

The thermal model in CitySim is based on analogy with an electrical circuit (Lefebvre *et al.* 1987), in which the heat flow between a wall and the outside air can be represented by an electric current through a resistor linking the two corresponding nodes (wall and outside air) and the wall's inertia can be represented by a capacitance linked at that node.

In our model (Kämpf and Robinson 2007), which is a refinement of that due to Nielsen (2005), an external air temperature node T_{ext} is connected with an outside surface temperature node T_{os} via an external film conductance K_e, which varies according to wind speed and direction (Figure 15.4). T_{os}, which also experiences heat fluxes due to shortwave and longwave exchange, is connected to a wall node T_w of capacitance C_w via a conductance defined by the external part of the wall. In fact this node resembles a mirror plane, so that we have similar connections to an internal air node T_a of capacitance C_i via an internal surface node T_{is}. T_{is} may also experience shortwave flux due to transmitted solar radiation and a longwave flux due to radiant heat gains from internal sources (people and appliances) and T_a may experience convective gains due to absorbed shortwave radiation, internal casual gains and heating/cooling systems. Finally, our internal air node may be connected with our external air temperature node via a resistance due to infiltration and ventilation.

For a whole building with many subspaces, the air nodes of each zone are linked via the separating wall conductance; interzonal airflow can also be handled through this conductance. To account for the corresponding inertia the capacitance of the separating wall is subdivided and allocated to the neighbouring zones' capacitances (Figure 15.5).

Predictions from this simplified model have been shown to compare well with those of ESP-r (Clarke 2001) for a range of monozone and multizone scenarios.

RADIATION MODELS

As with SUNtool, the Simplified Radiosity Algorithm (SRA) of Robinson and Stone (2004b) is used to solve for the shortwave irradiance incident on the surfaces defining our urban scene.

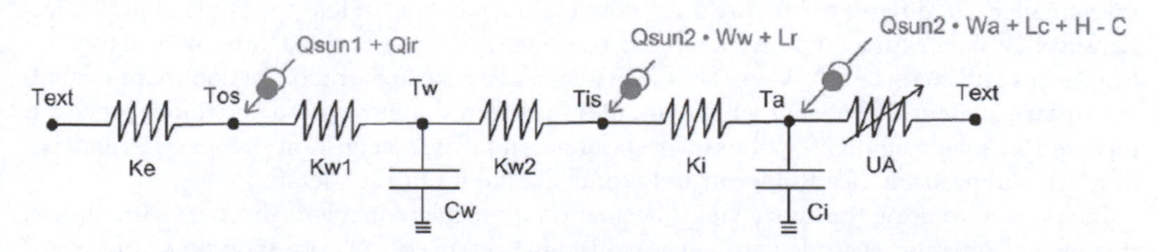

Figure 15.4 Monozone form of the CitySim thermal model.

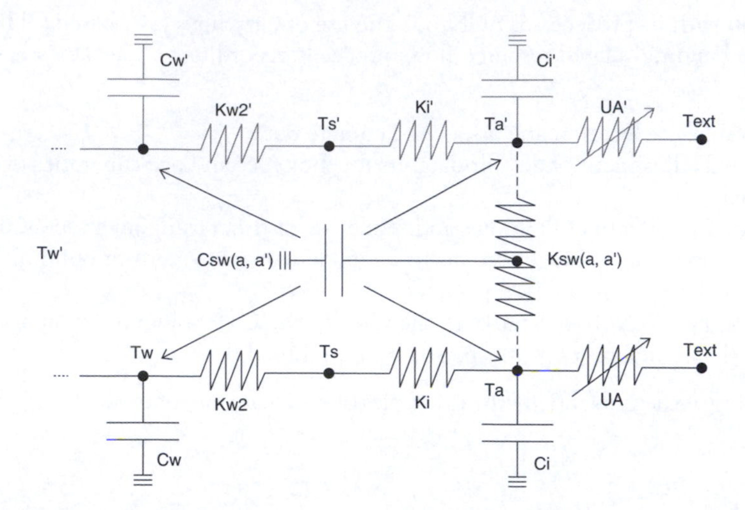

Figure 15.5 Interzonal connection between zones represented by the two-node thermal model.

For some set of p sky patches, each of which subtends a solid angle Φ (Sr) and has radiance R ($Wm^{-2}Sr^{-1}$) then, given the mean angle of incidence ξ (radians) between the patch and our receiving plane of slope β together with the proportion of the patch that can be seen σ ($0 \leq \sigma \leq 1$), the direct sky irradiance (Wm^{-2}) $I_{d\beta} = \sum_{i=1}^{p}(R\Phi\sigma \cos\xi)_i$.

For this the well known discretisation scheme due to Tregenza and Sharples (1993) is used to divide the sky vault into 145 patches of similar solid angle and the Perez all weather model (Perez *et al.* 1993) is used to calculate the radiance at the centroid of each of these patches. The direct beam irradiance $I_{b\beta}$ is calculated from the beam normal irradiance I_{bn} which is incident at an angle ξ to our surface of which some fraction ψ is visible from the sun, so that $I_{b\beta} = I_{bn}\psi \cos\xi$.

Now the direct sky and beam irradiance contributes to a given surface's radiance R which in turn influences the irradiance incident at other surfaces visible to it and vice versa. To solve for this a similar equation to that used for the sky contribution gives the reflected diffuse irradiance. In this case two discretised vaults are used, one above and one below the horizontal plane, so that $I_{p\beta} = \sum_{i=1}^{2p}(R^*\Phi\omega \cos\xi)_i$, where ω is the proportion of the patch which is obstructed by urban (reflecting) surfaces and R^* is the radiance of the surface which dominates the obstruction to this patch (in other words, that which contributes the most to ω^3). R^* itself depends on reflected diffuse irradiance as well as on the direct sky and beam irradiances. For this a set of simultaneous equations relating the beam and diffuse sky components to each surface's irradiance, which itself effects the reflected irradiance incident at other surfaces, may be formulated as a matrix and solved either by inversion or iteratively until a convergence criterion is passed (Robinson and Stone 2004b, 2005a).

The principle complication in the above algorithm lies in determining the necessary view factors. For obstruction view factors, views encapsulating the hemisphere are rendered from each surface centroid, with every surface having a unique colour. Each pixel is then translated into angular coordinates to identify the corresponding patch as well as the angle of incidence. For sky view factors then, $\Phi\sigma \cos\xi$ is treated as a single quantity obtained by numerical integration of $\cos\xi \, d\Phi$ across each sky patch. Likewise for $\Phi\omega \cos\xi$, for which the dominant occluding surface is identified as that which provides the greatest contribution. A similar process is repeated for solar visibility fractions for each surface, for which a constant size scene is rendered from the sun position. See Robinson and Stone (2005a) for further details.

In addition to using the above view information to calculate incident shortwave irradiance, this may also be used to calculate longwave irradiance; given the corresponding surface and sky temperatures (Robinson and Stone 2005a). Furthermore, additional renderings may be computed to determine the view information necessary to determine the direct and reflected contributions to internal illuminance (at known points) as well as the incoming luminous flux for internal reflection calculations (Robinson and Stone 2005a, 2006).

As Robinson and Stone (2005b) demonstrate, this simplified radiosity algorithm may be readily integrated into standard single building simulation programs to improve the accuracy of their treatment of radiation exchange in the urban context. For the interested reader, the range of urban radiation modelling techniques are presented and evaluated in detail in Robinson (2011).

BEHAVIOURAL MODELS

One of the key sources of uncertainty in building/urban simulation relates to occupants' behaviour, which is inherently stochastic in nature. Key types of behaviour and their main impacts on urban resource flows are as follows:

- Presence: metabolic heat gains and pollutants.
- Lights: heat gains and electrical power demand.
- Windows: infiltration rates.
- Blinds: illuminances and transmitted irradiance.
- Appliances (water/electrical): heat gains, electrical power demand, water consumption (and potential conversion of black water into biogas).
- Waste: production of combustible solids.

The central behavioural characteristic relates to occupants' presence, which of course determines whether they are available to exercise any other form of influence on resource flows. Based on the hypotheses that all occupants act independently and that their actions at time $t + 1$ depend only upon the immediate past (t), we model transitions in occupants' presence (present to absent (T_{10}) or present (T_{11}) and vice versa) based on the Markov condition that:

$$P\left(X_{t+1} = i \middle| X_t = j, X_{t-1} = k, \ldots, X_{t-N} = l\right) = P\left(X_{t+1} = i \middle| X_t = j\right) =: T_{ij}\left(t\right)$$

For this we take as input a profile of the probability of presence at each time t and a mobility parameter μ (the ratio of the probability of changing state to not changing state), so that the transition absent to present T_{01} is found from:

$$T_{01}(t) = \frac{\mu - 1}{\mu + 1} \cdot P(t) + P(t+1)$$

And that of present to present from:

$$T_{11} = \frac{1 - P(t)}{P(t)} \cdot T_{01} + \frac{P(t+1)}{P(t)} = \frac{1 - P(t)}{P(t)} \cdot \left[\frac{\mu - 1}{\mu + 1} \cdot P(t) + P(t+1)\right] + \frac{P(t+1)}{P(t)}$$

The remaining transitions are simply: $T_{10} = 1 - T_{11}$ and $T_{00} = 1 - T_{01}$. In addition to this we also need to consider long absences, for example due to illnesses or vacations. For this we use a daily profile of the probability of starting a long absence and a further profile for the cumulative probability of the duration of this absence. This takes precedence over the short time step transitions in presence, described above, and as such is computed first. For further details, and results from validation tests on this model, see Page *et al.* (2007).

Now, on entering a room, an occupant may experience significant changes in physical stimuli, such as the derivative of illuminance or temperature with time, as compared say with outside conditions. For this reason, as well as our acquired expectations of the conditions that we find comfortable for a given indoor environment, most interactions tend to take place at arrival. We tend to interact less frequently during occupation (this we refer to as intermediate times) and relatively more frequently at departure, for example to conserve resources or as a predictive strategy so that the envelope limits excess heating or cooling demands prior to the next time of occupation.

Following from this rationale, predicted interactions with lights are based on the same models as those integrated with Lightswitch-2002 (Reinhart 2004). In particular, the probability of switching on lights at arrival is expressed (Hunt 1980) as a logit function[4] of minimum internal workplane illuminance ($E_{i,min}$). Switching on lights at intermediate times is also predicted as a function of $E_{i,min}$ (Reinhart and Voss 2003) whereas switching off at departure is simply a function of expected duration of absence (Pigg *et al.* 1996): if we expect to be absent for a long time, we will tend to switch lights off.

In common with the model of switching on and off of lights, the window opening model (Haldi and Robinson 2009) also differentiates between actions at arrival, at intermediate times and upon departure. If at arrival (or rather 5 mins after arrival) the windows are closed, the probability of transition $P_{01}(t)$ is predicted and compared with a random number, r. If $P_{01}(t) > r$ then a transition is expected. A check is then made that the duration of this transition d, taken from a Weibull distribution, exceeds the simulation timestep, dt. If this is the case, there is indeed a transition in window status (the windows are opened). Irrespective of the outcome of this action at arrival we then enter into a period of intermediate presence. If windows are open we count down the opening duration until they are closed. If, on the other hand, windows are closed we repeat the above process to determine whether and for which duration they will be opened. Whilst windows are open, thermal simulations of the relevant zone are performed at a reduced timestep to solve for advective heat transfer. In this way our hybrid window opening algorithm avoids many potentially redundant simulations whilst maintaining a high degree of accuracy in the modelling of window openings and the associated responses. At departure, we simply predict the transition in status and, depending upon the value of a randomly drawn number, a transition is effected until the occupants' next arrival.

The model for blind usage (Haldi and Robinson 2010) also makes the distinction between actions at arrival and at intermediate times[5] the procedure is common for each of these occupancy types; only the probabilities differ). Firstly P_{lower} is compared with P_{raise}. The action of greatest probability is then compared with a random number and, as with windows, if $P(t) > r$ a transition in blinds status will take place. A similar procedure is then applied to determine whether the blinds will be fully raised (e.g. $P_{fully,raised}(t) > r$) or lowered. In the negative case a new shading fraction is computed from a Weibull distribution and the simulation proceeds to the next timestep.

The final model relates to the use of appliances, which may consume water, electrical energy or both. For this we envisage using the prototype model of Page (2007). Before launching the model at the first time-step an important pre-process will involve the stochastic allocation of appliances to the occupants. For residential buildings this is likely to depend upon household size and income and for non-residential uses the number of occupants and their function. The predicted use of these allocated appliances then depends on their category. Type 1 appliances are either in constant use (e.g. refrigerators and freezers) or follow a deterministic profile (e.g. hot water heaters). Type 2 appliances may be switched on by the user but are switched off automatically at the end of their programme (e.g. dishwashers). Occupants are responsible both for switching on and off Type 3 appliances. Type 4 appliances are a collection of appliances which are too small to be modelled individually and are thus modelled as an aggregate. In the case of Types 2 and 3, the probability of switching appliances on is first predicted. These may then be switched on upon comparison of this probability with a uniformly distributed random number. At each simulation timestep until predicted switch-off the inverse function method (IFM) is used to predict the power demand from a cumulative distribution function (CDF). A probably density function will be used to determine the duration of appliance use, from which the switch-off time may be deduced.

Waste production and processing is not currently handled in CitySim; likewise the modelling of water demand, storage and supply.

PLANT AND EQUIPMENT MODELS

This category of model includes both heating, ventilating and cooling (HVAC) systems and energy conversion systems (ECS).

The HVAC model (Kämpf 2009) simply computes the psychrometric state (temperature, moisture content and hence enthalpy) of the air at each stage in its supply (e.g. outside, heat recovered, cooled, re-heated, supplied). Given the required mass flow rate (which may be defined by the energy to be delivered or by the room fresh air requirement) the total delivered sensible and latent heating and cooling loads can then be calculated.

The family of ECS models (Kämpf 2009) comprise a range of technologies that provide/store heat and/or electricity to buildings. These include a thermal storage tank model for hot/cold fluids, boilers, heat pumps, cogeneration systems, combined cogeneration and heat pump systems, solar thermal collectors, photovoltaic cells and finally wind turbines. These ECS models are in general based on performance curve regression equations whereas a simplified thermal model simulates the change of state of the sensible/latent heat store (so that phase change materials are also accounted for).

If the ECS models have insufficient capacity to satisfy the HVAC demands then the supply state is adjusted and the predicted room thermal state is corrected (using the thermal model).

An example application of CitySim, in conjunction with an evolutionary optimisation algorithm, is described briefly on page 458. For the interested reader, the structure and theoretical basis of CitySim is described in considerably more detail in Robinson (2011).

Future research

Substantial progress has been made since the initial developments in urban environmental modelling began in the late 1990s, such that the latest generation of integrated resource flow models resolve building-related energy demand and supply with high resolution and include basic models of flows of water and waste. They are nevertheless far from complete. There is at present no model to predict the spatiotemporal temperature distribution in a given urban settlement and the associated consequences for buildings' space conditioning demands. These models also focus exclusively on resource flows associated with the operation of buildings and their services, so that neither monetary flows nor the financial and environmental consequences of the use of materials to construct and furnish buildings are considered. This latter precludes analyses of urban settlements throughout their life cycle. There has also been a tendency to continue to treat buildings as independent entities (albeit accounting for the urban radiant environment), so that potential synergetic energy and mass exchanges between buildings and between resources are not currently fully modelled. Finally, the use of parametric studies as a tool to help users optimise their proposals is both unsophisticated and highly unlikely to lead designers towards identifying the optimal configuration of the key variables which are available for manipulation.

The current generation of integrated resource flow models have also been developed, at least in the first instance, for relatively small scale analyses: the neighbourhood for SUNtool and both this and the district for CitySim. It would of course be a relatively straightforward matter to predict building-related resource flows at the scale of the entire city by extrapolating from results of statistical significant building typologies, in a similar approach to that of Jones *et al.* (2007), Shimoda *et al.* (2003) and Yamaguchi *et al.* (2007b). But this would fail to consider the influences of the urban radiant and thermal microclimate as well as potential local synergetic exchanges. Not having an accurate spatial representation of the city also leads to a lost opportunity to integrate the modelling of transportation;[6] which is not only the other dominant consumer of energy in cities but is also intrinsically linked to buildings' locations – in terms of the generation of demand for transportation.[7]

Finally, it would be especially interesting to model this building-transport dynamic within a modelling framework which also simulates the mechanisms of urban land and building use change. This would then enable us to simulate responses to planning and governance (such as fiscal interventions, regulations and educational programmes) strategies to improve the way in which our cities evolve with time: in particular their socio-economic and environmental sustainability.

Whether or not urban spatial dynamics are simulated, the near-term possibility for municipalities to prepare and update a model of the resource flows of their whole city is highly attractive. In this way it will be both straightforward and efficient to add-in new development proposals to test the possibilities for engaging in local synergetic energy and matter exchanges, to examine the implications for local transport and energy supply infrastructure and the environmental consequences of new buildings on those already existing and vice versa.

In the following sections we shall discuss how the above gaps in modelling capability might be filled.

Urban climate

Urban terrain, in which built and natural forms are extruded from the ground, is often highly geometrically complex, having greater rugosity than is found in rural settings where the climatic measurements that form the basis of standard meteorological databases tend to be made. Due to inter-reflections between the surfaces constituting this terrain its albedo tends to be lower, so that shortwave radiation is more efficiently absorbed relative to rural terrain. Furthermore, obstructions between surfaces diminish views to the sky, so that heat is less efficiently transferred at night by longwave exchange to the relatively cold sky. In consequence, more heat is absorbed and retained in urban than in rural terrain. Added to this is the heat gain due to anthropogenic sources, for example due to transport systems and heat sources in buildings, although this may be partially offset by evapotranspiration from vegetation. In consequence the mean temperature tends to be higher in urban than in rural terrain, with the intensity of the "urban heat island (UHI) effect" tending towards a maxima at the centre of a city, as the air entrained into the buoyant plume (which may be both deflected and diluted by wind) is successively heated. This type of behaviour has been observed in numerous experimental field studies (Graves *et al.* 2001; Palmer *et al.* 2000; Santamouris 1988). However, urban surfaces also tend to have greater thermal inertia than do rural surfaces. Indeed, Swaid and Hoffmann (1989) estimate the difference in time constant to be six to eight hours. The urban temperature profile is thus shifted in phase with respect to the rural profile so that the UHI profile (the difference between the two: $T_{urban}(t) - T_{rural}(t)$) tends to reach a minima in the late afternoon/early evening (to the extent that this instantaneous UHI may be negative) and a maxima during the night. Due to the above mechanisms, the measured intensity of the London UHI varies from –2 °C to +6 °C with corresponding implications for buildings' energy performance.

However the current practice in building simulation is to directly use the air velocity and temperature from a climate file, which is normally based on rural measurements. The influence of the UHI on the building energy balance is therefore ignored.

But modelling the urban climate to resolve this shortfall is complicated by the diverse time and length scales involved in atmospheric flow. The length scale for example can vary from a few meters (buildings) to a few kilometers (mountains). It is practically impossible to resolve all the scales in a single model with the present computational power available. A way to overcome this problem is to couple different models capable of resolving different scales.

A range of global numerical weather forecasting models having a resolution of a few hundred kilometers have been developed, such as the Global Forecast System "GFS" (http://www.emc.ncep.noaa.gov/modelinfo) or the Global Environmental Multiscale model "GEM" (http://collaboration.cmc.ec.gc.ca/science/rpn/gef_html_public/index.html). These global models may provide boundary conditions to a mesoscale model having a resolution of a few hundred meters, so that an entire city and its local topographic context may be represented. Examples include the finite volume model "FVM" developed at the Ecole Polytechnique Fédérale de Lausanne (EPFL) (http://lpas.epfl.ch/) and the open source models MM5 (http://www.mmm.ucar.edu/mm5/) and WRF (http://www.wrf-model.org). The exchange of mass, momentum and energy with the urban texture may then be implicitly represented by embedding an urban canopy model (Brown and Williams 1997; Kusaka *et al.* 2001; Martilli *et al.* 2002; Masson 2000; Rasheed 2009) within the mesoscale model. Given the necessary computing resources and an adequate calibration of the urban canyon model for each mesoscale cell, such a nested (global to urban mesoscale) model could be used to provide the necessary climate input data to both building and urban resource flow simulation software, to improve the accuracy of their energy predictions.

As a related example, using the urban parameterisation scheme of Martilli *et al.* (2002) embedded within the EPFL model "FVM", Rasheed (2009) has simulated the mesoclimate of Basel in Switzerland for a 24-hour period. From these results (Figure 15.6) it was possible to estimate the relative cooling demands within the city, suggesting an increase of up to 70%, due to Basel's UHI.

But when investigating natural ventilation strategies for buildings or the comfort of pedestrians in a city it is also important to account for the local temperature and velocity field in an accurate way, which depends on local as well as on large scale phenomena. The effects of buildings may be explicitly resolved using computational fluid dynamics software in conjunction with a reasonably fine mesh, whereas the large scale phenomena may be implicitly forced through the boundary conditions to this CFD model using results from the mesoscale model, so that the two models are coupled in one direction (the meso-model is essentially a pre-process for the micro-model). This is interesting because results from city mesoclimate simulations could be made publicly available (e.g. via an internet site) to support later simulations of the local microclimate, as a post-process.

Such a resource would be valuable not only for the improved prediction of urban resource flows, but also of pedestrians' comfort. For further details regarding urban climate modelling we refer the reader to Chapter 3 in Robinson (2011).

Pedestrian comfort

As we strive to improve the sustainability of urban developments, so there will be increasing pressure on urbanites to intensify their use of personal (walking, cycling) and public (buses, trains) means of transport. In consequence, a greater fraction of the urban realm can be expected to be designed to favour pedestrians and cyclists as they walk or cycle to and from their destinations, via modes of public transport. They will thus be expected to spend more time outdoors. It is important then that this outdoor experience be pleasant, if this strategy is to succeed.

But there are myriad factors which influence pedestrians' perception of and thus satisfaction with the outdoor environment. Broadly, we may think of these as being *environmental*,

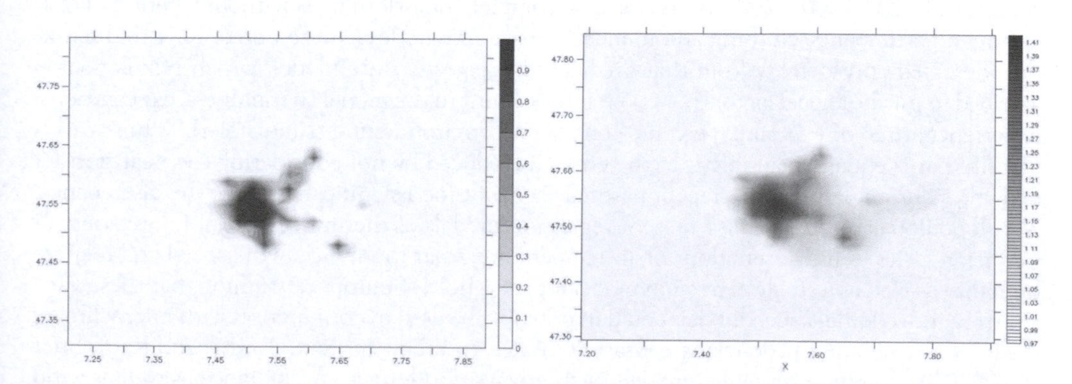

Figure 15.6 Fraction of land use classed as urban (left) and cooling energy demands (right) for the city of Basel: these latter relative to its local rural context (quotient of urban and rural cooling degree-hours).

Source: Rasheed 2009.

influencing our thermal, visual and aural comfort as well as our desire to remain dry and to perform our tasks in the presence of wind. We are also influenced by *sociological* factors. For example do we feel safe? Is the route crowded? Do we like the appearance of this environment? Ideally we would have some means for evaluating these diverse factors to arrive at an overall judgement regarding pedestrians' satisfaction. At present we are some way from achieving this, but progress has nevertheless been made in respect of some of the environmental factors mentioned above.

Based on field surveys, Lawson and Penwarden (1975) first proposed thresholds of time for which wind speeds may be exceeded for different pedestrian activities (sitting, standing, walking). For example when seated outdoors we may tolerate a Beaufort scale of three (~4–6m/s) for one percent of the time.

Such criteria are widely used in conjunction with CFD predictions of the urban environment to determine pedestrian comfort.[8] But they are a little too simple; pedestrians' comfort depends not only on their ability to conduct mechanical tasks without difficulty but also on achieving this within a comfortable range of environmental stimuli.

To support the design of *comfortable* urban spaces Arens and Bosselmann (1989) used simplified physical modelling techniques with which to adapt rural climate data to the urban context. The modified hourly data were then input, along with assumptions of activity and clothing level, to Gagge's two-node dynamic model of the human thermoregulatory system, to predict the corresponding thermal comfort. Hourly results were then processed to determine, for each point in an urban scene, the proportion of the year that pedestrians would be comfortable. Höppe (2002) also employed a dynamic human thermoregulatory model to study pedestrians' comfort, on the basis that people tend to be outside for relatively short periods of time, so that steady state conditions are seldom reached. In such cases a steady state model would give a false result; typically a pessimistic one, since external conditions tend to be more extreme than those experienced indoors. This he supported by predicting the time for skin and core temperature to reach steady-state in winter and summer.

But the majority of pedestrian comfort studies (Ahmed 2003; Givoni *et al.* 2003; Metje *et al.* 2008; Nagara *et al.* 1996; Nicolopoulou *et al.* 2001; Nicolopoulou and Steemers 2003; Nicolopoulou and Lykoudis 2006, 2007; Stathopoulos *et al.* 2004; Tacken 1989) have relied on field survey observations to develop regression models. In the studies of Tacken (1989) and Givoni *et al.* (2003) the study participants were placed at rest in different external conditions for several minutes and then asked to complete questionnaires while the corresponding microclimate conditions were recorded. In both cases multivariate linear regression analysis was used to derive expressions for thermal sensation.

In something of a departure from this approach respondents of the surveys of Nikolopoulou *et al.* (2001, 2003, 2006, 2007) were asked to complete questionnaires without a period of rest; although participants' clothing and immediate past activity level were noted. Reported sensation diverged considerably from that predicted by a steady state model, which Nikolopoulou ascribed to psychological factors not taken into account in the model (naturalness, expectations, experience, time of exposure, perceive control and environmental stimulation). Whilst this is plausible, it is equally plausible that the errors introduced by not considering the heat transfer dynamics during participants' recent past may account for a significant part of the discrepancy. Nikolopoulou *et al.* (2004) also present regression models of thermal sensation for a variety of European cities as linear functions of air temperature, solar irradiance, wind speed and relative humidity as well as a single regression model for the whole of Europe. Assuming that these variables are indeed linear such models could in principle be used in conjunction with microclimate predictions to predict pedestrians' sensation in a given urban context. To this end Katzschner *et al.* (2004) combine simplified models of shortwave and longwave irradiance as well as wind speed to predict, using Nikolopoulou's regression equations, the spatial distribution of pedestrians' sensation. Provided that the models used are valid, this is a potentially powerful tool.

But as Höppe points out, regression models are confined to the climatic conditions from which their data are derived. Furthermore these models tend to ignore the transient heat transfer between pedestrians and their environment for their recent activities and the environments in which these took place. To resolve these problems a more coherent approach might be to simulate the movement of pedestrians (as multiple agents) whilst simultaneously simulating the transient microclimate conditions encountered by them. Environmental, activity and clothing characteristics could then be parsed to a transient multi-segment human thermoregulatory model to predict our agent-pedestrians' thermal sensation. In such a way we may also be able to simulate agents' preferred routes and activities within their urban landscape.

Bruse (2007) has made a good first step in this direction by coupling a model of pedestrian agents with a simplified two-node dynamic human thermoregulatory model. These agents are placed within a domain of his urban microclimate modelling software ENVI-met (Bruse 1999). Thus, agents sample their local microclimate and choose their routes, from origin to destination, according to their predicted thermal sensation. To compensate for uncertainties in microclimate and comfort evaluation fuzzy-logic rules are used for the agents' routing decisions.

Despite this promising start, some further work would seem to be merited to relate skin and core temperature to pedestrians' thermal sensation and to compare predicted with observed movement, to refine the associated agents' rules. But these rules should also depend on agents' overall perception of outdoor space: visual, acoustic and protection from wind and precipitation as well as more complex sociological factors, such as those mentioned above. It would seem that considerably more field work is required here.

Comprehensive resource flow modelling

In the section on pages 447–453 two integrated resource flow modelling tools were introduced: *SUNtool* and its more comprehensive successor *CitySim*. Whilst undoubtedly useful in support of sustainable urban design projects, the initial versions of both of these tools concentrate almost exclusively on predicting buildings' operational energy consumption. To facilitate really comprehensive consideration of urban environmental sustainability, throughout the life cycle of a given development, considerable further progress is required.

- Attributes of physical objects in CitySim such as walls and HVAC plant currently only support operational energy use predictions (due to their influence on building energy demand). Their attribution should be extended to include cost and embodied energy content (and/or emissions resulting from their fabrication process) as well as replacement frequency, to support life cycle cost and environmental assessment.
- As noted in the introduction, net urban resource flows could be considerably reduced by improving synergetic exchanges amongst complementary components of the system. This might involve the derivation of energy from waste or the coupling of industrial processes, where certain waste products of one process are resources for another. This latter should probably be based on simple black-box models of the processes under consideration.
- Irrespective of the resource demands involved in water treatment and distribution, there are concerns that water is becoming an increasingly scarce resource. Indeed this is already the case in developing countries. There is a need therefore to model the set of water demands (not just based on water appliances, but also considering industrial processes, cleaning and irrigation) and means of water supply (rainwater, mains and coupling waste water from some processes with the supply to others, i.e. a cascading approach, with or without treatment).

In the future it would also be useful to couple urban resource flow modelling software with specialist tools for simulating surface water run-off, so that the efficiency of sustainable urban draining strategies can be examined.

One final core development, which also impacts upon resource flow predictions, relates to uncertainties, which for urban scale simulation can be considerable. Even for existing buildings we make assumptions about the internal layout of buildings, about constructional properties, the efficiency of energy supply systems and the behaviour of occupants. For new developments even the final building geometry is uncertain, let alone all of the parameters that depend on it. It is important therefore that our predictions reflect these uncertainties – that our performance indicators/variables are expressed as distributions rather than as single values.

Resource optimisation

For a new urban development, even with a relatively limited number of variables (geometry, type of use, occupancy and constructional characteristics, plant and energy supply technologies), the number of permutations is extremely large. Although the parameter space is smaller for a refurbishment exercise – in principle the geometry and some constructions will be retained – it remains very large. The probability of identifying an optimal configuration of these variables by manual trial and error or simple parametric studies is thus correspondingly small. It is appropriate then to use computational methods to efficiently explore this parameter space in search of the optimal solution.

Candidate methods include direct, indirect and heuristic search. Direct methods search for the optimal configuration in a random (e.g. Monte Carlo simulation) or in an algorithmic (e.g. Hooke-Jeeves) way which is not based on derivatives of the function being evaluated. Although more reliable than manual trial and error, random methods are relatively inefficient and algorithmic methods may become trapped in the region of a local optima. Indirect methods use mathematical techniques to identify an optimum in the parameter space. For example, by moving in a direction of steep gradient (steepest descent) where the solution should lie, but as with direct algorithmic methods this optimum may also be a local and not a global one. Improved efficiency is thus contrasted by uncertainty. Heuristic methods on the other hand adapt, using stochastic operators, according to what they have learnt about a given system. Such methods are relatively robust and efficient and can be applied to a wide variety of problems. They are thus our most promising candidate.Following from a review of available evolutionary algorithms, Kämpf and Robinson (2009a) have developed a hybrid (Covariance Matrix Adaptation – Evolutionary Strategy (CMA-ES) and Hybrid Differential Evolution (HDE)) algorithm which offers improved robustness over its individual counterparts for a larger range of optimisation problems.

As a starting point this algorithm has been applied to the relatively simple problem of optimising the layout and geometry of buildings for the utilisation of available solar radiation. First, candidate geometries are converted into RADIANCE format, a set of (*r*-trace) calculation points are then defined along with a cumulative annual sky radiance distribution for the location of interest. The simulations are then performed and the product of local solar irradiation (Wh.m^{-2}) and surface area (m^2) are cumulated for the set of *r*-trace points to determine total solar irradiation (Wh) – this being the objective function to maximise (Kämpf and Robinson 2010).

One example application is the problem of optimising tower height for a regular grid of buildings (Figure 15.7: right). Cheng *et al.* (2006) investigated a similar problem, but based on manual trial and error of a relatively small sample. Their conclusion that rather randomly dispersed tall buildings tends to optimise solar utilisation is contrasted with that of Kämpf and Robinson (2010): a well organised layout, consisting of a high perimeter with a south-facing gap and a low core (the gap providing access to the southern facades of rear buildings) is optimal (Figure 15.7: left).

More recently (Kämpf and Robinson 2009b) this new hybrid EA has been coupled with CitySim and applied to optimise (minimise) the primary energy demand, according to plausible renovation measures, of a block of 26 buildings in the district of Matthäus in Basel, Switzerland (Figure 15.8). For this block a combination of census data and visual observations was used to attribute the buildings according to façade glazing ratio, construction characteristics and embedded energy conversion systems. Buildings were also associated with different optimisation parameters depending on their protection status (three zones of different protection status

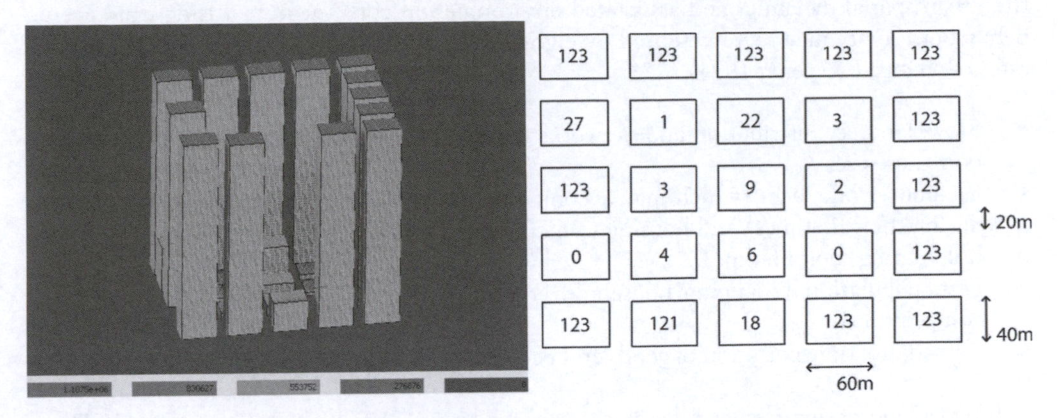

Figure 15.7 Optimisation of tower height within a regular grid for solar irradiation utilisation.

are identified in Figure 15.8): (i) internal insulation and ECS only, (ii) as (i) plus internal insulation, (iii) as (ii) plus modified façade glazing ratio and roof-integrated photovoltaic panels. The total allowed cost of these renovation measures was constrained to be within 2 million Swiss Francs. An XML template was then defined to describe this scene, in which the optimisation parameters' names were replaced by particular character combinations.

The results from this optimisation problem conform to expectations. Given the low primary energy conversion system heat pumps are always the ECS of choice. Where possible both photovoltaic panel area and insulation thickness are maximised (within their allowable limits). Any remaining money is then invested in substituting the glazing. This latter is less of a priority since glazing is relatively expensive and has a less marked impact on primary energy than the other options. We refer the interested reader to Chapter 8 in Robinson (2011) for further details regarding the hybrid CMA-ES/HDE optimisation algorithm and its applications.

Integrated land use and transport interaction (LUTI) modelling

Cities are complex systems which exhibit macroscopic emergent behaviour based on interactions between the actors accommodated (firms, individuals), who react to economic, governance, technological and educational stimuli and to the actions of their peers. To simulate

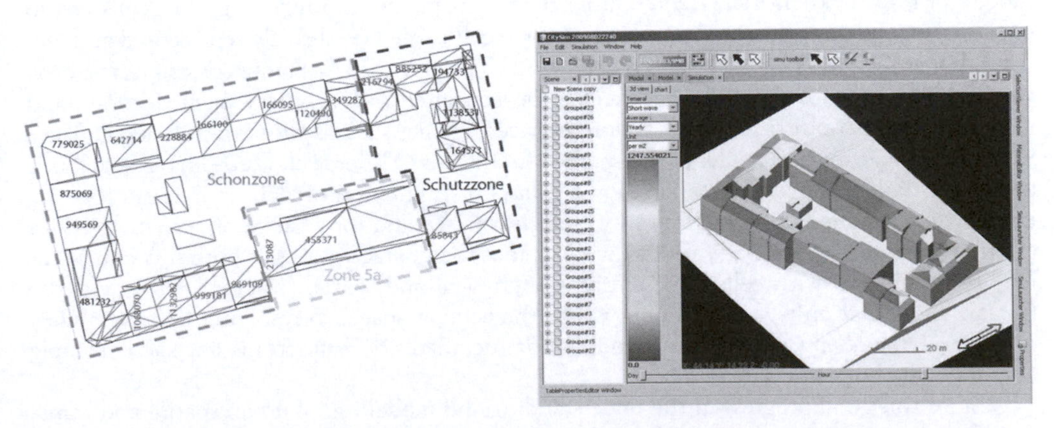

Figure 15.8 (a) An urban block in the district of Matthäus in Basel (Switzerland) distinguishing between zones of different protection status, and (b) falsecoloured according to incident annual shortwave irradiation (right).

the urban spatial dynamics and associated environmental consequences arising from agents' behaviours, a city may be represented by eight sub-systems, distinguished according to their rates of change (Wegener 1994):

- Slow: land use distribution and networks (utility, transport and communication) to support these uses.
- Medium: workplaces (which may accommodate several changes in type of employment during their lifetimes) and housing (which may accommodate several changes in household during their lifetime).
- Fast: population (changing household composition) and employment (changing firm composition).
- Immediate: transportation of goods and people.

A broad range of approaches have been employed to model these processes (Waddell and Ulfarsson 2004). The more popular early models were based on *spatial interaction* or *spatial input-output*. Spatial interaction models use a gravitational analogy to model travel between origin and destination zones which increases with the amount of activity within them and decreases with the square of the distance between them. Putman (1983) has applied this to model choices of trips, residential relocation and employment location; but this type of model lacks spatial resolution and many of the behavioural factors influencing the choices made. Spatial input-output models define regional zones which are treated as sub-economies that engage in production, consumption, import and export within and between them. In this way real estate and labour markets as well as travel demand are modelled (de la Barra 1989; Echenique *et al.* 1990), but the spatial aggregation level is high relative to that used in conventional transport modelling.

More recent models are based on much lower levels of spatial aggregation, for example using *cellular automata* (CA) or *multi agent systems* (MAS). Cellular automata have emerged from the study of complex systems whose emergent properties are simulated in response to simple local behavioural rules. These rules are applied to cells within a grid (Wolfram 1994), whereby the future state of a given cell depends upon its current state and that of its neighbours. CA have been used to study a broad range of phenomena ranging from the formation of star clusters to the behaviour of bacteria and insect populations. In the present context they have also been applied widely to the study of land cover or land use change (Batty 1997; Torrens and O'Sullivan 2001; White and Engelen 1993, 2000). But as Benensen *et al.* (2005) point out these models are limited by the immobility of the cells, whereas housing and transport dynamics necessitate mobile or non-fixed entities. Representing the urban milieu as a regular grid of cells is also constraining; features that are larger than a single cell cannot be directly represented and features that are smaller need to be aggregated. MAS have emerged as a more consistent response to these limitations, as individual agents may be associated with arbitrary shape, location and respond to more sophisticated interactions between agents (such as between individual, firms, workplaces, housing and modes of transport). On this basis Miller *et al.* (2004) have developed a highly detailed MAS-based land use and transport interaction model, resolving real estate market demand-supply interactions as well as travel demand for a range of activities. As an alternative to full MAS, Batty (2005) suggests that the characteristics of both CA and MAS may be combined, with cells representing the physical and spatial structure and agents the human and social units that function within the cellular space, so resolving Benensen's key criticism. *Urban-Sim* (an alternative and likewise detailed LUTI model) is one such example (Waddell and Ulfarsson 2004).

Despite this good progress in the understanding and modelling of urban spatial and transport dynamics, no attempt has been made to couple these LUTI models with urban resource flow models, and so provide a comprehensive basis for modelling sustainability and how this may evolve with time. Such models could provide powerful support to urban planners to guide the development of urban settlements along more sustainable trajectories. Achieving this in a computational tractable way is perhaps the next great challenge for urban environmental simulation!

As an intermediate step, work is underway to couple *CitySim* with *MATsim* (multi-agent transport simulation, see www.matsim.org), to resolve for the key building and transport related urban resource flows of the city of Zurich (see Chapter 6 in Robinson (2011)). The objective of this effort is to identify strategies for reducing average per capita energy demand (buildings, industry and transport; both operational and embedded) from its current value of around 5.8kW to 2kW, by the year 2050.

The form of this coupling will be via the exchange of occupants. A coherent basis for achieving this is to represent city inhabitants as agents, whose characteristics (e.g. income, household membership) and preferences (e.g. comfort, transport choice and consumption) are carried with them as they are transported between buildings.

Synopsis

The vast majority of buildings are constructed in urban environments. And this urban context has a direct impact on buildings' resource flows. Urban temperatures are on average warmer, the air is more contaminated and its velocity is lower. Adjacent buildings also reduce the availability of daylight and solar radiation. So whether we are concerned with designing just one or many buildings, this urban context needs to be considered. The urban context also offers great opportunities for reducing the growth in urban resource consumption and the associated adverse environmental consequences. For example, complementary building activities can be located adjacent to one another to reduce transport needs and/or to promote synergetic resource exchanges. The siting of buildings can also be arranged to maximise natural resource availability, subject to local density constraints. Energy generation can be locally centralised to benefit from the more stable aggregate resource demand profile (to improve the quality of fit between demand and supply) as well as economies of scale compared with the building-embedded case . . . etc. In order to maximise the utilisation of these opportunities there is a compelling case for the use of computer modelling; both for detailed masterplanning and to support the formulation of sustainable urban planning policy and guidance. In this chapter we have attempted to give a flavour of the progress that has been made in this respect.

Until now efforts have concentrated on the use of lighting simulation tools to evaluate the potential for the use of passive and active solar technologies as well as on energy modelling tools to assist municipalities to manage their stock of public buildings. More recently good progress has been made in the more general modelling of building-related urban resource flows. Indeed such tools are starting to be used to support really informed decision making in sustainable masterplanning. But there are many exciting challenges awaiting the urban modelling community.

Some further work is required to develop a comprehensive platform for the *modelling and optimisation* of urban resource flows for the spatially stationary case; that is, for a particular annual snapshot in time (i.e. land use does not change). But of course cities are not stationary. They are complex systems which exhibit macroscopic emergent behaviour based on the actions of the firms and individuals accommodated within them. To understand whether particular governmental (e.g. taxation, statutory, educational, subsidy) or urban planning interventions are likely to have a positive impact on the ways in which a city develops, the actions of firms and individuals and how they are influenced needs to be taken on board. Likewise the processes accommodated within buildings and the transport of people and goods between them. If we are to succeed in this there is a need for interdisciplinary research: integrating sociologists, climatologists, transport specialists, geographers, computer scientists, engineers and physicists. This work will keep us busy for quite some time to come!

Acknowledgements

Financial support received from the Swiss National Science Foundation and the European Commission's DG-TREN for the development of City-Sim is gratefully acknowledged. The contributions to the work described in this chapter of the following members, either past or present, of the Sustainable Urban Development research group at LESO-PB/EPFL are also very gratefully acknowledged: Frédéric Haldi, Jérôme Kämpf, André Kostro, Philippe Leroux, Jessen Page, Diane Perez, Adil Rasheed and Urs Wilke.

Notes

1 In principle this should also be increased according to the ratio of the solid angle of the region of sky represented by the light source ($2\pi/n$) to the actual solid angle of the light source (Φ_s).
2 Primitive *stochastic models* of occupants' presence and related use of lights and blinds and windows (Page 2007) were also integrated.
3 Note that $\omega = 1 - \sigma - \sigma_{self}$ where σ_{self} represents self obstructions (the proportion of a patch that is not visible because it is behind the receiving surface).
4 The probability of a binary outcome Y, given a set of n independent variables, may be found using a logistic regression equation of the form $P_Y(x_1 \ldots x_n) = \exp\left(a + \sum_{i=1}^{n} b_i x_i\right) \Big/ \left(1 + \exp\left(a + \sum_{i=1}^{n} b_i x_i\right)\right)$, where the coefficients a, b_i may be found by maximum likelihood estimation.
5 Behaviour at departure is not special in this case, so it is simple treated as an intermediate time step.
6 One could associate buildings at their true location with pre-calculated results for the corresponding typology, but this would ignore their spatial (climatic) context.
7 The better the local diversity of building activities in a city district for example, the less should be the need to make journeys outside of this district, so reducing resource demands and improving inhabitants' quality of life whilst also improving the potential for synergetic energy and matter exchanges between complementary building uses.
8 For each sector of say 22.5° the ratio of the wind speed at a given point of interest to that entering the domain is calculated; based on a reference wind speed of say 5m/s. As a post-process we may then read hourly climate data and multiply the wind speed ratio for the relevant sector by the wind speed in the climate file to obtain a local wind speed. Repeating this for each hour we may determine which activities would be considered comfortable.
9 If you have downloaded *GenCumulativeSky* (http://leso.epfl.ch/e/research_urbdev.html) a global cumulative sky radiance distribution is created simply using the command: gencumulativesky +s1 –a 46.25 –o 6.06 –m 15 –G ./Geneva.txt > cumulativesky.cal A diffuse distribution will be obtained by omitting '+s1' in the above command.
10 Such an *r*-trace command should look something like the following (using default RADIANCE parameters): *r*-trace –I –ab 2 scene.oct < datapoint where datapoint is a text file containing a single line: 22.5 0.1 1.5 0 1 0.
11 Albeit for this relatively extreme case of a point at the centre of the glazing at the ground floor of a south facing façade which is within a relatively deep canyon.

References

Ahmed, S.A. (2003) "Comfort in Urban Spaces: Defining the Boundaries of Outdoor Thermal Comfort for the Tropical Urban Environments", *Energy and Buildings* 35(1): 103–119.

Arens, E. and Bosselmann, P. (1989) "Wind, Sun and Temperature – Predicting the Thermal Comfort of People in Outdoor Spaces", *Building and Environment* 24(4): 315–320.

de la Barra, T. (1989) *Integrated Land Use and Transport Modelling*, Cambridge: Cambridge University Press.

Barton, H. (ed.) (2000) *Sustainable Communities: The Potential for Eco-neighbourhoods*, London: Earthscan.

Batty, M. (1997) "Cellular Automata and Urban Form: A Primer", *Journal of the American Planning Association* 63(2): 266–274.

Batty, M. (2005) *Cities and Complexity: Understanding Cities with Cellular Automata, Agent-Based Models, and Fractals*, Cambridge, MA: The MIT Press.

Benensen, I., Aronovich, S. and Noam, S. (2005) "Let's Talk Objects: Generic Methodology for Urban High-resolution Simulation", *Computers, Environment and Urban Systems* 29(4): 425–453.

Brown, M. and Williams, M. (1998) "An Urban Canopy Parameterisation for Mesoscale Meteorological Models", in *Proceedings of the 2nd AMS Urban Environment Symposium*, Albuquerque, NM.

Bruse, M. (1999) "Simulating Microscale Climate Interactions in Complex Terrain with a High-resolution Numerical Model: A Case Study for the Sydney CBD Area", in *Proceedings of the International Conference on Urban Climatology and International Congress of Biometeorology*, Sydney, Australia, pp. 8–12.

Bruse, M. (2007) "Simulating Human Thermal Comfort and Resulting Usage Patterns of Urban Open Spaces with a Multi-agent System", in *Proceedings of the 24th International Conference on Passive and Low Energy Architecture* (PLEA), Singapore, pp. 699–706.

Cheng, V., Steemers, K., Montavon, M. and Compagnon, R. (2006) "Urban Form, Density and Solar Potential", in *Proc. 23rd Int. Conf. Passive and Low Energy Architecture* (PLEA), Geneva, Switzerland.

Clarke, J.A. (2001) *Energy Simulation in Building Design*, Oxford: Butterworth-Heinemann.

Compagnon, R. (2004) "Solar and Daylight Availability in the Urban Fabric", *Energy and Buildings* 36: 321–328.

Compagnon, R. and Raydan, D. (2000) "Irradiance and Illuminance Distributions in Urban Areas", in *Proceedings of the 17th International Conference on Passive and Low Energy Architecture* (PLEA), Cambridge, UK, pp. 436–441.

Déqué, F., Olivier, F. and Poblador, A. (2000) "Grey Boxes Used to Represent Buildings with a Minimum Number of Geometric and Thermal Parameters", *Energy and Buildings* 31(1): 29–35.

Echenique, M.H., Anthony, D.J.F., Hunt, D.J., Mayo, T.R., Skidmore, I.J. and Simmonds, D.C. (1990) "The MEPLAN Models of Bilbao, Leeds and Dortmund", *Transport Reviews* 10(4): 309–322.

Erkman, S. (1998) *Vers une écologie industrielle*, Paris: Charles Léopold Mayer.

Gadsden, S., Rylatt, M., Lomas, K. and Robinson, D. (2003) "Predicting the Urban Solar Fraction: A Methodology for Energy Advisers and Planners Based on GIS", *Energy and Buildings* 35(1): 37–48.

Girardet, H. (1999) *Creating Sustainable Cities, Schumacher Briefing 2*, Darlington, UK: Green Books Ltd.

Givoni, B., Noguchi, M., Saaroni, H., Pochter, O., Yaacov, Y., Feller, N. and Becker, S. (2003) "Outdoor Comfort Research Issues", *Energy and Buildings* 35(1): 77–86.

Graves, H., Watkins, R., Westbury, P. and Littlefair, P. (2001) "Cooling Building in London: Overcoming the Heat Island", BRE Report BR 431, London: CRC Ltd.

Haldi, F. and Robinson, D. (2009) "Interactions with Window Openings by Office Occupants", *Building and Environment* 44(12): 2378–2395.

Haldi, F. and Robinson, D. (2010) "Modelling Actions on Shading Devices", *Building and Environment* (in press).

Höppe, P. (2002) "Different Aspects of Assessing Indoor and Outdoor Thermal Comfort", *Energy and Buildings* 34(6): 661–665.

Hunt, D.R.G. (1980) "Predicting Artificial Lighting Use – A Method Based upon Observed Patterns of Behaviour", *Lighting Research and Technology* 12(1): 7–14.

Johnston, D., Lowe, R. and Bell, M. (2005) "An Exploration of the Technical Feasibility of Achieving Reductions in Excess of 60% within the UK Housing Stock by the Year 2050", *Energy Policy* 33(13): 1643–1659.

Jones, P., Williams, J. and Lannon, S. (1998) "An Energy and Environmental Prediction Tool for Planning Sustainability in Cities", in *Proceedings of the 2nd European Conference REBUILD*, Florence, Italy.

Jones, P., Patterson, J., Lannon, S. and Prasad, D. (1999) "An Energy and Environmental Prediction Tool for Planning Sustainable Cities", in *Proceedings of the 16th International Conference on Passive and Low Energy Architecture* (PLEA) Brisbane, Australia, pp. 789–794.

Jones, P., Patterson, J. and Lannon, S. (2007) "Modelling the Built Environment at an Urban Scale – Energy and Health Impacts in Relation to Housing", *Landscape and Urban Planning* 83(1): 39–49.

Kämpf, J., (2009) "The Modelling and Optimisation of Urban Resource Flows", Unpublished PhD Thesis, EPFL, Lausanne, Switzerland.

Kämpf, J. and Robinson, D. (2007) "A Simplified Thermal Model to Support Analysis of Urban Resource Flows", *Energy and Buildings* 39(4): 445–453.

Kämpf, J. and Robinson, D. (2009a) "A Hybrid CMA-ES and DE Optimisation Algorithm with Application to Solar Energy Potential", *Applied Soft Computing* 2(9): 738–745.

Kämpf, J. and Robinson, D. (2009b) "Optimisation of Urban Energy Demand Using an Evolutionary Algorithm", in *Proceedings of Building Simulation 2009*, Glasgow, Scotland, pp. 668–673.

Kämpf, J. and Robinson, D. (2010) "Optimisation of Building Form for Solar Energy Utilisation Using Constrained Evolutionary Algorithms", *Energy and Buildings* (in press)

Katzschner, L., Bosch, U. and Roettgen, M. (2004) "Thermal Comfort Mapping and Zoning", in Nikolopoulou (ed), *Designing Open Spaces in the Urban Environment: A Bioclimatic Approach*, Attiki, Greece: Centre for Renewable Energy Sources.

Kusaka, H., Kondo, H., Kikegawa, Y. and Kimura, F. (2001) "A Simple Single-layer Urban Canopy Model for Atmospheric Models: Comparison with Multi-layer and SLAB Models", *Boundary Layer Meteorology* 101(3): 329–358.

Lawson, T.V. and Penwarden, A.D. (1975) "The Effects of Wind on People in the Vicinity of Buildings", in *Proceedings of the 4th International Conference on Wind Effects on Buildings and Structures*, Heathrow, UK.

Lefebvre, G., Bransier, J. and Neveu, A. (1987) "Simulation of the Thermal Behaviour of a Room by Reduced Order Numerical Methods" [simulation du comportement thermique d'un local par des methodes numeriques d'ordre reduit], *Revue Generale de Thermique* 26(302): 106–114.

Mardaljevic, J. (1999) "Daylight Simulation: Validation, Sky Models and Daylight Coefficients", Unpublished PhD Thesis, DeMontfort University, Leicester, UK.

Mardaljevic, J. and Rylatt, M. (2000) "An Image-based Analysis of Solar Radiation for Urban Settings", in *Proceedings of the 17th International Conference on Passive and Low Energy Architecture* (PLEA), Cambridge, UK, pp. 442–447.

Mardaljevic, J. and Rylatt, M. (2003) "Irradiation Mapping of Complex Urban Environments: An Image-based Approach", *Energy and Buildings* 35(1): 27–35.

Martilli, A., Clappier, A. and Rotach, M.W. (2002) "An Urban Surfaces Exchange Parameterisation for Mesoscale Models", *Boundary Layer Meteorology* 104(2): 261–304.

Masson, V. (2000) "A Physically-based Scheme for the Urban Energy Budget in Atmospheric Models", *Boundary Layer Meteorology* 94(3): 357–397.

Metje, N., Sterling, M. and Baker, C.J. (2008) "Pedestrian Comfort Using Clothing Values and Body Temperature", *Journal of Wind Engineering and Industrial Aerodynamics* 96(4): 412–435.

Miller, E.J., Hunt, J.D., Abraham, J.E. and Salvini, P.A. (2004) "Microsimulating Urban Systems", *Computers, Environment and Urban Systems* 28(1–2): 9–44.

Nagara, K., Shimoda, Y. and Mizuno, M. (1996) "Evaluation of the Thermal Environment in an Outdoor Pedestrian Space", *Atmospheric Environment* 30(3): 497–505.

Natarjan, S. and Levermore, G.J. (2007) "Predicting Future UK Housing Stock and Carbon Emissions", *Energy Policy* 35(11): 5719–5727.

Nicolopoulou, M. (2004) "Thermal Comfort Models for Open Urban Spaces", in Nikolopoulou (ed.), *Designing Open Spaces in the Urban Environment: A Bioclimatic Approach*, Attiki, Greece: Centre for Renewable Energy Sources.

Nicolopoulou, M. and Lykoudis, S. (2006) "Thermal Comfort in Outdoor Urban Spaces: Analysis across Different European Countries", *Building and Environment* 41(11): 1455–1470.

Nicolopoulou, M. and Lykoudis, S. (2007) "Use of Outdoor Spaces and Microclimate in a Mediterranean Urban Area", *Building and Environment* 42(10): 3691–3707.

Nicolopoulou, M. and Steemers, K. (2003) "Thermal Comfort and Psychological Adaptation as a Guide for Designing Urban Spaces", *Energy and Buildings* 35(1): 95–101.

Nicolopoulou, M., Baker, N. and Steemers, K. (2001) "Thermal Comfort in Outdoor Urban Spaces: Understanding the Human Parameter", *Solar Energy* 70(3): 227–235.

Nielsen, T.R. (2005) "Simple Tool to Evaluate Energy Demand and Indoor Environment in the Early Stages of Building Design", *Solar Energy* 78(1): 73–83.

Page, J. (2007) "Simulating Occupant Presence and Behaviour in Buildings". PhD Thesis, EPFL, Lausanne, Switzerland.

Page, J., Robinson, D., Morel, N. and Scartezzini, J.-L. (2007) "A Generalised Stochastic Model for the Prediction of Occupant Presence", *Energy and Buildings* 40(2): 83–98.

Palmer, J., Littlefair, P., Watkins, R. and Kolokotroni, M. (2000) "Urban Heat Islands", *Building Services Journal* 2000(5): 55–56.

Perez, R., Seals, R. and Michalsky, J. (1993) "All-weather Model for Sky Luminance Distribution – Preliminary Configuration and Validation", *Solar Energy* 50(3): 235–243.

Pigg, S., Eilers, M. and Reed, J. (1996) "Behavioral Aspects of Lighting and Occupancy Sensors in Private Offices: A Case Study of a University Office Building", in *Proceedings of the 1996 ACEEE Summer Study on Energy Efficiency in Buildings*, pp. 8161–8171.

Pout, C. "N-DEEM: the National Nondomestic Buildings Energy and Emissions Model", in Palmer, J., Littlefair, P., Watkins, R.and Kolokotroni, M. (2000) "Urban Heat Islands", *Building Services Journal* 2000(5): 55–56.

Putman, S.H. (1983) *Integrated Urban Models*, London: Pion.

Rasheed, A. (2009) "Multiscale Modelling of Urban Climate", Unpublished PhD Thesis, EPFL, Lausanne, Switzerland.

Ratti, C., Robinson, D., Baker, N. and Steemers, K. (2000) "LT Urban – The Energy Modelling of Urban Form", in *Proceedings of the 17th International Conference on Passive and Low Energy Architecture* (PLEA), Cambridge, UK.

Ratti, C., Baker, N. and Steemers, K. (2005) "Energy Consumption and Urban Texture", *Energy and Buildings* 37(7): 762–776.

Reinhart, C. (2004) "Lightswitch-2002: A Model for Manual and Automated Control of Electric Lighting and Blinds", *Solar Energy* 77(1): 15–28.

Reinhart, C.F. and Voss, K. (2003) "Monitoring Manual Control of Electric Lighting and Blinds", *Lighting Research and Technology* 35(3): 243–260.

Robinson, D. (2005) "Decision Support for Environmental Master Planning by Integrated Flux Modelling", in *Proceedings of CISBAT 2005*, Lausanne, Switzerland.

Robinson, D. (2011) "Building Modelling", in Robinson, D. (ed.), *Computer Modelling for Sustainable Urban Design*, London: Earthscan Press.

Robinson, D. and Baker, N. (2000) "Simplified Modelling – Recent Developments in the LT Method", *Building Performance* 3(1): 14–19.

Robinson, D. and Stone, A. (2004a) "Irradiation Modelling Made Simple – The Cumulative Sky Approach and Its Applications", *Proceedings of the 21st International Conference on Passive and Low Energy Architecture* (PLEA) Eindhoven, The Netherlands.

Robinson, D. and Stone, A. (2004b) "Solar Radiation Modelling in the Urban Context", *Solar Energy* (77)3: 295–309.

Robinson, D. and Stone, A. (2005a) "A Simplified Radiosity Algorithm for General Urban Radiation Exchange", *Building Services Engineering Research and Technology* 26(4): 271–284.

Robinson, D. and Stone, A. (2005b) "Holistic Radiation Modelling with a Fast Simplified Radiosity Algorithm", in *Proceedings of Building Simulation 2005*, Montreal, Canada.

Robinson, D. and Stone, A. (2006) "Internal Illumination Prediction Based on a Simplified Radiosity Algorithm", *Solar Energy* 80(3): 260–267.

Robinson, D., Stankovic, S., Morel, N., Deque, F., Rylatt, M., Kabele, K., Manolakaki, E. and Nieminen, J. (2003) "Integrated Resource Flow Modelling of Urban Neighbourhoods: Project SUNtool", in *Proceedings of Building Simulation 2003*, Eindhoven, The Netherlands, pp. 1117–1122.

Robinson, D., Scartezzini, J.-L., Montavon, M. and Compagnon, R. (2005) "Solurban: Solar Utilisation Potential of Urban Sites", Final Report, Swiss Federal Office of Energy. Online. Available at: http://www.bfe.admin.ch/php/modules/enet/streamfile.php?file=000000008944.pdf&name=000000250027.pdf

Robinson, D., Campbell, N., Gaiser, W., Kabel, K., Le-Mouele, A., Morel, N., Page, J., Stankovic, S. and Stone, A. (2007) "SUNtool – A New Modelling Paradigm for Simulating and Optimising Urban Sustainability", *Solar Energy* 81(9): 1196–1211.

Robinson, D., Haldi, F., Kämpf, J., Leroux, P., Perez, D., Rasheed, A. and Wilke, U. (2009) "CitySim – Comprehensive Micro-simulation of Resource Flows for Sustainable Urban Planning", in *Proceedings of Building Simulation 2009*, Glasgow, Scotland, pp. 1083–1090.

Rylatt, M., Gadsden, S. and Lomas, K. (2001) "GIS-based Decision Support for Solar Energy Planning in Urban Environments", *Computers, Environment and Urban Systems*, 25(6): 579–603.

Santamouris, M. (1998) "The Athens Urban Climate Experiment", *Proceedings of the 15th International Conference on Passive and Low Energy Architecture* (PLEA), Lisbon, Portugal.

Shimoda, Y., Fujii, T., Morikawa, T. and Mizuno, M. (2003) "Development of Residential Energy End-use Simulation Model at City Scale", in *Proceedings of Building Simulation 2003*, Eindhoven, The Netherlands, pp. 1201–1208.

Shimoda, Y., Yamaguchi, Y., Kawamoto, K., Ueshige, J., Iwai, Y. and Mizuno, M. (2007) "Effect of Telecommuting on Energy Consumption in Residential and Non-residential Sectors", in *Proceedings of Building Simulation 2007*, Beijing, China, pp. 1361–1368.
Shorrock, L.D. and Dunster, J.E. (1997) "The Physically-based Model BREHOMES and Its Use in Deriving Scenarios for Energy Use and Carbon Dioxide Emissions of the UK Housing Stock", *Energy Policy* 25(12): 1027–1037.
Stathopoulos, T., Wu, H. and Zacharias, J. (2004) "Outdoor Human Comfort in an Urban Climate", *Building and Environment* 39(3): 297–305.
Swaid, H. and Hoffman, M.E. (1989) "The Prediction of Impervious Ground Surface Temperature by the Surface Thermal Time Constant (STTC) Model", *Energy and Buildings* 13(2): 149–157.
Tacken, M. (1989) "A Comfortable Wind Climate for Outdoor Relaxation in Urban Areas", *Building and Environment* 24(4): 321–324.
Taniguchi, A., Shimoda, Y., Asahi, T., Yamaguchi, Y. and Mizuno, M. (2007) "Effectiveness of Energy Conservation Measures in Residential Sector of Japanese Buildings", in *Proceedings of Building Simulation 2007*, Beijing, China, pp. 1645–1652.
Torrens, P.M. and O'Sullivan, D. (2001) "Cellular Automata and Urban Simulation: Where Do We Go From Here?", *Environment and Planning B* 28(2): 163–168.
Tregenza, P. and Sharples, S. (1993) "Daylighting Algorithms", ETSU S 1350–1993, UK.
United Nations (2004) "World Urbanization Prospects: The 2003 Revision", Department of Economic and Social Affairs/Population Division, New York: United Nations, p. 3.
Waddell, P. and Ulfarsson, G.F. (2004) "Introduction to Urban Simulation: Design and Development of Operational Models", in Haynes, K., Stopher, P., Button, K. and Hensher, D. (eds), *Handbook 5: Transport Geography and Spatial Systems*, Oxford, UK: Pergamon Press.
Ward Larsen, G. and Shakespeare, R. (1997) *Rendering with Radiance – The Art and Science of Lighting Visualisation*, San Francisco: Morgan Kauffmann.
Wegener, M. (1994) "Operational Urban Models", *Journal of the American Planning Association* 60(1): 17–29.
White, R. and Engelen, G. (1993) "Cellular Automata and Fractal Urban Form: A Cellular Modelling Approach to the Evolution of Land Use Patterns", *Environment and Planning* A, 25(8): 1175–1199.
White, R. and Engelen, G. (2000) "High-resolution Integrated Modelling of the Spatial Dynamics of Urban and Regional Systems", *Computers, Environment and Urban Systems* 24(5): 383–400.
Wolfram, S. (1994) "Cellular Automata as Models of Complexity", *Nature* 311: 419–424.
Yamaguchi, Y., Shimoda, Y. and Mizuno, M. (2003) "Development of District Energy Simulation Model Based on Detailed Energy Demand Model", in *Proceedings of Building Simulation 2003*, Eindhoven, The Netherlands, pp. 1443–1450.
Yamaguchi, Y., Shimoda, Y. and Mizuno, M. (2007a) "Transition to a Sustainable Urban Energy System from a Long-term Perspective: Case Study in a Japanese Business District", *Energy and Buildings* 39(1): 1–12.
Yamaguchi, Y., Shimoda, Y. and Mizuno, M. (2007b) "Proposal of a Modelling Approach Considering Urban Form for Evaluation of City Level Energy Management", *Energy and Buildings* 39(5): 580–592.

Activities

For the present assignment we are going to use the public domain ray tracing program RADIANCE to perform a series of solar irradiation simulations. The purpose of this exercise is both to demonstrate the simplicity with which such simulations may be performed, but also to demonstrate the importance of properly accounting for diffuse irradiation in simulations of both resource availability and of resource flows.

Part 1: creating the sky

Using *GenCumulativeSky*, a module developed to describe a cumulative sky radiance distribution for solar irradiation simulations using RADIANCE, annual cumulative skies, both global and diffuse, have been created for Geneva (16.3° N, 6.1° E): to download these files and/or *GenCumulativeSky* visit http://leso.epfl.ch/e/research_urbdev.html. In each case the file *cumulativesky.cal* file[9] is reference from a file *cumulativesky.rad* as follows:

```
void brightfunc skyfunc
2 skybright ./cumulativesky.cal
0
0
skyfunc glow sky_glow
0
0
4 1 1 1 0
sky_glow source sky
0
0
4 0 0 1 180

skyfunc glow ground_glow
0
0
4 1 1 1 0

ground_glow source ground
0
0
4 0 0 -1 180
```

In this way any outgoing ray is associated with the cumulative radiance ($Wh/m^2.Sr$) of the corresponding sky patch. Performing a simulation using the program *r*-trace to sample a hemisphere along a vertical vector gives a predicted global horizontal solar irradiation of 1.129 MWh/m^2, which closely corresponds to the cumulative total irradiation from the initial climate file (which was 1.147 MWh/m^2). The actual cumulative global and diffuse radiance distributions are shown in Figure 15.9.

Part 2: creating a simple scene of a street canyon

Create a primitive description of a street canyon of height 22.5m, width 15m and depth 45m. For this you can use genbox:

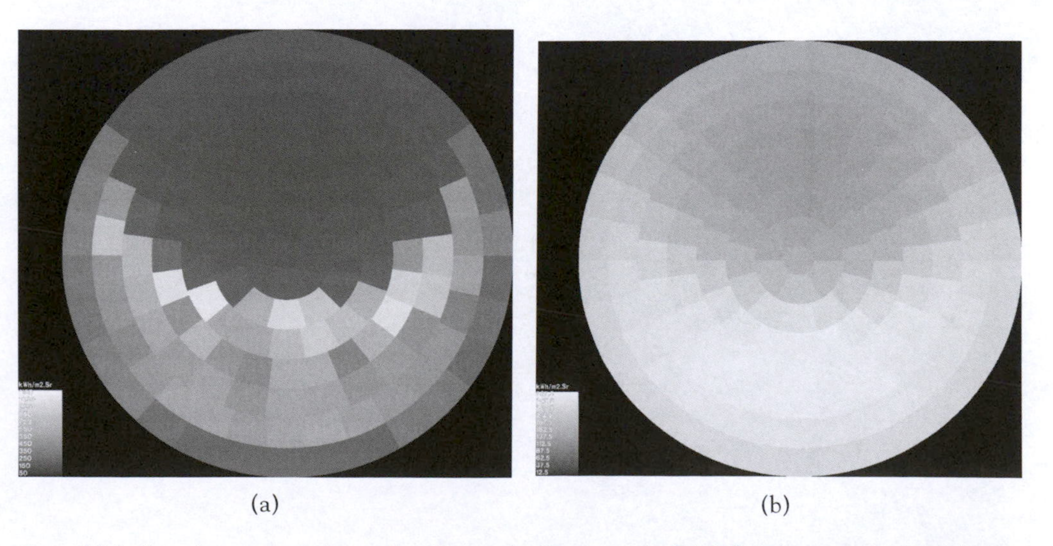

(a) (b)

Figure 15.9 Cumulative sky radiance distribution for Geneva: (a) global (left; lower scale), and (b) diffuse (right; upper scale).

```
genbox canyonwalls canyon 45 15 22.5 >> canyon.rad
```

Then add the following to the top of the file, to define the surface greyscale reflectances:

```
Void plastic canyonwalls
0
0
5 .2 .2 .2 0 0
```

And delete surfaces .5137, .4620, .6457.

Part 3: irradiation simulations

Now perform three simulations using the *r*-trace program,[10] from the viewpoint 22.5, 0.1, 1.5 looking in the view direction 0 1 0:

1. using the global cumulative sky radiance distribution (call this $I_g(obs)$),
2. using the diffuse cumulative sky radiance distribution (call this $I_d(obs)$),
3. using the diffuse cumulative sky radiance distribution, but with surface 3,2,6,7 removed (call this $I_d(unobs)$).

Now, most dynamic thermal simulation programs ignore obstructions to the sky vault due to adjacent buildings as well as reflections of diffuse radiation from them. Using our three simulation results we can calculate a good approximation of the relative error in incident annual solar irradiation due to this simplification:[11] $r.e = \varepsilon / I_g(obs)$, *where* $\varepsilon = I_d(unobs) - I_d(obs)$. Thus:

$$r.e. = \frac{I_d(unobs) - I_d(obs)}{I_g(obs)} = \frac{256.18 - 100.96}{145.54} = 1.07$$

or 107%. This is a significant over-estimation! Why not explore the sensitivity of this relative error to the proportions of the canyon or indeed to the position of the datapoint within the canyon, by performing some additional simulations?

16 Building simulation for policy support

Drury B. Crawley

Scope

This chapter discusses the use of simulation in support of building policy decision-making. Examples of building simulation in use for policy setting in this chapter include development and evaluation of building energy efficiency standards, setting building performance levels for utility incentives, evaluating potential impacts and direction for energy efficiency building retrofit programs, and sector-wide technology performance studies. The critical role of building simulation in policy making is emphasized.

Learning objectives

The reader is introduced to the use of building simulation to embody the performance of a portion of the building stock to evaluate options and directions for building policy. The reader should understand the key components of multi-building model policy studies including energy, environmental and financial decision-making.

Key words

Policy analyses/building stock/building energy standards/R&D planning/building technology assessments/policy analysis methodologies/sample design

Introduction

Building performance simulation is regularly used to support decision making in the design or retrofit of individual buildings. Yet, one of its most powerful uses lies beyond the performance of individual buildings in supporting building policy setting and decision making: to develop minimum standard regulations, assess the value of improved building performance for utilities or governments, or support high-level, public decision-making. Examples of policy targets include:

- defining a cost-effective performance level for various aspects of a new building energy efficiency standard,
- evaluating the performance of an existing or proposed building energy standard,
- establishing financial incentives for improved building energy performance, i.e., the value to the utility or political entity,
- evaluating the potential impact of and direction for voluntary programs encouraging energy-efficient new building design or existing building retrofit,
- evaluating the applicability of specific technologies or systems for the new building design or existing building retrofit markets,
- evaluating the potential for introduction of renewable energy technologies at the building, community or regional level,
- evaluating the potential impact of changes in regional, national or international policy,

- evaluating requirements for existing or new energy supply at a regional or national level, and
- evaluating the potential impact on building performance of changes relating to environment.

Coupling simulation with building models, which represent a range of building types and locations, can embody a portion of a building stock (existing or new, domestic, public, commercial, industrial, large, medium, or small) or the entire stock, which allows discrete modeling of building policy direction. Simulation provides policy decision makers a means of assessing what-if scenarios across a spectrum of buildings, ensuring that regulations and policy are set at the most financially and environmentally beneficial levels for individuals and the public.

In the context of building policy, performance simulation has been in wide use for more than 30 years in North America. These include evaluating the performance of new or proposed building energy standards, evaluating the value and design of energy savings programs for a utility or government entity, and evaluating the potential impact on the built environment of actions relating to climate change. Swan and Ugursal (2009) provide a comprehensive review of the various modeling techniques used for modeling residential sector energy consumption and they note:

> Two distinct approaches are identified: top-down and bottom-up. The top-down approach treats the residential sector as an energy sink and is not concerned with individual end-uses. It utilizes historic aggregate energy values and regresses the energy consumption of the housing stock as a function of top-level variables such as macroeconomic indicators (e.g. gross domestic product, unemployment, and inflation), energy price, and general climate. The bottom-up approach extrapolates the estimated energy consumption of a representative set of individual houses to regional and national levels, and consists of two distinct methodologies: the statistical method and the engineering method. Each technique relies on different levels of input information, different calculation or simulation techniques, and provides results with different applicability.

Swan and Ugursal go further to say that the most appropriate use of these three approaches in modeling residential buildings corresponds to its strongest attribute:

- top-down approaches for supply analysis based on projections of energy demand,
- bottom-up statistical techniques for determining energy end-uses including behaviour based on energy bills and simple surveys, and
- bottom-up engineering techniques for explicitly calculating energy end-uses based on detailed characteristics, enabling impact of new (or alternative) technologies.

This chapter reviews related commercial building research in support of building policy where comparison of alternative technologies, systems, or climatic response is required – a bottom-up engineering approach – building energy standards, building stock models, utility incentives, and climate modeling.

Evaluation of building energy standards

Building energy standards are a set of definitions of minimum performance of building components and equipment. They can be as simple as a table of minimum efficiency levels for a packaged air-conditioning unit, minimum insulation levels for walls and roofs, or maximum lighting power density. Building energy standards also can be performance based, requiring that a new or existing building has energy consumption no higher than a prescribed value. Where does building performance simulation fit into building energy standards? The primary aim of performance simulation in standards has been to evaluate the energy savings associated with a proposed standard (as compared to typical building practice or existing standards).

National building energy standards date back to the 1970s. For example, ASHRAE developed Standard 90-75 (ASHRAE 1975) to improve the efficiency of buildings in direct response to the oil embargo of the early 1970s. In the late 1970s, the US DOE proposed Building Energy Performance Standards (BEPS) (US DOE 1979) that set maximum energy performance levels for both residential and commercial buildings. A key feature of both the development and the proposed standards was the innovative use, at that time, of building performance simulation. Characteristics of typical buildings were collected throughout the United States, and baseline simulation models were constructed for more than 20 building types. Simulations were then performed and target levels established. Designers could then use building simulation models to demonstrate that their building performed at an energy level less than the maximum for that building type. Unfortunately, BEPS was too far ahead of the market and was buried under a landslide of adverse public comments.

Subsequent research by Pacific Northwest Laboratory (PNL 1983) used building simulation to propose aggressive updates to the then current Standard 90A-1980 in ASHRAE Special Project 41 (SP41) (Pacific Northwest Laboratory 1983). The SP41 proposals were based on life-cycle cost evaluation of the improved energy performance, evaluating ten buildings in six locations. This research was used as the starting point for the development of the next major update of Standard 90.1 in 1989 (ASHRAE 1989). Many of the public review drafts were tested by Pacific Northwest Laboratory for US DOE from 1985 through 1989, using five building prototypes and five locations (weather data) to evaluate energy savings levels. Interim energy simulation results were reported by Crawley and Briggs (1985a, 1985b), US DOE (1987), and Crawley and Boulin (1989), among others. A regional evaluation of state building energy codes in comparison with Standard 90.1-1989 was reported by Johnson *et al.* (1988). The five building prototypes were developed to be typical of commercial buildings rather than representative of the entire stock of buildings. With only five building prototypes and five weather locations, the results from these studies could not extrapolated to represent the value of potential savings for the new standard across the entire commercial sector.

Carlo *et al.* (2003) reported on an analysis of 12 prototype buildings to establish minimum requirements for a building energy-efficiency code in Brazil. While the code was based on Standard 90.1-1999, simulation allowed the researchers to customize the requirements to the climate and economic situation in Brazil. The goal was to create a multi-variable regression equation from the simulation results for calculating the annual energy consumption of a building in that location. The regression equation was tuned to that location with 10 coefficients. With 12 widely varying prototype buildings (in floor area and number of storeys), this required 1,616 combinations of envelope characteristics to be simulated. Carlo proposed to replicate this analysis for other Brazilian locations – requiring a similar number of simulations for each. Subsequently, Carlo and Lamberts (2008) reported on the impacts on electricity consumption of various envelope characteristics using 6 prototype buildings. Carlo developed a series of equations from simulation results which define the envelope efficiency level through a comparison of the proposed building envelope with a low and high efficiency level. These equations are the basis of new building energy efficiency regulations in Brazil.

Recently, the European Union (EU) has established a Directive on the Energy Performance of Buildings (Commission of the European Communities 2001), which requires member states to create and deploy calculation methods for rating energy performance for buildings larger than 1,000 m^2. This rating has a scale of A to H, where A is the lowest-energy and H the highest-energy buildings. Throughout Europe, this directive is requiring the use of building performance modeling of many more buildings than in the past. In the United Kingdom, The Netherlands, and a few other countries, the building performance modeling required to rate non-domestic buildings has been implemented in a simple set of equations for predicting energy use, known as Simplified Building Energy Model (SBEM) (British Research Establishment 2008a, 2008b). Rather than a model of energy use of a specific building, the SBEM is a comparative model for rating the relative carbon emissions of the proposed building. The

SBEM uses monthly utilization factor calculations and is limited in the technologies that it can evaluate.

Building models, building stock and utility incentives

In a policy framework relating to building energy performance, the targets are individual buildings, but there is a need to establish how the proposed policy or standard will perform in a broader context – of new or existing buildings within a region or country. This requires an understanding of the building stock:

- the number and size of buildings being built and their operating and thermo-physical characteristics;
- the number and size of existing buildings and their operation schedules, energy consumption, energy costs, and thermo-physical characteristics.

Internationally, detailed data relating new or existing commercial building stocks to actual building energy performance are relatively scarce in comparison to what Swan and Ugursal found for the residential sector. Macmillan and Köhler (2004) found that national residential and commercial sector-level energy use data were generally available throughout the world. But they also found that detailed data on the building stock and its energy use was very limited – citing studies in North America and Europe. Barrett (2009) supports this in a review of recent building stock research – saying that papers "concentrate on domestic buildings in detail with less about the non-domestic buildings." Barrett goes on to say that the non-domestic sector accounts for a large and growing proportion of energy use (as evidence in Figure 16.1) but the

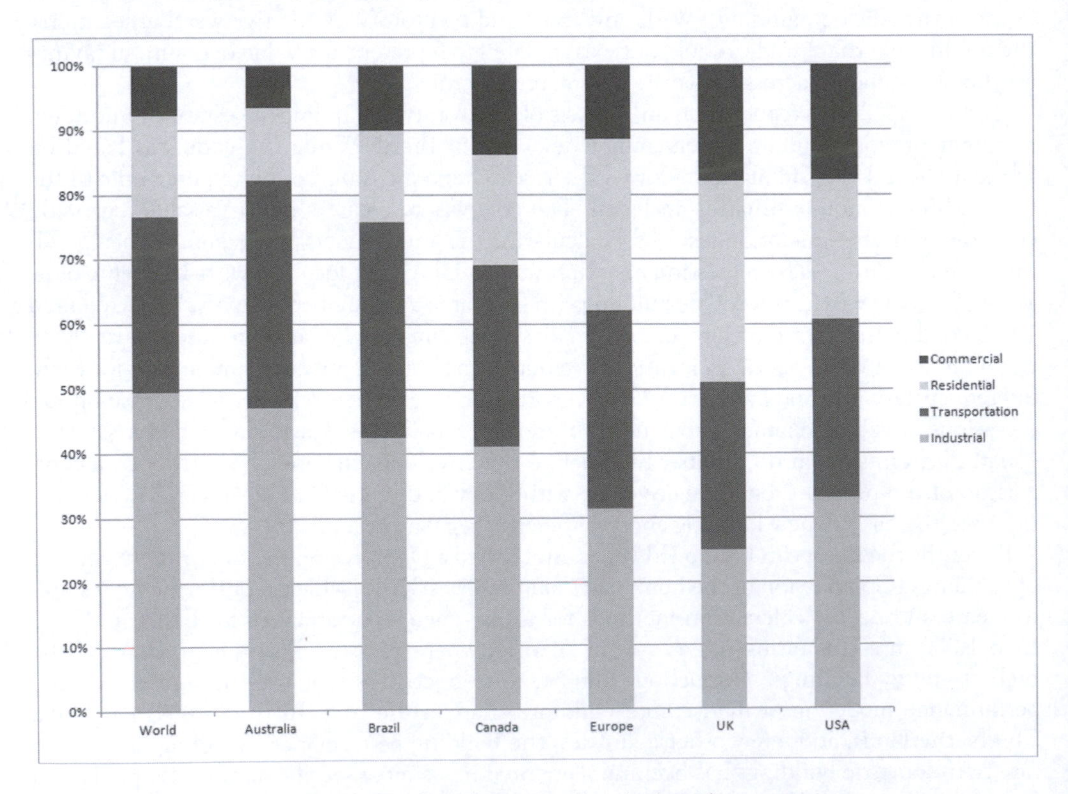

Figure 16.1 Percentage energy use by sector for the world and selected countries.

Source: UK data from Department of Trade and Industry (2002) and Department for Business Enterprise & Regulatory Reform (2008); other data from Energy Information Administration (2008).

literature he reviewed provides few details of this growth. He asserts that this is due to the more heterogeneous nature of the commercial building stock and that empirical data is sparse.

In the United States, the Energy Information Administration has conducted quadrennial energy consumption surveys since 1979 for commercial, industrial, and residential buildings. These surveys provide a wealth of information about the numbers and consumption of the entire building stock based on a statistical representation of the sector and the representativeness of several thousand buildings for each sector. The Commercial Building Energy Consumption Survey (CBECS) comprises survey data on more than 4,000 U.S. buildings (Energy Information Administration 2002, 2007). But these data are not complete thermo-physical models of each building. For example, there is little detail on the energy consuming systems or the thermal characteristics of the building envelope. Further, EIA masks certain data such as number of workers, number of floors and floor area in large buildings (more than 20 storeys and/or 45,000 m^2) so that individual buildings cannot be identified. Griffith *et al.* (2008) discusses the limitations of using the survey data in the context of a bottom-up research study – requiring supplementing the data with assumptions, defaults, data from other literature and probabilistic assignments.

Recently, Natural Resources Canada has published a similar commercial building survey for Canada (NRCan 2006), the Commercial and Institutional Consumption of Energy Survey (CICES). CICES includes information on 7,349 buildings, but it focuses on the floor area, building type, and the forms and amounts of energy used in the survey year. Key information such as the number of floors or any information about the building envelope or heating and cooling systems is not collected. This limits the use of the CICES as a source of input for building simulation.

California has created a similar survey with more details on the surveyed buildings (PG&E 1999). This survey was recently updated and extended state-wide as reported by Kinney and Piette (2002). The updated survey included the collection of sufficient building characteristics so that calibrated simulation models of each of the more than 2,800 commercial buildings could be created (Ramirez *et al.* 2005).

The Building Owners and Managers Association (BOMA) publishes an annual report on U.S. office markets (BOMA 2007). BOMA publishes information obtained from more than 3,000 office buildings including income, expenses, energy consumption, rent, and occupancy rate. But similar to the surveys noted above, the minimum geometric and other key information required for simulating the performance of those buildings is not collected by BOMA.

In Scotland, the Energy Systems Research Unit (ESRU) Domestic Energy Model (EDEM) (Clarke *et al.* 2003; Clarke *et al.* 2008; ESRU 2008) supports energy policy formulation for the residential sector. It provides representations of the entire domestic sector for Scotland, which allows policy makers to quickly apply a wide range of improvements analytically and evaluate which will provide the best energy performance or reduction in carbon. Analyses performed include national housing stock fabric upgrade strategies, local community upgrade strategy and carbon roadmap formulation, carbon and energy performance, and energy labelling in compliance with the EU Energy Performance of Buildings Directive. But the EDEM is limited to the domestic sector. An equivalent model for the non-domestic sector is not available.

The International Energy Agency (IEA) also publishes energy use data for buildings in member countries (IEA 2008), but this is a statistically disaggregated estimate, a top-down inference based on energy production and supply data. Little information is available on the average energy performance of buildings by type or climate zone.

While survey data are extremely useful for a building stock snapshot, they usually are limited by the information collected. Only rarely are there enough data to create a building model in energy simulation software. Several projects have worked to use existing survey data to create prototypes that represent large portions or building types and regions. Briggs *et al.* (1987, 1992) created 20 existing office building prototypes, and Crawley and Schliesing (1992) created 10 new office building prototypes for use by the Gas Research Institute in research and market assessment. The existing building models were based on the 1979 Nonresidential Buildings Energy Consumption Survey (NBECS), the first survey in the CBECS. The new building models were based on Standard 90.1. In both cases, substantially more input data were required for the models, and sources are documented in their reports. Huang *et al.* (1991) extended the prototypes to the entire commercial sector for 20 urban markets in the United States. Huang and Franconi (1999) updated the prototypes to evaluate the contributing components of commercial building loads. In all cases, these prototypes were limited by the assumptions that the authors had to make to create complete simulation models. Often these assumptions were not well documented.

Rather than producing prototypes to represent multiple buildings, Griffith *et al.* (2007, 2008) created building simulation models for each of the more than 4,000 buildings in the 2003 CBECS. These building models were developed first as a baseline with energy features and performance consistent with Standard 90.1-2004 (ASHRAE 2004). Then low-energy technologies and renewable energy systems anticipated to be available by 2025 were applied to the models. The result was an assessment of the technical potential for achieving zero-energy buildings throughout the commercial building sector in the United States. The conclusion was that the zero-energy building goal could be, on average, achieved in the commercial building sector. Building types which can reach the zero-energy target easily: offices, warehouses, schools, and retail. Restaurants, hospitals, and other energy-intensive building types will be the hardest to bring to net zero-energy. While not creating prototypes improved the breadth and representativeness of the analysis, it also created a logistic and quality assurance issue – how to deal with multiple thousands of building models. Griffith used an XML schema associated with a database of inputs to define each model. Since the inputs were derived directly from the CBECS data or related assumptions, the inputs could be automatically verified before a series of simulations were started.

Recently, Deru *et al.* (2008) published a set of 16 benchmarks for new commercial buildings based on the 2003 CBECS and ASHRAE Standard 90.1-2004. The benchmark buildings descriptions include a scorecard comprising detailed documentation of all inputs and assumptions as well as information and data sources. Each benchmark building also has a corresponding EnergyPlus input files (US DOE 2010).

For utilities, the economic benefit of incentives paid for improved energy efficiency is often obscured by their regulated environment and the complex valuation of equitable sharing of net benefits, cost capitalization, and risks with their stakeholders. Rather than evaluating individual incentives, they aggregate the energy efficiency measures into a portfolio, essentially a top-down approach to the utility sector. For example, Cappers *et al.* (2009) present an analysis of energy efficiency incentives in the context of a prototypical investor-owned utility. This analysis adapts a spreadsheet-based financial model (known as the Benefits Calculator) developed in support of the National Action Plan for Energy Efficiency (Jensen 2007).

The literature is full of similar portfolio-centric approaches to energy efficiency with a focus on program costs, costs and benefits to ratepayers, and similar utility rate case calculations. Many consultants offer services to develop and evaluate incentives for utilities but few publish in the peer-reviewed literature, which may indicate that they see their utility incentive calculation methodologies and procedures as a business advantage. Instead, utilities focus on supporting their customers and administering the incentive programs. Ter-Martirosyan (2003) says that this is due to changes which began in the late 1980s, with public utility commissions and other regulators focusing on regulation of the incentives themselves rather than the rate of return the utility could earn. For utilities, this has a two-fold result – decreased quality and cuts in service (as described by Ter-Martirosyan) and less interest in detailed engineering approaches that are overwhelmed by their heavily regulated environment.

Another critical issue for simulation for building policy is which climate data to use. Climate data issues and resources are discussed in Chapter 3.

Data issues in multi-building studies

The largest challenge of research involving large numbers of simulation models is dealing with the immense amount of data that building performance simulation programs can create. For example, a single EnergyPlus simulation with 10-minute time step output can produce more than 600 megabytes of data. Handling the data from hundreds or thousands of simulations can overwhelm a researcher.

In their studies of Standard 90.1 in the 1980s, Crawley *et al.* (1985a, 1985b, 1989) created five building type DOE-2.1E models for the commercial sector. They were based on the earlier models created for ASHRAE Special Project 41 and included variants tied to five climate regions throughout the United States. Three variations on the base model were created – Standard 90-75, Standard 90A-1980 (ASHRAE 1980), and the proposed draft standards, which became Standard 90.1-1989. This large number of simulations (5 buildings × 5 climates × 3 standards) required that the structure of the analysis be carefully designed, including a systematic file-name convention that included abbreviations for building type, climate zone, and variant of the standard. The researchers also created batch scripts to automate running the simulations and extracting summary energy results. When the ASHRAE committee developing the standard created three drafts in three years, this structure and automation proved invaluable. When Johnson *et al.* (1988) studied the potential for upgrading building energy standards in the Pacific Northwest; they were able to quickly adapt the 90.1 scripts and input files and complete the analysis in less than two months.

In the study for the Gas Research Institute described above, Briggs *et al.* (1992) chose to create a spreadsheet that included the key characteristics definitions for the building prototypes and all the parts of the simulation input files. Invoking a macro in the spreadsheet automatically generated the input files for the existing office building prototypes. This provided the added advantage of having all the input in tabular form, ready for insertion into the research report. A further advantage was that the prototypes could be modified easily and the data checked carefully.

When Huang and Franconi (1999) created new prototypes, they constructed these as snippets of simulation input files. The user simply invoked a batch script that assembled the correct parts into a complete input file.

Griffith *et al.* (2007, 2008) took a more robust approach in studying the technical potential for achieving zero-energy commercial buildings. Because this study involved creating a building simulation model for each of the more than 5,000 buildings in the 2003 CBECS, it necessitated the development of a method for automatically creating the input files. Griffith created a set of high-level XML definitions for all the key inputs, such as floor area, weather data, floor-to-floor height, lighting power intensity, and HVAC system. He created a specific part of the input file for each of these XML definitions. When the XML scripting tool was passed an XML key, a specific input file could be created. Further scripting tools were created to automatically submit simulations on multi-core computer clusters and extract summary results from the completed simulations.

Creating a structured approach to studies involving multi-building simulations is also vital to quality assurance. Checking the values of individual inputs is not easily accomplished if several hundred or thousand individual input files are involved. Summary tables of the values of key inputs that can be checked easily and then automatically transferred to input files are critical.

Donn (2007) studied the role of Quality Assurance (QA) in environment design decision support tools for architecture, including a variety of building performance simulation tools for daylighting, thermal design, and acoustics. He found that not only was it critical to have QA procedures as a standard part of any building analysis, it was critical to have QA measures that are codified and incorporated into the simulation tools themselves. These QA measures would be reality tests to examine whether outcomes from the design decision tools behave in a believable manner – like a real building. Donn also proposed the establishment of a shared database on QA performance data available to all design decision tool users.

Synopsis

This chapter has presented an overview of simulation in building policy studies in a number of areas:

- development and evaluation of building energy standards
- buildings models and building stock
- utility incentives
- QA and data volume issues in multi-building studies.

For the development and evaluation of building energy standards, having a number of prototype buildings which could represent the wider building stock was important. Several of the studies (US DOE 1987, Johnson *et al.* 1988) had few buildings and the results could not be extrapolated to the wider commercial building sector. Similarly, a large number of buildings raised the level of complexity as seen in the work by Carlo *et al.* (2003) which has implications for data management and QA. On the other extreme, a simple model such as the SBEM can inhibit the adoption of new technologies. It is important to be able to adequately cover the range of characteristics expected for the buildings to be simulated.

In setting policy for commercial buildings, the ultimate targets are individual buildings. But how the proposed policy will perform in a broader context of a region or country or a building type or entire building sector is important as well. Policy makers need to be able to relate individual decisions to their impacts. In that context, this requires an understanding of the target building stock including the size and number of buildings as well as their operating and physical characteristics. Few sources of detailed data below sector-level energy use are available. A few building characteristics surveys were available in North America and Europe but they all had limitations in the depth or quality of data available for building modeling. The exception was in California where the commercial building survey was coupled with calibrated building simulation models. Further review found work that filled in the data missing from the surveys with assumptions, rules of thumb, and other data. Often these supplementary data were not described in the reports or other documentation, making it difficult for potential users to fully understand the models. More recent benchmark buildings (Deru *et al.* 2008) are working to overcome these limitations by fully documenting all input assumptions and sources of data in the building models.

Interestingly, there were very few published studies on setting the level and technical attributes for utility incentives in the peer-reviewed literature. Most utility incentives literature focused instead on the delivery and administration of incentive programs for energy efficiency in commercial buildings.

Several studies noted that despite a substantial increase in the number of locations for which climate data are available over the last few years, limited availability and temporal resolution in many areas throughout the world is still an important issue.

Finally, studies that dealt with large number of building simulations found a number of means of automating the creation of building models. Some used spreadsheets, others scripts, but in all cases, it was critical to having a structured approach for the validation and error checking of individual inputs. This structure forms a critical part of the framework necessary for building research support policy makers. Crawley (2008) proposed a generalized framework for policy analysis using building performance simulation. This is shown in Table 16.1.

Table 16.1 Generalized framework for policy analysis using building performance simulation

Parameters			*Policy Study*			
			Minimum Building Energy Standard Development	*Utility Incentives Beyond Minimum Standards Study*	*Upgrading Existing Buildings*	*Sector-Wide Technology or Performance*
Research and Policy Focus						
Research Focus	Existing Building	Single building			•	
		Multiple buildings or stock			•	•
	New building	Single building	•	•		
		Multiple buildings or stock				•
Research Parameters	Building envelope (walls, roofs, fenestration)		•	•	•	•
	Lighting		•	•	•	•
	Internal loads		•	•	•	•
	HVAC		•	•	•	•
	Renewable technologies			•	•	•
Building Model						
Climate Data	Climatic design conditions		•	•	•	•
	Typical meteorological data		•	•	•	•
	Observed hourly data				•	•
Baseline Model	Simple prototype thermo-physical model		•	•		
	Existing building prototypes based on building stock				•	•
	New building prototypes based on minimum standards and building stock					•
	Low-energy prototypes including renewable technologies					•

Source: Crawley 2008.

References

ASHRAE (1975) *ASHRAE/IES Standard 90-75*, 'Energy Conservation in New Building Design', ASHRAE: New York.

ASHRAE (1980) *ANSI/ASHRAE/IES Standard 90A-1980*, 'Energy Efficiency Design of New Buildings', ASHRAE: Atlanta.

ASHRAE (1989) *ANSI/ASHRAE/IES Standard 90.1-1989*, 'Energy Efficient Design of New Buildings Except Low-Rise Residential Buildings', ASHRAE: Atlanta.

ASHRAE (2004) *ANSI/ASHRAE/IESNA Standard 90.1-2004*, 'Energy Efficient Design of New Buildings Except Low-Rise Residential Buildings', Atlanta: ASHRAE.

Barrett, M. (2009) 'Policy Challenges for Building Stocks', *Building Research and Information* 32(2): 201–205.

Briggs, R.S., Crawley, D.B. and Belzer, D.B. (1987) *Analyses and Categorization of the Office Building Stock*, GRI-87/0244. Chicago: Gas Research Institute.

Briggs, R.S., Crawley, D.B. and Schliesing, J.S. (1992) *Energy Requirements for Office Buildings, Volume 1, Existing Buildings*, GRI-90/0236.1, Chicago: Gas Research Institute.

British Research Establishment (2008a) 'National Calculation Method'. Online. Available at: http://www.ncm.bre.co.uk/

British Research Establishment (2008b) *Simplifed Building Energy Model Technical Manual*, Garston, UK: British Research Establishment.

Building Owners and Managers Association International (2007) *2007 Experience Exchange Report*. BOMA, Washington, DC.

Cappers, P., Goldman, C., Chait, M. Edgar, G., Schlegel, J. and Shirley, W. (2009) *Financial Analysis of Incentive Mechanisms to Promote Energy Efficiency: Case Study of a Prototypical Southwest Utility*, Berkeley, CA: Lawrence Berkeley National Laboratory.

Carlo, J. and Lamberts, R. (2008) 'Development of Envelope Efficiency Labels for Commercial Buildings: Effect of Different Variables on Electricity Consumption', *Energy and Buildings* 40(11): 2002–2008.

Carlo, J., Ghisi, E. and Lamberts, R. (2003) 'The Use of Computer Simulation to Establish Energy Efficiency Parameters for a Building Code of a City in Brazil', in *Proceedings of Building Simulation 2003*, Eindhoven, The Netherlands, pp. 131–138.

Clarke, J.A., Johnstone, C.M., Lever, M., McElroy, L.B., McKenzie, F., Peart, G., Prazeres, L. and Strachan, P.A. (2003) 'Simulation Support for the Formulation of Domestic Sector Upgrading Strategies', in *Proceeedings of Building Simulation 2003*, Eindhoven, Yhe Netherlands, pp. 219–226.

Clarke, J.A., Ghauri, S., Johnstone, C.M., Kim, J.M. and Tuohy, P.G. (2008) 'The EDEM Methodology for Housing Upgrade Analysis, Carbon and Energy Labelling and National Policy Development', in *Proceedings of the eSim 2008 Conference*, Quebec City, Canada, pp. 135–142.

Commission of the European Communities (2001) *Directive on the Energy Peformance of Buildings*, Brussels, Belgium.

Crawley, D.B. (2008) 'Building Performance Simulation: A Tool for Policymaking', PhD Thesis, University of Strathclyde, Glasgow, Scotland.

Crawley, D.B. and Boulin, J.J. (1989) 'Standard 90.1's ENVSTD: Both a Compliance Program and an Envelope Design Tool', in *Proceedings of Thermal Performance of the Exterior Envelopes of Buildings IV*, pp. 332–333. ASHRAE/DOE/BTECC/CIBSE, Orlando, Florida, pp. 332–333.

Crawley, D. B., and Briggs R.S. (1985a) 'Standard 90: The Value' *ASHRAE Journal* 27(11): 18–23.

Crawley, D.B. and Briggs R.S. (1985b) 'Envelope Impacts of ASHRAE Standard 90.1P: A Case Study View', in *Proceedings of Thermal Performance of the Exterior Envelopes of Buildings III*, ASHRAE/DOE/BTECC, Clearwater Beach, Florida, pp. 1128–1140.

Crawley, D.B. and Schliesing, J.S. (1992) *Energy Requirements for Office Buildings, Volume 2, Recent and Future Buildings*, GRI-90/0236.2. Chicago: Gas Research Institute.

Department of Trade and Industry (2002) *Energy Consumption in the United Kingdom*, London: Department of Trade and Industry.

Department for Business Enterprise & Regulatory Reform (2008) *Energy Consumption in the United Kingdom, 2008 Update*, London: Department for Business Enterprise & Regulatory Reform. Online. Available at: http://www.berr.gov.uk/whatwedo/energy/statistics/publications/ecuk/page17658.html

Deru, M., Crawley, D., Lui, B. and Huang, J (2008) *Commercial Building Benchmarks*, Washington: US DOE.

Donn, M.R. (2007) 'Simulation of Imagined Realities: Environmental Design Decision Support Tools in Architecture', PhD thesis, Victoria University of Wellington, New Zealand.

Energy Information Administration (2002) *Commercial Buildings Energy Consumption Survey – Commercial Buildings Characteristics 1999*. Washington: Energy Information Administration, US Department of Energy.

Energy Information Administration (2007) *Commercial Buildings Energy Consumption Survey – Consumption and Expenditures 2003*. Washington: Energy Information Administration. Online. Available at: http://www.eia.doe.gov/emeu/cbecs/cbecs2003/

Energy Information Administration (2008) *International Energy Outlook 2008*, DOE/EIA-0484 (2008) Washington: Energy Information Administration. Online. Available at: http://www.eia.doe.gov/oiaf/ieo/

Energy Systems Research Unit (2008) *ESRU Domestic Energy Model*, Glasgow: Energy Systems Research Unit, University of Strathclyde. Online. Available at: http://www.esru.strath.ac.uk/Programs/EEff/index.htm

Griffith, B., Long, N., Torcellini, P., Judkoff, R., Crawley, D. and Ryan, J. (2007) *Assessment of the Technical Potential for Achieving Net Zero-Energy Buildings in the Commercial Sector,* NREL/TP-550-41957, Golden, CO: National Renewable Energy Laboratory.

Griffith, B., Long, N., Torcellini, P., Judkoff, R., Crawley, D. and Ryan, J. (2008) *Methodology for Modeling Building Energy Performance Across the Commercial Sector*, NREL/TP-550-41956. Golden, CO: National Renewable Energy Laboratory.

Huang, Y.J. and Franconi, E.M. (1999) *Commercial Heating and Cooling Loads Component Analysis*, LBNL-37208. Berkeley, CA: Lawrence Berkeley National Laboratory.

Huang, Y.J., Akbari, H., Rainer, L. and Ritschard, R. (1991) *481 Prototypical Commercial Buildings for Twenty Urban Market Areas*, GRI-90/0326, Chicago IL: Gas Research Institute.

International Energy Agency (2008) *Key World Energy Statistics 2008*, Paris: International Energy Agency.

Jensen, V. (2007) *Aligning Utility Incentives with Investment in Energy Efficiency: A Product of the National Action Plan for Energy Efficiency*, Fairfax, VA: ICF International, Inc.

Johnson, B.K., Crawley, D.B. and Perez, R.C. (1988) 'Comparison of Energy Conservation Standards for New Commercial Buildings in the Pacific Northwest', in *Proceedings of the Pacific Northwest Regional Economic Conference*, Boise, Idaho.

Kinney, S. and Piette, M.A. (2002) *Development of a California Commercial Building Energy Benchmarking Database*, LBNL-50676, Berkeley, CA: Lawrence Berkeley National Laboratory.

Macmillan, S. and Kohler, J. (2004) *Modelling Energy Use in the Global Building Stock: A Pilot Survey to Identify Available Data Sources*, Norwich: Tyndall Centre for Climate Change Research, Technical Report 6.

Natural Resources Canada (2006) *2004 Commercial and Institution Consumption of Energy Survey*, Ottawa, Ontario: Natural Resources Canada.

Pacific Gas and Electric Company (1999) *Commercial Building Survey Report*, San Francisco.

Pacific Northwest Laboratory (1983) *Recommendations for Energy Conservation Standards and Guidelines for New Commercial Buildings*, PNL-4870. October 1983, Richland, WA: Pacific Northwest Laboratory.

Ramirez, R., Sebold F., Mayer, F.T., Ciminelli, M. and Abrishamim, M. (2005) 'A Building Simulation Palooza: The California CEUS Project and DrCEUS', in *Proceedings of Building Simulation 2005*, Montreal, Canada, pp. 1003–1010.

Swan, L.G. and Ugursal, V.I. (2009) 'Modeling of End-use Energy Consumption in the Residential Sector: A Review of Modeling Techniques', *Renewable and Sustainable Energy Reviews*, in press.

Ter-Martirosyan, A. (2003) 'The Effects of Incentive Regulation on Quality of Service in Electricity Markets', Working paper, Washington, DC: George Washington University.

US DOE (1979) *Notice of Proposed Rulemaking for Building Energy Performance Standards*, Washington, DC: USDOE.

US DOE (1987) *Notice of Proposed Rulemaking, Interim Standard, Voluntary Performance Standards for New Commercial and Multi-Family High Rise Residential Buildings*, Washington, DC: USDOE.

US DOE (2010) *EnergyPlus Version 6.0*. Online. Available at: http://www.energyplus.gov/.

Activities

Find out how the performance requirements for the applicable building energy standards in your area were determined using the categories listed in Table 16.1.

Learn what representations of building stock energy performance are available for your region from surveys or prototype buildings.

17 A view on future building system modeling and simulation

Michael Wetter

Scope

This chapter presents what a future environment for building system modeling and simulation may look like. As buildings continue to require increased performance and better comfort, their energy and control systems are becoming more integrated and complex. We therefore focus in this chapter on the modeling, simulation and analysis of building energy and control systems. Such systems can be classified as heterogeneous systems because they involve multiple domains, such as thermodynamics, fluid dynamics, heat and mass transfer, electrical systems, control systems and communication systems. Also, they typically involve multiple temporal and spatial scales, and their evolution can be described by coupled differential equations, discrete equations and events. Modeling and simulating such systems requires a higher level of abstraction and modularisation to manage the increased complexity compared to what is used in today's building simulation programs. Therefore, the trend towards more integrated building systems is likely to be a driving force for changing the status quo of today's building simulation programs. This chapter discusses evolving modeling requirements and outlines a path toward a future environment for modeling and simulation of heterogeneous building systems.

A range of topics that would require many additional pages of discussion has been omitted. Examples include computational fluid dynamics for air and particle flow in and around buildings, people movement, daylight simulation, uncertainty propagation and optimisation methods for building design and controls. For different discussions and perspectives on the future of building modeling and simulation, we refer to Sahlin (2000), Augenbroe (2001) and Malkawi and Augenbroe (2004).

Learning objectives

The learning objective of this chapter is to better understand the limitations of existing building simulation programs to model and simulate the performance of integrated building energy and control systems and how recent advances can overcome these limitations. In this chapter, the reader will be exposed to (i) motivating factors that may change the way we model and simulate building energy and control systems, (ii) technologies that enable such a change, and (iii) a path for realising a next generation modeling and simulation environment. While an in-depth discussion of the various enabling technologies is not possible in a single chapter, the reader will be exposed to the main concepts, and references to further literature are provided.

Key words

Requirements for building system simulation/modeling versus simulation/modular environments/object-oriented equation-based modeling/agent-based modeling/multi-domain modeling/management of complexity/Building Information Model

Introduction

Imagine an integrated design team sketching a building envelope in an object-oriented CAD system. A dashboard provides immediate feedback about the environmental impact and life cycle costs as different design variants are tested. After developing an initial building design, the HVAC engineer downloads the building information model (BIM), i.e., an electronic representation of the building parameters, draws the HVAC schematic diagram and drags standard supervisory control sequences from a library into the schematic diagram. At the press of a button, the HVAC engineer runs a whole building energy analysis and sizes the components while taking into account the reduced component size due to exploitation of free cooling and active façade systems. Next, the HVAC engineer replaces the simplified component models with refined equipment models downloaded from an electronic catalogue. The models in this catalogue have been used by the manufacturer during the product development and can represent the physical behaviour at different levels of temporal and spatial resolution; simplified performance models are used for annual energy analysis while detailed models may later be used during operation to determine a chiller's refrigerant charge in a fault detection and diagnostics program. After the HVAC engineer specifies the equipment, she/he uploads the data to the BIM that will be submitted to code authorities for performance-based code compliance checking within minutes and to generate bidding materials and construction plans. During building operation, the BIM data will be integrated with models and algorithms for optimisation-based control, fault detection, diagnostics and preventive maintenance.

What building simulation program will support such a life-cycle approach to simulation from product R&D to building operation? This is the focus of this chapter. Central to our discussion is the belief that the next generation tools that interface with the workflow of simulationists are not building simulation programs in the conventional sense. To explain this, we use the illustration from Cellier and Kofman (2006), shown in Figure 17.1, which distinguishes between the *mathematical model* and the *simulation program*. Cellier and Kofman state that the whole purpose of a mathematical model is to provide the human user of the modeling and simulation environment with a means to present knowledge about the physical system in a way that is as convenient to the user as possible. This may be for a building envelope an object-oriented CAD system, for an HVAC system a graphical schematic editor and for control systems a block diagram and finite state machines. Thus, the mathematical model represents the user interface. The mathematical model does not depend on how it is going to be used by the simulation program. The primary purpose of the simulation program is to compute the system response, typically in the form of time trajectories. The simulation program is automatically generated from the mathematical model. It typically provides a run-time interface with which the user can control the simulation process. Such a separation between mathematical model and simulation program is a critical underlying principle for a future modeling and simulation environment for building systems as we will see in this chapter. By following this principle, it is possible to structure the problem more naturally in the way a human thinks, and not how one computes a solution.

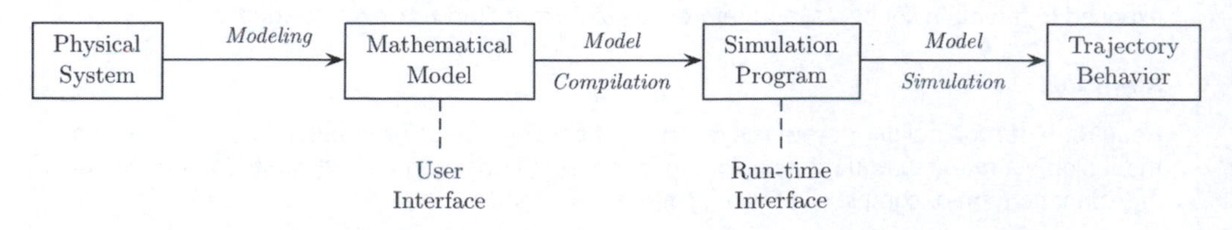

Figure 17.1 Modeling versus simulation.

Source: Adapted from Cellier and Kofman (2006), with kind permission of Springer Science+Business Media.

Motivating factors for a future modeling and simulation environment

To increase performance and improve comfort, building systems are becoming increasingly integrated. Integration is achieved physically by using systems that allow better energy recovery, energy storage and increased use of ambient energy sources and sinks for heating, cooling, ventilation and lighting. Integration is also achieved operationally by means of integrating controls for façade, lighting, HVAC, combined heat and power, solar systems, electrical grid and security. The characteristics of these systems are that they involve multiple functional domains (heat and mass transfer; fluid dynamics; daylighting; electricity generation and conversion; controls; communication across networks), and their temporal evolution may be described in the continuous time domain, in the discrete time domain and in a discrete event domain. Typical time-scales range from the order of subseconds (communication networks) to years (geothermal heat sources). This heterogeneity poses new challenges for modeling and simulation, which are used to reduce time-to-market during research and development, and to improve performance and cost-effectiveness during the design of innovative building energy systems and associated algorithms for controls, fault detection and diagnostics.

Due to the increased importance of controls to improve building performance, the control representation in building simulation programs needs to resemble more closely how actual control algorithms are implemented. Given the different topologies in which building systems can be configured, such as using a double façade as a ventilation air exchange path or using a thermally activated ceiling slab for thermal storage, a future modeling environment needs to allow a more intuitive representation of such complex systems that allows a user to quickly add new component models and reconfigure components to form new systems. Furthermore, the current separation between model developer and model user is not conducive to stimulate innovation.

Given the increased integration of different domain disciplines, a future environment for modeling and simulation should also enable the concurrent collaboration of multi-disciplinary teams, as opposed to a single person, and the sharing of models to integrate different domains. This may become increasingly important because much of the innovation in building science is likely to happen at the interface between different disciplines; for example, building physics for passive energy storage, mechanical systems for efficient energy conversion, controls systems for reducing irreversibilities (fluid flow friction and heat transfer), communication systems for propagation of sensor and actuator signals and mathematics to extract better quality information from the vast amount of data measured by building automation systems. Addressing the need for different discipline experts to collaborate more effectively will require different design of future tools.

Because embedded computing is becoming cheaper and more powerful and the performance of low-energy buildings is sensitive to controls and equipment degradation, it is likely that computational models will be used for commissioning, fault detection, diagnostics and model-based controls. In this context, the use of models can range from an individual piece of equipment, such as for a chiller fault detection, or to whole communities or estates that may be connected by a micro-grid for electricity, heating and cooling. Since sustainable buildings will have a lifespan that exceeds multiple generations of programming languages, operating systems and building simulation programs, models used during operation should be expressed at a high enough level of abstraction that ensures maintainability, portability, and facilitates upgrading to newer technology as buildings are retrofitted.

With respect to using building simulation for training of service technicians, commissioning agents and building operators, today's building simulation programs fall short in teaching how mechanical systems and their control systems operate. At present, models of these systems are often so simplified that they do not capture the dynamic behaviour and part-load operation of the mechanical system or the response of feedback control systems. Also, in many building simulation programs, the governing physical equations cannot be readily inspected by the user, which renders them as a black-box. Immersive simulation has the potential to improve the knowledge about the operational behaviour of buildings by presenting results of building simulation in a form that is closer to the human than a graph on a computer screen. Work to date has been primarily focused on computation and visualisation of spatial distribution of temperature, flow and comfort in rooms (Malkawi 2004; Wenisch *et al.* 2007). A versatile environment that would allow a realistic modeling and simulation of the mechanical and control system, and that could be reconfigured by the user to examine a variety of systems, would extend the promise of immersive simulation to training professionals in the operation of building systems.

The transition towards integrated systems introduces new challenges. Effective modeling strategies need to manage the increased complexity, the increased fragmentation of the supply chain and the need for working in multidisciplinary design teams using shared models. In the electronics design automation community, which faces challenges that are similar to the ones of the building industry, there are calls for a new design science to address system-level design challenges (Sangiovanni-Vincentelli 2007). The basic tenet of the advocated design flow is the identification of design as a meet-in-the-middle process, where successive refinements of specifications meet with abstractions of potential implementations. The refinement and abstraction process takes place at precisely-defined layers. It remains to be shown how this design approach can best benefit the building industry and what its implications are for building system models. The embedded systems community also called for a new system science that is jointly physical and computational in order to address challenges posed by networking physical devices that contain embedded control systems (Lee 2006). In this community, the semantic of a model is expressed at a high enough level of abstraction to allow formal analysis of system properties and to generate code automatically that can be used for design, synthesis and operation. It would be a surprise if advances in these related disciplines would not also cause the field of building simulation to embrace a higher level of abstraction that better addresses challenges in research, design and operation of integrated building systems.

A closer look at the software architecture of today's building simulation programs reveals that they are ill-suited to address the above needs. Many building simulation programs do not exploit advances in computer science that allow systems and their evolution to be expressed at a higher level of abstraction than that of imperative Fortran subroutines. Most of today's building simulation programs are large, monolithic programs that intertwine physics equations, statements that control the program flow logic, numerical solution methods, and data input and output routines. This makes extending these programs for applications that were not envisioned by their developers difficult. Lack of modularity has forced developers of building simulation programs to try to catch up by adding modeling capabilities that allow simulating new technologies. However, it may be more beneficial to develop a modular platform that can rapidly be extended and customised by users to drive the innovation of new systems and to integrate simulation better into building design and operation. In view of the gap between current building simulation programs and modern modeling and simulation environments, we will now take a step back to describe some basic needs that motivate a new approach to building system modeling and simulation.

Needs for more natural system representations

"Object programs produced by Fortran will be nearly as efficient as those written by good programmers." This statement from the Fortran manual (IBM 1956), released in 1956, shows that as early as fifty years ago, there was a trend towards languages that were closer to the problem at hand. Even then, experts questioned whether the "high-level" programming language would be adequately efficient. Today, most building simulation programs are still written using Fortran to encapsulate causal assignments in programming procedures. In these programs, procedures call other procedures, thereby transferring the locus of control from one procedure to another. Inside these procedures are imperative statements for algebraic equations, differential equations, difference equations and numerical solution algorithms. This syntax for formulating physical phenomena was motivated by the programming languages that were available decades ago. However, the syntax is far removed from how a physicist would describe the laws that relate physical quantities to each other or how an electrical engineer would describe communication in a network.

Physical systems are more conveniently modeled by (i) defining an object, (ii) exposing its boundary conditions and (iii) encapsulating inside the object mathematical constraints between the boundary conditions, the state variables and their derivatives. Consider for example a model for a heat conducting rod that stores no energy. It is not possible to determine whether heat flows through the thermal conductor because its ends are at different temperatures, or whether the temperatures at its ends are different because of the heat flow. So why should a user have to pick one or the other view when writing a model for a thermal conductor? It is more natural to create an object that exposes heat flow rate and temperature, i.e., (Q_1, T_1) and (Q_2, T_2), and encapsulates the constraints $Q_1 + Q_2 = 0$ and $Q_1 = K\,(T_1 - T_2)$, where K is the heat conductance (see Figure 17.2). How these equations are solved should be left entirely to an algorithm that generates a simulation program.

Such an equation-based, object-oriented model formulation is convenient to model physical systems that can be described by differential algebraic systems of equations. But is it also convenient to model *cyber-physical* systems? Cyber-physical systems are systems in which embedded computers and networks monitor and control physical devices with feedback loops, and the physical processes affect computations and vice versa (Lee 2006). Cyber-physical systems are becoming increasingly prevalent in buildings, and their size, in terms of nodes that send and receive signals, is growing rapidly. Components in cyber-physical systems exchange a sequence of discrete messages with each other through wired or wireless networks, which may lose or delay some messages. The sequence of discrete messages describes the evolution of each component's state variables. Such communication networks can be conveniently modeled by *finite state machines*. Finite state machines are objects that, if given the current state and an input, return the new state and an output. For computer modeling, it is convenient to represent such systems with actor-oriented component models.

Figure 17.2 illustrates the disconnect between how building simulation programs are written and how the evolution of physical systems and communication systems can be described. In procedural programs, a procedure that belongs to a class is called. This procedure manipulates data, may call other procedures, and then returns to transfer the locus of control back to the

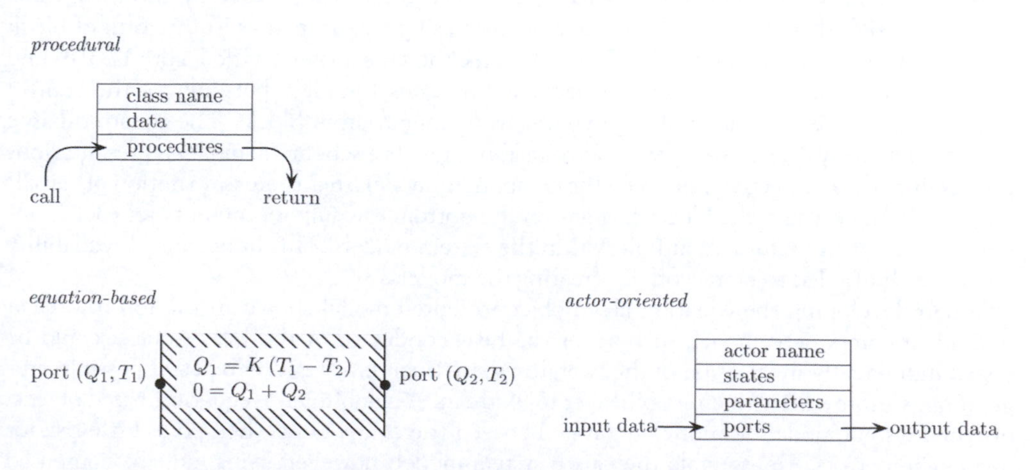

Figure 17.2 Comparison between procedural, equation-based and actor-oriented component models.

Source: Procedural and actor-oriented illustration adapted with permission from Lee (2003).

calling procedure. This is how typical building simulation programs are written. In equation-based programs, classes are typically encapsulated based on thermodynamic system boundaries. Ports are used to expose potential and flow variables (e.g., temperature and heat flow rate) without specifying what is input and output as this is not needed to describe physics. Acausal relations define constraints between the port data and between the thermodynamic states of the system. In actor-oriented programs, actors contain ports that receive and send tokens. A *model of computation* (Lee 2003) specifies operational rules that determine when actors perform internal computations, update their states and perform external communications, which can, for example, be at fixed time intervals or whenever new tokens arrive. The ports are connected to other components' ports in order to communicate with each other.

Needs to reduce software cost and development time

Cellier (1996) states the main factors that reduce cost and development time of software as follows:

- *Reusability:* A software design methodology that ensures optimal reusability of software components is the most essential factor in keeping the software development and maintenance cost down.
- *Abstraction:* Higher abstraction levels at the user interface help reduce the time of software development as well as debugging. The conceptual distance between the user interface and the final production code needs to be enlarged. Software translators can perform considerably more tasks than they traditionally did.

The reusability of code can be significantly increased through the use of equation-based object-oriented models that separate acausal relations between model variables from their solution algorithms. For example, Figure 17.3 shows models of a heat conducting wall in which the heat input is regulated by a feed-back control loop. Only half of the wall depth has been modeled because the wall is assumed to be symmetric and exposed to the same boundary conditions. In Figure 17.3(a), heat is added from the object called "source" to the wall's axis of symmetry, i.e., to the object called "mass1" which is at the centre of the wall. In Figure 17.3(b), heat is added to the surface of the wall. The top row shows the implementation using equation-based object-oriented models, where each icon encapsulates a model for an individual physical phenomenon (such as heat convection, heat conduction and heat storage). With this representation, the system being modeled is easily recognised. In contrast, the bottom row shows the same physical models, but they have been implemented using causal models in the form of block diagrams. Figure 17.3(a) and Figure 17.3(c) describe the same system, as do Figure 17.3(b) and Figure 17.3(d). Note that the top row also contains causal models, but only for the control system since this is most naturally represented using input/output blocks. The equation-based object-oriented system models are easier to read because the schematic model representations only defines the connectivity between the component models, but there is no notion of causality needed. In contrast, the block-diagrams in the bottom row impose causality for each component and for the information flow within the system models. This limits model readability and reuse. It also led to more work for creating the models.

When developing the equation-based object-oriented model shown in the top row, each individual component model, such as for the heat conductor or the heat storage, could be tested individually in isolation of the overall system. Such unit tests help detect possible programming errors earlier when it is cheaper to fix them. Assembling the equation-based object-oriented system model by connecting lines between the components then poses little risk for introducing errors. To assemble the causal system model, however, gains must be computed and heat flows must be added and connected to the corresponding integrator, which is time consuming and presents opportunities to make mistakes. The left and right columns of the

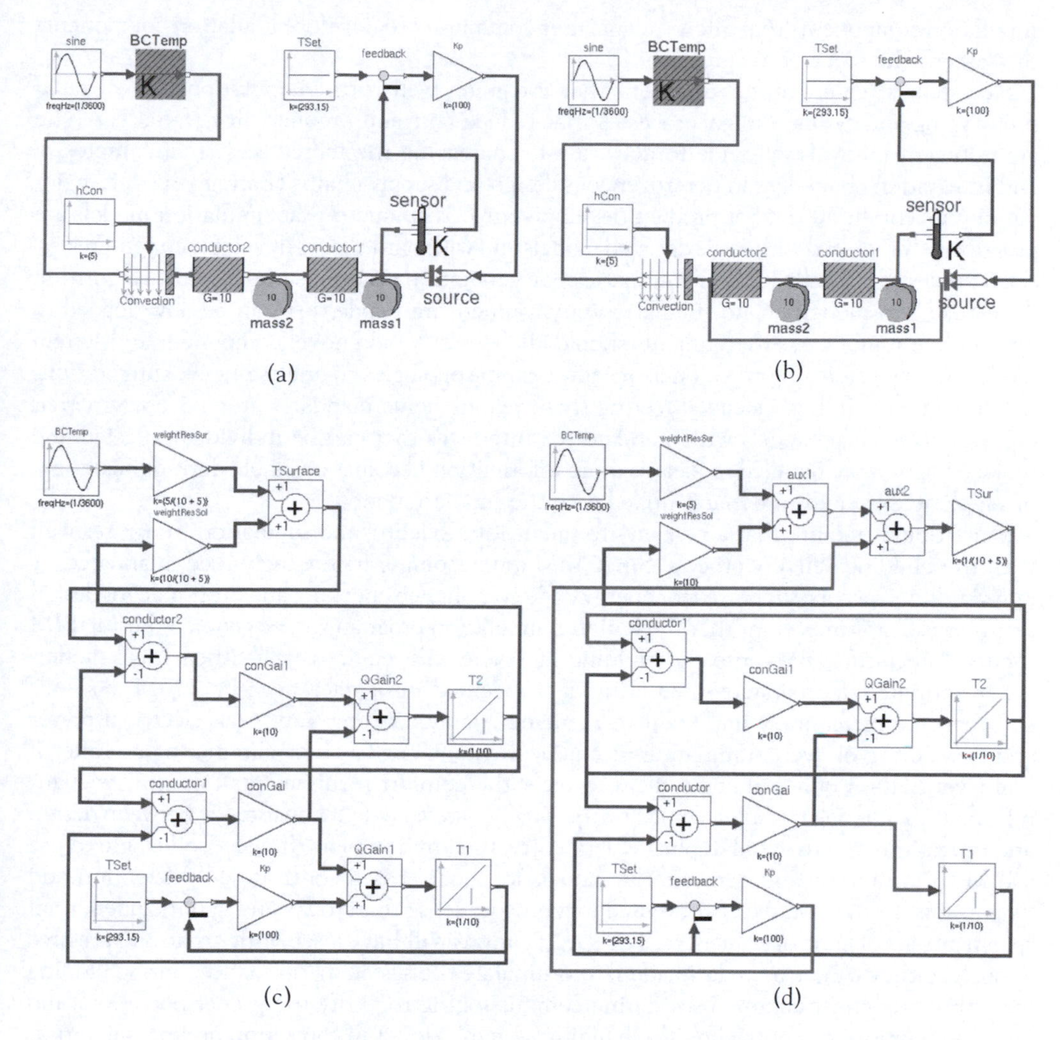

Figure 17.3 Representation of heat flow models with feedback control.

figure differ only in the location where heat is added to the construction. For the equation-based object-oriented model, all a modeler must do is reconnect one line; for the causal model the modeler must add and delete blocks and recompute gains, thus reusing a significantly smaller part of the model and possibly introducing mistakes. The example illustrates that the equation-based object-oriented system representation not only reduces development time but also allows a more natural system representation as discussed in the previous section.

Needs to integrate tools in a coherent workflow

Researchers and practitioners represent different user groups who use modeling and simulation within different workflows and therefore have different requirements that are best addressed by a modular environment. With a modular approach, the same underlying core of the environment can be used for both user groups while customised interfaces can address the individual requirements. An example of a modular software environment is Linux, which can be customised to run in embedded systems such as cell phones, in graphical desktop environments, in server farms and in super computers. Modularisation also makes it significantly easier for users to participate in program development; an example is the modular design of TRNSYS' simulation components that allows a large user community to develop simulation components that extend the scope of the program.

Researchers require an environment that integrates well with computational-based rapid prototyping and system design processes that reduce cost and product time-to-market. The environment must also provide formal ways for managing the increased system complexity and integration challenges in heterogeneous design and supply chains (Barnard 2005; Sangiovanni-Vincentelli 2007). For product development it is common that simulation models are progressively replaced with more detailed models and with hardware as development progresses. Such processes typically include automatic code generation that translates a control algorithm developed in a modeling and simulation environment into code that can be downloaded to control hardware. New environments should allow users to add novel component and system models rapidly, and combine models in new system topologies to analyse new configurations for integrated building systems. Meeting these requirements mandates an environment that consists of modular models with standardised interfaces that can be individually tested and replaced with more detailed models as more information becomes available during the design, or supplanted by hardware that is linked with the simulation model.

Practitioners require that a new environment for modeling and simulation integrates well with the building delivery process. Since BIM are becoming increasingly used, mandated by large building owners, supported for energy code compliance checking and profitable for design teams to use, it is important that the building simulation program can be generated using BIM inputs. Integrating BIMs into the building life cycle will require that mathematical models are instantiated and parameterised using data specified in or referenced by a BIM. As controls become increasingly important in reducing energy consumption, peak electrical power and downsizing or even eliminating mechanical equipment (such as eliminating a chiller if night-ventilation of a building suffices to meet the comfort requirements), control-relevant information also needs to be specified by the BIM. To avoid data inconsistency between design and operation, and to avoid duplicate data entry, the environment should also be able to use a BIM to instantiate and parameterise models for model-based controls, fault detection and diagnostics. In this context, BIM is used in a broader sense than today and will include formal specifications of control sequences. The specifications will have a semantics that is expressive enough so they can be used during design to simulate the system performance, during bidding for estimating construction costs, during commissioning to verify the proper operation, and during operation for controlling the building system. Model instantiation and parameterisation is easiest to achieve if the components specified in the BIM and in the simulation model have the same modularisation and if the connectivity between components is only constrained by how physical laws permit real components to be connected. To meet these requirements, a more general approach for component connectivity rules is needed than what can be found in many building simulation programs.

Enabling technologies

The previous section showed that building systems are becoming increasingly integrated across different functional domains. System integration increases design and operational complexity. A future modeling and simulation environment for building systems will address this challenge by using technologies that enable an intuitive representation of the system topology, the intrinsic behaviour of individual components and the interaction among the components once they become part of a system. This section discusses these technologies.

We will make a distinction between *modeling* and *simulation* as introduced by Cellier and Kofman (2006) and illustrated in Figure 17.1. Since building energy analysis requires large amounts of input data, it is convenient for our discussion to further separate mathematical modeling into behavioural modeling and data modeling. By *behavioural modeling* we mean specifying the mathematical equations that model a phenomenon, which Polderman and Willems

(2007) call the behavioural equations. Behavioural models may be based on physical laws, performance curves or by functions that update states and outputs of a finite state machine. By *data modeling* we mean representing the static data of a system without specifying the mathematical model that describes how the system evolves in time.

The purpose of modeling is to create a mathematical model from a physical system. From the mathematical model, a simulation program can be generated by using symbolic manipulations to sort equations, invert equations and replace derivative operators with numerical discretisation schemes. Generating a simulation program can be done manually by a programmer, as is the current practice in most building simulation programs such as in TRNSYS or EnergyPlus. Alternatively, it can be automated, as done in equation-based modeling environments such as Dymola, OpenModelica, SPARK, IDA or EES. The manual approach requires a model developer to write a simulation program (or subroutine), interface it with numerical solution methods, integrate it into the program kernel and implement methods to retrieve parameters and time-varying inputs and to write outputs. Manually generated programs are error prone and take considerably more time to write than specifying the actual physical relations that govern the object to be modeled. Equation-based modeling, in contrast, allows a model builder to concentrate the effort on describing the physical constraints and state transitions that define the evolution of the system. Generating efficient code for numerical simulation is left to a code generator.

Behavioural models

In the building simulation community, the need for more flexible means of constructing simulation models was expressed two decades ago by Sowell *et al.* (1986). A draft research plan was assembled by an international team working at the Lawrence Berkeley National Laboratory in 1985 for the collaborative development of a next generation building simulation program called the Energy Kernel System (EKS) for building simulation. A prototype software system called SPANK (Simulation Problem Analysis Kernel) that eventually became SPARK was created that contained some of the features envisioned for EKS (Sowell *et al.* 1986, 1989). Another development that embraced equation-based model formulation for building simulation is the Neutral Model Format (NMF) (Sahlin 1996; Sahlin and Sowell 1989). NMF is the primary modeling language used in IDA, an equation-based modeling and simulation environment that is used by researchers and practitioners and thas recently been used to translate and simulate models written in the Modelica language described below (Sahlin and Grozman 2003). The Engineering Equation Solver (EES) is another equation-based modeling and simulation environment with an interface to thermophysical property functions, such as for steam and refrigerants (Klein and Alvarado 1992).

In the meantime, a significantly larger industry-driven effort has been started that includes a wider range of industrial sectors and academic partners. In 1996, a consortium formed to develop Modelica, a freely-available, equation-based object-oriented modeling language that is designed for component-oriented, multi-domain modeling of dynamic systems. The goal of the consortium is to combine the benefits of existing modeling languages to define a new uniform language for model representation (Fritzson and Engelson 1998; Mattsson and Elmqvist 1997). Over the past decade, the Modelica language has gained significant adoption in various industrial sectors and has been used in demanding industrial applications. It is well positioned to become the de-facto standard for modeling complex physical or communication systems. The rise of Modelica provides an opportunity to revive the efforts of the International Energy Agency Task 22 to create a model library for building simulation, which was then implemented in NMF, that can be shared among different modeling and simulation environments (Vuolle *et al.* 1999). Modelica libraries for multi-domain physics include models for control, thermal, electrical and mechanical systems, as well as for fluid systems and different media (Casella *et al.* 2006; Elmqvist *et al.* 2003). Modelica models can contain differential equations, algebraic equations and discrete equations. Using standardised interfaces, a model's mathematical relations among its interface variables are encapsulated, and the model can be represented graphically by an icon. The encapsulation facilitates model reuse and exchange and allows connecting component models to system models using a graphical or textual editor. The tenets of Modelica models are that each component represents a physical device with physical interface ports, such as heat flow rate and temperature for solid ports and pressure, species flow and enthalpy for fluid ports. Defining a device- and physics-oriented system architecture enables intuitive model construction. The Modelica approach is realised by following the *object-oriented modeling paradigm*, a term that was coined by Elmqvist (1978) and summarised by Cellier *et al.* (1995) as follows:

- *Encapsulation of knowledge:* The modeler must be able to encode all knowledge related to a particular object in a compact fashion in one place with well-defined interface points to the outside.
- *Topological interconnection capability:* The modeler should be able to interconnect objects in a topological fashion, plugging together component models in the same way as an experimenter would plug together real equipment in a laboratory. This requirement entails that the equations describing the models must be declarative in nature, i.e., they must be acausal.
- *Hierarchical modeling:* The modeler should be able to declare interconnected models as new objects, making them indistinguishable from the outside from the basic equation models. Models can then be built up in a hierarchical fashion.
- *Object instantiation:* The modeler should have the possibility to describe generic object classes, and instantiate actual objects from these class definitions by a mechanism of model invocation.
- *Class inheritance:* A useful feature is class inheritance, since it allows encapsulation of knowledge even below the level of a physical object. The so encapsulated knowledge can then be distributed through the model by an inheritance mechanism, which ensures that the same knowledge will not have to be encoded several times in different places of the model separately.
- *Generalised networking capability:* A useful feature of a modeling environment is the capability to interconnect models through *nodes*. Nodes are different from regular models (objects) in that they offer a variable number of connections to them. This feature mandates the availability of *across* and *through variables*, so that power continuity across the nodes can be guaranteed.

Figure 17.4 illustrates how a multizone building model was created by graphically connecting physical ports for heat flow rate, air flow rate and control signals of thermal zone models, airflow models and HVAC system models. Each icon encapsulates a model that may encapsulate other models, thus allowing a model builder to manage the complexity of large system models through hierarchical modeling. Hierarchical model building also facilitates model reuse and testing of submodels before they are assembled as a large system model that may be difficult to debug. Object-oriented model construction with clearly defined encapsulations accelerates model development by allowing concurrent construction of multiple process models by their respective experts. For example, the HVAC and controls engineer can construct their models simultaneously and combine them later to a system model. The concurrent workflow with subsequent model combination is illustrated by Figure 17.4(a). On the left side of the figure, the three separate models represent the HVAC systems, the interzonal air flow and the building heat transfer (Wetter 2006a, b). The different domain models are integrated with each other by drawing lines between the model ports. The lines automatically generate equality constraints for across variables (temperature, pressure) and conservation equations for through variables (heat flow rate, mass flow rate).

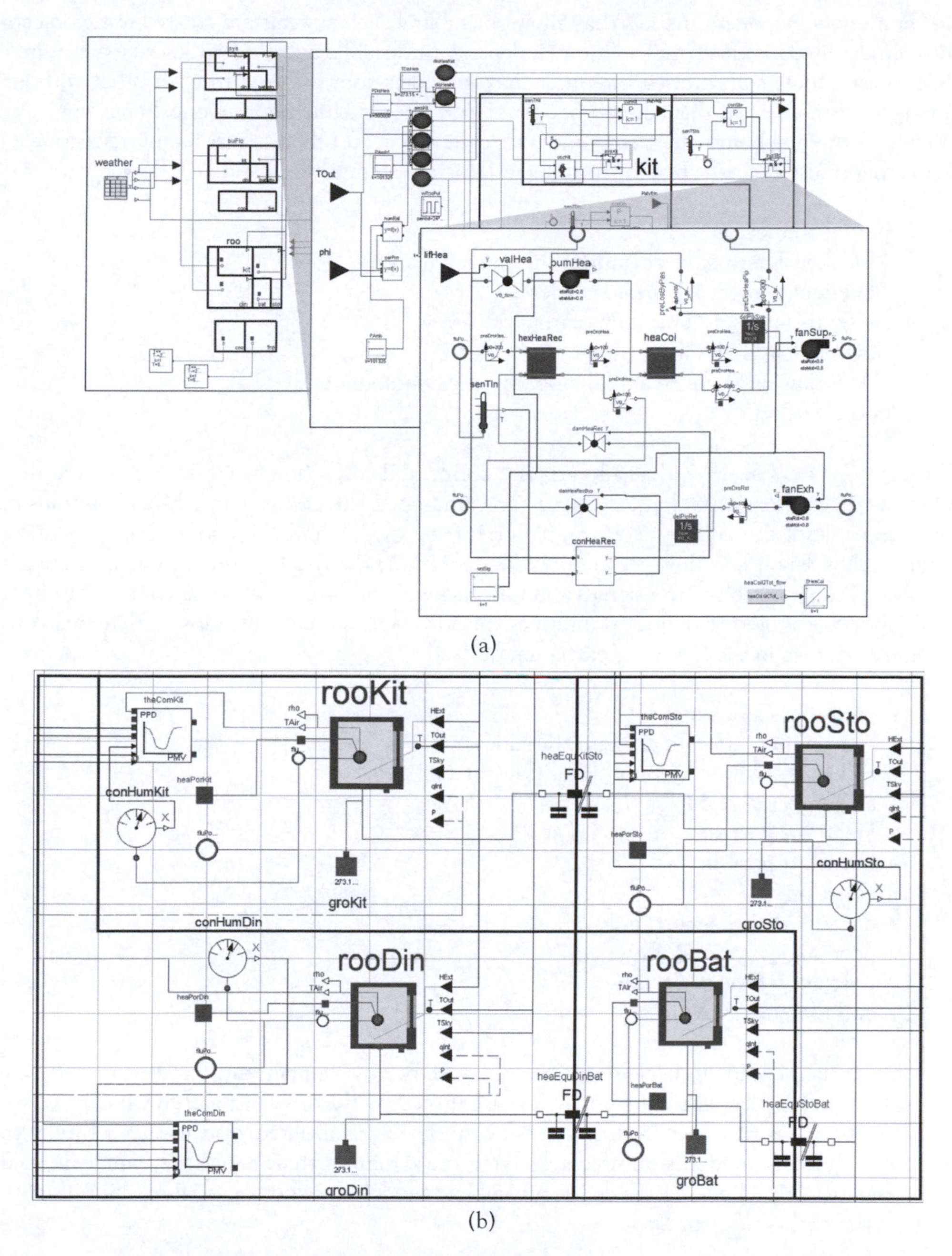

Figure 17.4 Schematic view of a building system model in Modelica that was constructed using different hierarchical layers to manage the model complexity, and that decomposes different domains. In the top figure (a), on the left, there are models for the HVAC system (top), for the interzonal air flow (middle) and for the building envelope heat transfer (bottom), which are connected to each other by fluid loops and control signals to form a coupled system of equations. The top right shows the inside of the HVAC system definition for each thermal zone, with an icon for the HVAC secondary loop, for the controller and for the user that may adjust the set points to maintain comfort. The bottom right shows the actual implementation of the HVAC secondary loop with the heat exchanges, the fans and pumps. The bottom figure (b), shows a section of the multizone building model with models for the rooms that are connected to each other with finite difference heat transfer models. Also shown are models for the thermal comfort.

Libraries of component models that allow assembling such systems are created using objects with standardised interfaces that define all independent variables necessary to describe the desired effect. Using these standardised interfaces makes models compatible with each other without having to add converters that convert one set of interface data into another one. For example, in Modelica's thermal library there is a connector called HeatPort that defines an interface for elements that transfer or store heat by introducing the tuple temperature and heat flow rate as:

```
1 partial connector
2 Modelica.Thermal.HeatTransfer.Interfaces.HeatPort
3 "Thermal port for 1-D heat transfer";
4 SI.Temperature T "Port temperature";
5 flow SI.HeatFlowRate Q_flow;
6 "Heat flow rate (positive if flowing into the component)";
7 end HeatPort.
```

The keyword flow declares that all variables connected to Q_flow need to sum to zero. For example, if two HeatPorts that are called port1 and port2 are connected, a Modelica translator will generate the equality constraint port1.T = port2.T and the conservation equation port1.Q_flow + port2.Q_flow = 0. Similar connectors are defined for other physical interfaces including fluid ports, electrical signals and translational and rotational bodies. The above heat port connector is used to define an interface for one-dimensional heat transfer elements with no energy storage in the following partial model:

```
1 partial model Element1D
2 "Partial heat transfer element that does not store energy"
3 SI.HeatFlowRate Q_flow "Heat flow rate";
4 SI.Temperature dT "port_a.T-port_b.T";
5 HeatPort port_a;
6 HeatPort port_b;
7 equation
8 dT = port_a.T – port_b.T;
9 port_a.Q_flow = Q_flow;
10 port_b.Q_flow = -Q_flow;
11 end Element1D.
```

This model is incomplete because it does not specify how temperatures and heat flow rate depend on each other. This incompleteness is denoted by the keyword partial which has two purposes: It shows to a user that the model cannot be instantiated, and it tells a Modelica translator that the model is allowed to have more variables than equations, which is used for automatic error checking. To define a complete model, the above partial model is used to define a thermal conductor as:

```
1 model ThermalConductor;
2 "Thermal element transporting heat without storage";
3 extends Interfaces.Element1D;
4 parameter SI.ThermalConductance G "Thermal conductance";
5 equation;
6 Q_flow = G*dT;
7 end ThermalConductor.
```

The thermal conductor model is graphically represented by an icon that is used, for example, in the system model shown in Figure 17.3. Note that by replacing the parameter declaration on line 4 and the equation on line 6, the semantics can be changed to represent other one-dimensional heat transfer elements, such as a model for convective heat transfer.

Equation-based modeling languages have been shown to significantly reduce model development time compared to conventional programming languages such as C/C++ or Fortran, as shown in Figure 17.5. A comparison of the development time for the BuildOpt multizone thermal building model which is implemented in C/C++ and a comparable implementation in Modelica showed a development time five to ten times faster when using Modelica (Wetter 2005, 2006a; Wetter and Haugstetter 2006). The Modelica development is faster primarily because Modelica is an acausal, equation-based, object-oriented language. This automated the work for writing routines for data input and output, sorting equations and procedure calls, implementing numerical solution methods and managing the large amount of data that is involved in building simulation. The shorter development time also manifests itself in a code size four times smaller (6,000 lines instead of 24,000 lines of code, not counting freely available libraries). In another experiment, the author asked three engineers familiar with Fortran and Modelica to implement the system shown in Figure 17.3(a) in Fortran and in the Modelica modeling and simulation environment Dymola. Even for such a simple problem where data input, output and storage are easy to implement, the model in Modelica was built 2.5 to 5 times faster than in Fortran. Similar increase in productivity has also been reported by Sahlin (2000), who was three times faster in implementing a thermal building simulator using the equation-based language NMF. These benchmarks indicate significant time-savings for modeling when using equation-based modeling languages.

Data models

A BIM is an instance of a data model that describes a specific building unambiguously. Several data models are used by the construction industry. The only open, object-oriented data model that covers the whole building life cycle are the Industry Foundation Classes (IFC), developed by the International Alliance for Interoperability and currently under consideration at the International Organization for Standardization (ISO) to become an international standard (ISO/PAS 16739:2005). Economic motivations for using BIMs include an increase in productivity through data interoperability among the participants of the building delivery process as well as data integrity during project planning, cost analysis, energy analysis and automatic code compliance checking. This section focuses upon how a future modeling and simulation environment for building systems can make BIM use throughout the building life cycle more seamless. Therefore, our discussion is tool-oriented to show how building performance assessment tools can best address requirements that originate from the use of BIM. We refer the reader to

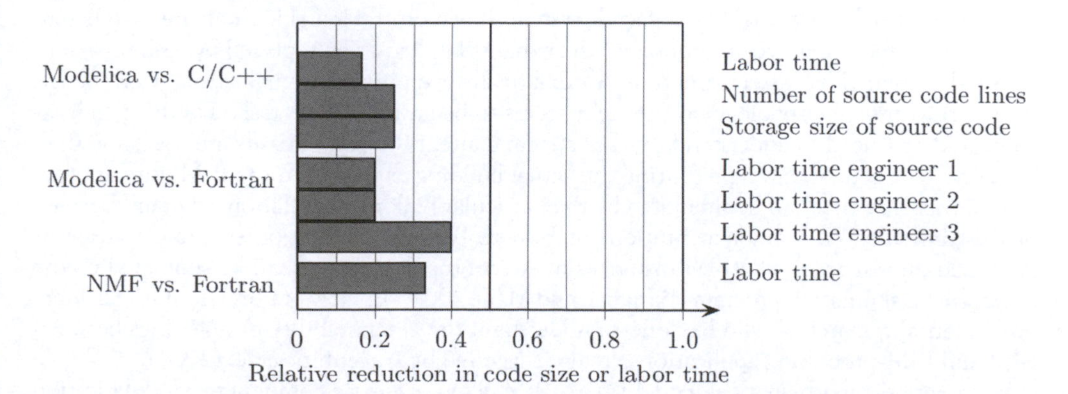

Figure 17.5 Comparison of relative reduction in labor time and code size for equation-based versus procedural programming languages.

Augenbroe *et al.* (2004) and Lam *et al.* (2004) for a process-oriented discussion of simulation models within the building process and to Bazjanac (2008) for a discussion of semi-automated data transformation between BIM authoring tools and building simulation tools.

This discussion assumes that future BIMs will contain specifications for building controls that will be used during design for building simulation as well as during operation for model-based algorithms for controls, fault detection and diagnostics. Since controls are an essential tool for increasing building system efficiency, they would be a natural extension to a BIM, and the IFC2x platform specification indeed mentions building automation as one discipline that it aims to support. To enable a seamless combination of data models represented by a BIM and behavioural models represented in an equation-based object-oriented language that can generate a building simulation program, the component modularisation and the variables that connect components should be identical in the data model and the behavioural model. Furthermore, data and behavioural models need to allow connecting components in the same way as one would connect actual components. Otherwise the set of possible HVAC systems is larger than the set of HVAC systems that can be simulated, which makes seamless BIM use difficult if not impossible for some HVAC systems.

Many building simulation programs are based on software principles that do not match how buildings and energy systems are actually built and controlled. Also, their modeling language is not expressive enough to allow computing the evolution of heterogeneous systems in a way that properly accounts for concurrency of event-driven systems. This can make the use of models for simulation-based controls and commissioning difficult. In addition, the model modularisation and model connectivity rule is imposed by the program developers. They decide what should be lumped in a monolithic subsystem model, what causality should be imposed on the equations to translate them into procedural code and what connections should be required for an object. These decisions are not motivated by how a BIM represents an HVAC system and they do not necessarily reflect real HVAC systems. Rather, the selection of component modularisation, component causality and connection variables is motivated by accommodating the data flow and solution algorithms imposed by the program architecture. For example, the room model of EnergyPlus computes a heating or cooling load in units of power. This load is then sent to zone equipment units, such as a fan coil, that try to meet the load. Thus, the control input of the fan coil is a load in watts, as opposed to a signal from a thermostat. This type of component connectivity is different from the connectivity of actual components. Furthermore, in EnergyPlus, a valve that adjusts a cooling coil's water flow rate is implicitly assumed to be part of the coil, which represents a component modularisation that does not correspond with that of actual systems. Further difficulties in generating a simulation model from a BIM can arise in non-conventional systems that do not follow the fluid loop structure of some building simulation programs. For example, because EnergyPlus (as well as DOE-2) does not allow equipment between thermal zones, the double façade system shown on the left side in Figure 17.6 had to be implemented using two air loops, and the exhaust fan had to be replaced by a supply fan to satisfy the EnergyPlus system structure. Automatically mapping the actual topology on the left from a BIM into the topology on the right seems to be an impossible task. The BIM translator would first need to understand in what system context the components are used, and then map one system topology, representing the actual building expressed in the BIM, into another system topology that can be simulated by the particular building simulation program. Because of this difficulty, early implementations of BIM to building simulation program translators are based on restricting the BIM instances to system topologies that can be simulated by the downstream simulation program (Bazjanac and Maile 2004). In view of these limitations, more fundamental research should be done to understand the computability of mappings between BIM and BIM-processing applications, as also suggested by Augenbroe *et al.* (2004).

To generate a mathematical model from a BIM, a more generic component modularisation and connectivity rule than what is employed in today's building simulation programs needs to be used. It is most natural to base the modularisation rule on how components are modularised

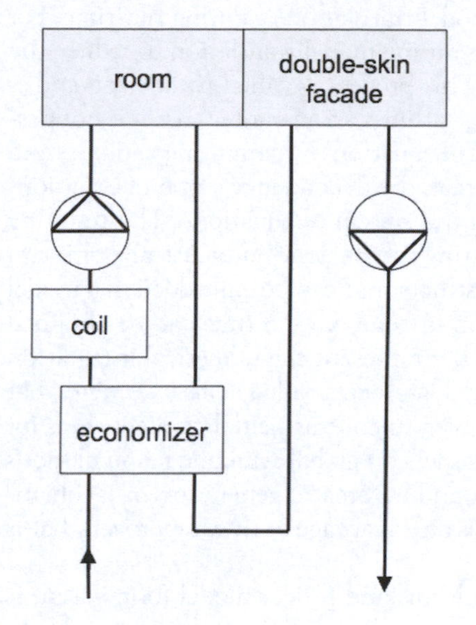

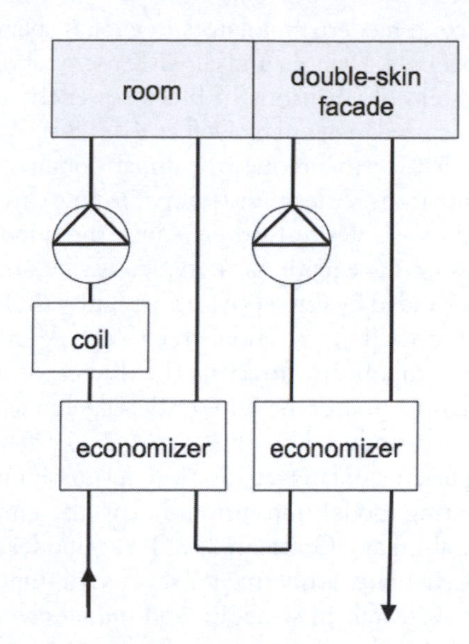

Figure 17.6 Schematic diagram of an HVAC system that uses a double-skin facade for the exhaust air (left side) and implementation in EnergyPlus that required two air loops (right-side).

Source: Figure with permission from Tobias Maile, Lawrence Berkeley National Laboratory.

in reality; typically modular units are defined by how they are packaged when shipped to the construction site. Furthermore, the connectivity rules of models should follow the same rules in which components can be connected in reality. This is easiest to realise by using acausal models that follow the object-oriented modeling paradigm and whose interface variables are sufficient to describe the interaction of connected objects. Models with these traits can be arbitrarily assembled to build systems as long as their connections have the same physical properties, such as temperature and heat flow rate for a solid-to-solid interface. Such a connection framework is indeed realised in the Modelica language.

A further difficulty arises when modeling cyber-physical systems. Cyber-physical systems will become more common as it becomes cheaper to densely instrument buildings with networks of sensor and actuators that control the building system. In cyber-physical systems, physical processes can affect when computations are performed and messages are sent across networks. For some systems, event detection, event ordering and other techniques for modeling and simulation of heterogeneous systems that properly model concurrency are needed. These needs may be easiest to meet by generating an actor-oriented system model for the communication network in which a model of computation controls the interactions between the different actors. The model may then be embedded into equation-based models for the physical components to form a hybrid system, which then can be used for designing and commissioning such networks.

Model translation and simulation

The translation of mathematical models into a simulation program can be fully automated. When translating a model, symbolic processing is important to reduce computing time since many building system simulation problems lead to large, sparse differential algebraic equation systems (DAE systems). Symbolic processing is typically used to reduce the index of the DAE system and to exploit sparsity. Pantelides (1988) presents an algorithm that reduces the index of DAE systems so that the equations can be solved using ordinary differential equation solvers. This algorithm is, for example, used by Dymola (Mattsson *et al.* 2000). It is current practice in modern simulators to exploit sparsity during model translation or during run-time. For example, Dymola and SPARK use symbolic processors during model translation to reduce the system of equations to block lower-triangular form. This process is called *partitioning* and is discussed in detail by Duff *et al.* (1989). Partitioning algorithms can guarantee that it is not possible to further reduce the dimensionality of the system of equations by permuting variables and equations. After partitioning, *tearing* can be used to break the dependency graph of equations and variables to further reduce the dimensionality of the system of equations. The resulting systems of equations are typically small but dense. Tearing can be done automatically or it can be guided by using physical insight with language constructs that can be embedded in a model library (Elmqvist and Otter 1994). A further reduction in computation time can be obtained by symbolically inserting the discretisation formulae that represent the numerical integration algorithm into the differential-algebraic equation model, a process called *inline integration* that was introduced by Elmqvist *et al.* (1995). The IDA solver, in contrast, employs algorithms for sparse Jacobian factorisation during simulation. It relies less on global symbolic manipulations during model translation, but it also employs tearing and automatic generation of Jacobians (Sahlin and Grozman 2003). Fast model compilation is an advantage of IDA's approach that is particularly attractive for short-term simulations.

Research in symbolic and numerical solution algorithms and proper model formulation is critically important for making equation-based building modeling and simulation accessible to a large audience. It is not realistic to demand that the typical simulation user have knowledge in these mathematical fields. Therefore it is important that robust solution algorithms and model libraries be put in place so that they are available for use by typical simulation users without experience in fixing numerical problems. While these topics have received little attention from the building simulation community, the fact that IDA is used by HVAC engineers indicates that this problem is not insurmountable.

There are several technologies in place that decrease the computation time and lead to increased numerical robustness. But how do they compare to conventional building simulation programs? In general, comparing the run-time efficiency of different simulators is a challenging endeavour because of the different temporal and physical resolution of the tools, and because solvers can be tuned to be more efficient for certain classes of problems at the expense of robustness. Nevertheless, the following comparison shows that commercial general purpose simulators can be as efficient in performing a building simulation as commercial special tailored building simulation programs. Sahlin *et al.* (2004) compared the computation time of IDA ICE with EnergyPlus. In their numerical experiments, IDA ICE required approximately half the computation time that was required by EnergyPlus for a three zone building with natural ventilation and twice the computation time in initial experiments for a 53 zone building without natural ventilation. Sowell and Haves (2001) compared the computation time of SPARK and HVACSIM+ for a variable air volume flow system that serves six thermal zones. They report that SPARK computes about 15 to 20 times faster than HVACSIM+, and they attribute the decrease in computation times to SPARK's symbolic processing. Another variable air volume flow network model with 26 rooms shows that the computation time of SPARK and Dymola are similar for this class of problems (Wetter *et al.* 2008). Wetter and Haugstetter (2006) compared the simulation time of a multizone building model written in Modelica with a building model using TRNSYS' TYPE 56 that gives comparable energy use and transient response. The Modelica model was simulated using the solver of Dymola and Simulink. They report that Dymola computed four times slower and Simulink three times slower than TRNSYS. In these experiments, TRNSYS computed the heat conduction with Conduction Transfer Functions, which are considered to be computationally faster than the finite difference scheme used in the Modelica implementation.

One should keep in mind that these benchmarks have been simulated using a single processor. The recent switch to parallel microcomputers requires, however, that programs are written differently to take advantage of the new hardware architecture. For example, the conventional wisdom is that multiplications are slow but data loading and storing is fast. This is reversed by new hardware, because modern microprocessors can take 200 clock cycles to access dynamic random access memory (DRAM), but a floating-point multiplication may take only four clock cycles (Asanovic *et al.* 2006). Although how to best address this change in programming paradigm is an open research field, it is reasonable to expect that models formulated in equation-based, high-level programming languages, as opposed to existing simulation programs, are in a better position to take advantage of this shift towards parallel hardware. One reason for this expectation is that equation-based simulation environments naturally separate the algebraic and numerical solvers from the model equations. This facilitates replacing algebraic and numerical solvers, as well as model translators, with newer versions that exploit the new hardware architecture.

Collaborative development

Over the past decade, Web 2.0 applications that enable distributed, concurrent, collaborative development of code and documentation have become a commodity. A characteristic of many successful open-source projects is that they employ a modular design that consists of small units with standardised interfaces. These units are good at doing one particular task, thereby providing a core of program functionalities around which further features can be built by a community of developers. Structuring the code into small units allows units to be developed and tested individually and then integrated into a larger system. It also reduces barriers to participation since only a small part of the program architecture needs to be understood by most contributors. This modularisation facilitates reusing code, distributing development efforts and using experts for the tasks that they are best at, such as developing numerical solvers, mathematical models or user interfaces. Involving different experts with a variety of backgrounds brings diverse points of view for solving a problem, which can further increase the quality of the code.

Many building simulation programs are developed by teams that do not always involve experts from other disciplines such as computer science, scientific computation and applied mathematics. Therefore, many advances in these fields are transferred to the building simulation community at a rather slow rate. To facilitate a better transfer of science and technologies, future developers of an environment for modeling and simulation should integrate, and where necessary, further develop existing technologies with other communities that are active in system modeling and simulation. This should enable sharing of models, solvers and interfaces to accelerate the development of more advanced tools. In such a model, significantly bigger investments for tool and library development can be leveraged.

A path towards a next generation environment for building simulation and computational engineering

Equation-based object-oriented modeling and simulation environments have been applied successfully in numerous demanding industrial applications. However, the building simulation community still rarely uses them, and little investment has been made to advance equation-based object-oriented building energy system modeling and simulation. If a new modular environment were to be used for whole building energy analysis, it would likely be measured against tools that have been developed for this purpose over several decades, often involving hundreds of man years of development time. Clearly, it does not make sense to target markets with heavy prior investment in existing technologies to help develop and mature a new environment for modeling and simulation. Instead, market segments with unmet needs should be pursued. Unmet needs are encountered primarily in the analysis and synthesis of algorithms for controls, fault detection and diagnostics, in research for building systems that integrate different domains such as HVAC, refrigeration, active façades and communication networks, and in the use of models during operation.

Specific applications that a new modular environment for building modeling and simulation should target include:

- *Building systems for Net Zero Energy Buildings*. To achieve cost-effective Net Zero Energy Buildings, new system approaches are needed that require analysis and synthesis of coupled multi-physics systems that may include active façade, lighting controls, solar-assisted cooling and dehumidification, natural ventilation, storage of heat in the building structure and storage of humidity in desiccants. Such systems pose new challenges and opportunities for multi-variate control, optimal energy storage management and reduction of system-internal irreversibilities, i.e., destroyed exergy.
- *Design optimisation of innovative building energy and control systems*. Innovative building energy and control systems are typically harder to design since there are not many design rules available and because they often involve complex systems that interact dynamically. However, once simulation input files have been created, optimisation algorithms can help find the optimal building and energy system design with little user effort. In optimisation, a user selects a vector of design parameters x and defines a cost function $f(x)$ that needs to be minimised, possibly subject to constraints. Nonlinear programming algorithms, a class of computationally efficient optimisation algorithms, require the cost function to be differentiable in the design parameters. However, building simulation programs contain iterative solution algorithms and therefore compute a numerical approximation $f^*(\varepsilon, x)$ to the cost function, where ε is the tolerance of the numerical solvers. The iterative solution algorithms introduce discontinuities in the numerical approximation to the cost function because a change in design parameters can cause a change in the number of solver iterations or in the grid on which a differential equation is discretised. Tightening the solver tolerance ε is possible in equation-based simulation programs, but it is often impractical in traditional building simulation programs since their solvers may be embedded in different component models and spread throughout hundreds of thousands of lines of code. Unfortunately, in building simulation programs, these discontinuities can be large enough to cause optimisation algorithms to fail far from an optimal solution (Polak and Wetter 2006; Wetter and Wright 2004). To illustrate this problem, we show in Figure 17.7 a section of a surface plot of normalised building energy consumption as a function of the width of the west and east windows, w_{west} and w_{east}. The plot was generated by sampling about 20,000 simulations on a two-dimensional grid. Globally, energy is minimised by increasing the width of both windows within the domain shown in the figure, i.e., the energy is minimal at the front corner in the figure. However, locally, due to the loose solver tolerance, the cost function exhibits "smooth" regions that are separated from each other by discontinuities. These "smooth" regions are visible in the graph as flat plateaus. The white lines that are visible on these plateaus are contours of equal function value. The problem is that in these "smooth regions", decreasing the west window width reduces energy. This caused the Hooke-Jeeves optimisation algorithm to generate a sequence of iterates that starts at the arrow on the right side in the figure and traverses towards the arrow in the left side in the figure where it jams at a discontinuity.
- *Model predictive control*. Model predictive control allows optimal equipment scheduling while taking into account the dynamics of the building energy system and the anticipated future building energy load. It allows one to minimise a user-specified cost function such as energy use, subject to state constraints like thermal comfort and indoor air quality. When solving the optimal control problem in a model predictive control algorithm, all state variables of the simulation model need to be reset to initial values prior to evaluating the performance of different control actions. However, in many traditional building simulation programs, resetting state variables is difficult if not impossible without significant code changes. Furthermore, for computational efficiency and to prove that the optimisation algorithm finds the optimal control sequence, the differential equations need to be smooth

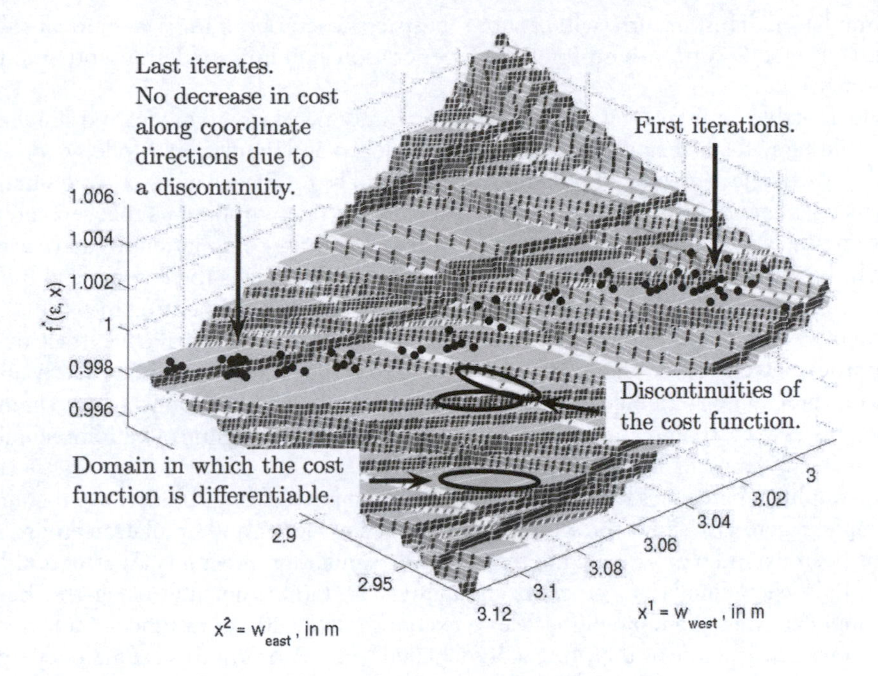

Figure 17.7 Parametric plot of the normalised source energy consumption for cooling and lighting as a function of the width of the west and east-facing window. The dots are iterates of the Hooke-Jeeves algorithm.

Source: Figure adapted from Wetter and Polak (2004).

in the control input, which is typically not the case in building simulation programs. Lack of smoothness, in addition to the loose solver tolerance mentioned above, can cause the optimisation to fail for the same reasons as discussed in the previous item.

- *System linearisation for development of control algorithms*. Many design methods for control algorithms of nonlinear systems require linearisation of the open loop response with respect to controls or disturbances, sometimes at multiple operating points. Such linearisations are difficult to perform with large monolithic building simulation programs. The reason is that system linearisation requires isolating the subsystem that is part of the feedback control loop, initialising its state variables and simulating its open loop response subject to different control inputs and disturbances. An additional difficulty posed by large monolithic building simulation programs is that their numerical solvers frequently fail to converge. This can render a numerical approximation to a derivative useless for functions such as the one shown in Figure 17.7.
- *Fault detection and diagnostics*. For fault detection and diagnostics of components and subsystems, only a subset of the building system needs to be simulated. Realistic modeling of the component dynamics can be important to discern between faults and transient effects. Such applications require modeling of realistic controls as opposed to the idealised controls that are found in many building simulation programs.

Theses applications do not in general require an annual whole building simulation. Rather they require a flexible modeling environment that allows a user to rapidly add new models, symbolically invert models and extract a subsystem of a model for further analysis, such as for input-output feedback linearisation or model reduction for controls design. By targeting these applications, the new environment can be developed with users who have the technical skills needed to mature the environment in terms of model scope, validation and formulation. This, in turn, will ensure robustness when obtaining a numerical solution and will ultimately form a base for a next generation building modeling and simulation environment.

While desirable for numerical efficiency it is often not necessary to model and simulate the whole building system in a new environment for modeling and simulation. Models expressed in equation-based object-oriented modeling languages can be interfaced with existing simulation programs to extend the scope of systems they can analyse. In environments that separate modeling from simulation, different models can be interfaced during compilation in such a way that they will share the same differential equation solver. An example is the commercial interface between Simulink and Dymola in which a Modelica model can be imported into Simulink in the form of an S-function. Monolithic simulation programs that do not allow extracting models from their solvers can be changed in such a way that they interface either directly another simulation program or a middleware that synchronises the data exchange as the simulation time progresses in each simulation program. A discussion about coupling of building simulation programs can be found in Trcka *et al.* (2006, 2009).

Both coupling approaches have been realised with many building simulation programs. As an example, Figure 17.8 shows the schematic editor of a modeling and simulation environment that has been extended to couple different building simulation programs (Wetter and Haves 2008). Here, each simulation program communicates during run-time using the Berkeley socket application programming interface to exchange data with an instance of a Java object that is shown in the figure by the white icons labelled "Simulator". In this example, EnergyPlus simulates the building heat flow and natural ventilation and MATLAB/Simulink simulates the controller for the window opening. A model of computation controls the interactions between the models. In this modeling environment, different domains such as discrete event, continuous time and synchronous data flow can be combined hierarchically. This enables the simultaneous simulation of building energy systems, control systems and communication networks.

Synopsis

In a future environment for modeling and simulation, models will be expressed at a higher level of abstraction, which will allow a modeler to formulate the system closer to how the

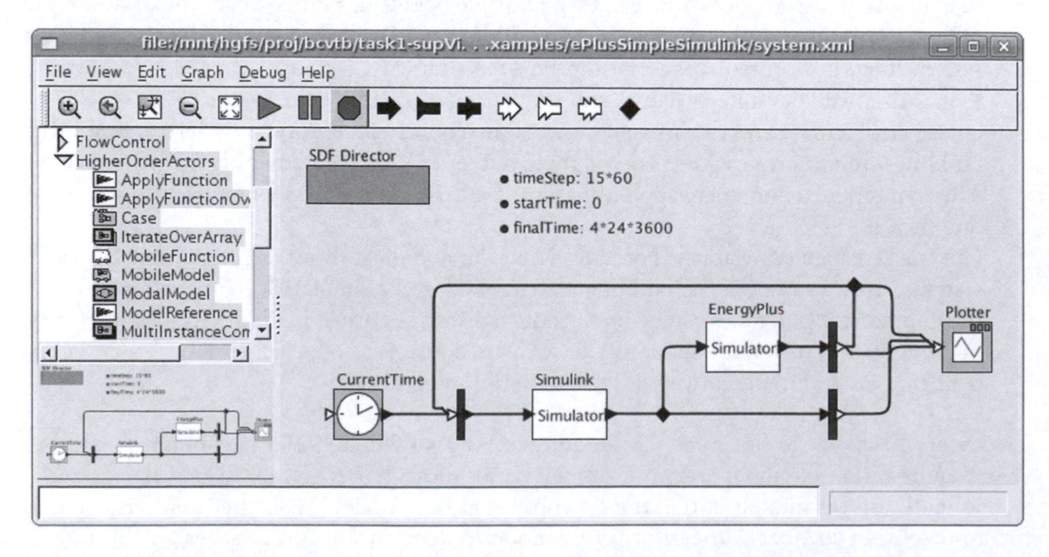

Figure 17.8 Ptolemy II graphical user interface with extensions to couple simulation programs, in this example EnergyPlus and MATLAB/Simulink®, for run-time data exchange.

actual system behaves. Mathematical models and simulation programs will be distinct system representations; the first accommodates a human's way of expressing and structuring systems and the second enables efficient computation of numerical solutions. Separating mathematical models from their solution processes will also facilitate assembling system models that combine different domains. Combining such models will be realised by formulating equality and conservation equations that link shared interface variables on an equation-level, as opposed to by integrating different programs, each with its individual solver, data management and input/output routines. Combining models on an equation-level is easier from the point of view of software engineering, mathematics (to ensure a consistent solution), and usability. Following the object-oriented modeling paradigm enables managing the complexity of large systems through hierarchical modeling. It also enables model assembly the same way as an experimenter would connect real components. This, in turn, allows creating a modeling environment where equation-based object-oriented models have the same modularity and connectivity rules as a modular BIM, facilitating the use of BIM to create simulation programs at different stages of the building life cycle. In addition, the higher level of abstraction allows sharing not only models, but also modeling and simulation environments and their associated technologies, such as numerical and symbolic solvers, with other industrial sectors.

References

Asanovic, K., Bodik, R., Catanzaro, B.C., Gebis, J.J., Husbands, P., Keutzer, K., Patterson, D.A., Plishker, W.L., Shalf, J., Williams, S.W. and Yelick, K.A. (2006) "The Landscape of Parallel Computing Research: A View from Berkeley", *Technical Report UCB/EECS-2006–183*, EECS Department, University of California, Berkeley.

Augenbroe, G. (2001) "Building Simulation Trends Going into the New Millennium", in Lamberts, R., Negrão, C.O.R. and Hensen, J. (eds), *Proceedings of the 7th International IBPSA Conference*, Rio de Janeiro, Brazil, pp. 15–27.

Augenbroe, G., de Wilde, P., Moon, H.J. and Malkawi, A. (2004) "An Interoperability Workbench for Design Analysis Integration", *Energy and Buildings* 36(8): 737–748.

Barnard, P.A. (2005) "Software Development Principles Applied to Graphical Model Development", *Modeling and Simulation Technologies Conference and Exhibit*, American Institute of Aeronautics and Astronautics, San Francisco, CA.

Bazjanac, V. (2008) "IFC BIM-Based Methodology for Semi-Automated Building Energy Performance Simulation", in *Proceedings of CIB-W78, the 25th International Conference on Information Technology in Construction*, Santiago de Chile, Chile.

Bazjanac, V. and Maile, T. (2004) "IFC HVAC Interface to EnergyPlus – A Case of Expanded Interoperability for Energy Simulation", *Proceedings of SimBuild 2004*, Boulder, CO.

Casella, F., Otter, M., Proelss, K., Richter, C. and Tummescheit H. (2006) "The Modelica Fluid and Media Library for Modeling of Incompressible and Compressible Thermo-fluid Pipe Networks", in Kral C. and Haumer A. (eds), *Proceedings of the 5th International Modelica Conference*, Vienna, Austria, pp. 631–640.

Cellier, F.E. (1996) "Object-oriented Modeling: Means for Dealing with System Complexity", in *Proceedings of the 15th Benelux Systems and Control Conference*, Mierlo, The Netherlands, pp. 53–64.

Cellier, F.E. and Kofman, E. (2006) *Continuous System Simulation*, Berlin: Springer.

Cellier, F.E., Elmqvist, H. and Otter, M. (1995) "Modeling from Physical Principles", in Levine W.S. (ed.) *The Control Handbook*, Boca Raton, FL: CRC Press.

Clarke, J.A. (2001) *Energy Simulation in Building Design*, 2nd ed., Oxford, UK: Butterworth-Heinemann.

Duff, I.S., Erisman, A.M. and Reid, J.K. (1989) *Direct Methods for Sparse Matrices – Monographs on Numerical Analysis*, Oxford: Oxford Science Publications.

Eastman, C., Liston, K., Sacks, R. and Teicholz, P. (2008) *BIM Handbook: A Guide to Building Information Modeling for Owners, Managers, Designers, Engineers and Contractors*, New York: John Wiley & Sons.

Elmqvist, H. (1978) "A Structured Model Language for Large Continuous Systems", PhD Thesis, Lund Institute of Technology, Sweden.

Elmqvist, H. and Otter, M. (1994) "Methods for Tearing Systems of Equations in Object-oriented Modelling", in *Proceedings of European Simulation Multiconference*, Barcelona, Spain, pp. 326–332.

Elmqvist, H., Otter, M. and Cellier, F.E. (1995) "Inline Integration: A New Mixed Symbolic/Numeric Approach for Solving Differential-algebraic Equation Systems", in *Proceedings of European Simulation Multiconference*, Prague, Czech Republic, pp. 23–34.

Elmqvist, H., Tummescheit, H. and Otter, M. (2003) "Object-oriented Modeling of Thermo-fluid Systems", in Fritzson, P. (ed.) *Proceedings of the 3rd Modelica Conference*, Linköping, Sweden, pp. 269–286.

Fritzson, P. (2004) *Principles of Object-Oriented Modeling and Simulation with Modelica 2.1*, New York: John Wiley & Sons.

Fritzson, P. and Engelson V. (1998) "Modelica – A Unified Object-oriented Language for System Modeling and Simulation", *Lecture Notes in Computer Science* 1445: 67–90, London: Springer.

IBM Corporation (1956) "The Fortran Automatic Coding System for the IBM 704 EDPM", *IBM Applied Science Division and Programming Research Department*, New York.

Klein, S.A. and Alvarado F.L. (1992) "Engineering Equation Solver (EES)", *F-Chart Software*, Madison, WI.

Lam, K.P., Wong, N.H., Mahdavi, A., Chan, K.K., Kang, Z. and Gupta, S. (2004) "SEMPER-II: An Internet-based Multi-domain Building Performance Simulation Environment for Early Design Support", *Automation in Construction* 13(5): 651–663.

Lee, E.A. (2003) "Model-driven Development – From Object-oriented Design to Actor-oriented Design", *Workshop on Software Engineering for Embedded Systems: From Requirements to Implementation (a.k.a. The Monterey Workshop)*, Monterey, CA.

Lee, E.A. (2006) "Cyber-physical Systems – Are Computing Foundations Adequate?", *Position Paper for NSF Workshop on Cyber-Physical Systems: Research Motivation, Techniques and Roadmap*, Berkeley, CA.

Lee, E.A. and Varaiya P.P. (2002) *Structure and Interpretation of Signals and Systems*, Addison Wesley.

Malkawi, A.M. (2004) "Immersive Building Simulation", in Malkawi, A.M. and Augenbroe, G. (eds), *Advanced Building Simulation*, New York: Spon Press.

Malkawi, A.M. and Augenbroe, G. (2004) *Advanced Building Simulation*, New York: Spon Press.

Mattsson, S.E. and Elmqvist, H. (1997) "Modelica – An International Effort to Design the Next Generation Modeling Language", in Boullart, L., Loccufier, M. and Mattsson S.E. (eds), *7th IFAC Symposium on Computer Aided Control Systems Design*, Gent, Belgium.

Mattsson, S.E., Olsson, H. and Elmqvist, H. (2000) "Dynamics Selection of States in Dymola", *Modelica Workshop*, Lund, Sweden, pp. 61–67.

Pantelides, C.C. (1988) "The Consistent Initialization of Differential-algebraic Systems", *SIAM Journal on Scientific and Statistical Computing* 9(2): 213–231.

Polak, E. and Wetter, M. (2006) "Precision Control for Generalized Pattern Search Algorithms with Adaptive Precision Function Evaluations", *SIAM Journal on Optimization* 16(3): 650–669.

Polderman, J.W. and Willems J.C. (2007) "Introduction to Mathematical System Theory", *Texts in Applied Mathematics*, Vol. 26, New York: Springer.

Sahlin, P. (1996) "Modeling and Simulation Methods for Modular Continuous Systems in Buildings", PhD Thesis, KTH, Stockholm, Sweden.

Sahlin, P. (2000) "The Methods of 2020 for Building Envelope and HVAC Systems Simulation – Will the Present Tools Survive?", in *Proceedings of the ASHRAE/CIBSE Conference*, Dublin, Ireland.

Sahlin, P. and Grozman P. (2003) "IDA Simulation Environment – A Tool for Modelica Based End-user Application Development", in Fritzson, P (ed.), *Proceedings of the 3rd Modelica Conference*, Linköping, Sweden, pp. 105–114.

Sahlin, P. and Sowell, E.F. (1989), "A Neutral Format for Building Simulation Models", in *Proceedings of the Second International IBPSA Conference*, Vancouver, Canada, pp. 147–154.

Sahlin, P., Eriksson, L., Grozman, P., Johnsson, H., Shapovalov, A. and Vuolle, M. (2004) "Whole-building Simulation with Symbolic DAE Equations and General Purpose Solvers", *Building and Environment* 39(8): 949–958.

Sangiovanni-Vincentelli, A. (2007) "Quo Vadis, SLD? Reasoning about the Trends and Challenges of System Level Design", *Proceedings of the IEEE* 95(3): 467–506.

Sowell, E.F. and Haves, P. (2001) "Efficient Solution Strategies for Building Energy System Simulation", *Energy and Buildings* 33(4): 309–317.

Sowell, E.F., Buhl, W.F., Erdem, A.E. and Winkelmann, F.C. (1986) "A Prototype Object-based System for HVAC Simulation", *Technical Report LBL-22106*, Lawrence Berkeley National Laboratory.

Sowell, E.F., Buhl, W.F. and Nataf, J.M. (1989) "Object-oriented Programming, Equation-based Submodels, and System Reduction in SPANK", in *Proceedings of the Second International IBPSA Conference*, Vancouver, Canada, pp. 141–146.

Tiller, M.M. (2001) *Introduction to Physical Modeling with Modelica*, Norwell, MA: Kluwer Academic Publisher.

Trcka, M., Hensen, J.L.M. and Wijsman, A.J.T.M. (2006) "Distributed Building Performance Simulation – A Novel Approach to Overcome Legacy Code Limitations", *International Journal of HVAC&R Research* 12(3a): 621–640.

Trcka, M., Hensen, J.L.M. and Wetter, M. (2009) "Co-simulation of innovative Integrated HVAC Systems in Buildings", *Journal of Building Performance Simulation* 2(3): 209–230.

Vuolle, M., Bring, A. and Sahlin, P. (1999) "An NMF Based Model Library for Building Thermal Simulation", in Nakahara, N., Yoshida, H., Udagawa, M. and Hensen, J. (eds), *Proceedings of the 6th IBPSA Conference*, Kyoto, Japan.

Wenisch, P., van Treeck, C., Borrmann, A., Rank, E., and Wenisch, O. (2007) "Computational Steering on Distributed Systems: Indoor Comfort Simulations as a Case Study of Interactive CFD on Supercomputers", *International Journal of Parallel, Emergent and Distributed Systems* 22(4): 275–291.

Wetter, M. (2005) "BuildOpt – A New Building Energy Simulation Program that Is Built on Smooth Models", *Building and Environment* 40(8): 1085–1092.

Wetter, M. (2006a) "Multizone Building Model for Thermal Building Simulation in Modelica", in Kral, C. and Haumer, A. (eds), *Proceedings of the 5th International Modelica Conference*, Vienna, Austria, pp. 517–526.

Wetter, M. (2006b) "Multizone Airflow Model in Modelica", in Kral, C. and Haumer, A. (eds), *Proceedings of the 5th International Modelica Conference*, Vienna, Austria, pp. 431–440.

Wetter, M. and Haugstetter, C. (2006) "Modelica Versus TRNSYS – A Comparison between an Equation-based and a Procedural Modeling Language for Building Energy Simulation", *Proceedings of SimBuild 2006*, Cambridge, MA.

Wetter, M. and Haves, P. (2008), "A Modular Building Controls Virtual Test Bed for the Integration of Heterogeneous Systems", *Proceedings of SimBuild 2008*, Berkeley, CA.

Wetter, M. and Polak, E. (2004) "A Convergent Optimization Method Using Pattern Search Algorithms with Adaptive Precision Simulation", *Building Services Engineering Research and Technology* 25(4): 327–338.

Wetter, M. and Wright, J. (2004), "A Comparison of Deterministic and Probabilistic Optimization Algorithms for Nonsmooth Simulation-based Optimization", *Building and Environment* 39(8): 989–999.

Wetter, M., Haves, P., Moshier, M.A. and Sowell, E.F. (2008) "Using SPARK as a Solver for Modelica", *Proceedings of SimBuild 2008*, Berkeley, CA.

Recommended reading

For a computational approach to heterogeneous systems, the reader is referred to the textbook of Lee and Varaiya (2002). For a discussion about trends and challenges of system level design we refer to Sangiovanni-Vincentelli (2007). The prerequisites for generating efficient simulation code from a general-purpose solver are discussed by Elmqvist *et al.* (1995). Cellier *et al.* (1996) discusses common pitfalls and misconceptions about the modeling of physical systems. Tiller (2001) and Fritzson (2004) discuss different equation-based modeling approaches for dynamical systems. Cellier and Kofman (2006) present in a rigorous way methods for transforming mathematical models into computational code. The textbook has been written for an engineering audience. Mattsson *et al.* (2000) describe how symbolic and numerical methods are combined in Dymola to solve reliably and efficiently high-index DAE systems. For a detailed discussion of BIM see for example Eastman *et al.* (2008). Clarke (2001) gives a detailed exposition of models that are typically found in building simulation programs. Malkawi and Augenbroe (2004) present recent trends in building simulation with a focus on combined air and heat flow, uncertainty propagation and integration of simulation into the design process.

Activities

1. Implement two versions of the system model shown in Figure 17.3(a). For the first version, use causal block-diagrams and for the second version, use acausal models for the physical components. As parameters, use for each heat capacitor *10 J/K*, for each heat conductor *10 W/K*, for the convective heat transfer *5 W/K*, for the setpoint and the initial temperatures 20°C and for the control gain *100 W/K*. For the boundary condition, use $T(t) = 20 + 5\ sin(2\ \pi\ t/3600)$, where t denotes time in seconds and the temperature is in degree Celsius. Compute the maximum control error. Next, reconfigure the system to the topology shown in Figure 17.3(b). Compute again the maximum control error. Discuss the time it took to develop and reconfigure the two model representations.

Hint: You may use Modelica or MATLAB/Simulink with the SimScape™ extension to create acausal and causal models.

2. How do you think a next generation building energy analysis tool will look like?

Answers

For configuration (a), the maximum control error is *0.12 K* and for configuration (b), it is *0.24 K*.

Index